Springer Series on Polymer and Composite Materials

Series editor

More information about this series at http://www.springer.com/series/13173

Mostafizur Rahaman · Dipak Khastgir
Ali Kanakhir Aldalbahi
Editors

Carbon-Containing Polymer Composites

Springer

Editors
Mostafizur Rahaman
Department of Chemistry, College of Science
King Saud University
Riyadh, Saudi Arabia

Ali Kanakhir Aldalbahi
Department of Chemistry, College of Science
King Saud University
Riyadh, Saudi Arabia

Dipak Khastgir
Rubber Technology Centre
Indian Institute of Technology Kharagpur
Kharagpur, West Bengal, India

ISSN 2364-1878 ISSN 2364-1886 (electronic)
Springer Series on Polymer and Composite Materials
ISBN 978-981-13-2687-5 ISBN 978-981-13-2688-2 (eBook)
https://doi.org/10.1007/978-981-13-2688-2

Library of Congress Control Number: 2018955443

This Springer imprint is published by the registered company Springer Nature Singapore Pte Ltd.
The registered company address is: 152 Beach Road, #21-01/04 Gateway East, Singapore 189721, Singapore

Preface

In this book, the methods of synthesis/preparation of different carbon materials like graphite, carbon blacks, carbon fibers, carbon nanotubes, and graphene have been discussed. The different processes/methods of functionalizations/modifications of these carbon materials for improving their composites' properties are reported in detail. The preparation techniques of polymer/carbon composites are shown pictorially to have a better idea about their preparation methods and have been discussed by focussing on methods that are more advantageous for getting improvement in properties. The merits and demerits associated with some preparation methods are mentioned in the respective chapter.

The different properties of the polymer composites like mechanical, electrical, dielectric, thermal, rheological, morphologic, spectroscopic, electronic, sensing are discussed elaborately on the basis of their geometrical/structural points of view in separate chapters. With the incorporation of carbon filler within the polymer matrix, the mechanical properties and the thermal stability of the composites improve up to a certain level and then decrease after further addition. However, the electrical conductivity, dielectric permittivity, thermal conductivity, and electromagnetic interference shielding effectiveness of the polymer composites increase regularly with the addition of conductive carbons. Good dispersion of carbons within the polymer matrices is observed at their certain loading. The properties are strongly dependent on the type and nature of both polymers and carbons.

Applicability of the composites is focussed on electrical and electronic fields, structural field, sensing and catalysis, fuel cell, solar energy, etc. It is well known that all polymers are not suitable for a particular application or a particular polymer is not suitable for all types of applications, and hence, the best choice of polymer for a particular application is reported herein. Similarly, it can be said that all carbon materials are not suitable for a particular application or a particular carbon material is not suitable for all types of applications. Hence, the readers can find out suitable carbon material for specific applications according to their choice. Moreover, the geometry/structure of carbons also affects different properties of the polymer composites. So, the effect of carbon types and their geometry/structure on the properties and applications of composites is discussed in detail in the chapters, and this is the speciality of this book and will be beneficial to the readers.

This book covers the synthesis/preparation of carbon materials, their functionalization, processing of composites, and almost all properties and applications of polymer/carbon composites. Hence, this book can be used as a supplementary reading/textbook in the future for the proposed branches of studies like polymer/carbon composites and polymer/carbon materials in nanotechnology suitable for the department of materials/polymer science and engineering, and center/department of polymer nanotechnology, respectively.

Riyadh, Saudi Arabia — Mostafizur Rahaman
Kharagpur, India — Dipak Khastgir
Riyadh, Saudi Arabia — Ali Kanakhir Aldalbahi
September 2018

Contents

About the Editors

Dr. Mostafizur Rahaman is Assistant Professor in the Department of Chemistry at the College of Science, King Saud University, Riyadh 11451, Saudi Arabia. He obtained his M.Sc. (chemistry) from Tilka Manjhi Bhagalpur University, India, and Ph.D. (chemical/polymer chemistry) from the Indian Institute of Technology Kharagpur, India. He completed his M.Tech. in plastics engineering at the Central Institute of Plastics Engineering and Technology (CIPET), Bhubaneswar, Orissa, India. He has published 60 papers and 5 communicated manuscripts in international journals and 15 research articles in international conference proceedings. He has also published one patent and one book. He has nine years of teaching and ten years of research experience. He has completed six research projects and attended/presented at various international conferences/seminars. He has been an active reviewer for various international journals and member of journal advisory boards and is an expert in handling sophisticated instruments. His research interests include polymer nanotechnology/nanocomposites; polymer membrane; polymer thin film; polymer-based sensors; catalytic synthesis of polymers and conducting polymers; polymer-based coating for corrosion protection; polymer fuel cell and solar energy and biopolymers for biomedical applications.

Dr. Dipak Khastgir is Professor of Rubber Technology Centre at the Indian Institute of Technology (IIT), Kharagpur, India. He obtained his M.Sc. (physical chemistry) and Ph.D. (polymer composites) at IIT Kharagpur in 1975 and 1984, respectively. His research areas include conductive polymer (polyaniline, synthesis, and characterization); rubber–carbon conductive composite for EMI shielding application and pressure-sensitive conductive rubber; high-voltage polymeric insulators and piezoelectric polymer composites; characterization of rubber and rubber-like materials; textile reinforcement of rubber products; and industrial rubber product design and technology. He has published over 180 papers in refereed international journals, 70 seminar and conference proceedings, and 6 books.

Dr. Ali Kanakhir Aldalbahi is Vice Dean (Research) at King Abdullah Institute for Nanotechnology; Associate Professor in the Department of Chemistry, College of Science, King Saud University, Riyadh 11451, Saudi Arabia; and Director of the Distinguished Student Program at King Saud University. He obtained his Ph.D. and master's in chemistry (nanotechnology) at Wollongong University in 2013 and 2008, respectively. He is Member of Australian Nanotechnology Network (ANN), Australian Student Leadership Association (ASLA), Saudi Chemical Society (SCS), and Material Research Society (MRS). He has published more than 100 papers in national and international journals.

Synthesis/Preparation of Carbon Materials

Purabi Bhagabati, Mostafizur Rahaman, Subhendu Bhandari, Indranil Roy, Ayan Dey, Prashant Gupta, M. A. Ansari, Aastha Dutta and Dipankar Chattopadhyay

Abstract This chapter focuses on the synthesis/preparation of different carbon materials namely, diamond, fullerene, graphite, carbon black, carbon fiber, carbon nanofiber, carbon nanotube, and graphene. Though there are several methods of synthesis/preparation of carbons, the most common and important ones are discussed herein. The synthesis of diamond has been majorly focussed by chemical vapor deposition (CVD) method, whereas the other ion beam sputtering method and an ionized deposition method has also been mentioned herein. The preparation of carbon black by various processes has been reported. Moreover, carbon black is also prepared by pyrolysis of polymer and hydrocarbon, plasma synthesis, and by hydrolysis of natural resources. Carbon fibers of high performance derived from different precursors are discussed in details. Carbon nanofiber is synthesized by

P. Bhagabati
Chemical Engineering Department, Indian Institute of Technology Guwahati, Guwahati 781039, India
e-mail: purabi.bhagabati08@gmail.com

M. Rahaman (✉)
Department of Chemistry, College of Science, King Saud University, Riyadh 11451, Saudi Arabia
e-mail: mrahaman@ksu.edu.sa

S. Bhandari (✉) · A. Dey · P. Gupta · M. A. Ansari · A. Dutta
Department of Plastic and Polymer Engineering, Maharashtra Institute of Technology, Aurangabad 431010, Maharashtra, India
e-mail: subhenduonly@gmail.com

A. Dey
e-mail: deyayanchem@yahoo.co.in

P. Gupta
e-mail: prashant.gupta@mit.asia

M. A. Ansari
e-mail: mujahid.ansari@mit.asia

A. Dutta
e-mail: aastha2006@gmail.com

M. Rahaman et al. (eds.), *Carbon-Containing Polymer Composites*, Springer Series on Polymer and Composite Materials,
https://doi.org/10.1007/978-981-13-2688-2_1

CVD as well as electrospinning and template methods. Physical methods like laser ablation, arc discharge, and chemical method like CVD are used to prepared carbon nanotube effectively. Two approaches namely top-down and bottom-up are considered to discuss the synthesis of graphene by different methods. The methods are discussed pictorially or diagrammatically where necessary.

Keywords Diamond · Fullerene · Graphite · Carbon black · Carbon fiber Carbon nanofiber · Carbon nanotube · Graphene

1 Synthesis of Diamonds

1.1 History of Diamond Synthesis

Diamond is among the most expensive and nonreactive allotropes of carbon that is naturally available on earth's mantle. The production process of diamond from carbon needs extreme pressure, temperature, and also quite a long span of time of 1–3 million years. Hence, diamond costs a lot until recent years, where humans started preparing diamonds artificially using various techniques. In this subchapter, a brief history about artificial diamond synthesis, different techniques, and its probable mechanism of conversion from its precursor will be discussed. Basically, there are four major processes followed for the synthesis of diamond: (1) conversion of carbonaceous material mostly graphite into diamond by using high-pressure–high-temperature (HT-HP) conditions; (2) direct conversion of graphite into diamond using strong shock waves; (3) Conversion of carbon-containing material like graphite into diamond using metal solvents; and (4) continuous deposition of diamond onto a substrate using a seed material from hydrocarbon gases.

Synthesis of diamond by artificial method by man requires high temperature at least of approximately 1150 °C and pressure of 45 kilo bars. The synthesis of diamond was first attempted in the year of 1880 by Hannay [1]. Later in the year of 1894, Moissan tried to prepare diamond. In the year of 1955, General Electric Company Research Laboratory synthesized substantiate amount of diamond [2]. In the same year, a Swedish farm Allmanna Svenska Elektriska Aktiebolage reported to have synthesized diamond in bulk quantity. Meanwhile, in the year of 1960, U. S. Army Electronic Research and Development Laboratory independently

I. Roy · D. Chattopadhyay
Department of Polymer Science and Technology, University of Calcutta, 92 A.P.C. Road, Kolkata 700009, India
e-mail: indranil.cit6@gmail.com

D. Chattopadhyay
e-mail: dipankar.chattopadhyay@gmail.com

published a method of successful synthesis of diamond [3]. Basically, two distinct thermodynamically feasible physicochemical processes were found to be proven successful. (1) Under the condition of high pressure and temperature, a reaction between carbon and an intermediate material produced during the process of synthesis mentioned above. (2) Direct conversion of graphite into diamond by subjecting the graphite to sudden and instantaneous high stress under explosive shock. This process was first reported by the Stanford Research Institute of Menlo Park, California [4]. An attempt reported by Union Carbide Corporation of Kenmore in 1962 involved high-temperature and low-pressure technique [5].

1.2 Different Methods for the Synthesis of Diamonds

In the subsequent sections of this chapter, a brief discussion on the instrumentation part of diamond synthesis will be included.

1. BELT apparatus: Fig. 1 illustrates a rough idea about the most conventional "BELT" apparatus of diamond synthesis by General Electric Company [6]. The min design principal of the apparatus features high pressure generated by the compressible preformed conical gasket. As mentioned, GE was capable of synthesizing diamond artificially using this apparatus. Still, GE holds the credibility of possessing the most effective technique to synthesize diamonds.
2. Supported stepped piston–cylinder: The cross-sectional view of the apparatus in Fig. 2 was designed and used by the United States Army Electronic Research

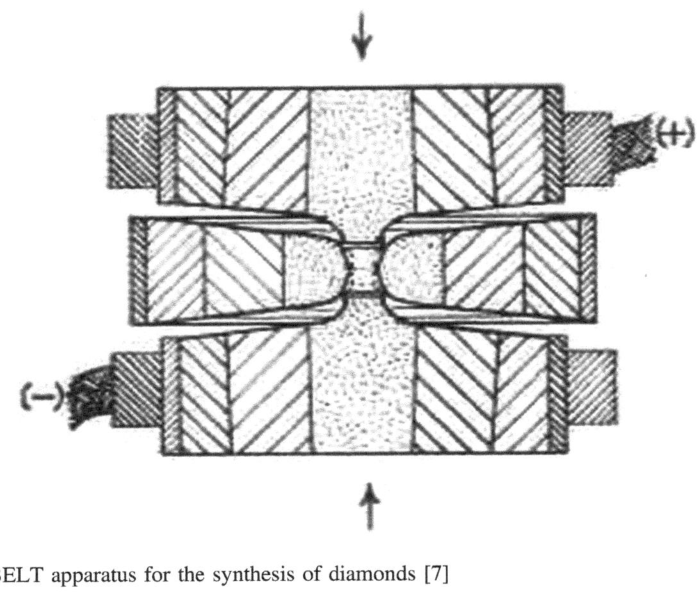

Fig. 1 BELT apparatus for the synthesis of diamonds [7]

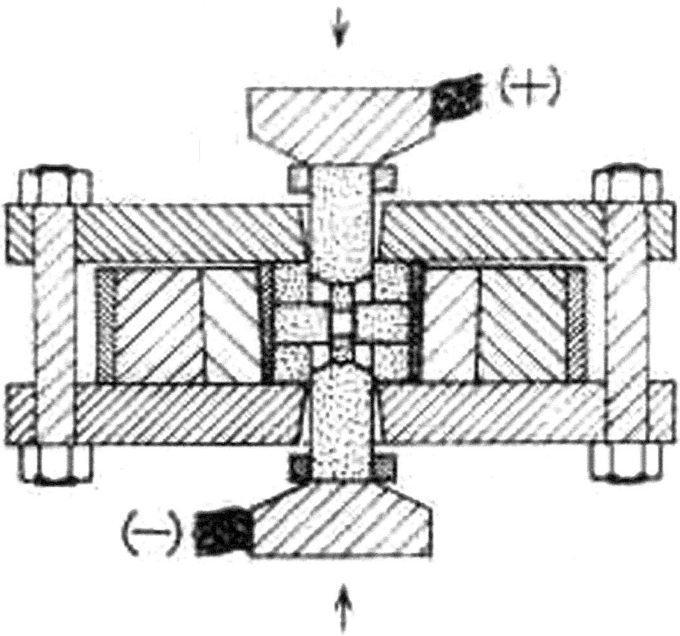

Fig. 2 Supported stepped piston–cylinder apparatus for the synthesis of diamond [7]

and Development laboratory. This device was capable of generating and sustaining high pressure and temperature of up to 100 kilo bars and 2500 °C.

3. Multi-anvil apparatus: Fig. 3 indicates a rough diagram of the H. T. Hall's "multi-anvil" instrument of tetrahedral design. The prime design principal of the apparatus is an extruded compressible gasket, which provides superior performance. The major advantage of this apparatus includes minimized sample deformation which is due to reduced displacement of compressible solid materials from the pressurized cavity.
4. "Gridle" chamber: Figure 4 represents a cross-sectional view of the "Gridle" chamber designed and developed by the "Battelle Memorial Institute". The geometrical appearance of this apparatus is almost comparable to the BELT apparatus, with the difference of utilization of multiple steel binding rings replacing the compressible gasket principle.
5. Explosive shock apparatus: The cross-sectional schematic view of the explosive shock apparatus is shown in Fig. 5. The apparatus was first developed and used by the Stanford Research Institute. An explosive charge is detonated against a piston in order to develop a shock wave onto a solid cylindrical specimen at a pressure of 400–500 kilo bars, and temperatures of 1000–1500 °C. This shock wave was generated at a time interval of one microsecond. This is the process successful in converting graphite into diamond.

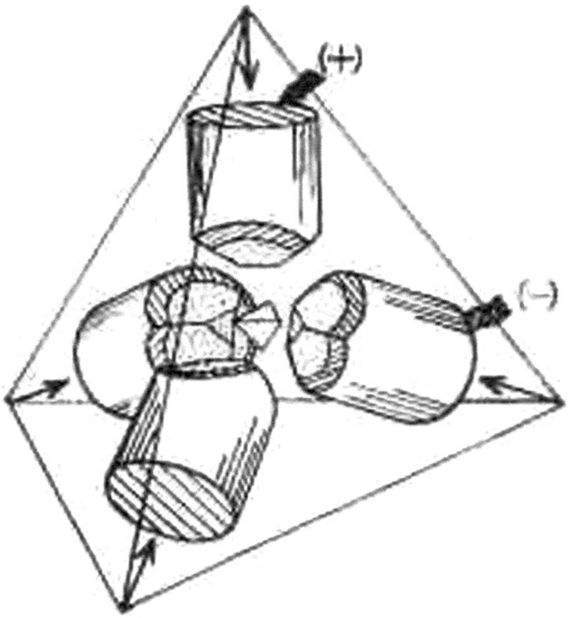

Fig. 3 Tetrahedral multi-anvil apparatus for the synthesis of diamond [7]

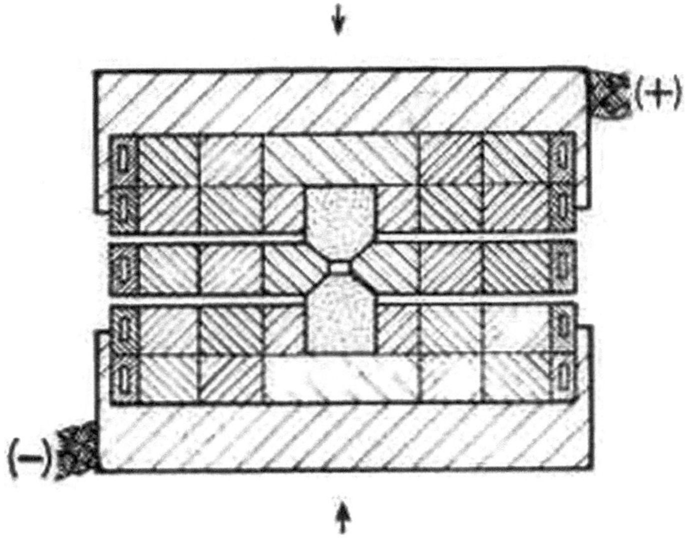

Fig. 4 "Gridle" chamber for the synthesis of diamond [7]

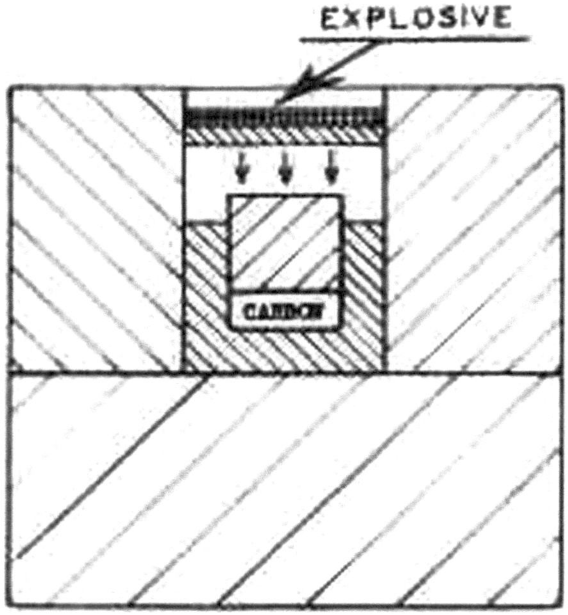

Fig. 5 Explosive shock apparatus for the synthesis of diamond [7]

1.3 Chemical Vapor Deposition (CVD) Technique

Other than abovementioned techniques, nowadays, some new techniques are becoming more popular and efficient in synthesizing diamonds. Chemical vapor deposition (CVD) techniques are more sophisticated and require precise controlling of various parameters, and proved to be very efficient.

Chemical vapor deposition (CVD) technique: The CVD technique is also well known as “Gas phase synthesis” of diamonds. Diamonds can also be synthesized from hydrocarbon gases over a substrate and the process is favorable for graphite. Hence, thermodynamically the process is in total contrast to high-pressure–high-temperature technique where diamonds are more stable than graphite during the synthesis process. Here, the co-deposition of graphite was dominated by the hydrogen atoms present in excess quantity within the reaction chamber.

CVD or gas phase synthesis of diamond was for the first time reported by Eversole, where a methyl group containing hydrocarbon gases like various alkanes (methane and higher alkanes)methyl chloride, methyl mercaptane, and acetone was passed over diamond powder at a temperature of 1000 °C and at a nominal pressure of 50 torr [8, 9]. The diamond particles of the powder act as seed material in the formation of diamonds. However, the deposition of black color carbon over the surface hampers the diamond accumulation process, and the carbon accumulated over the diamond was removed by heating in the presence of hydrogen gas. Hence, for continuous diamond formation process, the repeated accumulation–cleaning has to be carried out until our desired growth was obtained. In a nutshell, the diamond growth process takes place at high temperature of 1000 °C, where methyl radical formed by dissociating from its gaseous hydrocarbon reactant, which causes diamond growth on the seed, and finally cleaning of the obtained diamond is done using hydrogen gas at 1000 °C.

The three different CVD methods for the synthesis of diamond are used: (1) hot-filament; (2) RF-wave; and (3) microwave. Figure 6 shows the schematic diagram of hot-filament CVD apparatus for diamond synthesis. In the apparatus, a tungsten filament is placed above the substrate where the diamond is supposed to grow, and the distance between the substrate and the filament is almost 10 mm. The filament is then heated up to about 2000 °C, and both the substrate and the filament are inside a quartz tube of temperature around 600 °C that is placed inside a furnace. The temperature of the substrate eventually goes as high as 1000 °C due to the thermal radiation from the hot filaments and generates atomic hydrogen and chemically active species like radicals and ions by dissociating the hydrocarbon gases present inside the reactor. The chemically active species have long lifetime sufficient enough to travel up to the substrate and cause continuous growth of diamonds onto it. Figures 7 and 8 represent the CVD method of diamond synthesis using RF-wave and microwave, respectively. Here, the diamond grows onto the substrate which is submerged inside a mixture of hydrocarbon-hydrogen gas plasma generated by RF irradiation and microwave irradiation.

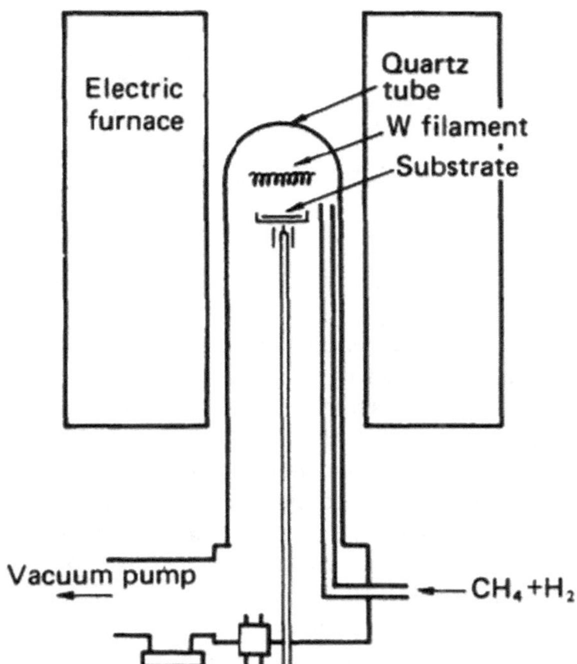

Fig. 6 CVD methods for the synthesis of diamond by hot-filament technique [10]

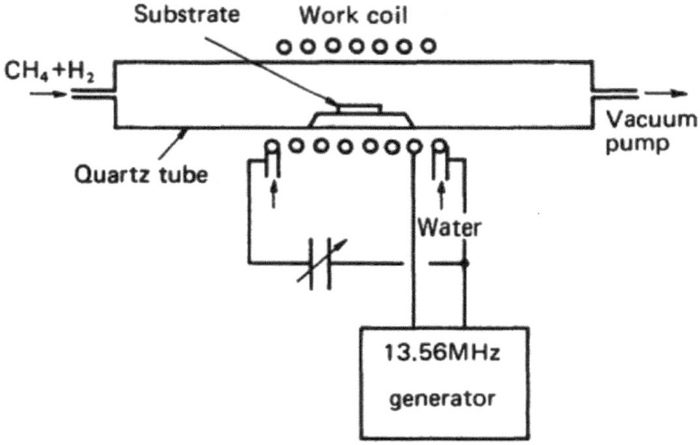

Fig. 7 CVD methods for the synthesis of diamond by RF-wave irradiation technique [10]

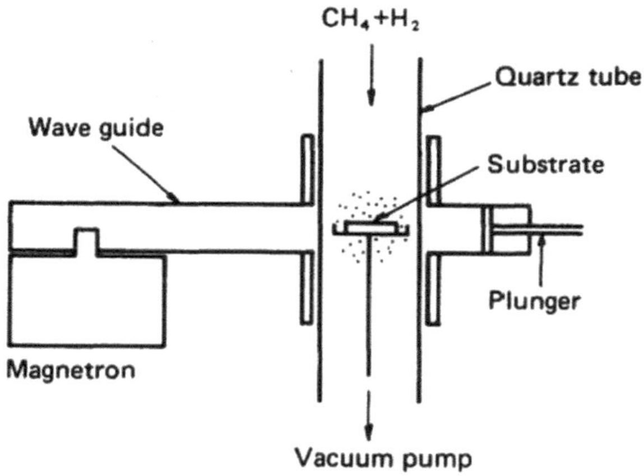

Fig. 8 CVD methods for the synthesis of diamond by microwave irradiation technique [10]

A generalized mechanism for the formation of diamond in the CVD technique by microwave technique as proposed by Seteka is shown in Fig. 9 [10, 11]. Though the exact mechanism of diamond synthesis is unclear, according to Seteka and his group, the hydrocarbons and hydrogen gas mixture form active radicals inside the chamber under the condition of plasma and these active hydrocarbon species diffuses and gets adsorbed onto the substrate and subsequently deposits onto it.

Recently, ion beam sputtering method and an ionized deposition method were reported to be a successful technique for the synthesis of diamond films over substrate [12, 13]. However, the efficiency of these two methods was found to be lesser than the CVD method. Further, Luo et al. in the year of 2015 reported synthesis of diamond from carbon nanofiber under the condition of low pressure and temperature [14]. Here, they have developed a new route to synthesize diamond

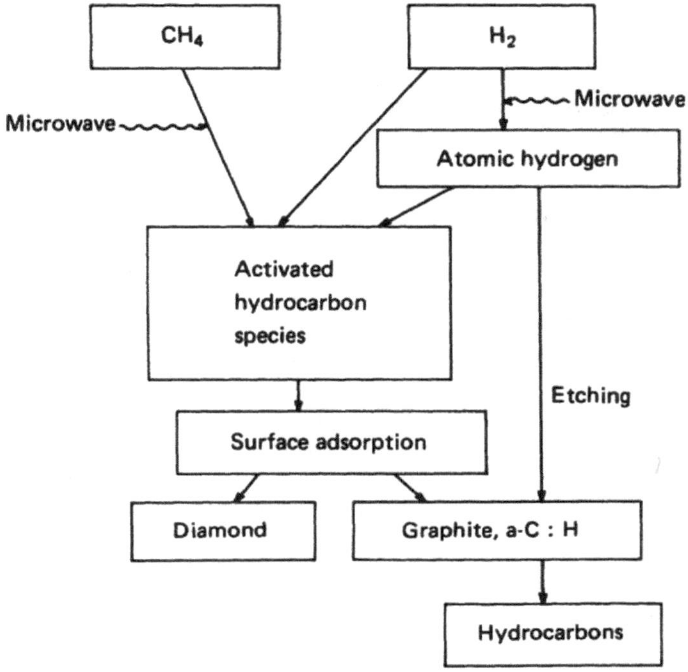

Fig. 9 Plausible mechanism of diamond formation as proposed by Setaka [10]

from carbon nanofibers in "Sperk Plasma Sintering" (SPS) system under the condition of atmospheric pressure and low temperature.

1.4 Nanodiamonds (NDs)

Nanodiamonds were discovered much back in the 1960s by American and Russian scientists, while the major exploration of NDs was started from the 90s. Due to its excellent binding capacity with biological molecules like peptides, proteins, nucleic acid, etc., it can be used in clinical applications. There are numerous ways for the synthesis of NDs: (1) Detonation technique, (2) Laser ablation technique, (3) High energy ball milling of high-pressure–high-temperature technique, (4) Chemical vapor deposition (CVD) technique, (5) Autoclave synthesis from supercritical fluids, (6) Ultrasound cavitation, and (7) Ion beam irradiation, etc. However, the first three methods, i.e., detonation technique, laser ablation technique, high energy ball milling of high-pressure–high-temperature technique are commercially viable techniques for the synthesis of NDs [15, 16]. Detonation method is the most extensively employed industrial method for the synthesis of NDs. In this method, strong explosives are detonated in a closed vessel under negative oxygen balance.

The average crystallite size of detonation nanodiamonds (DNDs) is 3–5 nm [17]. In the laser ablation technique, pulsed laser ablation of a solid target results in the synthesis of NDs. More detail about the mentioned synthesis techniques of NDs does not come under the scope of this chapter.

2 Fullerene Synthesis

Fullerenes commonly known with the chemical formula C_{60} was traced by Smalley and coworkers in 1985 in mass spectra of carbon soot prepared by laser vaporization of graphite under constant helium gas flow [18–20]. It was noticed that the amount of C_{60} species was quite significant. They proposed the structure of C_{60} species consisting of 20 numbers of hexagonal and 12 numbers of pentagonal rings, which was very much similar to the previously developed architectural structures known as "geodesic domes" as designed by R. Buckminster Fuller [21, 22]. The C_{60} species was later named as Buckminster fullerene or in short Bucky-Ball. However, the laser vaporization technique that was used for the preparation of C_{60} species and its other allotropes is far below its capacity of production in mass. A well-defined study of fullerene requires macroscopic quantity of fullerene species. In the subsequent section, a detailed discussion of the concepts of fullerene synthesis will be incurred. Formation of fullerene clusters with varying chemical structure C_{60}, C_{70}, etc., during the synthesis process is inevitable. Hence, the separation method and purification of these clusters will also be included in order to encapsulate complete knowledge on the production of fullerene.

2.1 *Different Methods of Fullerene Synthesis*

The different methods of synthesis of fullerene with varying yield % are mentioned below.

1. Arc vaporization of graphite

This is the most basic and simple process for the synthesis of fullerene species using graphite rods as the source of carbon. The procedure was designed by Kratschmer and Huffman in the year of 1990, and hence known as Kratschmer–Huffman method for macroscopic synthesis of fullerene [23]. In the apparatus, carbon soot collected consists of good quantity of fullerene species. A simplified schematic of the arc vaporization apparatus for fullerene synthesis is shown in Fig. 10. Three major components of the apparatus are: (1) High current electric feedthrough, (2) one movable electrode, and (3) Cool surface as carbon soot collector. The three connected components of the apparatus are encased within a vacuum chamber up to a pressure of $\leq 10^{-3}$ torr.

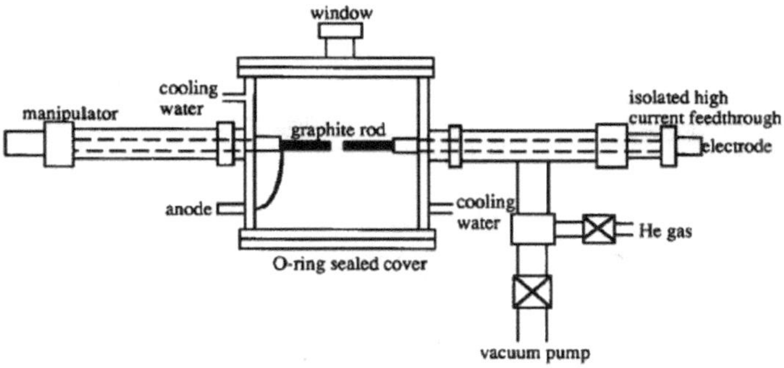

Fig. 10 Schematic illustration of arc vaporization apparatus for fullerene synthesis [24]

In a typical experiment, two pure graphite electrodes are placed facing each other and other ends of each rod are connected to high current feedthrough. Vacuum of $\leq 10^{-3}$ torr is applied to the chamber, followed by filling it with helium gas to a pressure of 150–250 torr in order to make a suitable inert environment to carry out the process. It is important to mention that both oxygen and moisture significantly inhibit the formation of fullerene. Then, the high current of around 100–200 A is passed through the two graphite rods. As optimized by the research group, it was mentioned that efficient fullerene formation was possible for graphite rods of 6.25 mm diameter with current of 100–200 A, where the rate of consumption of the graphite rods are 5–10 mm/min, and the carbon soot are collected onto a water or air-cooled surface and are stored in containers by easily brushing off the surface. This carbon soot contains a mixture of a variety of carbon species including fullerene species of chemical formula C_{60} and larger. Hence, the separation and the purification of our desired fullerene components are crucial and needs vast attention. Detailed processes of isolation and purification of different fullerene components from the carbon soot has been discussed in Sect. 2.2

2. Laser ablation of graphite

Another technique for macroscopic level fullerene synthesis is laser ablation of graphite under the inert environment of helium. This technique of fullerene production was initially ineffective due to the faster cooling of dense carbon plasma formed during the laser vaporization process. Such rapid cooling of carbon plasma could not provide enough time to rearrange the carbon fragments into stable and closed structures fullerene. However, Smalley and coworkers again made excellent modification through carrying out laser ablation of graphite at elevated temperature, which led to the efficient production of fullerene species at macroscopic quantity. A typical laser ablation apparatus for macroscopic production of fullerene is illustrated with a schematic diagram in Fig. 11.

In this technique, the graphite rod is ablated by the application of laser under high temperature of 1200 °C inside a furnace. The ablation of graphite at elevated temperature lets the carbon plasma to cool down slowly and thereby sets enough

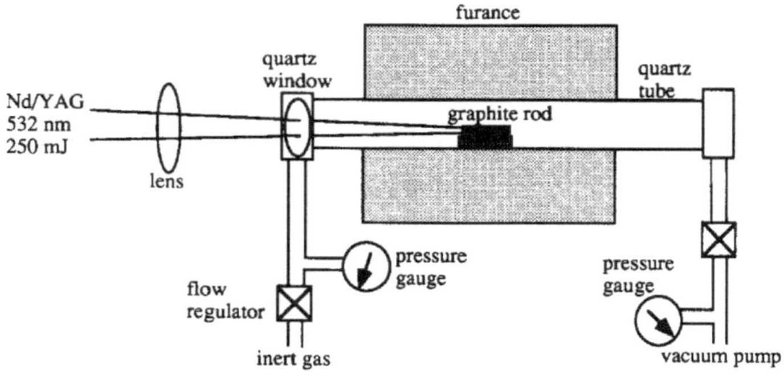

Fig. 11 Typical schematic illustration of laser ablation apparatus for macroscopic fullerene synthesis [24]

time for the carbon fragments to form fullerenes. Other benefits of laser ablation technique include the capability to change the characteristics of carbon plasma through varying the laser pulse energy and wavelength. Besides, varying the furnace temperature as well as the buffer gas pressure, it is possible to control the growth of the fullerene species.

3. Hydrocarbon combustion technique

In the year 1991, the macroscopic synthesis of fullerene by combustion of hydrocarbons was first reported by Scott [25]. Combusting a mixture of benzene and oxygen of ratio 1:1 using a laminar flow flame led to production of significant quantity (0.3%) of fullerene species in the obtained carbon soot. However, the yield of this technique is lower than the yield of the previously mentioned processes. The most important characteristic of this technique includes the ability of continuous operation by scaling-up the hydrocarbon combustion. Besides, varying the flame characteristic influences the size distribution of the fullerene species. Commercial plant for the production of fullerene by hydrocarbon combustion was recently been started in Japan that possesses a total capacity of 40 metric ton production of C_{60} per annum, which paved new era in the production of low-cost fullerene production [26]. United States (U.S.) is also establishing such low-cost fullerene production plant through improved combustion technique of hydrocarbon combustion [27]. Hydrocarbon combustion technique unlike electric arc discharge and laser ablation technique is a continuous process; hence it has tremendous potential for fullerene synthesis commercially. In 1998, Alford et al. developed a combustion apparatus industrially viable for the production of fullerene [28]. The major component of this apparatus that improves the efficiency of the typical combustion technique is a porous refractory plate burner. This technique was capable of producing multi-tons of fullerenes per year. Though the hydrocarbon combustion technique of fullerene synthesis is very efficient among all other processes, the actual mechanism of fullerene formation in the flames is not yet clearly been understood.

2.2 Extraction of Fullerene

The crude carbon soot obtained during the synthesis process of fullerene contained mixture of different types of fullerenes (e.g., C_{60}, C_{70} and higher) and carbon allotropes. Hence, extraction and separation of fullerene species is an essential step to be performed in obtaining pure fullerenes. The commonly followed technique to extract fullerenes from the crude carbon soot is solvent extraction technique. The technique involves exposing of the carbon soot to a liquid organic solvent for a long period of time in order to transfer the soluble material into the liquid. Initially, benzene was used as the solvent to extract fullerene species from the carbon soot. Besides benzene, hexane, heptane, pyridine, and tetramethylbenzene were all used as organic solvent in the extraction process. Nowadays, toluene is extensively used as the extracting solvent for the safety purpose. The soot obtained in the production process is extracted in toluene using Soxhlet apparatus. The continuity of the extraction process provides great benefit to the output of the process. The completion of the extraction process of fullerene species from the carbon soot can be determined as and when the toluene condensed in the extractor becomes colorless. The fullerene species extracted from the soot contains a mixture of C_{60}, $C_{70,}$ and other higher allotropes. In that case, the separation of individual fullerenes from each other can be done by chromatographic technique. It is to mention that hexane and heptane are capable of extracting C_{60}, C_{70} while tetramethylbenzene recovers higher molecular weight fullerenes. Further extraction using hexane or heptane followed by tetramethylbenzene provides the chance of fractionating the carbon soot into C_{60}, $C_{70,}$ and its higher molecular weight homologues.

Another method of fullerene extraction from carbon soot is sublimation technique. The process involves evaporation of volatile materials under vacuum or under inert atmosphere into gaseous phase at a constant temperature that is above sublimation temperature of the desired fullerene species. It is reported that depending on the type of carbon soot and the purity of carbon the sublimation temperature of C_{60} is within the range of 400–600 °C. After sublimation, the fullerene species are collected in the individual chambers of air-cooled collectors.

Recently, supercritical fluid extraction (SFE) technique has also effectively been introduced in order to obtain maximum purity of obtained fullerene species. Though liquid extraction technique is the most conventional and efficient technique registered till date in extracting fullerenes in bulk quantity, the time-consuming nature and requirement of large amount of solvents and its subsequent evaporation and fractionation process show the negative side of the process. SFE is reported to be more selective towards the desired fullerene species in the extraction process. In the SFE technique, a homogeneously dilute mixture of a strong organic solvent like toluene with a poor solvent power medium like supercritical CO_2 fluid under high pressure is employed.

Saim et al. published their work of investigation on using SFE technique to extract fullerene species, where they have adjusted the recovery and selectivity of SFE for fullerene species from carbon soot using different solvent mixtures temperatures and pressures. Quantitative recovery of C_{60} fullerene was reported by the same group by employing moderate concentrations of aromatic solvent modifier like toluene [29].

3 Synthesis of Graphite

Graphite is a carbonaceous material consisting of numbers of parallel graphene layers bound together by weak van der Waals force. Apart from the mere application of pristine graphite, contemporary researches involve modification of graphitic layers, improvement of processing and retaining electrical properties in composites, separation of graphene layers, coating of graphitic layers on to other materials, graphite encapsulated metal nanoparticles [29], etc.

3.1 Graphite Electrode

The idea of using graphite as electrode material has started more than a century back. Graphite electrodes are widely used in industrial manufacturing of aluminum. Mixture of petroleum coke with ca. 20% binder pitch is extruded in the desired shape of electrode. For anodes, ca. 15% pitch is added as binder. The extrudate is then baked at 850 °C to transform binder pitch to pitch coke, which is further baked at 3000 °C to graphitize the extrudate to form graphite electrode [30, 31].

3.2 Water-Soluble Graphite

Several efforts have been made by the researchers to solubilize graphite in water. Graphite is oxidized with strong acids and exfoliated ultrasonically [32, 33] or by means of thermal exfoliation [34] to yield graphene layers. The functionalized (–OH, –O–, –COOH, >C=O) graphitic layers have been studied by the researchers extensively to achieve water-soluble graphite which can be beneficial for easy processing to prepare different composites. The >C=O and –COOH groups exist in the periphery of the layers, whereas the –OH and –O– groups exist in the interior of stacked multilayers which enhances interlayer spacing from 0.34 nm (graphite) to >0.6 nm (graphite oxide) [35]. Irrespective of the exfoliation method, the graphene oxide layers prepared from exfoliated graphite oxide consists of defects and oxygen functionality. Therefore, apart from achieving water solubility, the oxygen functionality and the defects should be reduced to recover the electronic properties of graphene sheets by rigorous reduction process [36]. Moreover, graphitic nanoplatelets are dispersed into polymeric matrices using suitable dispersing agents, which exhibit the adverse effect for some applications [33, 37]. Depending on the

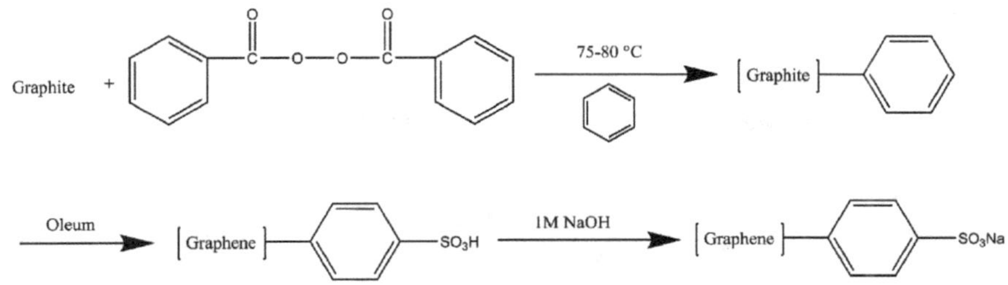

Fig. 12 Schematic representation of the synthesis of water-soluble graphite nanoplatelets by oleum exfoliation of graphite [39]

reduction process, conditions and reagents water solubility may differ widely. For example, very less water solubility (~ 0.5 mg/ml) is achieved while graphite oxide is reduced in presence of ammonia [38]. Efforts have been made to overcome such drawbacks. Si et al. [36] suggested a three-step method to separate graphitic layers: (i) removal of the majority of the oxygen functionality by treating graphene oxide with $NaBH_4$ for 1 h at 80 °C, (ii) sulfonation with aryl diazonium salt of sulfanilic acid for 2 h using ice bath, and (iii) removal of remaining oxygen functionality by postreduction with hydrazine for 24 h at 100 °C. A reasonable solubility to the extent of 2 mg/ml was achieved at broad pH range of 3–10.

Water-soluble graphite nanoplatelets may also be synthesized by oleum exfoliation of graphite. According to the method followed by Mukherjee et al. [39], benzoyl peroxide was added to the dispersion of defect-free graphite in benzene (Fig. 12) and the mixture was then heated for 2 h at 80 °C yield phenylated graphite. The dispersion of the phenylated graphite in oleum, i.e., H_2SO_4 with 20% free SO_3, was again heated for 4 h at 80 °C in inert atmosphere (argon) to yield phenyl sulfonated graphite which was then reacted to 1 M NaOH in similar temperature and inert condition to yield phenyl sulfonated salt of graphite. The solubility of the modified graphite salt was found to be 2.1 mg/ml. The sheet resistance of the phenyl sulfonated graphite platelets was found to be 212 Ω/sq in comparison to 960 Ω/sq as exhibited by the pristine graphite.

3.3 *Expanded Graphite*

Expanded graphite (EG) is a foam-like network structured graphite which can be used as adsorbent material to remove dye [40], pesticide [41], heavy oil [42], etc., as well as to prepare lightweight carbon-based articles. A typical example of the porous network structure of EG [43] has been shown in Fig. 13. Inside the layers of graphite, different atoms, ions or molecules may be inserted to produce graphite intercalation compounds (GICs). Modified EG with highly crystalline structure may be obtained by using H_2SO_4 as intercalator and H_2O_2 as oxidant which exhibits

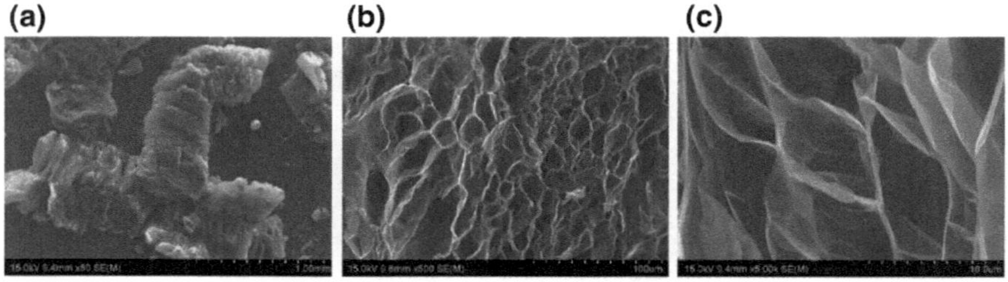

Fig. 13 Porous morphology of expanded graphite at different magnifications [43]

high adsorption efficiency towards oils. Sorption capacity increases with the increase in viscosity of oil following the trend: gasoline < kerosene < diesel oil < machine oil < crude oil < gear oil. For the modified EG with expanded volume of 320 ml/g, the sorption capacity for gear oil may be as high as 84.681 g/g [43]. Porous structure of EG is obtained after vaporization or degradation of GICs by rapid heating. EG obtained by thermal treatment of GIC is known as thermally expanded graphite (TEG). While being heated, the layers of graphite experience outward thrust along *c*-axis direction and the interlayer spacing increases. Graphite layers intercalated with anions inside the interlayer spacing are known as graphite salts. Penetration of bisulphate (HSO_4^-) ion into graphite lattice produces graphite bisulphate through different stages [44]:

(i) transport of a substance to the graphite host;
(ii) its adsorption by the surface along with separation of the transporting agent;
(iii) intercalation into the host;
(iv) diffusion within the host; and
(v) reaction between different intercalation stages.

The first stage is known as "blue graphite" because of its bluish color, whereas its color in subsequent stages is black. The lattice spacing for pristine graphite is 3.354 Å, whereas that of graphite bisulfate is 7.98 Å. During the first stage (heating at 200 °C), the lattice spacing increases to 11.33 Å, which further increases to 21.46 Å at the end of the fifth stage [45]. Graphite bisulfate can be prepared by swelling graphite rod in oleum (fuming sulphuric acid with 30% free SO_3). Cataldo et al. [46] prepared EG from carbon arc submerged in oleum at low (3 V, 5 A) and high (25 V, 10 A) current. The threshold concentrations of intercalation of weak acids via chemical method are very high, e.g., nearly 100% H_3PO_4 and $H_4P_2O_7$, whereas the threshold may be attained at relatively less concentrations for strong acids, e.g., 75% for HNO_3 and 63% for H_2SO_4 [47]. Although intercalation of weak organic acid, e.g., formic acid is reported [48], it may not be possible to intercalate many weak protonic acids like H_3BO_3 inside graphitic layers. Graphite bisulfate can be synthesized by either chemical [49] or electrochemical process. Intercalation of inorganic acids into graphitic layers in electrochemical method involves the following four stages [50]:

(i) induction period, i.e., oxidation and stress build-up within carbon network,
(ii) introduction of intercalates, i.e., diffusion of anions inside carbon macro-cations formed,
(iii) over oxidation of graphite salts with partial deformation of graphitic layers, and
(iv) oxidation of water yielding O_2 which is released as O_2, CO, and CO_2.

Thermally expanded graphite may be manufactured using "cyclone type", "opposite type" or other modified reactors [51]. Apart from conventional chemical or electrochemical processes, EG may also be synthesized via other nonconventional routes. Microwave-assisted synthesis route is very fast where an exposure of 30 s of mixture of graphite flakes, H_2SO_4, $KMnO_4$ and H_3PO_4 to 350 W yields EG [52]. Graphitic layers can be exfoliated by using ultrasound within wide range of frequency viz. 20–500 kHz [53]. Time of ultrasonication, type of solvent, centrifugation parameters, and processing temperature do not have significant influence on the type of graphite flake size distribution; however, mean size of graphite flake is greatly affected by the increase of ultrasonication and centrifugation time [54]. Sonochemical synthesis of potassium-graphite (KC_8) GIC at 20 kHz frequency and acoustic power density of 17 W/cm^2 reduces time of synthesis to less than 5 min in comparison to the solid-state synthesis at 150–200 °C requiring 1–8 h [55]. Exfoliation of graphitic layers may be improved after providing thermal shock at 1050 °C for 15 s by exposing it to ultrasonic irradiation at 100 W [56].

3.4 *Hydrothermal Synthesis of Graphite*

Hydrothermal treatment of different polymers, metal carbides, etc. in the form of chips, powders, or fibers may result in graphitization. Formation of graphitic carbon is highly dependent on synthesis conditions, i.e., temperature and pressure as well as on the reagents. For example, graphitization of poly(vinyl chloride) is possible at 600 °C temperature and 100 MPa pressure while reacted with $CaCO_3$ and water; whereas in the absence of these two reagents, graphitization is not possible even at 1400 °C and 1 GPa. Graphitization of carbon can be improved at the water pressure of 1 GPa at 900 °C [57]. Hydrothermal corrosion of SiC yields methane and SiO_2. However, graphitic carbon with layer thickness ranging from 10 nm to 2 μm may also be synthesized hydrothermally from SiC at 100 MPa for 25 h within the temperature range of 300–600 °C following the reaction [58]:

$$SiC + 2H_2O \rightarrow SiO_2 + C + 2H_2$$

Hydrothermal synthesis of carbonaceous materials may yield different shapes. Formation of different types of carbonaceous compounds including graphitic layers using poly(ethylene) (PE) and ethylene glycol (EG) as precursors of hydrothermal synthesis at different temperatures and pressures has been shown in Table 1, as

Table 1 Different types of carbonaceous deposits including graphitic structures prepared from hydrothermal synthesis [59]

Precursor formulation ratio (w/w)	Temperature (°C)	Phase 1 Pressure (MPa)	Phase 2 Pressure (MPa)	Carbonaceous deposits
Polyethylene + H_2O	750	100	n/a	Carbon films and spheres of amorphous characteristics
Ethylene glycol + H_2O	400–800	100	n/a	Carbon films and spheres of amorphous characteristics
Polyethylene chips + H_2O + Ni PE: H2O = 0.68–1.21: 1	740–760	79	100	Carbon films and spheres of amorphous characteristics Graphitic depsosits + tubes
Polyethylene chips + Ni	760	67	103	Carbon films and spheres of amorphous characteristics
Ethylene glycol + H_2O + Ni Ethylene glycol: H_2O = 7.85:1	730–800	52	90	Graphitic deposits + tubes

investigated by Libera et al. [59]. Flake-like graphite or tubular shaped carbon (diameter up to 1300 nm) with graphitic layers can be produced using this method.

3.5 Graphite Encapsulated Metal (GEM) Nanoparticles

In recent days, core–shell structured graphite encapsulated metal (GEM) nanoparticles have been drawing the attention of researchers because of its potential application in drug delivery, hydrogen storage [60], absorption of heavy metal ions [61], catalyst [62], etc. These particles exhibit immunity towards environmental degradation and strong corrosive medium. In its structure, nanocrystalline metals exist in the core, and the shell constituted of several layers of graphite/graphene. GEM nanoparticles were first discovered by Tomita et al. [63] and Ruoff et al. [64] where the syntheses of LaC_2 were carried out using the Krätschmer–Huffman arc discharge method [65]. Modified tungsten arc discharge method improves the encapsulation efficiency by reducing the amount of carbon deposition [66]. In this method, metal particles (e.g., iron, nickel, cobalt etc.) are kept in a graphite crucible which acts as anode, whereas a tungsten rod is used as cathode. Wide spatial distribution of arc plasma facilitates encapsulation of molten metals to form core–shell structured nanoparticles. During the synthesis of GEM nanoparticles in arc discharge method, the growth mechanism is based on coalescence. A two-step mechanism of GEM synthesis was suggested by Elliot et al. [67] where the metal block was separated from the graphite crucible by using a graphite foil, and mean particle size ranging between 7 and 14 nm was obtained by controlling the velocity of helium jet within the range of 56–20 m/s. Encapsulation efficiency can be tuned effectively by controlling the evaporated carbon:metal ratio, specifically in the thin coalescence region. The discharge may be carried out at the voltage of 30 V, current of 100 A and pressure of 200 Torr in helium atmosphere. The use of ethanol yields improved encapsulation efficiency as carbon source in comparison to diamond powder [29]. Formation of monolayer graphite in stepped surface of thermally annealed Ni has been reported by Shikin et al. by cracking of propylene at 500 °C temperature and 1×10^{-6} mbar pressure for 5 min duration [68]. Apart from ordinary GEM particles, magnetic carbon spheres (MCSs) containing magnetic inorganic particles in core and surrounded by graphitic layers may also be synthesized via different processes, viz., chemical vapor deposition [69], pyrolysis of organometallic polymers [70], magnetron and ion beam co-sputtering [71], self-generated template method [72], arc discharge [73], pulsed laser decomposition of organometallic species [74], hydrothermal carbonization [75], etc.

3.6 Miscellaneous Synthesis of Graphite

Apart from the conventional synthesis methods, several literatures reveal synthesis of graphite in nonconventional routes. Graphite ribbon can be synthesized by fusing benzenoid hydrocarbons via electrophile-induced cyclization route [76]. Graphite can also be manufactured by using high dose of carbon ion beam at 1×10^{18} C^+/cm^2 and 60 keV at the temperature of 600 °C [77]. Although graphite exists with layered morphology, helical structured graphite can be synthesized carbonizing helical polyacetylene (H-PA). Axially chiral biphenyl derivatives may be used as chiral dopant to prepare chiral nematic liquid crystals from nematic liquid crystals because it exhibits improved helical twisting power in comparison to asymmetric carbon-containing chiral dopants. Helical polyacetylene can be synthesized at 5–25 °C from the mixture of liquid crystals in presence of Ziegler–Natta catalyst and triethyl aluminum cocatalyst. Helical carbon film is obtained by carbonizing helical polyacetylene at 800 °C [78]. Application of direct current of +15 V to carbon electrode in aqueous H_2SO_4 medium followed by ultrasonication in dimethyl sulfoxide medium leads to exfoliation of graphite of the carbon electrode to produce graphite nanoplatelets of thickness 2.8–5.8 nm [79]. Synthesis of graphite has also been reported from the reaction of calcite with hydrogen which initiates at high temperature (500 °C) and high pressure [80]. Room temperature synthesis of graphite from mercuric carbide may be accomplished at very high pressure (ca. 15,000 atm) using a hydraulically operated tungsten carbide piston in a hollow steel cylinder [81]. Chlorination of ferric carbide in argon atmosphere yields graphite while heated at 600–100 °C [82]. Rod or pin-like polyhedral crystals of graphite may be prepared at 1000–1300 °C by using oxy-acetylene torch maintaining oxygen: acetylene ratio (v/v) at 0.9 and acetylene flow rate at 2 l/min [83]. Nevertheless, graphite can also be prepared by thermal decomposition of inorganic carbides by using laser. Ohkawara et al. [84] prepared graphite by irradiating sintered α-SiC by using pulse infrared laser (λ = 1064 nm, energy density of 150 J/cm^2, shot time of 5 ms) in the presence and also in the absence of pitch and boron carbide as binder.

4 Preparation of Carbon Black

The ultrafine structured carbon black is manufactured by the combustion of hydrocarbons in the limited presence of air. Since long, black colored materials as coloring agent have been obtained by charring various carbonaceous materials like wood, oils, etc., without any specified conditions. The blacks obtained this way were impure with variation in chemical composition in comparison to currently available blacks that are available as per required purity. The common areas of application of such carbon blacks are for making inks, cosmetics, tires, and other areas. Carbon black, manufactured in the present days is virtually pure elemental carbon, which is obtained by the partial combustion or controlled thermal decomposition of gaseous or liquid hydrocarbons. It appears in the form of pellet or black fluffy powder. It is made of carbon atoms arranged in plane layers of aromatic rings, where electrons can move easily along the layers which lead to its electrically conductive behavior. Carbon black is used in polymeric products, paper, pigment, etc. for diversified applications. Its important applications include reinforcing filler for rubber products like tire, belt, automotive parts, etc. Moreover, the UV resistance property of carbon blacks introduced its widespread applications in plastics where UV resistance is a necessary criterion like in outdoor water storage tanks etc., and for printing inks for newspapers and magazines.

4.1 Carbon Black Manufacturing Processes

Various manufacturing processes are followed to produce carbon black. Most of the production process involves partial or controlled combustion of hydrocarbons such as oil or natural gas [85]:

- Channel black process
- Gas black process
- Thermal black process
- Acetylene black process
- Lamp black process
- Furnace black process.

4.2 Channel Black Process

The channel black process involves combustion of natural gases is burnt with a continuous small wide thin luminous fan-shaped flame [85]. Air in restricted amount for complete combustion is supplied to the natural gas, and the produced carbon black is deposited through impingement upon the flat undersurface of moving channels [86]. The deposited carbon black is scrapped off, collected, and stored in storage silos. The channel process produces carbon black with a particle size of around 10–30 nm. The channel process is not being utilized nowadays because of pollution hazards and economic issues.

4.3 *Gas Black Process*

The major principle of the process is thermal oxidative decomposition, where the black is formed in a diffusion flame. The particle size of carbon black manufactured in this process ranges between 10 and 30 nm. The properties of this black are very close to that of channel black and known as the reinforcing part of tires [85]. Here coal tar is the main feedstock, which is vaporized using a hydrogen-rich gas burner. The formed flame is generally put in contact with a water-cooled roller, where the carbon blacks get deposited. These blacks are then filtered as per its size and pelletized as per requirement.

4.4 *Thermal Black Process*

The process involves thermal combustion of natural gas into elemental carbon and hydrogen under the condition of elevated temperature. The process is a cyclic operation, where natural gas is fed to combust the material into carbon of high purity. The endothermic process of decomposition of natural gas led to cooling off the furnace inside of which the whole process took place, and the furnace id reheated by utilizing the hydrogen gas evolved during the process (Fig. 14) [85]. The particle size of the carbon black synthesized in thermal black process is generally higher than 180 nm.

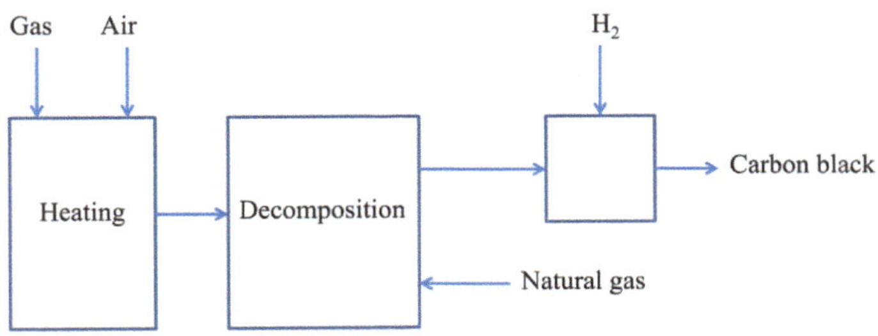

Fig. 14 Schematic representation of the thermal black process

4.5 Acetylene Black Process

Acetylene black is a commonly applicable additive in conductive rubbers and plastic materials and for dry cell as well. The production process involved continuous thermal decomposition of acetylene gas in the absence of oxygen gas. In the procedure of formation of this black, the reactor containing acetylene gas is heated to 600 °C by the combustion of the gas, followed by continuing acetylene gas combustion in absence of oxygen flow. The rapid combustion of acetylene gas led to formation of carbon and hydrogen gas, which is used further for heating. The cooling of the reactor is made by the water circulating cooling coils wrapped all around the reactor. In the process, the carbon black is formed in aerosols which are gradually extracted by removing the moisture [85].

4.6 Lamp Black Process

Lamp black is probably the oldest pigment known for use as black pigment in printing inks, shoe polishes, carbon paper, ceramics, and cements. The process involves uneven combustion of carbonaceous materials rich in carbon content like various resins, fatty oils and acids, coal tar, etc. the feedstock of the process is mostly burnt in presence of limited air supply, leading to deposition of carbon in the form of sot at considerable quantity. Lamp blacks are generally of large particle size (110–120 nm) [85] (Fig. 15).

4.7 Furnace Black Process

Carbon black production from furnace black is the current trend globally. This process utilizes heavy aromatic oil as feedstock. The process involves exposing the primary raw material to a hot gas stream by burning fuel like natural gas or oil (Fig. 16). The vaporized primary feed then undergoes pyrolysis and form carbon

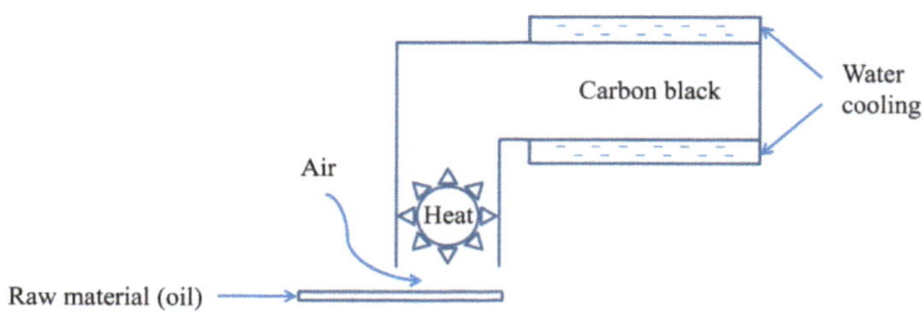

Fig. 15 Schematic representation of the lamp black process

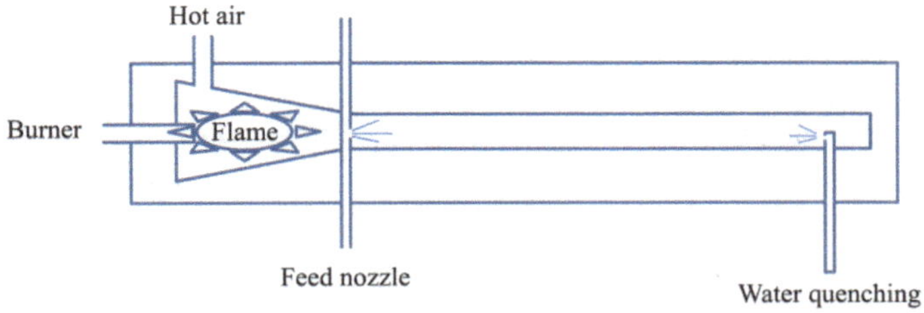

Fig. 16 Schematic representation of the furnace black process

black and gases [85]. This process yields carbon blacks with particle size ranging from 20 to 100 nm.

4.8 Miscellaneous Synthesis Methods of Carbon Black

4.8.1 Pyrolysis of Hydrocarbons

In the simplest way, carbon black may be obtained from the soot produced during thermal degradation of hydrocarbons. Mohan et al. [87] used kerosene oil, paraffin wax, lubricant oil, and diesel as precursors to prepare carbon black. The use of diesel as precursor yielded the product with the least disorder in crystallinity, whereas the use of lubricant oil yielded carbon black with the highest crystalline disorder. Nevertheless, carbon black obtained from diesel soot was found to have the smallest diameter (ca. 48–57 nm) as well as high purity (ca. 97.17% carbon content) (Fig. 17).

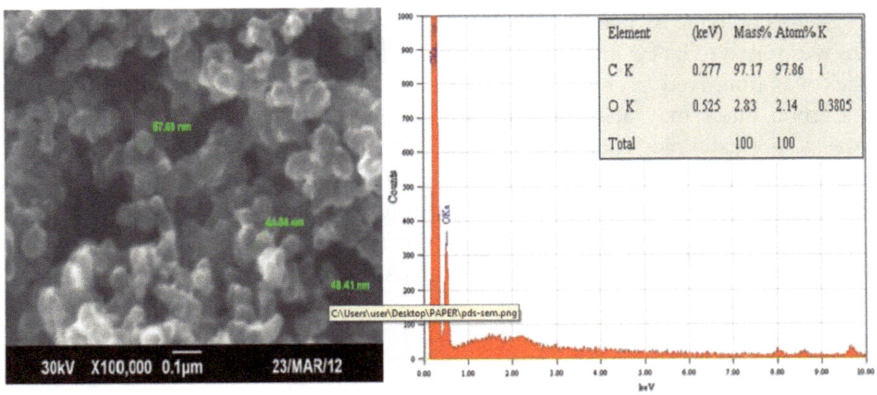

Fig. 17 SEM micrograph and electron dispersive spectroscopy (EDS) analysis of soot produced from the atmospheric combustion of diesel [87]

4.8.2 Pyrolysis of Polymer

Pyrolysis of polymers may also be carried out to produce carbon black. Aromatic fuels have higher tendency in comparison to aliphatic ones to produce carbon black at lower fuel to air ratio [88]. Ring-fracturing reaction leading to hydrocarbons is attached on to phenyl radicals, which lead to the formation of aromatic ring networks. Initial nucleation for carbon black synthesis takes place at these aromatic rings [89]. Vie et al. [90] synthesized carbon particles of diameter ranging between 50 and 300 nm by pyrolysis of resorcinol-formaldehyde copolymer. The pore size distribution was found to be ca. 38 nm (N_2 adsorption isotherm) with micropore volume of 0.4 $cm^3\ g^{-1}$ which was much higher than that of commercially available grade Vulcan XC 72 (0.06 $cm^3\ g^{-1}$).

4.8.3 Plasma Synthesis

Guo et al. [91] synthesized ultrafine carbon black by pyrolizing high density polyethylene (HDPE) and polystyrene (PS) by using direct current thermal plasma process. In that process, initially, HDPE and PS were catalytically or thermally pyrolized in a graphite chamber. For catalytic pyrolysis, catalyst pellets containing 30 wt% BaO/Silica were used. After pyrolysis, the produced hydrocarbon gases liquids and gases were passed through DC plasma jet of maximum power of 20 kW with torch efficiency 90%. Nitrogen was used in this process as carrier gas. The average diameter of carbon black was found to be as low as 56 nm. In this process, a maximum of 85% of recovery from PS was achieved. In this process, carbon black, manufactured from waste PS by pyrolizing at 900 °C, exhibited comparable surface area of ca. 142.5 $m^2\ g^{-1}$ (BET isotherm) in contrast to that obtained from commercially available PS (155.9 $m^2\ g^{-1}$).

Another preparation method for carbon black involving AC plasma arc was reported by Yuan et al. [92]. Carbon black nanoparticles were produced by cracking of gaseous propane by using plasma arc with power of 15 kW. The average particle size was found to be 80 nm. However, the BET surface area of the produced carbon black was improved from 96.7 to 113.2 $m^2\ g^{-1}$ by treating with nitric acid.

Carbon black may also be synthesized from solvents by liquid phase plasma process. For example, Thai et al. [93] exposed benzene precursor to AC plasma of 250 V at 30 kHz for 30 μs pulse time. BET surface area of the synthesized carbon black (163 $m^2\ g^{-1}$) was higher in comparison to that of commercial carbon black (127 $m^2\ g^{-1}$). Subsequent activation of synthesized carbon black by treating with KOH followed by heat treatment in inert atmosphere resulted in BET surface area to significantly high value of 3562 $m^2\ g^{-1}$.

4.8.4 Hydrolysis of Natural Resources

Carbon black may also be produced by thermal treatment of natural resources. Wang et al. [94] produced carbon black from rice husk by hydrolysis, carbonization, and pyrolysis processes. As per the claim, thus prepared carbon black contained the high surface area and pore volume and also exhibit graphitic structure.

4.9 Synthesis of Activated Carbon

Activated or conductive carbon also known as graphitized carbons possessing high specific surface area is prepared by graphitization, followed by ozone treatment and then chemical activation process. This process includes two activation steps, where the first step produces higher specific surface area, with poor structural order than the process involving graphitization followed by second step of activation. The step of ozone treatment leads to attachment of large number of C–O functional groups onto the surface of carbon black. This treatment facilitates activation [95]. High-performance carbon black with high surface area (2700 $m^2\ g^{-1}$) and more pore volume (1.98 $cm^3\ g^{-1}$) carbon blacks may also be prepared by method facile method of using phosphoric acid [96]. Activated carbons can also be formed via combustion of olive stones in an inert atmosphere followed by activation with carbon dioxide or zinc chloride, which lead to the formation of high surface area. Such active carbons have very low ash content and complete absence of sulfur [97]. Activated carbons with high surface area can also be prepared by KOH treated coconut shell. Greater degree of KOH treatment of coconut shell at elevated temperature results formation of highly activated carbons [98]. Similarly, macadamia nutshell was also investigated by researchers to produce activated carbon by carbonization, oxidization at stepped heating followed by activation in inert atmosphere at elevated temperature [99]. However, physical activation process may also be followed to produce activated carbon black by using steam [100], CO_2 [101], etc., in conventional heating as well as microwave heating [102].

Apart from the abovementioned types, a specific terminology is used as white carbon black (WCB) which, however, is basically composed of $SiO_2{\cdot}nH_2O$. It has properties similar to carbon black, and may be used as its substitute in different fields of application like coating, medicine, paper making etc. [103].

5 Synthesis of Carbon Fiber

5.1 Introduction

Carbon fibers have several micrometer-sized diameters. Filamentous carbon fibers have aspect ratio higher than 100. Bamboo-char filaments are the first ever known carbon fiber. They were used in the incandescent light bulb in 1880 by the famous scientist Thomas Alva Edison. As time progressed, carbon fibers were replaced by a tough element tungsten in light bulb applications. In the mid-period of the nineteenth century around 1950, necessity came from aerospace field to develop high-performance lightweight material. The importance of carbon fibers was felt once again. Now it was introduced as a reinforcing fiber in conjunction with plastics, metals, ceramics, and bulk carbons. High strength of these composite materials makes it one of the most demanding engineering materials [104–106].

PAN-based carbon fiber is stronger as compared to fibers obtained from other sources because of its greater carbon yield (>50% of the original precursor mass) [107]. Carbon fiber synthesis is mostly done in liquid phase and at temperature of 4100 K and pressure of 123 Kilo bar. The major precursor of carbon fiber is organic and is completed its preparation in three steps: stabilization of precursor at 300 °C, carbonization of the stabilized material at 1100 °C, followed by graphitization at above 2500 °C. The fibers up to the second step are called as carbon fiber and up to the third step are called as graphite fiber [105, 108].

Activated carbon fibers (ACF) offer several advantages compared to granular activated carbon (GAC) and powdered activated carbon (PAC) [109].

5.2 Synthesis Method

High-performance carbon fibers are derived from mesophase pitch or polyacrylonitrile (PAN). The higher strength of PAN-based fibers is due to the presence of turbostratic crystallites that restricts development of any flaw due to its complex interlinked structure [110]. Carbon fibers study has started in the late 1950s and the industrial production started in 1963. Nowadays, they have gained technological and commercial importance in the fields of aerospace, construction, sports, electronic device, and automobile industries [106, 111, 112]. Carbon-fiber-reinforced plastics (CFRPs) are considered as an alternative to heavy, iron-based parts in the automotive industry [113]. Use of carbon fiber based cement composites is attaining a lot of interest due to its capability to sense structural and electrical actuation and are able to sense both the compressive as well as tensile stress [114]. For producing carbon fibers, the carbon atoms are bonded together in crystals in such a manner that they get aligned more or less parallel to the long axis of the fiber. This aspect is specially taken care of because the crystal alignment in parallel direction gives the fiber high strength-to-volume ratio. Bundling large number of carbon fibers make then useful as woven fabric [110].

The most common method for synthesizing carbon filament or fiber is from polymer such as polyacrylonitrile (PAN), rayon, or petroleum pitch, known as a precursor. While general purpose carbon fibers are produced by melt blowing of isotrope pitch, but the high-performance carbon fibers are developed by melt spinning of mesophase pitch [115]. As PAN or Rayon are synthetic polymers, so an additional step is incorporated here, that is aligning the polymer molecules through spinning the polymers into filaments, and the process may vary based on the manufacturer. This is followed by heating the filaments in order to remove all the non-carbon elements, which may further be treated differently to improvise the properties of final carbon fiber. The entire process can be categorized into five steps, viz., spinning, stabilizing, carbonizing, treating the surface, and sizing [104, 110, 116, 117]. Each step is described below.

5.2.1 Spinning

PAN monomer that is acrylonitrile is taken in powder form and is mixed with another plastic, like methyl acrylate or methyl methacrylate, and the reaction is carried out under catalytic conditions. The catalysts used are itaconic acid, sulfur dioxide acid, sulfuric acid. The reaction is carried out using conventional suspension or solution polymerization process.

The PAN obtained is spun into fibers by following different methods. In a method, the reaction mixture is mixed with certain chemicals, drawn through a tiny orifice into a batch of chemicals and then quenched with water to obtain fibers. This method is the same as the one used for obtaining polyacrylic textile fibers.

In every case, the fibers are further cleaned with water, and stretched to obtain the required diameter of the fiber. It is also important to mention that the stretching of the fibers led to alignment of the molecules.

5.2.2 Stabilizing

The stabilization is carried out to convert the linear atomic bonding of the carbon fiber into thermally stable structure. Heating the fibers in the presence of oxygen to about 390–590° F (200–300 °C) for 30–120 min leads to breakage of several bonds thereby helping the fibers to catch up larger amount of oxygen molecules and cause rearrangement of atoms in molecular level. The chemical reaction involving in the stabilization is endothermic and needs proper precaution to keep the fibers unaffected by the heat. Commercially, the stabilization process uses a variety of equipment and techniques. Some methods make use of series of heated chambers for drawing out fibers. The other alternative is passing the fibers over hot rollers and through beds of loose materials held in suspension by a flow of hot air. Another option is using heated air mixed with certain gases that chemically accelerate the stabilization.

5.2.3 Carbonization

Carbonization is followed by the stabilization process of the fibers, heating under anaerobic condition at 1000–3000 °C for specified period of time. The inert gas pressure is maintained higher than the atmospheric pressure, and the absence of oxygen resists the fibers to get burnt. Under such condition, the carbon-to-carbon chains bond side-to-side leading formation of graphene sheets that subsequently develop a single columnar filament. Hence, PAN produces high-quality fiber, while pitch and rayon produces poor quality carbon fibers. Heating of fibers at high-temperature results in the expulsion of non-carbonaceous elements and also small amount of carbon atoms results in the form of some by-products like carbon dioxide, carbon monoxide, etc. Finally, the remaining carbon atoms are tightly packed with each other and get aligned along the *z*-axis of the fibers.

5.2.4 Treating the Surface

After carbonizing, the fibers surface properties are not activated, and the result is that they show poor bonding properties when bonded with resins such as epoxies for composite applications. The drawback is removed by minor oxidation of the surface. The oxidation leads to etching of surface, resulting in roughness in the surface, which can enhance the mechanical interlocking as well as chemical bonding. The various options for oxidizing the carbon fibers are: Gaseous phase chemicals such as air, carbon dioxide, or ozone; Liquid phase chemicals used are sodium hypochlorite or nitric acid. In both cases, the fibers are immersed in them for oxidizing. Ammonium sulfate is a commonly used commercially viable electrolyte to cause surface treatment of these materials. The fibers can be made conductive by electrolytically coating it in the electrolyte of conductive materials. The surface treatment process must be carefully controlled to avoid forming tiny surface defects.

5.2.5 Sizing

Followed by the surface treatment, is the sizing process where the fibers are coated to protect them from damage during winding or weaving. Before sizing the carbon fibers are pre-dried either in presence of air or rollers. In all cases, the coating of fibers is done with the help of compatibles like epoxy, urethane, etc., that may create bonding with the base matrix. The final step is winding of the fibers onto bobbins that can finalize the final size of fibers. The process explained above is depicted below by means of a flow sheet as shown in Fig. 18.

6 Synthesis of Carbon Nanofibers

6.1 Introduction

Carbon nanofibers with diameters in nanometric range emerge as a most prestigious material due to its excellent reinforcing ability, superior chemical and electrical

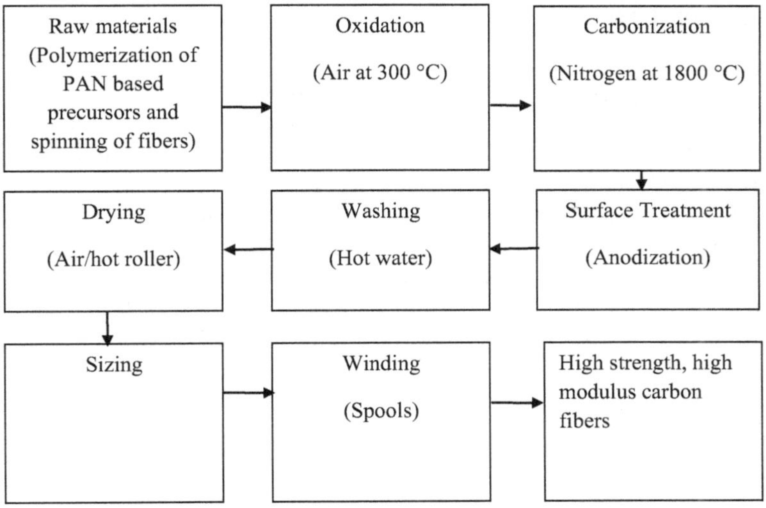

Fig. 18 Flow sheet for the manufacturing process of carbon fibers from PAN-based precursors [104]

property [118] and high surface reactivity [119]. Catalytic conversion of carbon-containing gases is one of the old techniques that produce such as fibers having diameters range 3–100 nm and length range 0.1–1000 μm [120]. However, it was not recognized until a recent outburst report of its structural similarity to fullerene and carbon nanotubes. Carbon nanofibers are also known as graphite nanofibers or carbon filaments [121]. Such materials can be developed by catalytic decomposition of various hydrocarbons over traditional metal particles such as iron, cobalt, nickel, etc. Alloys of various transition elements also gain immense interest for synthesizing CNF.

Various scientists describe numerous procedures to synthesis CNF. However, the main challenge relies on controlling length to diameter ratio, cross-sectional diameter and staple length of the fiber. It is noteworthy to mention that difficulty can be experienced to achieve narrow fiber diameter distribution during synthesis. In this chapter, an approach has been made to brief the various synthetic procedure for synthesizing CNF having various architecture and property.

6.2 *Synthesis Methods*

According to synthetic procedure, carbon nanofiber can be mainly classified into two categories: (i) Vapor grown [122, 123] and (ii) catalytically grown nanofiber. However, the synthetic procedure varies with the required morphology, structural composition, and the end use of such nanofibers.

6.3 Vapor Grown Nanofiber

Kong et al. reported a well-defined strategy to synthesize high-quality carbon nanofiber by chemical vapor deposition of gaseous methane over a specifically patterned surface [122]. Here, electron beam lithography was used to develop regularly spaced catalytic islands of $Fe(NO_3)_3 \times 9H_2O$, $MoO_2(acac)_2$ and alumina nanoparticles on the silicon surface. The surface was further heated until 1000 °C under inert environment of argon gas. Then, the argon flow was replaced by methane. Fan et al. used ferrocene as catalyst precursor to develop CNF by similar technique [124]. Here, the horizontal tubular reactor was used as the reaction vessel. Such vapor grown carbon nanofibers (VGCNF) are of highly crystalline nature [123].

6.4 Catalytically Grown Nanofibers

Synthesis of catalytically grown nanofibers includes two steps: (i) synthesis of catalysts and (ii) synthesis of nanofibers.

Synthesis of catalysts: Mainly Fe–Cu and Ni–Cu bimetallic catalysts are in use to synthesize carbon nanofibers. Marella and Tomaselli explain four different procedures to synthesize such catalyst systems [121]: (A) coprecipitation, (B) reduction–precipitation, (C) reverse microemulsion, and (D) high wetness of the material in the silica. Hence, the synthesized catalysts by the above process are exemplified in Table 2.

The flowchart for each process for catalytic synthesis is described in Fig. 19. Such bimetallic catalyst systems have a high effect on the overall morphology in the nanofibers. However, the best result can be achieved by the incipient wetness method.

Table 2 Synthetic procedures of various catalysts used for carbon nanofiber synthesis

S. No.	Catalyst	Procedure	Method	References
1	Fe–Cu (70:30 at.%)	A	Coprecipitation	[121]
2	Fe–Cu (70:30 at.%) doped with K	B		
3	Ni–Cu (70:30 at.%) doped with K	C		
4	Ni–Cu (70:30 at.%)	D	Reduction–precipitation	[125]
5	Fe–Cu (80.7:19.3 at.%)	E	Reverse microemulsion	[126]
6	Fe–Cu (70:30 at.%)	F	high wetness of the material in the silica	[121]
7	Ni–Cu (70:30 at.%)			

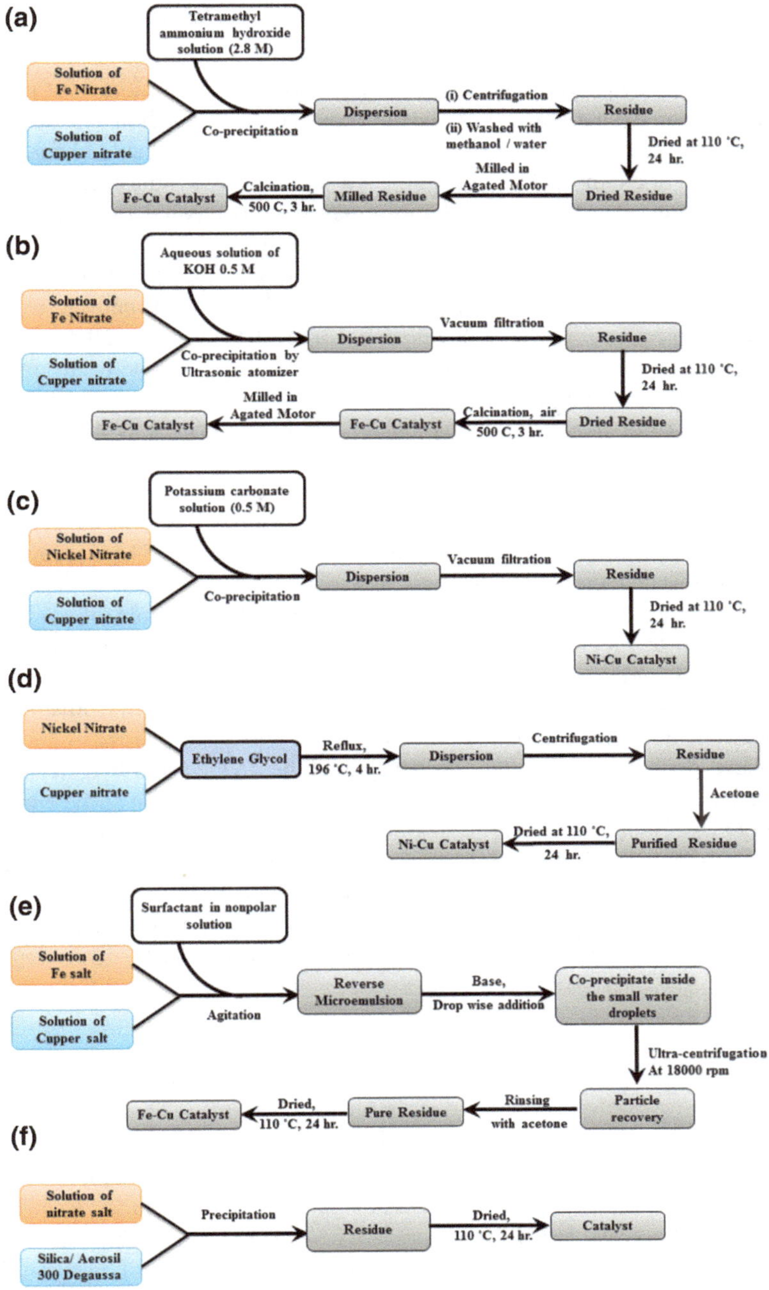

Fig. 19 Synthesis of catalyst used in developing carbon nanofiber

6.5 Procedure of Synthesizing Carbon Nanofibers

There are many synthetic routes for synthesizing carbon nanofibers out of which chemical vapor deposition (CVD) is the most common methods in developing CNF. The choice of route varies widely depending on its probable characteristics and end use. For example, Marella et al. have successfully synthesized the CNF for hydrogen storage data [121]. In addition, they have compared multiple synthetic routes and measured their hydrogen storage property.

Here, many catalyst systems (produced by [A] to [D]) produce different types of CNFs. Baker et al. introduced the growth mechanism of CNF and the review of Jong and Geus describes the synthetic procedure of CNF using Nickel catalysts [120, 127]. Generally, it is known that metal surfaces are capable of decomposing hydrogen molecules and produce carbon, which diffuses into the bulk of the metal and precipitate out in the form of graphite. However, later it is discovered that intermediate metal carbide has formed which then undergoes decomposition into graphite and metal. Magnetic measurements also evidence the presence of Ni_3C as an intermediate during the steady-state growth of carbon fiber on metal interface [120]. The carbon nanofibers were deposited from methane at 843 K. However, at first, a steady-state increase in C/Ni ratio has been observed but no fiber is produced with a significant drop in a magnetic moment. But after 12 min of exposure of nickel surface to methane, a sudden increase in the magnetic moment has been experienced. It indicates that ferromagnetic nickel is first converted into nonferromagnetic nickel carbide during the first 12 min of exposure but ferromagnetism is restored after this with the formation of CNF.

In view of such evidence, the mechanism can be divided into two steps: (i) the gaseous hydrocarbon methane undergoes decomposition over the surface of nickel and produces carbon and hydrogen atoms; (ii) Hydrogen molecule desorbs and carbon diffuses through the metal face forming a sub-stoichiometric nickel carbide and (iii) the metastable nickel carbide decomposes into metallic nickel and carbon fiber. It was also proven that Ni(110) and Ni(100) surfaces of the crystal lattice are more active during this process rather than Ni(111) surface [128, 129].

6.6 Electrospinning

Electrostatic spinning or electrospinning is established as a process which is used to fabricate nonwoven and ultrafine nanoscale fibers possessing diameters ranging from 3 nm to several microns. The technology was first coined by Reneker and coworkers in the 1990s [130]. High popularity of such process is beneath under its simplicity, affordability, flexibility in selection of wide range of materials, high surface-to-volume ratio, and tunable porosity. A standard design of electrospinning apparatus includes four major components: (i) spinneret (or needle), (ii) high voltage power supply, (iii) a glass syringe and (iv) metal collector. The staple length

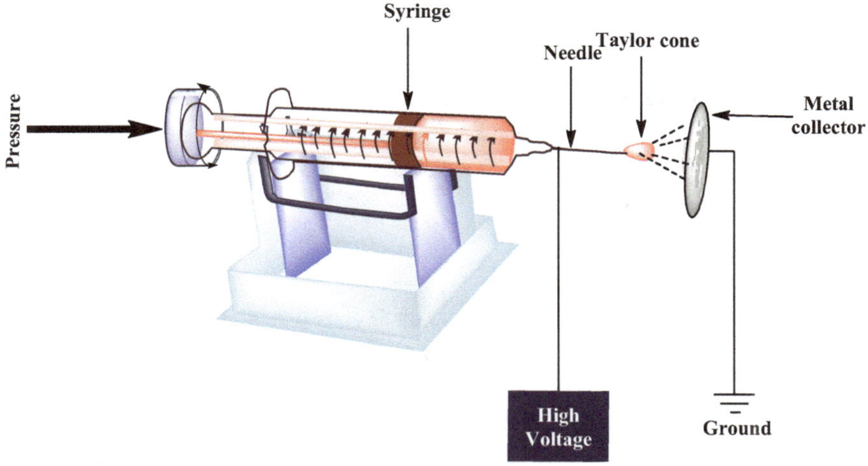

Fig. 20 Schematic diagram of electrospinning process

of Nanofibers depends on the electrical potential applied to the polymer solution and flow rate. Schematic diagram for electrospinning is shown in Fig. 20.

After loading the polymer solution in the syringe, it is first charged by applying voltage. After that, a small electrically charged droplet is formed. A repulsive force that is generated due to repulsion between similarly charged liquid surface and electric field results in the deformation of the droplet into conical shape (Taylor cone). The cone becomes unstable after a threshold charge and allows the exudation of liquid jet. The jet then travels through air towards the ground and finally it is collected on metal cylinder (used as collector) [130]. Required parameters and final characteristics of the electrospun CNFs from various materials line polyacrylonitrile, lignin, polyacrylic acid, polylactic acid, etc., are well described in the review of Zhang et al. [131]. The diameter of electrospun fiber has a proportional relationship with the feed rate of polymer solution.

6.7 Template Method

Template method is widely used to produce CNF. It is basically used to produce one-dimensional carbon material. Anodic aluminum oxide, mesoporous silica, and Te nanowires are generally used in this technique. The synthesis of fibers and submicron tubes of carbon is reported via the pyrolytic decomposition of propylene and produces carbon over the porous wall of aluminum oxide film. Here, removal of the template has been carried out through hydrofluoric acid washing. Another advancement of such process is used by treating precursor mixture of HCl, formaldehyde, F127, and phloroglucinol on the porous alumina substrate and subsequently the carbonization was carried out at a temperature of 500 °C. A self-template strategy is also reported where ethylene glycol (EG) is used as the carbon precursor and $Zn(CH_3COO)_2$ as the structural constructor as well as the porogen [132]. Pt-catalyzed pyrolysis of cellulose diacetate (CDA) nanofibrils has also been attempted using the same technique as well [133].

6.8 Carbon Film Casting

Such process develops carboneous material having two-dimensional architecture. Such process is generally used to develop carbon nanotubes and more specifically single-walled carbon nanotube. It involves four steps:

(i) film casting through using block type copolymers,
(ii) refining the structure through solvent annealing,
(iii) carrying out polymerization of the carbonaceous precursor
(iv) final carbonization

Recently, Feng et al. reported a preparation method of free-standing multi-wall carbon-based ordered thin films through conventional coating-etching method. Here, Resol/pluoronics are used as precursor which is carbonized at 600 °C [134].

7 Carbon Nanotube Synthesis

Various techniques have been employed for the successful production of carbon nanotubes (CNT) such as laser ablation, arc discharge, chemical vapor deposition, high-pressure carbon monoxide disproportionation (catalyzed using iron oxide/mixture of cobalt and molybdenum), electrolysis, flame synthesis, pyrolysis [135–141], etc. We can further group these processes into physical, chemical and miscellaneous processes based on the principles of carbon conversion into nanotubes as shown in Fig. 21.

To study different methods of carbon nanotube synthesis, the growth mechanism is of great importance. A three-step process generally distinguishes between extrusion/base/root growth and tip growth. It can be observed in Fig. 22, a precursor is formed on the surface of the metal catalyst which is generally pear or round in shape. In the second step, diffusion of carbon on the sides occurs which thereby leaves the top of the precursor free enabling in the hollow core structure of the nanotube resulting in the formation of rod-like structure of carbon. The base growth method enables the nanotube to grow in the upward direction from the metal particle which is attached with the substrate. In tip growth mechanism, in spite of detachment of particle from the substrate, it still manages to stay on top of building nanotube. Single-walled (single tube of graphene, commonly known as

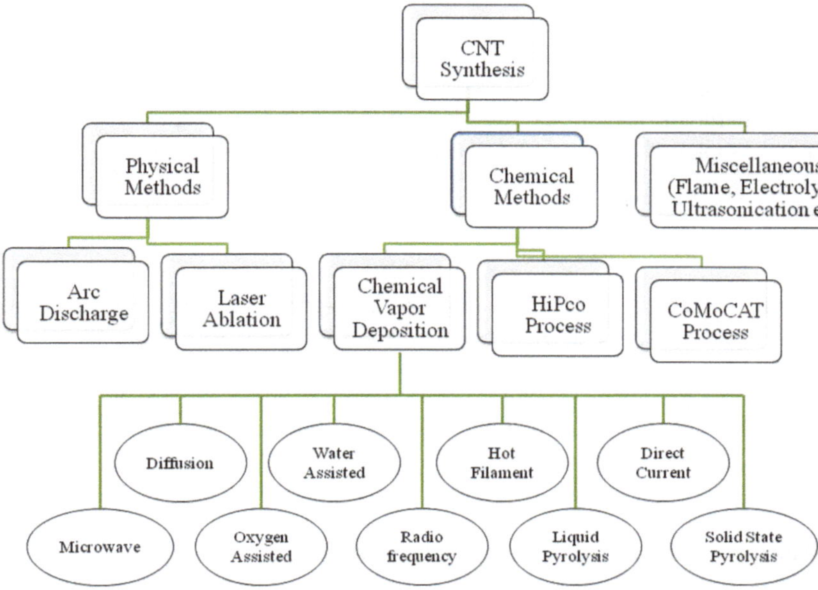

Fig. 21 Classification of carbon nanotubes synthesis methods

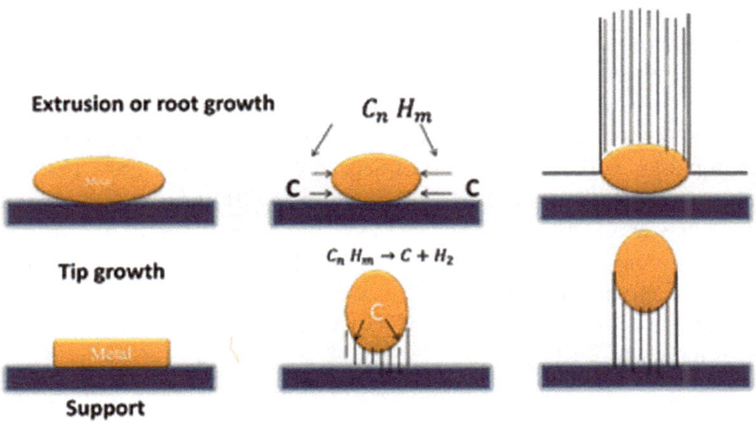

Fig. 22 General growth mechanism of carbon nanotubes [142]

SWNT) or multi-walled (several concentric tubes of graphene adjusted inside one another, commonly known as MWNT) nanotubes grow depending on the particulate size of catalyst employed for synthesis [142].

7.1 Physical Methods

As shown in Fig. 21, the physical techniques of CNT synthesis are laser ablation and arc discharge methods. They are high-temperature processes quite similar in the working principle, advantages, and disadvantages.

7.1.1 Laser Ablation

Laser ablation method is a high-quality and high-purity method for synthesis of SWNTs. A pulsated laser ablates the carbon pellet (graphite) containing catalyst materials (Cobalt or nickel) kept inside a furnace of tubular shape at temperatures ~ 1200 °C. The process is carried out in presence of inert gases such as helium/argon thereby vaporizing the graphite target. The carbon nanotubes basically grow over the relatively chiller surfaces in the reactor by condensation of vaporized carbon. However, there are instances where water-cooled surfaces are employed to collect the nanotubes. The by-products such as fullerene and amorphous carbon overcoating on the reactor sidewalls can be collected.

However, the properties of CNTs prepared using this process depend upon several parameters like laser properties (peak power, repetition rate, energy fluence, cw versus pulse, wavelength of oscillation), the chemical and structural composition of the target pellet used, the reactor pressure as well as the chemical composition of its atmosphere, pressure and flow of gas used as buffer and distance between the target and substrate.

In one of the very first experiments, high yields (>70%) of CNTs by the above process were achieved. In a two-step method, laser vaporization as the initial step was followed by the second laser pulse to increase the rate of target vaporization. It enables to minimize the amount of soot deposition. The CNTs grow in the form of mat of ropes with a diameter of 10–20 nm and up to 100 micron or more in length.

While Lebel et al. have prepared CNTs via utilization of ultraviolet laser [143] in another study done by Kusaba and Tsunawaki [144], XeCl Excimer laser with a wavelength of 308 nm was used and it was observed that the process carried out at 1350 °C produced highest yield of CNTs with a diameter of 1.2–1.7 nm with minimum length of 2 μm.

7.1.2 Arc Discharge

It is one among the oldest technique known for finite structured carbon production possessing needle-like tubes. The process is very much like a synthesis of fullerene, in which a graphite target is placed in an enclosed space such as a high-temperature reactor with inert gas atmosphere such as helium or argon. The pressures maintained are moderate (50–700 mbar). Two carbon rods placed vertically over each other are installed in the center of the reactor at different potentials. The positively charged anode is moved towards a negatively charged cathode until the formation of an arc [145]. A total process time of approximately 60 s with a distance of 1 mm between the electrodes results in the formation of nanotubes which can be collected along with its by-products after depressurization and cooling of the reactor.

Bethune et al. [146] reported the formation of SWNT with diameters as small as 1.2 nm by co-evaporation of carbon atoms and cobalt metal atoms by the arc discharge process. The usage of Helium atmosphere for SWNT synthesis with cocatalyst is also reported by Ajayan et al. [147] resulting in SWNTs within the size range of 1–2 nm diameter. The role of catalyst in the formation of carbon nanoclusters was studied by Seraphin et al. [148], where Ni, Pd, and Pt catalysts were used in presence of DC arc discharge, operated at 28 V and 70 A under He atmosphere at the pressure of 550 Torr. In the observations, it was reported that nickel-filled anode spurred on the growth of SWNTs. Saito et al. [149] also reported radial growth of SWNTs from fine Ni particles. Zhou et al. [150] used yttrium carbide filled anode and reported radial growth of SWNTs. The synthesis of SWNTs by the arc discharge method has also been reported with platinum group metals [151] such as Ru, Rh, Pd, Os, Ir, Pt, and also with Fe, Co, Ni, F/Ni, La, and Ce catalysts [152].

In one of the modern day arc discharge approach reported by Chen et al. [153] known as Ferrum Hydrogen (FH) arc discharge method, Fe (1% wt) filled anode is used in Hydrogen–Argon mixture gas atmosphere. The SWNTs synthesized are highly crystalline and are subjected to oxidation purification process with H_2O_2 for the removal of Fe catalyst nanoparticles which coexist with SWNTs. A purity of up to 90% has been attained using the modifications in the traditional arc discharge process [154]. The methods are still costly due to the raw materials involved. An effort of looking for a cheaper raw material of synthesis of SWNTs was taken up by Fan et al. [155]. They were able to achieve success in the synthesis of the same with argon gas environment and DC arc discharge of various charcoal materials which is a rich source of carbon using catalyst FeS (20 wt%). The analysis reported suggests that the high purity SWNTs synthesized were of diameter of 1.2 nm.

The problem with CNT remains with the presence of impurities which have an effect on the properties of the nanotubes. Ando et al. [156] reported that CNTs are synthesized by arc discharge of carbon-rich graphite in hydrogen gas that may be carried out by infrared irradiation at 500 °C for 30 min in air. It ensures the removal of coexisting carbon nanoparticles.

The study of calcination from 300 to 600 °C on CNTs synthesized by DC arc discharge was done by Pillai et al. [157]. A lot of research has been targeting to solve the long-lasting issue or CNT purification and several experiments, reviews, and research have been conducted over the same [158–160].

7.2 *Chemical Methods*

The chemical methods of synthesis include chemical vapor deposition (CVD), high-pressure carbon monoxide reaction (HiPco), and CoMoCAT Process.

7.2.1 Chemical Vapor Deposition

Chemical vapor deposition process is an effective process for control of growth directions on the substrate and efficient due to the quantity of CNT production which is large. A mixture of hydrocarbon gas and process gas (ethylene/methane/acetylene and ammonia/nitrogen/hydrogen) react in a chamber at a metal substrate heated at a temperature of around 700–900 °C at atm pressure. The hydrocarbon gas decomposes to result in the formation of CNTs growing on metal (catalyst) substrate. The substrate preparation can be done by evaporation of catalyst by electron beam, physical sputtering, and solution deposition. The growth of self-oriented nanotubes on large surfaces can be ideally done with the porous silicon substrate.

Chemical vapor deposition techniques are majorly classified into plasma [161] and thermal [162] process wherein the assistance modes for the CNT synthesis can be hot-filament, [163, 164] water assisted [165, 166], and oxygen assisted [167–170]. Catalytic chemical vapor deposition enhanced by either plasma or thermal process is economically feasible for large-scale synthesis of quite pure CNTs in comparison to physical methods such as laser ablation. Some of the few advantages offered by CVD have better control over the progressive course of the reaction and the purity of the product is high [171].

The role of catalyst in the decomposition of carbon source either by plasma or heat is vastly studied [161]. In plasma-assisted CVD, plasma assists in the generation of vertically grown carbon nanotubes with precise reactor geometry control. The synthesis of specified double-walled carbon nanotubes using CVD technique has been recently reported [172–174].

Low-temperature synthesis of CNTs on different metallic substrates such as NiV, Ir, Ag, Pt, W, and Ta and Fe/Al dual layer as catalyst using diffusion plasma enhanced chemical vapor deposition is reported by Kim and Gandloff [175]. Wang and Moore [176] experimented on preparation of CNTs with vertical alignment using glass substrate sputtered with FeNi or Fe by RF-PECVD or DC-PECVD. The difference between RF and DC-PECVD is the reactive radical concentration which is higher in the RF-PECVD. On the contrary, the vertical structural growth of CNTs was very well aligned. Better wettability and higher reactivity were observed with FeNi thin film catalyst in comparison with Fe thin film catalysts.

The effect of the morphology and composition of catalyst nanoparticles on the growth of CNTs is reported by Flahaut et al. [177], in which they investigated the effect of preparation conditions for the catalyst for CCVD synthesis of CNTs. The catalysts were prepared by combustion method fuelled by citric acid/urea. They found that mild conditions observed in citric acid fuelled process can either restrict the formation or increase the selectivity towards synthesis of CNTs with fewer walls depending upon the purity and composition of the catalyst. A mixture of catalysts used for PECVD synthesis is reported by Xiang et al. [178] in which acetylene was used over a variety of catalysts. The yield of synthesized CNTs increased with an increase in the Co content due to better dispersion characteristics of active Co moiety in increasing numbers. CNTs with smaller diameters and more structural order were also observed with an increase in the Co content.

As a catalyst, the use of spherical Ni nanoparticles film $Ni(NO_3)_2$ is reported by Luais et al. [179] for the PECVD synthesis of CVD. Mixture of gaseous C_2H_2/NH_3 along with temperatures of 520 °C was employed resulting in average thickness of 1 mm and diameter of around 50 nm. The synthesis was followed by surface treatments were carried out using microwave plasma for potential application in biosensors. The use of ferritin nanoparticles (Fe_2O_3 compound core enclosed by protein shell) as a catalyst for oxygen-assisted PCVD is also reported. Upon reaction with oxygen, amino acids present in proteins around the ferritin core were removed and separated from each other. It subsequently results in dense vertically aligned CNT structures growing well within the iron cores separated [180].

High-efficiency water-assisted CVD synthesis was demonstrated by Yamada et al. [181] using Fe catalyst. The achieved product was vertically aligned CNTs with heights of up to 2.2 mm with a carbon purity of 99.95%.

Metal-free CNTs on glass substrates have been successfully demonstrated by Seo et al. [182] with a microwave PECVD (MPECVD) process using CH_4 and H_2. Radiofrequency magnetron sputtering was used for carbon layer deposition acting as catalyst. The layer was pretreated with hydrogen plasma at 600 °C for 180 s. The CNTs grew for around 30 min at temperatures of 600 °C. Other CVD methodologies, i.e., liquid pyrolysis [183–185], and solid-state pyrolysis [186, 187] are less frequently used for the synthesis of CNTs.

7.2.2 High-Pressure Carbon Monoxide Reaction (HiPco®)

The gas phase continuous flow production of CNTs involves passing a mixture of source gas (carbon) and catalyst precursor (organometallic) molecules through a heated furnace. The catalyst disintegrates and reacts, forming congregates on which CNTs grow by nucleation. Some early reports [188, 189] have been for production of smaller quantities of SWNTs in addition to multi-wall carbon nanotubes.

A high-pressure reactor employed for the process will be a thin-walled quartz tube reactor with electrical heating elements on the surface further enclosed by thick-walled aluminum housing. A continuous flow of CO and $Fe(CO)_5$ through the quartz tube is used to carry out the process. The spraying of preheated CO through the showerhead is carried out. The showerhead CO preheater composes of thick-walled graphite tube which is resistively heated as rapid heating of gas mixture will enhance the synthesis of CNTs. As CO is passed through six channels longitudinally bored in the graphite tube having a small diameter graphite tube, it gets heated due to heat exchange with hot graphite rods [190, 191]. This method is suited for large-scale synthesis of CNTs, as the process is continuous and the product is free from catalyst supports. Also, due to continuous mode, it allows the reuse of carbon monoxide [192].

Smalley et al. [193] demonstrated the synthesis of CNTs by using continues flow gas phase carbon monoxide as a feedstock along with iron pentacarbonyl $Fe(CO)_5$

catalyst via HiPco method. The CNTs obtained were very thin, high-quality product with high intrinsic selectivity. In another work [194], the yield of the process was found to be as high as 97% for the synthesis and 90% for the purification method. HiPco process can also be modified in which a mixture of ferrocene ($Fe(C_5H_5)_2$) and benzene is reacted in continues flow of hydrogen gas to form CNT. The nanosize catalyst is however formed in the same manner, i.e., by the thermal disintegration of organomettalic compounds such as iron pentacarbonyl in the former and ferrocene in the latter.

7.2.3 CoMoCAT® Process

An alternative gas phase continues flow process used for the synthesis of CNTs using a unique mixture of Cobalt and molybdenum is known as CoMoCAT process [195]. The growth of CNTs is facilitated by disproportionation of pure CO into carbon and carbon dioxide at 700–950 °C and 1–10 atm pressure in the presence of Co and Mo catalysts. The growth rate with respect to the catalyst is reported to be about 0.25 gm CNT/gm of catalyst in 1–2 h. The combined effect of Co and Mo catalyst is only effective with low Co:Mo ratio together at the silica substrate. If they are separate, they are either inactive (only Mo) or unselective (only Co). The selectivity of the mixture of catalyst towards CNT synthesis by this process majorly depends upon the stabilization of Co^{2+} species by Mo oxide species [196]. CNTs obtained by CoMoCAT process have high selectivity, high quality, and a narrower distribution of diameters as the process yields much lesser number of bands.

CoMoCAT process can be scaled up as it operates at less severe parameters and atmosphere than other processes, and provides control in the structural characteristics of CNTs. Due to the simplicity and its similarity with other standard catalytic processes employed in the chemical industry, the CoMoCAT process has the potential for the commercial industrial production of CNTs at low cost.

7.3 *Miscellaneous Methods*

7.3.1 Flame Method

The flame method of CNT synthesis is in a restricted flame environment, using inexpensive hydrocarbons as fuel forming tiny aerosols of metal catalyst islands. The metal catalyst island formation can be attributed to any of the three routes. The first route consists of metal catalyst (Co) that is coated on a mesh over which the islands were formed through arc discharge and laser ablation methods [197]. As soon as they are exposed to flame, they become aerosol. The second method deals with creating aerosol tiny metal particles via burning a filter paper dipped through a metal ion (e.g., Fe Nitrate) solution. The third route deals with a thermal evaporation process in which a metal powder (Fe/Ni) is put in a trough and heated [198].

The fuel gas is burned partially to achieve temperatures of ~800 °C for CNT production.

Liu et al. demonstrated V-type flame method based on simple equipment and lesser complex experimental parameters. The process employed use of carbon source as CO and the resultant product yield was high with lesser impurities [199]. The size of CNT diameter was approximately in the range of 10–20 nm with the length going up to several microns [200].

7.3.2 Electrolysis

The use of electrolysis for to deposit atomic carbon was first demonstrated using methanol as well as carbon dioxide as carbon source [201]. In the electrochemical method, alkali metals were deposited on cathode made up of graphite from a molten salt system kept at high temperature. The intercalation of metallic atoms into the vacant space between the graphite sheets, thereby allowing it to diffuse through the bulk of cathode, results in formation of certain mechanical stresses inside graphite. This induced stress results in the ablation of individual graphitic sheets, eventually turning into CNTs because of interfacial forces in the effort to recombining of disintegrated C–C bonds. This method was demonstrated at the University of Miskolc by Kaptay and Sytchev [141]. This technique produces CNTs of high quality, but the method is not scalable to commercial large-scale production. Shawky et al. [202] reported the synthesis of CNTs by application of minor negative potential acetic acid solution over a Ni nanocatalyst supported by silver surface. The catalysts (Ni nanoparticles) were grown via electrochemical route on silver with a control over the deposition time for particle size. The effect of catalyst structure on CNT diameter distribution was studied using Scanning Electron Microscopy and Raman spectroscopy [203]. The growth of CNT is imitated by the transition metal catalysts, such as Fe/Ni nanomaterials. SEM and TEM characterization expressed the hollow structured CNT diameter in the range of 100 nm, and length to around 20 μm [204]. The process leads to reduced thermal fluctuations since synthesis progress in the liquid phase and normal room temperature conditions.

7.3.3 Ultrasonication

Sonication of carbon sources in liquid medium is another process for synthesis of carbon nanotubes. The sound waves are responsible for agitating particles in a sample thereby yielding CNTs. The conversion of two-dimensional graphene oxide nanosheets into CNTs is reported by sonication of the same in a solution comprising of 70% nitric acid without a catalyst. They reported that during the acid ultrasonication, graphene oxide can be easily disintegrated into polyaromatic hydrocarbons (PAHs). It was followed by condensation promoted by cavitation of

PAHs bringing in molecular rearrangement in the precursor. They eventually form folded carbon nanostructures which show magnetic properties [205, 206].

8 Synthesis of Graphene

8.1 Introduction

From the past few decades, many attempts have been made to synthesize large-scale defect-free graphene sheets. However in 2004, Prof. Geim and Novoselov successfully synthesized single-layer graphene sheets via repeated peeling of graphite crystals using strong adhesive tape and then they are transferred and extracted graphite layer on to the silicon wafer [207, 208]. This amazing discovery opened up a new horizon for researchers from all over the world owing to the presence of a wide range of extraordinary properties of graphene. High-quality mass scale production of graphene, many approaches have been taken that includes top-down and bottom-up techniques. Mechanical breakage, in contact probe type sonication, electrochemical exfoliation, super acid dissolution, chemical reduction of colloidal graphene oxide (GO), and thermal exfoliation of GO are included in the top-down category. The bottom-up approach consists of epitaxial growth using metal substrate, chemical vapor deposition (CVD), arc discharge, unzipping of carbon nanotube, and reduction of carbon monoxide [209].

8.2 Top-Down Approaches

8.2.1 Mechanical Breakage

Mechanical breakage method results in the formation of graphene using a scotch tape [207]. This technique is able to form sheets of bigger size, higher purity with the limitation of poor productivity and is only appropriate for basic researches and electronic applications [207]. Exfoliation of graphite is very much easy due to the poor interlayer van der walls force of graphite with inter-action energy of only $\sim$2 eV/nm [210]. In this method, graphene is detached from HOPG by using an adhesive tape. Multiple layers of graphene get adhered on the adhesive tape after peeling of the graphite crystal or HOPG. After repeated peeling, the obtained flakes are released in the acetone. In between, some of the flakes get deposited on the SiO_2/Si (300 nm) substrates which were already dipped in acetone. In this way, thin graphene sheets could be obtained by mechanical cleavage or micromechanical exfoliation or peeling of method or scotch tape method.

8.2.2 Direct Sonication

Sonication in the probe type is a very effective method of synthesis of graphene sheets through the process of exfoliation in the presence of a diluent liquid. Direct sonication method consists of three steps: (1) initially the graphite gets dispersed within the solvent uniformly, (2) under the effect of probe sonicator, the graphite sheet exfoliates or delaminates from each other, and (3) purification of the synthesized exfoliated graphite sheets. During ultrasonication, micrometer-sized bubbles or voids growth and collapse in the liquids are due to the fluctuation of pressure resulting shear force and cavitation which act on the bulk material to exfoliate [211]. The solvent–graphene interactions require balancing the inter-sheet attractive forces after exfoliation in the liquid. But the main question is which will be the perfect solvent for direct sonication method? Any solvent which minimizes the interfacial tension (mN m^{-1}) as well as the contact surface area of the liquid and graphite could serve the purpose [212]. If interfacial tension becomes high in graphite dispersion, the flakes got adhere with each other [212]. However, solvents with the surface tension value in the range of $\sim$40 mJ m^{-2} are the best fit for the method [213]. The major disadvantage in this method is toxicity of solvents used commonly in this technique. However, this method possesses simple steps: add graphite in the solvent and sonicated for few hours (more than 2 h) followed by the centrifugation of the graphite dispersion at the speed of 3000 rpm over the time span of 30 min in order to remove larger unexfoliated graphites. In the later stage, the supernatant of the dispersion was carefully decanted and repeatedly filtered with deionized water to collect graphene sheets [214]. However, the yield of graphene by this technique is very much low to meet the industrial needs.

8.2.3 Electrochemical Exfoliation

Recently, electrochemical exfoliation of graphite is drawing high attention for synthesis of high-quality graphene. This technique is very much easy, fast and environment friendly [215–218]. Electrochemical exfoliation process of graphite is carried out in two different types of electrolyte solution such as ionic liquids [217, 219] and aqueous acids (H_2SO_4 or H_3PO_4) [215, 218, 220]. However, using ionic liquids in exfoliation leads to the functionalization of the graphene sheets which results in a low conductivity, small lateral size and low yield [221, 222]. In case of acidic electrolytes, high yield of graphene with larger lateral size is obtained from graphite but functionalization of these graphite sheets is inevitable because of electrochemical oxidation process taking place onto the surface of exposed sheets [215, 218, 220]. Electrochemical exfoliation is performed using two electrodes, one of them is graphite. When direct current is applied to graphite electrode, graphite layers from graphite electrode begin to separate and get disperse into the electrolyte solution After that, the exfoliated products in the electrolyte solution are collected by vacuum filtration and purified to get graphene [223].

8.2.4 Super Acid Dissolution

The large-scale synthesis of graphene requires the development of new techniques for successful exfoliation of graphite. However, the separation and purification of the graphene from graphite needs much attention. The process of dissoluting graphites in chlorosulfonic acid has the ability to produce mass amount of graphene but hazardous chemicals and separation process may create some difficulties. Dispersibility of graphite strongly depends on the acid strength which is controlled through the mixing of chlorosulfonic acid and concentrated sulfuric acid (98%) with varying proportion [224]. Matteo Pasquali et al. showed that dispersibility of graphite significantly decreases when the proportion of the chlorosulfonic acid was lowered to 80% [225].

8.2.5 Chemical Exfoliation of GO

Chemical exfoliation involves two main steps; Step 1 creates interlayer gap between nearby layers and subsequently reduces the van der walls force of attraction between them. Step 2 involves the exfoliation of intercalated graphite by breaking the weak van der walls forces to form single-layer graphene or graphene oxide. Graphite oxide was first prepared by Bodie in 1859 [226]. The oxidation of graphite was done by $KClO_3$ and fuming HNO_3. After 40 years of interval, Staudenmaier modified the above method by using H_2SO_4, fuming HNO_3, and $KClO_3$ [227]. However, the most common technique to synthesize GO is the Hummer's method, which was reported in 1958 [228]. In this method, graphite powder is mixed with $NaNO_3$ and was oxidized with 98% concentrated H_2SO_4 in an ice bath. Then, $KMnO_4$ was added slowly into the black slurry while keeping the overall reaction temperature below 20 °C. Then in 4 h, the reaction was stopped, which is followed by removing the reaction system out of the ice bath, and keep stirring for 2 h at room temperature. The GO sample was collected after washing for several times and dried at 60 °C for 24 h. After oxidation, the interlayer spacing increases from 3.7 to 9.5 Å which depends on the nature of intercalants. The obtained GO possesses a huge number of oxygen-containing functional groups such as hydroxyl, carboxyl, carbonyl, and epoxy on basal plane. Still now, many approaches are going on to modify the Hummers method known as improved hummers method or modified Hummers method [229, 230]. Reduced graphene oxide (RGO) can be obtained by chemical reduction of the exfoliated GO in protic solvents like water, alcohol in presence heating and stirring. Further, GO can also be exfoliated with isocyantaes, [231] octadeclamine, [232] and as well as functionalizing with surfactants [232–234]. Various reducing agents are used for the synthesis of RGO such as hydrazine monohydrate, [235] sodium borohydrate, [236] hydroquinone, [237] strongly alkaline [238], and amines [239]. Between these reducing agents, hydrazine hydrate is mostly used because of strong reducing activity in order to remove oxygen-containing functional groups of GO. However, hydrazine is highly toxic and explosive therefore, several environment friendly reductants [240] such as vitamin C, [241] amino acid, [242] reducing sugar, [243] alcohol, [244] sodium citrate, [245] tea [246], and lactulose [247, 248] have been developed for synthesis of RGO. Oxidation of graphite destroys the conductivity of GO but after being reduced it restored the sp^2 hybridized network as well as electrical conductivity.

8.2.6 Thermal Exfoliation of GO

Thermally reduced graphene oxide (TRG) can be obtained through heating of GO at high temperature (1000 °C) under inert atmosphere [249–252]. Oxygen-containing functional groups of GO are decomposed at high temperature and forms mass amount of CO_2 gas which generates high pressure. This pressure helps to break down the van der wall forces of GO resulting massive exfoliation of GO. Volume expansion of graphene is 100–300 times. 30% weight loss occurred due to the decomposition of oxygen-containing functional groups and removal of water. TRG sheets contained single layer to few layers graphene with the average size of 500 nm [251]. All top-down approaches are shortly presented in Table 3.

8.3 Bottom-Up Approaches

8.3.1 Epitaxial Growth

Another attractive alternative process for the synthesis of graphene is epitaxial growth over hexagonal shaped substrate. For epitaxial growth of graphene, generally silicon carbide (SiC) and closely packed metal atoms are used as substrate. In 1961 [255], Badami first reported that the graphitization of hexagonal SiC crystal done during annealing at high temperature (1273–1773 K) in ultrahigh vacuum system resulting in sub-lamination of silicon from the surface of SiC while the remaining carbon atoms form a carbonic surface over the entire surface of SiC and arrange themselves as a graphene layers by controlling the growth conditions. Epitaxial technique can generate single-layer or bilayer graphene on the substrate which has a versatile range of applications such as nano or microelectronics and polymer nanocomposites [207]. The results are highly dependent on the various parameters such as temperature, heating rate, and pressure. Similarly, many approaches have been used such as graphene synthesized on Ruthenium (Ru) substrate by epitaxial growth method [256, 257] and so on.

8.3.2 Arc Discharge

Graphene is also synthesized by the arc discharge method using graphite as an electrode in the presence of hydrogen gas in a closed chamber [258, 259]. To prepare high-quality graphene, two graphite rods are placed in water-cooled

Table 3 Representation of top-down approaches

Method	Figure/illustration	Typical dimension	Advantages	Disadvantages	References
Mechanical cleavage	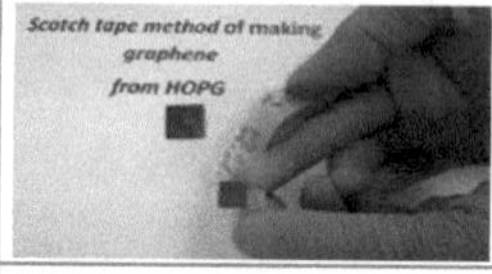	Single layer to few layers (µm–cm)	Directly from graphite and less defects	Mass production is not possible	[207]
Direct sonication	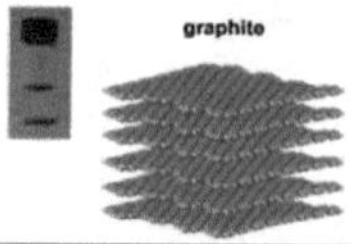	Multiple layers (µm–sub-µm)	Unmodified graphene and inexpensive	Low yield and difficult to separate	[214]
Electrochemical exfoliation	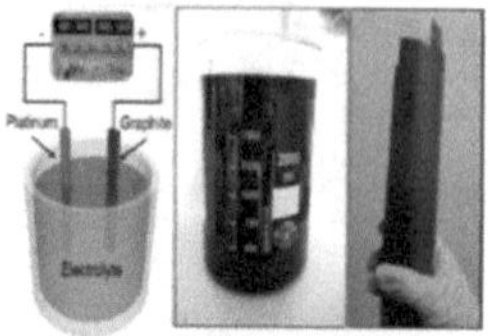	Single and few layers (500–700 nm)	Single step functionalization and exfoliation	Cost of ionic liquids	[223]
Superacid dissolution	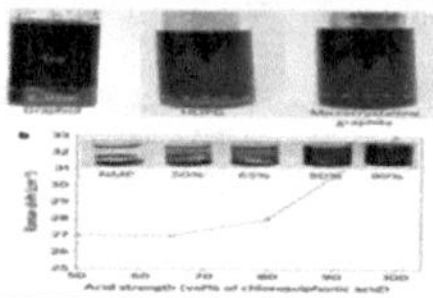	Mostly single layer (300–900 nm)	Unmodified graphene and scalable	Use of hazardous chlorosulfonic acid and cost of acid removal	[225]
Chemical reduction of colloidal GO	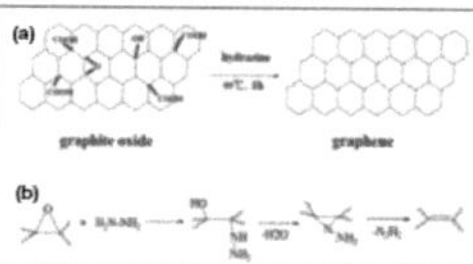	Single and multiple layers CRG (µm–sub-µm)	Large sheet size, low cost and mass production is possible	Hazardous chemicals and CRG dispersed in hydrophilic polymer only	[231]

(continued)

Table 3 (continued)

Method	Figure/illustration	Typical dimension	Advantages	Disadvantages	References
Chemical reduction of organically treated GO	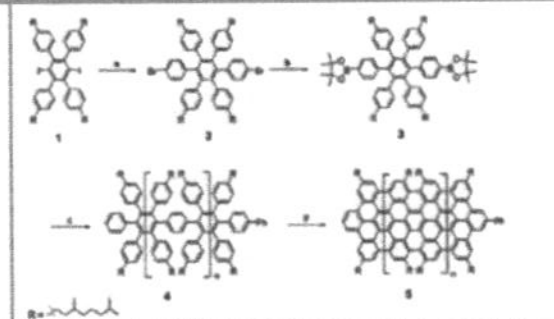	Mostly single layer CRG (few 100 nm–μm)	Mass production, colloidal stability in organic solvents and better exfoliation	Many defects	[253]
Thermally exfoliation of GO	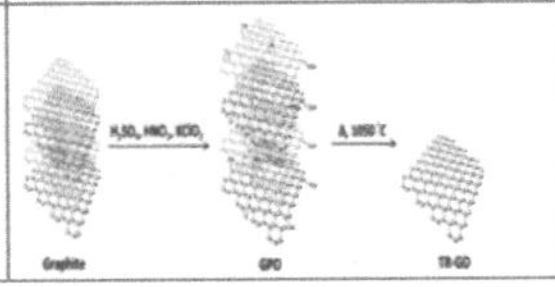	Single and few layers TRG ($\sim$500 nm)	One step exfoliation and short heating time	High heating temperature and short sheet size	[254]

stainless steel chamber in H_2 atmosphere. In typical experiment, high current (above 100 A), high voltage (>50 V), and high pressure (above 200 tor) are necessary for the synthesis of graphene. Graphene sheets with two or three layers are deposited on the inner walls of the arc chamber. To maintain the arc, the distances between both electrodes are well maintained. Arc discharge does not work at lower current and lower voltage even in the presence of hydrogen atmosphere. If the amount of the hydrogen content in the closed chamber is decreased, polyhedral content increases in the end product. This method is generally used for doping of graphene with boron and nitrogen [260]. The yield of the obtained graphene from arc discharge is very low around 10–20% [259].

8.3.3 Chemical Vapor Deposition

Chemical vapor deposition is a most promising and well-known method where graphene is directly grown on the metal substrate such as Ni, [261] Pd, [262] Ru, [256] Ir [263] and Cu [264] when exposed to hydrocarbon gas. Hydrocarbons like methane, ethylene, acetylene, and benzene were decomposed on the surface of metal at high temperature ($\sim$1000 °C). Conventionally, heating is done by filament in CVD process. Recently, graphene has been prepared through radio frequency enhanced chemical vapor deposition method (PECVD) with the advantage of energy saving characteristics and high purity in the production [265]. For better understanding, metal substrates like Ni film directly exposed to the gaseous mixture of H_2, CH_4, and Ar and heated at 1000 °C. At this high temperature, methane decomposes on the metal surface and hydrogen evaporates. Carbon diffuses on the Ni film. Graphene layer on the Ni film is obtained after cooling down in Ar atmosphere. The number of layers strongly depends on the behavior of hydrocarbon, the thickness of metal substrate, and reaction parameters. Graphene layers can be transferred to the other substrate through polymer support by etching the underlying transition metal.

8.3.4 Unzipping of Carbon Nanotube

Graphene can also be obtained from multi-walled carbon nanotube (MWCNT) through oxidative cutting [266]. MWCNT was dispersed in distilled water and strong oxidizers like potassium permanganate and sulfuric acid are added in it and then placed in autoclave reactor and heated at 180 °C for more than 18 h. After that black solution was purified through repeated washing. The conductivity of obtained graphene is very poor due to oxidation of basal plane [267]. This is a very expensive technique for the synthesis of graphene as CNT is utilized as a starting material.

8.3.5 Reduction of CO

Pristine graphenes are produced through reduction of CO by self-propagating high-temperature synthesis (SHS) [268]. Magnesium acts as a reducer of carbon for the synthesis of expanded graphite and multiple layers graphene. Magnesium reduced CO, [269] and CO_2 (dry ice) [270] to form graphene sheets. In conventional calcination method, magnesium reduced $CaCO_3$ is used to produce few layers of graphene. The mixture of Mg and $CaCO_3$ was placed in crucible then put in a closed steel container under CO_2 atmosphere. The reaction mixture is then ignited by electrical device to form black product. The obtained product is then purified to get pristine graphene. The prime drawback of this method is the presence of impurities which are not easy to separate from graphene [271]. All the down–up approaches are shortly presented in Table 4.

9 Summary

Diamond and fullerene are two natural allotropes of carbon with significant importance in the area of materials engineering. Synthesis of these two form of carbon by artificial means holds massive importance in developing high-performance materials. The unique morphology and inertness in structure towards various environments open up a new avenue of its engineering application. The high electrical and thermal conductivity are essentially relating its importance in various electrical equipment like solar cell, etc. Three types of graphite are mentioned herein, namely graphite electrode, water-soluble graphite, and expanded graphite. The synthesis of graphite was mentioned by hydrothermal process. Moreover, the graphite was also synthesized by nonconventional method. Graphite ribbon was synthesized by fusing benzenoid hydrocarbons via electrophile-induced cyclization route. Synthesis of carbon black involves the use of gas, liquid or solid precursors which may include synthetic as well as naturally occurring materials. Different synthesis methods and manufacturing procedures offer the scope of synthesis of carbon black with wide range of structural dimensions. However, the use of naturally occurring or synthetic waste materials as the necessary resources is of great interest for the future progress in the related field. Moreover, carbon black is also prepared by pyrolysis of polymer and hydrocarbon, plasma synthesis, and by hydrolysis of natural resources. The synthesis of high-performance carbon fibers was carried out from mesophase pitch or polyacrylonitrile (PAN). This entire process was categorized into five steps namely spinning, stabilizing, carbonizing, treating the surface, and sizing. Carbon nanofiber was synthesized by CVD as well as electrospinning and template methods. High-quality carbon nanofiber was synthesized by chemical vapor deposition of methane on the patterned substrates. Electrospinning process was used to fabricate nonwoven and ultrafine nanoscale fibers with diameters ranging from 3 nm to several microns. This process is highly popular because of its simplicity, affordability, and flexibility in selection of wide

Table 4 Representation of down–up approaches

Method	Figure/illustration	Typical dimension	Advantages	Disadvantages	References
Epitaxial growth	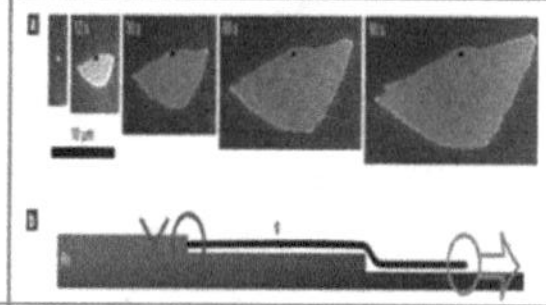	Few layers (up to cm size)	High quality and large surface area	Low production scale	[256]
Arc discharge	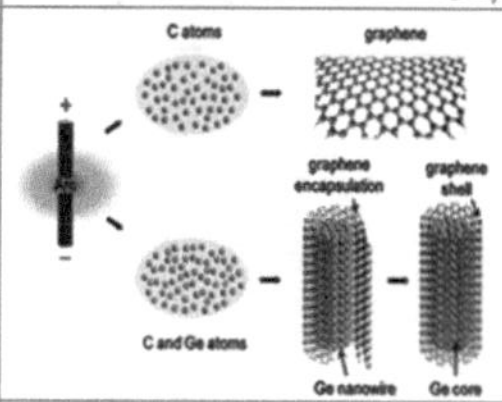	Single layer, bilayer and few layers (few 100 nm—few μm)	Produce 10 gm/h of graphene	Impurities	[272]
CVD technique	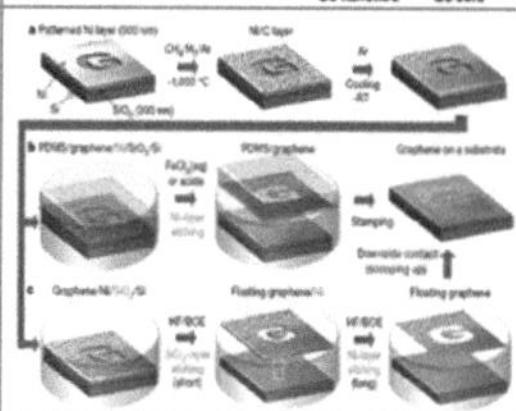	Single layer to few layers (100 nm–cm)	Thickness control, large size, and high quality-	High cost and small scale production	[261]
Unzipping of CNT	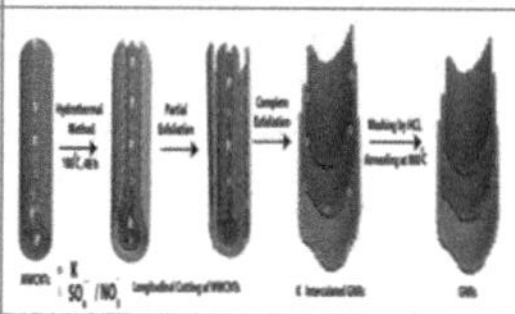	Multiple layers (few μm long nanoribbons)	Size of graphene sheets could be controlled by the selection of nanotube	High cost of starting material and oxidized graphene	[266]

(continued)

Table 4 (continued)

Method	Figure/illustration	Typical dimension	Advantages	Disadvantages	References
Reduction of CO and CO_2	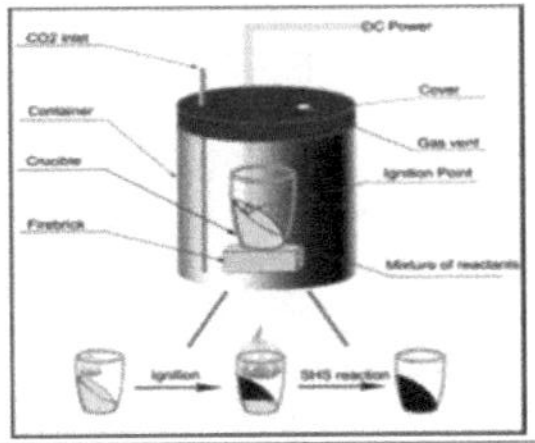	Multiple layers (Sub—μm)	Unoxidized sheets	Contamination and washing is difficult	[271, 273]

range of materials, high surface-to-volume ratio, and tunable porosity. Template method is widely and basically used to produce one-dimensional carbon material. Anodic aluminum oxide, mesoporous silica, and Te nanowires are generally used in this technique. Synthesis of fibers and submicron tubes of carbon is reported through the pyrolytic carbon decomposition of propylene on the pore wall of aluminum oxide film. Conventional techniques for synthesis of CNTs are performed at high temperatures to enable the catalyst and disintegrate carbon source for synthesis of CNTs. Such higher temperatures can lead to thermal fluctuations with the catalyst which may lead to deformation and aggregation of catalyst nanoparticles. Furthermore, it can lead to particle aggregation onto substrates leading to formation of CNTs with wider chirality distribution. Despite yielding high-quality CNTs, arc discharge and laser ablation methods are expensive method of synthesis of CNTs on large scale. CVD is the best-suited, economically feasible method of production of carbon nanotubes on large scale but the application is more suited towards the production of SWNT. CoMoCAT® & HiPco® have a tremendous potential of scaling up to large-scale process for the manufacture of CNTs. Pyrolysis electrolysis, flame method, and ultrasonication are some other processes which are well documented and can be worked upon to increase yield/purity of CNTs. Some other reported techniques for CNT production using mechanisms such as capillary micro fluids and Dip pen technology at room temperatures which can counter high-temperature processes such as CVD are the processes for the future. Among different synthesis methods, epitaxial growth and CVD techniques are suitable for the production of high-quality graphene and colloidal suspension method stands out for the high yield production of chemically modified graphene and used for the broad area of applications. However, all the methods have some advantages and disadvantage.

References

1. Hannay JB (1880) On the artificial formation of the diamond. J Franklin Inst 110(2):123–135
2. Bundy FP, Bovenkerk HP, Strong HM, Wentorf RH Jr (1961) Diamond-graphite equilibrium line from growth and graphitization of diamond. J Chem Phys 35(2):383–391
3. Giardini AA, Kohn JA, Eckart DW, Tydings JE (1961) The formation of coesite and kyanite from pyrophyllite at very high pressures and high temperatures. Am Mineral 46:976–983
4. De Carli PS, Milton DJ (1965) Stishovite: synthesis by shock wave. Science 147(3654):144–145
5. Eversole WG (1962) US Patent No. 3030 187 Washington DC: US Patent and Trademark Office
6. Hall HT (1960) Ultra-high-pressure, high-temperature apparatus: the"belt". Rev Sci Instrum 31(2):125–131
7. Giardini AA, Tydings JE (1962) Diamond synthesis-observations on mechanism of formation. Am Mineral 47(11–1):1393
8. Eversole WG (1962) US Patent No. 3030188 Washington DC: US Patent and Trademark Office

9. Eversole WG (1958) Diamond synthesis. US patents (3030187) 3030188
10. Setaka N (1984) Low pressure gas phase synthesis of diamonds. Hyomen (Surface) 22:110–117
11. Kamo M, Sato Y, Matsumoto S, Setaka N (1983) Diamond synthesis from gas phase in microwave plasma. J Cryst Growth 62(3):642–644
12. Mōri T, Namba Y (1983) Hard diamondlike carbon films deposited by ionized deposition of methane gas. J Vac Sci Technol, A 1(1):23–27
13. Kitabatake M, Wasa K (1985) Growth of diamond at room temperature by an ion-beam sputter deposition under hydrogen-ion bombardment. J Appl Phys 58(4):1693–1695
14. Luo C, Qi X, Pan C, Yang W (2015) Diamond synthesis from carbon nanofibers at low temperature and low pressure. Sci Rep 5:13879
15. Kazi S (2014) A review article on nanodiamonds discussing their properties and applications. Int J Pharm Sci Inv 3:40–45
16. Fang L, Ohfuji H, Irifune T (2013) A novel technique for the synthesis of nanodiamond powder. J Nanomater 2013:41
17. Baidakova M (2007) New prospects and frontiers of nanodiamond clusters. J Phys D Appl Phys 40(20):6300
18. Kroto HW, Heath JR, O'brien SC, Curl RF, Smalley REC (1985) This week's citation classic®. Nature 318:162–163
19. Guo T, Jin C, Smalley RE (1991) Doping bucky: formation and properties of boron-doped buckminsterfullerene. J Phys Chem 95(13):4948–4950
20. Heath JR, Curl RF, Smalley RE (1987) The UV absorption spectrum of C60 (buckminsterfullerene): a narrow band at 3860 Å. J Chem Phys 87(7):4236–4238
21. Rohlfing EA, Cox DM, Kaldor A (1984) Production and characterization of supersonic carbon cluster beams. J Chem Phys 81(7):3322–3330
22. Kroto H (1988) Space stars C60 and soot. Science 242(4882):1139–1145
23. Haufler RE, Conceicao J, Chibante LPF, Chai Y, Byrne NE, Flanagan S, Flanagan S, Haley MM, O'Brien SC, Pan C (1990) Efficient production of C_{60} (buckminsterfullerene), $C_{60}H_{36}$, and the solvated buckide ion. J Phys Chem 94(24):8634–8636
24. Howard JB, McKinnon JT, Makarovsky Y, Lafleur AL, Johnson ME (1991) Fullerenes C_{60} and C_{70} in flames. Nature 352(6331):139
25. Scott LT (2004) Methods for the chemical synthesis of fullerenes. Angew Chem Int Ed 43 (38):4994–5007
26. Howard JB, Kronholm DF, Modestino AJ, Richter H (2008) U.S. Patent No. 7,396,520. Washington, DC: U.S. Patent and Trademark Office
27. Takehara H, Fujiwara M, Arikawa M, Diener MD, Alford JM (2005) Experimental study of industrial scale fullerene production by combustion synthesis. Carbon 43(2):311–319
28. Saim S, Kuo KC, Stalling DL (1993) Supercritical fluid extraction of fullerenes C_{60} and C_{70} from carbon soot. Sep Sci Technol 28(8):1509–1525
29. Chiu CC, Lo JC, Teng MH (2012) A novel high efficiency method for the synthesis of graphite encapsulated metal (GEM) nanoparticles. Diam Relat Mater 24:179–183
30. Panaitescu C, Predeanu G (2010) Petrographic research applied to carbon materials. Rev Roum Chim 55:301–310
31. Predeanu G, Panaitescu C, Bălănescu M, Bieg G, Borrego AG, Diez MA, Hackley PB, Kwiecińska M, Marques M, Mastalerz M, Misz-Kennan S, Pusz I, Suárez Ruiz S, Rodrigues A, SinghK A, Varma K, Zdravkov A, Životić D (2015) Microscopical characterization of carbon materials derived from coal and petroleum and their interaction phenomena in making steel electrodes, anodes and cathode blocks for the Microscopy of Carbon Materials Working Group of the ICCP. Int J Coal Geol 139(l):63–79
32. Niyogi S, Bekyarova E, Itikis ME, McWilliams JL, Hammon MA, Haddon RC (2006) Solution properties of graphite and graphene. J Am Chem Soc 128:7720–7721
33. Stankovich S, Piner RD, Chen X, Wu N, Nguyen ST, Ruoff RS (2006) Stable aqueous dispersions of graphitic nanoplatelets via the reduction of exfoliated graphite oxide in the presence of poly (sodium 4-styrenesulfonate). J Mater Chem 16:155–158

34. Schniepp HC, Li JL, Mc Allister MJ, Sai H, Herrera-Alonso M, Adamson DH, Prud'homme, RK, Car R, Saville DA, Aksay IA (2006) Functionalized single graphene sheets derived from splitting graphite oxide. J Phys Chem B 110:8535–8539
35. Hontoria-Lucas C, Lopez-Peinado AJ, opez-Gonzalez JDL, Rojas-Cervantes ML, Martin-Aranda RM(1995) Study of oxygen-containing groups in a series of graphite oxides: physical and chemical characterization. Carbon 33:1585–1592
36. Si Y, Samulski ET (2008) Synthesis of water soluble graphene. Nano Lett 8:1679–1682
37. Stankovich S, Dikin DA, Dommmett GHB, Kohlhaas KM, Zimney EJ, Stach EA, Piner RDS, Nguyen T, Ruoff RS (2006) Graphene-based composite materials. Nature 442:282–286
38. Li D, Muller MB, Gilje S, Kaner RB, Wallace GG (2008) Processable aqueous dispersions of graphene nanosheets. Nat Nanotechnol 2:101–105
39. Mukherjee A, Kang JH, Kuznetsov O, Sun Y, Thaner R, Bratt AS, Lomeda JR, Kelly KF, Billups WE (2011) Water-soluble graphite nanoplatelets formed by oleum exfoliation of graphite. Chem Mater 23:9–13
40. Zhao MF, Liu P (2009) Adsorption of methylene blue from aqueous solutions by modified expanded graphite powder. Desalination 249:331–336
41. Zhou YY, Wang SW, Kim KN, Li JH, Yan XP (2006) Evaluation of expanded graphite as on-line solid-phase extraction sorbent for high performance liquid chromatographic determination of trace levels of DDTs in water samples. Talanta 69:970–975
42. Toyoda M, Inagaki M (2000) Heavy oil sorption using exfoliated graphite: new application of exfoliated graphite to protect heavy oil pollution. Carbon 38:199–210
43. Wang L, Fu X, Chang E, Wu H, Zhang K, Lei X, Zhang R, Qi X, Yang Y (2014) Preparation and its adsorptive property of modified expanded graphite nanomaterials. J Chem 2014:678151
44. Anderson Axdal SH, Chung DDL (1987) A theory for the kinetics of intercalation of graphite. Carbon 25:377–389
45. Holliday AK, Hughes G, Walker Carbon SM (1973) In: Bailar HJ, Emeleus R, Nyholm AF Trotman-Dickenson JC (eds Comprehensive inorganic chemistry. Pergamon Press, Oxford
46. Cataldo F, Valentini F, Cherubini V, Ursini O, Angelini G (2012) Synthesis of expanded graphite flakes by the submerged carbon arc in oleum. Fullerenes Nanotubes Carbon Nanostruct 20:152–162
47. Rudorff W (1939) Kristallstruktur der Säureverbindungen des Graphits. Z Phys Chem 45:42–69
48. Kang F, Leng Y, Zhang TY (1997) Electrochemical synthesis and characterization of formic acid-graphite intercalation compound. Carbon 35:1089–1096
49. Avdeev VV, Monyakina LA, Nikol'skaya IV, Sorokina NE, Semenenko KN (1992) Finaenov AI Chemical synthesis of graphite hydrogenosulfate: calorimetry and potentiometry studies. Carbon 30:825–827
50. Yakovlev AV, Finaenov AI, Zabud'kov SL, Yakovleva EV (2006) Thermally expanded graphite: synthesis, properties, and prospects for use. Russ J Appl Chem 79:1741–1751
51. Strativnov EV (2015) Design of modern reactors for synthesis of thermally expanded graphite. Nanoscale Res Lett 10:245
52. Zhang F, Zhao Q, Yan X, Li H, Zhang P, Wang L, Zhou T, Li Y, Ding L (2016) Rapid preparation of expanded graphite by microwave irradiation for the extraction of triazine herbicides in milk samples. Food Chem 197:943–949
53. Łoś S, Duclaux L, Alvarez L, Hawełek Ł, Duber S, Kempiński W (2013) Cleavage and size reduction of graphite crystal using ultrasound radiation. Carbon 55:53–61
54. Alaferdov AV, Gholamipour-Shirazi A, Canesqui MA, Danilov YA, Moshkalev SA (2014) Size-controlled synthesis of graphite nanoflakes and multi-layer graphene by liquid phase exfoliation of natural graphite. Carbon 69:525–535
55. Jess E, Jones Michael C, Cheshire Dominick J, Jr Casadonte, Carol C (2004) Phifer facile sonochemical synthesis of graphite intercalation compounds. Org Lett 6:1915–1917

56. Chen G, Weng W, Wu D, Wu C, Lu J, Wang P, Chen X (2004) Preparation and characterization of graphite nanosheets from ultrasonic powdering technique. Carbon 42:753–759
57. Hirano SI, Nakamura K, Somiya S (1989) Graphitization of carbon in the presence of calcium compounds under hydrothermal conditions by use of high gas pressure apparatus. In: Somiya S (eds) Hydrothermal reactions for materials science and engineering. Elsevier, London, pp 331–336
58. Gogotsi YG, Yoshimura M (1994) Formation of carbon films on carbides under hydrothermal conditions. Nature 30:367–628
59. Libera J, Gogotsi Y (2001) Hydrothermal synthesis of graphite tubes using Ni catalyst. Carbon 39:1307–1318
60. Chung SR, Wang KW, Teng MH, Perng TP (2009) Electrochemical hydrogenation of nanocrystalline face-centered cubic Co powder. Int J Hydrog Energy 34:1383–1388
61. Bystrzejewski M, Pyrzyńska K, Huczko A, Lange H (2009) Carbon-encapsulated magnetic nanoparticles as separable and mobile sorbents of heavy metal ions from aqueous solutions. Carbon 47:1189–1206
62. Saraswati TE, Prasiwi ODI, Masykur A, Anwar M (2017) Bifunctional catalyst of graphite-encapsulated iron compound nanoparticle for magnetic carbon nanotubes growth by chemical vapor deposition. In AIP conference proceedings 1788:030029
63. Tomita M, Saito Y, Hayashi T (1993) LaC_2 encapsulated in graphite nano-particle. Jpn J Appl Phys 32:L280–L282
64. Ruoff RS, Lorents DC, Chan B, Malhotra R, Subramoney S (1993) Single crystal metals encapsulated in carbon nanoparticles. Science 259:346–348
65. Kräschmer W, Huffman DR (1992) Fullerites: new forms of crystalline carbon. Carbon 30:1143–1147
66. Dravid VP, Host JJ, Teng MH, Elliott BR, Hwang JH, Johnson DL, Mason TO, Weertman JR (1995) Controlled-size nanocapsules. Nature 374–602
67. Elliot BR, Host JJ, Dravid VP, Teng MH, Hwang JH (1997) A descriptive model linking possible formation mechanisms for graphite-encapsulated nanocrystals to processing parameters. J Mater Res 12:3328–3344
68. Shikin AM, Prudnikova GV, Adamchuk VK, Soe WH, Rieder KH, Molodtsov SL, Laubschat C (2002) Synthesis of graphite monolayer stripes on a stepped Ni(771) surface. Phys Solid State 44:677–680
69. Miao JY, Hwang DW, Narasimhulu KV, Lin PI, Chen YT, Lin SH, Hwang LP (2004) Synthesis and properties of carbon nanospheres grown by CVD using Kaolin supported transition metal catalysts. Carbon 4:813–822
70. Koltypin Y, Fernandez A, Rojas TC, Campora J, Palma P, Prozorov R, Gedanken A (1999) Encapsulation of nickel nanoparticles in carbon obtained by the sonochemical decomposition of $Ni(C_8H_{12})_2$. Chem Mater 11:1331–1335
71. Hayashi T, Hirono S, Tomita M, Umemura S (1996) Magnetic thin films of cobalt nanocrystals encapsulated in graphite-like carbon. Nature 381:772–774
72. Liu B, Jia D, Rao J, Zuo P, Shao Y (2009) A self-assembly template approach for preparing hollow carbon microspheres. J Solid State Electrochem 13:497–501
73. Seraphin S, Zhou D, Jiao J (1996) Filling the carbon nanocages. J Appl Phys 80:2097–2104
74. Ye E, Liu B, Fan WY (2007) Preparation of graphite-coated iron nanoparticles using pulsed laser decomposition of $Fe_3(CO)_{12}$ and PPh_3 in hexane. Chem Mater 19:3845–3849
75. Liu Y, Ren Z, Wei Y, Jiang B, Feng S, Zhang L, Zhang W, Fu H (2010) Synthesis and applications of graphite carbon sphere with uniformly distributed magnetic Fe_3O_4 nanoparticles (MGCSs) and MGCS@Ag, MGCS@TiO_2. J Mater Chem 20:4802–4808
76. Goldfinger MB, Swager TM (1994) Fused polycyclic aromatics via electrophile-induced cyclization reactions: application to the synthesis of graphite ribbons. J Am Chem Soc 116:7895–7896
77. Heera V, Skorupa W, Pe´cz B, Dobos L (2000) Ion beam synthesis of graphite and diamond in silicon carbide. Appl Phys Lett 76:2847–2849

78. Goh M, Matsushita S, Akagi K (2010) From helical polyacetylene to helical graphite synthesis in the chiral nematic liquid crystal field and morphology-retaining carbonization. Chem Soc Rev 39:2466–2476
79. Thu TV, Tanizawa Y, Phuc NHH, Ko PJ, Sandhu A (2013) Synthesis and characterization of graphite nanoplatelets. J Phys: Conf Ser 433:012003
80. Giardini AA, Salotti CA, Lakner JF (1968) Synthesis of graphite and hydrocarbons by reaction between calcite and hydrogen. Science 159:317–319
81. Slawson CB (1953) Synthesis of graphite at room temperatures. Am Mineral 38:50–55
82. Dimovski S, Nikitin A, Ye H, Gogotsi Y (2004) Synthesis of graphite by chlorination of iron carbide at moderate temperatures. J Mater Chem 14:238–243
83. Okuno H, Palnichenko A, Despres JF, Issi JP, Charlier JC (2005) Synthesis of graphite polyhedral crystals using a combustion flame method. Carbon 43:692–697
84. Ohkawara Y, Shinada T, Fukada Y, Ohshio S, Saitoh H (2003) Synthesis of graphite using laser decomposition of SiC. J Mater Sci 30:2447–2453
85. Parkash S (2010) Petroleum fuels manufacturing handbook: including specialty products and sustainable manufacturing techniques. The McGraw-Hill Companies, New York
86. Drogin I (2012) Carbon black. J Air Pollut Control Assoc (TI-2 Chemical committee informative report No.09)
87. Mohan AN, Manoj B (2012) Synthesis and characterization of carbon nanospheres from hydrocarbon soot. Int J Electrochem Sci 7:9537–9549
88. Ivie JJ, Forney LJ (1988) A numerical model of the synthesis of carbon black by benzene pyrolysis. AIChE J 34:1813–1820
89. Abrahamson J (1977) Saturated platelets are new intermediates in hydrocarbon pyrolysis and carbon formation. Nature 266:323–327
90. Vié R, Drahi E, Baudino O, Blayac S, Berthon-Fabry S (2016) Synthesis of carbon nanospheres for the development of inkjet-printed resistive layers and sensors. Flex Print Electron 1:015003
91. Guo XF, Kim GJ (2010) Synthesis of ultrafine carbon black by pyrolysis of polymers using a direct current thermal plasma process. Plasma Chem Plasma Process 30:75–90
92. Yuan JJ, Hong RY, Wang YQ, Feng WG (2016) Plasma preparation of carbon black used in conductive coatings. Polym Compos 37:1078–1084
93. Nguyen M, Kimz TSJ (2015) Nano carbon black powder synthesized via liquid phase plasma process as a supercapacitor active material. J Electrochem Soc 162:A1445–A1450
94. Wang L, Wang X, Zou B, Ma X, Qu Y, Rong C, Li Y, Su Y, Wang Z (2011) Preparation of carbon black from rice husk by hydrolysis, carbonization and pyrolysis. Bioresour Technol 102:8220–8224
95. Lu W, Chung DDL (2001) Preparation of conductive carbons with high surface area. Carbon 39:39–44
96. Wang L, GuoY Zou B, Rong C, Ma X, Qu Y, Li Y, Wang Z (2011) High surface area porous carbons prepared from hydrochars by phosphoric acid activation. Bioresour Technol 102:1947–1950
97. De J, Lopez-Gonzalez D, Martinez-Vilchez F, Rodriguez-Reinoso F (1980) Preparation and characterization of active carbons from olive stones. Carbon 18:413–418
98. Hu Z, Srinivasan M (1999) Preparation of high-surface-area activated carbons from coconut shell. Microporous Mesoporous Mater 27:11–18
99. Tam MS, Antal MJ (1999) Preparation of activated carbons from macadamia nut shell and coconut shell by air activation. Ind Eng Chem Res 38:4268–4276
100. Sarkar SC, Bose A (1997) Role of activated carbon pellets in carbon dioxide removal. Energy Convers Manage 38:S105–S110
101. Su W, Zhou Y, Wei L, Sun Y, Zhou L (2007) Effect of microstructure and surface modification on the hydrogen adsorption active carbons. New Carbon Mater 22:135–140
102. Yang K, Peng J, Srinivasakanna C, Zhang L, Xia H (2010) Preparation of high surface area activated carbon from coconut shells using microwave heating. Bioresour Technol 101:6163–6169

103. Gao H, Song Z, Yang L, Wu H (2016) Synthesis method of white carbon black utilizing water-quenching blast furnace slag. Energy Fuels 30:9645–9651
104. Park SJ, Heo GY (2015) Precursors and manufacturing of carbon fibers. Springer, Berlin, pp 31–66
105. Huang X (2009) Fabrication and properties of carbon fibers. Materials 2:2369–2403
106. Edie DD (1998) The effect of processing on the structure and properties of carbon fibers. Carbon 36(4):345–362
107. Hanna SB, Yehia AA, Ismail MN, Khalaf AI (2012) Preparation and characterization of carbon fibers from polyacrylonitrile precursors. J Appl Polym Sci 123:2074–2083
108. Kim YA, Hayashi T, Endo M, Dresselhaus MS (2011) Carbon nanofibers. Springer, Belin, pp 2–27
109. Giraldo L, Ladino Y, Pirajánc JCM, Rodríguez MP (2007) Synthesis and characterization of activated carbon fibers from Kevlar. Eclet Quím 32(4):55–62
110. Sahin K, Nicholas AF, Chasiotis I, Lyons KM, Newcomb BA, Kamath MG, Chae HG, Kumar S (2014) High strength micron size carbon fibers from polyacrylonitrile–carbon nanotube precursors. Carbon 77:442–453
111. Rosas RR, Bedia J, Lallave M, Loscertales IG, Barrero A, Mirasol JR, Cordero T (2010) The production of submicron diameter carbon fibers by the electrospinning of lignin. Carbon 48:696–705
112. Maradur SP, Kim CH, Kim SY, Kim BH, Kim WC, Yang KS (2012) Preparation of carbon fibers from a lignin copolymer with polyacrylonitrile. Synth Met 162:453–459
113. Kim BJ, Eom Y, Kato O, Miyawaki J, Kim BC, Mochida I, Yoon SH (2014) Preparation of carbon fibers with excellent mechanical properties from isotropic pitches. Carbon 77:747–755
114. Sun M, Li J, Wang Y, Zhang X (2015) Preparation of carbon fiber reinforced cement-based composites using self-made carbon fiber mat. Constr Build Mater 79:283–289
115. Derbyshire F, Andrews R, Berkovich A, Jacques D, Jagtoyen M, Rantell T (2001) Synthesis of isotropic carbon fibers from pitch precursors. Fuel 80(3):345–356
116. Wazir AH, Lutfullah K (2009) Preparation and characterization of pitch-based carbon fibers. New Carbon Mater 24(1):83–88
117. Hao KT, Shiang LJ (2006) The effect of a chemical vapor deposited carbon film from acetylene on the properties of graphitized PAN-based carbon fibers. New Carbon Mater 21 (4):296–301
118. Zhang L, Aboagye A, Kelkar A, Lai C, Fong H (2014) A review: carbon nanofibers from electrospun polyacrylonitrile and their applications. J Mater Sci 49(2):463–480
119. Price RL, Ellison K, Haberstroh KM, Webster TJ (2004) Nanometer surface roughness increases select osteoblast adhesion on carbon nanofiber compacts. J Biomed Mater Res, Part A 70(1):129–138
120. Jong KPD, Geus JW (2000) Carbon nanofibers: catalytic synthesis and applications. Catal Rev 42(4):481–510
121. Marella M, Tomaselli M (2006) Synthesis of carbon nanofibers and measurements of hydrogen storage. Carbon 44(8):140–1413
122. Kong J, Soh HT, Cassell AM, Quate CF, Dai H (1998) Synthesis of individual single-walled carbon nanotubes on patterned silicon wafers. Nature 395(6705):878–881
123. Yang S, Taha-Tijerina J, Serrato-Diaz V, Hernandez K, Lozano K (2007) Dynamic mechanical and thermal analysis of aligned vapor grown carbon nanofiber reinforced polyethylene. Compos Part B Eng 38(2):228–235
124. Fan YY, Cheng HM, Wei YL, Su G, Shen ZH (2000) The influence of preparation parameters on the mass production of vapor-grown carbon nanofibers. Carbon 38(6):78–795
125. Bonet F, Grugeon S, Dupont L, Herrera Urbina R, Gue´ry C, Tarascon JM (2003) Synthesis and characterization of bimetallic Ni–Cu particles. J Solid State Chem 172(1):111–5
126. Zarur AJ, Ying JY (2000) Reverse microemulsion synthesis of nanostructured complex oxides for catalytic combustion. Nature 403(6765):65–67

127. Baker RTK, Barber MA, Harris PS, Feates FS, Waite RJ (1972) Nucleation and growth of carbon deposits from the nickel catalyzed decomposition of acetylene. J Catal 26(1):51–62
128. Schouten FC, Kaleveld EW, Bootsma GA (1977) AES-LEED-ellipsometry study of the kinetics of the interaction of methane with Ni(110). Surf Sci 63:460–474
129. Schouten FC, Gijzeman OLJ, Bootsma GA (1979) Interaction of methane with Ni(111) and Ni(100); diffusion of carbon into nickel through the (100) surface; an aes-leed study. Surf Sci 87:1–12
130. Eatemadi A, Daraee H, Zarghami N, Melat YH, Akbarzadeh A (2016) Nanofiber: synthesis and biomedical applications. Artif Cells Nanomed Biotechnol 44(1):111–121
131. Zhang B, Kang F, Tarascon JM, Kim JK (2016) Recent advances in electrospun carbon nanofibers and their application in electrochemical energy storage. Prog Mater Sci 76:319–380
132. Li W, Zhang F, Dou YQ, Wu ZX, Liu HJ, Qian XF, Gu D, Xia YY, Tu B, Zhao DY (2011) A self-template strategy for the synthesis of mesoporous carbon nanofibers as advanced supercapacitor electrodes. Adv Energy Mater 1:382–386
133. Luong ND, Lee Y, Nam JD (2008) Facile transformation of nanofibrillar polymer aerogel to carbon nanorods catalyzed by platinum nanoparticles. J Mater Chem 18:4254–4259
134. Feng D, Lv YY, Wu ZX, Dou YQ, Han L, Sun ZK, Xia YY, Zheng GF, Zhao DY (2011) Free-standing mesoporous carbon thin films with highly ordered pore architectures for nanodevices. J Am Chem Soc 133:15148–15156
135. Chrzanowska J, Hoffman J, Melolepszy A, Mazurkiewicz M, Kowalewski TA, Szymanski Z, Stobinski L (2015) Synthesis of carbon nanotubes by the laser ablation method: effect of laser wavelength. Phys Status Solidi B 252:1860–1867
136. Arora N, Sharma NN (2014) Arc discharge synthesis of carbon nanotubes: comprehensive review. Diamond Relat Mater 50:135–150
137. Andrews R, Jacques D, Qian D, Rantell T (2002) Multiwall carbon nanotubes: synthesis and application. Acc Chem Res 35:1008–1017
138. Nikolaev P, Bronikowski M, Bradley J, Kelley R, Rohmund F, Colbert D, Smith T, Smalley RE (1999) Gas-phase catalytic growth of single-walled carbon nanotubes from carbon monoxide. Chem Phys Let 313:91–97
139. Resasco DE, Alvarez WE, Pompeo F, Balzano L, Herrera JE, Kitiyanan B, Borgna A (2002) A scalable process for production of single-walled carbon nanotubes (SWNTS) by catalytic disproportionation of Co on a solid catalyst. J Nanopart Res 4:131–136
140. Prasek J, Drbohlavova J, Chomoucka J, Hubalek J, Jasek O, Adam V, Kizek R (2011) Methods for carbon nanotubes synthesis—review. Mater Chem 21:15872–15884
141. Rafique MMA, Iqbal J (2011) Production of carbon nanotubes by different routes—a review. JEAS 1:29–34
142. Kumar U, Sikarwar S, Sonekar RK, Yadav BC (2016) Carbon nanotube: synthesis and application in solar cell. JIOPM 26:1231–1242
143. Lebel LL, Aissa B, Khakani MAE, Therriault D (2010) Compos Sci Technol 70:518–524
144. Kusaba M, Tsunawaki Y (2006) Production of single-wall carbon nanotubes by a XeCl excimer laser ablation. Thin Solid Films 506:255–258
145. Iijima S (1991) Helical microtubules of graphitic carbon. Nature 354:56–58
146. Bethune DS, Kiang CH, Devries MS, Gorman G, Savoy R, Vazquez J, Beyers R (1993) Cobalt-catalysed growth of carbon nanotubes with single-atomic-layer walls. Nature 363:605–607
147. Ajayan PM, Lambert JM, Bernier P, Barbedette L, Colliex C, Planeix JM (1993) Growth morphologies during cobalt-catalyzed single-shell carbon nanotube synthesis. Chem Phys Lett 215:509–517
148. Seraphin S, Zhou D, Jiao J, Minke MA, Wang S, Yadav T, Withers JC (1994) Catalytic role of nickel, palladium, and platinum in the formation of carbon nanoclusters. Chem Phys Lett 217:191–198
149. Saito Y, Okuda M, Fujimoto N, Yoshikawa T, Tomita M, Hayashi T (1994) Single wall carbon nanotubes growing radially from Ni fine particles formed by arc evaporation. Jpn J Appl Phys Part 2(33):L526–L529

150. Zhou D, Seraphin S, Wang S (1994) Single-walled carbon nanotubes growing radially from YC_2 particles. Appl Phys Lett 65:1593–1595
151. Saito Y, Nishikubo K, Kawabata K, Matsumoto T (1996) Carbon nanocapsules and single layered nanotubes produced with platinum group metals (Ru, Rh, Pd, Os, Ir, Pt) by arc discharge. J Appl Phys 80:3062–3067
152. Saito Y, Okuda M, Koyama T (1996) Carbon nanocapsules and single-wall carbon nanotubes formed by arc evaporation. Surf Rev Lett 3:863–867
153. Chen B, Zhao X, Inoue S, Ando Y (2010) Fabrication and dispersion evaluation of single-wall carbon nanotubes produced by FH-arc discharge method. J Nanosci Nanotechnol 10:3973–3977
154. Chen B, Inoue S, Ando Y (2009) Raman spectroscopic and thermogravimetric studies of high-crystallinity SWNTs synthesized by FH-arc discharge method. Diamond Relat Mater 18:975–978
155. Fan WW, Zhao J, Lv YK, Bao WR, Liu XG (2010) Synthesis of SWNTs from charcoal by arc-discharging. J Wuhan Univ Technol-Mater Sci Ed 25:194–196
156. Ando Y, Zhao X, Kataura H, Achiba Y, Kaneto K, Tsuruta M, Uemura S, Iijima S (2000) Multiwalled carbon nanotubes prepared by hydrogen arc. Diamond Relat Mater 9:847–851
157. Pillai SK, Augustyn WG, Rossouw MH, McCrindle RI (2008) The effect of calcination on multi-walled carbon nanotubes produced by Dc-arc discharge. J Nanosci Nanotechnol 8:3539–3544
158. Chaturvedi P, Verma P, Singh A, Chaudhary PK, Harsh, Basu PK (2008) Carbon nanotube–purification and sorting protocols. Def Sci J 58:591–599
159. Saeed K (2010) Review on properties, dispersion and toxicology of carbon nanotubes. J Chem Soc Pak 32:559–564
160. Pillai SK, Ray SS, Moodley M (2007) Purification of single-walled carbon nanotubes. J Nanosci Nanotechnol 7:3011–3047
161. MacKenzie K, Dunens O, Harris AT (2009) A review of carbon nano-tube purification by microwave assisted acid digestion. Sep Purif Technol 66:209–222
162. Xu Y, Dervishi E, Biris AR, Biris AS (2011) Chirality-enriched semiconducting carbon nanotubes synthesized on high surface area MgO-supported catalyst. Mater Lett 65:1878–1881
163. Okazaki T, Shinohara H (2003) Synthesis and characterization of single-wall carbon nanotubes by hot-filament assisted chemical vapor deposition. Chem Phys Lett 5–6:606–611
164. Chen YM, Zhang HZ (2011) In: Bu JL, Jiang ZY, Jiao S (eds) Advances in composites. Trans Tech Publications Ltd, Stafa-Zurich, Pts 1 and 2, pp 1560–1563
165. Varshney D, Weiner BR, Morell G (2010) Growth and field emission study of a monolithic carbon nanotube/diamond composite. Carbon 48:3353–3358
166. Patole SP, Alegaonkar PS, Lee HC, Yoo YB (2008) Optimization of water assisted chemical vapor deposition parameters for super growth of carbon nanotubes. Carbon 46:1987–1993
167. Tempel H, Joshi R, Schneider JJ (2010) Ink jet printing of ferritin as method for selective catalyst patterning and growth of multiwalled carbon nanotubes. Mater Chem Phys 121:178–183
168. Byon HR, Lim H, Song HJ, Choi HC (2007) A synthesis of high purity single-walled carbon nanotubes from small diameters of cobalt nanoparticles by using oxygen-assisted chemical vapor deposition process. Bull Korean Chem Soc 28:2056–2060
169. Brown B, Parker CB, Stoner BR, Glass JT (2011) Growth of vertically aligned bamboo-like carbon nanotubes from ammonia/methane precursors using a platinum catalyst. Carbon 49:266–274
170. Kim HD, Lee Choi WS (2011) direct growth of carbon nanotubes with a catalyst of nickel nanoparticle-coated alumina powders. J Korean Phys Soc 58:112–115
171. Steiner SA, Baumann TF, Bayer BC, Blume R, Worsley MA, MoberlyChan WJ, Shaw EL, Schlogl R, Hart AJ, Hofmann S, Wardle BL (2009) Nanoscale Zirconia as a nonmetallic catalyst for graphitization of carbon and growth of single- and multiwall carbon nanotubes. J Am Chem Soc 131:12144–1215

172. Zhu YJ, Lin TJ, Liu QX, Chen YL, Zhang GF, Xiong HF, Zhang HY (2006) The effect of nickel content of composite catalysts synthesized by hydrothermal method on the preparation of carbon nanotubes. Mater Sci Eng B 127:198–202
173. Cumings J, Mickelson W, Zettl A (2003) Simplified synthesis of double wall carbon nanotubes. Solid State Commun 126:359–362
174. Deck CP, Mckee GSB, Vecchio KS (2006) Synthesis optimization & characterization of multi walled carbon nanotubes. J Electron Mater 35:211–223
175. Kim SM, Gangloff L (2009) Growth of carbon nanotubes (CNTs) on metallic underlayers by diffusion plasma-enhanced chemical vapour deposition (DPECVD). Physica E Low Dimens Syst Nanostruct 41:1763–1766
176. Wang H, Moore JJ (2008) Different growth mechanisms of vertical carbon nanotubes by RF- or dc-plasma enhanced chemical vapor deposition at low temperature. J Vac Sci Technol B 28:1081–1085
177. Flahaut E, Laurent C, Peigney A (2005) Catalytic CVD synthesis of double and triple-walled carbon nanotubes by the control of the catalyst preparation. Carbon 43:375–383
178. Xiang X, Zhang L, HimaHI Li F, Evans DG (2009) Co-based catalysts from Co/Fe/Al layered double hydroxides for preparation of carbon nanotubes. Appl Clay Sci 42:405–409
179. Luais E, Thobie-Gautier C, Tailleur A, Djouadi MA, Granier A, Tessier PY, Debarnot D, Poncin-Epaillard F, Boujtita M (2010) Preparation and modification of carbon nanotubes electrodes by cold plasmas processes toward the preparation of amperometric biosensors. Electrochim Acta 55:7916–7922
180. Haffner M, Schneider K, Schuster BE, Stamm B, Latteyer Fleischer M, Burkhardt Chasse T, Stett A, Kern DP (2010) Plasma enhanced chemical vapor deposition grown carbon nanotubes from ferritin catalyst for neural stimulation microelectrodes. Microelectron Eng 87:734–737
181. Yamada T, Namai T, Hata K, Futaba DN, Mizuno K, Fan J, Yudasaka M, Yumura M, Iijima S (2006) Size-selective growth of double-walled carbon nanotube forests from engineered iron catalysts. Nat Nanotechnol 1:131–136
182. Seo JK, Jung H, Lee JH, Deok SY, Young JJ, Choi WS (2010) Metal-free CNTs grown on glass substrate by microwave PECVD. Curr Appl Phys 10:S447–S450
183. Byeon H, Kim SY, Koh KH, Lee S (2010) Growth of ultra long multiwall carbon nanotube arrays by aerosol-assisted chemical vapor deposition. J Nanosci Nanotechnol 10:6116–6119
184. Jeong N, Seo Y, Lee J (2007) vertically aligned carbon nanotubes synthesized by the thermal pyrolysis with an ultrasonic evaporator. Diamond Relat Mater 16:600–608
185. Liu J, Zhang Y, Ionescu MI, Li R, Sun X (2011) Nitrogen-doped carbon nanotubes with tunable structure and high yield produced by ultrasonic spray pyrolysis. Appl Surf Sci 257:7837–7844
186. Ionescu MI, Zhang Y, Li R, Sun X, Abou-Rachid H, Lussier L-S (2011) Hydrogen-free spray pyrolysis chemical vapor deposition method for the carbon nanotube growth: parametric studies. Appl Surf Sci 257:6843–6849
187. Kucukayan G, Ovali R, Ilday S, Baykal B, Yurdakul H, Turan S, Gulseren O, Bengu E (2011) An experimental and theoretical examination of the effect of sulphur on the pyrolytically grown carbon nanotubes from sucrose-Based Solid State Precursors. Carbon 49:508–517
188. Endo M, Takeuchi K, Igarashi S, Kobori K, Shiraishi M, Kroto HW (1993) The production and structure of pyrolytic carbon nanotubes (PCNTs). J Phys Chem Solids 541:841–1848
189. Dresselhaus MS, Dresselhaus G, Eklund PC (1996) Science of fullerenes and carbon nanotubes. Academic Press, San Diego, pp 763–764
190. Isaacs JA, Tanwani A, Healy ML, Dahlben LJ (2010) Economic assessment of single-walled carbonnanotube processes. J Nanopart Res 12:551–562
191. Nikolaev P, Bronikowski MJ, Bradley RK, Rohmund F, Colbert DT, Smith KA, Smalley RE (1999) Gas-phase catalytic growth of single-walled carbon nanotubes from carbon monoxide. Chem Phys Lett 313:91

192. Bronikowski MJ, Willis PA, Colbert DT, Smith KA, Smalley RE (2001) Gas-phase production of carbon single-walled nanotubes from carbon monoxide via the HiPco process: a parametric study. J Vac Sci Technol A 19:1800–1805
193. Smalley RE, Yakobson B (1998) the future of the fullerenes. Solid State Com 107:597–606
194. Tang ZK, Zhang L, Wang N, Zhang XX, Wen GH, Li GD (2001) Superconductivity in 4 Å single walled carbon nanotubes. Science 292:2462–2465
195. Lolli G, Zhang LA, Balzano L, Sakulchaicharoen N, Tan YQ, Resasco DE (2006) Tailoring (n, m) structure of single-walled carbon nanotubes by modifying reaction conditions and the nature of the support of CoMo catalysts. J Phys Chem B 110:2108–2115
196. Alvarez WE, Kitiyanan B, Borgna A, Resasco DE (2001) Synergism of Co and Mo in the catalytic production of single-wall carbon nanotubes by decomposition of CO. Carbon 39:547–558
197. Wal V, Randall L, Hall LJ, Berger GM (2002) Optimization of flame synthesis for carbon nanotubes using supported catalyst. J Phys Chem B 106:13122–13132
198. Wal V, Randall L, Ticich TM (2001) Flame and furnace synthesis of single-walled and multi-walled carbon nanotubes and nanofibers. J Phys Chem B 105:10249–10256
199. Liu YC, Zheng NN, Huang JD, Sun BM (2011) In: Wang CH, Ma LX, Yang W (eds) Advanced polymer science and engineering. Trans Tech Publications Ltd, Stafa-Zurich, pp 99–103
200. Liu YC, Sun BM, Ding ZY (2011) In: Wang CH, Ma LX Yang W (eds) Advanced polymer science and engineering, pp 235–239
201. Dewulf DW, Jin T, Bard AJ (1989) Electrochemical and surface studies of carbon dioxide reduction to methane and ethylene at copper electrodes in aqueous solutions. J Electrochem Soc 136:1686–1691
202. Shawky A, Yasuda S, Murakoshi K (2012) Room-temperature synthesis of single-wall carbon nanotubes by an electrochemical process. Carbon 11:4184–4191
203. Zhou D, Chow L (2004) electrochemical deposition of carbon nanoparticles from organic solutions. US Patent 6,758,957 B1
204. Zhou D, Anoshkina EV, Chow L, Chai G (2006) Synthesis of carbon nanotubes by electrochemical deposition at room temperature. Carbon 44:1013–1024
205. Stankovich S, Piner RD, Nguyen ST, Ruoff RS (2006) Synthesis and exfoliation of isocyanate-treated graphene oxide nanoplatelets. Carbon 44:3342–3347
206. Wang S, Tang LAI, Bao Q, Lin M, Deng S, Goh BM Loh KP (2009) Room-temperature synthesis of soluble carbon nanotubes by the sonication of graphene oxide nanosheets. J Am Chem Soc 131:16832–16837
207. Singh V, Joung D, Zhai L, Das S, Khondaker SI, Seal S (2011) Graphene based materials: past, present and future. Prog Mater Sci 56:1178–1271
208. Soldano C, Mahmood A, Dujardin E (2010) Production properties and potential of graphene. Carbon 48:2127–2150
209. Kim H, Abdala AA, Macosko CW (2010) Graphene/polymer nanocomposites. Macromolecules 43:6515–6530
210. Fernandez-Moran H (1960) Single crystal of graphite and mica as specimen support for electron microscopy. J Appl Phys 31:1840
211. Mason TJ, Lorimer JP (2002) Applied sonochemistry: the uses of power ultrasound in chemistry and processing. Wiley-VCH, Weinheim
212. Israelachvili JN (2011) Intermolecular and surface forces. Academic press, New York (revised 3rd edition)
213. Hernandez Y, Nicolosi V, Lotya M, Blighe FM, Sun Z, De S, McGovern IT, Holland B, Byrne M, Gun'Ko YK, Boland JJ, Niraj P, Duesberg G, Krishnamurthy S, Goodhue R, Hutchison J, Scardaci V, Ferrari AC, Coleman JN (2008) High-yield production of graphene by liquid-phase exfoliation of graphite. Nat Nanotechnol 3:563–568
214. Ciesielski A, Samori P (2014) Graphene via sonication assisted liquid-phase exfoliation. Chem Soc Rev 43:381–398

215. Parvez K, Li RJ, Puniredd SR, Hernandez Y, Hinkel F, Wang S, Feng X, Müllen K (2013) Electrochemically exfoliated graphene as solution-processable, highly conductive electrodes for organic electronics. ACS Nano 7:3598–3606
216. Lee JH, Shin DW, Makotchenko VG, Nazarov AS, Fedorov VE, Kim YH, Choi JY, Kim JM, Yoo JB (2009) One-step exfoliation synthesis of easily soluble graphite and transparent conducting graphene sheets. Adv Mater 21:4383–4387
217. Liu BN, Luo F, Wu H, Liu Y, Zhang C, Chen J (2008) One-step ionic-liquid-assisted electrochemical synthesis of ionic-liquid-functionalized graphene sheets directly from graphite. Adv Funct Mater 18:1518–1525
218. Su CY, Lu AY, Xu YP, Chen FR, Khlobystov AN, Li LJ (2011) High-quality thin graphene films from fast electrochemical exfoliation. ACS Nano 5:2332–2339
219. Lu J, Yang JX, Wang JZ, Lim AL, Wang S, Loh KP (2009) One-pot synthesis of fluorescent carbon nanoribbons, nanoparticles, and graphene by the exfoliation of graphite in ionic liquids. ACS Nano 3:2367–2375
220. Liu JL, Yang HP, Zhen SG, Poh CK, Chaurasia A, Luo JS, Wu XY, Yeow EKL, Sahoo NG, Lin JY, Shen ZX (2013) A green approach to the synthesis of high-quality graphene oxide flakes via electrochemical exfoliation of pencil core. RSC Adv 3:11745–11750
221. Eda G, Fanchini G, Chhowalla M (2008) Large-area ultrathin films of reduced graphene oxide as a transparent and flexible electronic material. Nat Nanotechnol 3:270–274
222. Shang NG, Papakonstantinou P, Sharma S, Lubarsky G, Li MX, McNeill DW, Quinn AJ, Zhou WZ, Blackley R (2012) Controllable selective exfoliation of high-quality graphene nanosheets and nanodots by ionic liquid assisted grinding. Chem Commun 48:1877–1879
223. Parvez K, Wu ZS, Li R, Liu X, Graf R, Feng X, Müllen K (2014) Exfoliation of graphite into graphene in aqueous solutions of inorganic salts. J Am Chem Soc 136:6083–6091
224. Rai PK, Pinnick RA, Parra-Vasquez ANG, Davis VA, Schmidt HK, Hauge RH, Smalley RE, Pasquali M (2006) Isotropic-nematic phase transition of single-walled carbon nanotubes in strong acids. J Am Chem Soc 128:591–595
225. Behabtu N, Lomeda JR, Green MJ, Higginbotham AL, Sinitskii A, Kosynkin DV, Tsentalovich D, Parra-Vasquez ANG, Schmidt J, Kesselman E, Cohen Y, Talmon Y, Tour JM, Pasquali M (2010) Spontaneous high-concentration dispersions and liquid crystals of graphene. Nat Nanotechnol 5:406–411
226. Brodie BC (1859) On the atomic weight of graphite. Philos Trans R Soc Lond 149:249–259
227. Staudenmaier L (1893) Verfahren zur darstellung der graphitsäure. Ber Dtsch Chem Ges 31:1481–1487
228. Hummers WS, Offeman RE (1958) Preparation of graphitic oxide. J Am Chem Soc 80:1339–1339
229. Marcano DC, Kosynkin DV, Berlin JM, Sinitskii A, Sun Z, Slesarev A, Alemany LB, Lu W, Tour JM (2010) Improved synthesis of graphene oxide. ACS Nano 4:4806–4814
230. Li W, Xu Z, Chen L, Shan M, Tian X, Yang C, Lv H, Qian X (2014) A facile method to produce graphene oxide-g-poly(L-lactic acid) as an promising reinforcement for PLLA nanocomposites. Chem Eng J 237:291–299
231. Stankovich S, Dikin DA, Piner RD, Kohlhaas KA, Kleinhammes A, Jia Y, Wu Y, Nguyen ST, Ruoff RS (2007) Synthesis of graphene-based nanosheets via chemical reduction of exfoliated graphite oxide. Carbon 45:1558–1565
232. Wang G, Shen X, Wang B, Yao J, Park J (2009) Synthesis and characterisation of hydrophilic and organophilic graphene nanosheets. Carbon 47:1359–1364
233. Stankovich S, Piner RD, Chen X, Wu N, Nguyen ST, Ruoff RS (2006) Stable aqueous dispersions of graphitic nanoplatelets via the reduction of exfoliated graphite oxide in the presence of poly (sodium 4-styrenesulfonate). J Mater Chem 16:155–158
234. Lomeda JR, Doyle CD, Kosynkin DV, Hwang WF, Tour JM (2008) Diazonium functionalization of surfactant-wrapped chemically converted graphene sheets. J Am Chem Soc 130:16201–16206
235. Tung VC, Allen MJ, Yang Y, Kaner RB (2009) High-throughput solution processing of large-scale graphene. Nat Nanotechnol 4:25–29

236. Shen JF, Hu YZ, Shi M, Lu X, Qin C, Li C, Ye MX (2009) Fast and facile preparation of graphene oxide and reduced graphene oxide nanoplatelets. Chem Mater 21:3514–3520
237. Wang GX, Yang J, Park J, Gou XL, Wang B, Liu H, Yao J (2008) Facile synthesis and characterization of graphene nanosheets. J Phys Chem C 112:8192–8195
238. Fan BX, Peng W, Li Y, Li X, Wang S, Zhang G, Zhang F (2008) Deoxygenation of exfoliated graphite oxide under alkaline conditions: a green route to graphene preparation. Adv Mater 20:4490–4493
239. Shen X, Jiang L, Ji Z, Wu J, Zhou H, Zhu G (2011) Stable aqueous dispersions of graphene prepared with hexamethylenetetramine as a reductant. J Colloid Interface Sci 354:493–497
240. Bai S, Shen X (2012) Graphene–inorganic nanocomposites. RSC Adv 2:64–98
241. Fernandez-Merino MJ, Guardia L, Paredes JI, Villar-Rodil S, Solis-Fernandez P, Martinez-Alonso A, Tascon JMD (2010) Vitamin c is an ideal substitute for hydrazine in the reduction of graphene oxide suspensions. J Phys Chem C 114:6426–6432
242. Gao J, Liu F, Liu YL, Ma N, Wang ZQ, Zhang X (2010) Environment-friendly method to produce graphene that employs vitamin C and amino acid. Chem Mater 22:2213–2218
243. Zhu C, Guo S, Fang Y, Dong S (2010) Reducing sugar: new functional molecules for the green synthesis of graphene nanosheets. ACS Nano 4:2429–2437
244. Dreyer DR, Murali S, Zhu Y, Ruoff RS, Bielawski CW (2011) Reduction of graphite oxide using alcohols. J Mater Chem 21:3443–3447
245. Zhang Z, Cheng H, Xing C, Guo M, Xu F, Wang X, Gruber HJ, Zhang B, Tang J (2011) Sodium citrate: a universal reducing agent for reduction/ decoration of graphene oxide with au nanoparticles. Nano Res 4:599–611
246. Wang Y, Shi ZX, Yin J (2011) facile synthesis of soluble graphene via a green reduction of graphene oxide in tea solution and its biocomposites. ACS Appl Mater Interfaces 3:1127–1133
247. Roy I, Bhattacharyya A, Sarkar G, Saha NR, Rana D, Ghosh PP, Palit M, Das AR, Chattopadhyay D (2014) In situ synthesis of a reduced graphene oxide/cuprous oxide nanocomposite: a reusable catalyst. RSC Adv 4:52044–52052
248. Roy I, Rana D, Sarkar G, Bhattacharyya A, Saha NR, Mondal S, Pattanayak S, Chattopadhyay S, Chattopadhyay D (2015) Physical and electrochemical characterization of reduced graphene oxide/silver nanocomposites synthesized by adopting a green approach. RSC Adv 5:25357
249. Chen W, Yan L, Bangal PR (2010) Preparation of graphene by the rapid and mild thermal reduction of graphene oxide induced by microwaves. Carbon 48:1146–1152
250. Zhou M, Wang Y, Zhai Y, Zhai J, Ren W, Wang F, Dong S (2009) Controlled synthesis of large-area and patterned electrochemically reduced graphene oxide films. Chem Eur J 15:6116–6120
251. An SJ, Zhu Y, Lee SH, Stoller MD, Emilsson T, Park S, Velamakanni A, An J, Ruoff RS (2010) Thin film fabrication and simultaneous anodic reduction of deposited graphene oxide platelets by electrophoretic deposition. J Phys Chem Lett 1:1259–1263
252. Shao Y, Wang J, Engelhard M, Wang C Lin Y (2010) Facile and controllable electrochemical reduction of graphene oxide and its applications. J Mater Chem 20:743–748
253. Yang X, Dou X, Rouhanipour A, Zhi L, Rader HJ, Mullen K (2008) Two-dimensional graphene nanoribbons. J Am Chem Soc 130:4216–4217
254. Chee SY, Poh HL, Chua CK, Šaněk F, Sofer Z, Pumera M (2012) Influence of parent graphite particle size on the electrochemistry of thermally reduced graphene oxide. Phys Chem Chem Phys 14:12794–12799
255. Badami DV (1962) Graphitization of α-silicon carbide. Nature 193:569–570
256. Sutter PW, Flege JI, Sutter EA (2008) Epitaxial graphene on ruthenium. Nat Mater 7:406–411
257. De Parga ALV, Calleja F, Borca B, Passeggi MCG, Hinarejos JJ, Guinea F, Miranda R (2008) Periodically rippled graphene: growth and spatially resolved electronic structure. Phys Rev Lett 100:056807

258. Rao CNR, Subrahmanyam KS, Ramakrishna Matte HSS, Abdulhakeem B, Govindaraj A, Das B, Kumar P, Ghosh A, Late DJ (2010) A study of the synthetic methods and properties of graphenes. Sci Technol Adv Mater 11:054502
259. Subrahmanyam KS, Panchakarla LS, Govindaraj A, Rao CNR (2009) Simple method of preparing graphene flakes by an arc-discharge method. J Phys Chem C 113:4257–4259
260. Panchakarla LS, Govindaraj A, Rao CNR (2009) Boron- and nitrogen-doped carbon nanotubes and graphene. Inorg Chim Acta 363:4163–4174
261. Kim KS, Zhao Y, Jang H, Lee SY, Kim JM, Kim KS, Ahn JH, Kim P, Choi JY, Hong BH (2009) Large-scale pattern growth of graphene films for stretchable transparent electrodes. Nature 457:706–710
262. Kwon SY, Ciobanu CV, Petrova V, Shenoy VB, Bareno J, Gambin V, Petrov I, Kodambaka S (2009) Growth of Semiconducting Graphene on Palladium. Nano Lett 9:3985–3990
263. Coraux J, N'Diaye AT, Busse C, Michely T (2008) Structural coherency of graphene on Ir (111). Nano Lett 8:565–570
264. Li X, Cai W, An J, Kim S, Nah J, Yang D, Piner R, Velamakanni A, Jung I, Tutuc E, Banerjee SK, Colombo L, Ruoff RS (2009) Large-area synthesis of high-quality and uniform graphene films on copper foils. Science 324:1312–1314
265. Wang JJ, Zhu MY, Outlaw RA, Zhao X, Manos DM, Holloway BC, Mammana VP (2004) Free-standing subnanometer graphite sheets. Appl Phys Lett 85:1265–1267
266. Shinde DB, Majumder M, Pillai VK (2014) Counter-ion dependent, longitudinal unzipping of multi-walled carbon nanotubes to highly conductive and transparent graphene nanoribbons. Sci Rep 4:4363
267. Kosynkin DV, Higginbotham AL, Sinitskii A, Lomeda JR, Dimiev A, Price BK, Tour JM (2009) Longitudinal unzipping of carbon nanotubes to form graphene nanoribbons. Nature 458:872–876
268. Dabrowska A, Huczko A, Soszyński M, Bendjemil B, Micciulla F, Sacco I, Coderoni L, Bellucci S (2011) Ultra-fast efficient synthesis of one-dimensional nanostructures. Phys Status Solidi B 248:2704–2707
269. Kim CD, Min BK, Jung WS (2009) Preparation of graphene sheets by the reduction of carbon monoxide. Carbon 47:1610–1612
270. Chakrabarti A, Lu J, Skrabutenas JC, Xu T, Xiao Z, Maguireb JA, Hosmane NS (2011) Conversion of carbon dioxide to few-layer graphene. J Mater Chem 21:9491–9493
271. Wang C, Ju J, Yang Y, Tang Y, Lin J, Shi Z, Han RPS, Huang F (2013) In situ grown graphene-encapsulated germanium nanowires for superior lithium-ion storage properties. J Mater Chem A 1:8897–8902
272. Li N, Wang Z, Zhao K, Shi Z, Gu Z, Xu S (2009) Large scale synthesis of N-doped multi-layered graphene sheets by simple arc-discharge method. Carbon 48:255–259
273. Miao Q, Wang L, Liu Z, Wei B, Xu F, Fei W (2016) Magnetic properties of N-doped graphene with high Curie temperature. Sci Rep 6:21832

Surface Modification/Functionalization of Carbon Materials by Different Techniques: An Overview

Lalatendu Nayak, Mostafizur Rahaman and Radhashyam Giri

Abstract Carbon materials have agglomeration tendency because of high van der Wall force of interaction among the carbon particles. This agglomeration tendency has been an obstacle for their application in different fields. In order to reduce this agglomeration tendency and to explore their application areas, different surface modification/functionalization processes have been successfully developed by researchers. Surface functionalization reduces the agglomerating tendency of carbon materials and increases the carbon–polymer interfacial adhesion through covalent or ionic bonds. This chapter aims to depict an overview on the different types of surface functionalization techniques applied to different carbon materials like carbon blacks (CB), carbon nanotubes (CNTs), carbon nanofibers (CNFs), graphenes, and fullerenes. The methodology like wet oxidation (oxidation using nitric acid, sulfuric acid, hydrogen peroxide, potassium permanganate, etc.), dry oxidation (oxidation with air, ozone, plasma, etc.), amidation, silanization, silylation, polymer grafting, polymer wrapping, surfactant adsorption, and encapsulation have been presented with different examples. All the functionalization processes have been highlighted with their specific application. The gathering of different functionalization processes in this chapter will provide deep understanding regarding the selection of a particular technique for specific application.

L. Nayak (✉)
Phillips Carbon Black Ltd., Kolkata, West Bengal, India
e-mail: litupolymer@gmail.com

M. Rahaman
Department of Chemistry, College of Science, King Saud University, Riyadh 11451, Saudi Arabia
e-mail: mrahaman@ksu.edu.sa

R. Giri
Centre for Biopolymer Science and Technology (CBPST) a Unit of CIPET, Old JNM Campus, FACT Township, Eloor, Udyogamandal P.O., Kochi 683501, India
e-mail: giripolymer@gmail.com

M. Rahaman et al. (eds.), *Carbon-Containing Polymer Composites*, Springer Series on Polymer and Composite Materials,
https://doi.org/10.1007/978-981-13-2688-2_2

Keywords Functionalization · Carbon nanotube · Carbon nanofiber
Ozone · Plasma · Polymer grafting · Encapsulation

1 Introduction

This chapter discusses the surface modification of different carbon materials like CB, CNTs, CNFs, fullerene, and graphene. These materials are generally hydrophobic in nature and have low surface energy, which make these fillers incompatible with polar solvents and some polymer matrices. The strong interparticle van der Waals force of interaction and low surface energy tend to agglomerate these materials into macro-visible clusters. These clusters reduce the wetting properties with polymer matrices and act as barriers toward the effective load transfer from the polymer matrix to fillers. The carbon fillers with good wetting properties can be dispersed easily in polymer matrices and in different dispersion medium. The wetting properties of carbon materials are directly related to its degree of surface functionality. The uniform dispersion of fillers in the polymer matrix and desirable filler–polymer interaction are the major factors in order to produce composites with enhanced properties. The agglomerating tendency, poor dispersion, and weak interfacial bonding in carbon filler–polymer composite can be overcome by suitably modifying the surface of filler. This surface modification/functionalization reduces the agglomerating tendency of fillers by increasing filler–polymer interfacial adhesion through chemical bonding. Therefore, significant efforts have been directed in order to develop different methods to functionalize carbon fillers [1–5].

Functionalization methods can be conveniently divided into chemical functionalization (or covalent functionalization) and physical modification (or noncovalent functionalization) method based on the type of interactions involved between the active molecules and carbon atoms on the fillers. During chemical functionalization, different functional groups are attached to the surface of filler by covalent bonds. The chemically functionalized filler can produce strong interfacial bonds with polar or semipolar mediums/polymers and allow the composites to possess high mechanical strength and other functional properties. Functionalized fillers disperse easily in different polar media and enhance color properties in paint, costing, and ink.

In case of conductive applications, covalent functionalization of fillers is not required. During chemical treatment, $\pi-\pi$ conjugation system on the surface of filler which is responsible for electrical conduction is damaged with the change of hybridization from sp^2 to sp^3. Hence, there is a great interest in noncovalent functionalization in these areas. In this approach, the $\pi-\pi$ conjugation system in the surface graphene layer of CNTs/CNFs is not disrupted and their full original properties are retained. CNTs/CNFs undergo noncovalent functionalization by forming van der Wall force of interaction between the tube wall and the reacting molecules or by wrapping molecules helically round the tubes.

2 Different Techniques for the Modification/ Functionalization of Carbon Materials

Surface functionalization/modification of carbon materials is done basically by some oxidative processes. The oxidized materials again act as precursors for further chemical modifications, like silanation [6], polymer grafting [7], esterification [8], alkylation, arylation [9], etc. Oxidation takes place in wet condition using strong acids such as HNO_3, H_2SO_4 or a mixture of them [10], or with other strong oxidants such as $KMnO_4$ [11], and also in dry condition like ozone [6, 12], and reactive plasma [13, 14]. Nondestructive surface modification is done by wrapping and encapsulating processes [15, 16]. Taking the example of one carbon nanotube, possible reactive sites and functionalization processes of carbon materials are shown in Fig. 1.

2.1 *Surface Oxidation*

Carbon filler having highly polar surfaces is essential for high-quality coating, printing inks, and for reinforcing in polar polymers. Oxidized fillers improve

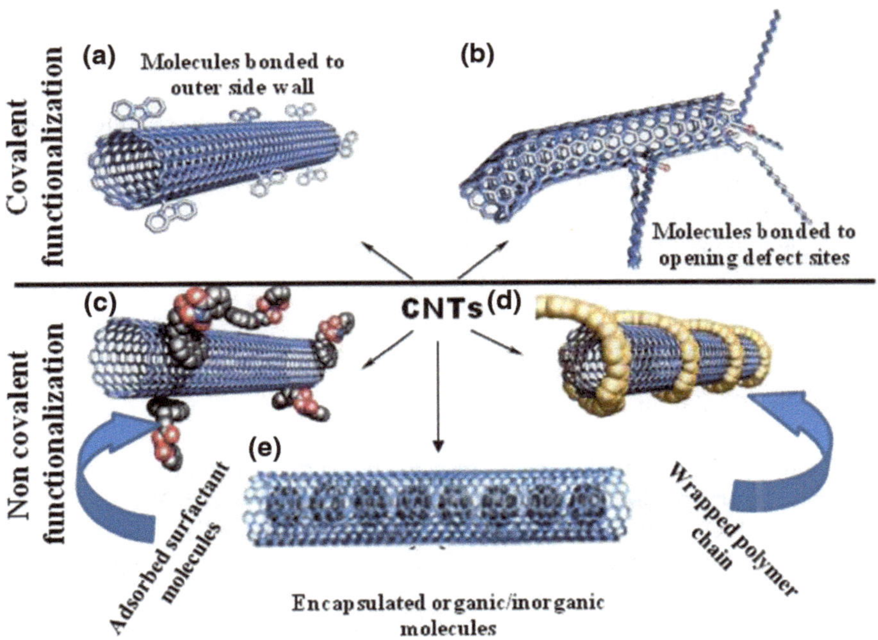

Fig. 1 Possible sites for functionalization and functionalization processes for CNTs: **a** covalent sidewall functionalization, **b** functionalization at defective sites, **c** noncovalent functionalization by surfactants adsorption, **d** wrapping of polymer chains, and **e** encapsulation of foreign materials [2]

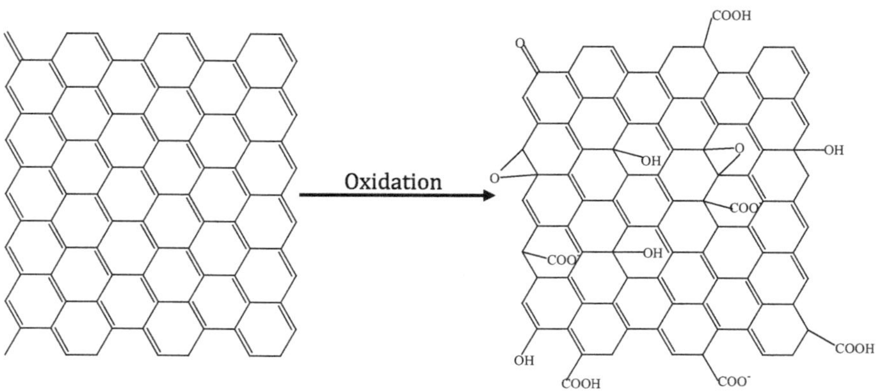

Fig. 2 Schematic presentation on the attachment of different functional groups after surface oxidation of carbon material

wettability and reinforcing efficiency in polymer compounds and enhance dispersion stability in the binder system (stability in storage) and the rheological behavior of the paint and printing ink. In case of carbon black system, furnace black possesses very small quantities of oxygen in the form of basic surface oxides. Similarly, CNTs and CNFs have also inert surfaces. A considerable amount of oxygen-containing functional groups can be attached on the surface of carbon materials by treating with strong oxidizing agents in a gas phase or an aqueous medium. Common oxidizing agents for large-scale industrial processes are ozone, air, nitric oxide/air mixtures, and nitric acid. Another highly efficient method for surface oxidation is plasma method, where powdery fillers are modified efficiently in a very short time. The attachment of functional groups to the carbon materials after oxidation is shown schematically in Fig. 2, where different functional groups like carboxylic, ketonic, quinine, hydroquinone, phenolic, or hydroxylic groups are attached to the surface of carbon materials surface by covalent bonding.

2.1.1 Wet Oxidation

Wet oxidation is a cost-effective process for the surface modification of carbon materials. Strong oxidizing agents such as HNO_3, $HNO_3 + H_2SO_4$ [10, 17, 18], $H_2SO_4 + KMnO_4$ [19], and $H_2SO_4 + H_2O_2$ [20, 21] are used to oxidize carbon fillers. By the surface oxidation, different oxygen-containing polar groups so-called volatile constituents are attached to the surface of carbon materials. The volatile constituents or polar groups have a crucial influence on the dispersion of carbon materials in aqueous systems. The greater the content of volatile constituents in the carbon materials, the greater is their hydrophilic character and the better is their dispersion in aqueous-based binder systems.

Wet oxidation of carbon blacks

From many years, channel carbon blacks have been exclusively used in printing inks and other coloring applications because of their excellent tint strength and high volatile contents. The volatile content of commercial channel blacks is of more than 5% by weight. The fluffy form and high volatile content readily disperse these blacks in solvents and contribute to excellent flow characteristics. Channel blacks are of high cost. Hence, maximum ink and color industries are using furnace black as a substitute of channel black. It is estimated that at least about 70% of all inks and color contain furnace blacks as pigments.

Furnace blacks have very low surface polar groups (volatile content less than 1 wt%) and hence adversely affect the dispersibility in pigment. To enhance the dispersibility and flow behavior, different surfactants are used. Without any surfactants, the flow behavior is improved by increasing the volatile content on the surface of furnace blacks by way of oxidation process. Many works have been carried out to prepare surface oxidized furnace blacks of excellent flow properties in ink vehicle and excellent tinting strength with a high degree of dispersibility, which will be suitable for use in high concentrations in different media. Out of them, nitric acid treatment is a very popular industrial scalable process in carbon black industries to convert low polar furnace carbon black to high polar color carbon black. Nitric acid is added to the pelletizing water and then the carbon black is oxidized at elevated temperatures during drying.

Jordan et al. [22] in their work developed flow behavior and color properties of a furnace black through nitric acid oxidation and fluid energy attrition process. In their study, furnace carbon black was mixed with nitric acid and rotated at 3 rpm in a drum drier at temperature 300 °F. The oxidized black was further subdivided by the action of a fluid energy mill. The so treated blacks showed superior PC fitness value, flow characteristics, and color properties than those of commercial channel black pigment.

In another study, Boonstra et al. [23] oxidized different grades of furnace carbon black with nitric acid and studied their reinforcing efficiency in butyl rubber. Oxidation was carried out taking different concentrations of nitric acid as shown in Table 1. The black was well mixed with nitric acid through stirring and pushed through an 8 mesh screen to break up any large lumps. The damp granules were

Table 1 The effect of nitric acid treatment on the surface properties of carbon black [23]

Carbon black	Concentration of HNO_3 (%) Black:HNO_3 = (1:1.1)	Volatile (%)	pH	N_2SA (m^2/g)
Furnace black	0	1.9	7.6	95.6
	5	3.8	3.3	
	10	4.9	2.9	
	15	5.7	3.2	
	25	6.6	3.2	96.4
	25[a]	7.6	2.4	105

[a]220 parts of 25%

then dried for 24 h at 250 °F (121 °C) in a carefully controlled forced air oven. They observed that the proportion of solution to black is not especially critical but the amount of solution should be at least enough to completely wet the black involved. The complete wetting of back was achieved by the use of approximately an equal amount of black to acid by weight. The polarity of black (% of volatile content) was increased with increasing concentration of nitric acid in solution. At 25% of nitric acid in solution, the percentage of volatile content was increased up to 6.6% at 1:1.1 black to nitric acid solution. At 5% nitric acid, a considerable amount (3.8%) of volatile content was achieved in black. Such chemically oxidized blacks improved tensile strength, modulus, rebound, hysteresis, and other such properties of the butyl vulcanizates. The black which has been treated with a solution of 15% HNO_3 has a higher tensile, higher percent elongation, lower modulus, lower hardness, lower hysteresis, and lower heat buildup than the corresponding composition made with untreated black.

In the surface oxidation process, the proportion of COOH groups is increased with the increase in oxidation time or concentration of oxidant; on the contrary, the proportion of OH-groups is increased to a specific amount and after that, it remains stationary, whatever the conditions of the oxidation process may be [24].

Oxidation of carbon black by hydrogen peroxide is another simple conventional process. Like nitric acid treatment, the carbon blacks are efficiently oxidized in industrial quantities by hydrogen peroxide treatment using standard carbon black production equipment and production rates. This treatment does not produce any hazardous emissions and does not require any special handling. Curtis et al. [25] used hydrogen peroxide solution for carbon black beading instead of water solution. Black was mixed with hydrogen peroxide solution at 1:1 weight ratio in pin beader. Wet black pellets were then dried for 1 h at 350 °F. Volatile content was increased from 2.6 to 3.8% by 30% hydrogen peroxide solution.

Water-based pigment ink is used in order to get a light and water resistance recorded image. Carbon black commercially available for color purpose is an acidic black. But, acidic black has no enough water dispersibility to be used as a colorant for water-based pigment ink. Therefore, carbon blacks are finely dispersed and stabilized in water medium by the use of different water-soluble dispersants like styrene-(meth)acrylic resin, styrene-maleic resin, etc. Resin-based ink is viscous. When it is used in ink jet ink printing or fine pen writing, it adheres to the orifice or pen tip. This adhered resin is difficult to dissolve again and orifice or pen tip is clogged. Another difficulty is its viscosity, which forms a resisting force for free flow of ink to the nozzle of ink jet printing head or tip of pen when continuous or high-speed printing or writing is conducted. To reduce the viscosity of this ink, low concentration of carbon black is used, which results in poor printing density. Hence, in order to overcome abovementioned problem, Nagasawa [26] prepared a water-based ink without using any dispersing aid, where an acidic CB was oxidized with a hypohalite salt (sodium hypohalite) in water. The resulted CB when applied to a printer showed high stability of ink jetting and printing was rapid and smooth. Water resistant of printed matter was good. In a similar study by Ito et al. [27] both channel black and acidic furnace black were treated with sodium hypochlorite.

These treated blacks also exhibited high print density and good storage stability even after storing the black for one month at 50 °C.

In 2014, Sekiyama et al. [28] described the modification of acidic carbon black which exhibited excellent blackness, dispersibility, and storage stability, and also excellent discharge capability in an aqueous black pigment ink. In their study, modified CB was prepared by neutralizing nearly 50% of acidic hydroxyl groups attached to CB with a polyvalent cation and remaining acidic hydroxyl groups with a monovalent cation.

Modification of CB for a nonaqueous ink was carried out by Adams and Belmont [29], where modified black was attached to at least one organic group and the organic group was attached to at least aromatic group and one ionic or ionizable group. CB was treated with salts of different organic materials, but the treatment of CB with sulfanilic acid and sodium nitrate exhibited more optically dense, jetter, bluer, and gloss properties to the coating compared to other treatments. In their study, modification was also done using both oxidized and non-oxidized CB. The oxidized CB product showed greater optical density, jetness, and bluer undertone than untreated CB.

Wet oxidation of CNTs and CNFs

Other different carbon materials like CNTs, CNFs, graphene, and graphite are also surface modified by the same processes as applied for carbon black modification. CNTs, CNFs, and graphene have received great attention in biomedical, electrical, electronics, and electromagnetic interference shielding applications due to their high stiffness and mechanical strength, high electrical and thermal conductivity (more conductive than copper), lightweight, and low density. The hydrophobic nature, high aspect ratio, high surface area, and strong van der Waal force of attraction tend CNTs and CNFs to agglomerates and consequently lead to inhomogeneous dispersion in different dispersing media. One of the important challenges in developing applications of these nanofillers is their homogeneous and reproducible dispersion in the applied medium. In order to realize their potential and attain the unique reinforcing and electronic conduction efficiency in applied medium, there is the necessity of homogeneous dispersion as well as a strong interfacial interaction between the filler particles and dispersing medium. Various types of chemical functionalizations [17–21] have been carried out by different researchers to increase their hydrophilicity for good dispersion and interfacial interaction with different media. Functional groups are attached on the sidewalls as well as at the open ends of CNTs/CNFs by different oxidative processes. The defects on CNTs/CNFs created by oxidants are stabilized by bonding with carboxylic acid (–COOH) or hydroxyl (–OH) groups.

Raw CNTs/CNFs have been oxidized by various types of oxidants at different processing conditions (varying oxidant concentration, temperature, treatment time and sonication time) to add functional groups on the surface. Oxidation of CNTs using concentrated acids is a popular functionalization method on the laboratory-scale in order to enhance the mechanical properties of polymer composites. Sahoo et al. oxidized raw MWNTs with H_2SO_4/HNO_3 (3:1) at 90 °C for

10 min with vigorous stirring and incorporated those in polyurathene. A 740% increase in modulus and 180% increase in tensile strength over pure PU were achieved at 20 wt% oxidized MWNT loading [17]. Chemical oxidation increases the dispersibility of CNTs and CNFs into the polymer matrix by covalent bonding and increases the mechanical and thermal properties of composites. But during the chemical functionalization, the ability of CNTs/CNFs to impart conductivity to the composite reduces because of the reduction in aspect ratio and damage of conjugated π bonds on the surface of CNTs/CNFs. So et al. [30] incorporated both oxidized and virgin CNTs into PI matrix and stated that unfunctionalized CNTs impart higher conductivity compared to functionalized one. A similar study was carried out Spitalsky et al. [31] introducing oxidized multi-walled carbon nanotubes (MWCNT) in epoxy, where a two fold increase in flexural modulus and two orders of magnitude decrease in DC conductivity were observed. In another study, Simsek et al. [32] reported that nanocomposites containing amino-functionalized carbon nanotubes exhibit relatively lower electrical conductivity compared to those with unfunctionalized ones.

2.1.2 Dry Oxidation (Gas Phase Oxidation)

Wet oxidation is unattractive for industrial scale functionalization because of its environment and health hazardous nature. Chemicals used in this process are highly corrosive and difficult to handle. During this process, different hazardous gases are eliminated, which may be subjected to regulatory scrutiny for environmental and health reason. This process reduces yield by overoxidation and also tend to form lumps or aggregates of carbon materials. Materials so treated obviously lose their fluffy nature so desired for pigment applications. In order to prepare loose, fluffy or if desired pelleted carbon materials without any effect on materials colloidal properties, different dry oxidation processes or called gas phase oxidation like air, ozone, steam, and mixture of air with some gases are used. Dry oxidation eliminates above mentioned problems during wet oxidation. It eliminates waste and drying steps.

Oxidation with Air

The oxidation of carbon materials using only air is a simple process, where surface hydrophilicity is increased along with the increase in surface area. Carbon materials in air start to oxidize at temperature above 300 °C. At temperature 300 °C, amorphous carbon starts to oxidize. Sufficient number of oxygen groups on the surface of carbon blacks can only be achieved at temperatures below 400 °C. Oxidation in the presence of both air and nitrogen dioxide provides more surface oxides compared to only air. The nitrogen dioxide acts as a catalyst in oxidizing the carbon surface. In this process, at first, the carbon black is oxidized with the mixture of nitrogen dioxide and air. Then in the second step, the oxidized carbon black is treated with hot air in order to desorb the nitrogen dioxide adsorbed on the carbon black surface at the time of oxidation [33]. Air oxidation technique is a time-consuming technique, which may take several hours, depending on the type of carbon black and the desired degree of oxidation. This type of aftertreatment is carried out preferentially in rotary drum or fluidized bed reactors.

Surface functionalization of CNTs/CNF/graphene by air oxidation takes place at higher temperature compared to carbon blacks. The amorphous carbons, surface layer with pentagons and heptagons are more reactive to air compared to hexagonal graphite layer of these fillers. At first, amorphous carbonaceous material is oxidized and etched out in the form of CO_2, then caped ends and curvature of these fillers are oxidized because these have a greater strain and contain pentagons or heptagons [34].

Li et al. [35] thermally oxidize multiwall carbon nanotubes in the presence of air in a tube type furnace. The quartz tube loaded with CNTs was placed in the oven and rotated at the rate of 25 rpm, so that sample was evenly exposed to the air in order to achieve uniform oxidation. They oxidized at different temperatures ranging from 480 to 750 °C for 20 min each and stated that air oxidation enhances the specific surface area and improved the intrinsic morphology of CNTs. Higher conductivity for air oxidized MWCNTs compared to acid oxidized ones at the same concentration was reported by Park et al. [36]. Similar results were also reported by Yuen et al. [37].

Oxidation with Ozone

Oxidation by O_3 is particularly a well suited industrially viable process to produce cost-effective modified carbon material. O_3 is now produced inexpensively in large quantity and also readily decomposed to O_2 in water or upon reacted with unsaturated molecules. Successful functionalization or modification of carbon materials using O_3 has been reported by many researchers [38–58].

Ozonation takes place basically at room temperature by properly exposing powder materials to ozone gas in a reactor. Ozone from ozone generator is passed through a reactor, where powder material is reacted with ozone. The remaining unreacted ozone is passed to the outside of reactor and is decomposed by an ozone decomposer. For the oxidation of all carbon particles, the carbon powders have to be kept in motion inside the reactor. Fluidized bed reactor, where carbon black is fluidized by incoming ozone gas, is particularly suitable for fine particles of high structure. Large particles of small structure are not fluidized by gas and these are treated with ozone in any stirred stationary reactor or rotary kilns. By the reaction of carbon materials with O_3, a variety of oxygen-bearing polar groups are generated. These functional groups permit carbon materials to disperse much more easily in different solvents.

Ozonation of Carbon Blacks

Carbon blacks are very reactive towards O_3 and their oxidation with gaseous ozone has been well studied. For pigment/ink purpose, long flow with high concentration of carbon black is required. To get a modified carbon black of good flow behavior and good color strength for ink or any color application, selection of suitable furnace black is very important; otherwise, the optimum combination of flow and color strength cannot be achieved. The furnace black subjected for modification should be of high surface area, small particle diameter, and very low structure. Carbon black having smaller particle diameter and smaller structure imparts higher flow characteristics and higher color in ink. Melore and Eckert [38] in their study showed a comparative statement on the effect of high color channel black and ozone oxidized furnace black in ink properties. An attempt to prepare ozone modified black at industrially scalable rate was taken. Carbon black was oxidized at different ratio of black to ozone concentration using one rotary drum reactor. On their modification condition, black to ozone ratio less than 17:1 showed flow behavior and blue tone much better than commercially available peerless channel black. They also observed ozonation at different temperature 20–400 °F and noticed that varying operating temperature over a range from 20 to 200 °F, no noticeable different in results, but at higher temperature around 300–400 °F, a pitting of surface of the black particles resulted.

In 1996, Sutherland et al. [39] studied the effect of ozone treatment on the surface of different furnace blacks using a fluidized bed reactor and examined the oxidation level and functional groups introduced on the surface by varying ozonation voltage. Ozone concentration in the O_3/O_2 mixture was increased and simultaneously the surface oxygen concentration was increased. A decrease in ozone concentration and increase in functionalization was observed with the increase in oxygen flow rate. When oxygen flow rate increases, O_3 concentration decreases, but the total volume of ozone increases, which increases the surface functionalization. Carbon black surface oxidation is also significantly affected by treatment time. Oxygen concentration on the surface of blacks is steadily increased with treatment time at constant voltage and at constant flow rate. From this study, it was also observed that under the same treatment conditions, surface oxygen concentration was decreased with increasing surface area as shown in Fig. 3a. They have also studied the effect of ozone oxidation on PH and stated that (Fig. 3b) the pH value initially decreases significantly with increasing oxygen concentration and reaches a limiting value around 2 above a surface oxygen concentration of around 6 at.%. The initial decrease may be attributed to the increase in acidic carboxylic groups on the surface of CB due to ozone oxidation. 6 wt% oxygen concentration may be related to the limiting monolayer oxidation level of CB. Further increase in surface oxygen concentration can be attributed to the increase in oxidation level below the first monolayer, and therefore it does not affect the pH value.

Fluidization of carbon blacks depends on their bulk density, surface area, structure, and particle size. The furnace black in the form of coarse particles do not form a stable fluidized bed. They often form channels through which carrier gases flow. For proper fluidization of carbon blacks with different carrier gases,

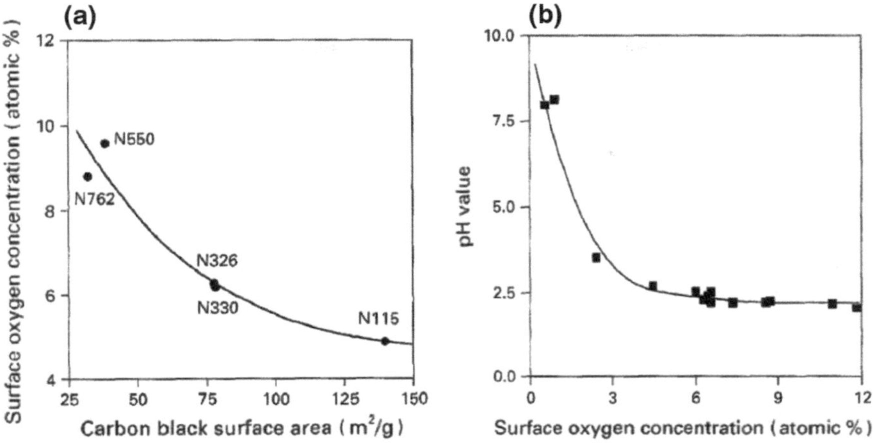

Fig. 3 **a** Effects of carbon black specific surface area on surface oxygen concentration by ozone treatment, **b** pH value of carbon black (N330) with various surface oxidation levels [39]

Stenger et al. [40] designed a fluidized reactor, where one additional gas stream was introduced to fluidized bed in the direction opposite to the main gas stream of the fluidized bed. The additional gas stream in counter direction creates complete fluidized bed for the carbon blacks having coarse particles. They exposed carbon black to ozone in this designed reactor and achieved 7.5% volatile content by supplying 50 gm/cm^3 ozone for 6.5 h.

Yeh et al. [41] ozonized carbon black in a solution phase. One channel carbon black was dispersed in water by one stirrer in a reactor vessel, then O_3 was supplied from the bottom side of reactor and carbon black was oxidized with O_3 dissolved in water. The reaction was carried out for 3 h and then adjusting pH of solution to 6.5–7 by adding sodium hydroxide, the treated black was recirculated through a microfluidizer for 5 h. The obtained product was self-dispersible without any particle sedimentation after 6 months.

Ozonation of CNTs/CNFs

Ozonation of other carbon materials like carbon nanotubes (MWCNTs and SWCNTs) have been successfully reported [42–46]. O_3 has been used for purification as well as surface modification CNTs and CNFs [47, 48]. By the ozone treatment, amorphous carbons are removed from the surface of nanotubes and a variety of functional groups, such as alcoholic, carboxylic, ketonic, esteric, and aldehydic moieties are generated on the surface of nanotubes. Functional groups are attached by a 1,3 dipolar cycloaddition of ozone to the double bonds present in CNTs following Criegee's mechanism [49]. These polar functional groups permit the carbon nanotubes to be dispersed easily in different solvents and in different polymers. Vennerberg et al. [42] used gas phase oxidation treatment with MWCNTs in a fluidized bed reactor. Significant level of oxidation (approx. 8 atomic % of O) was achieved with a very little damage to the sidewall of MWCNTs. They observed primarily hydroxyl functional groups are attached to the nanotubes surface during short exposure to O_3, but when oxidation takes place for long time, mainly carboxylic acid groups are formed. The defects present on the surface of nanotubes were found to play an important role in the oxidation mechanism.

The reaction of ozone with the carbon materials is basically carried out as a gas–solid process, but achieving of uniform oxidation of CNTs within a macroscopic agglomeration has been a great challenge, especially when the reaction size is large. Some reports are also available on the treatment of CNTs with ozone in solution phase, where CNTs are sonicated in water and O_3 is bubbled through the dispersion to oxidize CNTs to required level [19, 43–45, 47, 50–53]. In aqueous solution, O_3 can be decomposed into the hydroxyl radical, which is indeed stronger oxidant (standard reduction potential $E° = 3.06$ V) than the ozone molecule ($E° = 2.06$ V). This transformation of ozone into hydroxyl radicals is pH dependent [54, 55]. Morales-Lara et al. [56], in their study, compared the functionalization of multiwall carbon nanotubes by ozone at basic pH with oxygen plasma and ozone in gas phase and stated that oxidation efficiency of ozone in basic aqueous solution is less than that of ozone gas. At high pH, hydroxyl ions (OH^-) are adsorbed to the carbon atoms in CNTs, which kinetically restrict the hydroxyl radicals produced by O_3 in water to react with carbon atoms. Hence, an increase in basic pH in water reduces the oxidation efficiency of O_3.

Instead of water, Peng et al. [53] used water vapor with O_3 as a mixture to oxidize CNTs with low cost and mass production simultaneously. In their work, MWCNTs were exposed to the mixture of O_3 and water vapor at room temperature and more oxygenated functional groups were attached to the surface of CNTs compared to traditional approaches where high density of ozone is needed. In their study, the degree of oxidation using O_3-water vapor for the treatment time 0.5 h was more than five times to that of oxidation with only ozone for 6 h. Hydroxyl radicals might have been formed by the partial decomposition of O_3 in the presence of H_2O vapor [55, 57, 58]. These hydroxyl radicals are considered to be responsible for high degree of oxidation. The higher oxidation potential feature of hydroxyl radical compared to O_3 facilitates higher rate of electrophilic addition reaction with unsaturated "p" electrons present in CNTs [53, 59]. So, more amounts of oxygen-containing functional groups are introduced onto surface of CNTs.

Plasma

Plasma is an electrical conductive ionized gaseous form of a matter. When a gas is subjected to either electromagnetic field or high temperature, molecules in the gas are excited and variety of active species like electrons, ions, free radicals, and neutral atoms are formed. This ionized form of gas is called plasma. The electromagnetic energy is produced either by direct current (DC), or microwave (MW), or radiofrequency (RF) wave generator. It is a highly reactive chemical environment. Any substance subjected to this plasma environment is bombarded with the active species and bonding of new functional groups takes place on the surface of substance.

Compared to other oxidation processes of carbon materials, the plasma treatment process is of great interest because of its nonpolluting nature and very shorter reaction time. During plasma treatment, free radicals, ions, and metastable species cause ablation, cross-linking, and oxidation reactions depending on the nature of the gases used in the plasma zone.

Basically, films and flat materials are subjected to plasma treatment. Powder materials are not suitable for plasma treatment because of difficulties in their uniform exposure to plasma gas. This treatment is also extremely difficult for powdered carbon materials because of their highly agglomerating tendency and sensitive to vapor at the time of plasma treatment. The combination of plasma reactor geometry and plasma generator must be considered carefully for proper surface oxidation of a powder material.

Plasmas are of different types like high-pressure plasma (HPP), low-pressure plasma (LPP), and atmospheric pressure plasma (APP). In high HPP system, the plasma gas is generated in a high vacuum chamber. This type of plasma system is not suitable for the treatment of powdered materials. At the time of vacuuming the chamber, the powder materials taken for plasma treatment are sucked outside of the reactor chamber. Few literatures are available on the surface treatment of powdered carbon materials by low-pressure plasma system using different reactors like rotary drum reactor, downstream reactor, and circulating fluidized bed reactor [60–62].

Bruser et al. [63] modified the surface of carbon nanofibers by treating ammonia, O_2, CO_2, H_2O, and HCOOH using a rotary drum reactor under low-pressure plasma condition and observed improved wettability of fibers after treatment. Different plasma gases tested were found to show an increasing effectiveness in the following order: HCCO < H_2O < CO_2 < O_2. A study performed by Favia et al. [64] also used this type of reactor and observed an enhancement in hydrophillicity of carbon black granules by oxygen and ammonia RF plasma.

Heintze et al. [65] in their study used one fluidized bed reactor shown in Fig. 4 for the surface modification of CNFs in LPP condition. In this system, CNFs were fed to the plasma reactor from a reservoir by means of a screw system, and then the

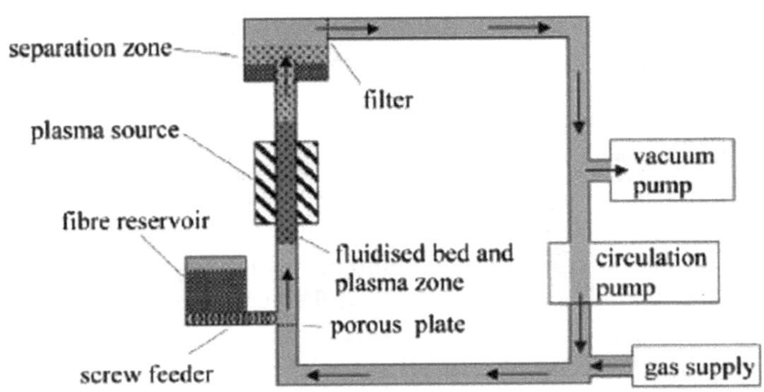

Fig. 4 Cold plasma reactor designed by Heintze et al. [65]

flow of gas forced the nanofibers to go throw the plasma and collected at the end of the reactor.

Although few literatures are available on plasma treatment of carbon materials at low and high pressure and carbon black surfaces are oxidized successfully, but the treatment of powder materials in high/low-pressure plasma condition is difficult, because in a vacuum condition it is difficult to stir, fluidize the powder materials for the homogeneous disposure to the plasma. High precautions are required during the pumping down time to avoid the powder traveling to the pump.

Treatment of carbon materials has been attempted using APP system. It can overcome some inconveniences in low/high-pressure plasma system but difficult for the uniform treatment of powdered material. The main advantages of APP are the elimination of vacuum systems, reduction of costs, and the possibility to be used in continuous systems. Compared to LPP, a very limited number of published papers are available in APP system [66, 67].

APP is generated by different ways like microwave, DC excitation (electric arc), AC excitation (corona discharge, dielectric barrier discharge). It is generated without chamber or a chamber at atmospheric pressure. Very limited works are available for the treatment of powder materials using APP, and basically, study is carried out by APP generated from dielectric barrier discharge (DBD), where one of the electrodes or both are covered by a dielectric layer in order to minimize electricity intensity and avoid arcing between the two electrodes.

First plasma treatment of carbon material using APP technique was done by Hanabasu et al. [68]; where activated carbon granules were coated with SiC using SiH_4–CH_4 and SiH_4–C_2H_4 as source gases. The reaction was carried out in a fluidized bed reactor coupled to plasma jet at atmospheric pressure and temperature. Erden et al. [69] in their study stated a continuous plasma oxidation of carbon fibers using APP technique. A schematic presentation of a continuous plasma modification setup is shown in Fig. 5, where plasma oxygen gas was ionized to form plasma jet. The reactor was designed using a long borosilicate glass. In order to maximize the fibers exposure to plasma oxygen, the carbon fiber roving was lined up by means of phenolic resin roller pins inside the reactor allowing the fibers to be looped 3 times through the plasma jet. Similar work has also been carried out by Bismark and his coworker [70]. They have introduced fluorine groups on the surface of carbon fibers by a continuous atmospheric plasma fluorination process to enhance the adhesion of fibers with fluoropolymers. Atmospheric plasma fluorination (APF) was conducted by ionizing chlorodifluoromethane in an open air Plasma Technology system (single rotating FLUME Jet RD1004, Plasmatreat, Steinhagen, Germany)

Surface treatment of CNTs and CNFs has also been extensively studied using air, oxygen, carbon dioxide, ammonia, nitrogen, fluorine, etc., as reactive plasma gages [67, 71–73]. Chirila et al. [74] studied the surface oxidation of CNFs using oxygen plasma. The effects of plasma power, chamber pressure, plasma frequency (radio frequency or microwave) as well as the effects of the treatment time on the quantity of functional groups attached to the surface of CNFs were examined.

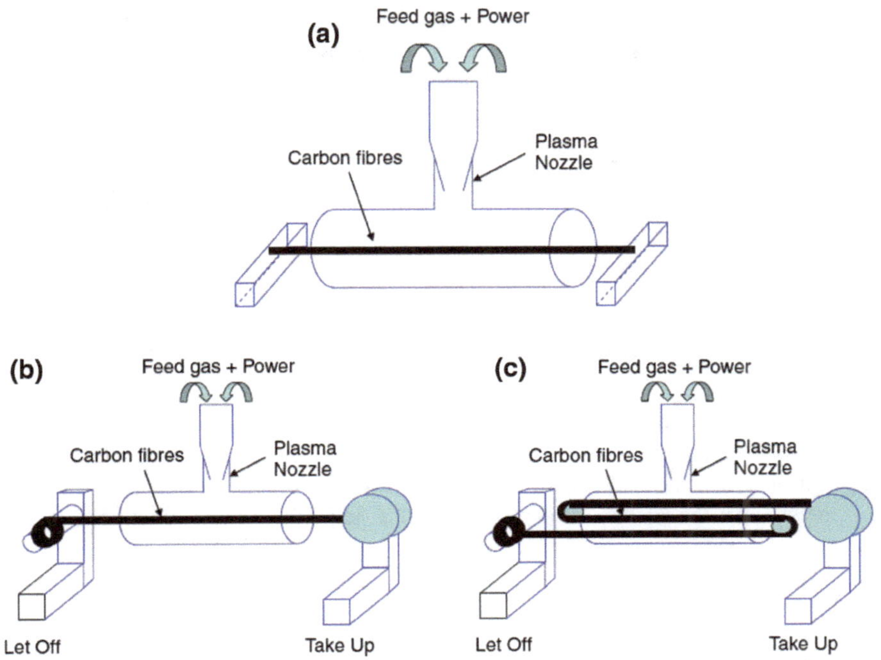

Fig. 5 Schematic presentation of different ways for carbon fibers fluorination by atmospheric plasma: **a** batch, **b** continuous single sided, and **c** continuous double-sided plasma treatment process [70]

In microwave plasma, the wettability of nanofibers was improved up to 68%, wherein RF plasma wettability was increased by 20%.

Bubert et al. [72] observed the influence of plasma power, treatment time, pressure of gas, and type of plasma gas on the surface functionality of CNTs using both vibrating as well as a fluidized bed reactor. Mixture of argon and oxygen or argon and carbon dioxide was used as plasma gas, treatment time was between 5 and 60 min, and plasma power was between 80 and 150 W. The efficiency of plasma treatment was gas pressure dependent. Oxygen concentration on CNTs surface was increased with the decrease in gas pressure. After some minutes of plasma treatment, the surface was saturated with oxygen-containing functional groups. Oxygen concentration was not increased by further treatment, but the distribution of different functional groups was influenced. The concentration of carboxylic group was increased with the decrease in concentration of hydroxyl groups. They also observed that the lower the plasma power, the higher the oxygen-containing functional groups.

The above mentioned plasma treatment of carbon materials has been able to treat very small amount of materials. No literature is available for industrial scale treatment. The main problem for continuous treatment of powder materials inproper fluidization which tends to inhomogeneous treatment of powder in plasma zone. Blockage of the system due to the adhesion of powder on the wall of plasma chamber is also the major problem for the treatment of powder materials in large scale. Especially for carbon nanomaterials like CNTs, CNFs, grapheme, and fulerene, plasma treatment is not suitable, because these materials are in agglomerated form and are not uniformly exposed to plasma. Hence, there is a need for further research to overcome the above mentioned problems for continuous operation in industrial scale.

2.2 *Covalent Coupling via the Oxidized Carbon Materials*

To get a stable and homogeneous dispersion of carbon materials in a specific solvent and their good interaction with specific polymer, and also for other special applications, the oxidized carbon materials are further functionalized. The carboxyl unit present on the surface of carbon materials helps for further functionalization.

2.2.1 Amidation

Oxidized carbon materials undergo amidation (attachment of nitrogen-containing molecules) in order to enhance their solubility with different solvents, to be used as catalyst, sensor, and to enhance the reinforcement in polymer composites. Various methods have been reported for the attachment of nitrogen-containing molecules on the surface of different carbon materials [75, 76]. Oxidized carbon materials undergo amidation either reacting directly with amine-containing molecules or through an intermediate reaction with thionyl chloride or a coupling agent. Active carboxyl groups first react with thionyl chloride, then this derivative reacts with different long-chain alkyl amines like octadecylamine (ODA) [77], glucosamine [78], glucosamine [79, 80], 4-dodecyl-aniline, and 4-$CH_3(CH2)_{13}C_6H_4NH_2$ [81]. To modify single-walled carbon nanotubes (SWCNTs) with the amide functionality, Haddon et al. [81, 82] in their work first stirred SWCNTs in $SOCl_2$ in dimethyl formamide (DMF) at 70 °C for 24 h, and after centrifugation, decantation, washing, and drying, the residual solid was mixed with octadecylamine (ODA) and heated at 90–100 °C for 96 h. This amine functionalized SWCNTs showed substantial solubility in chloroform, dichloromethane, aromatic solvents (benzene, toluene, chlorobenzene, 1,2-dichlorobenzene) CS_2, etc.

Oxidized carbon materials are also amine functionalized under mild and neutral conditions by using different coupling reagents like dicyclohexylcarbodiimide (DCC), 1-ethyl-3-(3-dimethylaminopropyl) carbodiimide hydrochloride (EDC), and *N*-hydroxyl sulfosuccinimide (sulfo-NHS) [83–86]. A schematic presentation

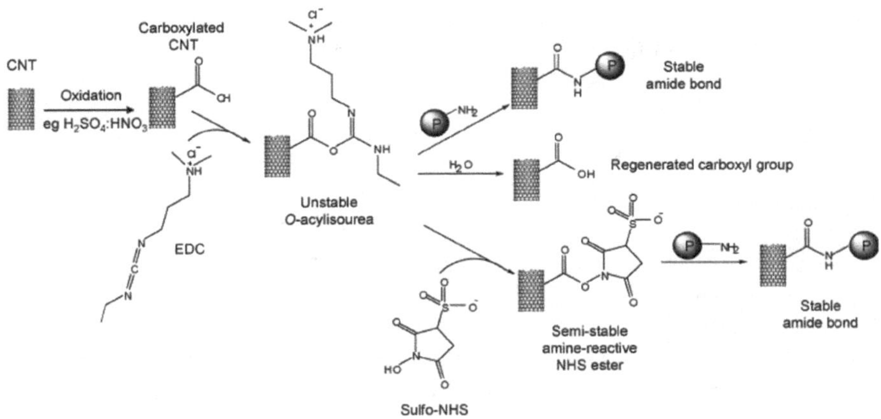

Fig. 6 Schematic presentation of CNT amidation using EDC as coupling agent [87]

of CNT amidation using EDC in order to prepare CNT-based biosensors by reacting carboxylated CNTs with amine terminated biomolecules like proteins is shown in Fig. 6.

In this study, at first EDC reacts with carboxyl group attached to CNT and an unstable amine-reactive *O*-acylisourea intermediate is formed, which subsequently reacts with an amine group and produces one stable amide bond. This intermediate *O*-acylisourea is very unstable and suddenly converted to carboxyl group by hydrolysis. This instability reduces the coupling efficiency. Hence, a water-soluble reagent like *N*-hydroxy succinimide (NHS) or *N*-hydroxy sulfosuccinimide (Sulfo-NHS) is added to the CNT dispersed solution before the addition of EDC. This NHS or sulfo-NHS stabilizes the intermediate by converting it to a semi-stable amine-reactive ester and increases the coupling efficiency.

Some researchers have also been able to attach amine-containing functional groups on the surface of carbon materials using microwave assisted functionalization technique, treating carbon materials with supercritical ammonia liquid, molten urea, and also treating carbon materials with ammonia at high temperature. Wang et al. [88] directly treated oxidized SWCNTs with 2,6-dinitroaniline in a microwave reactor for 10–15 min without any intermediate activated species and successfully attached amine-containing functional groups to oxidized SWCNTs. Amine modification is also done by exposing CNTs to a supercritical ammonia fluid. Shao et al. [89] prepared supercritical ammonia fluid by exposing ammonia at 308 °C temperature and 9.82 MPa pressure in the steel cell. They observed attachment of 5% atomic nitrogen by treating oxidized MWCNTs with this supercritical ammonia for 10 min. Kundu et al. [90] successfully introduced amine-containing functional groups on the surface of CNTs by directly heating oxidized multi-walled carbon nanotubes at different temperature for 6 h in 10 vol% NH_3 in He and observed 6% of nitrogen on CNT surface when CNT was treated with ammonia at 200 °C. In another study, Ford et al. [91] demonstrated a solvent-free amidation process for the preparation of excellent water-soluble material, where oxidized MWCNTs were treated with molten urea and ureido groups were attached to the MWCNTs surface and excellent water-soluble materials were developed.

2.2.2 Silylation

Silylation of carbon material is the attachment of a silyl group in order to enhance the interfacial adhesion of carbon materials with different polymer matrices or to disperse in specific solvent. A general silylation reaction mechanism is given in Fig. 7. Oxidized carbon materials undergo silylation by reacting with any silyl moiety in the presence of a base and are attached with a silyl group like $Si(R)_3$.

Different organosilanes like trimethoxysilane, hexaphenyldisilane, 3-mercaptopropyl-trimethoxysilane, 3-methacryl trimethoxysilane, *N* (*tert*-butyl dimethyl silyl)-*N*-methyl trifluoro acetamide and 1-(*tert*-butyl dimethyl silyl) imidazole [92], 3-glycidoxypropyl-trimethoxysilane, 3-isocyanatopropyl-triethoxysilane [93], vinyl triethoxysilane [94], 7-octenyltrichlorosilane, *n* octyl-trichlorosilane [95], octadecyltrimethoxysilane [96], and 3-aminopropyl triethoxysilane [97] are extensively used for the silylation of different carbon materials. Silanes act as coupling agent between carbon material and polymer molecules in a composite and effectively transfer mechanical load from polymer to the carbon filler and enhances the mechanical properties. Silylation increases the dispersibility and stability of carbon materials in different solvents.

Depending on type of silyl moieties, the reaction is carried out at different processing parameters. H. Gasper and his coworkers successfully introduced silyl groups to a surface oxidized MWCNTs by reacting with various types of organosilanes [98]. Silylation was carried out by two processes. In first process, oxidized CNTs were reacted with organosilane in the presence of a base material like triethylamine. First CNTs were dispersed in toluene and the reaction mixture having CNT dispersed toluene, organosilane, and triethylamine was refluxed for 21 h. After the reaction, CNTs were filtered and washed with toluene and dried in oven at 120 °C. In the second process, the reaction was carried out in the absence of a base material. The reaction mechanism of silylation carried out between oxidized CNTs

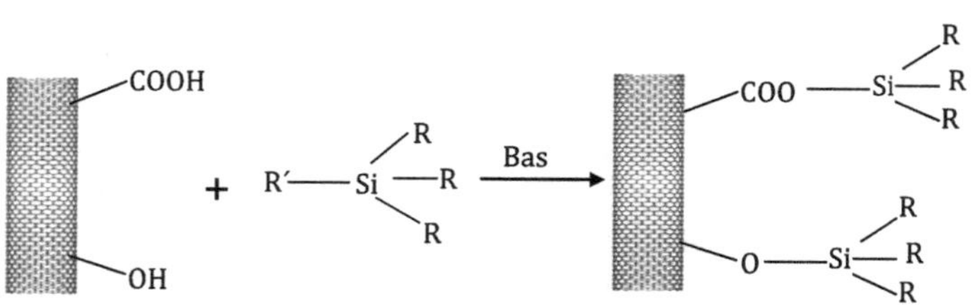

Fig. 7 Schematic presentation of silylation of oxidized CNTs. Three R may be the same or different chemical groups like alkyl, phenyl, alkoxy, or phenoxy groups. R′ is a leaving group

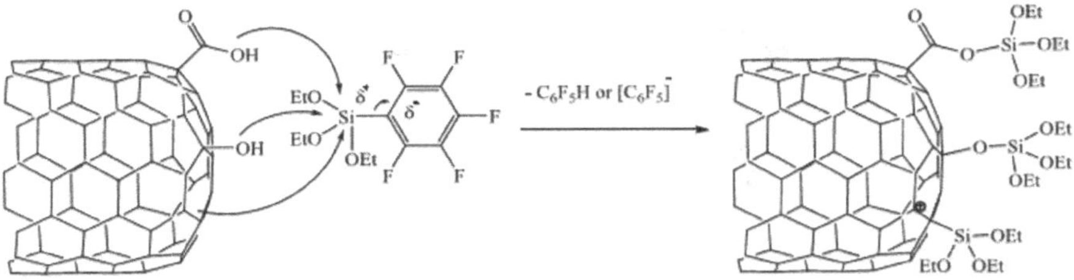

Fig. 8 Schematic presentation of silylation mechanism carried out between oxidized CNTs and (3-pentafluorophenyl) trimethoxysilane [99]

and (3-pentafluorophenyl) triethoxysilane are schematically shown in Fig. 8 where due to high electron-withdrawing capacity of phenyl ring, a strong electric dipole is created between Si (δ^+) and pentafluorophenyl ring (δ^-). A nucleophilic attack may be done on Si by the carboxyl or hydroxyl group attached to oxidized CNTs and silylated carbon nanotube is formed. Silyl group may be added to CNTs directly by the attack of π electrons in CNTs with the Si centre in alkyl moiety [99].

In another study, Masami Aizawa and Milo Shaffer refluxed different organosilanes (3-mercaptopropyl-trimethoxysilane and 3- methacryl trimethoxysilane) with oxidized MWCNTs at 1:1 weight ratio in an alcoholic medium and successfully attached the silyl group to the surface of MWCNTs [92]. Silylation has also been carried out in the presence of UV irradiation. Hemraj Benny and his coworker have successfully developed silane containing functional groups on the surface of acid functionalized single-wall carbon nanotubes by reacting with two organosilanes (trimethoxysilane and hexaphenyldisilane) in the presence of UV irradiation [100].

2.2.3 Silanization

Silanization of carbon materials is the process of attachment of organosilane coupling agent, which creates a bridge between carbon fillers and polymer matrices. It is basically a bifunctional molecule which reacts with both the fillers and the polymer. The basic structure of organosilane molecules used for silanization is

$$R_{(4-n)}\mathrm{Si}R'_n$$

where R is a nonhydrolyzable organic species that may possess a functional group. R may be amine, thiol, vinyl, aromatic ring, or epoxy ring, etc. R' is the hydrolyzable leaving group, which is readily attacked and replaced by the hydroxyl groups present on carbon material surface to form covalent bond between organosilane and carbon material. n is the number of hydrolyzable groups ($n \leq 3$) [101]. Silanization takes place through three steps as shown in Fig. 9. In first step (a), silane undergoes hydrolysis, where labile alkoxy groups are hydrolyzed and active silanol groups are formed. In second step (b), condensation reaction takes

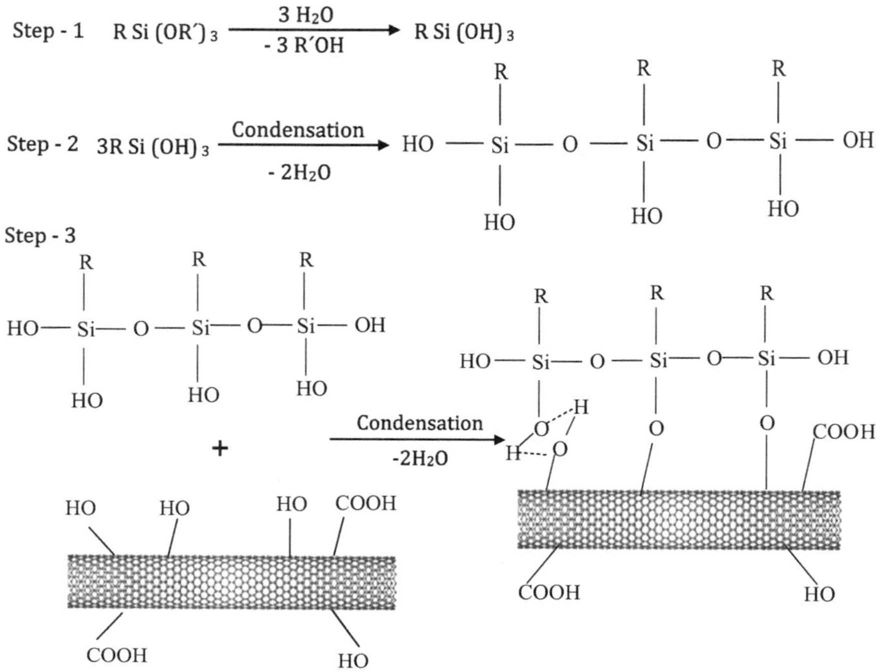

Fig. 9 Schematic presentation of silanization process for surface treatment of oxidized CNTs

place, where silanol groups condense each other to form siloxane oligomers and the silanol groups/oligomers create hydrogen bonding with the hydroxyl groups present on the surface of oxidized carbon materials. In final step called drying step or curing step, covalent bond is formed (C–O–Si) between substrate and siloxane linkage with the loss of water.

Silanization takes place in aqueous or alcoholic, or in any other solvents depending on the type of silane used. A number of works have been observed for the silanization of carbon nanotubes and carbon nanofibers. Bag et al. [102] in their study successfully attached organosilane moieties to the surface of MWCNTs by reacting with 3-methacryloxypropyltrimethoxysilane (MPTS) to improve their solubility in different organic solvents and for the copolymerization with any vinyl monomer. For silanization, oxidized MWCNT was mixed with MPTS by sonication in toluene. The reaction was carried out for 6 h at 100 °C with continuous stirring. The product was washed with DI water and acetone. Then, the product was separated by centrifuge and dried at 80 °C in vacuum oven.

Valasco-Santos et al. [103] also prepared silanized MWCNTs by reacting oxidized MWCNT with a silane coupling agent 3-mercaptopropyl-trimethoxysilane and achieved a good interaction between silanized MWCNTs and polymethyl methacrylate. Composites with silanized MWCNTs possessed higher storage modulus (*E*) than the composite filled with unfunctionalized MWCNTs. 1 wt% of silanized CNTs provided an increase in E of 38% at room temperature with respect to 1 wt% unmodified CNTs. Silanization was conducted in alcoholic medium (ethanol). Silane was added to the oxidized CNTs in 1:1 weight ratio. The mixture of silane and CNTs in ethanol was refluxed for 3 h at temperature around 70 °C under constant stirring. The silanized MWCNTs were dried at 100 °C for around 12 h to completely evaporate the ethanol and then washed with hot water and acetone to eliminate any non-reacted organosilane.

In order to enhance reinforcing efficiency of CNFs in epoxy, Zhu and his coworkers silanized oxidized CNF by reacting with 3-aminopropyl triethoxysilane (APTES) [104]. When incorporated these CNFs into epoxy polymers, a significant increase in tensile strength and storage modulus was achieved, but a decrease in electrical conductivity compared to the pristine CNFs/epoxy composites was observed. This decrease in electrical conductivity was due to the insulating silane layer on the surface of CNFs. They mixed oxidized CNFs with APTES (1 wt% of CNFs) in toluene by ultrasonically stirring for one hour. The reaction mixture was refluxed at 110 °C for 8 h. Then silanized CNFs were filtered and washed several times with DI water, ethanol, and acetone in sequence. The final product was dried in a vacuum oven at 80 °C overnight.

Hu et al. [105], in their study, functionalized graphene oxide (GO) with a silane coupling agent by reacting it with 3-methacryloxypropyltrimethoxysilane (3-MPTS) and observed 44% increases in tensile strength of silane functionalized graphene oxide/PMMA composite than that of the reduced graphene oxide (RGO)/PMMA composites. In their study, they first converted silane to silanol by adding 1 ml silane to 15 ml water, where pH value of solution was adjusted between 4 and 5 by adding acidic acid until the completion of hydrolysis. In another pot, 0.35 g GO was sonicated in 200 ml ethanol and 100 ml water solution for 1 h. Then, the prepared silanol was added to GO solution and stirred for 2 h at 60–65 °C for silylation. The silanized GO was filtered and washed with methanol and water sequentially and dried in a vacuum oven at 80 °C for 12 h (Table 2).

Table 2 Examples of different silanes used for the silanization of carbon fillers

Carbon material type	Type of silane	References
MWCNT	3-methacryloxy propyl trimethoxy silane	[102]
MWCNT	3-mercaptopropyl-trimethoxysilane	[103]
CNF	3-aminopropyl triethoxysilane	[104]
GO	3-methacryloxypropyltrimethoxysilane	[105]
MWCNT	3 aminopropyl triethoxysilane	[106, 107]
CNF	3-aminopropyltriethoxy silane, 3-aminopropyltrimetoxysilane, *N*-(2-aminoethyl)-3-(aminopropyl trimethoxy silane) and 3-glycidoxypropyltrimethoxy silane	[108]
MWCNTs	3-glycidoxypropyl-trimethoxysilane	[6]
MWCNT	3-aminopropyltriethoxysilane	[97]
CNF	Octadecyltrimethoxysilane	[96, 109]
MWCNT	3-isocyanatopropyl-triethoxysilane	[110]

2.2.4 Grafting of Polymers Chains

In order to achieve homogeneous dispersion of carbon fillers in a polymer matrix and to achieve enhanced mechanical properties in a composite, covalent attachment of polymer chains to the carbon fillers has been a great interest for researchers. Wang et al. [111] in their study grafted epoxy to vapor grown carbon nanofibers (VGCNF) and achieved a better dispersion in epoxy resin compared to pristine VGCNF. Both tensile strength and modulus of epoxy-modified VGCNF composites were higher than those of epoxy/pristine VGCNF composites. They first functionalize VGCNF with amine-containing pendants via a Friedel-Crafts acylation reaction with 4-(3-aminophenoxy) benzoic acid as shown in Fig. 10. The resulting amine functionalized CNF was treated with epichlorohydrine and *N,N*-diglycidyl modified CNF called epoxydized CNF was obtained.

For homogeneous dispersion of both SWCNTs and MWCNTs in polyimide matrix, Qu et al. [112] functionalize CNTs with a specially synthesized low molecular weight amine terminated polymide, which was structurally identical to the main polyimide matrix. The CNTs grafted with amine terminated PI were easily dispersible in many polar solvents like THF, DMF, DMAc, DMSO, NMP, etc. Similar polyimide grafting of CNTs was observed by Yuen et al. [113] and observed the tensile and modulus were significantly higher in PI-g-PI composite compared to those of pristine CNT filled PI composite. Polystyrene grafted SWCNT was prepared by Ajayan's group. They first treated CNTs with sec-butyl lithium and then used this as initiator for the anionic polymerization of styrene [114]. Malikov et al. [115] in their study prepared polyvinyl alcohol/MWCNT nanocomposite by grafting oxidized MWCNTs with polyvinyl alcohol by Fischer esterification process and observed higher thermal stability of nanocomposite compared to pristine PVA.

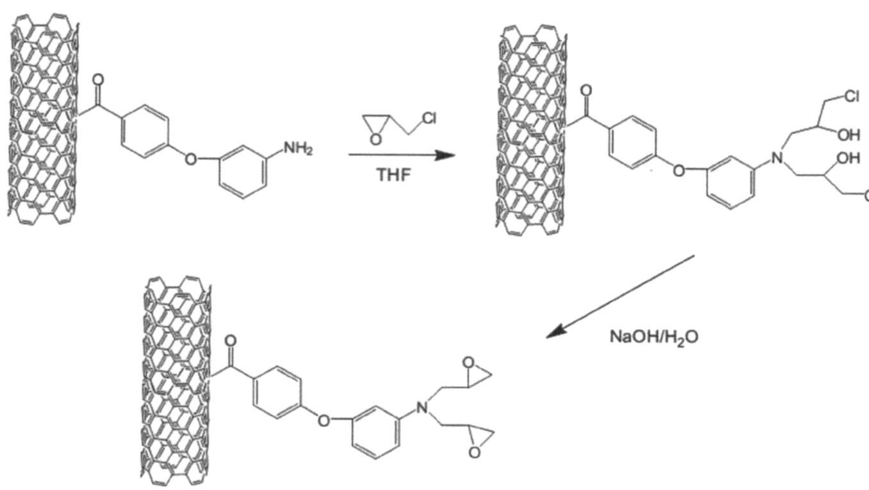

Fig. 10 Synthesis of epoxy grafted VGCNTs [111]

Water-dispersible MWCNTs ware developed by Shimou Chen and his coworkers by grafting CNTs with polyacrylic acid through a radiation-induced graft polymerization technique [116]. They first converted oxidized CNTs to hydroxyl ethylated CNTs by the γ radiation of CNTs in ethanol. Then, hydroxyl ethylated CNT, acrylic acid, and copper sulfate catalyst were mixed with water and radiation-induced graft polymerization was carried out in different doses of γ radiation.

2.3 Noncovalent Functionalization

Noncovalent functionalization is a nondestructive technique in order to tune the interfacial properties of carbon materials. In this technique, the original properties of carbon fillers like the electronic properties of CNTs, CNFs, carbon blacks, and fullerene are conserved by preventing the damage of Sp^2 conjugated structure. The noncovalent functionalization is carried out by basically following three ways.

1. Polymer wrapping
2. Surfactant adsorption
3. Encapsulation.

A typical example of noncovalent functionalization of CNTs is given in Fig. 11.

2.3.1 Polymer Wrapping

The polymer wrapping process is achieved through the van der Waals interactions and π–π stacking between carbon fillers and polymer chains containing aromatic

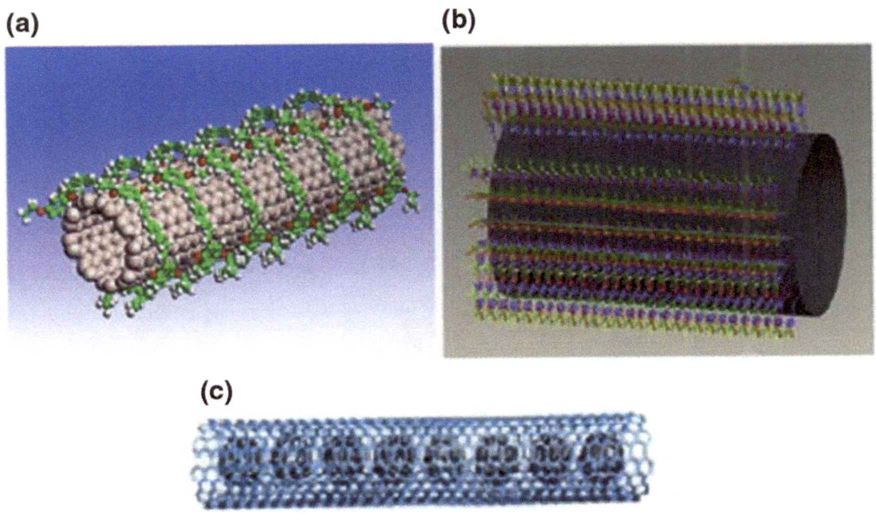

Fig. 11 Schematics of noncovalent functionalization of CNTs: **a** polymer wrapping **b** surfactant adsorption **c** encapsulation method [2, 117]

rings. In this technique, the extent of interaction of carbon filler with polymer matrix depends on the miscibility of polymer matrix with the wrapped polymer chain on the surface of carbon fillers. Extensive research work has been carried out on the physical wrapping of long-chain molecules on the surface of CNTs and CNFs. Polymers with high molecular weight, especially conjugated polymers are suitable for wrapping. The polymer chain wrapped to the CNTs reduces van der Wall interaction among CNTs and increases their dispersibility. The basic requirements of a polymer to be wrapped on the surface of carbon nanotubes/fibers are:

- One part of polymer should be hydrophobic in nature so that it can be easily compatible with the hydrophobic surface of CNT/CNF.
- Number of aromatic groups should be in the long polymer chains which will create interaction with CNT/CNF by $\pi-\pi$ stacking.

Polymers with helical structure are suitable for winding around CNTs/CNFs. Biopolymers like amylase contain peptide which has helical structure and are winded around CNTs/CNFs easily. In a study, it is observed that in situ synthesis of amphiphilic peptide on the presence of CNTs under ultrasonication has successfully wrapped the surface of CNTs by its long chain and the functionalized CNTs are easily dispersible in aqueous medium [118]. Nonpolar synthetic polymers like poly-m-phenylene vinylene [15] and polar polymers like polyvinyl pyrrolidone [119], and polystyrene sulfonate [120] are used for noncovalent functionalization of CNTs. A model for wrapping arrangements of polyvinyl pyrrolidone polymer around CNTs is presented in Fig. 12.

Another approach for noncovalent functionalization is the employment of small bifunctional molecules having planner groups like pyrenyl group. Chen et al. [122] in their study functionalized CNTs with a bifunctional molecule, 1-pyrenebutanoic acid-succinimidyl ester. The planner pyrenyl groups as shown in Fig. 13 are highly aromatic in nature; therefore, they are strongly interacted and irreversibly adsorbed to the basal graphitic plane of nanotube surface by π–π interaction. This pyrenyl adsorption helps CNTs to disperse completely in aqueous medium. Other end group, succinimidyl ester is highly reactive to nucleophilic substitution by amine groups present on the surface of the most protein. Hence, this type of functionalization has a great possibility to generate interaction of CNTs with different biomolecules.

2.3.2 Surfactant Adsorption

Apart from polymer wrapping, colloidal stabilization of carbon fillers using different surfactants has received great attention in biomedical and polymer composite filed. Different ionic and nonionic surfactants are used for stable colloidal suspension of carbon fillers. These surfactants are adsorbed on the surface of carbon filler and during sonication for dispersion; these surfactants keep separate carbon particles from one another by either electrostatic repulsion or steric hindrance.

Fig. 12 Different wrapping arrangements of PVP polymer around CNT [121]

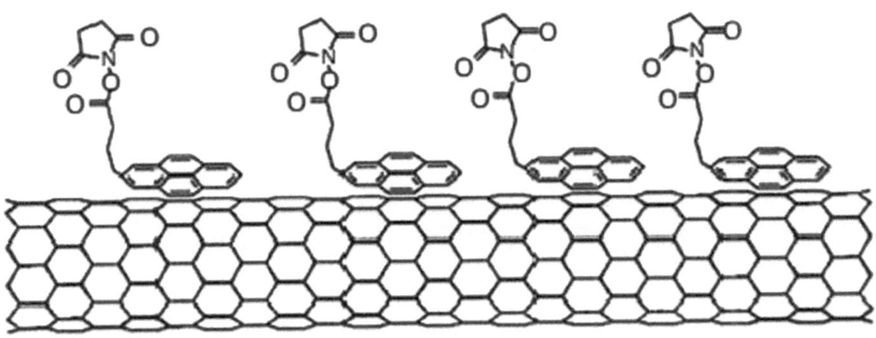

Fig. 13 Noncovalent functionalization of SWNT with 1-pyrenebutanoic acid-succinimidyl ester [122]

The repulsive force introduced by surfactant molecules overcomes the van der wall force of interaction among the carbon particles.

Anionic surfactant like sodium dodecyl sulfate (SDS), slithium dodecyl sulfate (LDS), sodium dodecylbenzene sulfate (SDBS), and nonionic surfactant like Triton

X-100 are widely used for colloidal stabilization of carbon materials. It has been reported that a stable black suspension is achieved by sonication of CNTs in a 1% aqueous solution of SDS [123, 124]. The hydrophilic part (alkyl group) of SDS is adsorbed to the graphitic surface of CNTs and the hydrophilic part is oriented towards the aqueous phase. This adsorption creates a negative charge environment around the CNTs that prevents the aggregation and creates a stable colloidal suspension in aqueous phase.

Colloidal stabilization of carbon materials depends on the structure of surfactant molecules. Presence of aromatic rings in surfactant molecules is more effective for stable dispersion of carbon materials. On a comparative study between the effect of SDS and its structurally related sodium dodecylbenzene sulfate (SDBS) in dispersion of CNTs, it was observed that SDBS is more effective for CNTs solubilization than SDS because of aromatic stacking of phenyl ring of SDBS with the CNTs within the micelle as shown in Fig. 14. CNTs dispersion in aqueous medium using nonionic surfactant like triton X-100 has also been extensively studied in the production of composites [126]. The colloidal suspension is governed by the steric hindrance created by bulkier hydrophilic group of surfactant.

2.3.3 Encapsulation

Encapsulation is the process of filling of any foreign material in the hollow cavity of another material. Carbon materials like fullerene and carbon nanotubes are used for encapsulation. This is a process to modify the properties of fullerene or CNTs by filling of any foreign material in their hollow cavity. Encapsulation of a foreign particle depends on the ratio of inner diameter of CNTs to the diameter of the foreign particle. The encapsulated particles are not removed out from the cavity by basically van der wall force of interaction. For optimum van der Wall interaction, the distance between inner surface of CNT and foreign particle should be close to van der Wall gap of ≈0.3 nm [127]. If the diameter of foreign particle is "*n*" nm,

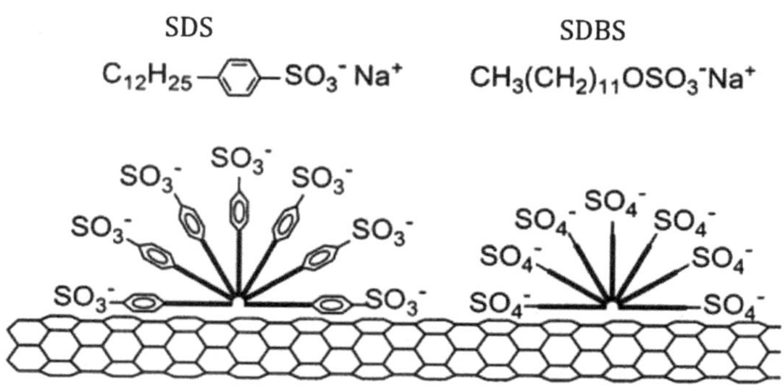

Fig. 14 Schematic representation of how surfactants may adsorb onto the nanotube surface. SDBS disperse the tubes better than SDS because of their benzene rings [125]

then optimum diameter of nanotube for an irreversible encapsulation should be $n + 2 \times 0.3$ nm. If inner diameter is more than this value, then there may be reversible encapsulation and the foreign particles will be removed out. If the inner diameter of CNTs will be less than this value, then foreign particles cannot be encapsulated. Liquid molecules are encapsulated in CNTs by the capillary filling process. According to Young's Laplas law, CNT immersion in a liquid with surface tension below 200 mN/m can be spontaneously filled by capillary action. The material which can be vaporized is encapsulated by putting the mixture of material and CNTs with the open end in a closed container and heating the container at temperature just above the sublimation point of foreign material.

A number of research papers are available for the encapsulation of different materials in CNTs for different applications like drugs microcarriers [128], high electrochemical performance in batteries [129], and for efficient photocatalysis [130], etc. Due to their nano needle-like shape, good cell membrane penetrability, and high drug-carrying capacities, CNTs can pass into the various cellular components without causing apparent cell damage and can be used as micro drug delivery vehicle [131]. Kostarelos et al. in their study observed the potential assessment of CNTs into the various cellular components [16] and in another study, Porter et al. stated a direct translocation of SWNTs into the cytoplasm of target cells without causing cell death [128]. Hence, any drugs or bioactive molecules like protein, DNA, or siRNA, which expert their therapeutic action only inside the nucleus can be transported using CNTs [132, 133].

Hampel and his coworker used MWCNTs as carrier of an anticancer agent for the treatment of cancer cells. Platinum-based anticancer agent, carboplatin was encapsulated inside the multi-walled nanotubes by wet chemical approach. They dispersed CNTs by sonication and stirring for 24 h in carboplatin solution. Carboplatin molecules were inserted into the CNTs by capillary action. This CNT encapsulated carboplatin molecules were used to study the growth of the cancer cells and observed that carboplatin-filled CNTs inhibited growth of bladder cancer cells [134]. Similarly, Li et al. [135] successfully encapsulated cisplatin (an FDA-approved chemotherapeutic drug) into multi-walled carbon nanotubes and demonstrated that encapsulated cisplatin could be delivered in living cells under physiological conditions to exert its pharmacological action.

Kim and coworker in their study encapsulated an organic active material, PTMA (poly (2,2,6,6-tetramethylpiperidinyloxy-4-vinylmethacrylate) in CNTs as shown in Fig. 15 and used these CNTs for the preparation of electrode for sodium ion batteries [129]. This PTMA encapsulated CNTs exhibited superior electrochemical performance in sodium ion battery compared to Li-ion battery. The encapsulation of PTMA in CNTs was done by dissolving PTMA in CNTs dispersed *N*-methyl-2-pyrrolidone, where PTMA was diffused into the free space of CNTs. The PTMA encapsulated CNTs were dried and washed with acetone to remove excess PTMA from outside of CNTs. These CNTs were blended with poly (vinylidene fluoride) to prepare the electrode for Na-ion batteries.

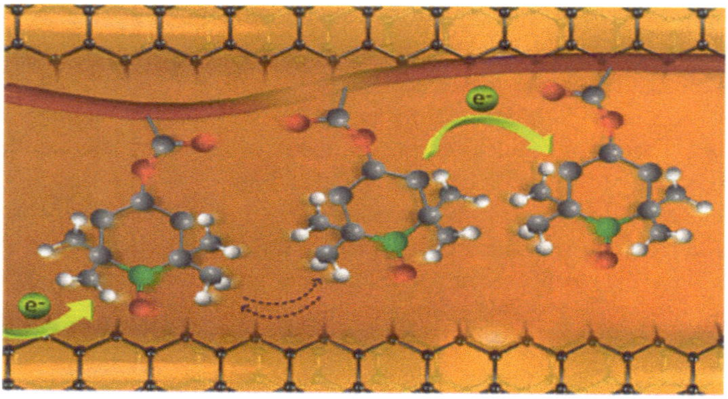

Fig. 15 Schematic presentation of encapsulation of organic material (PTMA) in carbon nanotube [129]

2.4 Conclusion

This chapter is the gathering of different functionalization techniques used for the surface modification of carbon materials for specific applications like polymer reinforcement, ink, coating, electronics, electrical, and biomedicals. On the basis of their application area, different functionalization techniques have presented in this chapter. A variety of covalent modification methods (acids, ozone, plasma treatment, amidation, and silylation) and noncovalent methods (polymer wrapping, polymer and other particle encapsulation, ionic liquid adsorption) have been depicted in this chapter. The different types of modification have extended versatile areas of application for carbon materials. The combination of different methodologies discussed in this chapter will provide extensive ideas to select the suitable modification technique for realistic applications of carbon materials.

References

1. Balasubramanian K, Burghard M (2005) Chemically functionalized carbon nanotubes. Small 1:180–192
2. Hirsch A (2002) Functionalization of single-walled carbon nanotubes. Angew Chem Int Ed 41:1853–1859
3. Hirsch A, Vostrowsky O (2005) Functionalization of carbon nanotubes. Top Curr Chem 245:193–237
4. Nayak L, Khastgir D, Chaki TK (2012) Influence of carbon nanofibers reinforcement on thermal and electrical behavior of polysulfone nanocomposites. Polym Eng Sci 52:2424–2434
5. Nayak L, Khastgir D, Chaki TK (2013) A mechanistic study on electromagnetic shielding effectiveness of polysulfone/carbon nanofibers nanocomposites. J Mater Sci 48:1492–1502

6. Ma PC, Kim JK, Tang BZ (2006) Functionalization of carbon nanotubes using a silane coupling agent. Carbon 44:3232–3238
7. Liu P (2005) Modifications of carbon nanotubes with polymers. Eur Polym J 41:2693–2703
8. Hamon MA, Hui H, Bhowmik P (2002) Ester-functionalized soluble single-walled carbon nanotubes. Appl Phys A 74:333–338
9. Stephenson JJ, Sadana AK, Higginbotham AL, Tour JM (2006) Highly functionalized and soluble multiwalled carbon nanotubes by reductive alkylation and arylation: the billups reaction. Chem Mater 18:4658–4661
10. Esumi K, Ishigami M, Nakajima A, Sawada K, Honda H (1996) Chemical treatment of carbon nanotubes. Carbon 34:279–281
11. Yu R, Chen L, Liu Q, Lin J, Tan KL, Ng SC, Chan HSO, Xu GQ, Andy Hor TS (1998) Platinum deposition on carbon nanotubes via chemical modification. Chem Mater 10:718–722
12. Sham ML, Kim JK (2006) Surface functionalities of multi-wall carbon nanotubes after UV/ozone and TETA treatments. Carbon 44:768–777
13. Ávila-Orta CA, Cruz-Delgado VJ, Neira-Velázquez MG, Hernández-Hernández E, Méndez-Padilla MG, Medellín-Rodríguez FJ (2009) Surface modification of carbon nanotubes with ethylene glycol plasma. Carbon 47:1916–1921
14. Wang SC, Chang KS, Yuan CJ (2009) Enhancement of electrochemical properties of screen-printed carbon electrodes by oxygen plasma treatment. Electrochim Acta 54:4937–4943
15. Star A, Stoddart JF, Steuerman D et al (2001) Preparation and properties of polymer-wrapped single-walled carbon nanotubes. Angew Chem Int Ed 40:1721–1725
16. Kostarelos K, Lacerda L, Pastorin G et al (2007) Cellular uptake of functionalized carbon nanotubes is independent of functional group and cell type. Nat Nanotechnol 2:108–113
17. Sahoo NG, Jung YC, Yoo HJ, Cho JW (2006) Effect of functionalized carbon nanotubes on molecular interaction and properties of polyurethane composites. Macromol Chem Phys 207:1773–1780
18. Wang Y, Wu J, Wei F (2003) A treatment method to give separated multi-walled carbon nanotubes with high purity, high crystallization and a large aspect ratio. Carbon 41:2939–2948
19. Wepasnick KA, Smith BA, Schrote KE, Wilson HK, Diegelmann SR, Fairbrother DH (2011) Surface and structural characterization of multi-walled carbon nanotubes following different oxidative treatments. Carbon 49:24–36
20. Bonifazi D, Nacci C, Marega R, Campidelli S, Ceballos G, Modesti S, Meneghetti M, Prato M (2006) Microscopic and spectroscopic characterization of paintbrush-like single-walled carbon nanotubes. Nano Lett 6:1408–1414
21. Ziegler KJ, Gu Z, Peng H, Flor EL, Hauge RH, Smalley RE (2005) Controlled oxidative cutting of single-walled carbon nanotubes. J Am Chem Soc 127:1541–1547
22. Jordan ME, Deery HJ, Hagopian E, Williams FR (1968) Process for producing furnace black pigments. Patent no. US3383232A
23. Boonstra BB, Dannenberg EM, Rossman RP, Williams FR (1971) Process for treating furnace carbon black. Patent no. US3565657A
24. Jean-Baptiste AD (1962) Water-soluble carbon black and production thereof. Patent no. US3023118A
25. Curtis JC, Taylor RL, Joyce GA (2000) Hydrogen peroxide oxidation of carbon black. Patent no. US6120594 A
26. Nagasawa T (1995) Water-based pigment ink and process for producing the same. Patent no. EP 0 688836 A2
27. Ito H, Momose M, Hayashi H, Ito S (2002) Aqueous pigment dispersion water base ink composition and recording method using the ink composition. Patent no. US6488753 B1
28. Sekiyama M, Saitoh T, Kirino T (2014) Method for producing aqueous dispersion of surface treated carbon black particles and aqueous dispersion of surface treated carbon black particles. Patent no. US20140000488A1

29. Adams CE, Belmont JA (1999) Modified carbon products and inks and coatings containing modified carbon products. Patent no. US5885335A
30. So HH, Cho JW, Sahoo NG (2007) Effect of carbon nanotubes on mechanical and electrical properties of polyimide/carbon nanotubes nanocomposites. Eur Polym J 43:3750–3756
31. Spitalsky Z, Krontiras CA, Georga SN, Galiotis C (2009) Effect of oxidation treatment of multiwalled carbon nanotubes on the mechanical and electrical properties of their epoxy composites. Compos Part A 40:778–783
32. Simsek Y, Ozyuzer L, Seyhan AT, Tanoglu M, Schulte K (2007) Temperature dependence of electrical conductivity in double-wall and multi-wall carbon nanotube/polyester nanocomposites. J Mater Sci 42:9689–9695
33. Karl A (2007) Method for producing post treated carbon black. Patent no. US 7217405B2
34. Park YS, Choi YC, Kim KS, Chung DC (2001) High yield purification of multiwalled carbon nanotubes by selective oxidation during thermal annealing. Carbon 39:655–661
35. Li C, Wang D, Liang Wang TX, Wu J, Hu X, Liang J (2004) Oxidation of multiwalled carbon nanotubes by air: benefits for electric double layer capacitors. Power technol 142:175–179
36. Park WK, Kim JH, Lee SS, Kim J, Lee GW, Park M (2005) Effect of carbon nanotube pre-treatment on dispersion and electrical properties of melt mixed multi-walled carbon nanotubes/poly(methyl methacrylate) composites. Macromol Res 13:206–211
37. Yuen SM, Ma CCM, Lin YY, Kuan HC (2007) Preparation, morphology and properties of acid and amine modified multiwalled carbon nanotube/polyimide composite. Compos Sci Technol 67:2564–2573
38. Melore PJ, Eckert FJ (1966) Preparation of long flow carbon black. Patent no. US3245820
39. Sutherland I, Sheng E, Bradley RH, Freakley PK (1996) Effects of ozone oxidation on carbon black Surfaces. J Mater Sci 31:5651–5655
40. Stenger F, Bergemann K, Nagel M (2013) Process for after treating carbon black. US patent no. US8574527 B2
41. Yeh AG, et al. (2005) Self dispersing pigment and process for making and use of same. Patent no. US6852156B2
42. Vennerberg DC, Quirino RL, Jang Y, Kessler MR (2014) Oxidation behavior of multiwalled carbon nanotubes fluidized with ozone. ACS Appl Mater Interfaces 6:1835–1842
43. Mawhinney DB, Naumenko V, Kuznetsova A, Yates JT, Liu J, Smalley RE (2000) Infrared spectral evidence for the etching of carbon nanotubes: ozone oxidation at 298 K. J Am Chem Soc 122:2383–2384
44. Cai L, Bahr JL, Yao Y, Tour JM (2002) Ozonation of single- walled carbon nanotubes and their assemblies on rigid self-assembled monolayers. Chem Mater 14:4235–4241
45. Chen Z, Ziegler KJ, Shaver J, Hauge RH, Smalley RE (2006) Cutting of single-walled carbon nanotubes by ozonolysis. J Phys Chem B 110:11624–11627
46. Lu X, Zhang L, Xu X, Wang N, Zhang QJ (2002) Can the sidewalls of single-wall carbon nanotubes be ozonized? J Phys Chem B 106:2136–2139
47. Banerjee S, Wong SS (2002) Rational sidewall functionalization and purification of single-walled carbon nanotubes by solution-phase ozonolysis. J Phys Chem B 106:12144–12151
48. Hernadi K, Siska A, Thien-Nga L, Forro L, Kiricsi I (2001) Reactivity of different kinds of carbon during oxidative purification of catalytically prepared carbon nanotubes. Solid State Ion 141–142:203–209
49. Criegee R (1975) Mechanism of ozonolysis. Angew Chem Int Ed 14(11):745–752
50. Simmons JM, Nichols BM, Baker SE, Marcus MS, Castellini OM, Lee CS, Hamers RJ, Eriksson MA (2006) Effect of ozone oxidation on single-walled carbon nanotubes. J Phys Chem 110:7113–7118
51. Byl O, Liu J, Yates JT (2005) Etching of carbon nanotubes by ozone: a surface area study. Langmuir 21:4200–4204

52. Hemraj-Benny T, Bandosz TJ, Wong SS (2008) Effect of ozonolysis on the pore structure, surface chemistry, and bundling of single-walled carbon nanotubes. J Colloid Interface Sci 317:375–382
53. Peng K, Liu LQ, Li H, Meyer H, Zhang Z (2011) Room temperature functionalization of carbon nanotubes using an ozone/water vapor mixture. Carbon 49:70–76
54. Staehelin J, Hoigne J (1982) Decomposition of ozone in water: rate of initiation by hydroxide ions and hydrogen peroxide. Environ Sci Technol 16:676–681
55. Sotelo JL, Beltran FJ, Benitez FJ, Beltran-Heredia J (1987) Ozone decomposition in water: kinetic study. Ind Eng Chem Res 26:39–43
56. Morales-Lara F, Perez-Mendoza MJ, Altmajer-Vaz D, Garca-Roman M, Melguizo M, Lopez-Garzon FJ, Domingo-Garca M (2013) Functionalization of multiwall carbon nanotubes by ozone at basic pH. Comparison with oxygen plasma and ozone in gas phase. J Phys Chem C 117:11647–11655
57. Alder MR, Hill GC (1950) The kinetics and mechanism of hydroxide ion catalyzed ozone decomposition in aqueous solution. J Am Chem Soc 72:1884–1886
58. Sehested K, Cotfltzen H, Holcman J, Flscher CH, Hart EJ (1991) The primary reaction in the decomposition of ozone in acidic aqueous solutions. Environ Sci Technol 25(9):1589–1596
59. Li W, Bai Y, Zhang YK, Sun ML, Cheng RM, Xu XC et al (2005) Effect of hydroxyl radical on the structure of multi-walled carbon nanotubes. Synth Met 155:509–515
60. Tressaud A, Shirasaki T, Nanse G, Papirer E (2002) Fluorinated carbon blacks: influence of the morphology of the starting material on the fluorination mechanism. Carbon 40:217–220
61. Schukin LI, Kornievich MV, Vartapetjan RS, Beznisko SI (2002) Low-temperature plasma oxidation of activated carbons. Carbon 40:2028–2030
62. Donnet JB, Wang WD, Vidal A (1994) Observation of plasma treated carbon black surfaces by scanning tunnelling microscopy. Carbon 32:199–2006
63. Bruser V, Heintze M, Brandl W, Marginean G, Bubert H (2004) Surface modification of carbon nanofibres in low temperature plasmas. Diamond Relat Mater 13:1177–1181
64. Favia P, Vietro ND, Mundo RD, Fracassi F, Agostino R (2006) Tuning the acid/base surface character of carbonaceous materials by means of cold plasma treatments. Plasma Processes Polym 3:66–74
65. Heintze M, Bruser V, Brandl W, Marginean G, Bubert H, Haiber S (2003) Surface functionalization of carbon nano-fibers in fluidized bed plasma. Surf Coat Technol 174–175:831–834
66. Sawada Y, Kogoma M (1997) Plasma-polymerized tetrafluoroethylene coatings on silica particles by atmospheric-pressure glow discharge. Powder Technol 90:245–250
67. Okpalugo TIT, Papakonstantinou P, Murphy H, Mclaughlin J, Brown NMD (2005) Oxidative functionalization of carbon nanotubes in atmospheric pressure filamentary dielectric barrier discharge (APDBD). Carbon 43:2951–2959
68. Hanabusa T, Uemiya S, Kojima T (1997) Surface modification of particles in a plasma jet fluidized bed reactor. Surf Coat Technol 88:226–231
69. Erden S, Ho KKC, Lamoriniere S, Lee AF, Yildiz H, Bismark A (2010) Continuous atmospheric plasma oxidation of carbon fibres: Influence on the fibre surface and bulk properties and adhesion to polyamide 12. Plasma Chem Plasma Process 30:471–487
70. Ho KKC, Lee AF, Lamoriniere S, Bismarck A (2008) Continuous atmospheric plasma fluorination of carbon fibres. Compos Part A Appl Sci Manuf 39:364–373
71. Utegulov ZN, Mast DB (2005) Functionalization of single walled carbon nanotubes using isotropic plasma treatment: resonant ruman spectroscopy study. J Appl Phys 97:104324
72. Bubert H, Haiber S, Brandl W, Marginean G, Heintze M, Bruser V (2003) Characterization of the uppermost layer of plasma-treated carbon nanotubes. Diamond Relat Mater 12:811–815
73. Haiber S et al (2003) Analysis of functional groups on the surface of plasma—treated carbon nanofibers. Bioanal Chem 375:875–883
74. Chirila V, Marginean TG, Brandl W (2005) Effect of the oxygen plasma treatment parameters on the carbon nanotubes surface properties. Surf Coat Technol 200:548–551

75. Chen J, Hamon MA, Hu H et al (1998) Solution properties of single-walled carbon nanotubes. Science 282:95–98
76. Li L, Lin R, He H, Sun M, Jiang L, Gao M (2014) Interaction of amidated single-walled carbon nanotubes with protein by multiple spectroscopic methods. J Lumin 145:125–131
77. Xu M, Huang Q, Chen Q, Guo P, Sun Z (2003) Synthesis and characterization of octadecylamine grafted multi-walled carbon nanotubes. Chem Phys Lett 375:598–604
78. Pompeo F, Resasco DE (2002) Water solubilization of single-walled carbon nanotubes by functionalization with glucosamine. Nano Lett 2:369–373
79. Liu L, Zhang S, Hu T, Guo ZX, Ye C, Dai L, Zhu D (2002) Solubilized multi-walled carbon nanotubes with broadband optical limiting effect. Chem Phys Lett 359:191–195
80. Wu W, Li J, Liu L, Yanga L, Guo ZX, Dai L, Zhu D (2002) The photoconductivity of PVK-carbon nanotube blends. Chem Phys Lett 364:196–199
81. Hamon MA, Chen J, Hu H, Chen Y, Itkis ME, Rao AM, Eklund PC, Haddon RC (1999) Dissolution of single-walled carbon nanotubes. Adv Mater 11:834–840
82. Hamon MA, Hu H, Bhowmik P, Niyogi S, Zhao B, Itkis ME, Haddon RC (2001) End-group and defect analysis of soluble single-walled carbon nanotubes. Chem Phys Lett 347:8–12
83. Gao Y, Kyratzis I (2008) Covalent immobilization of proteins on carbon nanotubes using the cross-linker 1-ethyl-3-(3-dimethylaminopropyl)carbodiimide—a critical assessment. Bioconjug Chem 19:1945–1950
84. Sano M, Kamino A, Okamura J, Shinkai S (2001) Ring closer of carbon nanotubes. Science 293:1299–1301
85. Huang WJ, Taylor S, Fu KF, Lin Y, Zhang DH, Hanks TW, Rao AM, Sun YP (2002) Attaching proteins to carbon nanotubes via diimide-activated amidation. Nano Lett 2:311–314
86. Jiang KY, Schadler LS, Siegel RW, Zhang XJ, Zhang HF, Terrones M (2004) Protein immobilization on carbon nanotubes via a two-step process of diimide-activated amidation. J Mater Chem 14:37–39
87. Jeykumari DRS, Ramaprabhu S, Narayanan SS (2007) A thionine functionalized multiwalled carbon nanotube modified electrode for the determination of hydrogen peroxide. Carbon 45:1340–1353
88. Wang Y, Iqbal Z, Mitra S (2005) Microwave-induced rapid chemical functionalization of single-walled carbon nanotubes. Carbon 43:1015–1020
89. Shao L, Bai Y, Huang X, Gao Z, Meng L, Huang Y, Ma J (2009) Multi-walled carbon nanotubes (MWCNTs) functionalized with amino groups by reacting with supercritical ammonia fluids. Mater Chem Phys 116:323–326
90. Kundu S, Xia W, Busser W, Becker M, Schmidt DA, Havenith M, Muhler M (2010) The formation of nitrogen-containing functional groups on carbon nanotube surfaces: a quantitative XPS and TPD study. Phys Chem Phys 12:4351–4359
91. Ford WE, Jung A, Hirsch A, Graupner R, Scholz F, Yasuda A, Wessels JM (2006) Urea-melt solubilization of single-walled carbon nanotubes. Adv Mater 18:1193–1197
92. Aizawa M, Shaffer MSP (2003) Silylation of multi-walled carbon nanotubes. Chem Phys Lett 368:121–124
93. Ma PC, Kim JK, Tang BZ (2007) Effects of silane functionalization on the properties of carbon nanotube/epoxy nanocompositos. Compos Sci Technol 67:2965–2972
94. Yuen SM, Ma CCM, Chiang CL, Teng CC, Yu YH (2008) Poly(vinyltriethoxysilane) modified MWCNT/polyimide nanocomposites—preparation, morphological, mechanical, and electrical properties. J Polym Sci Part A Polym Chem 46:803–816
95. Vast L, Lallemand F, Colomer JF, Van Tendeloo G, Fonseca A, Mekhalif Z, Delhalle J (2009) Multiwalled carbon nanotubes functionalized with 7-octenyltrichlorosilane and n-octyltrichlorosilane: dispersion in Sylgard®184 silicone and young's modulus. J Mater Sci 44:3476–3482
96. Wood W, Kumar S, Zhong WH (2010) Synthesis of organosilane-modified carbon nanofibers and influence of silane coating thickness on the performance of polyethylene nanocomposites. Macromol Mater Eng 295:1125–1135

97. Lee JH, Kathi J, Rhee KY, Lee JH (2010) Wear properties of 3-aminopropyl triethoxysilane-functionalized carbon nanotubes reinforced ultra high molecular weight polyethylene nanocomposites. Polym Eng Sci 50:1433–1439
98. Gaspar H, Pereira C, Rebelo SLH, Pereira MFR, Figueiredo JL, Freire C (2011) Understanding the silylation reaction of multi-walled carbon nanotubes. Carbon 49:3441–3453
99. Lin TW, Salzmann CG, Shao LD, Yu CH, Green MLH, Tsang SC (2009) Polyethylene glycol grafting and attachment of encapsulated magnetic iron oxide silica nanoparticles onto chlorosilanized single-wall carbon nanotubes. Carbon 47:1415–1420
100. Hemraj-Benny T, Wong SS (2006) Silylation of single-walled carbon nanotubes. Chem Mater 18:4827–4839
101. Velasco-Santos C, Martinez-Hernandez AL, Lozada-Cassou M, Alvarez-Castillo A, Castano VM (2002) Chemical funtionalization of carbon nanotubes through an organosilane. Nanotechnol 13:495
102. Bag DS, Dubey R, Zhang N, Xie J, Varadan VK, Lal D, Mathur GN (2004) Chemical functionalization of carbon nanotubes with 3-methacryloxypropyl trimethoxysilane (3-MPTS). Smart Mater Struct 13:1263
103. Velasco-Santos C, Martinez-Hernandez AL, Brostow W, Castano VM (2011) Influence of silanization treatment on thermomechanical properties of multiwalled carbon nanotubes: poly (methylmethacrylate) nanocomposites grafting of polymers chain. J Nanomater 2011:1–9
104. Zhu J, Wei S, Ryu J, Budhathoki M, Liang G, Guo Z (2010) In situ stabilized carbon nanofiber (CNF) reinforced epoxy nanocomposites. J Mater Chem 20:4937–4948
105. Hu X, Su E, Zhu B, Jia J, Yao P, Bai Y (2014) Preparation of silanized graphene/poly (methyl methacrylate) nanocomposites in situ copolymerization and its mechanical properties. Compos Sci Technol 97:6–11
106. Kathi J, Rhee KY (2008) Surface modification of multi-walled carbon nanotubes using 3-aminopropyltriethoxysilane. J Mater Sci 43:33–37
107. Scheibe B, Borowiak-Palen E, Kalenczuk RJ (2009) Effect of the silanization processes on the properties of oxidized multiwalled carbon nanotubes. Acta Phys Pol A 116:S150–S155
108. Palencia C, Rubio F, Merino C, Rubio J, Oteo JL (2009) Study of the silanization process in CNFs: time, temperature, silane type and concentration influence. J Nano Res 4:33–43
109. Liu T, Wood W, Zhong WH (2011) Sensitivity of dielectric properties to wear process on carbon nanofiber/high-density polyethylene composites. Nanoscale Res Lett 6:7
110. Yuen SM, Ma CCM, Chiang CL (2008) Silane grafted MWCNT/polyimide composites—preparation, morphological and electrical properties. Compos Sci Technol 68:2842–2848
111. Wang DH, Sihn S, Roy AK, Baek JB, Tan LS (2010) Nanocomposites based on vapor-grown carbon nanofibers and an epoxy: functionalization, preparation and characterization. Eur Polym J 46:1404–1416
112. Qu L, Lin Y, Hill DE, Zhou B (2004) Polyimide-functionalized carbon nanotubes: synthesis and dispersion in nanocomposite films. Macromolecules 37:6055–6060
113. Yuen SM, Ma CCM, Chiang CL, Lin YY, Teng CC (2007) Preparation and morphological, electrical, and mechanical properties of polyimide-grafted MWCNT/polyimide composite. J Polym Sci Part A Polym Chem 45:3349–3358
114. Viswanathan G, Chakrapani N, Yang H et al (2003) Single-step in situ synthesis of polymer-grafted single-wall nanotube composites. J Amer Chem Soc 125:9258–9259
115. Malikov EY, Muradov MB, Akperov OH, Eyvazova GM, Puskas R, Madarász D, Nagy L, Kukovecz A, Kónya Z (2014) Synthesis and characterization of polyvinyl alcohol based multiwalled carbon nanotu be nanocomposites. Phys E 61:129–134
116. Chen S, Wu G, Liu Y, Long D (2006) Preparation of poly(acrylic acid) grafted multiwalled carbon nanotubes by a two-step irradiation technique. Macromol 39:330–334
117. Geng Y, Liu MY, Li J, Shi XM, Kim JK (2008) Effects of surfactant treatment on mechanical and electrical properties of CNT/epoxy nanocomposites. Compos Part A 39:1876–1883

118. Mao J, Liu Q, Lv X et al (2007) A water-soluble hybrid material of single-walled carbon nanotubes with an amphiphilic poly(phenyleneethynylene): preparation, characterization, and photovoltaic properties. J Nanosci Nanotech 7:2709–2718
119. Herranz MA, Ehli C, Campidelli S, Guti Errez M, Hug GL, Ohkubo K, Fukuzumi S, Prato M, Martín N, Guldi DM (2008) Spectroscopic characterization of photolytically generated radical ion pairs in single-wall carbon nanotubes bearing surface-immobilized tetrathiafulvalenes. J Am Chem Soc 130:66–73
120. Kavakka JS, Heikkinen S, Kilpelainen I, Mattila M, Lipsanen H, Helaja J (2007) Noncovalent attachment of pyro-pheophorbide a to a carbon nanotube. Chem Commun 5:519–521
121. O'Connell MJ, Boul P, Ericson LM et al (2001) Reversible water-solubilization of single-walled carbon nanotubes by polymer wrapping. Chem Phys Lett 342:265–271
122. Chen RJ, Zhang Y, Wang D, Dai H (2001) Noncovalent sidewall functionalization of single-walled carbon nanotubes for protein immobilization. J Amer Chem Soc 123:3838–3839
123. Bonard JM et al (1997) Purification and size-selection of carbon nanotubes. Adv Mater 9:827–831
124. Richard C, Balavoine F, Schultz P, Ebbesen TW, Mioskowski C (2003) Supramolecular self-assembly of lipid derivatives on carbon nanotubes. Science 300:775–778
125. Islam MF, Rojas E, Bergey DM, Johnson AT, Yodh AG (2002) High weight fraction surfactant solubilization of single-wall carbon nanotubes in water. Nano Lett 3:269–273
126. Kang M, Myung SJ, Jin HJ (2006) Nylon 610 and carbon nanotube composite by in situ interfacial polymerization. Polymer 47:3961–3966
127. Guldi DM, Martin N (2010) Carbon nanotubes and related structures—synthesis, characterization, functionalization, and applications, p 351
128. Porter AE, Gass M, Muller K, Skepper JN, Midgley PA, Welland M (2007) Direct imaging of single-walled carbon nanotubes in cells. Nat Nanotechnol 2:713–717
129. Kim JK, Kim Y, Park S, Ko H, Kim Y (2016) Encapsulation of organic active materials in carbon nanotubes for application to high-electrochemical-performance sodium batteries. Energy Environ Sci 9:1264–1269
130. Cui X, Wang Y, Jiang G, Zhao Z, Xu C, Duan A, Liu J, Wei Y, Bai W (2014) The encapsulation of CdS in carbon nanotubes for stable and efficient photocatalysis. J Mater Chem A 2:20939–20946
131. Meng L, Zhang X, Lu Q, Fei Z, Dyson PJ (2012) Single walled carbon nanotubes as drug delivery vehicles: targeting doxorubicin to tumors. Biomaterials 33:1689–1698
132. Pantarotto D, Briand JP, Prato M, Bianco A (2004) Translocation of bioactive peptides across cell membranes by carbon nanotubes. Chem Commun 10:16–17
133. Kam NWS, Liu Z, Dai H (2005) Carbon nanotubes as intracellular transporters for proteins and DNA: an investigation of the uptake mechanism and pathway. Angew Chem 44:1–6
134. Hampel S, Kunze D, Haase D, Kramer K, Rauschenbach M, Ritschel M et al (2008) Carbon nanotubes filled with a chemotherapeutic agent: a nanocarrier mediates inhibition of tumor cell growth. Nanomedicine 3:175–178
135. Li J, Yap SQ, Yoong SL, Nayak TR, Chandra GW, Ang WH et al (2012) Carbon nanotube bottles for incorporation, release and enhanced cytotoxic effect of cisplatin. Carbon 50:1625–1634

Preparation/Processing of Polymer–Carbon Composites by Different Techniques

Mostafizur Rahaman, Ali Aldalbahi and Purabi Bhagabati

Abstract In this chapter, the discussion has been made on some important methodologies to prepare polymer/carbon composites. These procedures are solution mixing technique, melt mixing technique, in situ polymerization technique, dry mixing technique, powder mixing technique, and aqueous mixing technique. Solution mixing has been categorized into evaporative casting, vacuum filtration, 3D printing, and wet spinning. In the melt mixing process, the discussion has been focused on melt blending through internal mixer and melt spinning. Some diagrams have been drawn and discussed for better understanding of the composite preparation processes. The advantages and disadvantages associated with the composite preparation processes are mentioned herein where necessary.

Keywords Polymer/carbon composites · Composite processing
Solution mixing · Melt mixing · In situ polymerization · Dry mixing
Powder mixing · Aqueous mixing

M. Rahaman (✉) · A. Aldalbahi
Department of Chemistry, College of Science, King Saud University,
Riyadh 11451, Saudi Arabia
e-mail: mrahaman@ksu.edu.sa

A. Aldalbahi
e-mail: aaldalbahi@ksu.edu.sa

P. Bhagabati
Chemical Engineering Department, Indian Institute of Technology Guwahati,
Guwahati 781039, India
e-mail: purabi.bhagabati08@gmail.com

M. Rahaman et al. (eds.), *Carbon-Containing Polymer Composites*,
Springer Series on Polymer and Composite Materials,
https://doi.org/10.1007/978-981-13-2688-2_3

1 Processing of Conducting Composite Materials

Different processing methods have been employed to fabricate various forms of composite materials such as freestanding films, fibers, buckypapers, and printed patterns. Commonly used methods for preparing these composite materials include solution processing, melt mixing, in situ polymerization process, etc. [1–3].

1.1 Composites by Solution Processing

Solution mixing process is an important method for preparing polymer–carbon composite materials in both organic solvents and water [4–6]. In general, this method requires the carbons to be efficiently dispersed in a solvent that can dissolve the polymer. Then both components are mixed by magnetic/mechanical stirring or sonication [2, 3, 7]. Different types of solution processing have been used to produce composite materials including evaporative casting, vacuum filtration, fiber spinning, and printing. These process techniques involve the two important steps: initially dispersing carbons in the polymer matrix and then removal of solvent or dispersant from the mixture (Fig. 1).

1.1.1 Evaporative Casting

In the evaporative casting method, the carbons are dispersed in a solvent with or without dispersants, such as polymers, biopolymers, and surfactants. The dispersion

Fig. 1 Flowchart representing the steps of solution processing for preparing polymer–carbon composites

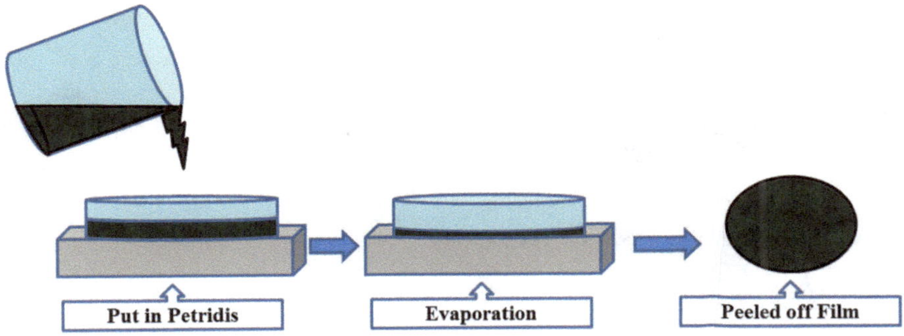

Fig. 2 Schematic of evaporative casting process in order to form freestanding composite film

is then transferred into a mold and the solvent is evaporated in a controlled manner, resulting in the formation of a composite film (Fig. 2) [8, 9]. These films may be either freestanding or attached to the substrate onto which they are cast. A similar well-known film formation method is drop casting [10, 11]. Polymer/carbon composites especially with CNTs have been prepared in the past using this evaporative casting method and discussed their different properties [12, 13]. It has been mentioned herein that the properties like mechanical and electrical are mostly dependent on the polymer and filler ratio that is the increment in filler concentration up to a certain level improve the electrical conductivity and mechanical strength [12].

Styrene butadiene rubber (SBR)/graphene nanocomposites were prepared by solution mixing and evaporative casting method [14]. Initially, graphite intercalated compound (GIC) was subjected to thermal sock at high temperature and then ultrasonicated in tetrahydrofuran (THF) to get graphene nanoplatelets (GNPs). The mixture was added with SBR solution, stirred mechanically at rpm 200, and sonicated for 1 h below 30 °C. The solvent was evaporated at 60 °C, where ethanol was used to precipitate, collect, wash, and dry the composites power. The authors compared their results with the composites prepared by melt mixing technique and reported better dispersion of GNPs within the SBR matrix when prepared by evaporative solution casting method. Graphite oxide (GO)/poly(propylene carbonate) (PPC) composites and poly(methyl methacrylate) (PMMA)/carbon black (CB) composites were prepared by solution mixing followed by evaporative casting technique [15, 16].

1.1.2 Vacuum Filtration

The vacuum filtration method is one of the simplest processes for manufacturing of electrically conductive ultrathin film made of densely packed and homogeneously dispersed carbon networks [17, 18]. These films are called buckypapers in which the carbon networks are entangled and are held together by van der Waals force of attraction at the carbon–carbon interface where it forms a pseudo-two-dimensional

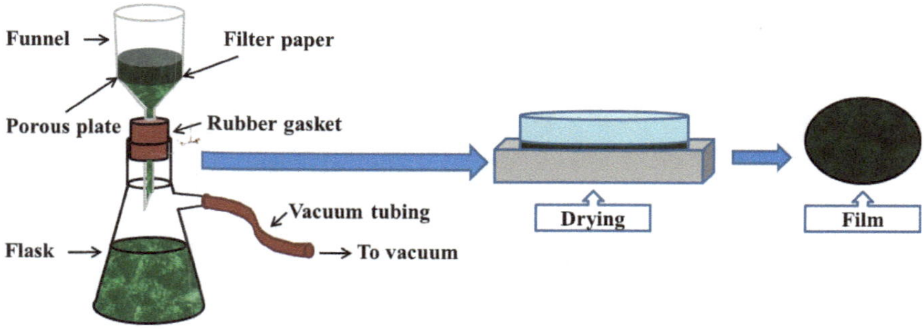

Fig. 3 Schematic for the preparation of composite films using the vacuum filtration method

structure [19]. This technique is having four simple steps: preparation of the carbon dispersions, vacuum filtering of this dispersion through a suitable membrane, washing the product with distilled water and methanol for making composite film, and drying in the oven to easily peel off the films from the filtration membrane [17, 20]. A schematic diagram of vacuum filtration technique is shown in Fig. 3.

This filtration technique is advantageous because of many reasons: (i) homogeneity as a result of flow dynamics. (ii) Maximum overlap with other nanotubes as they accumulate due to the strong vacuum, which causes the nanotubes to flatten and straighten out. This results in the increase in maximum electrical conductivity and mechanical strength throughout the composite film. (iii) Thickness control of the film by controlling the concentration of nanotubes and volume of the dispersion filtered [17].

It is shown in the literature that the properties of buckypapers composite films are influenced by the condition of preparation method and types of used dispersants [18, 21]. It can be exemplified that the in-plane Poisson's ratio of buckypapers films is tuned by the combination of single-walled carbon nanotubes (SWCNTs) and multi-walled carbon nanotubes (MWCNTs) [22]. The preparation of buckypapers with the change of MWCNTs length significantly affects the pore diameter present in the final material [18]. It has also been investigated that the sonication time, volume of solvent used, and type of used membrane filter affect the tensile property and surface morphology of SWCNT buckypapers [21]. In addition, it has been reported that the electrical and mechanical properties of buckypapers are shown to decrease with the increase in dispersant's molecular mass [21]. The flexibility and excellent electrical and dielectric properties make the carbon nanotube buckypapers an ideal material for different types of applications such as in solar cell, display system, touch screen, sensor, electronic paper, supercapacitor, and batteries [23–26].

The vacuum filtration technique was used for the fabrication of graphene oxide (GO)/polyvinyl alcohol (PVA) and GO/polymethyl methacrylate (PMMA) composite films with a wide range of filler loading [27]. In this typical process, graphite was pre-oxidized by stirring for 6 h within a mixture of concentrated sulfuric acid (15 mL), potassium persulfate (10 g), and phosphorous pentoxide (10 g). The

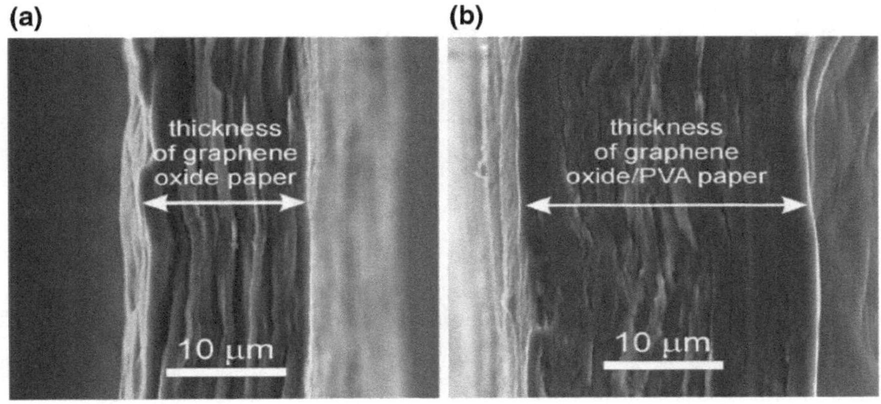

Fig. 4 SEM images of **a** GO paper and **b** GO/PVA composite. Reproduced after permission from Ref. [27]

extracted powder after filtering, washing, and drying was then stirred at 35 °C within the mixture of concentrated sulfuric acid and potassium per manganite to get oxidized graphite. The GO was further exfoliated by ultrasonication in a bath-type sonicator. The GO/PVA composites were obtained by aqueous dispersion of GO within the aqueous solution of PVA, whereas GO/PMMA composites were obtained by dispersing GO within *N,N*-dimethylformamide (DMF) and subsequent mixing with PMMA solution in DMF. The GO film and composite films were obtained by vacuum filtration and later on drying in a Petridis. Scanning electron microscopic images showed the macroscopic paper-like morphology of both GO film and composite film with paper thickness 10–30 μm as shown in Fig. 4.

1.1.3 Fiber Spinning

Spinning is the manufacturing process for preparing composite fibers. This is performed using a spinneret, which is fed by an extrusion process, to eject a thin continuous fiber. Different spinning techniques exist; these include wet spinning, melt spinning, and electrospinning [28, 29].

A common method used to produce composite fibers containing conducting fillers polymers, thermoplastic polymers, and carbons is wet spinning [30–34]. Wet spinning involves injecting of a composite's solution into a liquid bath that contains low molecular weight solution, ensuring that the solvent does not dissolve the composite solution (Fig. 5). Due to the exchange of solvent and non-solvent in the coagulation bath, a solid gel fiber is formed. This solid gel fiber is stretched and dried by passing it over spinning rollers [35]. Previously, it has been shown in the literature that, by adding a small quantity of CNTs, a successful reinforcement of the properties of the polymer fibers was achieved [35]. In another study, biopolymers (gellan gum and chitosan) were used in a wet-spinning technique where one

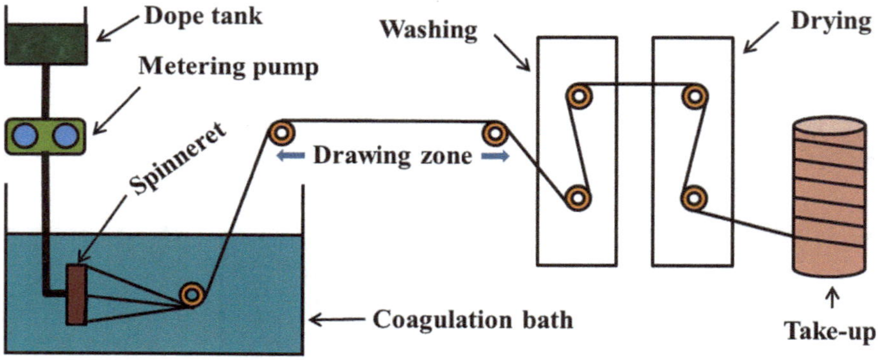

Fig. 5 Schematic illustration of the setup to form fibers using the wet-spinning process

biopolymer was used as CNT dispersant and the other as a coagulant medium for the continuously spun fibers. This was known as polyelectrolytic complexation, as gellan gum and chitosan are oppositely charged biopolymers [32].

1.1.4 Printing

In printing, a material, especially ink is deposited onto a suitable substrate for producing a particular structure. In the previously described processing methods (evaporative casting, vacuum filtration, and fiber spinning) composite materials are prepared in film and fiber forms. Printing methods are mainly classified into two main groups to fabricate composite material in pattern form: inkjet and extrusion printing [36–39].

Inkjet printing is a commonly used industrial printing method due to advantages such as precise and controlled pattern generation, solution saving effects, non-contact injection, high repeatability, and scalability [40, 41]. This method also allows for printing of various materials onto both rigid and flexible substrates [38, 42]. The ink consists of a solute, such as carbons, dispersed in a solvent, such as a polymer, and is jetted from a nozzle by deflection of piezoelectric materials (Fig. 6). This method usually requires low viscous inks and filtration of the ink prior to the jetting to prevent nozzle clogging [43, 44]. It has been demonstrated as a method for printing bio-scaffold and cell [45, 46]. Other applications of inkjet printing include components for drug screening and biosensors [47, 48]. Furthermore, it has a potential use in regenerative medicine and tissue engineering [49–51].

Different conducting materials such as polymers and CNTs were successfully printed using inkjet printing. For example, it has been reported that by inkjet printing SWCNTs and MWCNTs with conductive polymers onto flexible substrates by using a piezoelectric inkjet printer, very low sheet resistance 225 Ω/sq was reported [36]. Another study showed that inkjet printed patterns of SWNTs

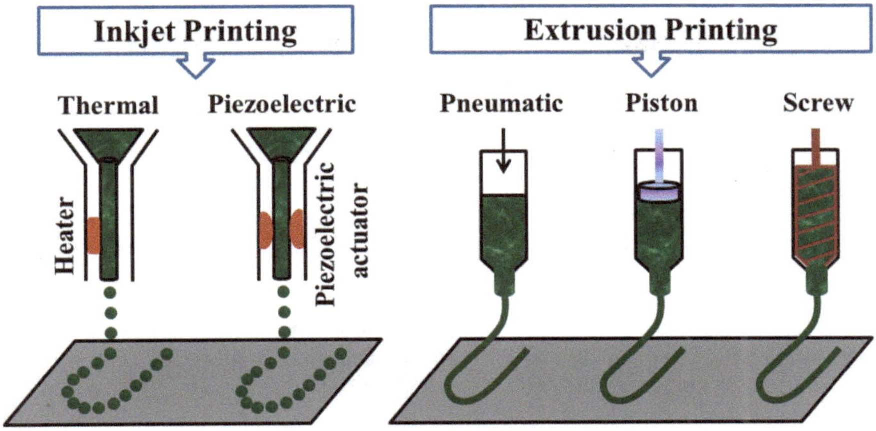

Fig. 6 Schematic illustrations of inkjet and extrusion printing methods

stabilized with poly (2-methoxyaniline-5-sulfonic acid) (PMAS) onto poly(ethylene terephthalate (PET) sheets displayed a sheet resistance 100 kΩ/sq and optical transparency 85% as well as sensitivity to alcohol vapors under static and flow conditions [52].

Extrusion printing is a process that involves pushing (using constant pressure) the inks through a small nozzle onto a moving substrate [38, 53]. Compared to inkjet printing, extrusion printing creates patterns which can reliably deposit large volumes through cheap and replaceable parts [54]. It has been shown diagrammatically in Fig. 6. This method can also be applied to fabricate three-dimensional structure and to embed material onto substrate [38, 55]. However, extrusion printing is slower and has poorer resolution and scalability compared to the inkjet process. Unlike inkjet printing, extrusion printing allows for the patterning of high-viscosity inks [38]. Very careful control of the flow properties, surface tensions, and wettability of the inks is required in order to successfully produce a high-quality extruded material [56].

In recent studies, it has been shown that both inkjet and extrusion printing methods were used to prepare conducting tracks by depositing a conducting polymer, poly(3,4-ethylenedioxythiophene)/poly(sodium 4-styrenesulfonate), onto a chitosan biopolymer substrate [38]. In that work, a highly viscous paste was used for the syringe extrusion method, and a more aqueous solution was used for inkjet printing. Using extrusion printing, the authors prepared embedded tracks, for potential application as electrodes, by inserting the extruding needle under the surface of the biopolymer solution. The authors reported a higher conductivity for the extruded patterns compared to the inkjet printed ones, due to the greater amount of deposited conducting inks.

Another study found that gellan gums are a good rheological modifier for MWCNT ink [39]. It was found that a single layer of ink printed onto a flexible substrate exhibited resistance value of 7–8 kΩ/cm, which was reduced by increasing the number of printed layers. Many polymer/carbon composites like ABS/carbon fiber [57], PS/carbon nanofiber and PS/graphite [58], nylon 12/carbon black [59], ABS/graphene and PLA/graphene [60], etc., were prepared by 3D printing technique, where mechanical, electrical, and some important properties were discussed in detailed.

1.2 Composites by Melt Processing

Instead of using a solvent, the polymer substrate can also be melted and then intermixed with carbons using different techniques such as shearing, extrusion, etc. This process is known as melt mixing process [2, 3].

The benefit of melt processing is that it does not use any solvent, simple, and it can be used as an industrial process to impart carbons into thermoplastic polymers such as polypropylene. This method uses elevated temperatures together with applied shear forces to form the polymer–carbon composites and requires equipment such as extruders and injection devices that are designed to withstand these conditions [2, 3]. While this method can also be used to generate composite fibers, the composites prepared through this synthesis route usually have low carbon filler content [6]. However, there are also reports of mixing high content of carbons within the polymer matrices using this methodology [61–65]. A flowchart diagram of this type of mixing process has been depicted in Fig. 7. In this process, polymer, carbon, and other necessary additives/ingredients are fed into mixing mill and processed approximately 20 °C higher than the melting point of the polymer. The mixing is done for a certain time at a specific rotor speed. The molten mass is then taken out and passes through the two-roll mill, and finally, the composite is sheeted out by using any compression mold. The mixing can be done in a batch or continuously, depending on the used mixing machine and our choice.

A previous study reported polycarbonate–CNT composites prepared by melt processing. It was found that by controlling the mixing condition, an electrical percolation threshold of about 1 wt% was obtained [66]. Other studies showed a threefold increase in the modulus and strength for polyamide containing 2 wt% MWNTs compared to the pure matrix [67].

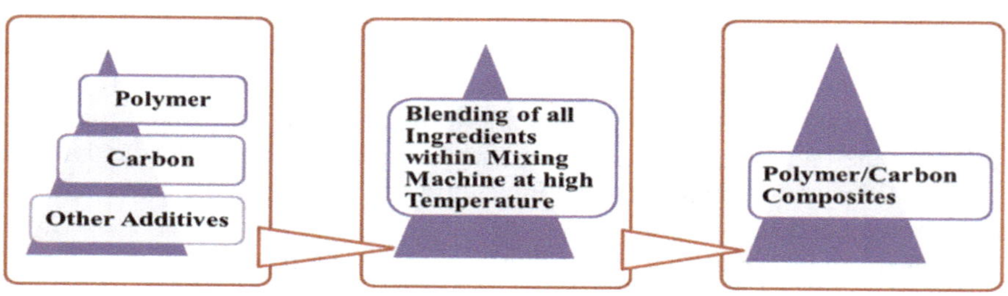

Fig. 7 Flowchart presenting the steps of the melt processing

Rahaman et al. have reported the melt mixing of carbon fiber with ethylene vinyl acetate (EVA) and acrylonitrile butadiene rubber (NBR) and their blend compositions mixed with other ingredients [68]. The mixing process was carried out within Haake Rheocord at temperature of 120 °C for 6 min at shear rate of 60 rpm. The composites were vulcanized by compression molding at 160 °C at 5 MPa pressure at their determined curing time. The polymer, carbon fiber, and other ingredients were added sequentially as follows: polymer → carbon fiber → other ingredients. The mechanical property, electrical conductivity, and electromagnetic interference shielding effectiveness (EMI SE) for these polymer carbon fiber composites were studied. It has been reported that elongation at break and tensile strength have been reduced after the addition of carbon fiber but the tensile modulus within the stress–strain proportionality limit, electrical conductivity, and EMI SE has been increased. More importantly, there is substantial breakage of fiber length within the polymer matrices and the extent of breakage has been increased with the increase in loading of carbon fiber when the authors used this melt mixing technique.

The preparation of polypropylene grafted maleic anhydride (PP-g-MA) and expandable graphite oxide (EGO) composites was carried out by melt mixing technique in Haake kneading mixer [69]. The mixing was conditioned at 200 °C for 30 min at 60 rpm. In this mixing process, PP is nonpolar polymer, and hence MA was used to increase its compatibility with EGO particles. It was reported through XRD analysis that there was not any change in the d-spacing of graphite layers within the PP-g-ME/EGO composites.

Masterbatches of polycaprolactum (PCL)/MWCNT composites were prepared by melt mixing technique using an internal mixer where the mixing was done for 10 min at 65 °C and 60 rpm as reported by Maiti et al. [70]. The masterbath thus again was melt blended with polycarbonate (PC) at 280 °C at the same time period and rpm. It was investigated through scanning electron microscopic (SEM) and transmission electron microscopic studies that the CNTs were homogeneously dispersed within the polymer matrices at their low loading as shown in Fig. 8. The obtained electrical percolation threshold was 0.14%, which suggested the formation of interconnected continuous conductive networks within the polymer matrices.

The thermoplastic polyurethane (TPU) elastomer was melt blended with high-structure carbon black and carbon nanofiber by a special miniature asymmetric batch mixer [71]. The authors ensured good dispersion of the particles within the polymer matrix. They investigated the thermal, mechanical, electrical, and flammability properties for this composite system.

In another study, the nylon 6/carbon black composite was prepared by masterbatch dilution and melt mixing technique [72]. The effect of carbon black size and their distribution on the mechanical and electrical properties of the composites was investigated. The electrical resistivity of nylon 6 was reduced from 10^{15} to 10^{7} Ω cm in case of 1 wt% diluted masterbatch and 6 wt% melt mixed nylon 6/carbon black composites, indicating that composite prepared by masterbatch dilution method gives better property. This is because of the formation of small carbon black clusters when the composite is prepared by masterbatch dilution

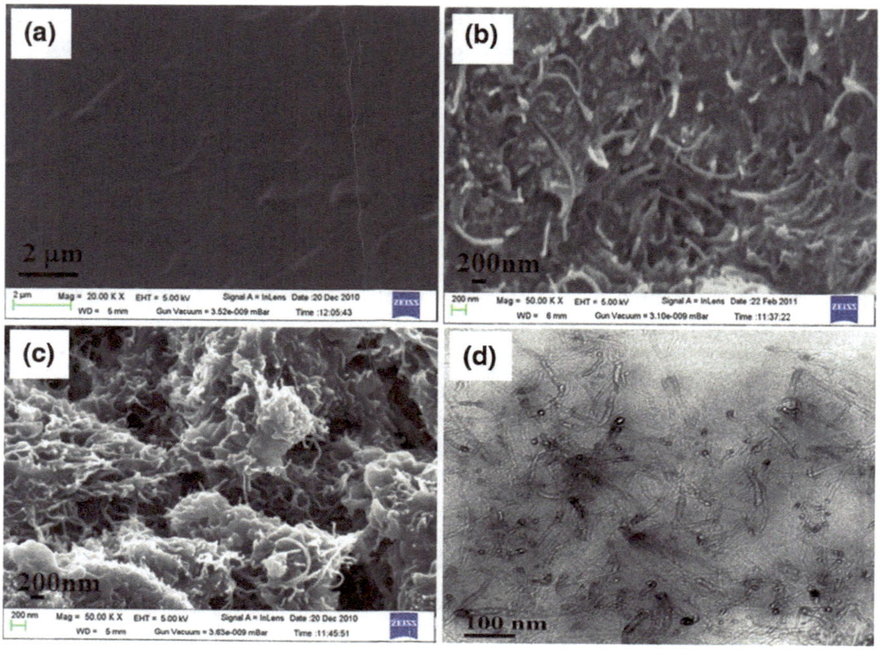

Fig. 8 FESEM image of **a** (90/10 w/w) PC/PCL miscible blend, **b** (90/10/0.35 w/w) PC/PCL/MWCNT composite, **c** (80/20/0.7 w/w) PC/PCL/MWCNT composite, and **d** TEM image of (90/10/0.35 w/w) PC/PCL/MWCNT composite. Reproduced after permission from Ref. [70]

method. The authors also reported that the increment in carbon black content decreases the elongation at break and tensile strength, but increases the tensile modulus. There are also many reports of preparation polymer/carbon black composites by melt mixing technique [73, 74].

Polymer/graphene composites are also prepared by this melt mixing technique. For example, polyvinyl chloride (PVC)/multilayer graphene composite was prepared by conventional melt mixing process for improving the toughness of rigid PVC [75]. Before melt mixing, the powder PVC, graphene, stabilizers, and lubricants were dry mixed at room temperature using a high-speed mixer. The mixture ingredients were then melt mixed using a torque rheometer at temperature of 165 °C for 5 min. The composite blends were then transferred into two-roll mill for further mixing at 170 °C for squeezing into thin sheet. These thin sheets were stacked together and placed into hot-press machine to get flat composite sheets. It was reported that only 0.36 wt% graphene loading into PVC greatly improves the toughness and impact strength of the composites. Besides that, many polymers like poly(lactic acid) (PLA), PP, PU, poly(ethylene terephthalate) (PET), polystyrene (PS), poly(ether ether ketone) (PEEK), styrene–ethylene/butylene–styrene triblock copolymer, etc., were blended with graphene for obtaining the composites by this method for investigating their mechanical, thermal, and electrical properties along with their morphology [76–82].

The polymer/carbon composites fiber is prepared by melt spinning technique, which is very convenient, economical and can be prepared industrially in large scale. In this typical process, the fiber-forming substances that are the mixed polymer and carbon or their prepared master batch are placed in hopper and extruded through spinneret (Fig. 9). The extruded mass is then solidified by cooling. The used polymer should not be volatile in nature or degradable at the processing temperature. Generally, the single-screw or twin-screw extruder is used depending on the nature of polymer and filler particles. The metering pump shown in the diagram, control the flow of molten mass to the spinneret head and filter any unmelted mass before extrusion. The extruded molten fiber mass is cooled by quench air, and then drawn into continuous filament, stretched, and finally take up/wind up in a roll. The strength of the resultant composite fiber depends on the winding speed. The advantage of using this method is that it is cost-effective, can be used for both continuous and staple filaments, no environmental pollution, no requirement of any solvent, nontoxic, and high production rate (1000–2000 m/min). The only disadvantage is that the input of heat is high and the high maintenance cost is needed.

Polycarbonate (PC)/MWCNT (2 wt%) conductive composite was prepared by melt spinning technique using a piston-type apparatus where take-up velocities were up to 800 m/min and draw down ratio 250 [83]. The composite was investigated through TEM, which revealed that the carbon nanotubes were aligned along the fiber axis but its curve shape still existed in the melt spun fibers. At high draw down ratio, the curvature was reduced. It was reported that with the increase in alignment, the electrical conductivity of the fiber was reduced. It was also mentioned that the tensile strength and elongation at break was less at the low take-up compared to neat PC, whereas it was increased at high take-up speed. There are also many reports of preparing polymer/carbon composites by this melt-spinning technique [84–89].

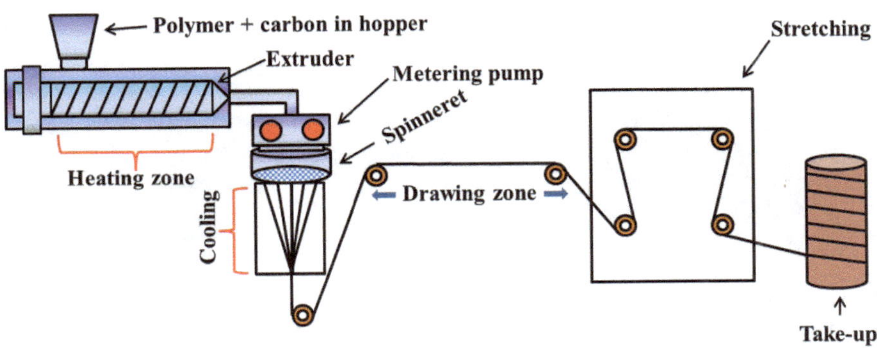

Fig. 9 Schematic illustration of the setup to form fibers using melt-spinning process

1.3 Composites by In Situ Polymerization Process

In situ polymerization is employed in order to uniformly dispersing the conducting filler, while retaining the aspect ratio and improve polymer–filler bonding strength. It is particularly useful for preparing polymers that cannot be produced by solution (insoluble polymer) and melt mixing (thermally unstable polymer) techniques [3].

In this method, the monomer is used as a starting material. It involves polymer and carbons being dispersed in a solvent. This allows for the carbons to act as a template for the polymer's growth. The mixture is then sonicated and stirred to allow for growth. These composite structures are then isolated from the solution by filtering or evaporation to dryness [90]. A general schematic diagram of this type of polymerization process is depicted in Fig. 10.

This method allows the preparation of composite with high carbon weight fraction, which can be diluted by other methods [2]. Composite materials produced by this method are more conductive. This may be because there is an increased interaction between the polymer and the carbons. It may also be due to the fact that the product has a more favorable morphology. Many polymer–CNTs composites were prepared using in situ polymerization, which included polystyrene–MWNT [91] and polyimide–SWNT [92]. In addition, epoxy–CNT [93] and polycarbonate–carbon

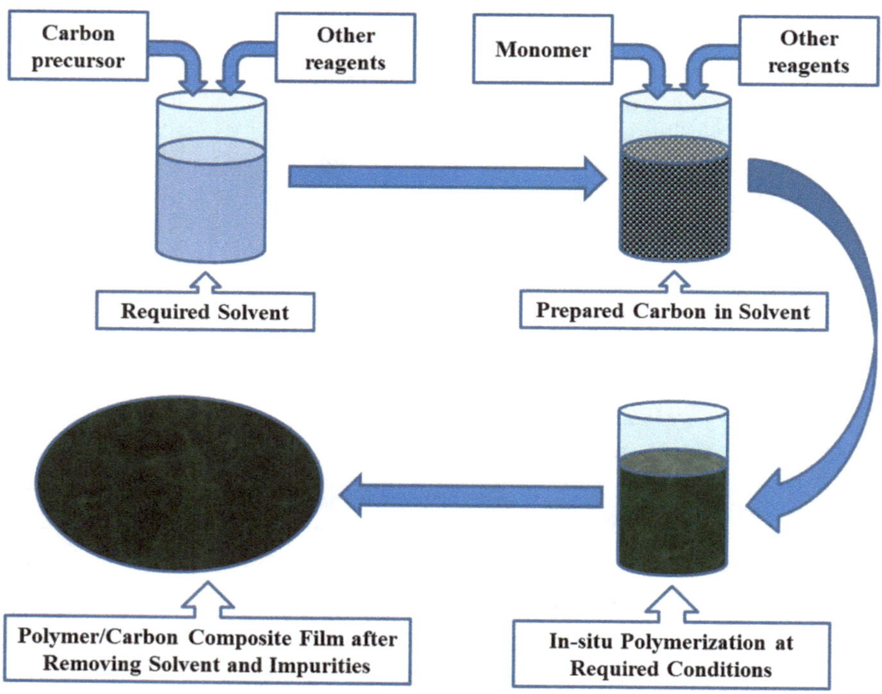

Fig. 10 Flowchart presenting the steps of the in situ polymerization processing

Fig. 11 Scanning (**a**) and transmission (**b**) electron microscopy images of polyaniline–MWNT composites prepared by in situ polymerization processing. Reproduced after permission from Ref. [95]

nanofiber [94] composites have been prepared using this method. Figure 11a, b shows the scanning and transmission electron microscopic image of polyaniline–MWCNT composite made by this in situ polymerization technique [95].

The synthesis of graphene, graphene oxide (GO), and functionalized graphene oxide (f-GO)—epoxy nanocomposites was carried out by in situ polymerization technique [96]. Initially, the filler, GO/f-GO was dispersed in acetone by means of ultrasonication and then added with epoxy matrix. The mixture was then placed within a vacuum oven at 50 °C. After evaporation of nearly 80% of solvent, m-phenylenediamine was added within it and stirred vigorously. It was then poured into a stainless steel mold and heating at 60 °C for removing the residual solvent followed by pre-cure at 80 °C for 2 h in an oven and post-cure at 120 °C for additional 2 h. The prepared composites were then characterized for several studies. It was reported that the better dispersion was achieved for epoxy/graphene and epoxy/f-GO composites compared to epoxy/GO composite.

Fim et al. reported the synthesis of polyethylene (PE)/graphite composites via in situ polymerization technique [97]. Intercalated graphite was prepared from the graphite flake by treating with H_2SO_4/HNO_3. It was then heated at 1000 °C in air for 30 s to get expanded graphite (EG), which on ultrasonication in 70% ethanol for 8 h produces graphite nanosheet (GNP). The GNP was surface treated with 15 wt% methylaluminoxane (MAO) by stirring in toluene for 30 min, and then, the solvent was removed under reduced pressure. The ethylene was polymerized using 100 ml PAAR reactor, toluene as solvent, MAO as cocatalyst, bis(cyclopentadienyl)zirconium dichloride (Cp_2ZrCl_2) as catalyst, at temperature 70 °C, ethylene pressure 2.8 bar, and for time duration 30 min. Different amount of MAO treated GNP was added to the reactor as filler before PE synthesis to get PE/graphite composites. It was mentioned that thermal treatment and ultrasonic agitation had expanded graphite layers along with the reduction of crystal size. The synthesis of polymer/graphene nanocomposites was also reported in some literature by this in situ polymerization technique [98, 99].

The polymer/carbon composites are also prepared by in situ emulsion polymerization technique. Polystyrene (PS)/carbon black (CB) composites were prepared by this in situ emulsion polymerization technique, where CB was manually mixed with styrene monomer at ambient temperature [100]. The absorption of monomer on the surface of CB made viscous paste, which was reduced by adding a surfactant within it. Azobisisobutyronitrile (AIBN) initiator was added to it for preparing emulsified monomer droplets. The surfactant solution was added to it and ultrasonicated to get well-dispersed system. The dispersion was transferred to the reactor for polymerization conditioned at reaction temperature of 60 °C, mixing speed of 350 rpm, and reaction time of 120 min. It was reported that the particle diameter was 50 nm and high polydispersity index for the polymer. The synthesis of PS/graphene nanocomposites by this in situ emulsion polymerization technique was also mentioned in literatures [101, 102].

1.4 Dry Mixing Technique

This method is most useful for rubbery materials where polymers are soft enough at room temperature. A schematic diagram for the preparation of polymer composites using this type of method is shown in Fig. 12. Initially, the conductive filler along with other ingredients are mixed with rubber in an open mixing mill or internal mixer. Generally, the mixing process starts at room temperature but during mixing the temperature may rise to 60 °C. The mix is then vulcanized at a specific temperature, pressure, and time duration to make the conductive composite.

Liu et al. have prepared styrene butadiene rubber (SBR)/carbon black (CB) composites, where carbon blacks with their different structure were used as conductive filler [103]. The concentration of carbon black within the composite was 50 phr. Other ingredients like aromatic oil (10 phr), zinc oxide (3 phr), stearic acid (2 phr), antioxidant (2 phr), wax (1.5 phr) accelerant DM (1.2 phr), accelerant D

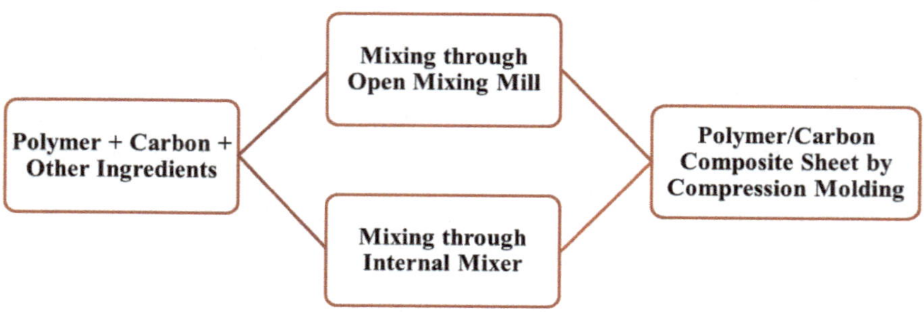

Fig. 12 Flowchart presenting the steps of dry mixing process

(0.6 phr), and sulfur (1.5 phr) were also used. The composites were prepared using two-roll mixing mill at ambient temperature via general compounding method. The composites were vulcanized at temperature of 150 °C and pressure 15 kPa to make sheet thickness of 2 mm. They reported that the mechanical properties of SBR/CB composites increase with the increase in CB loading but the composites loaded with low particle size carbon black exhibits high mechanical properties compared to high particle size ones.

Cardanol-grafted natural rubber (CGNR) and HAF carbon black composites were prepared by simple compounding method using two-roll mixing mill at ambient temperature according to standard ASTM D-3182-07 [104]. The aspect ratio of two-roll mill was length:diameter = 33:15 and friction ratio during mixing was 1:1.2. They reported that during mixing, due to friction, the temperature of two-roll raised up to 80–85 °C. The composites, later on, vulcanized at 150 °C using compression molding machine according to their optimum cure time determined by rheometer. The authors have compared the dispersion of carbon black in CGNR matrix and aromatic oil plasticized NR matrix along with their other mechanical properties. It has been reported that the dispersion of carbon black is better in CGNR matrix compared to the aromatic oil plasticized NR matrix. The mechanical strength of CGNR/CB composites is higher compared to oil plasticized NR/CB composites.

Sethi and his co-workers have reported the preparation of polychloroprene (CR) rubber and carbon black composites by this dry mixing technique [105]. The mixing was performed using two-roll mixing mill at ambient temperature by maintaining the friction ratio 1:1.1. All the mixing was done at the identical condition and the same sequence of adding the ingredients. Later on, the composites were cured at 150 °C under compression molding at hydraulic pressure 5 MPa according to optimum cure time determined from the rheometric curves. The authors investigated the effect of different mechanical deformation and temperature on electrical and dynamic mechanical properties of CR/CB composites. The electrical conductivity and dynamic mechanical properties dropped when both the number of bending cycles and compression flex cycles of deformation was increased. Figure 13 shows that the electrical conductivity of different carbon filled CR composite has reduced after bend flexing of the composites. The increase in temperature also resulted in the variation in electrical conductivity of the composites.

MWCNT based NR nanocomposites were also prepared by this dry mixing technique using an open two-roll mixing mill at the friction ratio 1:1.1, nip gap 1 mm, mixing time 19 min at room temperature [106]. In this study, MWCNT was surface functionalized with silane, 3-aminopropyltriethoxysilane and investigated its effect on the rheological and mechanical properties of NR vulcanizates. It was reported that the modulus and strength were increased due to the interaction between silane functionalized MWCNT and NR vulcanizates. There are also many reports of preparation of polymer/carbon composites by dry mixing technique in the past [107–110].

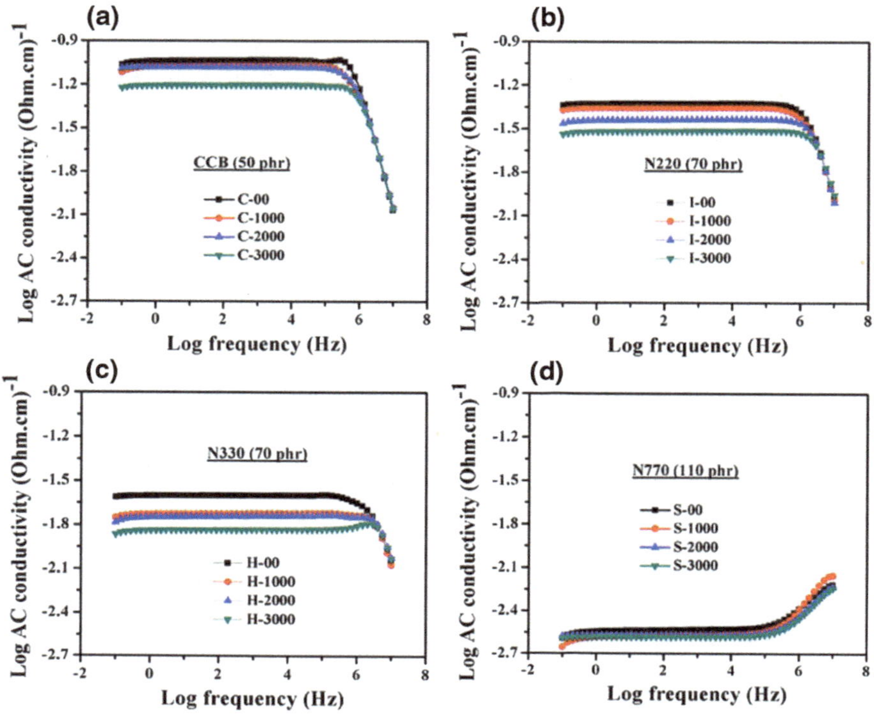

Fig. 13 AC conductivity of CR/CB composites measured before and after bend flexing at the highest loading of different CB **a** CCB, **b** N220, **c** N330, and **d** N770. Reproduced after permission from Ref. [105]

1.5 Powder Mixing Technique

Powder mixing is a very important technique that is widely used in many domestic areas and industrial fields. This is a type of dry mixing technique. In domestic areas, it is used in food making and gardening, whereas in industrial fields, it is used in the manufacturing process of pharmaceuticals, ceramics, plastics, fertilizers, detergents, cement, food production, powder metallurgy, animal feed, etc. [111–113]. In short, one can say that this technique is practiced when particulate materials are processed [111]. This technique can also be effectively used for both thermoplastic and thermosetting resins. A general diagram of this type of mixing processes is shown in Fig. 14. In this method, powdered polymer or oligomer is mixed with carbon filler along with other ingredients. The mixing is done using a blender normally at ambient temperature. The resultant mixture is then processed in batch-wise or continuously depending upon the nature of final product. The powdered polymer, in this case, is considered as continuous phase and the carbon particles as dispersed phase. In a batch-type process, the mixture is molded in either in hot or cold

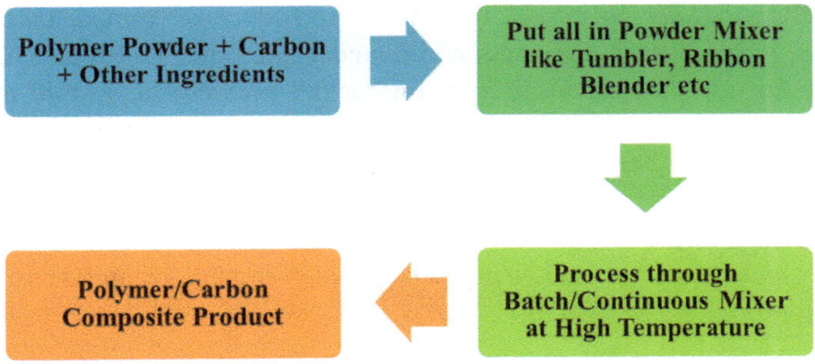

Fig. 14 Diagram representing powder mixing technique

condition to get polymer/carbon composite. The continuous mixing process is adopted for making any composite fiber or long tube/sheets, where the continuous removal of mixture is important.

Polyamide 11 (PA11)/graphene nanoplatelets (GNP) nanocomposites were prepared by this powder mixing technique aiming to be used as electrostatic discharge material [114]. They used Resodyn™ Resonant Acoustic and Thinky™ mixers for this powder–powder mixing and the mixture powder was pressed to make thin film for characterization. The SEM, TGA, Raman spectroscopy, and electrical conductivity were studied. It was revealed that the addition of GNP reduced the thermal stability of composites because of its poor dispersion. The electrical conductivity for all composites was within the electrostatic dissipation range.

Islam et al. have prepared the polyvinyl chloride (PVC)/carbon black (CB) composites in a blender, where both ingredients were in dry and powdered form [115]. The mixing was done for 1 min at 1000 rpm. Then, the mixture was compression molded at temperature of 185 °C, pressure of 200–250 kN, and mixing time for 10 min. An additional pressure of 50 kN was applied to make void-free specimen. All the molding was done at the identical condition. The compression-molded samples were taken out from the mold. The dimension of samples was 14 cm in diameter and 4 mm in thickness. Carbon content within the composites ranges from 5 to 30 wt%. The composites were tested for their electrical and mechanical properties. Electrical conductivity was increased with the addition of CB but the mechanical properties were improved only up to 15 wt% of CB and thereafter decreased. Microstructural analysis showed good dispersion of CB within the PVC matrix.

It is also specially applied for the modification of electrical conductivity of polytetrafluoroethylene (PTFE). Suh et al. have reported the electrical property and morphology of PTFE/FLG (few-layer graphene) composites prepared by this solid-state powder mixing technique and subsequent hot pressing [116]. Planetary milling process was used for the preparation of composites at 50 rpm for 10 min using stainless steel balls with diameter 10 mm and 4.5 g in weight. The weight ratio of ball to powder was 10:1. The mixture powder was compacted at ambient temperature within steel compaction die under pressure 15 MPa. The mixture powder within the compacted die was further hot pressed for 30 min at pressure 15 MPa and temperature 300 °C. A significant increase in electrical conductivity was observed by this technique using a small amount of conducting filler. It was shown that FLG was compacted on the surface of composite and overlapped without making any gaps in between them. The preparation of polymer/carbon composites was also reported using this powder mixing technique in literature [117–119].

1.6 Aqueous Mixing Technique

Biopolymer composites and conductive coatings can be prepared by this technique [13, 120, 121]. A schematic diagram of this process is shown in Fig. 15. In this process, the polymer in granules/powder form is dissolved in an aqueous medium either at room temperature or by heating in a hot plate with the help of a mechanical or magnetic stirring during a certain time period. Thereafter, carbon particle is added to this aqueous solution of polymer and disperse by mechanical stirring or sonication over a certain time period. The mixture is then evaporated by solvent casting method or vacuum filtrated to form a film of polymer/carbon composite. This methodology is same as solution mixing technique: the difference arises in the use of solvent only. In solution mixing technique, one can use any suitable solvent for dissolving the polymers, whereas in this case, only aqueous medium can be used. Hence, the composites based on only water-soluble polymer can be prepared by this technique. For the preparation of conductive coatings, the very small powdered particles of conductive fillers are added in an aqueous emulsion of polymers. The emulsion can be coagulated or used directly for the purposes.

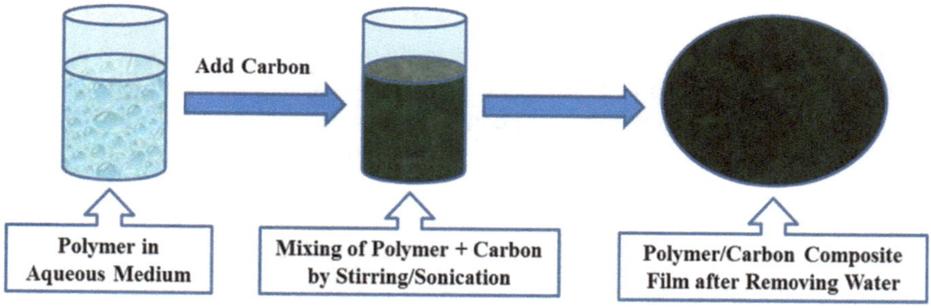

Fig. 15 Flowchart representing the steps of aqueous mixing technique

Zhao et al. have reported the preparation of poly(vinyl alcohol) (PVA)/graphene composites via facial aqueous solution mixing technique [122]. Initially, they prepared PVA/graphene oxide (GO) composites and then reduced GO to exfoliated graphene nanosheet. GO was first dispersed in aqueous solution using sonicator and stabilized by adding 0.5 wt% sodium dodecylbenzenesulfonate (SDBS). The GO was reduced to graphene using hydrazine and ammonia. The resultant mixture was then cast in a clean glass plate using a doctor blade and dried in an oven for film formation. They have shown that there was 150% increment in tensile strength and 10 times increment in tensile modulus for 1.8 vol.% PVA/graphene composite compared to neat PVA.

Preparation of chitosan/MWCNT nanocomposites was performed by simple solution mixing and evaporative casting method in aqueous medium [123]. In this typical procedure, the MWCNTs were first swelled in the required amount of distilled water (100 ml) and homogenized for 60 min in an ultrasonic bath. Acetic acid (1 ml) and chitosan (1 g) were then added with this already made MWCNT suspension and shaken for 1 h for dissolving chitosan. The solution mixture was stir mechanically at 18,000 rpm for 30 min and then sonicated for 20 min for removing any bubble present in the system. The composite solution was poured in a plastic Petridis and heated at 50 °C for evaporating water. The obtained film has an average thickness of 0.08 mm. The authors investigated the morphology of the composite system through SEM, TEM, optical microscopy, XRD, and mechanical properties like tensile and nanoindentation. Their results showed that the CNTs were homogeneously dispersed within the chitosan matrix. Moreover, 93 and 99% improvement in tensile modulus and strength, respectively, were reported for 0.8 wt % MWCNT loaded composite.

Our research group has investigated the electrical, dielectric, and impedance spectra behavior of PVA/starch (70/30) blend graphene nanocomposites prepared by solution aqueous mixing and evaporative casting technique [124]. In this procedure, 4.2 g of PVA was separately dissolved in 50 ml of deionized water in a beaker. In another beaker, starch (1.8 g) mixed with glycerol (3 g) was dispersed in 50 ml of deionized water. Different amounts of graphene were dispersed in deionized water by ultrasonication for 2 min at amplitude of 30%. Subsequently, the starch and graphene dispersions were added to the already dissolved PVA solution and continuously stirred for 4 h at 400 rpm. The mixture was degassed, decanted into flat glass plate, and evaporated overnight at 50 °C to get composite film. The polymer/carbon composites were also prepared by several authors in the past by this aqueous mixing technique [125–130]. This method is advantageous in the sense that it is economical and less hazardous because of the use of aqueous medium in the processing of composites.

1.7 Conclusions

The preparation of polymer/carbon composites by different techniques has been discussed in this chapter. Solution mixing method is suitable for the polymers that are soluble in some common solvents. Moreover, in some cases, solution mixing process gives a better dispersion of carbons within polymer matrices. However, for making buckypaper, vacuum filtration is a favorable process. Melt mixing technique is important because of its industrial viability and for those polymers, which are insoluble in common solvents. Wastage of solvent gives more cost to the solution mixing, whereas in melt mixing, energy loss due to heating produces more cost to the process. For making composite fiber, one has to follow either wet or melt spinning technique depending on the nature of polymer and carbons. Dry mixing is favorable for those polymers which are quite soft at ambient temperature and the other ingredients are in soft, liquid, or powder form. Rubbery materials are generally processed through this technique. Powder polymers are processed through the powder mixing technique when the other ingredients are either in powder or liquid form. Biopolymer composites are generally prepared through aqueous mixing technique as they are water-soluble in nature.

References

1. Harris P (2004) Carbon nanotube composites. Int Mater Rev 49(1):31–43
2. Spitalsky Z, Tasis D, Papagelis K, Galiotis C (2010) Carbon nanotube-polymer composites: chemistry, processing, mechanical and electrical properties. Prog Polym Sci 35(3):357–401
3. Al-Saleh MH, Sundararaj U (2009) A review of vapor grown carbon nanofiber/polymer conductive composites. Carbon 47(1):2–22
4. Cadek M, Coleman JN, Barron V, Hedicke K, Blau WJ (2002) Morphological and mechanical properties of carbon-nanotube-reinforced semicrystalline and amorphous polymer composites. Appl Phys Lett 81(27):5123–5125
5. Paiva MC, Zhou B, Fernando KAS, Lin Y, Kennedy JM, Sun YP (2004) Mechanical and morphological characterization of polymer-carbon nanocomposites from functionalized carbon nanotubes. Carbon 42(14):2849–2854
6. Moniruzzaman M, Winey KI (2006) Polymer nanocomposites containing carbon nanotubes. Macromolecules 39(16):5194–5205
7. Safadi B, Andrews R, Grulke EA (2002) Multiwalled carbon nanotube polymer composites: synthesis and characterization of thin films. J Appl Polym Sci 84(14):2660–2669
8. Songmee N, Singjai P, in het Panhuis M (2010) Gel-carbon nanotube materials: the relationship between nanotube network connectivity and conductivity. Nanoscale 2 (9):1740–1745
9. Ferris CJ, in het Panhuis M (2009) Gel-carbon nanotube composites: the effect of carbon nanotubes on gelation and conductivity behavior. Soft Matter 5(7):1466–1473
10. Bayer IS, Caramia V, Fragouli D, Spano F, Cingolani R, Athanassiou A (2012) Electrically conductive and high temperature resistant superhydrophobic composite films from colloidal graphite. J Mater Chem 22(5):2057–2062
11. Sreekumar TV, Liu T, Kumar S, Ericson LM, Hauge RH, Smalley RE (2002) Single-wall carbon nanotube films. Chem Mater 15(1):175–178

12. Bauhofer W, Kovacs JZA (2009) Review and analysis of electrical percolation in carbon nanotube polymer composites. Compos Sci Technol 69(10):1486–1498
13. Coleman JN, Cadek M, Ryan KP, Fonseca A, Nagy JB, Blau WJ, Ferreira MS (2006) Reinforcement of polymers with carbon nanotubes. The role of an ordered polymer interfacial region. Experiment and modeling. Polymer 47(26):8556–8561
14. Araby S, Meng Q, Zhang L, Kang H, Majewski P, Tang Y, Ma J (2014) Electrically and thermally conductive elastomer/graphene nanocomposites by solution mixing. Polymer 55:201–210
15. Bian J, Wei XW, Lin HL, Gong SJ, Zhang H, Guan ZP (2011) Preparation and characterization of modified graphite oxide/poly(propylene carbonate) composites by solution intercalation. Polym Degrad Stab 96:1833–1840
16. Ou R, Gupta S, Parker CA, Gerhardt RA (2006) Fabrication and electrical conductivity of poly(methyl methacrylate) (PMMA)/carbon black (CB) composites: comparison between an ordered carbon black nanowire-like segregated structure and a randomly dispersed carbon black nanostructure. J Phys Chem B 110:22365–22373
17. Wu Z, Chen Z, Du X, Logan JM, Sippel J, Nikolou M, Kamaras K, Reynolds JR, Tanner DB, Hebard AF, Rinzler AG (2004) Transparent, conductive carbon nanotube films. Science 305(5688):1273–1276
18. Kukovecz A, Smajda R, Konya Z, Kiricsi I (2007) Controlling the pore diameter distribution of multi-wall carbon nanotube buckypapers. Carbon 45(8):1696–1698
19. Whitby RLD, Fukuda T, Maekawa T, James SL, Mikhalovsky SV (2008) Geometric control and tuneable pore size distribution of buckypaper and buckydiscs. Carbon 46(6):949–956
20. Wang Q, Moriyama H (2011) Carbon nanotube-based thin films: synthesis and properties. In: Yellampalli S (ed) Carbon nanotubes—synthesis, characterization, applications. InTech, Rijeka, pp 488–514
21. Boge J, Sweetman LJ, in het Panhuis M, Ralph SF (2009) The effect of preparation conditions and biopolymer dispersants on the properties of SWNT buckypaper. J Mater Chem 19(48):9131–9140
22. Hall LJ, Coluci VR, Galvao DS, Kozlov ME, Zhang M, Dantas SO, Baughman RH (2008) Sign change of Poisson's ratio for carbon nanotube sheets. Science 320(5875): 504–507
23. Agnes SC, John EF, Chad BH, Andrew GR, Richard ES (2000) Solid-state electrochemistry of the Li single wall carbon nanotube system. J Electrochem Soc 147(8):2845–2852
24. Liu T, Sreekumar TV, Kumar S, Hauge RH, Smalley RE (2003) SWNT/PAN composite film-based supercapacitors. Carbon 41(12):2440–2442
25. Prokudina NA, Shishchenko ER, Joo OS, Hyung KH, Han SH (2005) A carbon nanotube film as a radio frequency filter. Carbon 43(9):1815–1819
26. Rein MD, Breuer O, Wagner HD (2011) Sensors and sensitivity: carbon nanotube buckypaper films as strain sensing devices. Compos Sci Technol 71(3):373–381
27. Putz KW, Compton OC, Palmeri MJ, Nguyen SBT, Brinson LC (2010) High-nanofiller-content graphene oxide–polymer nanocomposites via vacuum-assisted self-assembly. Adv Funct Mater 20:3322–3329
28. Ziabicki A (1976) Fundamentals of fibre formation: the science of fibre spinning and drawing. Wiley, New York
29. Nakajima T (1994) Advanced fiber spinning technology. Woodhead Publishing, Cambridge
30. Luo Y, Gong Z, He M, Wang X, Tang Z, Chen H (2011) Fabrication of high-quality carbon nanotube fibers for optoelectronic applications. Solar Energy Mater Solar Cells 97:78–82
31. Pomfret SJ, Adams PN, Comfort NP, Monkman AP (199) Advances in processing routes for conductive polyaniline fibres. Synth Met 101(1–3):724–725
32. Granero AJ, Razal JM, Wallace GG, in het Panhuis M (2008) Spinning carbon nanotube-gel fibers using polyelectrolyte complexation. Adv Funct Mater 18(23):3759–3764
33. Higgins T, Warren H, in het Panhuis M (2011) Film, buckypapers and fibers from clay chitosan and carbon nanotubes. Nanomaterials 1:3–19
34. Raza JM, Gilmore KJ, Wallace GG (2008) Carbon nanotube biofiber formation in a polymer-free coagulation bath. Adv Funct Mater 18(1):61–66

35. Mottaghitalab T (2006) Development characterisation of polyaniline-carbon nanotube conducting composite fibers. Ph.D. thesis, Mechanical and Materials Engineering, University of Wollongong, Wollongong
36. Denneulin A, Bras J, Blayo A, Khelifi B, Roussel-Dherbey F, Neuman C (2009) The influence of carbon nanotubes in inkjet printing of conductive polymer suspensions. Nanotechnology 20(38):385701
37. Kordás K, Mustonen T, Tóth G, Jantunen H, Lajunen M, Soldano C, Talapatra S, Kar S, Vajtai R, Ajayan P (2006) Inkjet printing of electrically conductive patterns of carbon nanotubes. Small 2(8–9):1021–1025
38. Mire CA, Agrawal A, Wallace GG, Calvert P, in het Panhuis M (2011) Inkjet and extrusion printing of conducting poly(3,4-ethylenedioxythiophene) tracks on and embedded in biopolymer materials. J Mater Chem 21:2671–2678
39. Pidcock G, in het Panhuis M (2010) Extrusion printing conducting gel-carbon nanotube structures upon flexible substrates. In: Optoelectronic and microelectronic materials and devices (COMMAD), IEEE, Canberra, pp 179–180
40. Jayesh B, Yang Y (1998) Polymer electroluminescent devices processed by inkjet printing: I. polymer light-emitting logo. Appl Phys Lett 72(21):2660–2662
41. Song JW, Kim J, Yoon YH, Choi BS, Kim JH, Han CS (2008) Inkjet printing of single-walled carbon nanotubes and electrical characterization of the line pattern. Nanotechnology 19(9):095702
42. Mustonen T, Kordás K, Saukko S, Tóth G, Penttilä JS, Helistö P, Seppä H, Jantunen H (2007) Inkjet printing of transparent and conductive patterns of single-walled carbon nanotubes and PEDOT-PSS composites. Physica Status Solidi (B) 244(11):4336–4340
43. McCallum D, Ferris C, Calvert P, Wallace G, in het Panhuis M (2010) Printed hydrogel materials. In: International conference on nanoscience and nanotechnology (ICONN), IEEE, Sydney, pp 257–260
44. Calvert P (2001) Inkjet printing for materials and devices. Chem Mater 13(10):3299–3305
45. Calvert P (2007) Printing cells. Science 318(5848):208–209
46. Sachlos E, Reis N, Ainsley C, Derby B, Czernuszka JT (2003) Novel collagen scaffolds with predefined internal morphology made by solid freeform fabrication. Biomaterials 24 (8):1487–1497
47. Lemmo AV, Rose DJ, Tisone TC (1998) Inkjet dispensing technology: applications in drug discovery. Curr Opin Biotechnol 9(6):615–617
48. Collier WA, Janssen D, Hart AL (1996) Measurement of soluble L-lactate in dairy products using screen-printed sensors in batch mode. Biosens Bioelectron 11(10):1041–1049
49. Nakamura M, Nishiyama Y, Henmi C, Yamaguchi K(2006) Inkjet bioprinting as an effective tool for tissue fabrication. Soc Imaging Sci Technol, 89–92
50. Yoo J, Wake T (2008) Inkjet printing technology for regenerative medicine. Soc Imaging Sci Technol, 7–9
51. Boland T, Xu T, Damon B, Cui X (2006) Application of inkjet printing to tissue engineering. Biotechnol J 1(9):910–917
52. Small WR, in het Panhuis M (2007) Inkjet printing of transparent, electrical conducting single-walled carbon-nanotube composites. Small 3(9):1500–1503
53. Calvert P, Crockett R (1997) Chemical solid free-form fabrication: making shapes without molds. Chem Mater 9(3):650–663
54. Lu X, Lee Y, Yang S, Hao Y, Evans JRG, Parini CG (2010) Solvent-based paste extrusion solid free forming. J Eur Ceram Soc 30(1):1–10
55. Leong KF, Cheah CM, Chua CK (2003) Solid freeform fabrication of three-dimensional scaffolds for engineering replacement tissues and organs. Biomaterials 24(13):2363–2378
56. Rauwendaal C (2001) Polymer extrusion. Hanser, Munich
57. Tekinalp HL, Kunc V, Velez-Garcia GM, Duty CE, Love LJ, Naskar AK, Blue CA, Ozcan S (2014) Highly oriented carbon fiber-polymer composites via additive manufacturing. Compos Sci Technol 105:144–150

58. Rymansaib Z, Iravani P, Emslie E, Medvidovic Kosanovic M, Sak Bosnar M, Verdejo R, Marken F (2016) All polystyrene 3D printed electrochemical device with embedded carbon nanofiber graphite polystyrene composite conductor. Electroanalysis 28(7):1517–1523
59. Athreya SR, Kalaitzidou K, Das S (2010) Processing and characterization of a carbon black-filled electrically conductive nylon-12 nanocomposite produced by selective laser sintering. Mater Sci Eng A 527(10):2637–2642
60. Wei X, Li D, Jiang W, Gu Z, Wang X, Zhang Z, Sun Z (2015) 3D printable graphene composite. Sci Rep 5:11181
61. Rahaman M, Chaki TK, Khastgir D (2011) Development of high performance EMI shielding material from EVA, NBR, and their blends: effect of carbon black structure. J Mater Sci 46 (11):3989–3999
62. Gubbels F, Jérôme R, Teyssie P, Vanlathem E, Deltour R, Calderone A, Parente V, Bredas JL (1994) Selective localization of carbon black in immiscible polymer blends- a useful tool to design electrical conductive composites. Macromolecules 27:1972–1974
63. Rahaman M, Chaki TK, Khastgir D (2013) Control of the temperature coefficient of the DC resistivity in polymer-based composites. J Mater Sci 48(21):7466–7475
64. Li Y, Shimizu H (2007) High shear processing induced homogenous dispersion of pristine multiwalled carbon nanotubes in a thermoplastic elastomer. Polymer 48:2203–2207
65. Rahaman M, Thomas SP, Hussein IA, De SK (2013) Dependence of electrical properties of polyethylene nanocomposites on aspect ratio of carbon nanotubes. Polym Compos 34 (4):494–499
66. Potschke P, Abdel-Goad M, Alig I, Dudkin S, Lellinger D (2004) Rheological and dielectrical characterization of melt mixed polycarbonate-multiwalled carbon nanotube composites. Polymer 45(26):8863–8870
67. Zhang WD, Shen L, Phang IY, Liu T (2003) Carbon nanotubes reinforced nylon-6 composite prepared by simple melt-compounding. Macromolecule 37(2):256–259
68. Rahaman M, Chaki TK, Khastgir D (2011) High-performance EMI shielding materials based on short carbon fiber-filled ethylene vinyl acetate copolymer, acrylonitrile butadiene copolymer, and their blends. Polym Compos 32(11):1790–1805
69. Shanks RA, Cerezo FT (2012) Preparation and properties of poly(propylene-g-maleic anhydride) filled with expanded graphite oxide. Compos Part A 43:1092–1100
70. Maiti S, Suin S, Shrivastava NK, Khatua BB (2014) Low percolation threshold and high electrical conductivity in melt-blended polycarbonate/multiwall carbon nanotube nanocomposites in the presence of poly(ε-caprolactone). Polym Eng Sci 54:646–659
71. Cruz SM, Viana JC (2015) Melt blending and characterization of carbon nanoparticles-filled thermoplastic polyurethane elastomers. J Elastom Plast 47(7):647–665
72. Koysuren O, Yesil S, Bayram G (2006) Effect of composite preparation techniques on electrical and mechanical properties and morphology of nylon 6 based conductive polymer composites. J Appl Polym Sci 102:2520–2526
73. Garzón C, Palza H (2014) Electrical behavior of polypropylene composites melt mixed with carbon-based particles: Effect of the kind of particle and annealing process. Compos Sci Technol 99:117–123
74. Zhang Q, Zhang BY, Guo ZX, Yu J (2017) Tunable electrical conductivity of carbon-black-filled ternary polymer blends by constructing a hierarchical structure. Polymers 9:404
75. Wang H, Xie GY, Yang C, Zheng YX, Ying Z (2017) Enhanced toughness of multilayer graphene-filled poly(vinyl chloride) composites prepared using melt-mixing method. Polym Compos 38:138–146
76. Verdejo R, Bernal MM, Romasanta LJ, Lopez-Manchado MA (2011) Graphene filled polymer nanocomposites. J Mater Chem 21:3301–3310
77. Kim H, Miura Y, Macosko CW (2010) Graphene/polyurethane nanocomposites for improved gas barrier and electrical conductivity. Chem Mater 22:3441–3450

78. Wakabayashi K, Pierre C, Dikin DA, Ruoff RS, Ramanathan T, Brinson LC, Torkelson JM (2008) Polymer−graphite nanocomposites: effective dispersion and major property enhancement via solid-state shear pulverization. Macromolecules 41:1905–1908
79. Zhang HB, Zheng WG, Yan Q, Yang Y, Wang JW, Lu ZH, Ji GY, Yu ZZ (2010) Electrically conductive polyethylene terephthalate/graphene nanocomposites prepared by melt compounding. Polymer 51:1191–1196
80. Al-Saleh MH, Jawad SA (2016) Graphene nanoplatelet–polystyrene nanocomposite: dielectric and charge storage behaviors. J Electro Mater 45:3532–3595
81. Arya T, Justin H, Dong Z, Taghon M, Tse S, Chiu G, Mayo WE, Kear B, Nosker T, Lynch J (2017) Characterization of melt-blended graphene–poly(ether ether ketone) nanocomposite. Mater Sci Eng B 216:41–49
82. You F, Wang D, Cao J, Li X, Dang ZM, Hu GH (2014) In situ thermal reduction of graphene oxide in a styrene–ethylene/butylene–styrene triblock copolymer via melt blending. Polym Int 63:93–99
83. Pötschke P, Brünig H, Janke A, Fischer D, Jehnichen D (2005) Orientation of multiwalled carbon nanotubes in composites with polycarbonate by melt spinning. Polymer 46:10355–10363
84. Johannsen I, Jaksik K, Wirch N, Pötschke P, Fiedler B, Schulte K (2016) Electrical conductivity of melt-spun thermoplastic poly(hydroxy ether of bisphenol A) fibres containing multi-wall carbon nanotubes. Polymer 97:80–94
85. Pötschke P, Andres T, Villmow T, Pegel S, Brünig H, Kobashi K, Fischer D, Häussler L (2010) Liquid sensing properties of fibres prepared by melt spinning from poly(lactic acid) containing multi-walled carbon nanotubes. Compos Sci Technol 70:343–349
86. Soroudi A, Skrifvars M (2010) Melt blending of carbon nanotubes/ polyaniline/ polypropylene compounds and their melt spinning to conductive fibres. Synth Met 160:1143–1147
87. Yanga Z, Xu D, Liu J, Liu J, Li L, Zhang L, Lv J (2015) Fabrication and characterization of poly(vinyl alcohol)/carbon nanotube melt-spinning composites fiber. Prog Natural Sci Mater Int 25:437–444
88. Weise B, Völkel L, Köppe G, Schriever S, Mroszczok J, Köhler J, Scheffler P, Wegener M, Seide G (2017) Melt- and wet-spinning of graphene-polymer nano-composite fibres for multifunctional textile applications. Mater Today Proc 4:S135–S145
89. Strååt M, Toll S, Boldizar A, Rigdahl M, Hagström B (2011) Melt spinning of conducting polymeric composites containing carbonaceous fillers. J Appl Polym Sci 119:3264–3272
90. Ghatak S, Chakraborty G, Meikap AK, Woods T, Babu R, Blau WJ (2010) Synthesis and characterization of polyaniline/carbon nanotube composites. J Appl Polym Sci 119(2):1016–1025
91. Shaffer MP, Koziol K (2002) Polystyrene grafted multi-walled carbon nanotubes. Chem Commun 18:2074–2075
92. Park C, Ounaies Z, Watson KA, Crooks RE, Smith J Jr, Lowther SE, Connell JW, Siochi EJ, Harrison JS, Clair TLS (2002) Dispersion of single wall carbon nanotubes by in situ polymerization under sonication. Chem Phys Lett 364(3–4):303–308
93. Spitalsky Z, Tsoukleri G, Tasis D, Krontiras C, Georga SN, Galiotis C (2009) High volume fraction carbon nanotube-epoxy composites. Nanotechnology 20(40):405–702
94. Higgins BA, Brittain WJ (2005) Polycarbonate carbon nanofiber composites. Eur Polym J 41(5):889–893
95. Konyushenko E, Stejskal J, Trchova M, Hradil J, Kovarova J, Prokes J, Cieslar M, Hwang J, Chen K, Sapurina I (2006) Multi-wall carbon nanotubes coated with polyaniline. Polymer 47 (16):5715–5723
96. Guo Y, Bao C, Song L, Yuan B, Hu Y (2011) In situ polymerization of graphene, graphite oxide, and functionalized graphite oxide into epoxy resin and comparison study of on the flame behavior. Ind Eng Chem Res 50:7772–7783
97. Fim FDC, Guterres JM, Basso NRS, Galland GB (2010) Polyethylene/graphite nanocomposites obtained by in situ polymerization. J Polym Sci Part A Polym Chem 48(3):692–698

98. Xu Z, Gao C (2010) In situ polymerization approach to graphene reinforced nylon-6 composites. Macromolecules 43(16):6716–6723
99. Fabbri P, Bassoli E, Bon SB, Valentini L (2012) Preparation and characterization of poly (butylene terephthalate)/graphene composites by in-situ polymerization of cyclic butylene terephthalate. Polymer 53:897–902
100. Zaragoza-Contreras EA, Hernández-Escobar CA, Navarrete-Fontes A, Flores-Gallardo SG (2011) Synthesis of carbon black/polystyrene conductive nanocomposite. Pickering emulsion effect characterized by TEM. Micron 42:263–270
101. Hassan M, Reddy KR, Haque E, Minett AI, Gomes VG (2013) High-yield aqueous phase exfoliation of graphene for facile nanocomposite synthesis via emulsion polymerization. J Colloid Interface Sci 410:43–51
102. Hu H, Wang X, Wang J, Wan L, Liu F, Zheng H, Chen R, Xu C (2010) Preparation and properties of graphene nanosheets–polystyrene nanocomposites via in situ emulsion polymerization. Chem Phys Lett 484(4–6):247–253
103. Liu C, Dong B, Zhang LQ, Zheng Q, Wu YP (2015) Influence of strain amplification near crack tip on the fracture resistance of carbon black–filled SBR. Rubber Chem Technol 88 (2):276–288
104. Mohapatra S, Nando GB (2015) Analysis of carbon black–reinforced cardanol-modified natural rubber compounds. Rubber Chem Technol 88(2):289–309
105. Sethi D, Ram R, Khastgir D (2017) Analysis of electrical and dynamic mechanical response of conductive elastomeric composites subjected to cyclic deformations and temperature. Polym Compos. https://doi.org/10.1002/pc.24429
106. Shanmugharaj AM, Bae JH, Lee KY, Noh WH, Lee SH, Ryu SH (2007) Physical and chemical characteristics of multiwalled carbon nanotubes functionalized with aminosilane and its influence on the properties of natural rubber composites. Compos Sci Technol 67:1813–1822
107. Sau KP, Chaki TK, Khastgir D (1998) The change in conductivity of a rubber-carbon black composite subjected to different modes of pre-strain. Compos Part A 29:363–370
108. Das A, Kasaliwal GR, Jurk R, Boldt R, Fischer D, Stöckelhuber KW, Heinrich G (2012) Rubber composites based on graphene nanoplatelets, expanded graphite, carbon nanotubes and their combination: a comparative study. Compos Sci Technol 72:1961–1967
109. Jovanović V, Samaržija-Jovanović S, Budinski-Simendić J, Marković G, Marinović-Cincovic M (2013) Composites based on carbon black reinforced NBR/EPDM rubber blends. Compos Part B 45:333–340
110. Sau KP, Khastgir D, Chaki TK (1998) Electrical conductivity of carbon black and carbon fibre filled silicone rubber composites. Macromol Mater Eng 258(1):11–17
111. Staniforth JN (1987) Order out of chaos. J Pharm Pharmacol 39:329–334
112. Poux M, Fayolle P, Bertrand J (1991) Powder mixing: some practical rules applied to agitated systems. Powder Technol 68:213–234
113. Fan LT, Chen YM (1990) Recent developments in solids mixing. Powder Technol 61:255–287
114. Chen DZ, Lao S, Koo JH, Londa M, Alabdullatif Z (2010) Powder processing and properties characterization of polyamide 11- graphene nanocomposites for selective laser sintering, pp 435–450. Paper presented at 21st annual international solid freeform fabrication symposium —an additive manufacturing conference, SFF 2010, Austin, TX, USA, 9–11 Aug 2010
115. Islam I, Sultana S, Ray SK, Nur HP, Hossain MT, Ajmotgir WM (2018) Electrical and tensile properties of carbon black reinforced polyvinyl chloride conductive composites. J Carbon Res 4:15
116. Suh JY, Shin SE, Bae DH (2017) Electrical properties of polytetrafluoroethylene/few-layer graphene composites fabricated by solid-state processing. J Compos Mater 51(18):2565–2573
117. Lee KM, Oh SM, Lee SM (2008) Electrochemical properties of carbon composites prepared by using graphite ball-milled in argon and air atmosphere. Bull Korean Chem Soc 29 (6):1121–1124

118. Chan CM, Cheng CL, Yuen MMF (1997) Electrical properties of polymer composites prepared by sintering a mixture of carbon black and ultra-high molecular weight polyethylene powder. Polym Eng Sci 37(7):1127–1136
119. Schulze M, Lorenz M, Kaz T (2002) XPS study of electrodes formed from a mixture of carbon black and PTFE powder. Surf Interface Anal 34:646–651
120. Ma CY, Huang SC, Chou PH, Den W, Hou CH (2016) Application of a multiwalled carbon nanotube-chitosan composite as an electrode in the electrosorption process for water purification. Chemosphere 146:113–120
121. Li C, Hou T, She X, Wei X, She F, Gao W, Kong L (2015) Decomposition properties of PVA/graphene composites during melting-crystallization. Polym Degrad Stab 119:178–189
122. Zhao X, Zhang Q, Chen D (2010) Enhanced mechanical properties of graphene-based poly (vinyl alcohol) composites. Macromolecules 43:2357–2363
123. Wang SF, Shen L, Zhang WD, Tong YJ (2005) Preparation and mechanical properties of chitosan/carbon nanotubes composites. Biomacromol 6:3067–3072
124. Bin-Dahman OA, Rahaman M, Khastgir D, Al-Harthi MA (2018) Electrical and dielectric properties of poly(vinyl alcohol)/starch/graphene nanocomposites. Can J Chem Eng 96 (4):903–911
125. Hasan M, Das SK, Islam JMM, Gafur MA, Hoque E, Khan MA (2013) Thermal properties of carbon nanotube (CNT) reinforced polyvinyl alcohol (PVA) composites. Int Lett Chem Phys Astron 12:59–66
126. Yang A, Yang P, Huang CP (2017) Preparation of graphene oxide–chitosan composite and adsorption performance for uranium. J Radioanal Nucl Chem 313:371–378
127. Lin JH, Lin ZI, Pan YJ, Hsieh CT, Huang CL, Lou CW (2016) Thermoplastic polyvinyl alcohol/multiwalled carbon nanotube composites: preparation, mechanical properties, thermal properties, and electromagnetic shielding effectiveness. J Appl Polym Sci 133:43474
128. Nasr GM, Abd El-Haleem AS, Klingner A, Alnozahy AM, Mourad MH (2015) The DC electrical properties of polyvinyl alcohol/multi-walled carbon nanotube composites. J Multidiscip Eng Sci Technol 2(5):884–889
129. Cheng-an T, Hao Z, Fang W, Hui Z, Xiaorong Z, Jianfang W (2017) Mechanical properties of graphene oxide/polyvinyl alcohol composite film. Polym Polym Compos 25(1):11–16
130. Simsek EB, Saloglu D, Ozcan N, Novak I, Berek D (2017) Carbon fiber embedded chitosan/ PVA composites for decontamination of endocrine disruptor bisphenol-A from water. J Taiwan Inst Chem Eng 70:291–301

Mechanical Properties of Carbon-Containing Polymer Composites

K. Sasikumar, N. R. Manoj, T. Mukundan, Mostafizur Rahaman and Dipak Khastgir

Abstract The carbon particles in micro- and nanometric size are the prominently used as particulate filler in polymer matrix to improve mechanical properties. The carbon black is the most widely used cost-effective filler, but require higher loading levels to get optimum mechanical properties. Researchers all over the world are investigating the various nanoforms of carbon like Fullerenes, Carbon nanofibres, Carbon nanotubes, Nano diamond and Graphene as fillers in polymer matrix to improve mechanical properties at low loading levels. These nano-sized carbon particles, impart additional properties like improved electrical conductivity, reduced friction, higher heat dissipation and improved hydrophobic nature, etc., which makes them preferred candidate despite of higher cost. New methods for functionalization and dispersion are attempted to capitalize the full potential of these nanomaterials. This chapter gives a bird's eye view of progress made in the preparation of these nanocomposites, advanced functionalization, and dispersion methods, and their impact on the mechanical properties of polymer nanocomposites. The obstacles in

K. Sasikumar (✉)
Advanced Technologies Division, Nano and Composites Group,
Combat Vehicles Research and Development Establishment,
DRDO, Avadi, Chennai 600054, India
e-mail: sasikumar.k@cvrde.drdo.in

N. R. Manoj · T. Mukundan
Materials and MEMS Group, Naval Physical and Oceanographic Laboratory,
DRDO, Kochi 682023, India
e-mail: nr.manoj@npol.drdo.in

T. Mukundan
e-mail: mukundant@npol.drdo.in

M. Rahaman
Department of Chemistry, College of Science, King Saud University,
Riyadh 11451, Saudi Arabia
e-mail: mrahaman@ksu.edu.sa

D. Khastgir
Rubber Technology Center, Indian Institute of Technology Kharagpur,
Kharagpur 721302, India
e-mail: khasdi@rtc.iitkgp.ernet.in

M. Rahaman et al. (eds.), *Carbon-Containing Polymer Composites*,
Springer Series on Polymer and Composite Materials,
https://doi.org/10.1007/978-981-13-2688-2_4

converting lab-scale achievements into large-scale applications also discussed and suitable suggestions are given.

Keywords Mechanical properties · Carbon · Polymers · Nanoparticles Graphene

1 Introduction

The particulate fillers improve the mechanical properties of polymeric materials due to their reinforcing nature. The degree of reinforcement depends on physical interactions between the filler and the polymer matrixes [1, 2]. The fillers such as Carbon particles, Silica, and clay show significant reinforcement due to the better adhesion with the polymer matrix. The presence of active sites, make these fillers more compatible with the polymer matrix [3–5]. Among the various reinforcing fillers, carbon particles are the most preferred fillers for improving the mechanical properties of polymers especially elastomers due excellent reinforcing ability. Carbon exists in different allotrophs, among them carbon black and graphite are conventionally used in polymer composites. However, the research in material technologies coupled with the emergence of electron microscopes has resulted in the discovery of newer forms of carbon such as valerian, nano diamond, carbon nanofibre, carbon nanotubes and graphene, which have at least one dimension in the nanometric range [6, 7]. The nanoforms of carbon become the best choice for polymer scientists to design lightweight composites, which possess high strength-to weight ratio, high elastic modulus and excellent mechanical strength [8–10]. In the present chapter, we will discuss about the various methods of preparation of the following carbon particle filled polymer composites and their impact on the mechanical properties.

1. Carbon Black Composites
2. Graphite Composites
3. Fullerene Composites
4. Nano Diamond Composites
5. Carbon Nanofibre (CNF) Composites
6. Carbon Nanotube Composites
7. Graphene Nanocomposites

2 Carbon Black Composites

Carbon black is the most widely used reinforcing filler in polymer matrixes due to the presence of active sites on its surface, smaller size, and lower cost. Almost all engineering rubber components contain large quantity of carbon black as reinforcing filler. There is plenty of research papers available published over last seven decades, which throws light on the reinforcing ability of carbon black in polymers especially elastomeric compounds. The surface of carbon black contains many types of functional groups like carboxyls, lactones, phenols, and Quinones [11], which dictate the adhesion with the polymer matrix and overall improvement in the strength. Further, carbon black forms physical bonding with the polymer chains and thereby enhances the strength [12]. Many types of carbon black are available based on the manufacturing method, and each type has a distinct structure, which governs its property. The reinforcing type carbon blacks are used to improve the mechanical properties and the conductive carbon blacks are used in electro-magnetic interference (EMI) shielding materials and antistatic surfaces. Generally, the mechanical property improvement becomes prominent when the concentration of carbon black is above 30% [13]. For example, in EPDM rubber the polymer-filler interaction becomes noticeable above 30 phr indicated by enhanced physical stiffening and it reaches saturation at about 60 phr loading level [14]. The addition of high abrasion furnace type carbon black (N330) into EPDM/PP blend show increase in tensile strength and decrease in elongation at break up to 60 phr loading levels, beyond which the properties start decreasing due to improper wetting of carbon black particles in the polymer matrix [15]. The gasification of carbon black particles in carbon dioxide gas, prior to the incorporation into polymer matrix was found to enhance the modulus due to formation of higher degree of primary aggregates. The mechanical strength of electro-spun fibres of carbon black impregnated polyurethane increases up to 120% at a loading level of 3.65% of carbon black, beyond which the composite loses its strength due to agglomeration and increase in the viscosity of the precursor solution [16]. The addition of 5% of silane treated carbon black results in an increase of tensile properties of HDPE [17]. The reinforcement due to carbon black found to be better in polar polymers like Nitrile rubber compared to the nonpolar polymers like EPDM [18]. Majority of the published literature show that carbon black as reinforcing filler has tremendous potential at medium to higher loading levels. However, the agglomeration of the carbon black at higher loading level tends to cause un-uniformity in the mechanical properties from batch to batch. Further, carbon black cannot be used to achieve multi-functional properties like strength and damping, strength and conductivity, strength and hydrophobic nature, etc., simultaneously. To get multi functional properties nanovariants of carbon black are the best choice.

3 Graphite Composites

The Graphite is one of the simplest natural forms of carbon and it has high electrical conductivity and most preferred for the preparation of polymer composites, which are lightweight and electrically conductive [19]. The layered structure of natural graphite has to be broken during the preparation of conductive polymer composites to get higher conductivity at low level of filler loading. To achieve this, natural graphite is converted into graphite oxide (GO) by oxidation. The GO contains various polar functional groups such as hydroxyl, epoxy and carbonyl groups [20]. These functional groups interact with the polar moiety of the polymer chains and there by improve the intercalation and exfoliation of the graphene layers [21, 22] into the hydrophilic polymers. However, dispersion of GO in hydrophobic polymers is difficult due to the absence of polar–polar interaction. Hence, to disperse the GO in hydrophobic polymers a novel method of reversible addition-fragmentation chain transfer (RAFT) reaction had been used [23]. Graphite oxide was immobilized with dodecyl isobutyric acid trithiocarbonate (DIBTC) by RAFT method (Fig. 1) and the resultant GO was used in the preparation of Polystyrene-Graphite nanocomposites. This has resulted in the enhancement of mechanical, dynamic mechanical properties and increase in the glass transition temperature from 74 to 105 °C for as little as 1% of GO.

Another interesting method for increasing dispersion is increasing the inter-laminar distance. The normal inter-laminar distance in natural graphite is about 0.66 nm, which can be increased to about 2 μm by treating the natural graphite with H_2SO_4–HNO_3 acid mixture followed by heating the Graphite Intercalated Compound (GIC) to 900 °C for short duration of 15 s [24]. The graphite formed by this method is known as exfoliated or expanded graphite (EG). The addition of EG into polypropylene matrix by in-situ processing method was found to improve the mechanical and thermal properties better than melt mixing due to high degree of uniform dispersion [25]. The incorporation of EG into PMMA matrix by direct solution blending method had resulted in the enhancement of

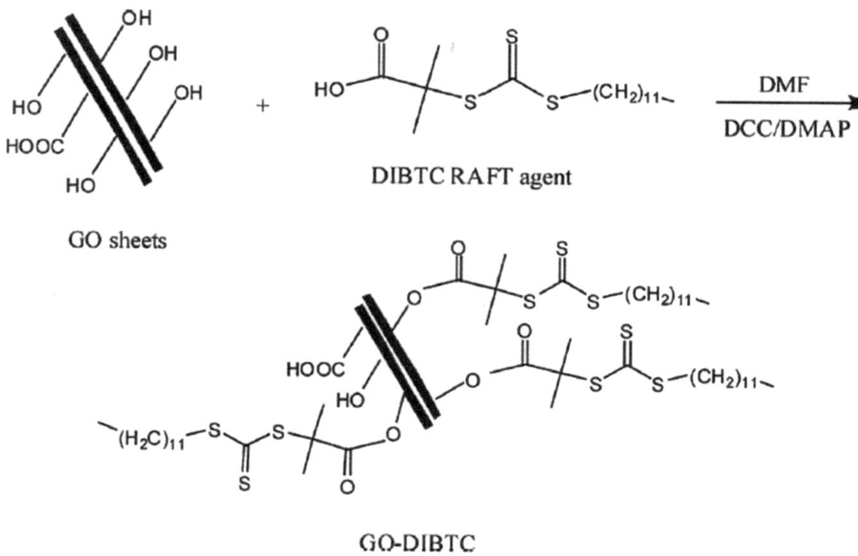

Fig. 1 The synthesis route for the preparation of RAFT immobilized GO nanosheets. Reprinted with permission from Ref. [23]

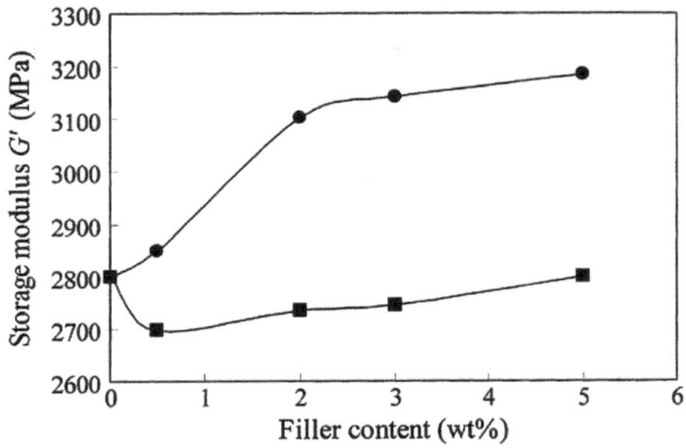

Fig. 2 G' of PMMA/graphite (filled square) and PMMA/EG (filled circle) plotted as a function of filler weight fractions at 25 °C. Reprinted with permission from Ref. [26]

storage modulus, loss modulus and tan delta at very low concentrations of EG [26]. The PMMA/graphite composite shows lesser modulus than the neat PMMA due to the formation of weak interface, whereas the PMMA/EG composites show excellent dynamic properties due to the better mechanical interlocking between the EG layers and PMMA polymer chains (Fig. 2).

The grafting of chemical functional groups in expanded graphite yields still better mechanical properties than EG. The addition of 3-aminopropoxyltriethoxy silane grafted EG into Epoxy resin improved the rubbery modulus from 160 to 240 MPa due to the covalent bonding between the functionalized EG and epoxy resin system [27]. The study also revealed that the addition of EG or functionalized EG beyond certain loading level leads to re-agglomeration of graphite platelets and thereby decreases the glass transition temperature. Further, at higher loading levels, the viscosity of the epoxy resin—graphite system reaches very high value and causes processing problems. The graphite nano platelets are a special class of expanded graphite in which the interlayer distance is around 200 nm. The addition of 2.5% of graphite nano platelets into epoxy resin found improve modulus tensile strength, dynamic modulus as well as glass transition temperature due to the better interfacial adhesion between the platelets and the resin matrix [28]. Though the expanded graphite and its functionalized variants show significant improvement in mechanical properties at lower loading levels, the preparation of EG and further functionalization is a tedious method and industrially not viable. Hence, the large-scale application of graphite is a limited one.

4 Fullerene Composites

Fullerene is an interesting form of molecular carbon, which contains 60 carbon atoms arranged in a spherical form consisting of 20 hexagons and 12 pentagons. The spherical shape of the Fullerene makes it a zero-dimensional material, in which structural rigidity occurs in all the three dimensions leading to very high mechanical strength. However, to capitalize this to the maximum in the composites, we need to do functionalize the Fullerene molecule according to the nature of matrix [29]. The addition of a little quantity of Fullerene (0.5%) in thermoplastic polyurethane matrix by melt mixing leads to 50% improvement in tensile strength, which further increases to 70% if the surface of Fullerene molecules are modified with nitric acid and Silane coupling agent [30]. Interestingly, the elongation at break also increases, which shows that the addition of Fullerene gives simultaneous strengthening as well as toughening. The addition of just 1% of Fullerene (C_{60}) into the epoxy matrix was found to improve the fracture toughness by 50% and tensile strength by 20% [31]. To achieve the same level of improvement in properties, it requires at least 5 to 10% of other nano fillers like nano SiO_2 or nano TiO_2 or Carbon Nanotubes (CNT). The formation of entanglements is very much low and dispersion is better than other nanofillers due to the spherical shape of Fullerene molecules, which results in the enhancement of mechanical properties.

The incorporation of about 0.1% of Fullerene increases the tensile strength and elongation at break by 20% due to the restriction offered by the nanofiller upon the mobility of epoxy polymer chains under load [32]. The simultaneous increase in strength and elongation at break makes the nanocomposite tougher with the three-fold increase in the impact strength. The addition of low concentration (1–3 wt%) of carbon soot containing Fullerene into epoxy matrix was found to change the brittle nature of epoxy compound into ductile nature [33] due to the enhancement in toughness by 1000–1800% (Fig. 3).

The significant change in the behavior of the nanocomposite is attributed to the crack inhibiting nature of Fullerenes. The change in the nature of the epoxy material has a tremendous advantage in the painting and coating applications, which undergo plastic deformation. The in-situ addition as well as addition of Fullerene during melt mixing of PA6 polymer favors the selective crystallization of the polymer in α form and thereby increases the mechanical strength at very low level of loading from 0.01 to 1% [34]. The addition of plasticizers like amide of oleic acid found to decrease the agglomeration and enhance the uniform dispersion of the

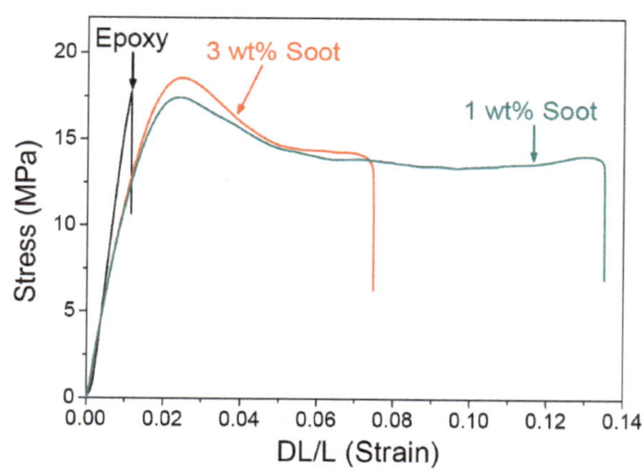

Fig. 3 Tensile testing results of epoxy resin, and the composites with 1 and 3 wt% Carbon soot. Reprinted with permission from Ref. [33]

nanofiller in the polymer matrix. The micro-hardness of Polyethylene thin films improves by 30% due to the addition of 2.5% of Fullerene particles through solution mixing process [35]. Annealing of these composites at 130 °C leads to increase in the micro hardness due to phase separation of Fullerene moieties within the nanocomposite matrix. The incorporation of small amount of Fullerene (0.1–0.5%) into the epoxy resin used in the fabrication of Carbon fibre reinforced plastics (CFRP) found to improve various mechanical properties. The tensile strength, compressive strength, and compression after impact (CAI) strength were increased by 15% and inter-laminar fracture toughness increased by 60% [36] due to the inter facial interaction between the polymer and the Fullerene molecules. Despite its high mechanical strength, Fullerenes has very limited application in industrial polymer composites, due to the variation in the purity and cost aspects.

5 Nano Diamond Composites

The nano diamond is a form of diamond, which has particle size in the range of 5 to 50 nm and near spherical shape. It has high mechanical strength like bulk diamond. The presence of additional functional groups on the surface of nano diamond makes it attractable one to the researchers in the field of nanocomposites [37]. The advantages of nano diamond over other nano forms of carbon are higher Young's modulus, hardness, thermal conductivity, electrical resistivity, chemical stability, and low coefficient of friction [38]. The near spherical shape and smaller size gives maximum possible interaction volume between the matrix and the nano diamond particles, which is equal in all directions unlike nanotubes [39, 40]. The air oxidation of nano diamond particles results in improved purity and leads to the formation of certain functional groups on the surface of the nano diamond particles [41, 42]. Nano diamond based composites are useful as wear resistant coatings, friction modifier, high performance grease, and as high strength composites [43, 44]. The conventional dispersion techniques like shear mixing, high power ultra-sonication would yield only marginal improvement in the dispersion due to the high affinity between spherical nano particles [45, 46]. Further, the casting and melting methods are not suitable for incorporating large quantity of nano diamond particles in the polymeric fibres. Hence, electro-spinning was tried to disperse the nano diamonds in polymeric fibres [47]. The nano diamond particles produced by detonation method were first air oxidized for 2 h at 425 °C to remove non-diamond carbon forms and then treated with concentrated HCl at 100 °C to remove metallic oxides. The acid treated nano diamond particles were washed to neutral pH and dried. These nano diamond particles dispersed in solvent composed of formic acid (FA) and dichloromethane (DCM) in a 1:1 volume ratio, and then suitable amount of polyamide (PA 11) powder was added. The mixed solution was used to develop electro-spun nano-fibres. The Young's modulus found to increase by four times and hardness by two times due to addition of 20% of pre treated nano diamond particles. Interestingly, the electro-spun fibre mats obtained by this method exhibit not only superior mechanical properties but also show improved protection from ultra violet rays, transparent and are scratch resistant. This multifunctional characteristic makes it suitable for a variety of applications. The addition of very low quantity of nano diamonds (0.5%) into vinyl trimethoxy silane grafted high-density polyethylene (g-HDPE), by ball milling followed by melt mixing had resulted in enhanced mechanical strength and modulus [48]. However, above 0.5% loading, the mechanical properties decrease gradually. The nanocomposite with 0.5% nano diamond show improved strength, toughness, and higher elongation at break, whereas the nanocomposites with 5% nano diamond show low toughness and higher brittleness (Fig. 4).

A novel method of grafting polyimide chains on the surface of functionalized nano diamonds has shown remarkable increase in the hardness of Polyimide-nano diamond nanocomposites [49]. The nano diamonds were synthesized by high temperature, high pressure detonation method to get nano diamond particles containing carboxylic functional groups. The nano particles were first treated with thionyl chloride and then with constituent monomers of polyimide namely 4,4′-(m-phenylene dioxy) dianiline (APB), 3,3′,4,4′-benzophenone tetracarboxylic dianhydride (BTDA) to get the polyimide grafted nano diamond particles (Fig. 5). This type of grafting of polymers from the surface of the nanoparticles has facilitated breaking the agglomerations and enhancing the dispersion. The hardness of nanocomposite containing 1% of polyimide grafted nano diamond is about 30% higher than the neat polyimide whereas that of nanocomposite containing 5% of unfunctionalized nano diamond particles is about 25% higher than the neat polymer. This shows that chemical grafting is one of the best methods to improve mechanical properties of nano diamond composites.

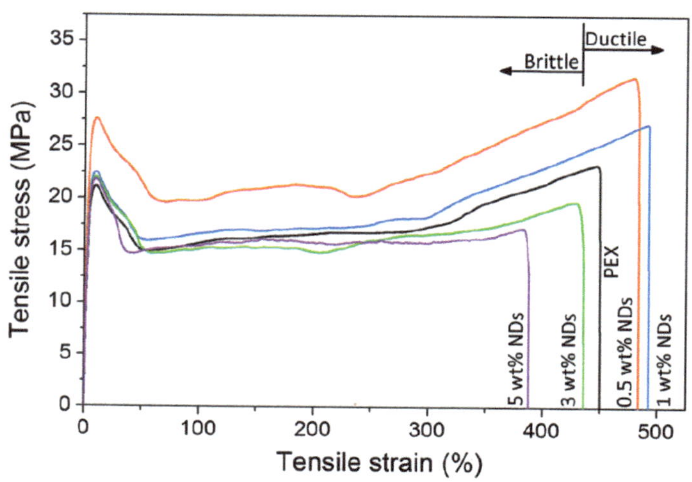

Fig. 4 Stress–strain curves of PEX and its nanocomposites with nanodiamonds. Reprinted with permission from Ref. [48]

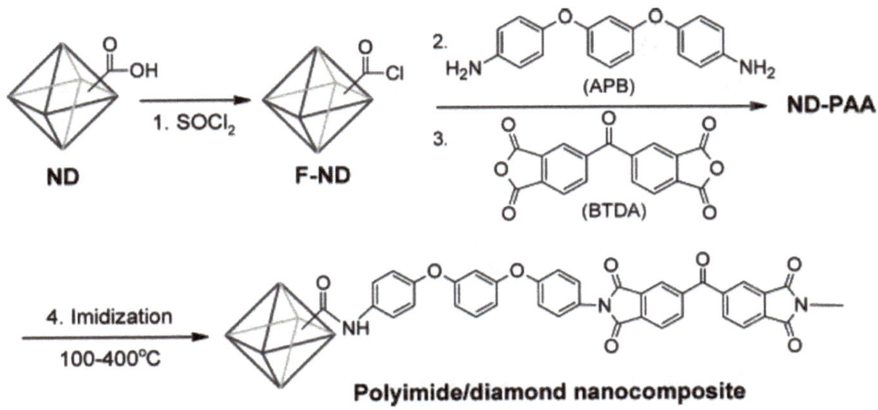

Fig. 5 Acyl chloride functionalization of nanodiamonds and grafting of polyimide from diamond surface. Reprinted with permission from Ref. [49]

The addition of 1% of aqueous dispersed nano diamond particles in Poly (vinyl alcohol), PVA matrix led to increase in the Young's modulus by 250% and tensile strength by 30% [50]. The increase in modulus was found to match well with the value predicted by rule of mixtures. The surface of nano diamond particles contain few carboxylic functional groups, which are expected to form hydrogen bonding with the hydroxyl group of the PVA, additionally the presence of water molecules also enhance the interaction and thereby demonstrate higher modulus and strength. The carboxylated nano diamond particles when added to epoxy resin undergo esterification reaction with the epoxy groups, resulting in the enhancement of tensile modulus and strength by 50% at very low concentration of 0.3% of nano diamond particles [51]. A novel method of chemical interaction was attempted by grafting amine groups on the nano diamond particles through Trimethylene tetra-amine (TETA), which in turn reacts with epoxy resin to form chemical cross-linking (Fig. 6). The chemical grafting aids to improve both storage modulus as well as fracture toughness [52].

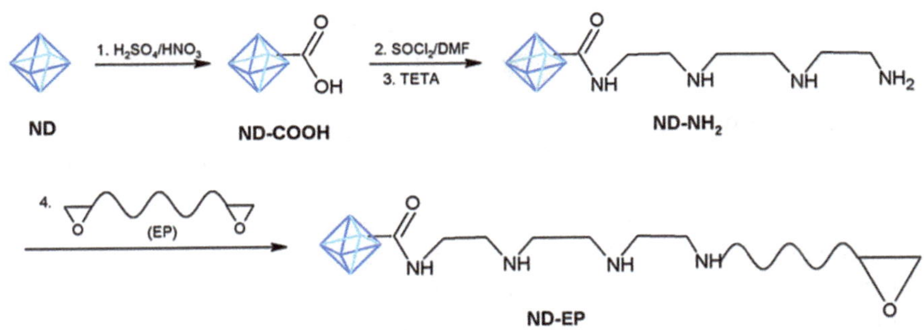

Fig. 6 Amino functionalization of ND with TETA and grafting epoxy from ND surface. Reprinted with permission from Ref. [52]

The addition of nano diamond in large quantity like 25% by solution mixing in tetra hydro furan and sonication along with resin part of epoxy polymer prior to polymerization reaction found to enhance the modulus by 470% and hardness by 300%. The scratch resistance and thermal conductivity also improved remarkably due to formation of interconnected diamond network at higher filler loading [53]. The combination of higher mechanical strength as well as solid lubrication properties of nano diamond makes it a potential material for the preparation of high pressure self lubricating polymeric seals for dynamic applications like suspension systems, hydraulic actuators and landing gears of aircrafts.

6 Carbon Nano Fibre (CNF) Composites

The carbon nano fibre is a potential replacement for the conventional short fibres due the good mechanical strength (7 GPa) and modulus (400 GPa) [54] combined with crack inhibiting property. The nano-sized fibres inhibit the nucleation and growth of crack at the nano scale length and stop their propagation into micro-scale, thereby improve the fracture toughness of polymer composites. The addition of 5% of carbon nano fibres into Poly (ethylene terephthalate) (PET) powder by ball milling and further melt processing has shown 50% improvement in the compressive modulus and torsion modulus, which help to arrest the kink propagation [55]. However, the improvement in tensile strength was very marginal, possibly due to the random distribution of nanofibres in the matrix. Likewise, in PP/CNF composites also the tensile strength is decreased beyond 2% CNF loading, though the modulus increase steadily as the concentration of CNF increased to 10% [56]. The addition of 1% CNF into Poly (lactic acid) by solution blending was found to increase the tensile modulus increases by 24 at 0.5% loading of CNF and reduces to 16 at 1% loading of CNF. This is due to the agglomeration of CNF at higher loading levels. However, the tensile strength increases steadily by 40% due to addition of 1% CNF, indicating alignment of CNF in axial directional under tensile load [57]. Dispersion of CNFs in acetone by sonication and then adding the mixture into epoxy resin was found to reduce the agglomeration. Further, slow curing of the epoxy-CNF nanocomposite under refrigeration at 4 °C for 8 days also reduces the degree of agglomeration by arresting the movement of CNFs during the curing process [58].

The addition of 0.5% CNF into epoxy resin by this procedure was found to enhance the bending modulus by 60% and hardness by 100%. The electrochemical surface treatment of CNFs using nitric acid as the electrolyte was found to improve the flexural modulus of epoxy-CNF nanocomposites significantly [59]. The increase in flexural modulus due to addition of 12% of untreated CNF was about 45%, which increases gradually with increase in the time of electrochemical surface treatment and reaches to a maximum of 250% enhancement for 12 min of surface treatment.

This is due to the oxidation of surface carbon atoms into various functional group containing C=O moieties, which form better bonding with the epoxy matrix. At still higher duration of surface treatment, the surface of nanofibers becomes more porous leading to decrease in the modulus. Another interesting method to disperse CNF in epoxy composites is synthesis of CNF/PVA nanocomposite mats through electro-spinning. A very low level of CNFs (0.015%) was added to PVA solution and the solution of electro-spun to get the nanofibre mats. These mats were used as intermediate layers in the fabrication of test coupons of epoxy-CNF composites. The flexural modulus of the nanocomposites got improved by 200% due to the addition of 0.015% of CNF in the matrix [60]. The PANI functionalized CNF nano-paper was found to have excellent electrical conductivity as well as mechanical strength. Incorporation of this nano-paper as protective layer over glass fibre reinforced epoxy composite improved the flexural strength by 24% and offer excellent electromagnetic shielding [61].

The vapor grown carbon nano-fibres (VGCNF), prepared by catalytic decomposition of hydrocarbons over metal catalysts are economical alternatives for CNTs due to the improved mechanical reinforcement, electrical conductivity, and high aspect ratio [55, 62, 63]. The incorporation of 3% of VGCNF into low-density polyethylene (LDPE) by melt mixing using single screw extruder was found to improve the modulus by more than 40% due to the excellent reinforcement [64]. The graphitization of CNFs by heat treatment at 2500 °C in nitrogen atmosphere was found to make the CNF surface more defects free, which led to increase in electrical conductivity of epoxy-CNF composites. However, the heat treatment was detrimental for the mechanical modulus and strength due to the decrease in the interfacial adhesion between the epoxy matrix and heat-treated CNFs [65]. A novel method to form semi aligned continuous carbon nano fibres, is electro-spinning of PAN nano-fibres over a rotating drum and then carbonization of the electro-spun mat at high temperature. The carbon nano fibre mats formed by this method can be used as intermediate layers in the fibre-reinforced polymer composites to achieve higher modulus, tensile strength, and inter-laminar shear strength. The addition of 2% of continuous CNF mats enhances the tensile strength and inter-laminar shear strength by 200%, due to the higher surface area of the CNFs and enhanced interaction between the nano fibres and the resin matrix [66]. The degree of agglomeration hampers the use of CNFs in polymer composites in bulk quantities. The electro-spinning of CNF precursors is one of the novel methods, which has huge potential in the manufacture of CFRP in large scale.

7 Carbon Nanotube Polymer Composites

Among all the allotropes of carbon, carbon nanotubes have become a prime candidate for many engineering applications, due to very high mechanical strength and the ease of manufacturing and functionalizing the CNTs [67–70]. The pristine MWNTs found to have tensile strength of 11–63 GPa and modulus of 270–950 GPa as determined by AFM studies [71]. The maximum strain that MWNTs can undergo before breaking was found to be about 12%. The addition of nanotubes into polymer matrixes is expected to result in huge increase in the modulus and strength as per the rule of mixture. However, in practical scenario, it is not possible to get the expected theoretical values. Nanotubes cannot act as stand-alone strength member due to the finite dimensions, especially length in few micrometer ranges. It needs other component for the tube-to-tube connectivity. In most of the cases, the matrixes in which the nanotubes are dispersed play the role of interconnecting medium. The effective stress transfer between the nanotubes and the matrix has to be ensured to get better strength and modulus. The stress transfer between the nanotubes and the matrix can be improved either by mechanical interlocking or by chemical or physical bonding. The shrinkage of the polymer matrix found to cause buckling and bending of nanotubes due to the compressive load exerted by the matrixes like epoxy and thermoplastic polymers [72].

The addition of pristine MWCNTs itself found to enhance the tensile strength, if the host polymer matrix contains polar groups. The addition of 7.5% of pristine MWCNT to hydroxylated nitrile-butadiene rubber (HNBR) found to enhance the tensile strength from 26.6 to 29.5 MPa due to the presence of ACN groups, which have affinity towards the nanotubes [73]. In the case of MWCNT-Natural Rubber composites, the addition of pristine MWCNT up to 10 phr found to increase the modulus but decrease the ultimate tensile strength [74]. The addition of nanotubes found to support the stress induced crystallization in natural rubber resulting in higher modulus and less elongation at break as well as strength (Fig. 7).

To improve the dispersion of nanotubes in the polymer matrix, researchers follow various methods like in-situ polymerization, solution mixing, surfactant-assisted processing, and melt compounding. The physical method of interaction achieved

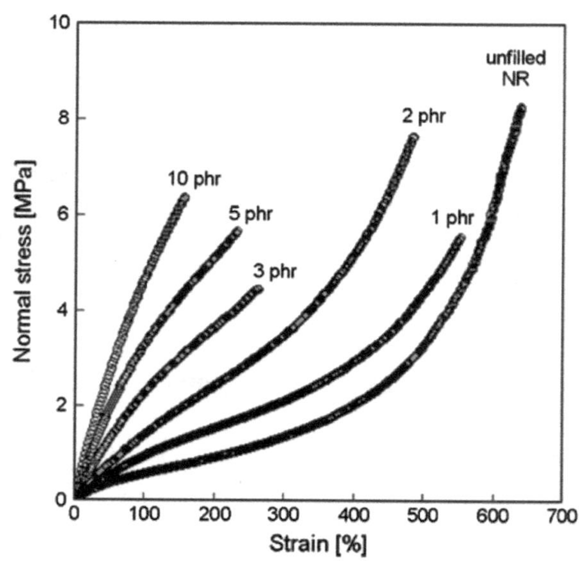

Fig. 7 Stress–strain plots of NR-MWCNT composites. Reprinted with permission from Ref. [74]

through surfactants like gum Arabic, sodium carboxymethyl cellulose, etc., Carbon nanotubes dispersed in poly vinyl alcohol using Sodium dodocyl sulphite (SDS) as surfactant (CNT: 0.4%; SDS: 0.5%) found to increase the modulus by 50% (from 6.6 to 9.4 GPa) in the electro-spun nano-fibres. The addition of SDS overcomes agglomeration of CNTs, and electro-spinning favors alignment of the nanotubes along the fibre axis [75]. The chemical bonding is achieved by functionalizing the nanotubes to get functional groups like carboxylic group, hydroxyl group, etc., the carboxylic functionalization of MWNTs found to enhance the tensile strength by 15% and modulus by 13% in MWNT-Epoxy composites. If the carboxylic functionalization is replaced with addition of gum Arabic, the tensile strength increases further up to 22% and modulus up to 29%. The comparison of Electron microscopic analysis of unfunctionalized MWNT-Epoxy, carboxylic functionalized MWNT-Epoxy and gum Arabic incorporated MWNT-Epoxy samples show that the functionalization methods decrease the agglomeration of nanotubes and enhance the degree of interaction between the nanotubes and the polymer matrix. The non-covalent functionalization of nanotubes by gum Arabic seems to be better option than the carboxylic functionalization [76]. The addition of amino functionalized MWCNTS and amino functionalized DWCNTs found to increase the tensile strength of isophalic acid based unsaturated polyester resin system by about 15–17%, due to the increase interfacial adhesion of the amino groups [77]. The increase in tensile strength is more in case of DWCNTs due to better load transfer characteristics originated from the higher surface area of DWCNTs (Fig. 8). Here the dispersion of nanotubes into the matrix was achieved by 3-roll mill mixing method.

The chemical functionalization of nanotubes leads to defect initiation in the nanotubes structure. To overcome this, a novel method of chemical functionalization using half-neutralized dibasic acids was attempted by Jagtap et al. [78]. Here the MWCNTs were treated with half-neutralized succinic acid (HNSCA), half-neutralized adipic acid (HNAA), and half-neutralized sebacic acid (HNSA).

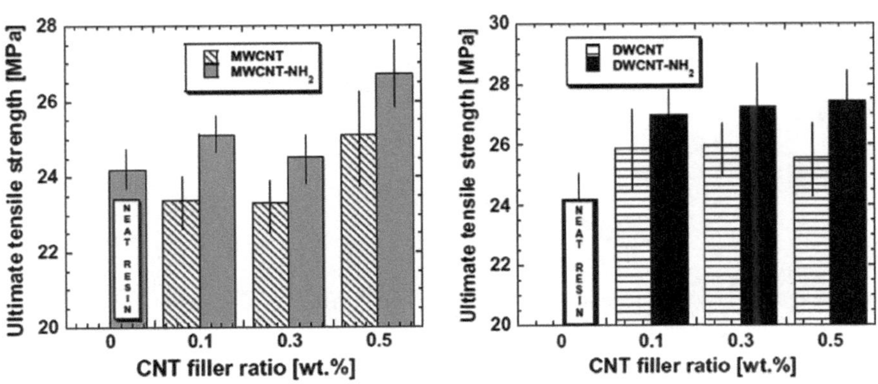

Fig. 8 Ultimate tensile strength (UTS) of the CNT/polyester nanocomposites (a) with MWCNT and MWCNT–NH_2 as a function of CNT filler ratio, (b) with DWCNT and DWCNT–NH_2 as a function of filler ratio. Reprinted with permission from Ref. [77]

Among the three, the adipic acid salt modified MWCNTs were found to improve the tensile strength and tensile modulus most effectively. The addition of 0.5 wt% HNAA modified MWCNTs showed 70 and 190% increase in the tensile strength and tensile modulus, respectively. The cation–Π interaction between the HNAA and MWCNT results in the decrease in agglomeration of nanotubes and chemical bonding between HNAA and epoxy chain results in better interfacial adhesion (Fig. 9).

In thermoplastic polymers, solution mixing and melt mixing methods are widely used to disperse nanotubes. In solution mixing, the polymer is dissolved in a solvent and nanotubes are added to the solution, ultra sonicated to disperse the nanotubes. However, this is a tedious method and requires large volume of solvent to dissolve the polymer and removal of the solvent after mixing is a cumbersome activity. In melt mixing, the polymer is melted at high temperature in a screw mixer and the nanotubes are added during mixing. The addition of MWCNT in LDPE matrix by melt mixing at 140 °C found to increase the modulus and strength gradually (Fig. 10). About 10% addition of MWNT led to increase in Young's modulus by

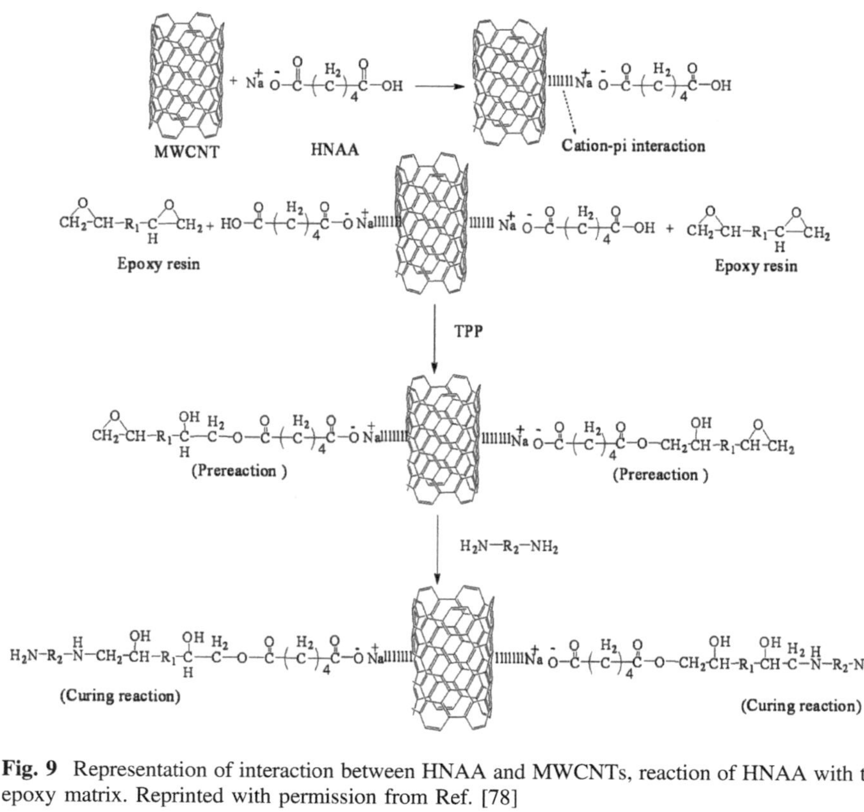

Fig. 9 Representation of interaction between HNAA and MWCNTs, reaction of HNAA with the epoxy matrix. Reprinted with permission from Ref. [78]

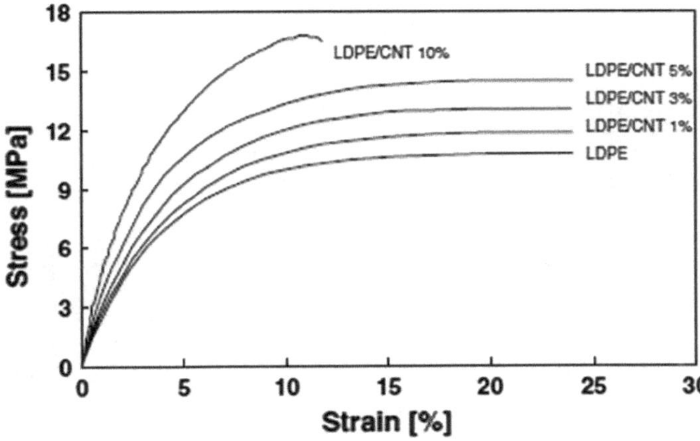

Fig. 10 Stress–strain curves of the LDPE composites with varying MWNT content. Reprinted with permission from Ref. [79]

89% and tensile strength by 56% [79]. The coiling nature of nanotubes makes more mechanical interlocking with the polymer chains and thereby improves the reinforcement in the matrix.

A recent study shows that melting mixing of already solution mixed sample of nanocomposites can give better properties. For example, the addition of MWCNT, into Ethyl Methyl Acrylate polymer by solution cum melt mixing found to increase the tensile strength from 9.61 to 12.9 MPa [80]. The increase in strength is supported by the enhanced dispersion of carbon nanotubes in the matrix as observed in the TEM images (Fig. 11).

Another method to enhance the reinforcement between nanotubes and polymer matrix is by grafting of nanotube walls with short chain polymer moieties like maleic anhydride under plasma conditions. The polyimide—mMWCNT nanocomposites found to exhibit superior mechanical strength and electrical properties at very low concentration of grafted MWCNTs [81]. The addition of pristine nanotubes in polyimide found to decrease the strength whereas the grafted MWCNTs found to increase the strength due to better interaction (Fig. 12).

The addition of MWCNTs by wet mixing method was found to give better strength in the case of Styrene-Butadiene and Butadiene Rubber blends [82]. The nanotubes were dispersed in ethanol using ultra-sonication, and this solution was added to the rubber compound during the mill mixing. The ethanol was evaporated during the mill mixing and the resulting rubber compound had well dispersed nanotubes. The stress–strain plot (Fig. 13) shows that the strength of rubber compound containing 5-phr wet mixed CNT is about double of the rubber compound containing 10 phr of dry mixed CNT. Further, the wet mixed neat, silane or hydroxyl functionalized CNT, all composites gave almost similar results. It shows that wet mixing is more effective for the rubber matrixes.

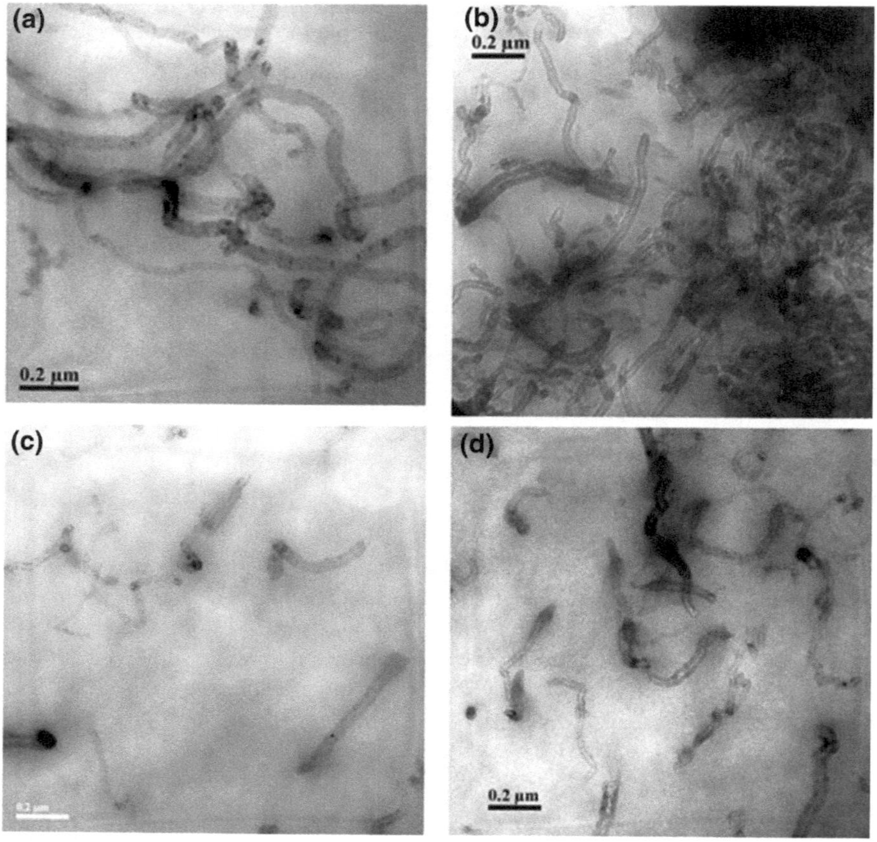

Fig. 11 TEM image of MWNT/EMA composites with different concentration of MWNT: **a** sMWNT 1.0%, **b** sMWNT 2.5%, **c** smMWNT 1.0%, **d** smMWNT 2.5%. Reprinted with permission from Ref. [80]

The sidewall functionalization of MWCNTs with triethylene tetraamine (TETA), was found to increase the impact strength tremendously [83]. The nanotubes were treated with sulphuric acid and nitric acid mixture to get carboxyl function groups. The carboxyl groups were converted into acid chloride by adding thionyl chloride. The acid chloride groups were treated with TETA to get the TETA functionalized MWCNTs (Fig. 14).

Addition of 0.6% of the TETA functionalized MWCNTs to the base polymer DGEBA, had resulted in increase of impact strength, bending strength, and bending modulus by 84, 29, and 22% respectively. The short chain moieties of TETA act a shell over the MWCNT core (Fig. 15), and thereby aids for efficient load transfer from the epoxy matrix to the nanotubes [83]. The surface modification of nanotubes found to enhance the interaction with the epoxy matrix significantly [84].

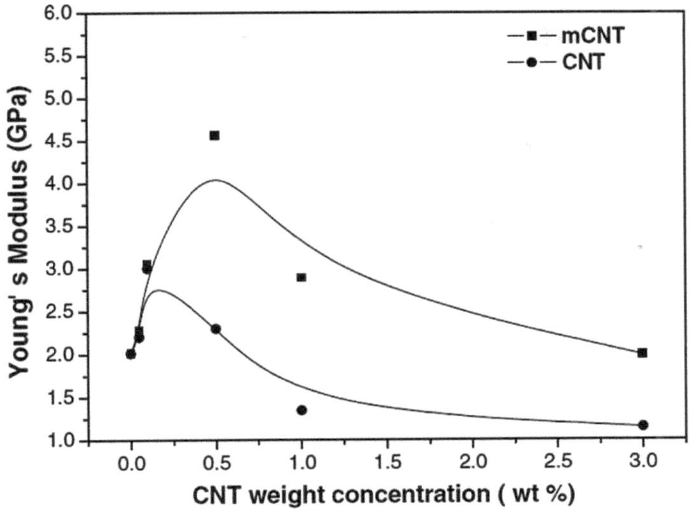

Fig. 12 Plots of the Young's modulus of the mCNT/polyimide and CNT/polyimide films against CNT content. Reprinted with permission from Ref. [81]

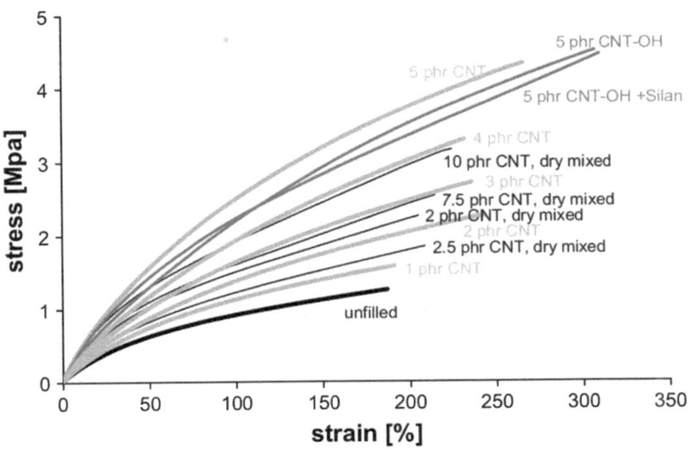

Fig. 13 Stress–strain diagram of CNT filled SBR-BR blend. Reprinted with permission from Ref. [82]

Comparing to the chemical methods like treatment with acid and amine, the plasma method was found to be better method to induce surface functionality in the nanotubes and thereby to improve the interaction with the epoxy matrix. The enhanced interaction led to effective stress transfer, which contributes to the huge increase in tensile strength, modulus, as well as elongation at break (Table 1).

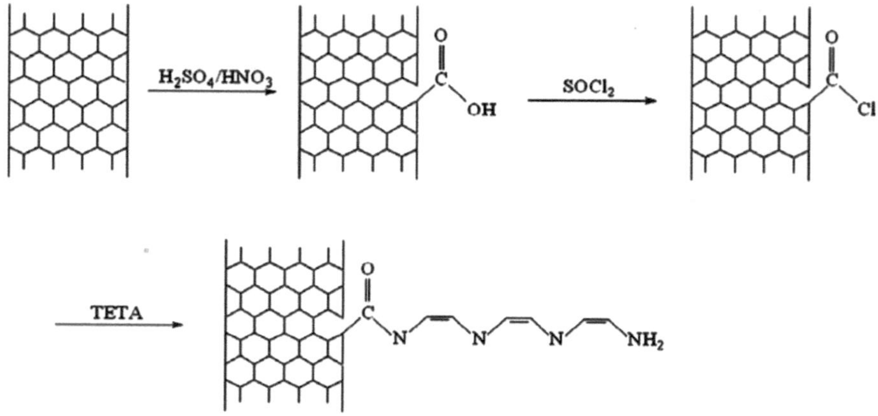

Fig. 14 Scheme for the process of grafting TETA onto the MWCNT surface. Reprinted with permission from Ref. [83]

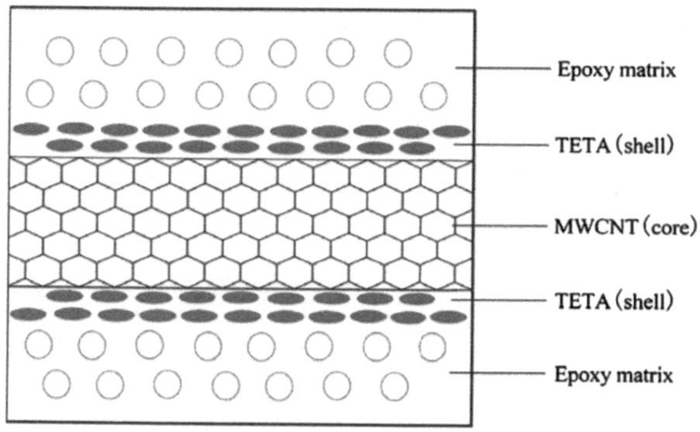

Fig. 15 Model structure of TETA functionalized MWCNT/epoxy composites. Reprinted with permission from Ref. [83]

Table 1 Mechanical properties of pure epoxy and CNT/epoxy composites. Reprinted with permission from Ref. [84]

Fillers	Tensile strength (MPa)	Young's modulus (GPa)	Elongation at break (%)
No Filler (Epoxy)	26	1.21	2.33
Untreated CNTs	42	1.38	3.83
Acid treated CNTs	44	1.22	4.94
Amine treated CNTs	47	1.23	4.72
Plasma treated CNTs	58	1.61	5.22

The method of in-situ dispersion of MWNCT in the synthesis of polyimide thin films was found to give better electrical conductivity even at very low concentration of 0.15% volume fraction [85]. About 11 orders of magnitude increase in electrical conductivity was observed. However, the increase in the mechanical properties was marginal in this method even at 2% volume fraction. It is due to inertness of the pristine MWCNT. The addition of acid treated MWCNTs (sulphuric acid and nitric acid in the ratio 3:1), had yielded significant improvement in tensile strength and fracture strength of epoxy matrix [86]. About 70% increase in tensile strength was observed by addition of 8% of acid treated MWCNT, due to the high specific strength as well as specific surface area of the MWCNTs. Interestingly, the modulus of the composites shown reverse trend due the increase in the toughness of the composites. The acid treatment cause decrease in length and aspect ratio of the nanotubes, which is believed to be the reason for the increase in the toughness of the composites.

The Young's modulus of polyimide could be improved by 90% by adding just 1% of nitric acid treated SWCNTs. The nanocomposite became stronger as well as brittle due to the presence of nanotubes [87]. Among the three types of nanotubes namely, SWCNT, MWCNT and DWCNT, the DWCNTs were found to offer better strength and fracture toughness [88, 89]. Addition of 0.5% of amino functionalized DWCNT led to improve the mechanical strength by 10% and fracture toughness by 43%. Usually, if the interfacial bonding between the nanotubes and matrix is more, then the fracture toughness should decrease, but in the case of DWCNT-epoxy composites, both increase simultaneously, This was explained by considering the fact that the elongation at break of nanotubes are higher than the epoxy matrix. The nanotubes act as bridge between the polymer chains and delay the crack initiation. A novel method of spray winging had been experimented in the preparation of Polyimide-MWNT nanocomposites [90]. The sheets of MWNT synthesized by CVD method were wound over a polymeric spool to required thickness. The polymer precursor, polyamic acid was sprayed over the MWNT sheets during the winding process, and the resulting nanocomposite prepregs were heated to high temperature under pressure to get the Polyimide–MWNT nanocomposite thin films. The tensile properties of the resultant films show significant improvement in strength (300%) as well as modulus (1300%). The modulus increase from 4.04 ± 0.72 GPa for the neat polyimide film to 53.73 ± 3.29 GPa for the MWNT composite (Fig. 16).

In the conventional nanocomposites, the MWNT form discontinuous channels and hence improvement in the modulus could be less, whereas here the continuous aligned structure of nanotubes result in very high modulus and strength. The addition of MWNT in polyamide (PA) matrix by twin screw compounding method has yielded enhancement of modulus tremendously [91]. The modulus of PA composite with 8% MWNT was more than the composite with 20% carbon black. Further, the composite containing carbon black became more conductive due to the formation of conductive channels at high loading fractions whereas the MWNT containing composites did not significant increase in the conductivity. The addition of 1% SWNT into RTV silicone resin system was found to increase the initial modulus significantly, but the modulus started decreasing at about 10–20% strain due to the breakdown of effective interface between the nanotubes and the matrix [92]. The in-situ Raman spectral analysis of

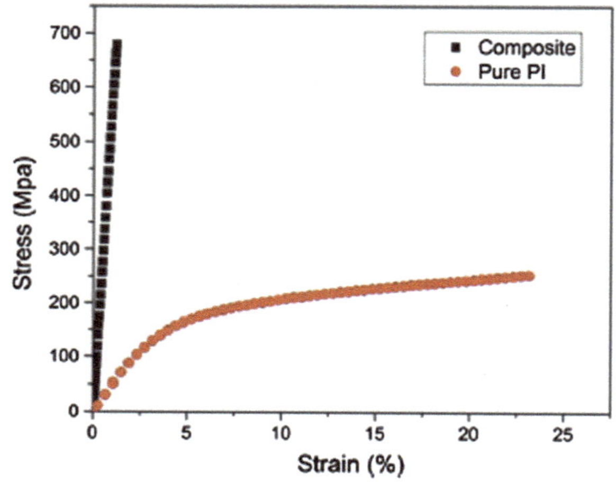

Fig. 16 The typical stress–strain curve comparison of MWNT/BPDA-PDA composite and pure BPDA-PDA polyimide. Reprinted with permission from Ref. [90]

samples during the tensile testing revealed that the nanotubes undergo reorientation under the applied stress, thereby decrease the effective stress transfer. However, in most of the applications, the rubber components undergo tensile strain less than 10%. Hence, the improved modulus at low strains makes the SWNT nanocomposites a better choice for low strain application.

The uniform distribution and dispersion of nanotubes in rubber matrix is a challenging one due to the high viscosity of the rubber matrix. The solvent method, which is widely used in thermoplastics like PVDF cannot be used in the case of rubber materials, because it requires large quantity of solvent and removal of solvent becomes a tedious one. A novel approach of dispersing nanotubes in ethanol in the ratio of 1:30 and adding ionic dispersing liquids like 1-methyl-3-octylimidazolium chloride (MOIC) and 1-allyl-3-methyl imidazolium chloride (AMIC) in the range of 1 mmol per gram of MWNT was experimented by Sahoo et al. [93]. This method had resulted in significant increase in the strength as modulus. About four times increase in strength was observed due to 6 phr MWNT (Fig. 17).

The gradual replacement of carbon black with MWCNT in natural rubber/butyl rubber blend was found to increase the modulus and strength at lower volume fractions of nanotubes, beyond which the strength started decreasing due to the agglomeration of nanotubes [94]. The addition of pristine MWCNT upto 10-phr level in carboxylated Nitrile rubber enhanced the tensile modulus and strength by 140 and 330% respectively [95]. Surprisingly, the elongation at break also increased in this case. This shows that the nanofiller can offer both strengthening as well as toughing simultaneously, which is not possible with micron-sized fillers. The addition of amine functionalized MWCNT into water borne polyurethane by covalent bonding of amine groups with the isocyanide group of polyurethane pre-polymer was found to enhance the tensile stress and tensile modulus significantly due to the formation of isocyanate linkage [96]. About 370% increase in tensile stress and 171% increase in tensile modulus were observed for the addition of 4 phr of amine functionalized MWCNT.

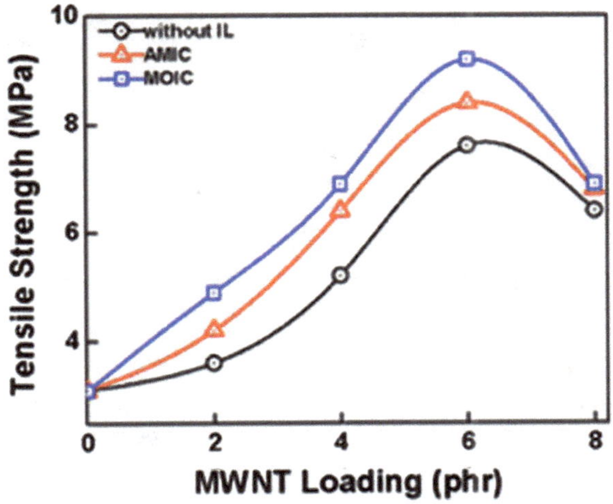

Fig. 17 Tensile strength of AEM/MWNT nanocomposites. Reprinted with permission from Ref. [93]

The addition of SWNTs into isotactic polypropylene in very low loading levels by melt mixing was found to increase the tensile modulus and strength. About 40% increase in modulus was observed due to the addition of 0.75% nanotubes, whereas the increase in modulus due to addition of same quantity of carbon black was about 12% only [97]. The difference in modulus is due to the difference in degree of interfacial adhesion between the filler and the matrix. The addition of MWCNT in very low concentrations (0.5 phr) was also found show the Payne effect due to the breakage of filler-filler networks [74]. The storage modulus of the composites decrease with increase in the strain amplitude, and effect was more pronounced at higher loading of MWCNT (Fig. 18).

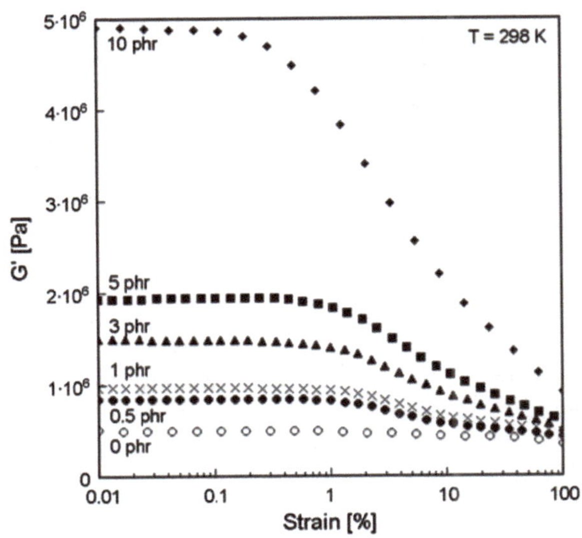

Fig. 18 Strain dependence of storage modulus of unfilled NR and of MWNTs/NR composites at room temperature. Reprinted with permission from Ref. [74]

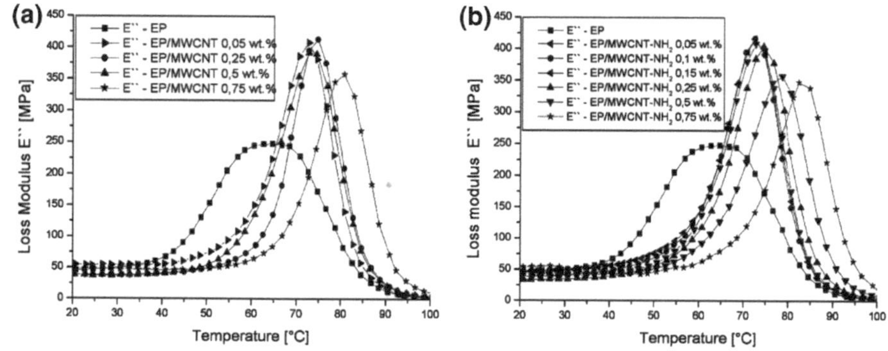

Fig. 19 Loss modulus of nanotube/epoxy nanocomposites with **a** non-functionalized nanotubes and **b** amino-functionalized nanotubes. Reprinted with permission from Ref. [88]

The addition of functionalized carbon nanotubes can increase the loss modulus and glass transition temperature (T_g) of epoxy composites (Fig. 19). The addition of about 0.75% of pristine nanotubes found to increase the T_g from 66 to 80 °C and if the nanotubes were functionalized by amino groups, then T_g increased to 83 °C. The increase in T_g is attributed to the enhanced interfacial interaction between the nanotubes and the host matrix, which restrict the mobility of the polymer chains [89]. Further, the damping nature of the composites also enhanced significantly even if the nanotubes quantity was very less (0.05%).

The increase in damping was due to the resistance offered by the nanotubes against the viscoelastic deformation of the epoxy matrix under mechanical stress. A slight decrease in loss modulus was observed in nanocomposites with higher concentration of nanotubes due to the presence of agglomeration. The gradual replacement of carbon black in natural rubber/ butyl rubber compound had resulting in increase in the storage modulus [94] due to increased reinforcement. The enhanced reinforcement also led to decrease in tan delta values at glass transition temperature (T_g). However, the peak position of loss tangent had shown a negative shift (Fig. 20). Usually, the increase in reinforcement should lead to positive shift in the T_g peak. The surface of pristine MWNCT has fewer tendencies to absorb the polymer chains due to the difference in the energy levels of nanotubes and polymer chains. Hence, the short range localized mobility of the polymer chains could not be disturbed by the nanotubes. Moreover, nanotubes act as voids in between the macromolecular chains and aid in the onset of segmental motion of the polymer chains at lesser temperature.

The addition of pristine MWCNT into carboxylated Nitrile rubber found to increase the storage modulus gradually due to hydrodynamic effect [95]. The effective aspect ratio of the nanotubes in the rubber matrix was calculated from the value of storage modulus at 25 °C by using Guth-Smallwood equation [98, 99]. The effective aspect ratio was found to vary in the range 75 for nanocomposite with 1 phr of MWNT and 20 for the nanocomposite with 10 phr of MWNT, whereas it

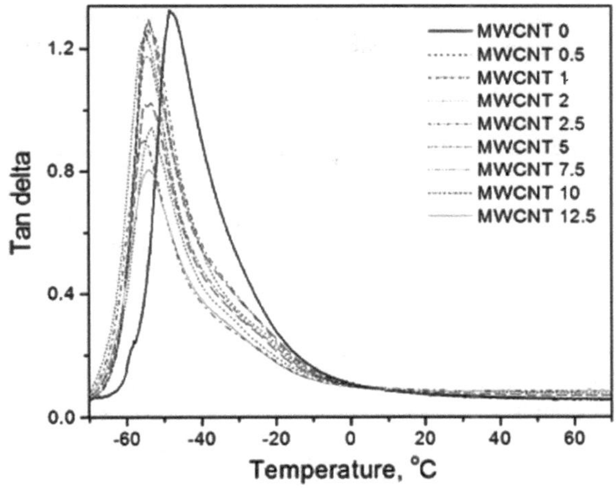

Fig. 20 Temperature dependence of loss tangent of the CB and MWNCT filled NR/IR rubber compounds. Reprinted with permission from Ref. [94]

should be in the range of 500–1000 based on the TEM and SEM images. The significant decrease in the aspect ratio of nanotubes in the composites was explained by considering the agglomeration of nanotubes due to Vander Waals force and possible breakage of nanotubes due to the shearing forces experienced by the nanotubes during mill mixing.

The majority of wok done in polymer nanocomposites discusses about CNT composites. The worldwide research has exploited all combinations of nanotubes and polymers and results are very encouraging. The cost of nanotubes is reducing day by day due to the ease of bulk production. Some of the CNT nanocomposites are already in use for specialized applications in defense and consumer products sectors. The need of the hour is the study of long-term stability and toxicology of the nanocomposites, to ensure that the CNT nanocomposites are stable and safer for use.

8 Graphene Polymer Nanocomposites

A single layer of graphitic form of carbon is known as graphene. The thickness of graphene sheets is in the range of 0.4–0.6 nm whereas the length (*X*) and breath (*Y*) are very large. The quantum confinement in the thickness direction and hexagonal ordered arrangement of carbon atoms in *X* and *Y* directions facilitate high electron mobility [15,000 $cm^2/(V\ sec)$], and high thermal conductivity (4840 −5300 W/mK), giant thermoelectric effect apart from reinforcing nature [100–103]. Graphene is the first material in the new class of materials known as two-dimensional crystalline materials. This new material has a number of unique properties, which make it interesting for fundamental studies and potential applications, especially as next generation electronic material [104]. Its unique properties are high current density, ballistic electron transport, chemical inertness, high thermal conductivity, optical transmittance, and super hydrophobicity. The covalent carbon bonds that make up the crystal lattice of graphene make it extremely resilient to loading. A pure sample of graphene has an intrinsic strength of 21 Nm^{-1} making it the strongest material in existence. A sheet of graphene stretched over a coffee cup could support the weight of a truck bearing down on a pencil point. The engineering applications of Graphene are significant, but limited to laboratories until a manufacturing process is scaled to produce large quantities of the material that is pure enough to yield competitive results.

The addition of graphene in poly (vinyl alcohol) has shown remarkable improvement in Young modulus [105]. However, the actual improvement depends on the degree of dispersion of the graphene particles in the polymer matrix. Graphene possess very high surface area (2600 m^2/g) [106], surface energy and undergo agglomeration and even restack together to form graphitic layers due to interlayer Vander Waals forces [107, 108]. Overcoming these interactions by conventional mixing methods is a challenging task in making graphene nanocomposites on industrial scale. Generally three methods are followed to disperse graphene in polymers, they are in-situ mixing during polymerization, solvent mixing and melt mixing [109]. In the in-situ polymerization, the polymerization reaction is carried out in presence of the graphene particles. The dispersion of graphene is better in both in-situ polymerization and solvent mixing. However, these two methods require large volume of organic solvents and the removal of solvent at the end becomes a tedious process. In melt mixing the dispersion is not as good as in the other two methods, but melt mixing is well-suited and economic method for the dispersion of graphene in thermoplastic polymers.

To improve the efficacy of melt mixing method, a novel method of master batch melt mixing was attempted for dispersing graphene in poly lactic acid matrix [110]. The master batch containing 20% graphene was prepared by mixing chemically reduced graphene with poly lactic acid in tetrahydrofuran as solvent. The mixture was ultra sonicated to ensure good dispersion and then the solvent was removed by thermal evaporation. The presence of polymer chains prevents restacking of graphene layers. To get the nanocomposites of desired composition, required quantity of this master batch had been melt mixed with the pristine PLA. The master batch method has resulted in good dispersion of graphene. Further, it is less solvent incentive and prevents re-aggregation of graphene particles. The master batch has high bulk density, which is desired for the melt mixing. The sequence of entire process is shown in Fig. 21.

The tensile strength and hardness of graphene nanocomposites increase with increase in the graphene content up to 0.08%, due to the reinforcing effect. Above 0.08%, the hardness increases but tensile strength decreases steadily and reaches even below the tensile strength of neat polymer. This is because the formation of graphene interlayer at higher concentration results in the obstruction of interaction between the polymer chains leading to decrease in reinforcement. However, the hardness and storage modulus, which dependent on the stiffness of the material increase steadily. The addition of 0.1% of graphene in Nylon-6 enhances the

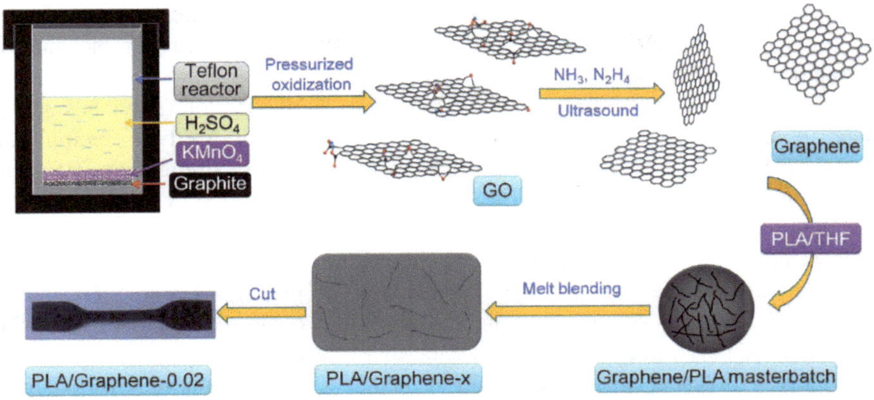

Fig. 21 Illustration for the preparation of graphene and PLA/graphene nanocomposites. The graphite is oxidized by pressurized oxidization and then reduced to single-atom-thick graphene by a multiplex reduction method. The graphene/PLA masterbatch (20% graphene) is prepared by solvent blending from PLA and graphene in THF media. The "x" in "PLA/Graphene-x" is the percentage of the graphene. The black sample is PLA/graphene-0.02 which contains 0.02% graphene. Reprinted with permission from Ref. [110]

Young's modulus by 140% and tensile strength by 110% due to better dispersion and hydrogen bonding [111]. Another novel method attempted for better dispersion of graphene in polymer matrix was chemical functionalization followed by polymerization with the host polymer matrix. The graphitic oxide was reacted with Octa decylamine (ODA) and methacryloyl chloride to get polymeriable –C=C– linkages embedded in the graphene sheets [112]. The functionalized graphene sheets were added at the time of polymerisation reaction of methyl mechacrylic acid (MMA) to get PMMA-Graphene covalently bonded nanocomposites. By this method, the glass transition temperature (T_g) of the nanocomposite increased from 119 to 131 °C with 0.5% of graphene and the storage modulus also improved by 60%.

The reduction of graphene oxide (GO) into graphene monolayer during in-situ polymerization of Poly Vinyl alcohol (PVA) found to improve the tensile strength and dynamic storage modulus and glass transition temperature significantly [113]. The addition of 0.5% GO improved the tensile strength of PVA/GO composite by 35% whereas the in-situ reduction of 0.5% GO into pure unfunctionalized graphene enhanced the tensile strength by 212%. The increase in reinforcement was confirmed by the simultaneous increase in the glass transition temperature from 22.7 °C for PVA to 27.2 °C for PVA/GO and 30.5 °C for PVA/Graphene composites. The tremendous improvement in mechanical properties achieved at the low loading level of graphene is explained based on the better dispersion, interaction of graphene with the PVA molecules and presence of oxide moieties to certain extent. Further investigation of PVA/GO and PVA/Graphene nanocomposites revealed that the enhancement in mechanical properties is due to the presence of hydrogen bonding between Graphene/GO moieties and PVA molecules [114]. Though GO forms more hydrogen bonds with PVA, the improvement in tensile properties found to better in the case of Graphene/PVA composites due to higher mechanical strength of Graphene/GO layers present.

The dispersion of 2% of graphene nanosheets (GNS) in Poly(butylenes succinate) by solvent processing using chloroform as solvent followed by precipitation using methanol had resulted in the enhancement of tensile strength by 21% and storage modulus by 24% [115]. The experimental values were in agreement with the theoretically predicted values for randomly oriented filler particles. The significant enhancement is due to the better interaction between the PBS matrix and graphene layers and reduced level of stacking of graphene sheets. The better interaction has reduced the damping nature of the composites also. The in-situ formation of polyimide chains in presence of isocyanate functionalized graphene oxide particles dispersed in DMF found to enhance the modulus tremendously at very low concentrations of graphene oxide [116]. The isocyanate functionalization enhances the interaction between the nano particles and the matrix, which was confirmed from the SEM images of tensile fractured surfaces. The incorporation of graphene in low concentrations like 0.5 to 5% found to give remarkable increase in the mechanical properties and the increase is isotropic in nature due to the smaller size of the graphene layers. The advantage of Graphene is that the mechanical properties of the polymer composite can be improved drastically by adding very low quantity of Graphene particles, but the reproducibility seems to be difficult due to the agglomeration effect. Hence, more research needs to be carried out to achieve perfection in the development of graphene nanocomposites.

9 Hybrid Nanocomposites

The latest trend in the development of polymer nanocomposites is the development of hybrid nanocomposites in which combination of two or more nano-sized fillers or mixture of micro sized and nano-sized filler are incorporated into the polymer matrix. The resulting nanocomposite has multi functionality and superior performance. The incorporation of nano silica and MWNTs into epoxy polymer by adopting suitable functionalization is one such development [117]. The nanoparticles of silica were attached to MWCNT chemically by using 3-Glycidoxy propyl triethoxy silane as coupling agent (Fig. 22). The addition of nano silica coupled MWNTs into epoxy matrix were found to increase the tensile strength from 3270 ± 32 MPa for the neat epoxy to 3955 ± 58 MPa for the hybrid nanocomposite. The addition of 0.5% amino functionalized SWCNT to the carbon fibres was found to increase the tensile strength by 10% in the carbon fibre reinforced epoxy composites [118]. The presence of amino-functionalized SWCNTs found to inhibit the interfacial cracking and delamination during the fatigue also due to the chemical bonding with the fiber. The research in hybrid nanocomposites is gaining momentum in recent years and has good potential for making multifunctional composites.

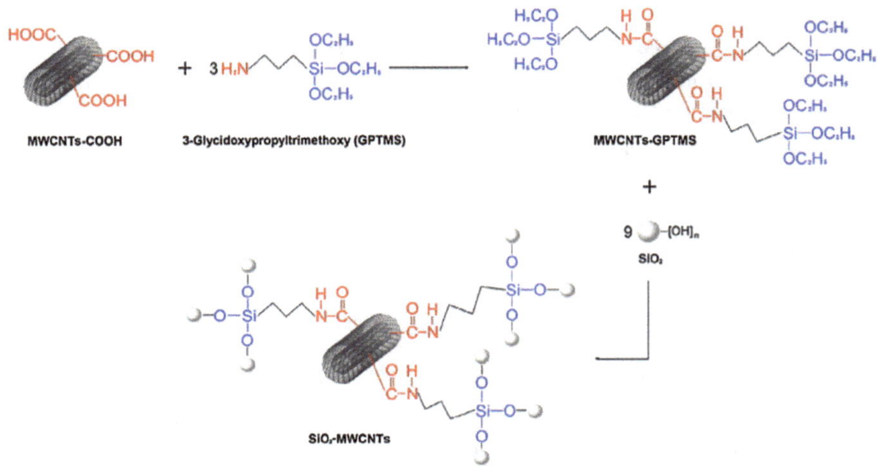

Fig. 22 Schematic of MWNT-SiO_2 preparation. Reprinted with permission from Ref. [117]

10 Conclusions

In this chapter, the recent research activities in the development of carbon-containing polymer composites were reviewed with special emphasis on the effect of various forms of micro- and nanometric carbon particles on the mechanical properties. The use of carbon black as reinforcing filler is found to be the most economic and easier method and this will continue for next few years until the large-scale production of nanomaterials and nanocomposites attains reasonable level of maturity. The advent of modern technologies in materials science research has resulted in the discovery of newer forms of carbon in nanometric size. The curiosity of researchers in the field of polymer technology has resulted in the emergence of polymer nanocomposites containing nano-sized carbon particles like fullerene, carbon nanofibres, carbon nanotubes, nano diamond particles, and graphene. The extensive literature review shows that all these nano particles have great potential for the development of multi functional lightweight polymer nanocomposites. However, the functionalization and dispersion of these nano particles in bulk scale for industrial applications are yet to stabilize and are costlier. Hence, the focus of next generation scientists should be on converting the lab-scale achievements into industrial scale applications in a cost effective manner. In this regard, the preparation of semi aligned carbon nano fibres through the development of electro-spun fibre mats of PAN is a promising method, to get high strength fibre reinforced composites required for various engineering applications. Likewise, the incorporation of carbon nano particles during the synthesis of raw polymers is another interesting and potential method to get uniform dispersion of the nano particles in the polymer matrix. This method can open up new vistas in the bulk synthesis and supply of various synthetic rubbers used in tire and specialty rubber components. The concept of hybrid nanocomposites can also be useful to get multi scale reinforcement and multi functionality in the composites. The parallel studies on the health hazards associated with the handling of various nano forms of Carbon and safe disposal methods will give more value addition for these promising next generation polymer-nanocarbon composites.

References

1. Donnet JB, Voet A, Dekker (1976) Carbon black, physics, chemistry, and elastomer reinforcement. New York
2. White JL (1995) Rubber processing: technology, materials, and principles. Hanser Publishers, Munich
3. Hajji P, David L, Gerard JF, Pascault JP, Vigier G (1999) Synthesis, structure, and morphology of polymer–silica hybrid nanocomposites based on hydroxyethyl methacrylate. J Polym Sci Part B Polym Phys 37(22):3172–3187
4. Park S J (1999). In: Hsu JP (ed) Interfacial forces and field: theory and applications, Dekker, New York, p 385
5. Zhang Z, Zhang G, Li D, Liu Z, Chen X (1999) Chlorohydrin water-swellable rubber compatibilized by an amphiphilic graft copolymer. II. Effects of PVA-*g*-PBA and CPA on water-swelling behaviors. J Appl Polym Sci 74:3145–3152
6. Perez LD, Giraldo LF, Brostow W, Lopez B (2007) Poly (methyl acrylate) plus mesoporous silica nanohybrids: mechanical and thermophysical properties. E-polymers, 29:324–334
7. Zhou XW, Zhu YF, Liang J (2007) Preparation and properties of powder styrene–butadiene rubber composites filled with carbon black and carbon nanotubes. Mater Res Bull 42:456–464
8. Ibarra L, Rodrıguez A, Mora I (2007) Ionic nanocomposites based on XNBR-OMg filled with layered nanoclays. Eur Polym J 43:753–761
9. Miyagawa H, Drzal LT (2004) Thermo-physical and impact properties of epoxy nanocomposites reinforced by single-wall carbon nanotubes. Polymer 45:5163–5170
10. Schadler LS, Giannaris SC, Ajayan PM (1999) Load transfer in carbon nanotube epoxy composites. Appl Phys Lett 73:3842–3844
11. Bansal RC, Donnet JB, Stoeckli F (1988) Active carbon. Marcel Dekker, New York and Basel
12. Wang MJ, Wolff S (1993) Carbon black science and technology. In: Donnet JB, Bansal RC, Wang MJ (eds), Dekker, NewYork, p 229
13. Mather PJ, Thomas KM (1997) Carbon black/ high density polyethylene conducting composite materials. J Mater Sci 32:401–407
14. Ghosh P, Chakrabarti A (2000) Conducting carbon black filled EPDM vulcanizates: assessment of dependence of physical and mechanical properties and conducting character on variation of filler loading. Eur Polym J 36:1043–1054
15. Katbab AA, Nazockdast H, Bazgir S (2000) Carbon black-reinforced dynamically cured EPDM/PPThermoplastic elastomers. I. Morphology, rheology, and dynamic mechanical properties. J Appl Polym Sci 75:1127–1137
16. Hwang J, Muth J, Ghosh T (2007) Electrical and mechanical properties of carbon-black-filled, electrospun nanocomposite fiber webs. J Appl Polym Sci 104:2410–2417
17. Liang JZ, Yang QQ (2009) Mechanical properties of carbon black-filled high-density polyethylene antistatic composites. J Reinf Plasto Compos 28–3:295–304

18. Sau KP, Chaki TK, Khastgir D (1999) Electrical and mechanical properties of conductingcarbon black filled composites based on rubberand rubber blends. J Appl Polym Sci 71:887–895
19. King JA, Tucker KW, Vogt BD, Weber EH, Quan CL (1999) Electrically and thermally conductive nylon 6,6. Polym Compos 20(5):643–654
20. Zheng G, Wu J, Wang W, Pan C (2004) Characterizations of expanded graphite/polymer composites prepared by in situ polymerization. Carbon 42:2839–2847
21. Hontoria-Lucas C, Lopez-Peinado AJ, Lopez-Gonzalez JdD, Rojas-Cervantes ML, Martın-Aranda RM (1995) Study of oxygen-containing groups in a series of graphite oxides: physical and chemical characterization. Carbon 33:1585–1592
22. Lerf A, He H, Forster M, Klinowski J (1998) Structure of graphite oxide revisited. J Phys Chem B 102:4477–4482
23. Etmimi HM, Tonge MP, Sanderson RD (2011) Synthesis and characterization of polystyrene-graphite nanocomposites via surface RAFT-mediated miniemulsion polymerization. J Polym Sci Part A Polym Chem 49:1621–1632
24. Cao NZ, Shen WC, Wen SZ, Liu YJ, Wang ZD, Inagaki M (1996) The factors influencing the porous structure of expanded graphite. Mater Sci Eng (Chin) 14:22–26
25. Kalaitzidou K, Fukushima H, Drzal LT (2007) A new compounding method for exfoliated graphite-polypropylene nanocomposites with enhanced flexural properties and lower percolation threshold. Compos Sci Technol 67:2045–2051
26. Zheng W, Wong SC (2003) Electrical conductivity and dielectric properties of PMMA/ expanded graphite composites. Compos Sci Technol 63:225–235
27. Ganguli S, Roy AK, Anderson DP (2008) Improved thermal conductivity for chemically functionalized exfoliated graphite/epoxy composites. Carbon 46:806–817
28. Yasmin A, Daniel IM (2004) Mechanical and thermal properties of graphite platelet/epoxy composites. Polymer 45:8211–8219
29. Nierengarten J-F, Guttierez-Nava M, Zhang S, Masson P, Oswald L, Bourgogne C (2004) Fullerene-containing macromolecules for materials science applications. Carbon 42:1077–1083
30. Tayfun U, Kanbur Y, Abaci U, Guney HY, Bayramli E (2015) Mechanical, flow and electrical properties of thermoplasticpolyurethane/fullerene composites: Effect of surface modification offullerene. Compos Part B Eng 80:101–107
31. Rafiee MA, Yavari F, Rafiee J, Koratkar N (2011) Fullerene–epoxy nanocomposites-enhanced mechanicalproperties at low nanofiller loading. J Nanopart Res 13:733–737
32. Pikhurov DV, Zuev VV (2013) The effect of Fullerene C60 on the mechanical and dielectrical behavior of epoxy resins at low loading. Nanosyst Phys Chem Math 4(6):834–843
33. Okonkwo AO, Jagadale P, Herrera JEG, Hadjiev VG, Saldana JM, Tagliaferro A, Hernandez FCR (2015) High-toughness/low-friction ductile epoxy coatings reinforced with carbon nanostructures. Polym Test 47:113–119
34. Pikhurov DV, Zuev VV (2016) The study of mechanical and tribological performance of fulleroid materials filled PA6 composites. Lubricants 4:13
35. Calleja FJB, Giri L (1996) Structure and mechanical properties ofpolyethylene-fullerene composites. J Mater Sci 31:5153–5157
36. Ogasawara T, Ishida Y, Kasai T (2009) Mechanical properties of carbon fiber/ fullerene-dispersed epoxy composites. Compos Sci Technol 69:2002–2007
37. Mochalin VN, Shenderova O, Ho D, Gogotsi Y (2012) The properties and applications of nanodiamonds. Nat Nanotechnol 7(1):11–23
38. Shenderova O, Tyler T, Cunningham G, Ray M, Walsh J, Casulli M, Hens S, McGuire G, Kuznetsov V, Lipa S (2007) Nanodiamond and onion-like carbon polymer nanocomposites. Diam Relat Mater 16:1213–1217
39. Shenderova OA, Gruen DM (2006) Ultrananocrystalline diamond: synthesis, properties and applications. William Andrew Publishing, New York

40. Chou CC, Lee SH (2010) Tribological behavior of nanodiamond-dispersed lubricants on carbon steels and aluminium alloy. Wear 269:757–762
41. Osswald S, Yushin G, Mochalin V, Kucheyev SO, Gogotsi Y (2006) Control of sp^2/sp^3 carbon ratio and surface chemistry of nanodiamond powders by selective oxidation in air. J Am Chem Soc 128(35):11635–11642
42. Osswald S, Havel M, Mochalin V, Yushin G, Gogotsi Y (2008) Increase of nano diamond crystal size by selective oxidation. Diamond Relat Mater 17(7–10):1122–1126
43. Steenackers M, Lud SQ, Niedermeier M, Bruno P, Gruen DM, Feulner P, Stutzmann M, Garrido JA, Jordan R (2007) Structured polymer grafts on diamond. J Am Chem Soc 129:15655–15661
44. Zhao YQ, Lau KT, Kim JK, Xu CL, Zhao DD, Li HL (2010) Nanodiamond/poly (lactic acid) nanocomposites: effect of nanodiamond on structure and properties of poly (lactic acid). Compos Part B Eng 41:646–653
45. Barnard AS, Sternberg M (2007) Crystallinity and surface electrostatics of diamond nanocrystals. J Mater Chem 17(45):4811–4819
46. Krüger A, Kataoka F, Ozawa M, Fujino T, Suzuki Y, Aleksenskii AE (2005) Unusually tight aggregation in detonation nanodiamond: identification and disintegration. Carbon 43 (8):1722–1730
47. Behler KD, Stravato A, Mochalin V, Korneva G, Yushin G, Gogotsi Y (2009) Nanodiamond-polymer composite fibers and coatings. ACS Nano 3–2:363–369
48. Roumeli E, Pavlidou E, Avgeropoulos A, Vourlias G, Bikiaris DN, Chrissafis K (2014) Factors controlling the enhanced mechanical and thermal properties of nanodiamond-reinforced cross-linked high density polyethylene. J Phys Chem B 118:11341–11352
49. Zhang Q, Naito K, Tanaka Y, Kagawa Y (2008) Grafting polyimides from nanodiamonds. Macromolecules 41:536–538
50. Morimune S, Kotera M, Nishino T, Goto K, Hata K (2011) Poly(vinyl alcohol) nanocomposites with nanodiamond. Macromolecules 44:4415–4421
51. Zhai YJ, Wang ZC, Huang W, Huang JJ, Wang YY, Zhao YQ (2011) Improved mechanical properties of epoxy reinforced by low content nano diamond powder. Mat Sci Eng A Struct 528:7295–7300
52. Guo H, Sheng H, Peng X, Yu X, Naito K, Qu X, Zhang Q (2014) Preparation and mechanical properties of epoxy/diamond nanocomposites. Polym Compos 35:2144–2149
53. Neitzel I, Mochalin V, Knoke I, Palmese GR, Gogotsi Y (2011) Mechanical properties of epoxy composites with high contents of nanodiamond. Compos Sci Technol 71:710–716
54. Zhou ZP (2009) Development of carbon nanofibers from aligned electrospun polyacrylonitrile nanofiber bundles and characterization of their microstructural, electrical, and mechanical properties. Polymer 50(13):2999–3006
55. Ma H, Zeng J, Mary LR, Kumar S, Schiraldi DA (2003) Processing, structure, and properties of fibers from polyester/carbon nanofiber composites. Compos Sci Technol 63:1617–1628
56. Tong X, Chen Y, Cheng H (2005) Influence of carbon nanofiber addition on mechanical properties and crystallization behavior of polypropylene. J Mater Sci Technol 21(5):686–690
57. Park JH, Lee SC (2016) Crystallization kinetics and mechanical properties of poly(lactic acid)/carbon nanofiber composites. Text Sci Eng 53:55–61
58. Bal S (2010) Experimental study of mechanical and electrical properties of carbon nanofiber/epoxy composites. J Mater Des 31(5):2406–2413. https://doi.org/10.1016/j.matdes.2009.11.058
59. Lafdi K, Fox W, Matzek M, Yildiz E (2008) Effect of carbon nanofiber-matrix adhesion on polymeric nanocomposite properties—Part II. J Nano mater, Article ID 310126, p 8. https://doi.org/10.1155/2008/310126
60. Danni N, Sasikumar T (2016) Characterization of electrospun carbon nanofiber mat reinforced polymer composites using ultra-sonic scanning method. Dig J Nanomater Biostruct 11(1):141–148

61. Zhang D, Zhao Y, Cabrera E, Zhao Z, Castro JM, Lee LJ (2016) Improved sand erosion resistance and mechanical properties of multifunctional carbon nanofiber nanopaper enhanced glass fiber/epoxy composites. SPE ANTEC Indianap, 411–415
62. Al-Saleh MH, Sundararaj U (2009) A review of vapor grown carbon nanofiber/polymer conductivecomposites. Carbon 47:2–22
63. Kingsuk M, Dipiti P, Dhannu L, Kanik R, Gyanesh NM (2004) Synthesis of coiled/straight carbon nanofibers by catalystic chemical vapor deposition. Carbon 42:3251–3272
64. Khattab A, Liu C, Chirdon W, Hebert C (2012) Mechanical and thermal characterization of carbon nanofiber reinforced low-density polyethylene composites. J Thermoplast Compos 26(7):954–967
65. Guadagno L, Raimondo M, Vittoria V, Vertuccio L, Lafdi K, De VB, Lamberti P, Spinelli G, Tucci V (2013) The role of carbon nanofiber defects onthe electrical and mechanical properties of CNF-based resins. Nanotechnology 24:305704
66. Dhakate SR, Chaudhary A, Gupta A, Pathak AK, Singh BP, Subhedara KM, Yokozekib T (2016) Excellent mechanical properties of carbon fiber semi-aligned electrospun carbon nanofiber hybrid polymer composites. RSC Adv 6:36715–36722
67. Meyyappan M (2005) Carbon nanotube: science and applications. CRC Press, Florida
68. O'Connell MJ (ed) (2006) Carbon nanotubes—properties and applications. Taylor and Francis, Boca Raton
69. Wong EW, Sheehan PE, Lieber CM (1997) Nanobeam mechanics: elasticity, strength, and toughness of nanorods and nanotubes. Science 277:1971–1975
70. Dresselhaus MS, Dresselhaus G, Eklund PC (1996) Science of fullerenes and carbon nanotubes. Academic Press, San Diego
71. Yu MF, Lourie O, Dyer M, Moloni K, Kelly TF, Ruoff RS (2000) Strength and breaking mechanism of multi-walled carbon nanotubes under tensile load". Science 287:637–639
72. Lau KT, Hui D (2002) The revolutionary creation of new advanced materials—carbon nanotube composites. Compos Part B Eng 33:263–277
73. Rahul MC, Ghassemieh E (2012) Optimized process for the inclusion of carbon nanotubes in elastomers with improved thermal and mechanical properties. J Appl Polym Sci 124:4993–5001
74. Bokobza L (2012) Multiwall carbon nanotube-filled natural rubber: electrical and mechanical properties. Express Polym Lett 6(3):213–223
75. Wang W, Ciselli P, Kuznetsov E, Peijs T, Barber AH (2008) Effective reinforcement in carbon nanotube–polymer composites. Philos Trans R Soc A 366:1613–1626
76. Kim MT, Park H, Hui D, Rhee KY (2011) Carbon nanotube modification using gum arabic and its effect on the dispersion and tensile properties of carbon nanotubes/epoxy nanocomposites. J Nanosci Nanotechnol 11:7369–7373
77. Tugrul Seyhan A, Gojny FH, Tanoglu M, Schulte K (2007) Critical aspects related to processing of carbon nanotube/unsaturated thermoset polyester nanocomposites. Eur Polym J 43:374–379
78. Jagtap BS, Ratna D (2013) Novel method of dispersion of multiwalled carbon nanotubes in a flexible epoxy matrix. J Appl Polym Sci 130:2610–2618
79. Xiao KQ, Zhang LC, Zarudi I (2007) Mechanical and rheological properties of carbon nanotube-reinforced polyethylene composites. Compos Sci Technol 67:177–182
80. Basuli U, Chaki TK, Chattopadhyay S, Sabharwal S (2010) Thermal and mechanical properties of polymer-nanocomposites based on ethylene methyl acrylate and multiwalled carbon nanotube. Polym Compos 31:1168–1178
81. Chou WJ, Wang CC, Chen CY (2008) Characteristics of polyimide-based nanocomposites containing plasma-modified multi-walled carbon nanotubes. Compos Sci Technol 68:2208–2213
82. Das A, Stockelhuber KW, Jurk R, Saphiannikova M, Fritzsche J, Lorenz H, Kluppel M, Heinrich G (2008) Modified and unmodified multiwalled carbon nanotubes in high performance solution-styrene–butadiene and butadiene rubber blends. Polymer 49:5276–5283

83. Yang K, Gu M, Guo Y, Pan X, Mu G (2009) Effects of carbon nanotube functionalization on the mechanical and thermal properties of epoxy composites. Carbon 47:1723–1737
84. Kim JA, GiSeong D, Kang TJ, Youn JR (2006) Effects of surface modification on rheological and mechanical properties of CNT/epoxy composites. Carbon 44:1898–1905
85. Jiang X, Bin Y, Matsuo M (2005) Electrical and mechanical properties of polyimide–carbon nanotubescomposites fabricated by in situ polymerization. Polymer 46:7418–7424
86. Guo P, Chen X, Gao X, Song H, Shen H (2007) Fabrication and mechanical properties of well-dispersed multiwalled carbon nanotubes/epoxy composites. Compos Sci Technol 67:3331–3337
87. Yu A, Hu H, Bekyarova E, Itkis ME, Gao J, Zhao B, Haddon RC (2006) Incorporation of highly dispersed single-walled carbon nanotubes in a polyimide matrix. Compos Sci Technol 66:1190–1197
88. Gojny FH, Schulte K (2004) Functionalisation effect on the thermo-mechanical behavior of multi-wall carbon nanotube/epoxy-composites. Compos Sci Technol 64:2303–2308
89. Gojny FH, Wichmann MHG, Fiedler B, Schulte K (2005) Influence of different carbon nanotubes on the mechanical properties of epoxy matrix composites—a comparative study. Compos Sci Technol 65:2300–2313
90. Jiang Q, Wang X, Zhu Y, Hui D, Qiu Y (2014) Mechanical, electrical and thermal properties of aligned carbon nanotube/polyimide composites. Compos Part B Eng 56:408–412
91. Meincke O, Kaempfer D, Weickmann H, Friedrich C, Vathauer M, Warth H (2004) Mechanical properties and electrical conductivity of carbon-nanotube filled polyamide-6 and its blends with acrylonitrile/butadiene/styrene. Polymer 45:739–748
92. Frogley MD, Ravich D, Wagner HD (2003) Mechanical properties of carbon nanoparticle-reinforced elastomers. Compos Sci Technol 63:1647–1654
93. Sahoo BP, Naskar K, Tripathy DK (2016) Multiwalled carbon nanotube-filled ethylene acrylic elastomer nanocomposites: influence of ionic liquids on the mechanical, dynamic mechanical, and dielectric properties. Polym Compos 37:2568–2580
94. Poikelispaa M, Das A, Dierkes W, Vuorinen J (2013) The effect of partial replacement of carbon black by carbon nanotubes on the properties of natural rubber/butadiene rubber compound. J Appl Polym Sci 130:3153–3160
95. Sasikumar K, Manoj NR, Mukundan T, Khastgir D (2014) Design of XNBR nanocomposites for underwater acoustic sensor applications: effect of MWNT on dynamic mechanical properties and morphology. J Appl Polym Sci 131:40752
96. Kuan HC, Ma C-CM, Chang W-P, Yuen S-M, Wu H-H, Lee T-M (2005) Synthesis, thermal, mechanical and rheological properties of multiwall carbon nanotube/waterborne polyurethane nanocomposite. Compos Sci Technol. 65:1703–1710
97. Lopez MMA, Valentini L, Biagiotti J, Kenny JM (2005) Thermal and mechanical properties of single-walled carbon nanotubes–polypropylene composites prepared by melt processing. Carbon 43:1499–1505
98. Guth E, Gold O (1938) On the hydro dynamical theory of the viscosity of suspensions. Phys Rev 53:322–325
99. Smallwood HM (1944) Limiting law of the reinforcement of rubber. J Appl Phys 15:758–766
100. Geim AK, Novoselov KS (2007) The rise of graphene. Nat Mater 6:183–191
101. Chen JY, Wen YG, Guo YL, Wu B, Huang LP, Xue YZ, Geng DC, Wang D, Yu G, Liu YQ (2011) Oxygen-aided synthesis of polycrystalline graphene on silicon dioxide substrates. J Am Chem Soc 133:17548–17551
102. Singh V, Joung D, Zhai L, Das S, Khondaker SI, Seal S (2011) Graphene based materials: past, present and future. Prog Mater Sci 56:1178–1271
103. Dragoman D, Dragoman M (2007) Negative differential resistance of electrons in graphene barrier. Appl Phys Lett 90:203116
104. Choi W, Lahiri I, Seelaboyina R, Kang YS (2010) Synthesis of graphene and its applications: a review. Crit Rev Solid State 35(1):52–71

105. Zhao X, Zhang QH, Chen DJ, Lu P (2010) Enhanced mechanical properties of graphene-based poly (vinyl alcohol) composites. Macromolecules 43:2357–2363
106. Novoselov KS, Geim AK, Morozov SV, Jiang D, Zhang Y, Dubonos SV, Grigorieva IV, FirsovA A (2004) Electric field effect in atomically thin carbon films. Science 5696:666–669
107. Wu H, Zhao W, Hu H, Chen G (2011) One-step in situ ball milling synthesis of polymer-functionalized graphene nanocomposites. J Mater Chem 21:8626–8632
108. Su Q, Pang SP, Alijani V, Li C, Feng XL, Mullen K (2009) Composites of graphene with large aromatic molecules. Adv Mater 21:3191–3195
109. Kim H, Kobayashi S, AbdurRahim MA, Zhang MLJ, Khusainova A, Hillmyer MA, Abdala AA, Macosko CW (2011) Graphite oxide, graphene, and metal-loaded graphene for fire safety applications of polystyrene. Polymer 52:1837–1846
110. Bao C, Song L, Xing W, Yuan B, Wilkie CA, Huang J, Guo Y, Hu Y (2012) Preparation of graphene by pressurized oxidation and multiplex reduction and its polymer nanocomposites by masterbatch-based melt blending. J Mater Chem 22:6088–6096
111. Xu Z, Gao C (2010) In situ polymerization approach to graphene-reinforced nylon-6 composites. Macromolecules 43:6716–6723
112. Pramoda KP, Hussain H, Koh HM, Tan HR, He CB (2010) Covalent bonded polymer-graphene nanocomposites. J Polym Sci Part A Polym Chem 48:4262–4267
113. Wang J, Wang X, Xu C, Zhang M, Shang X (2011) Preparation of graphene/poly(vinyl alcohol) nanocomposites with enhanced mechanical properties and water resistance. Polym Int 60:816–822
114. Bao C, Guo Y, Song L, Hu Y (2011) Poly(vinyl alcohol) nanocomposites based on graphene and graphite oxide: a comparative investigation of property and mechanism. J Mater Chem 21:13942–13950
115. Wang X, Yang H, Song L, Hu Y, Xing W, Lu H (2011) Morphology, mechanical and thermal properties of graphene-reinforced poly(butylene succinate) nanocomposites. Compos Sci Technol 72:1–6
116. Luong ND, Hippi U, Korhonen JT, Soininen AJ, Ruokolainen J, Johansson LS, Nam JD, Sinh LH, Seppälä J (2011) Enhanced mechanical and electrical properties of polyimide film by graphene sheets via in situ polymerization. Polymer 52:5237–5242
117. Jia X, Liu B, Huang L, Hui D, Yang X (2013) Numerical analysis of synergistic reinforcing effect of silica nanoparticle–MWCNT hybrid on epoxy-based composites. Compos Part B Eng 54:133–137
118. Davis DC, Wilkerson JW, Zhu J, Hadjiev VG (2011) A strategy for improving mechanical properties of a fiber reinforced epoxy composite using functionalized carbon nanotubes. Compos Sci Technol 71(8):1089–1097

Electrical Conductivity of Polymer–Carbon Composites: Effects of Different Factors

Mostafizur Rahaman, Ali Aldalbahi, Lalatendu Nayak and Radhashyam Giri

Abstract In this chapter, the electrical conductivity/resistivity of polymer–carbon composites has been discussed in detail. The types of electrical resistivity and their measurement procedure have been depicted pictorially. The electrical conductivity of different carbon materials like diamond, graphite, fullerene, carbon fiber, carbon black, carbon nanotubes, and graphene are noted and discussed. The different techniques of preparation/processing of conducting polymer/carbon composites are mentioned here within short. Moreover, how the geometry/structure of different carbons controls the electrical conductivity of polymer composites has been critically reviewed. The electrical percolation threshold and the conductivity of polymer/carbon composites that depends on many physical and chemical factors are investigated from different literature sources and reported in this chapter.

Keywords Polymer/carbon composites · Electrical conductivity
Percolation threshold · Measurement of conductivity · Dependency of conductivity

---

M. Rahaman (✉) · A. Aldalbahi
Department of Chemistry, College of Science, King Saud University, Riyadh 11451, Saudi Arabia
e-mail: mrahaman@ksu.edu.sa

A. Aldalbahi
e-mail: aaldalbahi@ksu.edu.sa

L. Nayak
Phillips Carbon Black Ltd., Kolkata, West Bengal, India
e-mail: litupolymer@gmail.com

R. Giri
Central Institute of Plastics Engineering and Technology, Ahmedabad 382445, India
e-mail: giripolymer@gmail.com

M. Rahaman et al. (eds.), *Carbon-Containing Polymer Composites*,
Springer Series on Polymer and Composite Materials,
https://doi.org/10.1007/978-981-13-2688-2_5

1 Introduction

1.1 Polymers on the Basis of Electrical Property

On the basis of electrical property, polymers are classified into three groups; electrically insulator, semiconductor, and conductor. Most of the polymers are inherently insulating in nature due to the nonavailability of delocalized electrons or holes in their backbone chain. However, with the technological revolution, the needs for electrically conducting polymers have gained a significant importance due to their various applications in the electrical and electronic industry [1–5]. Electrically conducting polymers are of three types; intrinsically/inherently conducting polymers, extrinsically conducting polymers also known as conducting polymer composites, and ionically conducting polymers.

1.2 Intrinsically Conducting Polymers

Intrinsically conducting organic polymers possess similar electrical, electronic, and magnetic properties to that of metals and are commonly known as synthetic metals [6]. The development of these intrinsic conducting polymers has become a promising field of research since the discovery of polyacetylene in 1977 by Alan J Heeger and his co-workers [7], for which they got the Nobel Prize in Chemistry, 2000. Some known intrinsically conducting polymers are polyacetylene, polyaniline, polypyrrole, polythiophene, poly (phenylene), poly (phenylene vinylene), etc. These conducting polymers are having delocalized π electrons in their main chain backbone. The essential characteristics of these polymers are their relatively small π electronic band gap ($\sim$1–3.5 eV) with corresponding low excitation energy; they can be easily oxidized or reduced using different dopant species; the mobility of the charged particles are high enough in their conductive state and the presence of quasiparticles, which move relatively more freely through the materials [8, 9]. The main disadvantages associated with these intrinsically conducting polymers are their processibility [10–13]. These polymers are insoluble in most of the solvents, infusible in nature, and mechanically very stiff. Thus, these polymers cannot be applied to those cases where some degree of flexibility and mechanical property are required.

1.3 Extrinsically Conducting Polymers

Extrinsically conducting polymer composite is prepared through the incorporation of different conducting fillers into an insulating polymer matrix. Generally, carbon black [14, 15], carbon fiber [16–19], carbon nanotubes [20–24], graphite [25], coke [26], metal-coated oxide [27], nonoxidizing metal powders/fibers like silver [28], aluminum [29], steel [30], iron [31, 32], copper [33], nickel and nickel-coated glass fibers [34, 35], and particles of conducting polymers [36, 37], etc., are used as conducting filler in insulating polymer matrix. In the present study, we will confine our discussion on the extrinsic conductive polymer.

1.4 Conductivity Mechanism

The mechanism of electrical conductivity, in general, can be carried by electrons, or through ionic transport. In the case of metallic conductors and semiconductors, electrons act as charge carriers, while in the case of ionic compounds such as sodium chloride, the conduction role is played by ions. Electronic transport depends on factors such as atomic bond strength and lattice imperfections, while ion transport is governed by ion diffusion rates [38]. Figure 1 shows this transport mechanism for three classes of conductors: metals, semiconductors, and ionic compounds. Metallic conductors have a “sea” of free electrons, one associated with affecting this conductivity such as temperature and type of metal [38, 39]. Semiconductors have a band gap in their density of states and hence, can alternate between insulating and conducting depending on the input voltage. Ionic conductivity is the movement of an ion from one site to another in the crystal lattice; it is dependent upon the concentration and presence of defects [40].

2 Different Types of Electrical Resistivity/Conductivity and Their Measurements

The electrical resistivity is a fundamental property of any insulating or conducting materials. Depending upon the application point of views, the electrical resistivity is of three types namely surface resistivity, volume/bulk resistivity, and contact resistivity. The resistivity of an insulating material is measured according to

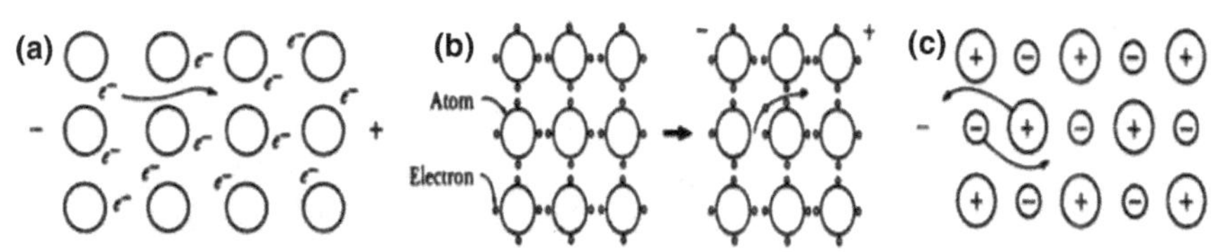

Fig. 1 The mechanism of electrical conductivity. **a** Valence electrons in metals flow easily, hence metals are good conductors. **b** Covalent bonds are disrupted in semiconductors, hence they are poorer conductors. **c** Ion diffusion is the charge carrying mechanism for ionic compounds. Reproduced from Ref. [38]

standard ASTM D 257-99 [41]. The electrical resistance of a material is measured by using Ohm's law, where a known voltage is passed through the material and the resultant current passing through the material is noted down. The resistivity is calculated from this resistance value depending upon the physical dimension of the test sample.

2.1 Surface Resistance and Resistivity

The definition of surface resistance and surface resistivity has been mentioned in many standards and books [41–44]. Surface resistance is defined as the ratio of DC voltage to the current flowing between two electrodes, where the electrodes are of specified configuration and are in contact with the same surface of the test sample (Fig. 2). In surface resistance measurement, the input is voltage and the output is current (Fig. 3). Mathematically, it can be expressed as (Eq. 1):

$$R_s = V/I \tag{1}$$

where R_s is the surface resistance of the material, V is the applied DC voltage and I is the resultant current passed through the sample.

Surface resistivity (ρ_s) is defined as (Eq. 2):

$$\rho_s = K_s R_s \tag{2}$$

where $K_s = P/L$. P is the effective perimeter of the guarded electrode in mm and L is the distance between the guarded electrode and the ring electrode in mm. The dimension L is determined from Fig. 3, that is, $L = (D_2 - D_1)/2$. The magnitude of P for the circular electrode is πD_0, where $D_0 = D_1 + L$ as determined from Fig. 4.

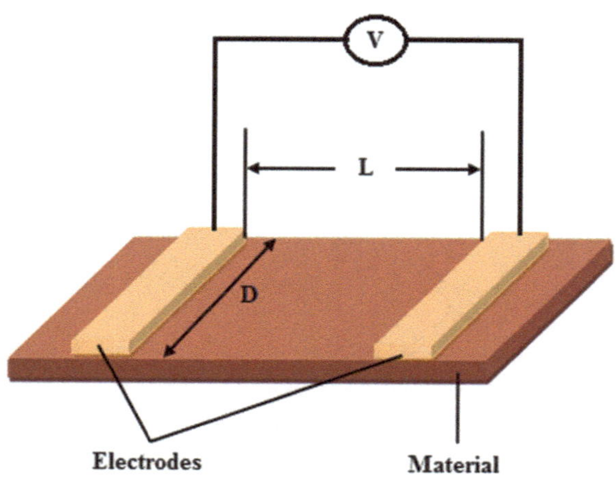

Fig. 2 Placement of electrodes on the same side of material for surface resistance measurement

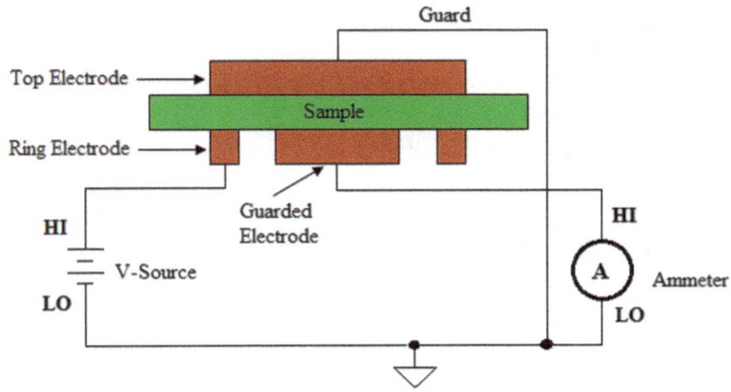

Fig. 3 Technique for the measurement of surface resistance

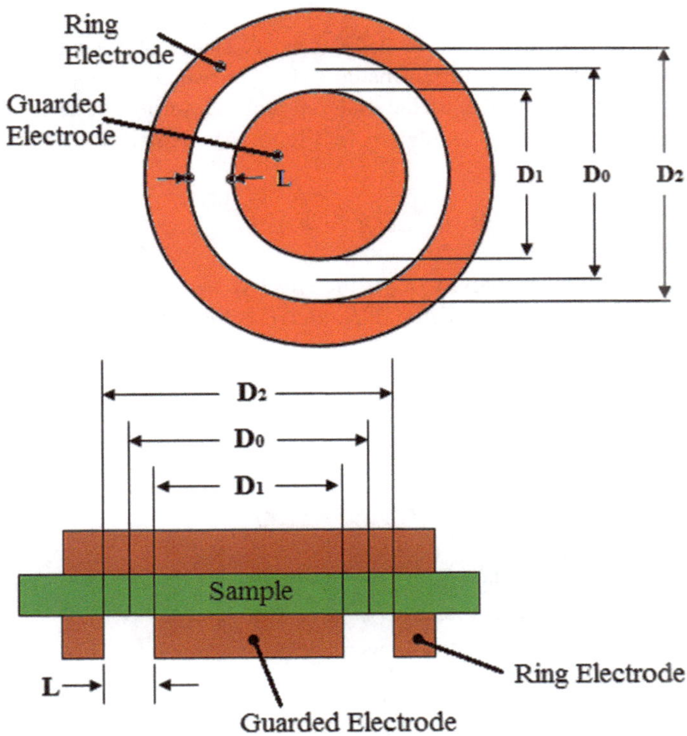

Fig. 4 Dimension of the circular electrode

The surface resistance depends on the characteristics of the material and the geometry of electrodes used during the measurement technique. The unit of electrical resistance/surface resistance is Ohm. However, the calculated physical unit of surface resistivity is also Ohm. Thus, the unit of surface resistance and surface resistivity is mixed up and makes confusion. Hence, to avoid any confusion, the unit of surface resistivity is often mentioned as Ohm/square, though it is not valid on the dimensional point of views.

Theoretically, it is assumed that during the measurement of surface resistance/resistivity of any material, the current flows only along the surface of the material and do not penetrate into the bulk of the material. But in reality, it is not true because a portion of the current always flows through the material. However, to study the surface property of different materials, it has been assumed that the surface current flows through the infinitesimally thin layer of the surface and this layer is so thin that its thickness can be neglected.

2.2 *Volume/Bulk Resistance and Resistivity*

Volume resistivity of a material is defined as its electrical resistance, when the current flow through a cube of that material. It is measured by applying a voltage across the opposite faces/sides of the material and measuring the resultant current flows through the materials as shown in Fig. 5. Mathematically, it is given as (Eq. 3):

$$\rho_{\text{v}} = (K_{\text{v}} \times R_{\text{v}})/T \tag{3}$$

where ρ_{v} is the volume resistivity, K_{v} is the effective area of the guarded electrode, R_{v} is the volume resistance in Ohm (*V*/*I*), and *T* is the thickness of the sample in mm. For circular electrode (Eq. 4):

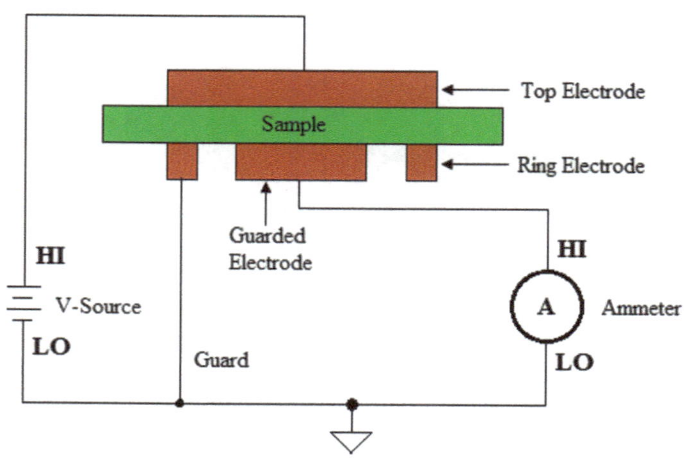

Fig. 5 Measurement technique of volume resistivity

$$K_v = \pi(D_1/2 + B * L/2)^2 \tag{4}$$

where D_1 is the outer diameter of the guarded electrode, L is the distance between the guarded electrode and ring electrode, and B is the effective area coefficient. The dimension D_1 and L can be calculated from Fig. 5. For the effective area, coefficient B for the measurement of volume resistivity is typically used as 0.

2.3 *Contact Resistance and Resistivity*

Contact is a necessary part of many electrical and electronic devices. So it is necessary to measure the contact resistance to get an idea that how it affects the performance of the devices. The contact resistance of a resistor can be measured by the extension of the resistance measurement technique. The measurement can be done by considering the geometry of a simple resistor as shown in Fig. 6. The resistor consists of a semiconducting material in contact with two metals at the end. The length of the material is L and the contact area at its metal interface is Ac. Thus, the total resistance can be given as (Eq. 5):

$$R_T = 2R_m + 2R_c + R_{semi} \tag{5}$$

where R_T is the total resistance of the resistor, R_m is the resistance of the metal, R_c is the contact resistance in between metal and semiconducting interface, and R_{semi} is the resistance of semiconducting materials. In most of the situations, the resistivity of metal in contact is very low compared to the semiconducting material, that is $R_m \ll R_c$, hence R_m can be neglected. So, the above equation can be written as (Eq. 6):

$$R_T = 2R_c + R_{semi} \tag{6}$$

The resistance of semiconducting material can be written as $R_{semi} = R_s * L/W$, where R_s is the sheet resistivity and W is the area of the sheet. Hence, the above equation can be written as (Eq. 7):

$$R_T = (R_s * L)/W + 2R_c \tag{7}$$

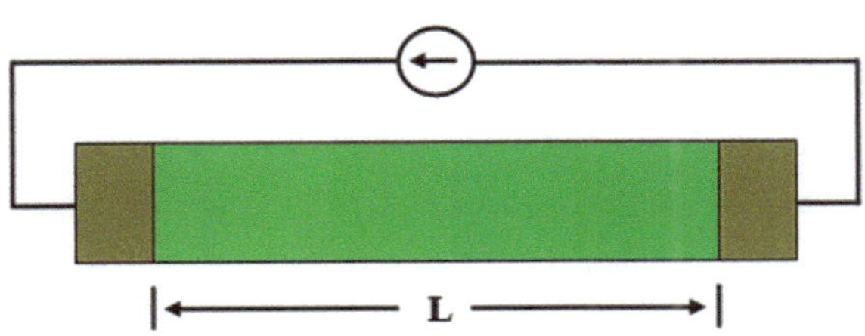

Fig. 6 Geometry of a simple resistor

Fig. 7 Plot of R_T against L for contact ratio measurement

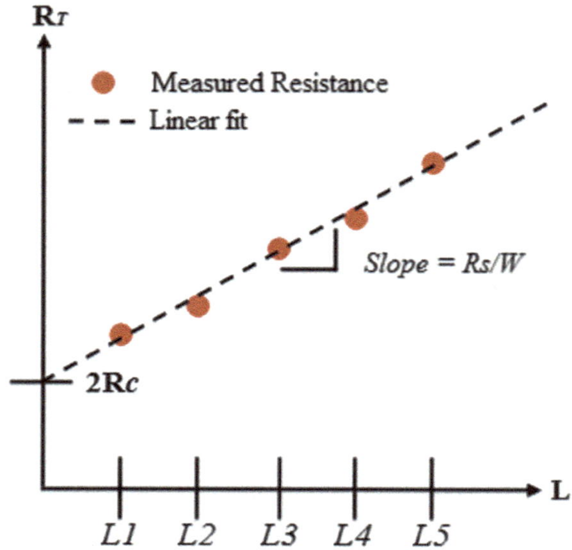

If we take the resistor of different lengths keeping all other parameters as constant, then the above equation will be the equation of the straight line with intercept $2R_c$ and slope R_s/W. Hence, R_T can be plotted against L as shown in Fig. 7. The value of total resistance R_T for each length of the resistor can be measured from the instrument. The contact resistance and resistor sheet resistivity can be calculated from the intercept and slope of the fitted straight line, respectively.

The contact resistance of any resistor depends on its contact area, and hence, on comparison point of view, it is not logical. Hence, it is better to define the term contact resistivity. To calculate the contact resistivity, we can consider a small region of contact with thickness Δx as shown in Fig. 8. Thus, the contact resistance by considering the contact thickness Δx can be given as (Eq. 8):

$$R_c = \rho'_c * (\Delta x / A_c) \tag{8}$$

where A_c is the area of contact and ρ'_c is the contact resistivity when contact thickness is Δx. But, in reality, the thickness of contact is very small and can be considered that it tends to zero. In this respect, the contact resistivity can be given as (Eq. 9):

$$\rho_c = \lim_{\Delta x \to 0} (\rho'_c * \Delta x) = R_c * A_c \tag{9}$$

So, the contact resistance has the unit of $\Omega\ cm^2$.

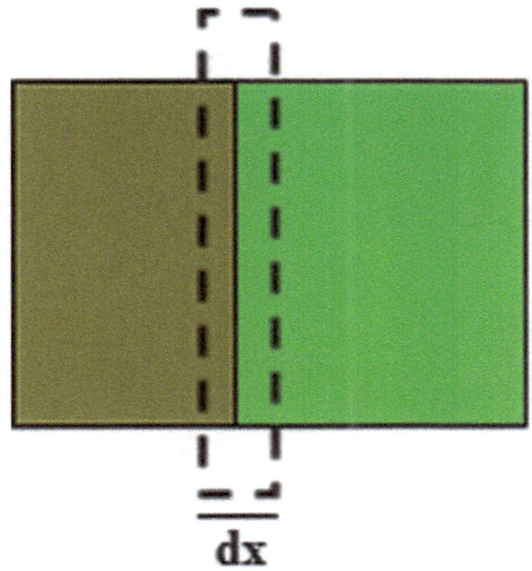

Fig. 8 Small contact region

2.4 *Van der Pauw Technique for Resistivity Measurement*

This four probes technique is used to measure the resistivity of a sample of any arbitrary shape [45–47]. The sample should be flat with uniform thickness, free from any isolated hole, homogeneous in composition and isotopic/symmetrical. The four contacts must be located at the perimeter of the sample and the area of any individual contact should be one order of magnitude smaller than the area of the sample for the accuracy of the measurement. The sample with their four contacts is shown in Fig. 9, where the contacts are numbered as 1, 2, 3, and 4. To measure the resistance, current is passed through the one edge of the sample and the voltage drop is measured across its opposite edge. The measurement procedure is improved by reciprocal measurement and reverse polarity measurement, where the effect of offset voltage such as thermoelectric potential is canceled out by taking total eight measurements as shown in Fig. 9. Four measurements are taken in a horizontal position and the rest four measurements are taken in the vertical position. The resistance is calculated from the relation $R = V/I$. Hence, the average resistance in the horizontal position (R_h) and vertical position (R_v) are $R_h = (R_1 + R_2 + R_5 + R_6)/4$ and $R_v = (R_3 + R_4 + R_7 + R_8)/4$, respectively. Thus, the sheet resistance (R_s) is calculated using the Van der Pauw formula (Eq. 10):

$$\text{Exp}\,(-\pi R_h/R_s) + \text{Exp}(-\pi R_v/R_s) = 1 \quad (10)$$

A numerical method is applied to calculate the sheet resistance value. The resistivity (ρ) is then calculated from the relation $\rho = R_s * t$, where t is the thickness of the sample.

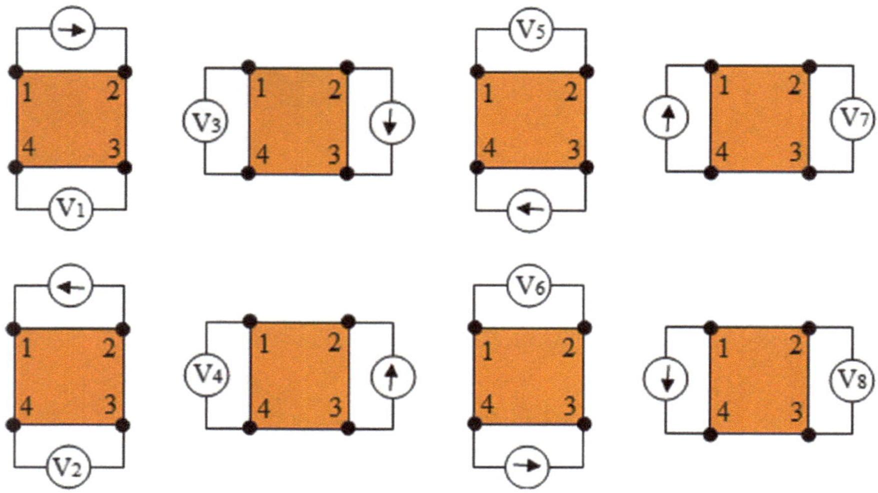

Fig. 9 Van der Pauw resistance configurations

3 Preparation Methods of Conductive Polymer Composites

Available literature reveal that conductive polymer composites can be prepared by several methods. These methodologies depend mainly on the types of polymer, their initial form, and the nature of conductive fillers. Different preparation techniques for polymer/carbon composites are melt mixing technique, solution mixing technique, dry mixing technique, powder mixing technique, aqueous mixing technique, etc. The details of the processing techniques are discussed in Chap. 3.

4 Percolation Theory

It has been mentioned earlier that most of the polymers are mostly insulating in nature. The electrical conductivity of these polymers is due to the addition of conductive filler into the system. The conductivity of this type of system increases drastically at a particular concentration of filler known as percolation threshold. Below this percolation threshold, the conductive system is considered an insulator where the conductive particles are separated from each other in the polymer matrix. At percolation concentration, the filler particles contact each other and continuous conducting networks are formed. The behavior of electrical resistivity with respect to filler concentration has been shown in Fig. 10. Many models and equations have been proposed to understand the behavior of the percolation threshold.

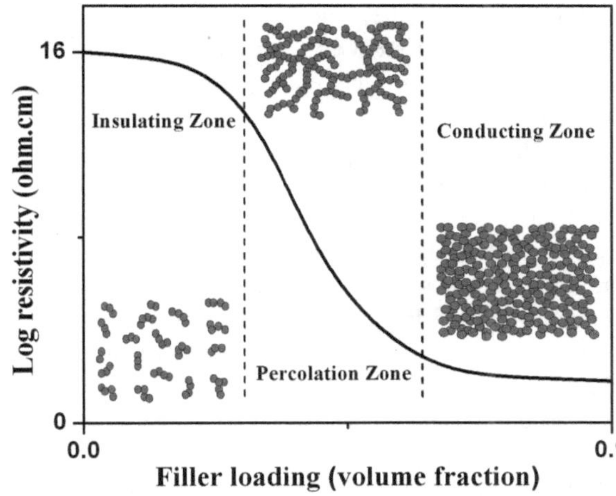

Fig. 10 Behavior of electrical resistivity with respect to filler concentration

The electrical percolation threshold of DC conductivity is predicted from the scaling/power law equation [48, 49]. This can be written as (Eq. 11).

$$\sigma_{\mathrm{dc}} = k(v_{\mathrm{f}} - v_{\mathrm{fc}})^t \quad \text{for} \quad v_{\mathrm{f}} > v_{\mathrm{fc}} \tag{11}$$

where k is a constant quantity known as a pre-exponential factor, σ_{dc} is DC conductivity of composite; v_{f} is volume fraction of filler, v_{fc} is volume fraction of filler at critical concentration/percolation threshold, and t is a critical exponent. The plot of log σ against log $(v_f - v_{fc})$ according to the power law equation should be a straight line. The value of critical exponent t is obtained from the slope of the straight line and the value of pre-exponential factor k is obtained from the intercept. The value of k depends on the conductivity of the fillers, their contact resistance, and the network topology [50], whereas, the value of t depends on the connectivity of the system. The universal value of t for two and three dimensions (2D and 3D) composite systems are $\approx$1.33 and $\approx$2, respectively [51, 52]. However, considerable variation in the experimental value of critical exponent has been observed as have been reported in many literature [48, 49]. The electrical percolation threshold of polymer carbon composites depend on many factors and some of these are mentioned herewith.

Tchoudakov et al. [53] reported the electrical behavior of polypropylene/CB and nylon/CB composites prepared through melt mixing technique. The percolation threshold for PP composites was 2.4 phr, while for nylon composite it was more than 13 phr. This large difference in the percolation threshold is due to the differences in interface energy between PP/CB and nylon/CB composites. Nylon is having better interaction with CB compared to PP. This leads to better dispersion of CB particles within the nylon matrix, which impede in the formation of aggregated structure. Thus, a higher amount of CB is required to reach percolation threshold. A variety of work had been reported on conductive composites prepared through melt mixing technique, where percolation threshold ranges from 2 to 27 vol.% [54–58]. This wide difference in percolation threshold had been attributed to polymer crystallinity and surface energy that largely influence the conductive network formation.

Percolation behavior of CB in epoxy resin was investigated by Schueler et al. [59]. They observed that the Coulombic force between the CB particles acted as a potential barrier that hindered in the formation of agglomeration structure in the liquid polymer. To overcome this, they applied external shear force before curing and added some salt to increase the ionic strength which accelerates the formation of agglomeration leading to a lower value of the percolation threshold. It has been mentioned that the percolation threshold was reduced to 0.06 vol.% of CB by the addition of chlorine–methanol solution with the maximum conductivity near 0.001 S/cm. Investigation on epoxy-based conductive composites was reported in the literature [60–62]. Some literature also reveal that composites made from solution mixing technique exhibited percolation threshold in the range 0.9–3 vol.% [63, 64].

The effect of crystallinity on the percolation threshold was reported on CB-filled PP and ethylene octane copolymer by Huang [65]. The crystallinity of PP and ethylene octane copolymer was 80% and 10–15%, respectively. As the CB cannot reside in the crystalline region, the percolation threshold for PP was found to be 2 vol.%; whereas, for ethylene octane copolymer it was found to be 7 vol.%. The effect of crystallinity on the percolation threshold was also investigated by several other researchers [53, 54, 66].

The percolation threshold is also influenced by the molecular weight of the polymer matrix. The percolation behavior of CB-filled composite with different polymer molecular weights was reported by Sumita et al. [56]. They concluded that the problem of dispersion increases with an increase in molecular weight, which leads to higher percolation threshold for the higher molecular weight of the polymer. The effect of melt flow index on three different EVAs was investigated by Huang and Wu [67]. They had reported that a higher melt flow index with higher crystallinity reduced the percolation threshold.

5 Effect of Different Factors on Electrical Conductivity of Polymer Composites

Electrically conductive composites can be made by incorporating of different conducting fillers in an insulating polymer matrix. However, the different electrical properties of these composites depend on the nature and type of polymer and filler, blend composition, filler concentration, filler dispersion, processing parameter, temperature, etc. The AC electrical conductivity is greatly affected by the frequency of the applied electrical field.

5.1 *Types of Fillers*

Different types of natural and synthetic carbon materials are available in the market such as diamond, fullerene molecules, graphite, carbon blacks, carbon fibers, nanotubes, graphene, etc. In the extrinsically conductive composites, where these conductive additives are incorporated into suitable polymer matrices, the types of these conductive fillers affect the electrical conductivity of such composites.

5.1.1 Diamond

In diamond, each carbon atom is bonded with other four carbon atoms by a strong sp^3 covalent bond as shown in Fig. 11. As there is no conjugated double bond in the backbone of diamond, so its electrical conductivity is very low because all the electrons in diamond are localized and there is no electron present for conduction of electricity. However, the diamond can be made electrically conductive by using dopants. The electrical conductivity of undoped and doped diamond falls in the range 10^{-16}–10^{-2} S/cm with a band gap 1–4 eV [68]. SEM image of the diamond film is shown in Fig. 12 [69].

5.1.2 Fullerene

Fullerene molecule C_{60} is having hexagonal and pentagonal carbon ring in its chemical structure (Fig. 13). The number of hexagonal rings is 20 and the

Fig. 11 Structure of diamond

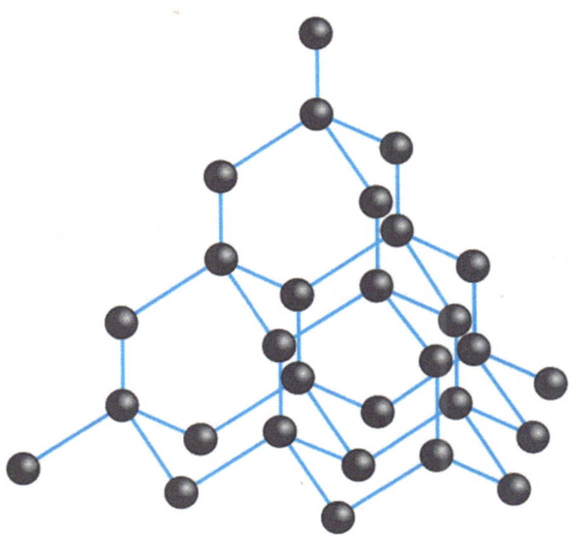

Fig. 12 SEM photo of the as-grown epitaxially oriented diamond film. Reproduced from Ref. [69]

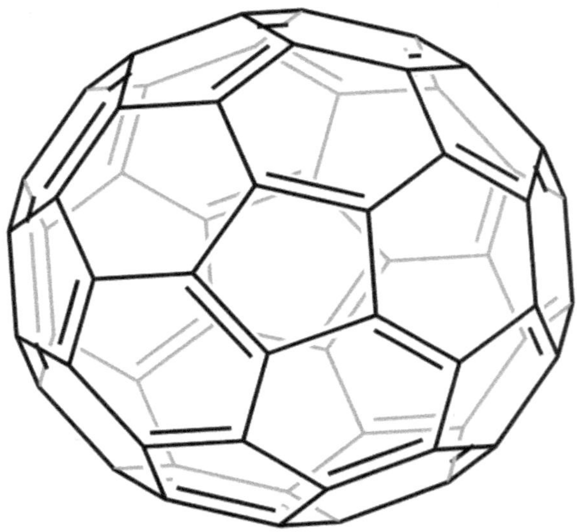

Fig. 13 Structure of fullerene C_{60}

pentagonal ring is 12. It is having icosahedral symmetry closed cage structure similar to football. So it is also called buckyball. There are two bond lengths in the structure of C_{60} molecule: the 6:6 and 6:5 ring bonds. The 6:6 ring bonds is considered as double bond and is shorter in length than 6:5 ring bond. The C_{60} molecule is not super aromatic as its pentagonal ring lacks in double bond. As a result, there is poor electron delocalization in C_{60} molecule and hence behave like an electron deficient alkene. So the electrical conductivity of C_{60} molecule falls within the antistatic/semiconducting range that is 10^{-6} S/cm with a band gap of 1.8 eV [70]. However, high conductivity of C_{60} molecule can be achieved by charge doping [68]. The smallest fullerene is C_{20}. Moreover, some common fullerenes are C_{70}, C_{72}, C_{76}, C_{84}, and even up to 100 carbon atoms are available. The HRTEM image of pure C_{60} fullerene is shown in Fig. 14 [71].

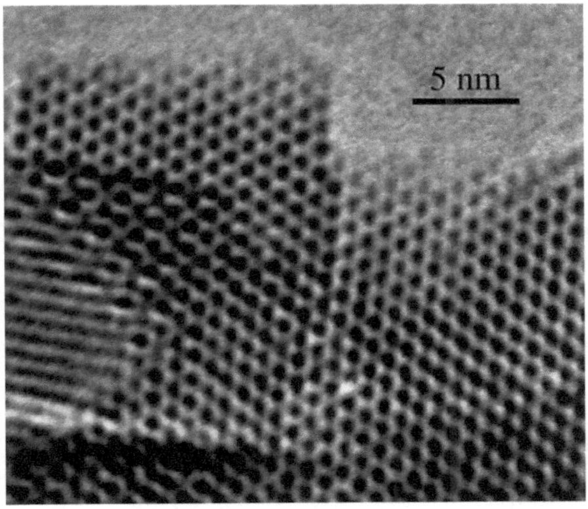

Fig. 14 HRTEM image of fullerene C_{60}. Reproduced from Ref. [71]

5.1.3 Graphite

In graphite, the carbon atoms in each layer are hexagonally linked together with sp^2 hybridization where the layers are stacked parallel to each other (Fig. 15). Each layer forms two-dimensional sheets having a thickness of one atom. The inter-atomic distance within a layer plane in the hexagonal lattice is 1.42 Å and the interlayer distance between the planes is 3.354 C35. Each carbon atom has one out-of-plane π-bond, which is responsible for the delocalization of electrons and accounts for its thermal and electrical conductivity. The adjacent layers are held together by weak Van der Waals force of attraction which accounts for its softness and lubricating nature. Based on the literature, the electrical conductivity of graphite and modified graphite is 2×10^4 S/cm [72], 17–28 S/cm [73], 1.8×10^4 to 6×10^5 S/cm [74], 6×10^4 S/m [75]. A SEM micrograph of exfoliated graphite is shown in Fig. 16 [72].

5.1.4 Carbon Black

Carbon blacks are the most widely used additives, which can be processed easily. Carbon black improves electrical and different physical properties of a composite [76–78]. The desired level of conductivity can be obtained for a black filled composite by simply controlling the amount of carbon black. Different grades of carbon blacks like channel black, thermal black, furnace black, acetylene black, etc., are available in the market according to the manufacturing process and raw material source. The main factors those governing the electrical conductivity of carbon black composites are the particle size of carbon black, structure (aggregating tendency), and degree of graphitization [79–82]. High structure and a higher degree

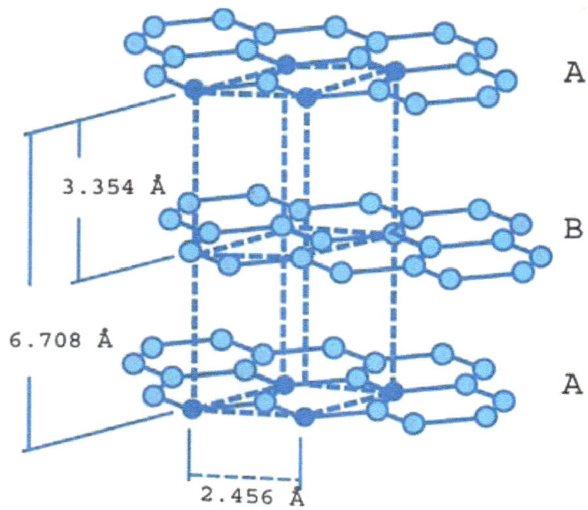

Fig. 15 Mosaic-like crystal particle structure of graphite where layers of stratified planes are hexagonal and parallel to each other

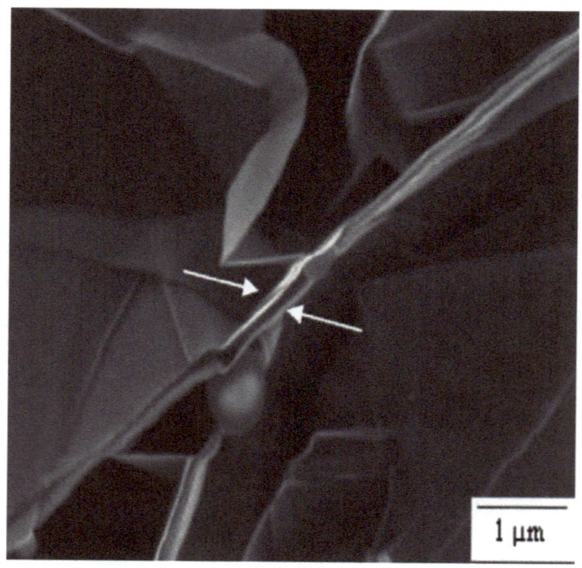

Fig. 16 High magnification SEM micrograph of exfoliated graphite nanoparticles. Reproduced from Ref. [72]

of graphitization favor electrical conductivity, whereas surface adsorbed groups containing oxygen and hydrogen affect the electrical properties somewhat adversely through hindering movement of charged species. However, the presence of such groups on the surface of carbon black is responsible for the enhancement of the physical property of the composites.

The electrical conductivity of carbon black depends on its grade and sources. Nannan Guo and Ming C. Leu reported the electrical conductivity of carbon black

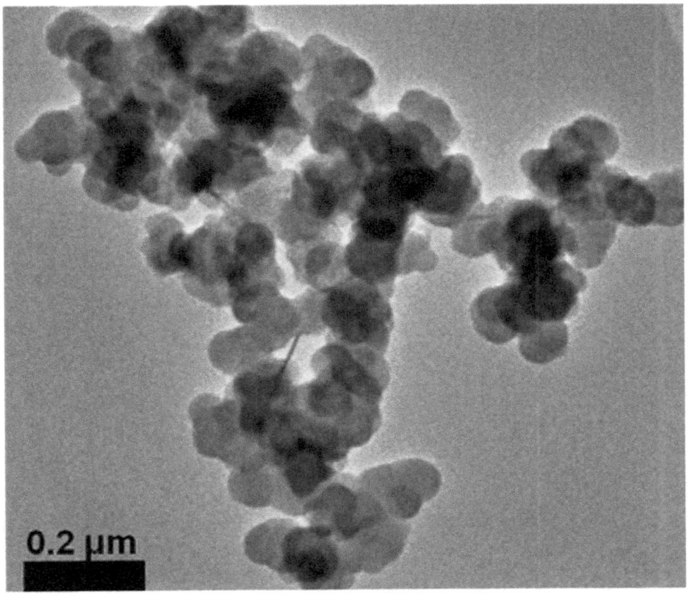

Fig. 17 TEM images (left side) reveal the grape-like aggregation of CB spheres. Reproduced from Ref. [83]

powder (Grade—5303, purchased from Asbury Graphite Mills, Inc., New Jersey, USA) 2.93 S/cm [73]. The intrinsic isolated single-particle conductivity of carbon black was reported as 1×10^6 S/cm [83]. Sánchez-González et al. reported the electrical conductivity of different grades of carbon blacks and showed that the conductivity of blacks varied with their types and compactness [84]. The conductivity increased with the increase in compactness. A TEM image of the carbon black structure is shown in Fig. 17 [83].

5.1.5 Carbon Fiber and Nanofiber

The physical and electrical properties of carbon fiber depend on the precursor and technology of preparation. Generally, these fibers are manufactured by the pyrolysis of organic compound at very high temperatures in an inert atmosphere. Conventional carbon fibers (CCF) are prepared from the precursor like polyacrylonitrile (PAN), pitch, viscous jute, etc. A carbon fiber may be considered as a macromolecule. It is composed of both ordered and disordered structure, for example, graphite-like layered structure with some less ordered polyene and cyclic defect structure. The idealized macromolecular structure of PAN-based carbon fiber obtained after the process of stabilization and subsequent carbonization has been proposed as shown in Fig. 18 [85]. The cross-linking of the chain, aromatization,

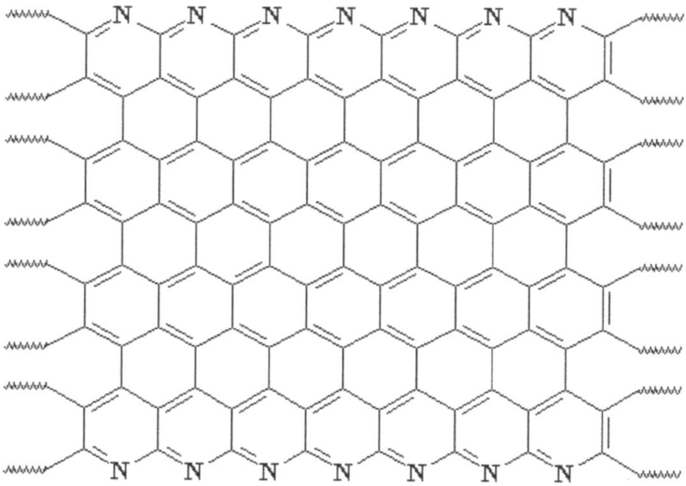

Fig. 18 An idealized macromolecular structure of PAN-based carbon fiber. Reproduced from Ref. [85]

and graphite-like structure of the element is obtained after carbonization of stabilized fiber. Carbon nanofiber (CNF) is prepared by catalytic thermal chemical vapor deposition growth and electrospinning followed by heat treatment from different precursor [86]. CCF has a diameter in the micrometer range whereas CNF has a diameter in the nanometer range. Its diameter is larger than CNTs. The length of CNF is comparable with CNTs but smaller than CCF. CNFs are hollow core nanofibers comprises of a single layer or double layer of graphite planes stacked parallel to each other or at a certain angle from the fiber axis. The structure, dimension, mechanical and electrical properties of carbon fibers depend on its production technique and posttreatment method. The electrical conductivity of carbon fiber and nanofiber are $\approx 6 \times 10^2$ S/cm and 1×10^4 S/cm, respectively [87]. In another review article, the electrical conductivity of CNF was mentioned as 5×10^2 S/cm at high temperature and 2×10^4 S/cm at room temperature [88]. A TEM image of vapor grown carbon nanofiber with a cylindrical hollow core structure is shown in Fig. 19 [88].

5.1.6 Carbon Nanotubes

CNTs can be thought of as graphene sheets rolled up into a tube-shaped (cylinder) structure. This graphene sheet is made up of planar parallel sheets of carbon atoms arranged in hexagonal aromatic rings, which are covalently bound by SP^2 hybridization [89]. CNTs can exist as single-walled nanotubes (SWNTs) or multi-walled nanotubes (MWNTs), depending on the number of rolled up

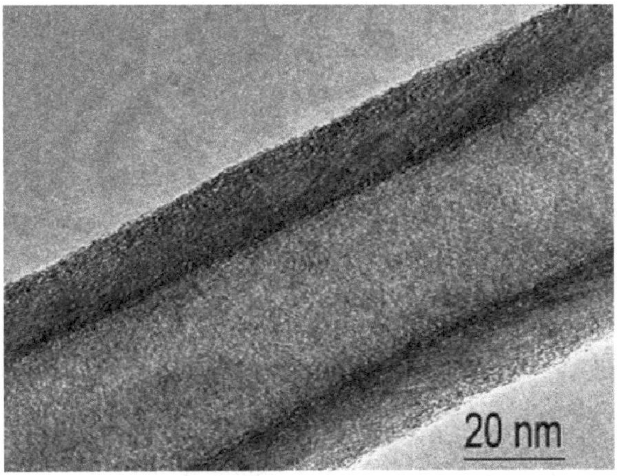

Fig. 19 TEM micrograph of VGCNF. Reproduced from Ref. [88]

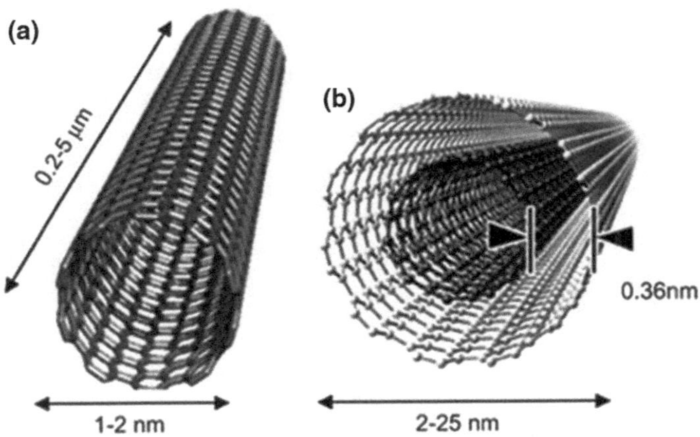

Fig. 20 Schematic diagram of **a** single wall carbon nanotube (SWNT) and **b** multi-walled carbon nanotube (MWNT). Reproduced from Ref. [92]

graphene sheets (Fig. 20). The diameter of the nanotubes can vary from 0.7 to 2 nm (SWNTs) to 10–300 nm (MWNTs) and the length of the nanotubes can reach into macroscale [90]. CNTs have many impressive properties such as electrical conductivity, mechanical strength, optical transparency, and thermal stability, which make them suitable for current and future applications in nanotechnology [91].

One of the reasons that CNTs have such good properties is because they have a symmetrical structure [93]. Individual CNTs vary in their conformational arrangements as a result of the orientation of the graphene sheets in the CNT structure. The most common description of this orientation is the chiral vector (C_h) joining two equivalent points on the original graphene lattice. The vectorial equation is given as (Eq. 12) [89]:

$$C_h = na_1 + ma_2 \tag{12}$$

where a_1 and a_2 are the distance between the nearest equivalent points on a graphene sheet and the integers n and m indicate the number of zigzag steps along the graphene sheet. There are three different orientations, depending on the values of n and m, of the graphene sheet within the structure of CNTs: armchair, zigzag, and chiral (Fig. 21). An armchair nanotube results when $n = m$; a zigzag nanotube when $m = 0$; and chiral nanotubes arise when $n \neq m \neq 0$ [89]. The length and direction of the chiral vector (the values of n and m) will clearly influence the diameter of nanotubes and hence influence their properties including conductivity and density [94]. When $(n - m)/3$ is a whole number the nanotube will be metallic in nature, i.e., electrically conducting. Otherwise, a semiconducting nanotube is resultant. Therefore, two-thirds of nanotubes will be semiconducting and one-third conducting [95].

The different value of electrical conductivity of CNTs was mentioned in different literature. The conductivity of MWCNTs on the basis of literature were 1×10^4 S/m [96], 23 S/m [97], 2×10^2 S/m [98], 2×10^4 S/m [99], and 572 S/m [100]; whereas, the conductivities of SWCNTs were 5.1×10^4 S/m [101], 4×10^3 S/m [102], 3×10^4 S/m [103].

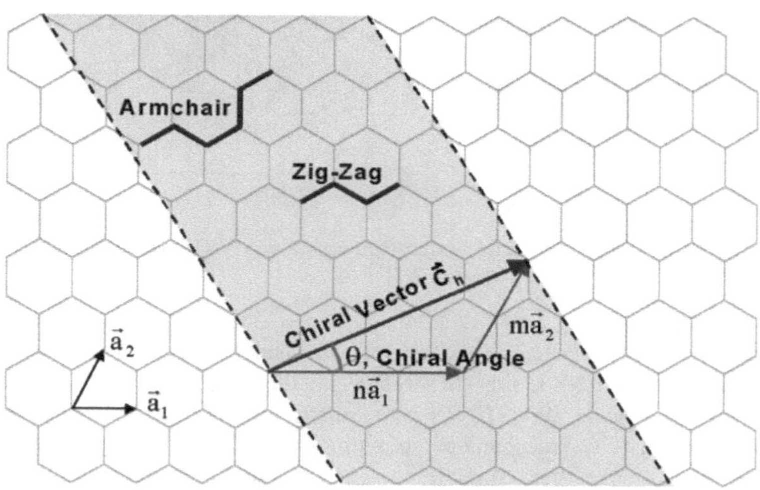

Fig. 21 Schematic diagram showing how a graphene sheet is rolled to form a CNTs. Reproduced from Ref. [93]

5.1.7 Graphene

Graphene is one-atom thick, monolayer, sp^2 hybridized, hexagonal planar carbon sheet (Fig. 22), which is believed as the thinnest and strongest material with exceptional physical and chemical properties. The carbon–carbon bond length in graphene is approximately 0.142 nm [104] and the layer thickness ranges from 0.35 to 1.0 nm [105]. The electron mobility in graphene at room temperature is 200,000 $cm^2\ V^{-1}\ s^{-1}$ [106]. The electrical conductivity of graphene depends on its precursors, method of synthesis, and measurements, the number of stacking layers, level of oxide etc. As for example, graphene oxide reduced chemically by hydrazine resulted in an electrical conductivity of 5–200 S/m [107]; whereas, thermally reduced graphene oxide gave much higher conductivity (8×10^3 to 2×10^5 S/m) [108]. The intrinsic single-particle conductivity of graphene was mentioned as 10^7–10^8 S/m [83]. A SEM image of graphene is shown in Fig. 23 [109].

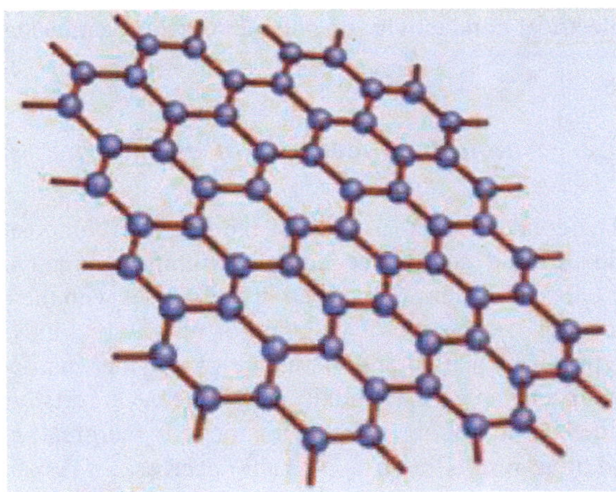

Fig. 22 Hexagonal planar monoatomic graphene sheet

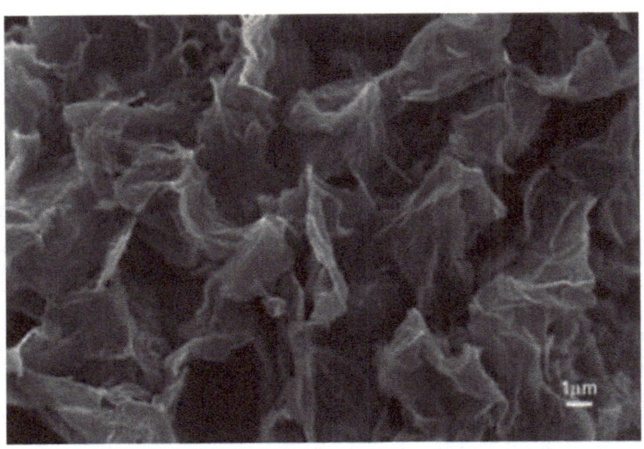

Fig. 23 SEM image of graphene sheets. Reproduced from Ref. [109]

The intrinsic individual inherent particle conductivity of these carbon materials depends on their purity, source of materials/precursors, shape, size, crystallinity, etc. However, their bulk conductivity is governed by their degree of compactness, alignment, the direction of measurement, etc. The bulk conductivity was reported to be always lower than their inherent conductivity because of contact resistance [83]. Compactness increases the bulk conductivity for all carbons [83]. Among the carbons, the lowest inherent conductivity was observed for diamond. The highest inherent conductivity was reported for graphene compared to CNTs [83, 110].

The electrical conductivity of polymer/carbon composites filled with fullerenes [111], graphite [72, 75], carbon blacks [112–114], carbon fibers [86–88], carbon nanotubes [115–117], and graphene [118, 119] were extensively reported in some review and research articles. It is revealed from these articles that the electrical conductivity of the composites not only depends on the inherent conductivity of the individual carbon but also on their types, functionality, geometry, dispersion in polymer matrix, nature of polymer, crystallinity in polymer, blending of polymers, processing methods of composites, etc. Hence, these affecting parameters on the electrical conductivity of polymer/carbon composites are discussed herewith.

5.2 Amount/Concentration of Fillers

The conductivity/resistivity of the polymer composites depends on the concentration of the conductive additive within the matrix polymer. Generally, for all composite systems, the resistivity decreases with the increase in conductive additive loading. The nature of variation of resistivity with filler concentration depends on types of filler used and the nature of polymer matrices. In most of the literature, it has been shown that initially, the decrease in resistivity with the addition of carbon fillers in insulating polymer matrices is marginal. Further addition of conductive additive results in an appreciable decrease in resistivity and then again the resistivity decrement becomes marginal. However, some other literature it was reported that initially there is a sharp decrease in resistivity is observed followed by a marginal decrease up to the highest loading of conductive additives. The first one is observed for low structured/low aspect ratio carbon fillers whereas the later one is observed for high structures/high aspect ratio carbon fillers. We have reported the variation of DC resistivity against carbon black concentration, where EVA and NBR are used as polymer matrices, and Conductex and Printex blacks as conductive additives as shown in Figs. 24 and 25, respectively [120]. Conductex carbon black filled composites show initially marginal decrease in resistivity up to 20 phr loading and then the decrement in resistivity become sharp up to 40 phr loading followed by an again marginal decrease in resistivity up to its highest concentration. However, for Printex black filled composites, initially, there is a sharp decrease in resistivity and beyond 20 phr of loading the decrement in resistivity becomes marginal. This decrement in electrical resistivity with the addition of conductive carbon fillers in insulating polymer matrices was explained

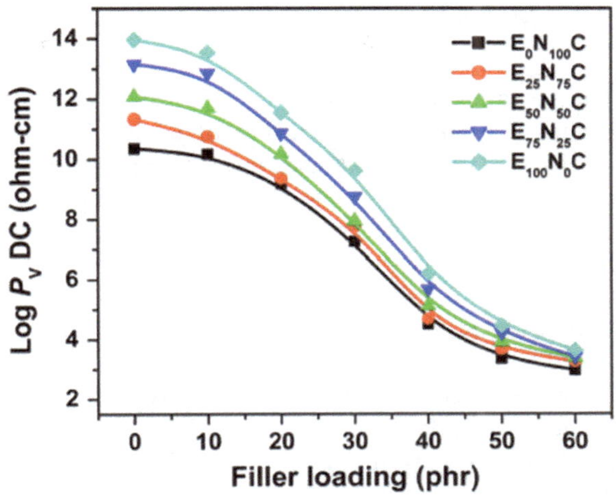

Fig. 24 DC resistivity versus filler loading of EVA/NBR blend composites filled with Conductex carbon black. Reproduced from Ref. [120]

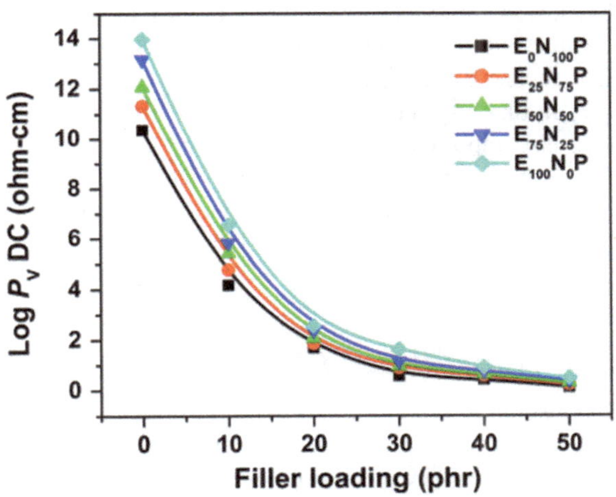

Fig. 25 DC resistivity versus filler loading of EVA/NBR blend composites filled with Printex XE2 carbon black. Reproduced from Ref. [120]

in terms of the general theory of electrical conduction mechanism. Initially, the addition of conductive additive results in the formation of discrete aggregates in the insulating polymer matrix. On further addition, these discrete aggregate grows in size and leads to the formation of conductive continuous networks at a certain concentration known as the percolation threshold of electrical conductivity. The addition of more conductive additive beyond this concentration only increases the number of such conductive continuous networks and forms the conductive mesh. So the increase in electrical conductivity becomes marginal. The effect of conductive additive loading on the electrical conductivity of insulating polymer matrices was also reported in many literature and observed almost the same phenomenon [121–124].

5.3 Structure of Fillers

The percolation threshold and electrical resistivity of the polymer/carbon composites depend on the structure of conductive additives, especially for carbon blacks. Carbon black particles form strong aggregates (primary structure) and these aggregates are loosely clustered into bigger aggregates (secondary structure). The aggregate with more complex shape, size, and internal voids are termed as the higher structure and vice versa. The structure can be indirectly measured through the quantity of dibutyl phthalate (DBP) absorbed by aggregated carbon black. Carbon black with higher structure can form more easily continuous conductive network in an insulating matrix. In our previous investigation mentioned earlier where we used two different conductive carbon blacks namely Conductex and Printex that differ from each other in terms of their structure values [120]. The matrix polymers were EVA, NBR, and their blends. It is shown that Printex black has a higher structure and exhibited a lower percolation threshold compared to those of Conductex black (Figs. 24 and 25). The percolation threshold for Printex black was around 25 phr and that of for Conductex black was 45 phr. The electrical conductivity of Printex black filled composites is also higher compared to Conductex black filled composites at their similar black loading. The effect of carbon black structure on the electrical conductivity of polymer carbon composites was also reported in the literature and shown that highly structured carbon black favors the electrical conductivity which means it exhibited lower percolation threshold with higher electrical conductivity at the same filler loading compared to low structured black one [113, 125, 126].

5.4 Aspect Ratio of Fillers

The aspect ratio of rod-like carbon fillers like fibers, nanofibers, nanotubes affects the electrical conductivity/resistivity of polymer/carbon composites. Aspect ratio is defined as the ratio of length to diameter of any rod-like carbon fillers. It has been reported that the electrical resistivity of any polymer/carbon composites decreases with an increase in the aspect ratio of carbon filler at their any particular loading [127]. It has been mentioned in the literature that high aspect ratio carbon filler is equivalent to the combination of several low aspect ratio carbon fillers. As a result, the average distance as well as the number of contacts between the fillers particles decreases with an increase in aspect ratio. Hence, the flow of electrons through hopping is facilitated as both the average distance and numbers of contacts among the particles are decreased, resulting in higher electrical conductivity of the composites filled with higher aspect ratio carbon particle.

We reported the effect of aspect ratio of carbon nanotubes on the electrical resistivity of polyethylene nanocomposites and the same is presented in Fig. 26 [127]. It is revealed that the composites with a high aspect ratio of CNTs

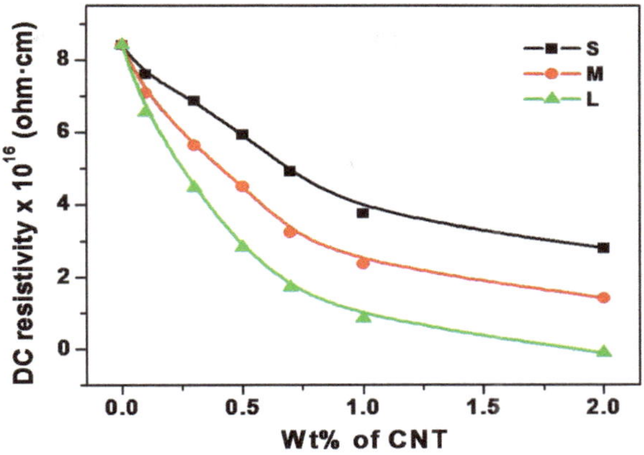

Fig. 26 Effect of aspect ratio on electrical resistivity of LDPE/CNT composites. Reproduced from Ref. [127]

(L) showed lower electrical resistivity at all CNT loading compared to composites with a low aspect ratio of CNTs. Similar observations have also been made by several authors [128, 129]. Guo et al. [129] reported that the higher is the aspect ratio of CNT, the lower is the percolation threshold and electrical resistivity. This is also true for other rod-like/tube-like carbon filled polymer composites [116].

5.5 *Functionalization of Carbons*

The functionalization of carbon materials affects the electrical conductivity of polymer/carbon composites. It has been mentioned earlier that the conductive carbon fillers are associated with a conjugated double bond in its backbone structure that assists in the flow of electrons. The functionalization of carbon materials results in the opening of double bonds. Thus, the number of double bonds in their conjugation is reduced and in some cases, the conjugation is lost. Thus, the flow of electrons is hindered in their path and results in lowering of electrical conductivity. Another important point is that the functionalization of carbons by any acid or amine fatty acids or by other means increases the distances among the carbon particles. It has been mentioned in the literature that the hopping of electrons from one conductive site to another is facilitated when the average distances among the particles $\approx$10 Å [120]. Thus, functionalization reduced the hopping of electrons by increasing the average distances among them and results in the decrease in electrical conductivity of polymer/carbon composites.

Selvin et al. have studied the effect of 1-octadecanol functionalization of carbon nanotubes (C18-CNT) on the electrical resistivity/conductivity of natural rubber (NR) nanocomposites as shown in Fig. 27 [130]. It is seen that the functionalized NR/C18-CNT composites exhibit higher electrical resistivity compared to NR/CNT composites. The effect of functionalization of carbons on the electrical resistivity of polymer/carbon composites was studied by several authors [109, 119, 131–133].

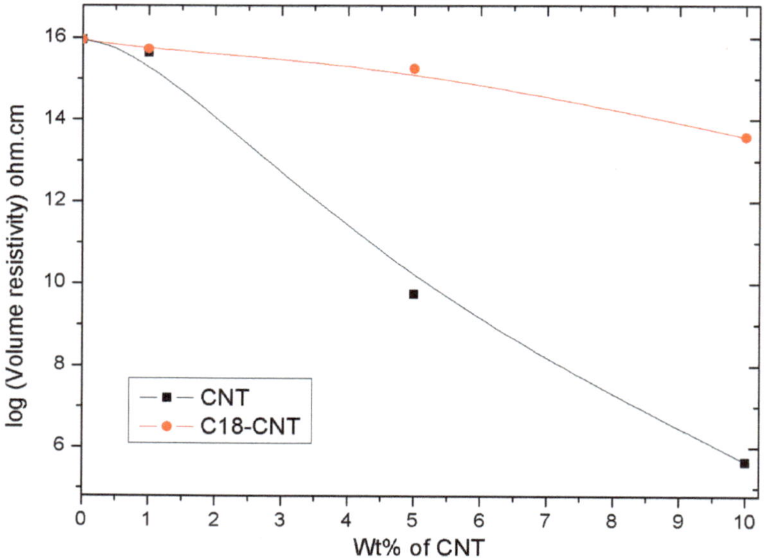

Fig. 27 DC volume resistivity versus filler loading of NR composites. Reproduced from Ref. [130]

5.6 *Filler Orientation and Waviness*

The orientation/alignment of rod-like that means anisotropic fillers greatly affects the electrical percolation threshold and electrical conductivity of polymer/carbon composites. It has been mentioned in the literature that the alignment of anisotropic fillers leads to the fewer contacts between them resulting in the destruction of conductive networks. Consequently, the conductivity of polymer composites is drastically reduced [116]. The low percolation threshold of electrical conductivity is obtained when the alignment of fillers is isotropic or slightly anisotropic in nature in the polymer matrices. However, this percolation threshold value increases when the alignment of fillers becomes more anisotropic which means it is uniaxial in nature. Thus, there is a critical alignment above which the electrical percolation threshold value increases even at volume fraction value of fillers well above the percolation threshold of the isotropically oriented network. Above this value, there is the destruction of percolative conductive networks within the polymer matrix. It has been reported that this critical alignment value increases with the increase in volume fraction and aspect ratio of fillers [134].

Khan et al. [135] studied the effect of multi-walled carbon nanotubes alignment on the electrical conductivity of epoxy nanocomposites as shown in Fig. 28. They measured the electrical conductivity of randomly oriented, parallelly aligned, and perpendicularly aligned composites. It was shown that the percolation threshold of parallelly aligned composites (0.0031 vol.%) was one order of magnitude lower

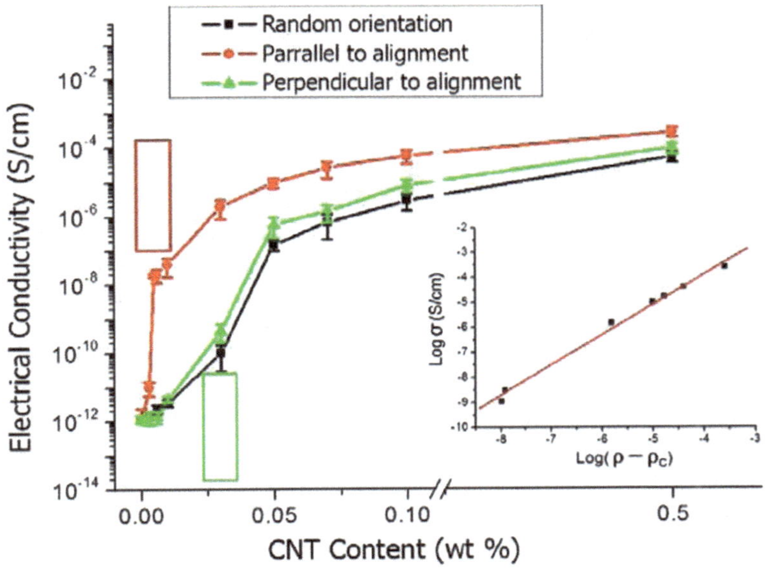

Fig. 28 Electrical conductivity (σ) of nanocomposites as a function of CNT content (ρ). Inset: log σ plotted against log ($\rho - \rho_c$) giving the percolation threshold of $\rho_c = 0.0048$ wt% (equivalent to 0.0031 vol.% assuming the densities of CNT and the matrix material to be 1.68 and 1.11 g/cm^3, respectively). Reproduced from Ref. [135]

than the perpendicular and randomly oriented composites (0.034 vol.%). This was due to preferential orientation of the parallelly aligned composites. This was attributed to the fact that upon the application of electric field, the CNTs aligned immediately and start the lateral migration to form bundles of even at their low concentration. This results in the formation of extensive conductive percolative networks in the direction of alignment and the similar networks are not formed in the perpendicular direction of alignment at the same CNT content. It can be noted that the percolative networks in the perpendicular direction start to form only when the percolative network in the parallel direction are developed. The migration of CNTs in the perpendicular direction increases with the increase of its concentration in the polymer matrix. Hence, the difference in absolute conductivity is diminished at a higher volume fraction of CNTs. The results also showed that the electrical percolation and conductivity of randomly oriented and perpendicularly aligned composites was almost the same; indicated that the significant portions of CNTs took part in the formation of the percolative network in the direction of perpendicular to the alignment.

A little bit different behavior is observed for polyurethane/reduced graphene oxide nanocomposites [136]. The percolation threshold and electrical conductivity increases in the parallel direction of alignment. However, in the perpendicular direction, initially, there is the increase in electrical conductivity up to the percolation threshold, and thereafter it falls abruptly above the percolation threshold. This was attributed to the hindrance of electron transport due to the presence of some oxides on the surface of graphene sheets.

In many theoretical studies of rod-like fillers, it was assumed that the geometry of fillers was rigid, straight cylindrical or sphero-cylindrical. This assumption is in good agreement with the nanowires for predicting any physical properties. However, this assumption is not logical for carbon nanotubes based polymer composites because the microscopic study was already revealed that the CNTs embedded in polymer composites are curved/wavy rather than straight as can be seen from Fig. 29 [137]. Several authors theoretically studied the effect of waviness on the electrical percolation and conductivity of polymer/CNTs nanocomposites [137–139]. They reported that the electrical percolation threshold of the composites increases with the increase in waviness of fillers and this effect is more pronounced for low aspect ratio fillers. Simulation study also proved that the electrical conductivity is decreased with the increase in waviness of the CNTs within the polymer matrices [137, 138].

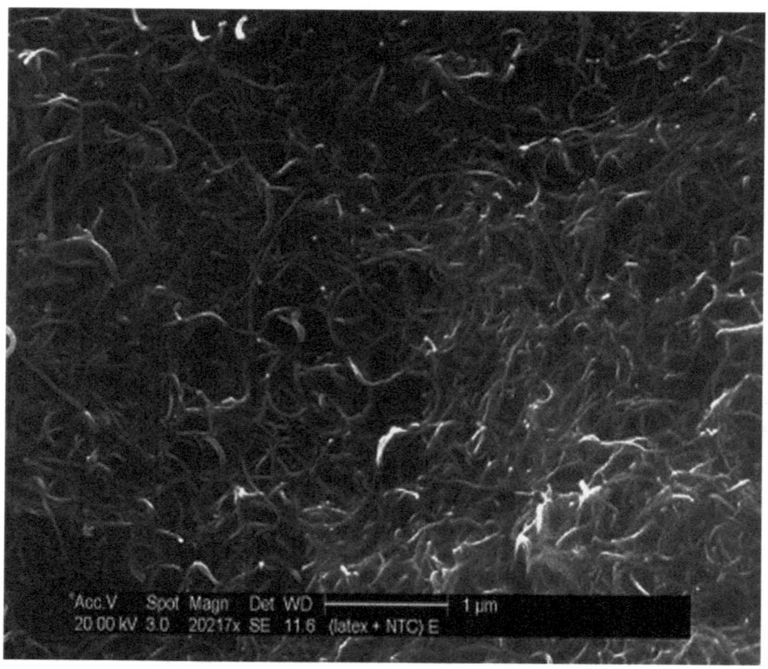

Fig. 29 FESEM image of a MWCNT/polymer nanocomposite showing the nanotube waviness. Reproduced from Ref. [137]

5.7 Polymer Matrix

The electrical properties of conductive composite systems depend on the type of polymer matrix. Studies on the electrical conductivity of different rubber vulcanizates filled with carbon filler have been carried out [140, 141]. It has been observed that different volume fraction of carbon filler is required to get the same level of conductivity for all the composite systems. A similar type of observations has also been made by other authors [54, 55]. They have pointed out that the surface tension of polymer plays an important role in the dispersion of fillers particles in the polymer matrix and hence is responsible for obtaining different values of electrical properties. It has been mentioned that more volume fraction of filler is required to get a desired level of conductivity if the surface tension of the polymer matrix is increased. The presence of the polar group in the backbone of the polymer matrix also affects the dispersion of filler particles in the polymer matrix. There is a report regarding the effect of polymer viscosity on electrical conductivity for the conductive composite system [142]. The high viscosity of polymer exerts a more shearing force on the filler particles during mixing which results in more breakdown of conducting filler. This reduces the electrical conductivity of the system. A researcher has reported the effect of MFI (melt flow index) and crystallinity on the percolation threshold of electrical properties of the polymer [67].

Pöschke et al. [143] have studied the electrical percolation behavior of polycarbonate (PC)/MWCNTs nanocomposites prepared by melt mixing technique. They reported the percolation threshold of electrical conductivity was reached at 1.44 wt% of CNTs. In another study, it was reported that the percolation threshold of electrical conductivity for polypropylene (PP)/MWCNTs nanocomposites prepared by melt mixing technique was 1.10 wt% of CNTs [144]. This lower value of percolation threshold for PP nanocomposites was due to higher crystallinity of PP, where the CNTs distribute more to the amorphous phase and results in lower percolation threshold. Thus, the electrical percolation thresholds vary according to the polymer matrix used.

5.8 Polymer Blend

The blending of two or more polymers is of great significance for achieving tailor-made properties [140–142, 145]. Plastic–plastic or plastic–rubber or rubber–rubber blending having different polarities of the constituents was proved to be important for achieving a high degree of conductivity at a relatively lower concentration of filler compared to the individual one [53, 146]. This type of blending provides a well-defined interface, where filler particles accumulate and facilitate the formation of conducting networks more easily. The conductivity of the black filled blending systems depends on the black concentration in the blend, morphology of the blend, and distribution of black particles within the blend [147, 148]. Polymer blends are classified into different categories; namely, miscible, immiscible, compatible, and incompatible. It is expected that polar–polar or non-polar–non-polar polymer blends are miscible or compatible; whereas polar–nonpolar blends are immiscible or incompatible. For mechanical property improvement, polar–polar or nonpolar–nonpolar (miscible/compatible) polymer blends are desirable where the formation of homogeneous interface takes place. On the other hand, for electrical property improvement, the generation of the heterogeneous interface is required and hence, immiscible/incompatible polymer blend can be preferred [149]. Generally, in the polymer blend, the filler particles tend to concentrate in one phase or in between the interface of two polymers. This finally leads to an increase in electrical conductivity [150].

We reported the effect of polymer blending at a different proportion on electrical resistivity of EVA and NBR filled with 60 phr Conductex (ENC_{60}) and 50 Printex (ENP_{50}) blacks as shown in Fig. 30 [120]. The results showed that NBR and NBR-rich composites exhibit lower resistivity compared to EVA and EVA-rich ones. This was attributed to the lower viscosity of NBR compared to EVA at their mixing temperature, which resulted in less breakdown of carbon particles and hence lower electrical conductivity.

5.9 *Effect of Temperature*

Generally, polymer-based conductive composites can exhibit two types of variation in resistivity when subjected to temperature variation known as positive temperature coefficient (PTC) of resistivity and negative temperature coefficient (NTC) of resistivity [151]. In NTC effect the resistivity decreases with the increase in

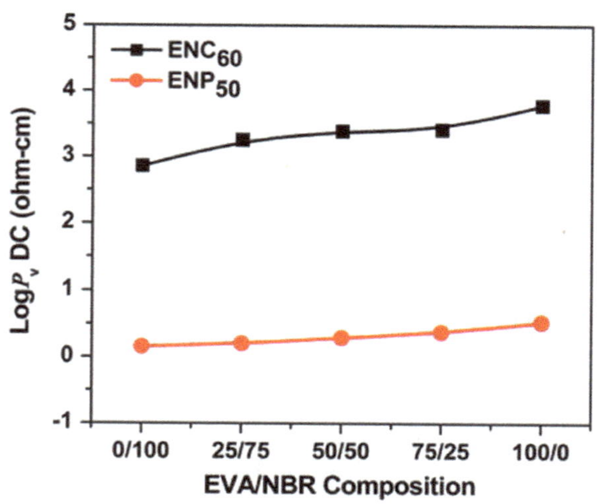

Fig. 30 DC resistivity versus blend compositions of ENC60 and ENP50 composites. Reproduced from Ref. [120]

temperature due to the increased effect of thermal emission of an electron between adjacent filler particles generally observed in the case of semiconducting materials [147, 152]. In PTC effect, the resistivity is increased with the increase in temperature due to the uneven thermal expansion of the polymer matrix and the filler particles. The uneven thermal expansion leads to an increase in the average gap between filler particles/filler aggregates making tunneling or hopping of electrons more difficult [148]. In fact, the temperature dependency of composites depends on various factors like the type of polymer and filler particles, concentration of filler, Tg of polymer, the crystallinity of polymer, etc. [153–156]. Sometimes, extremely high positive temperature coefficient (HPTC) in some composites has also been observed under certain conditions [157]. The crystalline and noncrystalline polymers filled with the same type of carbon black behave differently when subjected to a rise in temperature. The noncrystalline polymer composites exhibit either PTC or NTC effect depending on the nature of polymer and filler; whereas, crystalline polymer may exhibit both the PTC and NTC effect due to the melting of the crystalline phase. Literature regarding thermoplastic–carbon black composites reported a several fold increase in resistivity at and around Tg of the polymer [158, 159]. It is noteworthy to mention here that the mechanism of electrical conduction at low temperature is mainly due to tunneling or hopping of electrons. Carbon fiber filled polymer composites generally exhibit PTC effect due to the large difference between thermal expansion coefficients of polymer and fiber. Polymer/rubber exhibits a positive thermal expansion coefficient in the range 50–220 $\times 10^{-6}$; whereas carbon fiber exhibits negative thermal expansion coefficient (-1.45×10^{-6}) [160]. Metal filled polymer composites may show PTC effect (metallic type conduction) or NTC effect (semiconducting type) depending upon the concentration of filler particles and level of the oxide coating on the metal particles [161].

He et al. have observed the PTC effect of resistivity for 5.4 wt% MWCNT/HDPE composite and 16 wt% carbon black/HDPE composites as reported in Fig. 31, respectively [162]. It has been shown that the electrical resistivity of both composites increases up to the melting point of the composites (Fig. 31) with the increase in temperature. This PTC behavior has been attributed to the thermal expansion of HDPE due to its crystalline melting. Beyond the melting temperature, there is the NTC effect of resistivity for both the composites. However, this effect is more pronounced for HDPE/CB composite compared to HDPE/MWCNT composites. This NTC effect of resistivity is due to the agglomeration of conductive particles because of polymer melt at which the chain segmental motion of polymer takes place. As the geometrical size of CNT is larger than carbon black, hence more energy is required for the agglomeration of CNT compared to carbon black. This is why HDPE/MWCNT composite exhibits weak NTC effect compared to HDPE/CB composite.

The NTC effect of resistivity can be eliminated or minimized by cross-linking/vulcanization of the polymer composites. The authors irradiated the HDPE/MWCNT composites with gamma-ray at room temperature using the dose 80 kGy and the resultant plots are shown in Fig. 32 [162]. It is observed that the NTC effect

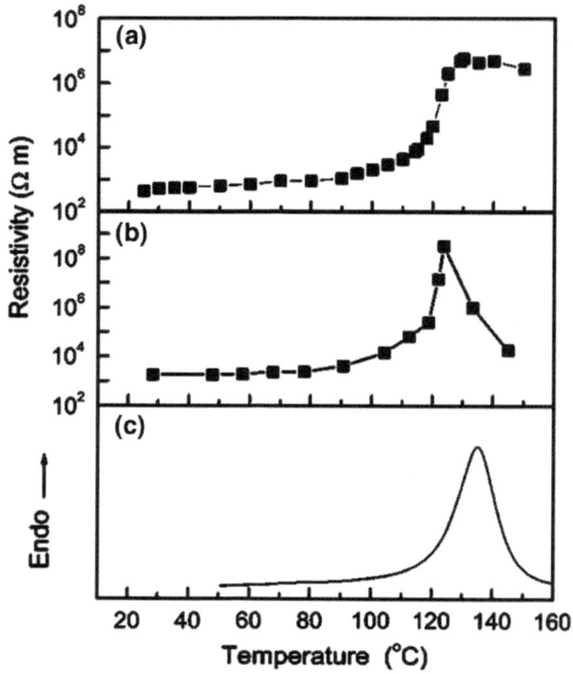

Fig. 31 Comparison of resistivity–temperature curve for **a** 5.4 wt% MWNT/HDPE composite and **b** 16 wt% CB/HDPE composite; **c** DSC curve of the 5.4 wt% MWNT/HDPE composite. Reproduced from Ref. [162]

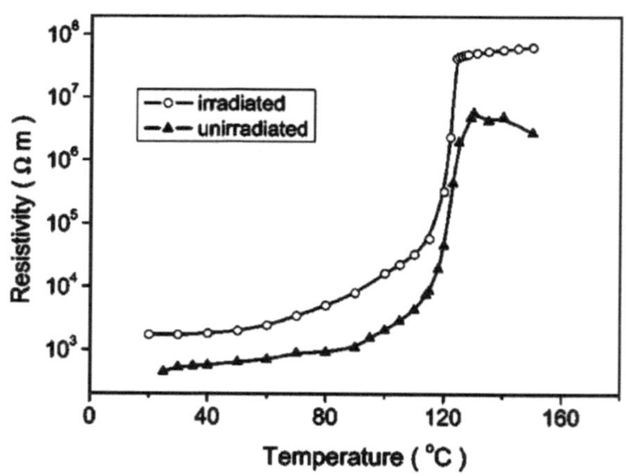

Fig. 32 Resistivity–temperature curves of the 5.4 wt% MWNT/HDPE composite before and after irradiation. Reproduced from Ref. [162]

of resistivity beyond the melting temperature has been eliminated after cross-linking the composite through irradiation. The cross-linking of polymer makes restriction in the movement of its chains and hence the further agglomeration of conductive particles becomes more difficult; as a result, NTC effect of resistivity is depressed.

The PTC and NTC behavior of resistivity of ethylene vinyl acetate copolymer (EVA), acrylonitrile butadiene rubber (NBR), and their blend composites filled with conductive carbon blacks and short carbon fiber are investigated by our group [163]. It has been shown that the PTC and NTC behavior of resistivity depends on the crystallinity of polymer, characteristics of polymer, their melting temperature, types and nature of carbons used, differential thermal expansion coefficient of polymer and fillers, and their curing characteristics. Double PTC behavior for EVA/ Carbon composites has been reported herewith. Vulcanization of polymer minimizes the NTC characteristics of the polymer composites beyond their melting points. The effect of heating and cooling, and repeated heating and cooling on resistivity are also reported. It was shown that the cooling path of resistivity does not follow its heating path and there is always an electrical hysteresis with the existence of an electrical set. This difference in electrical resistivity in the heating and cooling path is due to the formation of some additional conductive networks during a heating cycle with the simultaneous destruction of some existing conductive networks. If the formation of networks dominates the destruction of networks then resistivity in the cooling path will be lower compared to the heating path and vice versa. PTC and NTC behavior of resistivity was also reported for HDPE/ Carbon black composites [164]. Jiyun Feng and Chi-Ming Chan showed that double PTC behavior of resistivity can be achieved by blending of two different semicrystalline polymers [165, 166].

5.10 *Effect of Pressure*

The electrical properties of conductive composites are affected by the application of stress or strain. Conductivity may increase/decrease depending upon the nature of the polymer and conducting filler particles, and concentration of filler. It has been mentioned that the electrical resistivity of non-conducting rubber composites decreases with the increase in pressure where the electrical conduction is mainly due to the ionic charge carrier [157]. However, the polymer composites filled with any conductive particles where the electrical conduction is governed by the presence of electrons or hole, the resistivity is found to decrease with the increase in pressure. Similar investigations for metal/carbon black filled composites have been reported in literature [167, 168]. The development of pressure-sensitive polymer composites has also been a subject of interest for some authors [169, 170]. Generally, the variation of electrical conductivity against applied pressure is nonlinear in nature [171].

Das et al. have studied the effect of applied pressure on electrical resistivity of ethylene vinyl acetate copolymer, ethylene propylene diene monomer (EPDM), and their blend filled with carbon black and carbon fiber [172] as shown in Fig. 33 and 34, respectively. The plots show that the pressure effect of electrical resistivity is dependent on the magnitude of applied pressure, type of polymers, and nature of fillers. EPDM/carbon black composite (EP 0.50 V) showed an increase in electrical

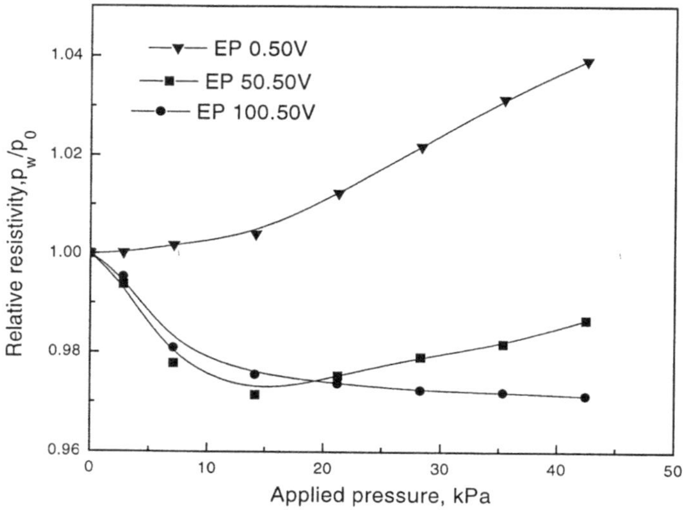

Fig. 33 Effect of pressure on the volume resistivity of carbon black filled EVA, EPDM, and its 50/50 blend composites. Reproduced from Ref. [172]

resistivity, EVA/carbon black composite (EP 100.50 V) showed decrease in electrical resistivity, and EVA/EPDM/carbon black 50/50 blend composite (EP 50.50 V) showed initially a decrease in electrical resistivity followed by an increase after certain pressure. On the contrary, all composites filled with carbon fiber showed an increase in electrical resistivity with the increase in applied pressure (Fig. 34). It has been mentioned in the literature that the application of pressure creates the movement of polymer chains, which, in turn, affect the conductive networks by the formation of some additional conductive networks and simultaneous destruction of some existing conductive networks [172–174]. If the formation of conductive networks is dominant over the destruction of some existing networks then the resistivity will decrease with the increase in applied pressure and vice versa. For carbon black filled composites (except EPDM one), the formation of conductive networks is the dominant factor and hence resistivity decrease with the increase in applied pressure. The effect of pressure loading and unloading on resistivity was reported in the literature [175]. The resistivity of the unloading path did not coincide with their loading path and there was an electrical hysteresis where there exists an electrical set.

5.11 Processing Conditions

Different processing parameters like mixing time, mixing temperature, rotor speed, mold pressure, forming process, and vulcanization affect the mechanical and

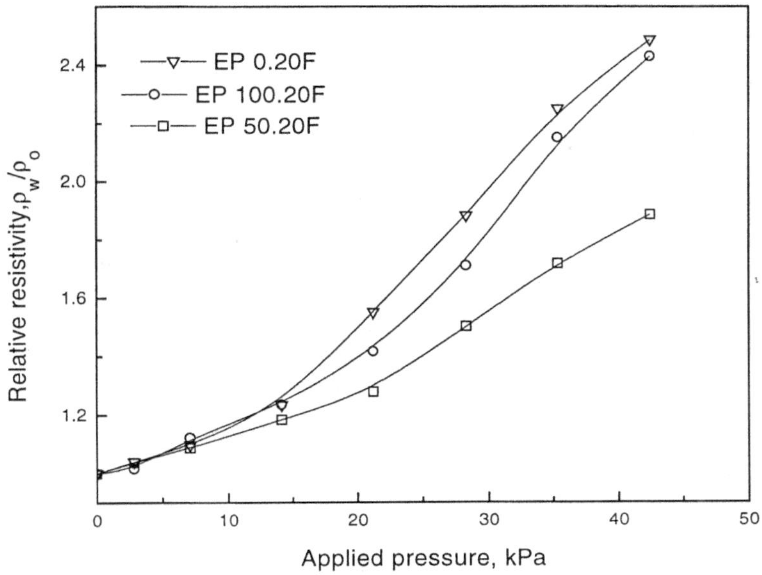

Fig. 34 Effect of pressure on the volume resistivity of SCF-filled EVA, EPDM and its 50/50 blend composites. Reproduced from Ref. [172]

electrical properties of elastomer-based conductive composites. Therefore, it is necessary to optimize these processing parameters to obtain reproducible results.

5.11.1 Dispersion of Conductive Particles

Dispersion of conductive particles in the polymer matrix greatly affects the electrical properties of polymer composites. The degree of dispersion of filler particles in polymer matrix depends on mixing time, mixing temperature, and shear rate during dry/melt mixing. However, better dispersion of filler particles requires higher filler concentration to make the composite enough conducting (i.e., higher percolation threshold). The effect of dispersion on the electrical conductivity of graphite-filled polyethylene composite had been reported in the literature [176]. The control of physical as well as electrical properties of the composites by the degree of dispersion has been investigated in details [177, 178]. However, there is a conflicting report on the effect of dispersion of particles on electrical conductivity. In some literature, it has been shown that uniform dispersion gives better conductivity [101, 179–181], whereas in other literature it has been reported that agglomeration of a particle gives higher conductivity with a lower value of percolation threshold [182–185]. In fact, preferential dispersion of the filler in polymer matrix leads to percolation at a lower loading of filler whereas more uniform mixing requires much higher loading of filler.

5.11.2 Mixing Time

Mixing time affects the electrical properties of polymer composites. Generally, the electrical resistivity is found to increase with the increase in dispersion time [186]. The filler aggregates and short carbon fiber undergo breakage due to the shearing force exerted by the polymer matrix on them during mixing. Consequently, the increase in mixing time cause more breakage of filler aggregate which adversely affects the electrical property. Pramanik et al. [187] have reported that the increase in mixing time increases the electrical resistivity for both carbon black and carbon fiber filled NBR composites. A similar observation has also been made for EVA, EPDM, and their blend composites loaded with 50 phr carbon black and 20 phr carbon fiber as shown in Fig. 35 [172]. The composites were processed by melt mixing technique. The effect of sonication time of electrical conductivity of polymer/carbon composites was also investigated by various authors [117, 188, 189]. They reported that the increase in sonication time reduces the electrical conductivity of the polymer/carbon composites at any particular filler loading.

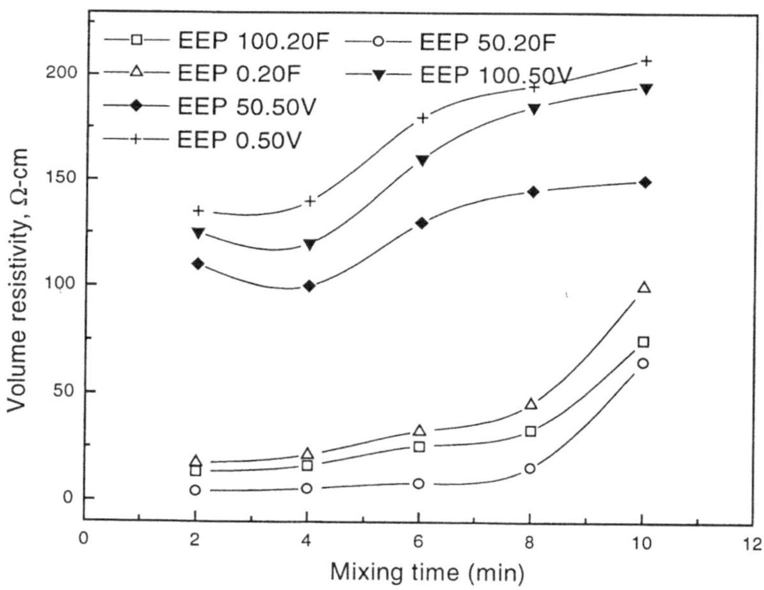

Fig. 35 Effect of mixing time on the volume resistivity of SCF and carbon black filled EVA, EPDM and its 50/50 blend composites (mixing temperature, 70 °C; rotor speed, 60 rev./min). Reproduced from Ref. [172]

5.11.3 Rotor Speed

The effect of shear rate on electrical resistivity has also been investigated on the same system [187]. They have found that the resistivity was increased with the increase in shear rate. Actually, the increase in shear rate increases the shear stress on the filler aggregate which results in the appreciable breakdown and more uniform dispersion of the filler in matrix polymer leading to the increase in electrical resistivity [190]. The effect of rotor speed from 20 to 100 rpm on electrical resistivity of EVA, EPDM, and their blend composites filled with carbon black and carbon fiber at the mixing time 5 min and mixing temperature 70 °C are presented in Fig. 36 [172]. The resistivity is seen to increase with the increase in rotor speed. The increment in electrical resistivity for carbon black filled composites at lower rotor speed is quite prominent. However, at high rotor speed, the increment in resistivity is marginal. Actually at high rotor speed, due to the high shearing, there is the rise in temperature of the polymer melt leading to the lowering of its viscosity. Thus, the shearing force on the filler aggregates is reduced which favor less breakdown of conductive particle carbon black at high rotor speed and consequently results in a lower increase in electrical resistivity. However, the carbon fiber filled composites exhibited opposite behavior. The increase in resistivity at low

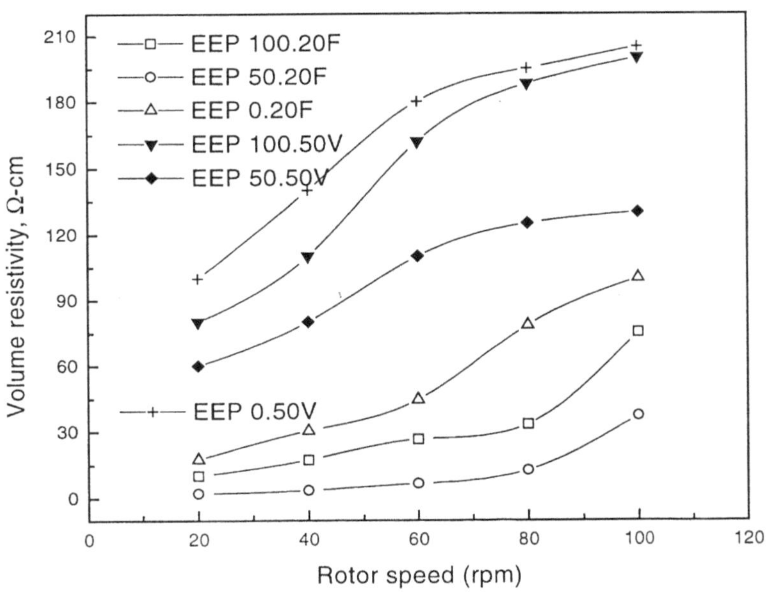

Fig. 36 Effect of rotor speed on the volume resistivity of SCF- filled and carbon black filled EVA, EPDM and its 50/50 blend composites (mixing temperature, 70 °C; mixing time, 5 min). Reproduced from Ref. [172]

rotor speed is less and at high rotor speed is high. The fiber is having low bending strength. As a result, it undergoes more breakdowns at high rotor speed and results in a high increase in electrical resistivity.

5.11.4 Mixing Temperature

The increase in mixing temperature reduces the matrix viscosity which in turn reduces the shear stress during mixing leading to reducing in the breakdown of particle structure, consequently resistivity decrease sharply when mixing temperature is increased [187]. Das et al. have also studied the effect of mixing temperature on electrical resistivity of EVA, EPDM, and their blend composites filled with carbon black and carbon fiber as can be seen in Fig. 37. The resistivity of melt-mixed composites decreases with the increase in mixing temperature as expected. The sonication temperature during solution mixing process also affects the electrical conductivity of the composites. Low mixing temperature during sonication leads to better dispersion of conductive particles. The electrical conductivity may decrease or increase depending on the preferential formation of conductive networks within the polymer matrix. A detailed study in this matter is required.

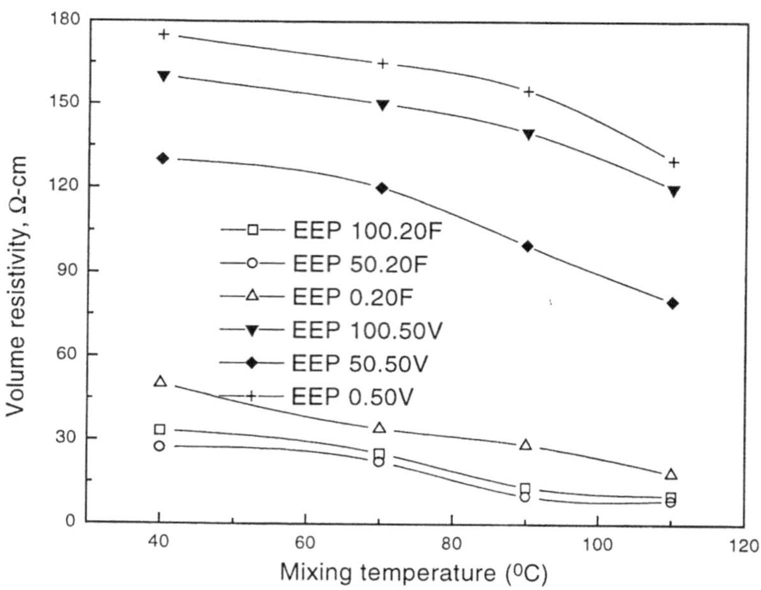

Fig. 37 Effect of mixing temperature on the volume resistivity of SCF-filled and carbon black filled EVA, EPDM and its 50/50 blend composites (mixing time, 5 min; rotor speed, 60 rev./min). Reproduced from Ref. [172]

5.11.5 Mold Pressure

The effect of mold pressure on electrical resistivity exhibits anomalous behavior. There was a report on the increase of electrical resistivity with the increase in mold pressure applied to the composite. This was due to the breakdown of filler aggregates caused by higher and fast expansion of rubber when the mold pressure was released [191]. However, the opposite observations were also reported. It had been observed that there was a decrease in resistivity of composites with an increase in mold pressure which helped in the formation of three-dimensional conductive network [191] and increased compactness [161] of conductive carbon black. Another investigation had revealed that the hydrostatic pressure during vulcanization tends to rearrange the filler particles inside the polymer matrix, which leads to the decrease in resistivity that is an increase in conductivity [192]. The results reported by Das et al. in Fig. 38 also shows the similar behavior that is the resistivity decreases with the increase in mold pressure [172]. Badawy and Nasr investigated the effect of mold pressure on resistivity for polyvinyl chloride (PVC) mixed NBR composites filled with high abrasion furnace black (HAF) [193]. They reported that the conductivity decreases when mold pressure is suddenly increased and then increases exponentially with time.

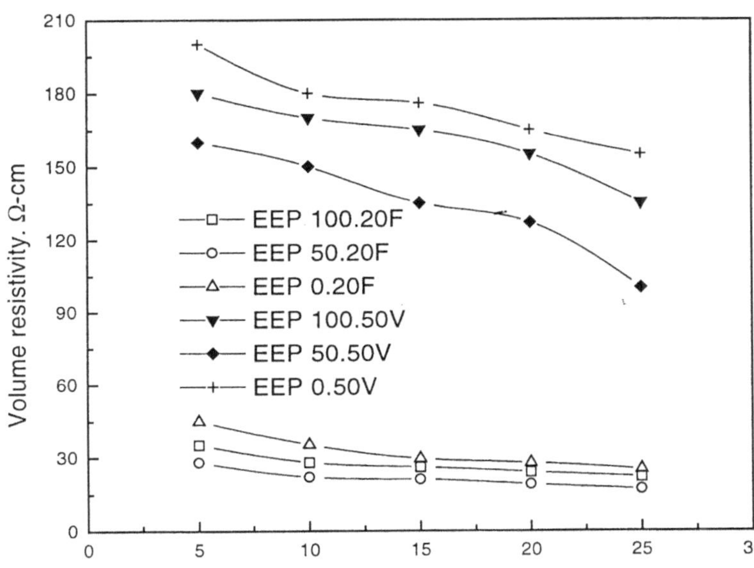

Fig. 38 Effect of vulcanization pressure on the volume resistivity of SCF-filled and carbon black filled EVA, EPDM and its 50/50 blend composites (mixing temperature, 70 °C; mixing time, 5 min; rotor speed, 60 rev./min). Reproduced from Ref. [172]

5.12 Forming Process and Vulcanization

Different manufacturing processes are available depending upon the demand of product and nature of materials. Plastics materials are processed by injection molding, extrusion process, calendaring, compression molding, thermoforming, etc. In the manufacture of rubbery materials, curing agents are required along with other ingredients. The filler dispersion and formation of conducting network vary with respect to the manufacturing process and vulcanization time. It has been observed that the degree of cross-linking reduces the resistivity due to the formation of three-dimensional networks in the vulcanizate. The effect of vulcanization on physicochemical and electrical properties at different mold pressure and temperature of siloxane composite has been investigated and reported a marginal change in resistivity with vulcanization system. Boonstra [194] observed that the resistivity decreases after one or two minute pressing of the sample. The effect of vulcanization time on electrical resistivity of EVA, EPDM, and their blend composites are presented in Fig. 39 [172]. The resistivity decreases with curing time for all cases. During the curing process, there is the formation of free ions that reduce the resistivity at the initial stage of vulcanization. However, at higher curing time there is the formation of three-dimensional networks within the polymer chains which binds the conductive networks compactly and results in a marginal decrease in resistivity.

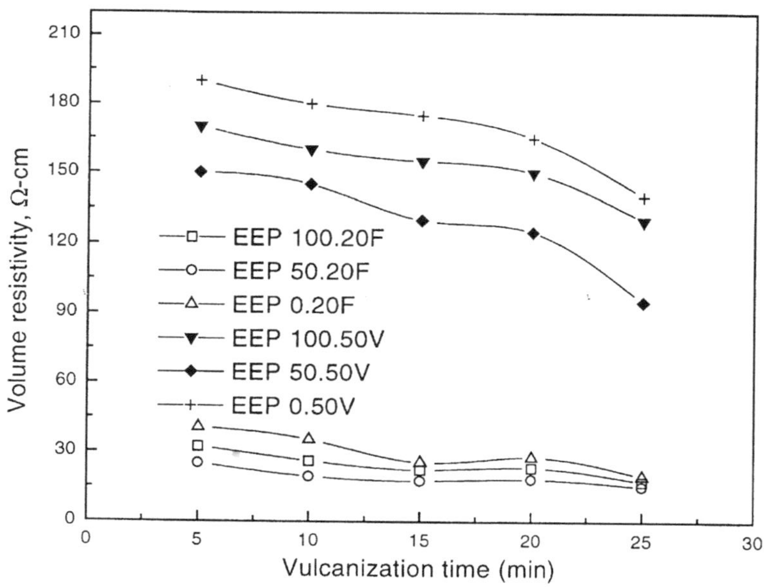

Fig. 39 Effect of vulcanization time on the volume resistivity of SCF-filled and carbon black filled EVA, EPDM and its 50/50 blend composites (mixing temperature, 708C; mixing time, 5 min; rotor speed, 60 rev./min). Reproduced from Ref. [172]

5.13 Mechanical Deformation

Physical deformation like stretching, bending, and compression affect the electrical properties of conductive rubber composite systems. The behavior of these physical deformations toward electrical properties depends on the nature of polymer, nature of additive, loading level, and degree of deformation [195–197]. An investigation regarding the effect of stretching on resistivity has been made based on the carbon black filled rubber composite [198]. It is revealed that the electrical resistivity is increasing while stretching and decreasing to a lesser extent during relaxation. Hashem et al. [199] have reported the effect of stretching on electrical resistivity on carbon black filled butyl rubber composite. He pointed out that the resistivity is increasing with stretching up to a certain limit due to the breakdown of some existing conducting networks. However, on further stretching the resistivity was found to decrease due to the reorientation of carbon black networks at high stretching. Thus, it can be mentioned here that stretching causes breakdown as well as reorientation of conductive networks. At low stretching, the breakdown of conductive networks gets priority over the orientation of networks; whereas, at high stretching reorientation process acts as predominant factor. Several types of studies regarding the effect of strain on the electrical conductivity of carbon black filled polymer composites have been reported by a number of authors [200, 201].

5.14 Effect of Frequency

The AC conductivity of polymer–carbon composites is greatly affected by the frequency of electric/electromagnetic field. AC conductivity was found to increase with the increase in the frequency of the electric field [19]. With the increase in frequency, the electrons/charge carriers present in the composite system gets energies, which, in turn, increase the hopping/tunneling of electrons/charge carriers resulting in the increase in electrical conductivity. However, the nature of the variation of electrical conductivity with frequency is dependent on the type of polymers, types of fillers, and the concentration of filler used. The neat insulating polymers exhibit strong frequency-dependent behavior. Moreover, this frequency dependency of conductivity decreases as the loading of the conductive additive in the insulating polymer matrices increases. After certain loading, the conductive composites exhibit frequency independent behavior and at this stage the AC conductivity coincides with the DC conductivity of the composite. Sophie et al. were reported the effect of frequency on the electrical conductivity of polyepoxy carbon nanotubes nanocomposites as presented in Fig. 40 [202]. They found that below 0.4 weight fraction of CNT loading, the composites exhibit frequency-dependent behavior and beyond this loading the composites exhibit frequency independent conductivity. The composite with 2.5 weight fraction of CNT loading shows

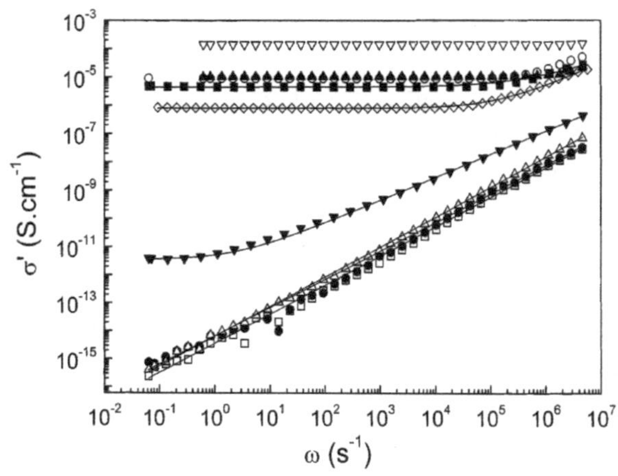

Fig. 40 Frequency dependence of the real part ó¢ of the complex conductivity for different carbon nanotubes weight fraction at 20 °C: (□) 0, (●) 0.04, (Δ) 0.2, (▼) 0.3, (◊) 0.4, (■) 0.6, (O) 0.8, (▲) 1.5, and (V) 2.5 wt%. Reproduced from Ref. [202]

complete frequency-independent conductivity. The study on the effect of frequency on polymer–carbon composites was investigated by many authors [203–205].

In addition to these factors discussed herewith, several others points also affect the electrical conductivity of polymer–carbon composites. The same composites prepared by different techniques like in situ, solution casting, or melt blending would exhibit different electrical percolation thresholds and conductivities. It has been reported that the composites prepared by in situ or solution casting technique shows lower electrical percolation threshold and higher electrical conductivity compared to the melt blending technique [19, 119]. Actually, in the melt blending technique, there is the extensive breakage of carbon particles, which results in the lowering of its structure/aspect ratio, and consequently the higher value of electrical percolation threshold and resistivity.

6 Conclusions

Polymer–carbon composites fall under extrinsically conducting polymers when conductive carbons are mixed with insulating polymers. Generally, we measure three types of resistances namely surface resistance, volume/bulk resistance, and contact resistance either in two-probe or four-probe technique. Among the carbons, diamond shows the lowest electrical conductivity and monolayer graphene shows the highest inherent electrical conductivity. However, the electrical conductivity of these carbons depends on their purity, types, methods of manufacturing, geometry/structure, aspect ratio, number of layers, surface area, etc. High purity and high structure/aspect ratio of carbons show high bulk conductivity due to low defects and low contact resistance, respectively. It is revealed that the composite with high polymer crystallinity, preferential distribution of carbon fillers, and low viscosity of polymer and molecular weight, high structure and aspect ratio of carbons exhibited low electrical percolation threshold and high electrical conductivity. Functionalizations of carbons and waviness reduce the electrical percolation threshold and conductivity. The increase in ac frequency increases the electrical conductivity. Polymer blending, and dispersion and orientation of carbons affect the conductivity of the composites but the conductivity may increase or decrease depending on the nature of blending, quality of dispersion, and nature of orientation. Vulcanization of composites increases the electrical conductivity up to certain curing time. Both the measurement and processing temperature, pressure, and time affect the electrical conductivity. The composites may exhibit PTC or NTC effect of resistivity depending on the nature of polymers and fillers. Hence, from the above-going discussion, it is clear that the electrical conductivity and percolation threshold of polymer–carbon composites depends on many factors/parameters. So, it can be mentioned/concluded herewith that to get a desired level of electrical conductivity of polymer–carbon composites, the choice of polymer, carbon filler, preparation method, and the control of other processing parameters/factors are necessary.

References

1. Kaynak A, Mohan S, Unsworth J, Clout R (1994) Plane-wave shielding effectiveness studies on conducting polypyrrole. J Mater Sci Lett 13:1121–1123
2. Kohlman RS, Min YG, MacDiarmid AG, Epstein AJ (1996) Tunability of high frequency shielding in electronic polymers. J Eng Appl Sci 2:1412–1416
3. Tan S, Zhang M, Zeng H (1998) Electro-conductive polymer composite for shielding EMI. Cailiao Gongcheng/J Mater Eng 5:6–9
4. Rupprecht L, Hawkinson C (1999) Conductive plastics for medical applications. Med Device Diagn Ind 21:8
5. Wen S, Chung DDL (2005) Pitch-matrix composites for electrical, electromagnetic and strain-sensing applications. J Mater Sci 40:3897–3903
6. MacDiarmid AG (2001) Nobel lecture. Angew Chem Int Ed 40:2581–2590
7. Shirakawa H, Louis EJ, MacDiarmid AG, Chiang CK, Heeger AJ (1977) Synthesis of electrically conducting organic polymers: halogen derivatives of polyacetylene, $(CH)_x$. J Chem Soc Chem Commun 578–580
8. Heeger AJ, Kivelson S, Schrieffer JR, Su WP (1988) Solitons in conducting polymers. Rev Mod Phys 60:781–850
9. Salaneck WR, Bredas JL (1994) Conjugated polymers. Solid State Comm 92:31–36
10. Unsworth J, Conn C, Jin Z, Kaynak A, Ediriweera R, Innis P, Booth N (1994) Conducting polymers: properties and applications. J Intell Mater Sys Struc 5:595–604
11. Sauerer W (1994) Intrinsically conducting polymers—from exploratory research to applications. Galvanotechnik 85:1467–1472
12. Borgmans APJH, Glaser RH (1995) Design considerations for EMI shielding conductive plastic compounds. Evaluat Eng 34:S32–S37
13. Chung DDL (2001) Electromagnetic interference shielding effectiveness of carbon materials. Carbon 39:279–285
14. Sau KP, Chaki TK, Chakraborty A, Khastgir D (1997) Electromagnetic interference shielding by carbon black and carbon fibre filled rubber composites. Plast Rub Compos Process Appl 26:291–297

15. Flandin L, Hiltner A, Baer E (2001) Interrelationships between electrical and mechanical properties of a carbon black-filled ethylene-ocetene elastomer. Polymer 42:827–838
16. Jana PB, Mallick AK, De SK (1993) Electromagnetic interference shielding effectiveness of short carbon fibre-filled polychloroprene vulcanized by barium ferrite. J Mater Sci 28:2097–2104
17. Xing L, Liu J, Ren S (1998) Study on electromagnetic property of short carbon fibers and its application to radar absorbing materials. Cailiao Gongcheng/J Mater Eng 1:19–21
18. Jimenez G, Jana SC (2007) Electrically conductive polymer nanocomposites of polymethylmethacrylate and carbon nanofibers prepared by chaotic mixing. Composites Part A Appl Sci Manuf 38:983–993
19. Sohi NJS, Bhadra S, Khastgir D (2011) The effect of different carbon fillers on the electrical conductivity of ethylene vinyl acetate copolymer-based composites and the applicability of different conductivity models. Carbon 41:1349–1361
20. Potschke P, Bhattacharyya AR, Janke A (2003) Morphology and electrical resistivity of melt mixed blends of polyethylene and carbon nanotube filled polycarbonate. Polymer 44:8061–8069
21. Potschke P, Bhattacharyya AR, Janke A (2004) Carbon nanotube-filled polycarbonate composites produced by melt mixing and their use in blends with polyethylene. Carbon 42:965–969
22. Ounaies Z, Park C, Wise KE, Siochi EJ, Harrison JS (2003) Electrical properties of single wall carbon nanotube reinforced polyimide composites. Compos Sci Technol 63:1637–1646
23. Yoon JK, Taek SS, Hyung DC, Jong HK, Yeon-Choon C, Ho GY (2005) Electrical conductivity of chemically modified multiwalled carbon nanotube/epoxy composites. Carbon 43:23–30
24. Agnihotri P, Basu S, Kar KK (2011) Effect of carbon nanotube length and density on the properties of carbon nanotube-coated carbon fiber/polyester composites. Carbon 49:3098–3106
25. Zou JF, Yu ZZ, Pan YX, Fang XP, Ou YC (2002) Conductive mechanism of polymer/graphite conducting composites with low percolation threshold. J Polym Sci Part B Polym Phys 40:954–963
26. Savvinova ME, Kovalenko NA (2009) Influence of technological parameters on the electrical conduction of carbon composites. Russ Eng Res 29:487–489
27. Sidhu A, Reike J, Michelsen U, Messinger R, Habiger E, Wolf J (1997) Metallization of plastics for shielding. In: IEEE international symposium on electromagnetic compatibility, Piscataway, New Jersey, USA: IEEE, pp 102–105
28. Jiang H, Moon KS, Li Y, Wong CP (2006) Surface functionalized silver nanoparticles for ultrahigh conductive polymer composites. Chem Mater 18:2969–2973
29. Gwinner D, Scheyrer P, Fernandez W (1996) Selective deposition of aluminum on plastic parts for EMI shielding. In: Proceedings of 39th annual technical conference. Society of Vacuum Coaters, Albuquerque, New Mexico, USA, p 336
30. Notingher PV, Panaitescu D, Paven H, Chipara M (2004) Some characteristics of conductive polymer composites containing stainless steel fibers. J Optoelectro Adv Mater 6:1081–1084
31. Francis G, Hanejko GW, Ellis Timothy JH (1998) Application of high performance material processing—electromagnetic products. International conference on powder metallurgy & particulate materials. 31 May–4 June, Las Vegas, NV, USA
32. Stabik J, Dybowska A, Chomiak M (2010) Polymer composites filled with powders as polymer graded materials. J Achiev Mater Manuf Eng 43(1):153–161
33. Pierre Deltour R, Perenboom JA, Van Bentum PJM (1990) Electrical-conduction mechanisms in polymer-copper-particle composites. 1. Temperature and high-magnetic-field dependence of the conductivity. Phys Rev B 42:3380–3385
34. Zhu M, Chung DDL (1991) Nickel fiber silicone–matrix composites as resilient electrical conductors. J Electron Packag 113:417–420
35. Shui X, Chung DDL (2000) Submicron diameter nickel filaments and their polymer-matrix composites. J Mater Sci 35:1773–1785

36. Fernando GSJ, Soares BG, Siddaramaiah Barra GMO, Herbst MH (2006) Influence of plasticizers (DOP and CNSL) on mechancial and electrical properties of SBS/polyaniline blends. Polymer 47:7548–7553
37. Lakshmi K, John H, Mathew KT, Joseph R, George KE (2009) Microwave absorption, reflection and EMI shielding of PU–PANI composite. Acta Mater 57:371–375
38. Askeland D (1988) The science and engineering of materials. Cengage Learning, Stamford
39. Lee PA, Stone AD, Fukuyama H (1987) Universal conductance fluctuations in metals: Effects of finite temperature, interactions, and magnetic field. Phys Rev B 35(3):1039–1070
40. Kumar P, Yashonath S (2006) Ionic conduction in the solid state. J Chem Soc 118(1):135–154
41. ASTM Standard D 257-99 (1999) Standard test methods for DC resistance or conductance of insulating materials
42. ESD STM 11.11-2001 Standard (2001) Surface resistance measurement of static dissipative planar materials
43. Heaney MB (1999) Electrical Conductivity and Resistivity. In: The measurement, instrumentation and sensors handbook. CRC Press, Boca Raton
44. IEC 61340-5-1 Standard (1998) Electrostatics—part 5-1: Protection of electronic devices from electrostatic phenomena—general requirements
45. Banaszczyk J, Schwarz A, De Mey G, Van Langenhove L (2010) The Van der Pauw method for sheet resistance measurements of polypyrrole-coated para-aramide woven fabrics. J Appl Polym Sci 117:2553–2558
46. Rietveld G, Koijmans ChV, Henderson LCA, Hall MJ, Harmon S, Warnecke P, Schumacher B (2003) DC conductivity measurements in the Van Der Pauw geometry. IEEE Trans Instrum Meas 52(2):449–453
47. Majid AJ (2011) Resistivity measurements of conductors and semiconductors of different geometrical shapes using Van der Pauw technique. Int J Sci Eng Res 2(10):1–5
48. Ram R, Rahaman M, Khastgir D (2015) Electrical properties of polyvinylidene fluoride (PVDF)/multi-walled carbon nanotube (MWCNT) composites: modelling of DC conductivity. Composites Part A Appl Sci Manuf 69:30–39
49. Rahaman M, Chaki TK, Khastgir D (2012) Modeling of DC conductivity for ethylene vinyl acetate (EVA)/polyaniline conductive composites prepared through in-situ polymerization of aniline in EVA matrix. Compos Sci Technol 72:1575–1580
50. Foygel M, Morris R, Anez D, French S, Sobolev V (2005) Theoretical and computational studies of carbon nanotube composites and sus-pensions: electrical and thermal conductivity. Phys Rev B 71:104201/1–104201/8
51. Stauffer D (1987) Introduction to percolation theory. Taylor and Francis, Inc., Philadelphia, p 181
52. Sahimi M (1994) Applications of percolation theory. Taylor and Francis, London, p 276
53. Tchoudakov R, Breuer O, Narkis M, Siegmann A (1996) Conductive polymer blends with low carbon black loading: polypropylene/polyamide. Polym Eng Sci 36:1336–1346
54. Miyasaka K, Watanabe K, Jojima E, Aida H, Sumita M, Ishikawa K (1982) Electrical conductivity of carbon-polymer composites as a function of carbon content. J Mater Sci 17:1610–1616
55. Sumita M, Abe H, Kayaki H, Miyasaka K (1986) Effect of melt viscosity and surface tension of polymers on the percolation threshold of conductive-particle-filled polymeric composites. J Macromol Sci Phys B25:171–184
56. Sumita M, Asai S, Miyadera N, Jojima E, Miyasaka K (1986) Electrical conductivity of carbon black filled ethylene-vinyl acetate copolymer as a function of vinyl acetate content. Colloid Polym Sci 264:212–217
57. Lee GJ, Suh KD, Im SS (1998) Study of electrical phenomena in carbon black–filled HDPE composite. Polym Eng Sci 38:471–477
58. Feller JF, Linossier I, Levesque G (2002) Conductive polymer composites (CPCs): comparison of electrical properties of poly(ethylene-co-ethyl acrylate)-carbon black with

poly(butylene terephthalate)/poly(ethylene-co-ethyl acrylate)-carbon black. Polym Advan Technol 13:714–724
59. Schueler R, Petermann J, Schulte K, Wentzel HP (1997) Agglomeration and electrical percolation behavior of carbon black dispersed in epoxy resin. J Appl Polym Sci 63:1741–1746
60. Fournier J, Boiteux G, Seytre G, Marichy G (1997) Percolation network of polypyrrole in conducting polymer composites. Synth Met 84:839–840
61. Boiteux G, Fournier J, Issotier D, Seytre G, Marichy G (1999) Conductive thermoset composites: PTC effect. Synth Met 102:1234–1235
62. Flandin L, Prasse T, Schueler R, Schulte K, Bauhofer W, Cavaille JY (1999) Anomalous percolation transition in carbon-black–epoxy composite materials. Phys Rev B 59:14349–14355
63. Tang H, Chen XF, Luo YX (1996) Electrical and dynamic mechanical behavior of carbon black filled polymer composites. Eur Polym J 32:963–966
64. Grunlan JC, Gerberich WW, Francis LF (1999) Electrical and mechanical property transitions in carbon-filled poly(vinylpyrrolidone). J Mater Res 14:4132–4135
65. Huang JC (2002) Carbon black filled conducting polymers and polymer blends. Adv Polym Tech 21:299–313
66. Zhang MY, Jia WT, Chen XF (1996) Influences of crystallization histories on PTC/NTC effects of PVDF/CB composites. J Appl Polym Sci 62:743–747
67. Huang JC, Wu CL (2000) Processability, mechanical properties, and electrical conductivities of carbon black-filled ethylene-vinyl acetate copolymers. Adv Polym Technol 19:132–139
68. Grill A (1999) Electrical and optical properties of diamond-like carbon. Thin Solid Films 355–356:189–193
69. Chen Q, Wang L-X, Zhang Z, Yang J, Lin Z (1996) Epitaxially oriented growth of diamond on silicon by hot filament chemical vapor deposition. Appl Phys Lett 68(2):176–178
70. Yadav BC, Kumar R (2008) Structure, properties and applications of fullerenes. Int J Nanotech Appl 2:15–24
71. Goel A, Howard JB, Sande JBV (2004) Size analysis of single fullerene molecules by electron microscopy. Carbon 42:1907–1915
72. Sengupta R, Bhattacharya M, Bandyopadhyay S, Bhowmick AK (2011) A review on the mechanical and electrical properties of graphite and modified graphite reinforced polymer composites. Prog Polym Sci 36:638–670
73. Guo N, Leu MC (2012) Effect of different graphite materials on the electrical conductivity and flexural strength of bipolar plates fabricated using selective laser sintering. Int J Hydrogen Energy 37:3558–3566
74. Matsumura K, Takahashi A, Tsukamoto J (1985) Structure and electrical conductivity of graphite fibers prepared by pyrolysis of cyanoacetylene. Synth Met 11:9–20
75. Sadasivuni KK, Ponnamma D, Thomas S, Grohens Y (2014) Evolution from graphite to graphene elastomer composites. Prog Polym Sci 39:749–780
76. Kraus G (1965) Interactions of elastomers and reinforcing fillers. Rubb Chem Technol 38:1070–1114
77. Gent AN (1978) Science and technology of rubber. In: Erich FR(ed). Academic Press, Inc., New York
78. Wolff S, Wang M (1993) In: Donnet JB, Bansal RC, Wang M(ed) Carbon black science and technology. Marcel Dekker Inc., p 289
79. Janzen J (1975) On the critical conductive filler loading in antistatic composites. J Appl Phys 46:966–969
80. Bigg DM, Bradhury JE (1981) Conducting polymers. In: Seymour RB (ed) Polymer science and technology, vol 15. Plenum, New York, p 13
81. Medalia AI (1986) Electrical conduction in carbon black composites. Rubb Chem Technol 59:432–454
82. Nelson JR (1986) Morphology of electrically conductive grades of carbon black. Carbon 24:115–121

83. Marinho B, Ghislandi M, Tkalya E, Koning CE, de With G (2012) Electrical conductivity of compacts of graphene, multi-wall carbon nanotubes, carbon black, and graphite powder. Powder Technol 221:351–358
84. Sánchez-González J, Macías-García A, Alexandre-Franco MF, Gómez-Serrano V (2005) Electrical conductivity of carbon blacks under compression. Carbon 43:741–747
85. Fitzer E, Frons W, Heine M (1986) Optimization of stabilization and carbonization treatment of PAN fibres and structural characterization of the resulting carbon fibres. Carbon 24:387–395
86. Feng L, Ning Xie, Zhong J (2014) Carbon nanofibers and their composites: a review of synthesizing, properties and applications. Materials 7:3919–3945
87. Al-Saleh MH, Sundararaj U (2009) A review of vapor grown carbon nanofiber/ polymer conductive composites. Carbon 47:2–22
88. Tibbetts GG, Lake ML, Strong KL, Rice BP (2007) A review of the fabrication and properties of vapor-grown carbon nanofiber/polymer composites. Compos Sci Technol 67:1709–1718
89. Rao C (2005) Natotubes and nanowires. Royal Society of Chemistry, Cambridge
90. Kang I, Heung YY, Kim JH, Lee JW, Gollapudi R, Subramaniam S, Narasimhadevara S, Hurd D, Kirikera GR, Shanov V, Schulz MJ, Shi D, Boerio J, Mall S, Ruggles-Wren M (2006) Introduction to carbon nanotube and nanofiber smart materials. Composites Part B Eng 37(6):382–394
91. Baughman RH, Zakhidov AA, Heer WAD (2002) Carbon nanotubes–the route toward applications. Science 297(5582):787–792
92. Iijima S (2002) Carbon nanotubes: past, present, and future. Phys B 323:1–5
93. Thostenson ET, Ren Z, Chou TW (2001) Advances in the science and technology of carbon nanotubes and their composites: a review. Compos Sci Technol 61(13):1899–1912
94. Dillon AC, Yudasaka M, Dresselhaus MS (2004) Employing raman spectroscopy to qualitatively evaluate the purity of carbon single-wall nanotube materials. J Nanosci Nanotechnol 4(7):691–703
95. Baddour C, Briens C (2005) Carbon nanotube synthesis: a review. Int J Chem React Eng 3 (1):1–20
96. Sundaray B, Subramanian V, Natarajan TS, Krishnamurthy K (2006) Electrical conductivity of a single electrospun fiber of poly(methyl methacrylate) and multiwalled carbon nanotube nanocomposite. Appl Phys Lett 88:143114
97. Deng J, Ding X, Zhang W, Peng Y, Wang J, Long X et al (2002) Carbon nanotube-polyaniline hybrid materials. Eur Polym J 38:2497–2501
98. Long Y, Chen Z, Zhang X, Zhang J, Liu Z (2004) Synthesis and electrical properties of carbon nanotube polyaniline composites. Appl Phys Lett 85:1796–1798
99. Wu TM, Lin SH (2006) Characterization and electrical properties of polypyrrole/multiwalled carbon nanotube composites synthesized by in situ chemical oxidative polymerization. J Polym Sci Part B Polym Phys 44:1413–1418
100. Karim MR, Lee CJ, Chowdhury AMS, Nahar N, Lee MS (2007) Radiolytic synthesis of conducting polypyrrole/carbon nanotube composites. Mater Lett 61:1688–1692
101. Ramasubramaniam R, Chen J, Liu H (2003) Homogeneous carbon nanotube/polymer composites for electrical applications. Appl Phys Lett 83:2928–2930
102. Karim MR, Lee CJ, Lee MS (2006) Synthesis and characterization of conducting polythiophene/carbon nanotubes composites. J Polym Sci Part A: Polym Chem 44:5283–5290
103. Grimes CA, Mungle C, Kouzoudis D, Fang S, Eklund PC (2000) The 500 MHz to 5.50 GHz complex permittivity spectra of single-wall carbon nanotube-loaded polymer composites. Chem Phys Lett 319:460–464
104. Andres PL, Ramirez R, Verges JA (2008) Strong covalent bonding between two graphene layers. Phys Rev B 77:045403/1–045403/15

105. Nemes-Incze P, Osvatha Z, Kamarasb K, Biro LP (2008) Anomalies in thickness measurements of graphene and few layer graphite crystals by tapping mode atomic force microscopy. Carbon 46:1435–1442
106. Bolotin KI, Sikes KJ, Jiang Z, Klima M, Fudenberg G, Hone J, Kim P, Stormer HL (2008) Ultrahigh electron mobility in suspended graphene. Solid State Commun 146:351–355
107. Gomez-Navarro C, Weitz RT, Bittner AM, Scolari M, Mews A, Burghard M, Kern K (2007) Electronic transport properties of individual chemically reduced graphene oxide sheets. Nano Lett 7:3499–3503
108. Wu ZS, Ren W, Gao L, Zhao J, Chen Z, Liu B, Tang D, Yu B, Jiang C, Cheng HM (2009) Synthesis of graphene sheets with high electrical conductivity and good thermal stability by hydrogen arc discharge exfoliation. ACS Nano 3:411–417
109. Ansari S, Giannelis EP (2009) Functionalized graphene sheet—poly(vinylidene fluoride) conductive nanocomposites. J Polym Sci Part B Polym Phys 47:888–897
110. Du J, Zhao L, Zeng Y, Zhang L, Li F, Liu P, Liu C (2011) Comparison of electrical properties between multi-walled carbon nanotube and graphene nanosheet/high density polyethylene composites with a segregated network structure. Carbon 49:1094–1100
111. Wang C, Guo Z-X, Fu S, Wu W, Zhu D (2004) Polymers containing fullerene or carbon nanotube structures. Prog Polym Sci 29:1079–1141
112. Huang J-C (2002) Carbon black filled conducting polymers and polymer blends. Adv Polym Technol 21(4):299–313
113. Spahr ME, Gilardi R, Bonacchi D (2013) Carbon black for electrically conductive polymer applications. Encyclopedia of polymers and composites. Springer, Berlin. https://doi.org/10.1007/978-3-642-37179-0_32-1
114. Sanjinés R, Abad MD, Vâju C, Smajda R, Mionić M, Magrez A (2011) Electrical properties and applications of carbon based nanocomposite materials: an overview. Surf Coat Technol 206:727–733
115. Spitalsky Z, Tasisb D, Papagelis K, Galiotis C (2010) Carbon nanotube–polymer composites: chemistry, processing, mechanical and electrical properties. Prog Polym Sci 35:357–401
116. Mutiso RM, Winey KI (2015) Electrical properties of polymer nanocomposites containingrod-like nanofillers. Prog Polym Sci 40:63–84
117. Ma P-C, Siddiqui NA, Marom G, Kim J-K (2010) Dispersion and functionalization of carbon nanotubes for polymer-based nanocomposites: a review. Compos A 41:1345–1367
118. Kuilla T, Bhadra S, Yao D, Kim NH, Bose S, Lee JH (2010) Recent advances in graphene based polymer composites. Prog Polym Sci 35:1350–1375
119. Khanam PN, Ponnamma D, AL-Madeed MA (2015) Electrical properties of graphene polymer nanocomposites. In: Sadasivuni KK et al (eds) Graphene-based polymer nanocomposites in electronics. Springer series on polymer and composite materials. Springer International Publishing, Switzerland, pp 25–47. https://doi.org/10.1007/978-3-319-13875-6_2
120. Rahaman M, Chaki TK, Khastgir D (2011) Development of high performance EMI shielding material from EVA, NBR, and their blends: effect of carbon black structure. J Mater Sci 46 (11):3989–3999
121. Sohi NJS, Rahaman M, Khastgir D (2011) Dielectric property and electromagnetic interference shielding effectiveness of ethylene vinyl acetate based conductive composites: effect of different type of carbon fillers. Polym Compos 32:1148–1154
122. Nayak L, Rahaman M, Khastgir D, Chaki TK (2011) Thermal and electrical properties of carbon nanotubes based polysulfone nanocomposites. Polym Bull 67:1029–1044
123. Rahaman M, Chaki TK, Khastgir D (2011) High performance EMI shielding materials based on short carbon fiber filled ethylene vinyl acetate copolymer, acrylonitrile butadiene copolymer, and their blends. Polym Compos 32(11):1790–1805
124. Ram R, Rahaman M, Khastgir D (2014) Mechanical, electrical and dielectric properties of polyvinylidene fluoride/short carbon fiber composites with low electrical percolation threshold. J Appl Polym Sci 131(3):39866

125. Huang J-C (2002) Carbon black filled conducting polymers and polymer blends. Adv Polym Technol 21:299–313
126. Li ZH, Zhang J, Chen SJ (2008) Effects of carbon blacks with various structures on vulcanization and reinforcement of filled ethylene-propylene-diene rubber. eXPRESS Polym Lett 2(10):695–704
127. Rahaman M, Thomas SP, Hussein IA, De SK (2013) Dependence of electrical properties of polyethylene nanocomposites on aspect ratio of carbon nanotubes. Polym Compos 34:494–499
128. Shehzad K, Dang Z-M, Ahmad MN, Sagar R-UrR, Farooq MU, Wang T-B (2013) Effects of carbon nanotubes aspect ratio on the qualitative and quantitative aspects of frequency response of electrical conductivity and dielectric permittivity in the carbon nanotube/polymer composites. Carbon 54:105–112
129. Guo J, Liu Y, Prada-Silvy R, Tan Y, Azad S, Krause B, Pötschke P, Grady BP (2014) Aspect ratio effects of multi-walled carbon nanotubes on electrical, mechanical, and thermal properties of polycarbonate/MWCNT composites. J Polym Sci Part B Polym Phys 52:73–83
130. Thomas SP, Abdullateef AA, Al-Harthi MA, Atieh MA, De SK, Rahaman M, Chaki TK, Khastgir D, Bandyopadhyay S (2012) Electrical properties of natural rubber nanocomposites: effect of 1-octadecanol functionalization of carbon nanotubes. J Mater Sci 47:3344–3349
131. Jiang Mei-Juan, Dang Zhi-Min, Yao Sheng-Hong, Bai Jinbo (2008) Effects of surface modification of carbon nanotubes on the microstructure and electrical properties of carbon nanotubes/rubber nanocomposites. Chem Phys Lett 457:352–356
132. Sulong AB, Muhamad N, Sahari J, Ramli R, Deros BM, Park J (2009) Electrical conductivity behaviour of chemical functionalized MWCNTs epoxy nanocomposites. Eur J Sci Res 29(1):13–21
133. Park O-K, Kim S-G, You N-H, Ku B-C, Hui D, Lee JH (2014) Synthesis and properties of iodo functionalized graphene oxide/polyimide nanocomposites. Compos B 56:365–371
134. White S, DiDonna B, Mu M, Lubensky T, Winey K (2009) Simulations and electrical conductivity of percolated networks of finite rods with various degrees of axial alignment. Phys Rev B 79:024301/1–024301/17
135. Khan SU, Pothnis JR, Kim J-K (2013) Effects of carbon nanotube alignment on electrical and mechanical properties of epoxy nanocomposites. Compos A 49:26–34
136. Yousefi N, Gudarzi MM, Zheng QB, Aboutalebi SH, Sharif F, Kim JK (2012) Self alignment and high electrical conductivity of ultra large graphene oxide/polyurethane nanocomposites. J Mater Chem 22:12709–12717
137. Dalmas F, Dendievel R, Chazeau L, Cavaille J, Gauthier C (2006) Carbon nanotube-filled polymer composites. Numerical simulation of electrical conductivity in three-dimensional entangled fibrous networks. Acta Mater 54:2923–2931
138. Li C, Thostenson E, Chou T (2008) Effect of nanotube waviness on the electrical conductivity of carbon nanotube-based composites. Compos Sci Technol 68:1445–1452
139. Kyrylyuk AV, van der Schoot P (2008) Continuum percolation of carbon nanotubes in polymeric and colloidal media. Proc Natl Acad Sci USA 105:8221–8226
140. Sumita M, Sakata A, Asai S, Miyasaka K, Nakgawa H (1991) Dispersion of fillers and the electrical conductivity of polymer blends filled with carbon black. Polym Bull 25:265–271
141. Sau KP, Chaki TK, Khastgir D (1998) Carbon fiber filled conductive composites based on nitrile rubber (NBR), ethylene propylene diene rubber (EPDM) and their blend. Polymer 39:6461–6471
142. Sau KP, Chaki TK, Khastgir D (1997) Conductive rubber composites from different blends of ethylene-propylene-diene rubber and nitrile rubber. J Mater Sci 32:5717–5724
143. Pöschke P, Dudkin SM, Alig I (2003) Dielectric spectroscopy on melt processed polycarbonate–multiwalled carbon nanotube composites. Polymer 44:5023–5030
144. Mĭcŭsik M, Omastova M, Krupa I, Prokĕs J, Pissis P, Logakis E et al (2009) A comparative study on the electrical and mechanical behaviour of multi-walled carbon nanotube

composites prepared by diluting a masterbatch with various types of polypropylene. J Appl Polym Sci 113:2536–2551
145. Paul DR, Newman S (eds) (1978) Polymer blends, vol 1, 2. Academic Press, Inc., New York
146. Sirkar AK, Lamond TG (1973) Carbon black transfer in blends of cis poly(butadiene) with other elastomers. Rubb Chem Technol 46:178–191
147. Meyer J (1973) Glass transition temperature as a guide to selection of polymers suitable for PTC materials. Polym Eng Sci 13:462–468
148. Sichel EK, Gittleman JI, Sheng P (1978) Transport properties of the composite material carbon-poly(vinyl chloride). Phys Rev B Condensed Mater B18:5712–5716
149. Jeuskens G, Gielens JL, Geshef D, Deltour R (1987) The electrical conductivity of polymer blends filled with carbon-black. Eur Polym J 23:993–995
150. Sirkar AK (1981) Softer conductive rubber compounds by elastomer blending. Rubb Chem Technol 54:820–834
151. Voet A (1980) Temperature effect of electrical resistivity of carbon black filled polymers. Rubb Chem Technol 54:42–50
152. Bhattacharya SK, Basu S, De SK (1980) Effect of temperature on the electrical conductivity of poly(vinyl chloride)–copper composites. J Appl Polym Sci 25:111–118
153. Amin M, Hassan HH, Abdel-Bary EM (1974) Conductivity of carbon black-loaded styrene–butadiene rubber. J Polym Sci Polym Chem 12:2651–2657
154. Abdel-Bary EM, Amin M, Hassan HH (1979) Factors affecting electrical conductivity of carbon black-loaded rubber. II. Effect of concentration and type of carbon black on electrical conductivity of SBR. J Polym Sci Polym Chem 17:2163–2172
155. Amin M, Hassan HH, Abdel-Bary EM (1989) Influence of solvent penetration on the electrical conductance of pre-extended FEF carbon black-loaded rubbers. J Appl Polym Sci 37:1209–1219
156. Allak HM, Brinkman AW, Woods J (1993) I-V characteristics of carbon black-loaded crystalline polyethylene. J Mater Sci 28:117–120
157. Aminabhavi TM, Cassidy PE, Thomson CM (1990) electrical resistivity of carbon-black-loaded rubbers. Rubb Chem Technol 63:451–471
158. Klason C, Kubat J (1975) Anomalous behavior of electrical conductivity and thermal noise in carbon black-containing polymers at T_g and T_m. J Appl Polym Sci 19:831–845
159. Ghofraniha M, Saovey R (1988) Electrical conductivity of polymers containing carbon black. Polym Eng Sci 28:58–63
160. Langley M (1973) Carbon fiber in engineering. McGraw Hill, London
161. Bhattacharya SK, Chaklader AC (1982) Review on metal-filled plastics. part1: electrical conductivity. Polym Plast Technol Eng 19:21–51
162. He XJ, Du JH, Ying Z, Cheng HM (2005) Positive temperature coefficient effect in multiwalled carbon nanotube/high-density polyethylene composites. Appl Phys Lett 86:062112
163. Rahaman M, Chaki TK, Khastgir D (2013) Control of the temperature coefficient of the DC resistivity in polymer-based composites. J Mater Sci 48:7466–7475
164. Park S-J, Seo M-K, Lee J-R (2001) PTC/NTC behaviors of nanostructured carbon black-filled HDPE polymer composites. Carbon Sci 2(3&4):159–164
165. Feng J, Chan C-M (2000) Positive and negative temperature coefficient effects of an alternating copolymer of tetrafluoroethylene–ethylene containing carbon black-filled HDPE particles. Polymer 41:7279–7282
166. Feng J, Chan C-M (2000) Double positive temperature coefficient effects of carbon black-filled polymer blends containing two semicrystalline polymers. Polymer 41:4559–4565
167. Saito S, Sasabe H, Nakajima T, Yada K (1968) Dielectric relaxation and electrical conduction of polymers as a function of pressure and temperature. J Polym Sci Part A-2 Polym Phys 6(7):1297–1315

168. Pramanik PK, De SK, Saha TN, Khastgir D (1990) Pressure-sensitive electrically conductive nitrile rubber composites filled with particulate carbon black and short carbon fiber. J Mater Sci 25:3848–3853
169. Akakabe M (1986) JP 61, 32, 913 [86, 32, 913] C1 HO 1B13/00 15 Feb. Appl 84/156, 099; 25 July 1984, 4 p
170. Bickley AC, Donnet G (1988) Eur. Pat. Appl. EP 283, 193 (C1 HO 1B1/24) 02 Nov. GB Appl 8719, 355, 21 April 1987, 10 p
171. Moshimo S, Nagayasu S, Yamaguch Y, Noguchi T, Nakajima M, Kakluchi H, Tanida K (1987) EP 207, 450 (CI HO1B1/24) 07 Jan. JP Appl 85/147, 160;03 July 185, 58 p
172. Das NC, Chaki TK, Khastgir D (2002) Effect of processing parameters, applied pressure and temperature on the electrical resistivity of rubber-based conductive composites. Carbon 40:807–816
173. Rahaman M, Chaki TK, Khastgir D (2013) Polyaniline, ethylene vinyl acetate semi-conductive composites as pressure sensitive sensor. J Appl Polym Sci 128:161–168
174. Rahaman M, Chaki TK, Khastgir D (2014) Polyaniline/ethylene vinyl acetate composites as dielectric sensor. Polym Eng Sci 54:1632–1639
175. Mahmoud WE, El-Lawindy AMY, El-Eraki MH, Hassan HH (2007) Butadiene acrylonitrile rubber loaded fast extrusion furnace black as a compressive strain and pressure sensors. Sens Actuators A 136:229–233
176. Agari Y, Ueda A, Nagai S (1991) Thermal conductivities of composites in several types of dispersion systems. J Appl Polym Sci 42:1665–1669
177. Cembrola RJ (1982) The relationship of carbon black dispersion to electrical resistivity and vulcanizate physical properties. Polym Eng Sci 22:601–609
178. Nakajima N, Harrell ER (1984) Contributions of elastomer behavior to mechanisms of carbon black dispersion. Rubb Chem Technol 57:153–167
179. Li J, Ma PC, Chow WS, To CK, Tang BZ, Kim JK (2007) Correlations between percolation threshold, dispersion state, and aspect ratio of carbon nanotubes. Adv Funct Mater 17:3207–3215
180. Du F, Fischer JE, Winey KI (2003) Coagulation method for preparing single-walled carbon nanotube/poly(methyl methacrylate) composites and their modulus, electrical conductivity, and thermal stability. J Polym Sci, Part B: Polym Phys 41:3333–3338
181. Kashiwagi T, Fagan J, Douglas J, Yamamoto K, Heckert A, Leigh S, Obrzut J, Du F, Lingibson S, Mu M, Winey KI, Haggenmueller R (2007) Relationship between dispersion metric and properties of PMMA/SWNT nanocomposites. Polymer 48:4855–4866
182. Bryning MB, Islam MF, Kikkawa JM, Yodh AG (2005) Very low conductivity threshold in bulk isotropic single-walled carbon nanotube–epoxy composites. Adv Mater 17:1186–1191
183. Martin CA, Sandler JKW, Shaffer MSP, Schwarz MK, Bauhofer W, Schulte K, Windle AH (2004) Formation of percolating networks in multi-wall carbon-nanotube? Epoxy composites. Compos Sci Technol 64:2309–2316
184. Sandler JKW, Kirk JE, Kinloch IA, Shaffer MSP, Windle AH (2003) Ultra-low electrical percolation threshold in carbon-nanotube–epoxy composites. Polymer 44:5893–5899
185. Pegel S, Pötschke P, Villmow T, Stoyan D, Heinrich G (2009) Spatial statistics of carbon nanotube polymer composites. Polymer 50:2123–2132
186. Abdel-Bary EM, Amin M, Hassan HH (1977) Factors affecting electrical conductivity of carbon black-loaded rubber I. Effect of milling conditions and thermal-oxidative aging on electrical conductivity of HAF carbon black-loaded styrene–butadiene rubber. J Polym Sci Polym Chem 15:197–201
187. Pramanik PK, Saha TN, Khastgir D (1992) Effect of some processing parameters on the resistivity of conductive nitrile rubber composites. Plast Rubb Compos Process Appl 17:179–185
188. Huang YY, Terentjev EM (2012) Dispersion of carbon nanotubes: mixing, sonication, stabilization, and composite properties. Polymers 4:275–295

189. Kim S, Lee JW, Hong I-K, Lee S (2014) Electrical conductivity enhancement of polycarbonate/poly(styrene-co-acrylonitrile)/carbon nanotube composites by high intensity ultrasound. Macromol Res 22(2):154–159
190. Fernandez DR, Marzocca AJ (1991) Analysis of resistivity in a rubber compound. Rubb Chem Technol 64:501–509
191. Thomson CM, Besuden TW, Beumel LL (1988) Resistivity of rubber as a function of mold pressure. Rubb Chem Technol 61:828–841
192. Nasr GM, Amin M, Osman HH, Badway MM (1989) Influence of hydrostatic pressure on the electrical properties of unvulcanized FEF-loaded SBR. J Appl Polym Sci 37:1327–1337
193. Badawy MM, Nasr GM (1997) Effect of molding pressure on the electrical conductivity of conductive NBR/PVC composites. Polym Test 16:155–164
194. Boonstra BB (1977) Resistivity of unvulcanized compounds of rubber and carbon black. Rubb Chem Technol 50:194–210
195. Voet A, Sirkar AK, Mullens TJ (1969) Electrical properties of stretched carbon black loaded vulcanizates. Rubb Chem Technol 42:874–891
196. Voet A, Morawaki JC (1974) Dynamic mechanical and electrical properties of vulcanizates at elongations up to sample rupture. Rubb Chem Technol 47:765–777
197. Burton LC, Hwang K, Zhang T (1989) Dynamic electrical and electromechanical properties of carbon-black loaded rubber. Rubb Chem Technol 62:838–849
198. Pramanik M, Saha TN, Khastagir D (1993) Effect of extensional strain on the resistivity of electrically conductive nitrile-rubber composites filled with carbon filler. J Mater Sci 28:3539–3546
199. Hashem AA, Ghani AA, Eatah AI (1991) Effect of preextension on electrical conductivity and physicomechanical properties of butyl rubber (IIR) loaded with different types of carbon black. J Appl Polym Sci 42:1081–1085
200. Amin M, Nasr GM, Hassan HH, Ei-Guiziri S, Abdem MA (1989) Investigation on the dependence of the electrical conductivity of FEF/SBR vulcanizates on the cyclic strain. Polym Bull 22:413–420
201. Hasan HH, Khairy SA, El-Guiziri S, Abdel-Moneim HM (1991) Effect of tensile deformation on the electrical conductivity of SRF black-loaded SBR blend. J Appl Polym Sci 42:2879–2883
202. Barrau S, Demont P, Peigney A, Laurent C, Lacabanne C (2003) DC and AC conductivity of carbon nanotubes-polyepoxy composites. Macromolecules 36:5187–5194
203. McLachlan DS, Chiteme C, Park C, Wise KE, Lowther SE, Lillehei PT, Siochi EJ, Harrison JS (2005) AC and DC percolative conductivity of single wall carbon nanotube polymer composites. J Polym Sci Part B Polym Phys 43:3273–3287
204. Elimat ZM (2006) AC electrical conductivity of poly(methyl methacrylate)/carbon black composite. J Phys D Appl Phys 39:2824–2828
205. Jäger KM, McQueen DH, Tchmutin IA, Ryvkina NG, Klüppel M (2001) Electron transport and ac electrical properties of carbon black polymer composites. J Phys D Appl Phys 34:2699–2707

Dielectric Properties of Polymer–Carbon Composites

Suryakanta Nayak

Abstract Polymer composites are widely used in recent days for various applications. Carbon-based polymer composites have great attention mainly towards electrical applications. This chapter is focused on different carbon fillers and their polymer-based composites/nanocomposites. The effect of various carbon fillers on different polymer composites has been thoroughly described. The dielectric properties of carbon–polymer composites can be affected by many factors. These factors are as follows: processing condition, composite morphology, frequency, concentration, temperature, electric field, and pressure. The effects of all the above factors on dielectric properties of carbon–polymer composites have been discussed.

Keywords Polymer · Carbon · Dielectric · Composite · Electrical

1 Introduction

Technologically, materials play an important role to make our lives more comfortable. There is always requirement of new materials to improve the properties and to satisfy customers. So, the properties of the final product can be tailored by mixing two or more materials. Polymeric materials are more desired due to their lightweight, low cost, and ease of processing. There is a substantial expansion of nanotechnology in the field of polymer composites to develop advanced materials for different applications [1].

In the twenty-first century, nanotechnology is one of the most promising research areas. Nanocomposites are hybrid materials which consist of dispersed

S. Nayak (✉)
Department of Electrical and Computer Engineering, National University of Singapore, Singapore 117582, Singapore
e-mail: suryakanta.nayak@nus.edu.sg

M. Rahaman et al. (eds.), *Carbon-Containing Polymer Composites*,
Springer Series on Polymer and Composite Materials,
https://doi.org/10.1007/978-981-13-2688-2_6

phase (size in nanometer scale) in a suitable matrix, which can enhance the existing properties as well as gives rise to new properties [2–4]. Polymeric materials have more importance among all other materials for their electronic and mechanical properties, whether they are used in pure form or reinforced with different fillers. For optoelectronic applications, polymers also have been used to prepare light-emitting diodes (LEDs) and flexible electronic papers [5–9]. Polymer nanocomposites (nanoparticles reinforced polymer matrix) have attracted great attention of many researchers because the addition of nanoparticles into matrix polymer enhances the mechanical, electrical, thermal, and barrier properties [4, 10]. Polymer composites/nanocomposites mostly contain nanoclay, carbon fillers, nanofiber, carbon nanotubes (CNTs), or inorganic oxide particles dispersed in the suitable polymer matrix [4, 10–16]. The addition of conducting polymers (polypyrrole, polythiophene, and polyaniline) to insulating matrix-polymer, produces composites which have high dielectric constant [16]. These composites have different applications in the field of electronics like electronic packaging, angular accelerometers, integrated decoupling capacitors, and acoustic emission sensors [4, 10, 17].

In the present chapter, we have focused on the effect of different factors on dielectric properties of carbon–polymer composites.

2 Types of Carbon Fillers and Their Use in Polymer Composites

Carbon fillers are available in different forms which are used for the reinforcement (mechanical/electrical) of matrix polymer. The various forms of carbon fillers are carbon black, graphite, carbon fiber, carbon nanotubes (CNTs), graphene, reduced graphene oxide, and fullerenes [18–20]. Polymer–carbon nanocomposites have attracted much attention to many researchers for their potential applications in different fields. The incorporation of carbon filler into polymer matrix enhances the electrical and mechanical properties, which are well-known reinforcing effects of carbon fillers. There are many studies on physical and structural properties of polymer composites containing several forms of carbon. These carbon particles have different geometry, size, and physical properties. The size of carbon fillers plays an important role to reinforce the composite material due to their higher surface area. The CNTs improve the mechanical and electrical properties significantly at relatively low loading compared to other carbon fillers but carbon nanotubes have high cost. The significant improvement in properties in case of CNTs is due to their high aspect ratio.

3 Factors Affecting the Dielectric Properties of Polymer–Carbon Composites

The dielectric properties of polymer–carbon composites can be affected by many factors such as geometry, size, surface area, physical and chemical properties of filler particles, interfacial interactions between the filler and matrix-polymer, filler–filler, concentration of filler, mixing method, temperature, shear effects, composite morphology, and frequency of operation [1, 18]. The fillers used for composite preparation can be one/two/three-dimensional. The carbonaceous nanofillers such as CNTs and graphene play an important role due to their superior functional and structural properties such as high mechanical strength, high aspect ratio, and high electrical properties [1]. We have discussed a few important factors in detail which affects the dielectric/electrical properties of carbon–polymer composites.

3.1 Effect of Processing Conditions

The effect of processing conditions on dielectric properties of different polymer–carbon composites is discussed here in detail. There are different processing conditions which affect the dielectric properties of the polymer composites are as follows: mixing method, amount of loading, shear effects, and temperature [18]. Ameli et al. have reported dielectric properties of polypropylene–MWCNT composites and they have done a comparative study of dielectric properties of different nanocomposites with and without foaming. The foaming is introduced to the composite with the help of supercritical carbon dioxide ($scCO_2$). Foaming can improve different factors such as (i) filler distribution and dispersion in the composite matrix, (ii) reduce the composite weight, and (iii) decrease the electrical percolation threshold [21]. Figure 1 describes the difference in dielectric properties of PP-MWCNT composites with and without foaming and also describes the variations of permittivity (ε') and dielectric loss (tan δ) against frequency for both solid/foamed composites at different loadings of MWCNT. The change of dielectric permittivity with frequency is marginal for solid composites at lower loadings (0 and 0.26%). But there is a change in dielectric permittivity with frequency for composites containing MWCNT amount beyond 0.26%. However, foaming composites show frequency-independent dielectric behavior for composites up to a loading of 0.77% MWCNT, which shows that the change in the property is due to the change in matrix density [21].

Dalmas et al. have reported the influence of processing conditions on electrical properties and viscoelastic behavior of the flexible polymer–nanofiber composites. They prepared the nanocomposites in two different ways/methods using poly(styreneco-butyl acrylate), P(S-BuA) as the base matrix and cellulose

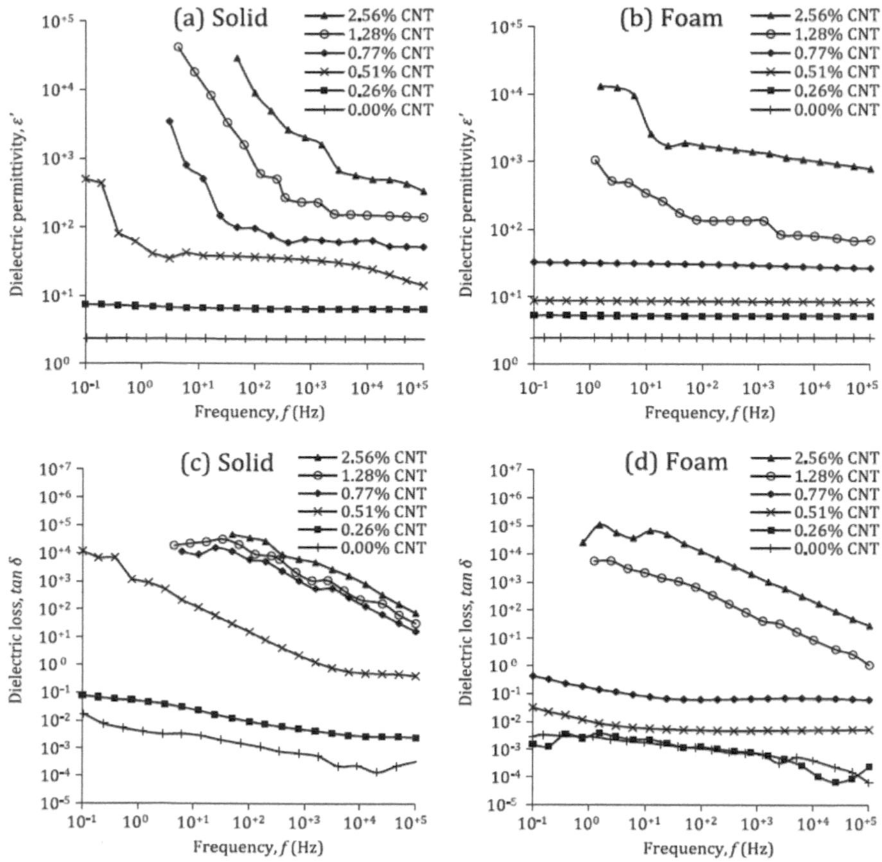

Fig. 1 Variation of **a**, **b** permittivity and **c**, **d** dielectric loss of solid/foamed PP-MWCNTs nanocomposites with frequency. Reproduced with permission of Elsevier, Ameli et al. Carbon [21]

nanofibrils/MWCNTs as filler. In both cases, suspensions of nanofillers were mixed with the above latex. In the first process, the composite mixtures were cast in an aluminum mold (Teflon coated) and dried at 35 °C under vacuum for 5 days to evaporate water slowly followed by film (named as *E*) formation. In the other method, the composite mixture was freeze-dried to make compact soft powder. Then, film (named FP) was prepared by pressing the above compact powder at 100 °C for 5 min under a pressure of 1 MPa after 45 min. Thermomechanical behavior of the above composites was studied with respect to cellulose nanofibrils/MWCNT concentration. Figure 2 describes the effect of temperature on the storage modulus of composites containing MWCNT filler of 1.5 and 3.0 vol%, which were prepared by two methods (E/FP) as described above. The pristine P(S-BuA) matrix is amorphous below glass transition temperature, T_α (281 K for *E* matrix). There is a drastic reduction in shear modulus once it reaches the glass transition temperature

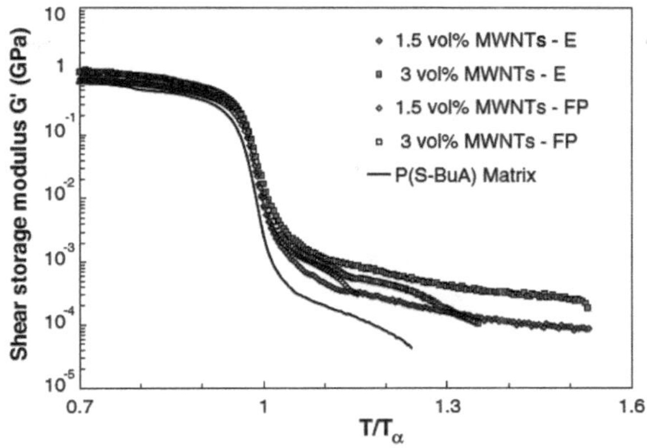

Fig. 2 Variation of shear storage modulus (G_0) with temperature (normalized) for unfilled matrix P(S-BuA) and composites filled with MWCNTs. Reproduced with permission of Elsevier, et al. Composites Science and Technology [22]

(T_α). There is mechanical reinforcement with MWCNT loading irrespective of the preparation methods and the effect of nanotube loading is moderate below T_α and higher above T_α. The increase in thermomechanical stability with MWCNT loading cannot be understood only from the interaction between polymer chains and nanotubes, but percolating nanotube network within composite should be taken into account. Figure 3 illustrates the difference between an entangled contact and a simple contact of two types of nanofibers. The elastic behavior of the structure depends on the above entanglements. The FP composites also show similar mechanical properties but these composites rapidly flow with temperature [22].

Scheme 1 explains the effect of processing conditions on thermomechanical behavior of the composites. In the first process (*E*), the nanofibers rearrangement is possible because of slow evaporation and presence of Brownian motions in suspension. The final structure is fully relaxed and has entanglements between the nanofibers. In case of second process (FP), the nanofiber movement in suspension was first frozen, and during hot pressing rearrangement of these fibers was not possible. In the final structure, there were very few entanglements leading to weak thermomechanical stability. The entanglements are less if the polymer is more viscous leading to weak strength. The increase in shear storage modulus is due to stress transfer between the nanotubes and matrix polymer [P(S-BuA)].

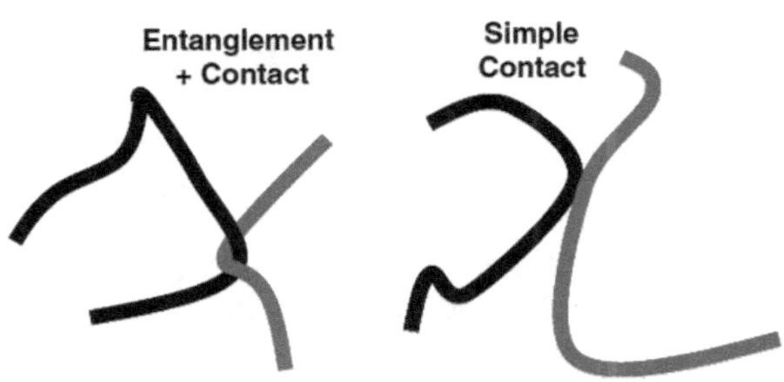

Fig. 3 Contact of two different fibers in the process of entanglement and simple fiber–fiber contact. Reproduced with permission of Elsevier, Dalmas et al. Composites Science and Technology [22]

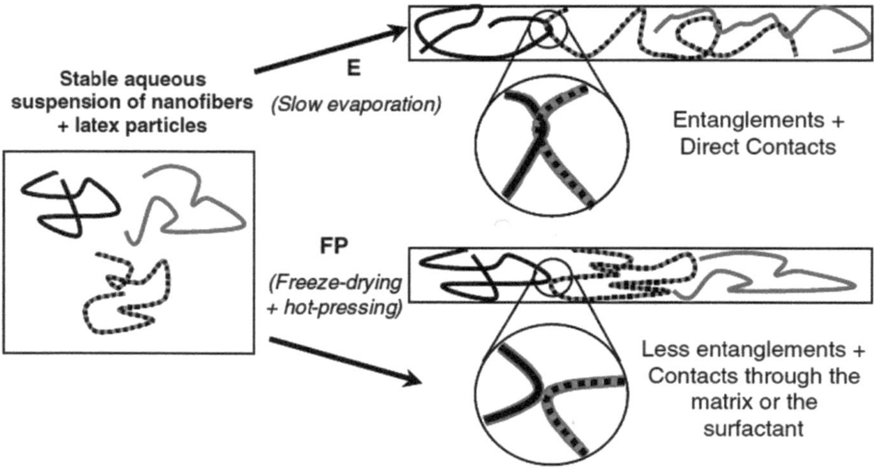

Scheme 1 Influence of processing conditions on nanofibers entanglements. Reproduced with permission of Elsevier, Dalmas et al. Composites Science and Technology [22]

3.2 *Effect of Morphology*

This section describes the effect of particle/composite morphology on dielectric properties. Particles shape and size affects both the electrical and mechanical properties of polymer composites/nanocmposites. There are many studies on the effect of particle shape and/or size on the electrical properties of polymer composites. Bai et al. have investigated the effect of MWCNTs length and aggregate size on electrical and mechanical properties of composites based on epoxy resin (Bisphenol A-epichlorhydrin) and aromatic hardener (triethylenetetramine) [23]. MWCNTs were prepared by CVD using benzene (source of carbon), ferrocene (catalyst), thiophene (growth promoter), and hydrogen (carrier gas). The prepared MWCNT nanotubes were treated chemically in three conditions (Treatment A: MWCNTs dispersed in methanol solution under magnetic agitation; Treatment B: MWCNTs passed through a 1 mm sieve without pressure followed by dispersion in magnetically agitated methanol solution; Treatment C: MWCNTs forced through a 0.5 mm sieve with pressure) to get different lengths of nanotubes. The length of nanotubes prepared through treatment 1, 2 and 3 are 50, 10, and 1 μm, respectively. The structures of the above nanotubes are shown in the SEM image (Fig. 4).

The change in conductivity as a function of nanotube loading and three treatment conditions are shown in Fig. 5. The effect of nanotubes length is clearly seen, both on the shape of curves and on the transition of insulator-to-conductor transition. Composites with type A MWNTs have better performance in terms of conductivity. It is observed that composite with type A nanotubes shows percolation threshold (conducting path formed due to nanotubes interconnection and aggregates) and at a very low loading whereas type C composite shows a marginal change in

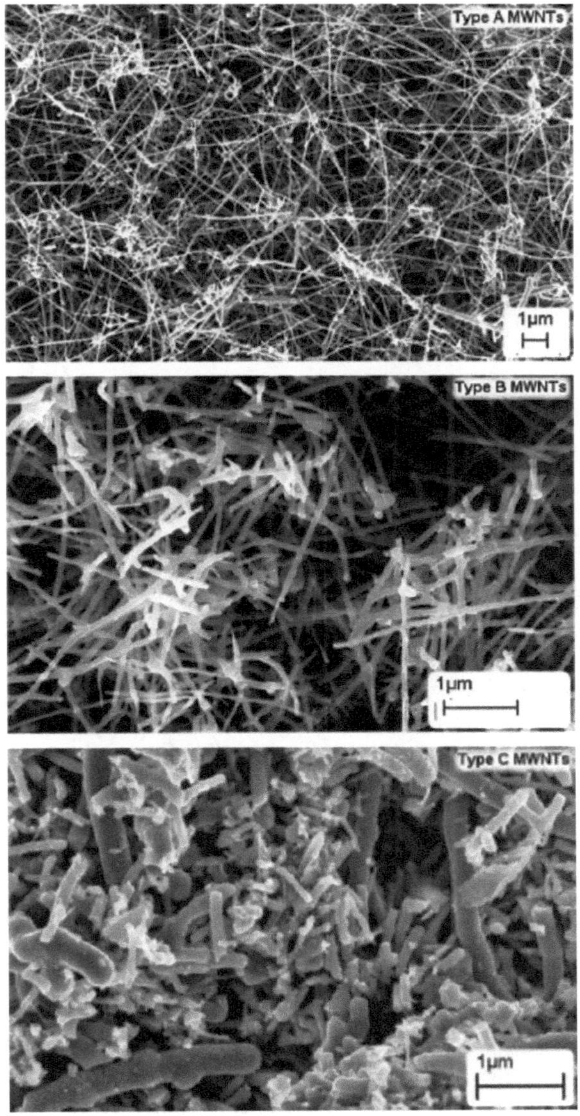

Fig. 4 SEM images of MWNTs after three different treatment conditions. Reproduced with permission of Elsevier, Bai et al. Composites Part A: Applied Science and Manufacturing [23]

conductivity up to 4 wt% loading. The length of MWCNTs has the measure role in improving the electrical conductivity of composites/nanocomposites [23].

The effect of alignment of single-walled nanotubes (SWNTs) on electrical properties is studied by Wang et al. [24]. The composites were prepared from SWNTs and epoxy as the base matrix where the composites were stretched many times (50–100) when it was half cured to get the SWNTs aligned. The electrical properties of the composites with aligned SWNTs are compared with composites

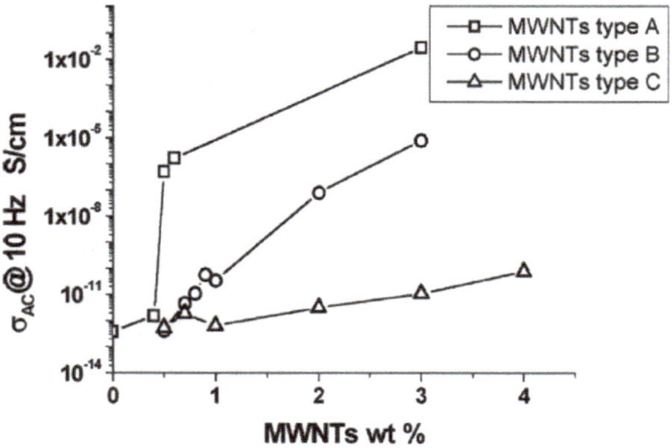

Fig. 5 Variation of conductivity of the composites with different concentrations of MWNTs prepared by three different treatment conditions. Reproduced with permission of Elsevier, Bai et al. Composites Part A: Applied Science and Manufacturing [23]

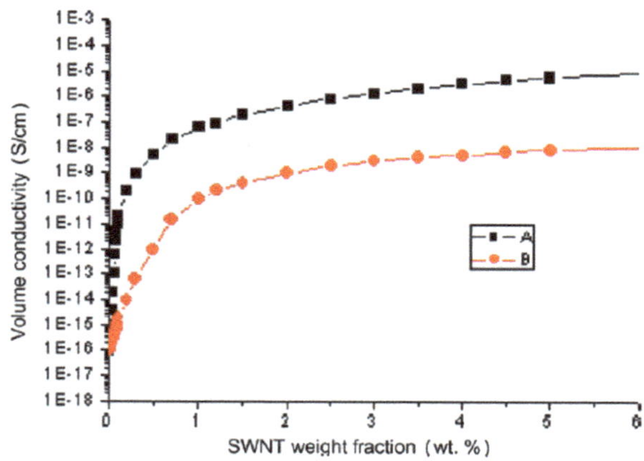

Fig. 6 Variation of DC volume conductivity with SWNT loading for epoxy/SWNT composites: (A) aligned SWNTs and (B) perpendicular SWNTs. Reproduced with permission of Elsevier, Wang et al. Composites Science and Technology [24]

with the perpendicular SWNTs. The composites with aligned SWNTs have more electrical conductivities and mechanical properties compared to the composites with perpendicular SWNTs. There is also an increase in electrical conductivity with the increase in the concentration of SWNTs. The variation of DC volume conductivity of both types of SWNT/epoxy resin composites as a function of SWNT concentration is presented in Fig. 6. The composite with aligned SWNTs loading of 0.5–3 wt%, conductivity changed from 5×10^{-9} to 1×10^{-6} S/cm, whereas the composite with perpendicular SWNTs, the conductivity changed from 1×10^{-12} to 2.8×10^{-9} S/cm. The percolation thresholds of the epoxy–SWNT composites can be determined from the first derivatives of logarithmic conductivity (electrical). Figure 7a, b shows the variation of derivative log (conductivity) with two types of SWNT loading. The percolation thresholds for epoxy–SWNT composite in the SWNT aligning direction is 0.05 wt%, whereas 0.5 wt% in case of perpendicularly aligned SWNT. The significant difference is due to the anisotropic electrical property of CNT and ultrahigh aspect ratio [24].

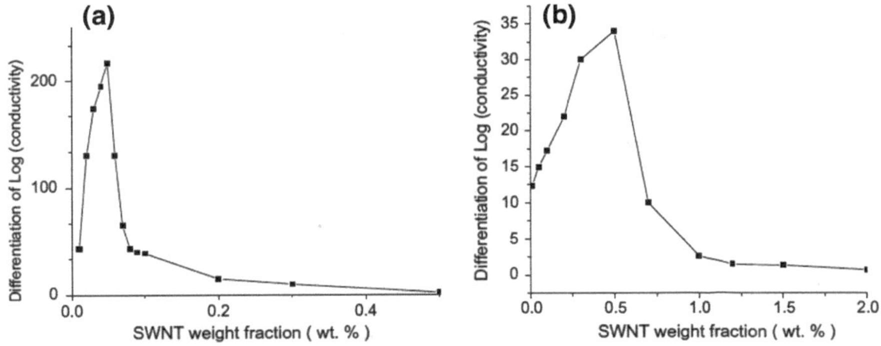

Fig. 7 Variation of derivative log (conductivity) with SWNTs (two types) loading for SWNT/epoxy composite: **a** aligned SWNTs and **b** perpendicular SWNTs. Reproduced with permission of Elsevier, Wang et al. Composites Science and Technology [24]

3.3 *Effect of Frequency and Filler Concentration*

The concentration of filler particles in polymer composite is an important parameter to control the bulk properties of polymer composites/nanocomposites. The electrical and mechanical properties can be tailored by varying the concentration of filler with respect to the base polymer matrix. Frequency is also another parameter which affects the dielectric properties of polymer composites. The effect of frequency on permittivity at room temperature has been studied by Dang et al. for PVDF-trifluorophenyl (TFP) functionalized MWNTs composites. They found that the value of dielectric permittivity is 900 at 10^2 Hz for PVDF-TFP-MWNT composites with loading of ca. 0.08 (percolation threshold) which is 90 times larger than neat PVDF. Moreover, it is observed that the maximum permittivity is more than 8000 at 100 Hz for composite with a loading of ca. 0.15, which is approximately four orders of magnitude larger than the value for neat PVDF. The loading of ca. 0.08 fulfills the percolation limit whereas the composite was still electrically insulating due to thin layer of insulating PVDF. There is significant dielectric relaxation observed with the increase in frequency between 10^2 and 10^4 Hz (Fig. 8). High permittivity at low frequency is due to the accumulated charge carriers at the interface (Maxwell–Wagner–Sillars effect) which may produce heterogeneous capacitors. However, the permittivity of PVDF-MWNT composites was dominated by PVDF, when the frequency was more than 10^5 Hz. Dang et al. have also studied the effect of TFP-MWNT concentration with dielectric permittivity at a frequency of 1 kHz and at room temperature (Fig. 9). It is observed that both conductivity and permittivity increased with the increase in the concentration of TFP-MWNT in PVDF matrix whereas there is a sharp increase in dielectric permittivity once it reached the threshold limit (ca. 0.08). The dielectric permittivity reaches a maximum at loading of ca. 0.15 beyond which there is a sharp decrease in the value of dielectric permittivity [25].

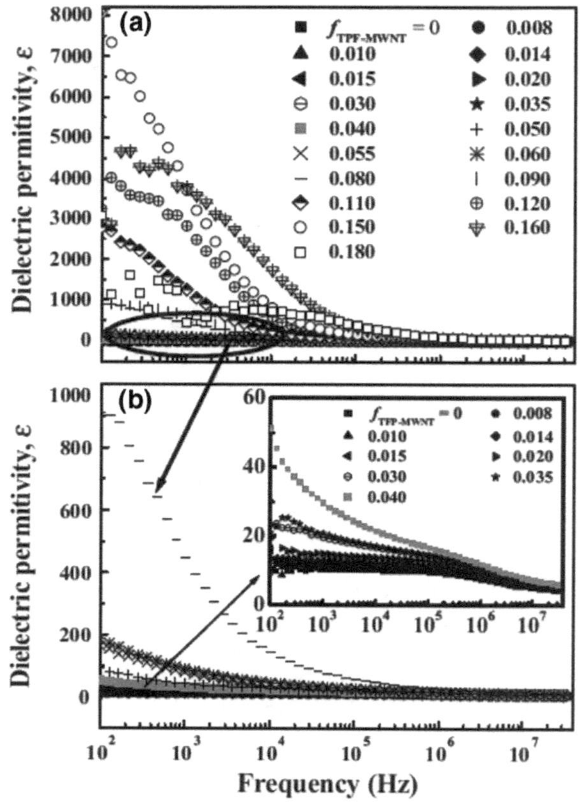

Fig. 8 Variation of permittivity with frequency for TFP-MWNT/PVDF composites at room temperature, when the volume fraction of TFP-MWNTs is **a** 0–0.15 and **b** 0–0.08. Inset in (**b**) shows variation of permittivity with frequency at room temperature when the volume fraction of TFP-MWNTs was 0–0.04. Reproduced with the permission of Wiley-VCH Verlag GmbH and Co. KGaA., Dang et al. Advanced Materials [25]

Seo et al. have studied the rheological behavior and electrical resistivity of CNT filled polypropylene composites. They found that there is a decrease in DC resistivity with an increase in the concentration of carbon nanotubes in polypropylene matrix (Fig. 10). However, there is a significant reduction in resistivity observed at 2 wt% of carbon nanotubes loading. The significant change in resistivity is due to the formation of MWNTS interconnects. This indicates that there is a high percentage of electron flow through the specimen due to the formation of an interconnecting conductive pathway at a loading in between 1 and 2 wt% of MWNT. There is a marginal decrease in resistivity which is observed with a further increase in the concentration of MWNTs in polypropylene matrix [26].

Song et al. have studied the influence of dispersion states of CNTs on the physical properties of epoxy composites. They have studied the dispersion of CNTs in epoxy resin matrix with/without solvent and found that composite with solvent shows better dispersion compared to the composite prepared without solvent (Figs. 11 and 12). Figure 11a shows the TEM image of epoxy–CNT composites prepared with solvent whereas Figs. 11b and 12 show the FESEM images of

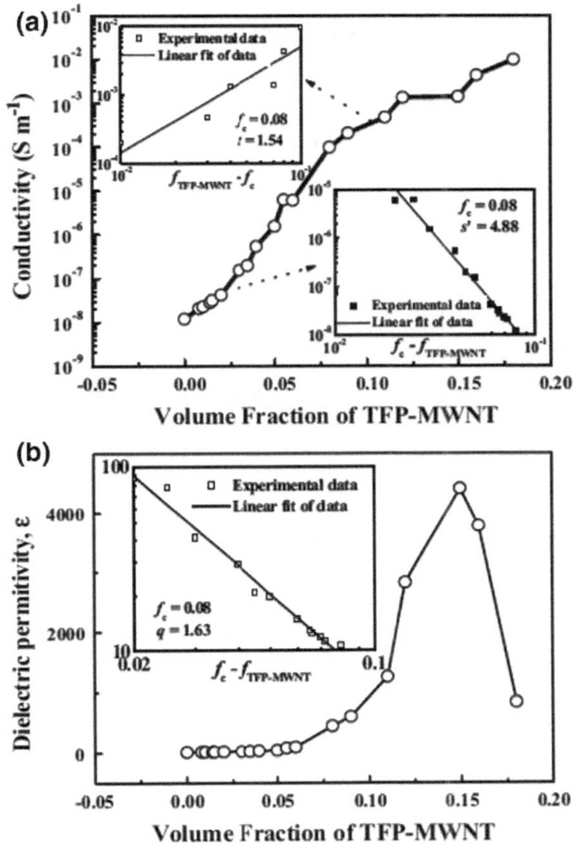

Fig. 9 Variation of conductivity and permittivity with TFP-MWNT volume fraction in TFP-MWNT/PVDF composites, measured at room temperature and 10^3 Hz. Reproduced with the permission of Wiley-VCH Verlag GmbH and Co. KGaA., Dang et al. Advanced Materials [25]

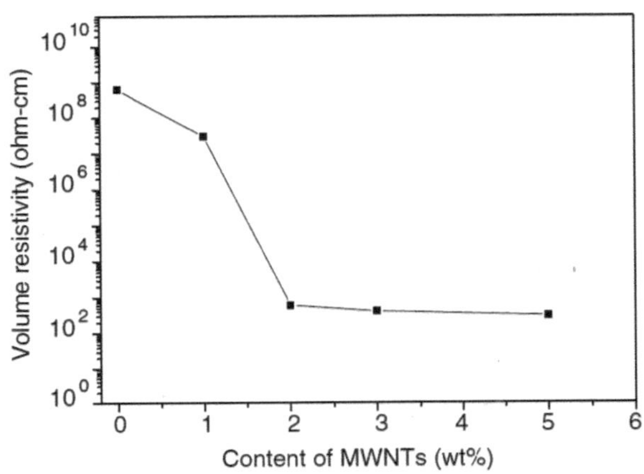

Fig. 10 Variation of volume resistivity with MWNT loading for MWNTs/PP composites. Reproduced with permission of Elsevier, Seo et al. Chemical Physics Letters [26]

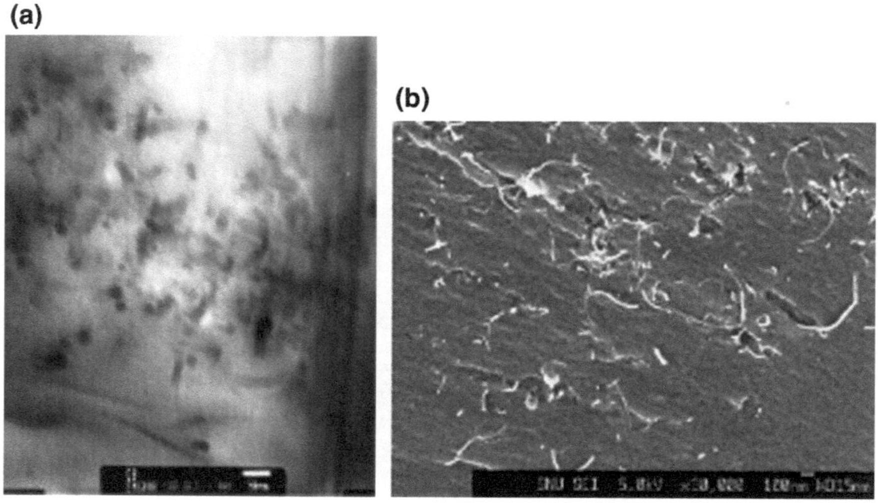

Fig. 11 **a** TEM and **b** FESEM images of fractured surface of composites prepared by using solvent. Reproduced with permission of Elsevier, Song et al. Carbon [27]

epoxy–CNT composites prepared with and without solvent, respectively. FESEM samples were chosen from the tensile fractured samples. Agglomerations are observed in case of composites prepared without solvent (Fig. 12). The white portion in Fig. 12a represents the agglomeration part and one part (rectangle in Fig. 12a) is enlarged and showed in Fig. 12b in order to verify the existence of the CNTs. The dispersion of CNTs in the epoxy matrix also affects the electrical conductivity of the composites. The well-dispersed composite shows better/more electrical conductivity compared to the poorly dispersed composites (Fig. 13) [27]. There are six orders of magnitude increase in conductivity with loading up to 1.5 wt % CNTs for both poor and well-dispersed composites. However, well-dispersed composites show one order more conductivity compared to the poorly dispersed composites. So, the well-dispersed CNTs form the conducting path due to their homogeneous dispersion than the poorly dispersed composites. It is observed that the percolation limit is seen at a lower loading of 0.5 wt% or less which is more lower than the conventional composites [27].

Potschke et al. studied the rheological and dielectric properties of melt mixed polycarbonate–MWCNT composites. They have prepared the composites of MWCNT and polycarbonate (PC) by melt mixing technique. The dielectric measurements were carried out at room temperature in the frequency range of 10^{-3}–10^{7} Hz and results presented in Fig. 14. The effect of both frequency and MWCNT concentration with the real part permittivity (ε') and real part of conductivity (δ') is described in Fig. 14. The PC-MWCNT composites show two types of behaviors which change with the concentration of MWCNT in PP matrix.

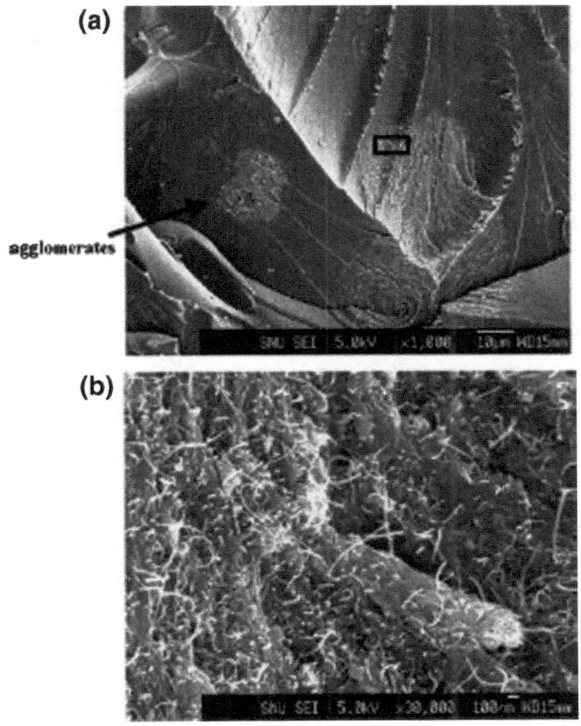

Fig. 12 FESEM images of CNTs/epoxy composites prepared without using solvent. Reproduced with permission of Elsevier, Song et al. Carbon [27]

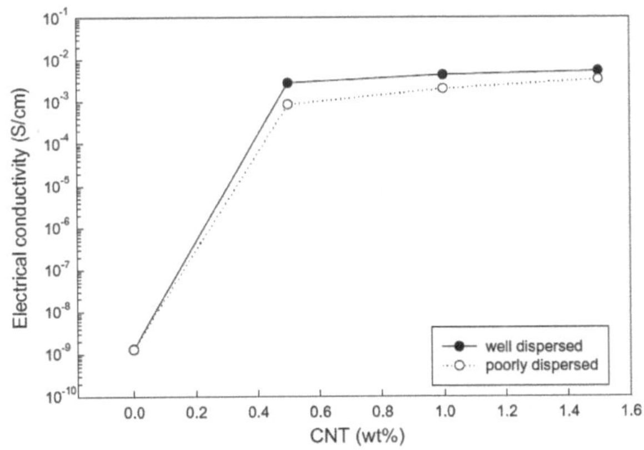

Fig. 13 Electrical conductivity of composites with different degrees of CNTs dispersion as a function of carbon content. Reproduced with permission of Elsevier, Song et al. Carbon [27]

The composites with less than 1.0 wt% of MWCNT show nearly constant ε′ value, whereas composites with 1.0 wt% or more loading show frequency-dependent behavior. The significant difference is also seen in conductivity spectra. The percolation in composites is observed in between 0.875 and 1.0 wt% [28].

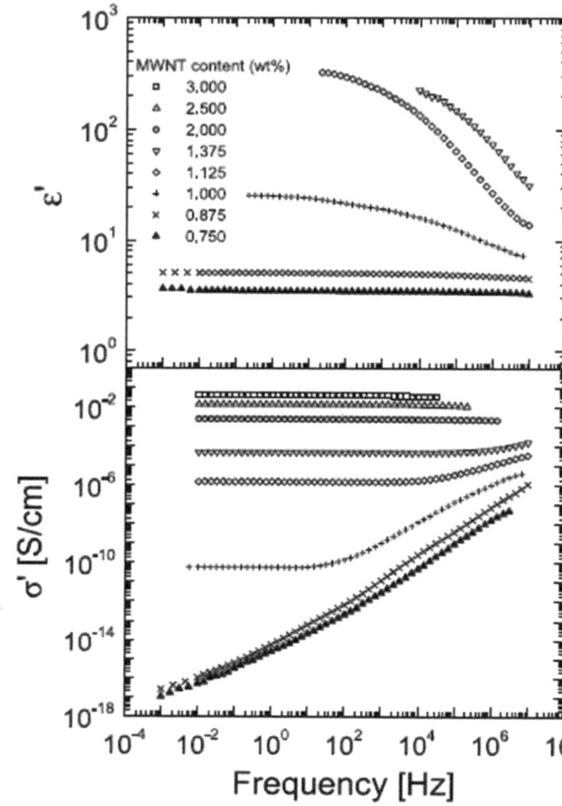

Fig. 14 Variation of permittivity and conductivity with frequency for PC-MWCNT composites at different concentrations of MWCNT. Reproduced with permission of Elsevier, Potschke et al. Polymer [28]

3.4 Effect of Temperature

The permittivity of any material depends upon the orientation of the permanent dipoles. The dipoles can be orientated by the change in temperature leading to a change in permittivity of the material. However, the change in permittivity with the increase in temperature does not follow the same path. The increase in permittivity with an increase in temperature is due to the increase in mobility of bound charges such as interfacial space charges and induced dipoles. The thermal excitation of these charges with an increase in temperature is the main reason for the increase in permittivity. The decrease in permittivity beyond a certain temperature is due to the effect of thermal expansion of matrix leading to a decrease in charge density. The electrical conductivity of polymer composites also increases with the increase in temperature but up to the certain limit as reported in the previous literature reports [29–32].

The effect of temperature on dielectric properties of the TFP-MWNT/PVDF nanocomposites has been studied at 1 kHz by Dang et al. There is an increase in

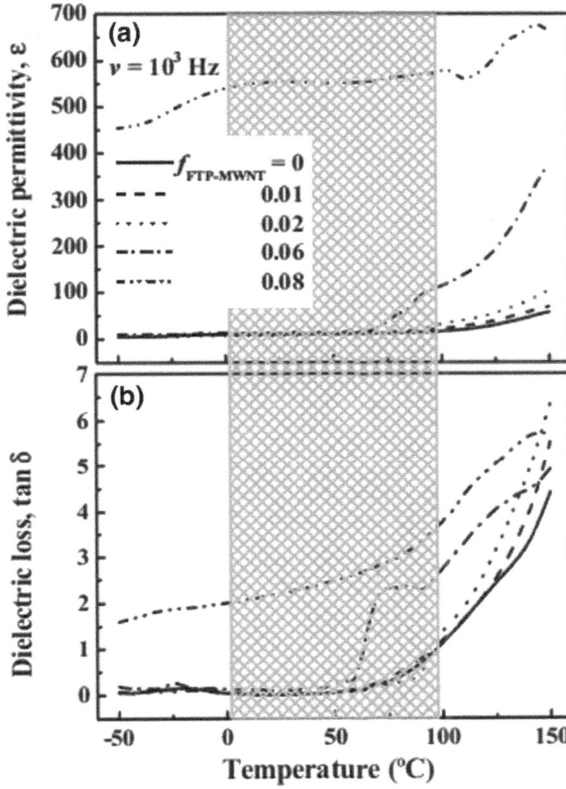

Fig. 15 Variation of permittivity and tan δ with temperature of TFP-MWNT/PVDF nanocomposites at 10^3 Hz. Reproduced with permission of Wiley-VCH Verlag GmbH & Co. KGaA., Dang et al. Advanced Materials [25]

both dielectric permittivity and loss with the increase in temperature, which is due to the induction of polarization at high temperatures (Fig. 15). The permittivity and loss of TFP-MWNT/PVDF composites with a loading of 0.06 (volume fraction) increased with the temperature above 70 °C. The composite with a loading of 0.08 (volume fraction) always shows higher permittivity and loss compared to other composites over a wide range of temperature, −50 to 150 °C. The dielectric permittivity at percolation threshold maintained a stable value of 550 with a temperature range of 0–100 °C [25]. Yousefi et al. have studied the effect of temperature on permittivity and conductivity of rGO/epoxy nanocomposites and represented in Fig. 16a, b, respectively. Neat epoxy and rGO/epoxy nanocomposites at a loading of 0.1 wt% of graphene have very low permittivity value (approx. 5) whereas there is a significant increase in the permittivity value at a loading of 0.5 wt% of graphene (Fig. 16a). The rGO/epoxy nanocomposites maintained high permittivity at all temperatures (Fig. 16b). There are peaks at about 70–80 °C, especially in composites with high rGO concentration, which coincides with the T_g, of composites. The permittivity of any polymer shows a peak at T_g as a result of molecular α-relaxation. The activation and orientation of the dipoles are responsible for the

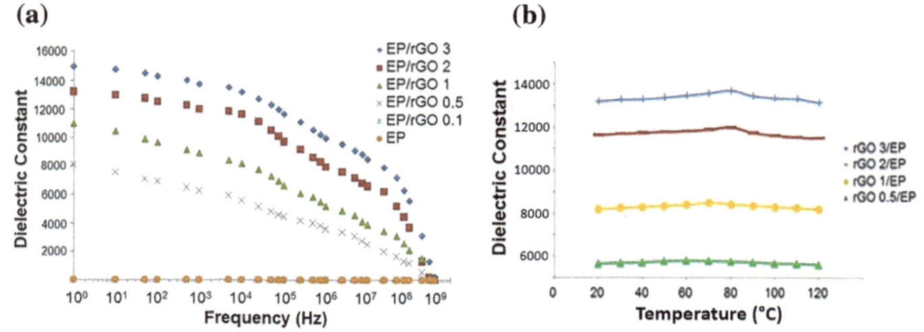

Fig. 16 Variation of dielectric constant with **a** frequency and **b** temperature for rGO/epoxy composites. Reproduced with permission of Wiley-VCH Verlag GmbH & Co. KGaA., Yousefi et al. Advanced Materials [20]

above peak when molecules start to vibrate vigorously. Figure 17 shows the effect of temperature on the electrical conductivity of rGO/epoxy nanocomposites. There is a significant increase in conductivity observed for composites with an increase in temperature (ambient to 100 °C) at low loading of graphene. However, there is a marginal change in conductivity with temperature for composites at higher loadings of 1.0 wt% or more [20].

3.5 *Effect of Applied Pressure*

The dielectric/electrical properties of polymer composites/nanocomposites are affected by external stress/pressure. The decrease/increase in conductivity/resistivity and dielectric properties may depend on type and nature of the matrix polymer and filler type. In case of conducting polymer composites, the electrical conduction mainly depends on the electrons/holes, where conductivity increases

Fig. 17 Variation of conductivity with temperature for rGO/epoxy composites. Reproduced with permission of Wiley-VCH Verlag GmbH & Co. KGaA., Yousefi et al. Advanced Materials [20]

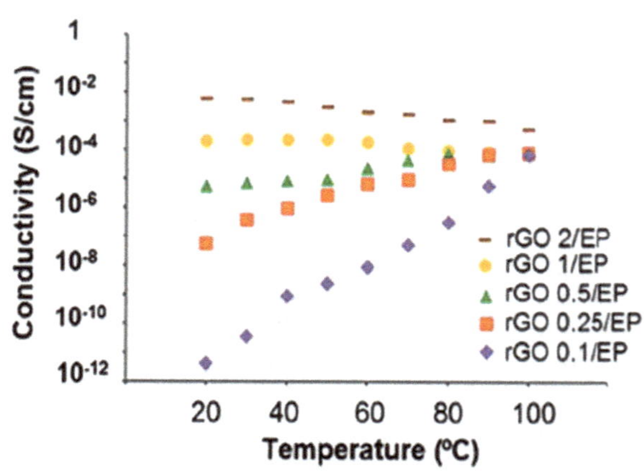

with the increase in the applied stress [33]. However, the electrical conductivity of insulating polymer composites increases with the external stress/pressure, where the electrical conduction is due to the ionic charge carriers [34]. The effect of pressure on the electrical properties of polymer composites has been studied by many researchers. Mohanraj et al. have studied the effect of pressure on AC conductivity (σ_{ac}) of SBR-particulate metal alloy nanocomposites and found the conductivity increased with increase in pressure. They have also studied the effect of constant stress for a certain time and found that after a particular time, there is a marginal change in conductivity [30]. A similar study on CNT-PEEK composites has been reported in the literature [31, 32].

Hu et al. have reported the effect of external pressure on the change in resistance of PDMS/MWNT composites. Figure 18 demonstrates the change in relative resistance of PDMS/MWNT nanocomposites with different MWCNT loadings. There are different responses to the change in relative resistance of different PDMS/MWNT composites by the applied stress/pressure. The aspect ratio of the used CNT was about 1000, so they form entanglement in the composites as seen in Fig. 19. After the application of external pressure, these CNTs inside composite may bend further and have larger curvatures (inset of Fig. 18) followed by change in resistance of the composites. It is observed that the MWCNT mat also shows a change in resistance with pressure (Fig. 19). The MWCNT mat was deposited from an ethanol suspension. So, there are gaps in between the nanotubes and these gaps can be compressed under applied pressure and create a better contact in between the nanotubes. The applied pressure causes the CNTs to bend further and leads to an increase in resistance up to 1 MPa but there is a marginal change in resistance

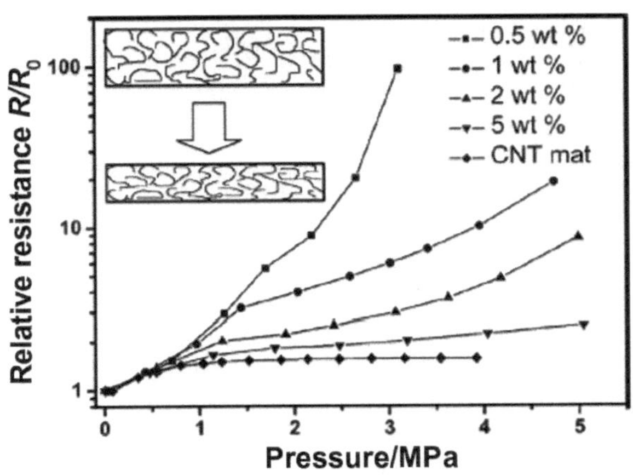

Fig. 18 Variation of relative resistance with applied pressure on MWCNT/PDMS composites with different concentrations of MWCNT, where R_0 is the ambient condition resistance. Inset: schematic representation of the MWCNT networks in PDMD composite before and after application of external pressure. Reproduced with permission of AIP Publishing, Hu et al. Applied Physics Letters [35]

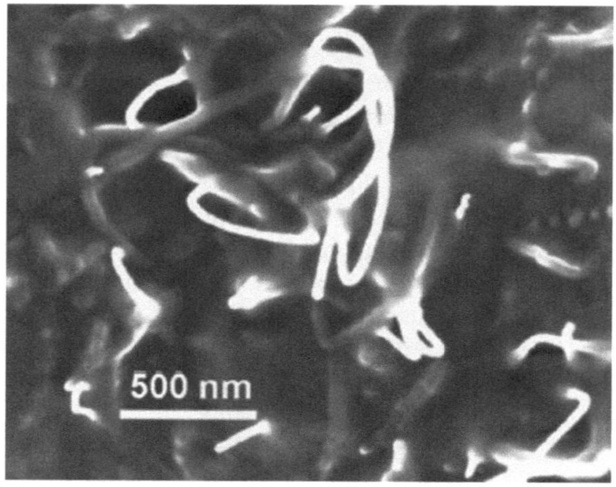

Fig. 19 SEM image of fractured surface of PDMS-MWCNT composite with 5 wt% CNT loading. Reproduced with permission of AIP Publishing, Hu et al. Applied Physics Letters [35]

beyond a pressure of 1 MPa. It is also observed that the composites with lower MWCNT loadings have more resistance change under the same applied pressure. The modulus of PDMS/MWNT composites is more for composites with higher loading. Therefore, composites with lower MWCNT loading experience greater curvatures than those in high loading composites. This is the reason why composites with low loading show more resistance change than the composites with high loading at same pressure [35]. Hwang et al. have studied the effect of applied pressure on the resistance change of Poly(3-hexylthiophene) wrapped CNT/poly (dimethylsiloxane) [P3HT-MWCNT/PDMS] composites which can be used as finger-sensing piezoresistive pressure sensors. They have studied the effect of pressure on the change in resistance of three different types of composites (Fig. 20b). These composites differ from each other by the difference in weight fraction of P3HT (1, 2, and 3 wt%) where the concentration of MWCNT remains constant (1 wt%) in all composites. Figure 20a shows the schematic representation of the experimental setup used for determination of piezoresistive properties. The resistance change with applied pressure can be explained by two factors: (i) formation of new conducting networks or (ii) destruction of existing conductive networks. The composites with spherical conducting fillers induce the formation of conducting networks upon applied pressure. Therefore, the relative resistance decreases with applied pressure in above composites. In contrast, composites with high aspect ratio fillers (e.g., MWCNT) destruct the existing conducting network, and hence, relative resistance increases upon the application of external pressure. In the present case, the filler particles are dispersed well and a low percolation threshold observed due to P3HT wrapping leading to a "PPC" effect over the pressure range of 0–0.12 MPa (Fig. 20b). The average distance between MWCNTs particles in composite with a 3:1 ratio of P3HT:MWCNT was larger than that of the 1:1 P3HT:MWCNT composite (Fig. 20c, d). Therefore, there is a significant change in resistance as seen in Fig. 20b due to the increase in the concentration of P3HT [36].

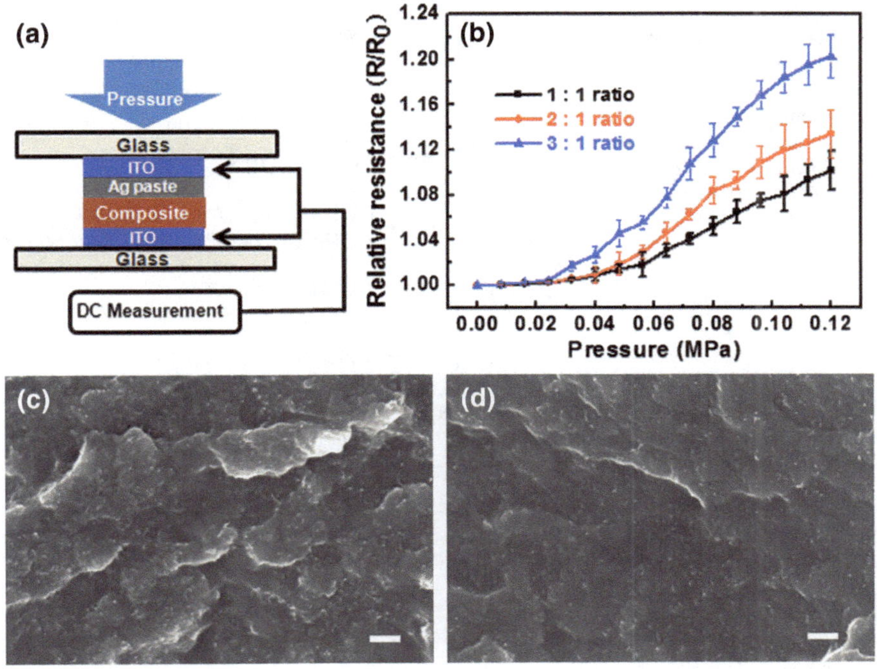

Fig. 20 **a** Schematic representation of experimental setup used for the determination of piezoresistive properties; **b** variation of relative resistances with applied pressure for P3HT-MWCNTs/PDMS composites at different ratios of P3HT:MWCNTs; and **c**, **d** SEM images of P3HT-MWCNTs/PDMS composites with two different ratios (1:1 and 3:1) of P3HT:MWCNTs. Reproduced with permission of Elsevier, Hwang et al. Carbon [36]

3.6 Effect of Electric Field (Poling)

The process in which dipoles are oriented in the same direction by the application of high voltage is known as poling. The direction along which the dipoles align is known as the poling direction. Sa-Gong et al. have studied the effect of poling on piezoelectric properties of epoxy (Ecogel)–PZT composites. The conducting particles (carbon/germanium/silicon) were added to the above composites to control the conductivity. Rapid poling at low voltages is possible by controlling the conductivity, resulting in properties comparable with the composites prepared without a conductive phase. Figure 21 shows the electrical resistivities of the carbon–polymer, germanium–polymer, and PZT–polymer composites with/without conductive phase where all measurements were done at a poling temperature of 100 °C. The resistivities of Ecogel polymer (epoxy) and PZT are also listed for comparison. The resistivity of Ecogel polymer is reduced by order of magnitude by the addition of a small amount of germanium or carbon. The resistivity of polymer–PZT composite also decreases with the addition of germanium or carbon. Figure 22

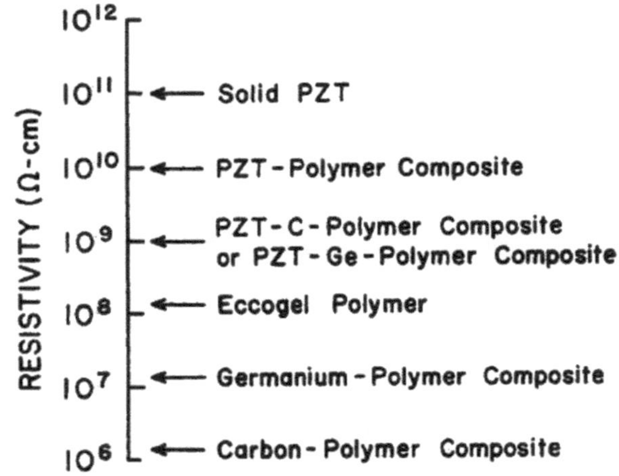

Fig. 21 Resistivity of the PZT, the Ecogel polymer (1365-0), and several other composites measured at poling temperature. Reproduced with permission of Taylor & Francis Group, Sa-Gong et al. Ferroelectrics Letters Section [37]

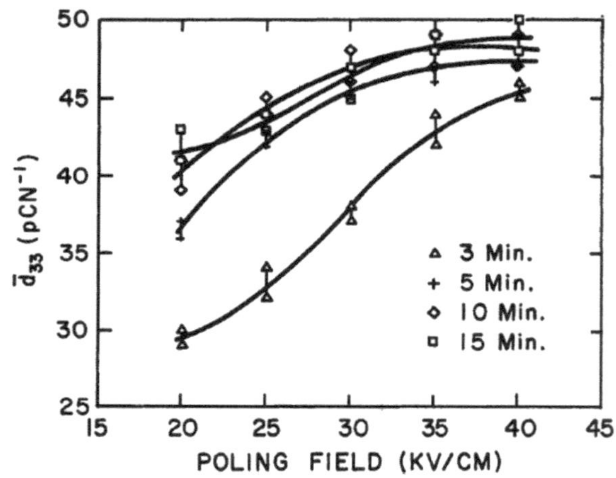

Fig. 22 Variation of piezoelectric coefficient of PZT–carbon–Ecogel composite with poling field. Reproduced with permission of Taylor & Francis Group, Sa-Gong et al. Ferroelectrics Letters Section [37]

describes the change in d33 coefficient as a function of poling voltage for composite containing 68.5 vol% PZT, 1.5 vol% carbon, and 30 vol% Ecogel 1365-0. All the composites were poled at 100 °C for different times and found that the d33 coefficient value increases with the poling voltage and saturated at about 35 kV/cm. It is also observed that 5 min poling duration is enough for a composite [37].

Park et al. have developed flexible and large-area nanocomposite generators (NCGs) based on PZT particles and CNTs. The NCGs were prepared from a composite of PDMS (Sylgard 184), PZT particles, and carbon nanotubes. The above NCG was sandwiched between two layers of PDMS dielectric followed by two layers of ITO coated PET substrates and subjected to poling at 140 °C by applying an external electric field from 0.5 to 2 kV for 12 h. Figure 23 shows the measured output voltages of NCG devices poled by an external electric field

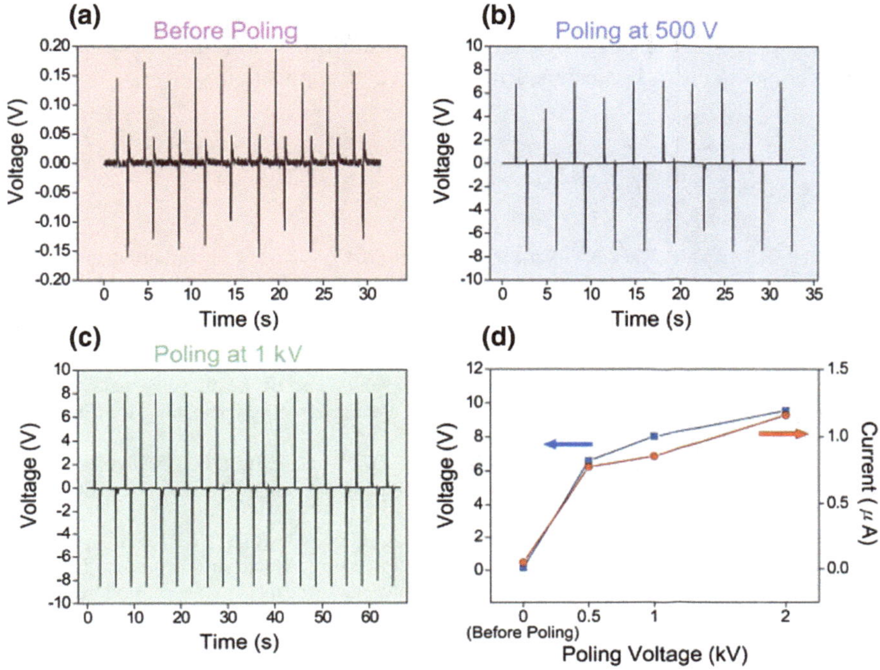

Fig. 23 **a** The measured output voltage before poling process. **b**, **c** After poling process at 500 V (**b**) and 1 kV (**c**), the converted output voltage from periodical bending/unbending deformation. **d** Output voltage and current signals measured from NCG device as a function of poling voltage. By increasing poling voltage, outputs also gradually increase up to ~10 V. Reproduced with permission of Wiley-VCH Verlag GmbH & Co. KGaA., Park et al. Advanced Energy Materials [38]

of 0, 500, and 1000 V. Low output voltage for the non-poled NCG device increases to 8.2 V, when the device is poled at 140 °C under 1000 V for 12 h. The amplitude of the current signal also increases as a function of poling voltage (Fig. 23d) [38].

4 Summary and Conclusions

Polymer–carbon composites are promising materials with emerging applications in various fields such as nanotechnology, materials science, and electronic applications (e.g., EMI shielding). However, different types of composites need to be prepared from various polymers and carbon fillers like carbon black, carbon nanotubes, and carbon fiber to fulfill the application requirement. This chapter focused on different types of carbon fillers and their polymer-based composites and nanocomposites. The above chapter also described many factors which affect the dielectric properties of polymer–carbon composites. Based on the above factors, a proper composite can be designed for a particular application.

References

1. Mittal G, Dhand V, Rhee KY, Park S-J, Lee WR (2015) A review on carbon nanotubes and graphene as fillers in reinforced polymer nanocomposites. J Ind Eng Chem 21:11–25
2. Youssef AM (2013) Polymer nanocomposites as a new trend for packaging applications. Polym-Plast Technol Eng 52:635–660
3. Manias E (2007) Nanocomposites stiffer by design. Nat Mater 6:9–11
4. Nayak S, Chaki TK, Khastgir D (2014) Development of flexible piezoelectric poly (dimethylsiloxane)–$BaTiO_3$ nanocomposites for electrical energy harvesting. Ind Eng Chem Res 53(39):14982–14992
5. Koulouridis S, Kiziltas G, Zhou Y, Hansford DJ, Volakis JL (2006) Polymer–ceramic composites for microwave applications: fabrication and performance assessment. IEEE Trans Microw Theory Tech 54(12):4202–4208
6. Rogers JA, Bao Z, Baldwin K, Dodabalapur A, Crone B, Raju VR, Kuck V, Katz H, Amundson K, Ewing J, Drzaic P (2001) Paper-like electronic displays: large-area rubber-stamped plastic sheets of electronics and microencapsulated electrophoretic inks. Proc Natl Acad Sci USA 98(9):4835–4840
7. Hwang DH, Chuah BS, Li XC, Kimt ST, Moratti SC, Holmes AB (1998) New luminescent polymers for LEDs and LECs. Macromol Symp 125(1):111–120
8. Li R, DeJean G, Tentzeris MM, Papapolymerou J, Laskar J (2005) Radiation-pattern improvement of patch antennas on a large-size substrate using a compact soft-surface structure and its realization on LTCC multilayer technology. IEEE Trans Antennas Propag 53 (1):200–208
9. Nayak S, Chaki TK, Khastgir D (2016) Dielectric relaxation and viscoelastic behavior of polyurethane–titania composites: dielectric mixing models to explain experimental results. Polym Bull 74(2):369–392
10. Nayak S, Kumar Chaki T, Khastgir D (2012) Development of poly(dimethylsiloxane)/$BaTiO_3$ nanocomposites as dielectric material. Adv Mater Res 622–623:897–900
11. Gam S, Meth JS, Zane SG, Chi C, Wood BA, Winey KI, Clarke N, Composto RJ (2012) Polymer diffusion in a polymer nanocomposite: effect of nanoparticle size and polydispersity. Soft Matter 8(24):6512
12. Liang GD, Tjong SC (2006) Electrical properties of low-density polyethylene/multiwalled carbon nanotube nanocomposites. Mater Chem Phys 100(1):132–137
13. Patel HA, Somani RS, Bajaj HC, Jasra RV (2006) Nanoclays for polymer nanocomposites, paints, inks, greases and cosmetics formulations, drug delivery vehicle and waste water treatment. Bull Mater Sci 29(2):133–145
14. Segal E, Tchoudakov R, Narkis M, Siegma A (2002) Thermoplastic polyurethane-carbon black compounds: structure, electrical conductivity and sensing of liquids. Polym Eng Sci 42 (12):2430–2439
15. Nayak S, Sahoo B, Kumar Chaki T, Khastgir D (2013) Development of polyurethane–titania nanocomposites as dielectric and piezoelectric material. RSC Adv 3(8):2620
16. Nayak S, Rahaman M, Pandey AK, Setua DK, Chaki TK, Khastgir D (2013) Development of poly(dimethylsiloxane)-titania nanocomposites with controlled dielectric properties: effect of heat treatment of titania on electrical properties. J Appl Polym Sci 127(1):784–796

17. Manna R, Nayak S, Rahaman M, Khastgir D (2014) Effect of annealed titania on dielectric and mechanical properties of ethylene propylene diene monomer-titania nanocomposites. e-Polymers 14(4):267–275
18. Kasgoz A, Akın D, Ayten AI, Durmus A (2014) Effect of different types of carbon fillers on mechanical and rheological properties of cyclic olefin copolymer (COC) composites. Compos B Eng 66:126–135
19. Zhou Z, Wang S, Zhang Y, Zhang Y (2006) Effect of different carbon fillers on the properties of PP composites: comparison of carbon black with multiwalled carbon nanotubes. J Appl Polym Sci 102(5):4823–4830
20. Yousefi N, Sun X, Lin X, Shen X, Jia J, Zhang B, Tang B, Chan M, Kim J-K (2014) Highly aligned graphene/polymer nanocomposites with excellent dielectric properties for high-performance electromagnetic interference shielding. Adv Mater 26:5480–5487
21. Ameli A, Nofar M, Park CB, Tschke PP, Rizvi G (2014) Polypropylene/carbon nanotube nano/microcellular structures with high dielectric permittivity, low dielectric loss, and low percolation threshold. Carbon 71:206–217
22. Dalmas F, Cavaille J-Y, Gauthier C, Chazeau L, Dendievel R (2007) Viscoelastic behavior and electrical properties of flexible nanofiber filled polymer nanocomposites. Influence of processing conditions. Compos Sci Technol 67:829–839
23. Bai JB, Allaoui A (2003) Effect of the length and the aggregate size of MWNTs on the improvement efficiency of the mechanical and electrical properties of nanocomposites-experimental investigation. Compos A Appl Sci Manuf 34:689–694
24. Wang Q, Dai J, Li W, Wei Z, Jiang J (2008) The effects of CNT alignment on electrical conductivity and mechanical properties of SWNT/epoxy nanocomposites. Compos Sci Technol 68:1644–1648
25. Dang Z-M, Wang L, Yin Y, Zhang Q, Lei Q-Q (2007) Giant dielectric permittivities in functionalized carbon-nanotube/electroactive-polymer nanocomposites. Adv Mater 19:852–857
26. Seo M-K, Park S-J (2004) Electrical resistivity and rheological behaviors of carbon nanotubes-filled polypropylene composites. Chem Phys Lett 395:44–48
27. Song YS, Youn JR (2005) Influence of dispersion states of carbon nanotubes on physical properties of epoxy nanocomposites. Carbon 43:1378–1385
28. Potschke P, Abdel-Goad M, Alig I, Dudkin S, Lellinger D (2004) Rheological and dielectrical characterization of melt mixed polycarbonate-multiwalled carbon nanotube composites. Polym Bull 45:8863–8870
29. Li Q, Xue QZ, Gao XL, Zheng QB (2009) Temperature dependence of the electrical properties of the carbon nanotube/polymer composites. eXPRESS Polym Lett 3(12):769–777
30. Mohanraj GT, Dey PK, Chaki TK, Chakraborty A, Khastgir D (2007) Effect of temperature, pressure, and composition on dc resistivity and ac conductivity of conductive styrene-butadiene rubber–particulate metal alloy nanocomposites. Polym Compos 28(5):696–704
31. Mohiuddin M, Hoa SV (2011) Electrical resistance of CNT-PEEK composites under compression at different temperatures. Nanoscale Res Lett 6:419–423
32. Nayak S (2015) Electroactive ceramic filled flexible poly(dimethylsiloxane) and polyurethane composites for dielectric and piezoelectric applications. PhD Thesis, Indian Institute of Technology Kharagpur
33. Rahaman M, Chaki TK, Khastgir D (2014) Polyaniline/ethylene vinyl acetate composites as dielectric sensor. Polym Eng Sci 54(7):1632–1639
34. Aminabhavi MT, Cassidy PE, Thompson CM (1990) Electrical resistivity of carbon-black-loaded rubbers. Rubber Chem Technol 63(3):451–471
35. Hu CH, Liu CH, Chen LZ, Peng YC, Fan SS (2008) Resistance-pressure sensitivity and a mechanism study of multiwall carbon nanotube networks/poly(dimethylsiloxane) composites. Appl Phys Lett 93:03310-1-033108-3

36. Hwang J, Jang J, Hong K, Kim KN, Han JH, Shin K, Park CE (2011) Poly(3-hexylthiophene) wrapped carbon nanotube/poly(dimethylsiloxane) composites for use in finger-sensing piezoresistive pressure sensors. Carbon 49:106–110
37. Sa-Gong G, Safari A, Jang SJ, Newnham RE (1986) Poling flexible piezoelectric composites. Ferroelectr Lett Sect 5(5):131–142
38. Park K-I, Jeong CK, Ryu J, Hwang G-T, Lee KJ (2013) Flexible and large-area nanocomposite generators based on lead zirconate titanate particles and carbon nanotubes. Adv Energy Mater 3:1539–1544

Thermal Properties of Polymer–Carbon Nanocomposites

Ayippadath Gopi Jineesh and Sunita Mohapatra

Abstract Carbon-based nanomaterials are widely used in polymer nanocomposites with enhanced properties. The dispersion of carbon nanomaterials and their interaction with the polymer can be improved by the surface functionalization of the carbon nanofillers, optimum processing conditions, and the methods of preparation of nanocomposites. Carbon nanofillers such as carbon nanotube (CNT), graphene, fullerene, and carbon nanofiber (CNF) can influence the thermal stability of the polymer matrix due to several mechanisms acting between the carbonaceous fillers and the basic polymer matrix. Optimum loading of carbon nanomaterials and proper interactions at the interphase between polymer and fillers can increase the thermal stability to a greater extent. Carbon nanomaterials can also impart changes in glass transition temperatures, crystallization temperatures as well as the extent of crystallinity of the polymer phase in the composites.

Keywords Carbon nanotube · Graphene · Carbon nanofiber · Fullerene
Nanocomposites · Thermal properties

1 Introduction

Polymer nanocomposites are "the combination of polymer matrix and fillers that have at least one dimension in the nanometer range". The fillers can be of one-dimensional—(nanotubes), two-dimensional—(nanoclay), or three-dimensional particles (spherical nanofillers). The large surface area to volume ratio of the nanofillers is one of the reasons behind the outstanding mechanical properties of the

A. G. Jineesh (✉)
Department of Chemistry, M.S. Ramaiah Univeristy of Applied Sciences,
470, IV Phase, Peenya Industrial Area, Bengaluru 560058, Karnataka, India
e-mail: jineesh.cy.mp@msruas.ac.in

S. Mohapatra
Raychem Innovation Centre, Raychem RPG (P) Ltd., Halol 390008, Gujarat, India
e-mail: Sunita_mohapatra@raychemrpg.com

M. Rahaman et al. (eds.), *Carbon-Containing Polymer Composites*,
Springer Series on Polymer and Composite Materials,
https://doi.org/10.1007/978-981-13-2688-2_7

polymer composites even with the incorporation of small amount of nanofillers. Nanofillers can also be used to improve thermal resistance, scratch resistance, barrier resistance, and flame resistance of the polymer matrix. Nanofillers can introduce significant changes in the optical, magnetic, and electrical characteristics of the polymer. The dimension of the nanoparticles, surface chemistry of nanoparticles, and the dispersion of the particles in the matrix, play major roles in deciding the properties of the polymer nanocomposites [1–3].

To formulate ideal nanocomposites for specific applications one should follow the procedures given below.

a. Appropriate selection of nanomaterials and their loadings in polymer matrix.
b. Optimize the level of dispersion of the nanomaterials in the polymer phase to get the desired phase morphology.
c. Appropriate surface modification of nanofillers which depends on the type of filler and the matrix.
d. Selection of proper processing/fabrication techniques for the polymer nanocomposites.

The discovery of fullerene buckyballs in 1985, the discovery of graphene, and surprised properties of carbon nanotubes attracted the attention of the scientists and engineers in the field of polymer technology. Polymer–carbon nanocomposites attracted a great interest in the electronic, automobile, space, and biomedical industries owing to their specific properties [4]. A slight change in electronic or in macroscopic structure of carbon leads to fascinating properties. Carbon thin-films can be used to prepare micro-arrays for microscopy and micro-electromechanical systems [5] and electrodes for sensor applications at high temperatures [6]. But agglomeration of carbon nanotubes, nanofibers, and graphene in the polymer matrix is a major concern in the fabrication/preparation of polymer composites. Chemical modifications/surface functionalization methods are used to address these problems and these modification leads to properly dispersed polymer–carbon nanocomposites with better polymer–filler interactions [7]. The different types of carbon nanofillers used in the preparation of polymer composites are zero-dimensional (fullerene), one-dimensional (carbon nanotubes and carbon nanofibers), two-dimensional (graphene), and three-dimensional (graphite) fillers. The graphical representations of the structures of these fillers are shown in Fig. 1.

2 Thermal Properties of Polymers and Their Nanocomposites

2.1 Thermal Stability of Polymers and Nanocomposites

Performance during the usage of materials decides the real fate of those products in indoor and outdoor applications. The durability is an important characteristic of the polymeric materials because it determines property retention during their lifetime,

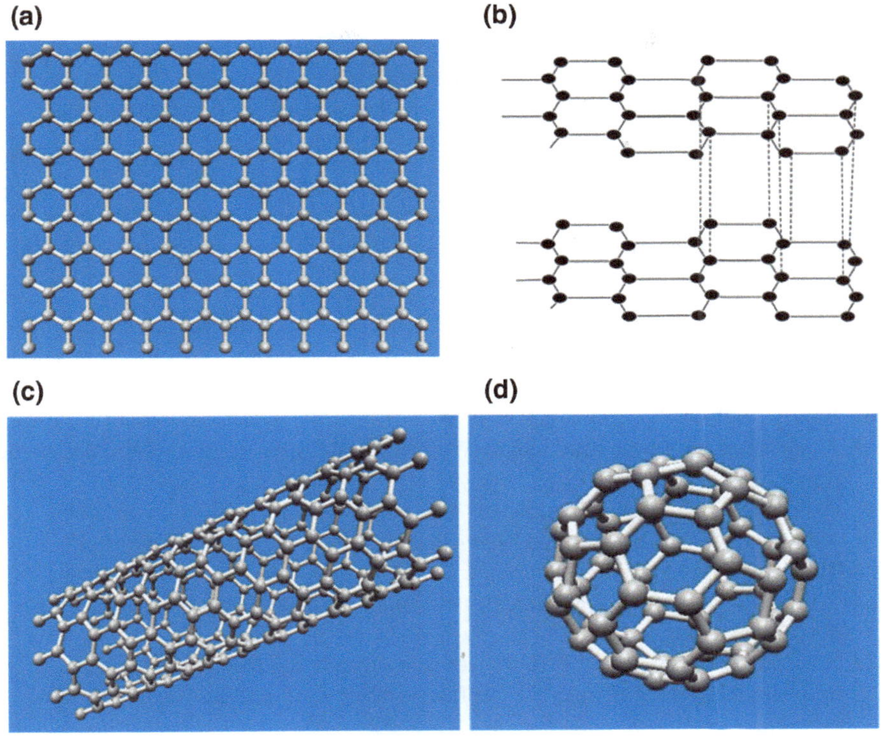

Fig. 1 Representation of **a** graphene **b** graphite **c** carbon nanotube (CNT) **d** buckminsterfullerene (The geometries were modeled using Nanotube Modeler, ©JCrystalSoft Trial version)

maintenance, and replacement of the polymeric materials during their life period, etc. Polymers degrade at low temperatures compared to ceramics and metals. This limits their applications in the high-temperature environment.

Thermal degradation of polymers occurs via different possible mechanisms such as random scission, depolymerisation, and side group elimination. Random scission occurs through the production of free radicals in the polymer, which causes fragmentation of the polymer molecules into smaller molecules with dissimilar chain lengths [8]. Side group elimination involves the elimination of atoms or molecules from the backbone of polymer chain, by leaving an unsaturated chain [9, 10]. Thermogravimetric analysis (TGA) is one of the popular methods used to learn the degradation behavior of polymers. The experimental indices used to characterize the thermal degradation/stability properties of neat polymers and their composites are

T_i (onset degradation temperature): The temperature at which the degradation starts and this has been demarcated by the temperature, at which the decline in the mass percentage of the material commences in the mass percentage versus temperature graph.

T_5: This is the temperature at which the 5% mass loss occurs at the early stage of the degradation of the sample.
T_{50}: The temperature at which half of the mass is lost.
T_d: the temperature at which maximum weight loss occur at different stages.

In recent years, it has been found that nanoparticles filled composites with less amount of nanofillers exhibit remarkable improvement of thermal and other properties compared with neat polymers and their micro-composites [11–14]. The major reasons behind the enhancement of thermal stability of polymer nanocomposites are given below

(a) The barrier effect of nanofillers to heat and mass transfer is one of the major mechanisms behind thermal stability of nanocomposites. Nanoparticles act as thermal barrier and this prevents the polymer from degradation due to heat.
(b) Nanoparticles which also act as a barricade to mass transfer reduces the diffusion of the gaseous products derived from the initial stage of degradation of polymer matrix.
(c) Polymer/nanoparticles interphase also reduces the molecular mobility of polymers which also leads to better thermal stability.
(d) Trapping of free radicals generated during the degradation of the matrix [15].
(e) Combined physical and chemical mechanisms enhance the thermal stability of polymer nanocomposites [12].
(f) Nanoparticles can also absorb the volatile degradation product either physically or chemically to enhance the thermal stability of polymer nanocomposites [16, 17].

2.2 *Transition Temperatures of Polymers and Nanocomposites*

The transition of any amorphous polymer from the glassy into the liquid state is escorted by sharp jump in the heat capacity. The temperature at which this transition takes place is called glass transition temperature (T_g). The macromolecular chains befall mobile, and the polymer can relax by changing toward its more stable shape in the vicinity of the T_g. The interfacial area between polymer and the nanofillers is huge in nanocomposites and the interaction of the polymer with the nanofillers induces changes in the chain mobility and relaxation dynamics. This results a change in T_g of the polymer [18, 19]. Confinement effect in the polymer nanocomposites due to the nanofillers induces changes in the transition temperature of a polymer and the changes can be even tens of degrees [20–22]. The type of interaction and the extent of interaction between the surface of the nanofiller and the polymer chain define the extent of mobility of the polymer segments. The nature of bulk polymer and the polymer–nanofiller interphase are quite different. The volume fraction of the interphase in the nanocomposites also increases with the increase in the interface area between polymer and nanofillers. These factors are the reasons behind the remarkable changes in the nanocomposite properties [23]. Recent reports showed that the T_g increased for the nanoparticles with attractive polymer–nanofiller interaction [24, 25], and decreased with repulsive interaction between matrix and fillers [26, 27]. Qiao et al. [28] studied the effect of agglomeration of particle on the glass transition temperature of polymer nanocomposites using finite elemental method. Volume fraction of the polymer–nanofiller is reduced with increase of the agglomeration of fillers. This may be the reason behind less change in the T_g of the nanocomposites with agglomerated nanoparticles compared to properly dispersed nanocomposites. Fundamental information on the influence of confinement and interfacial interaction between the nanofiller and the polymer chains on the characteristics of the nanocomposites is very vital in different industrial applications. For example, the differences in the dynamic modulus and T_g of the elastomer nanocomposites with respect to neat polymer matrix have significant effects upon the rolling resistance and wet skid traction of tire tread in tire applications [19, 29]. Molecular simulations and experiments proved that the T_g's of polymer nanocomposites can be increased or decreased by creating suitable polymer–nanofiller interactions [27]. To determine transition temperatures of polymers and nanocomposites, dynamic mechanical analysis (DMA), and differential scanning calorimetry (DSC) are used.

Crystallization of polymers involves nucleation and growth mechanisms. The incorporation of a second phase in the polymer matrix has an intense effect on the crystallization nature of the polymer. The effects include acceleration/deceleration of crystallization, changes in the crystal structure and spherulites. In nanocomposites, the size of the second phase matches the sizes of the polymer chains and even the small amount of nanofiller can significantly alter the crystalline behavior of polymers [30]. Techniques like DSC, X-ray diffraction, and nuclear magnetic resonance can be used to investigate the crystallization reactions and mechanisms under isothermal and non-isothermal conditions [30].

Avrami's theory of phase transformation can be used to learn and interpret the isothermal crystallization of polymers.

$$1 - X_t = \exp(-Kt^n), \tag{1}$$

where 'n' is the Avrami exponent, X_t is the fractional crystallization at time t and K is the rate constant [31, 32].

To study non-isothermal crystallization behavior of polymers, Ozawa equation [33] is being used.

$$1 - X_t = \exp(-K(T)/\varphi^m), \tag{2}$$

where X_t is the fractional crystallization at time t, φ is the cooling rate and $K(T)$ is the cooling rate function. The crystallization activation energy can be measured using Kissinger's method [34]. Crystallization of polymers depends not only on the crystallization environment but also the presence of other ingredients in the molecular phase.

Nanoparticles can act as heterogeneous nucleating agents for different types of semi-crystalline polymers [35, 36]. The large amount of nucleating effect of nano-sized fillers is mostly due to their large surface areas and aspect ratios. These conditions initiate the crystallization at elevated temperature by decreasing the thermodynamic barrier energy for nucleation [37, 38]. Nanoparticles can induce unique crystalline morphologies in the polymer due to their geometrical effects [39, 40]. For example, (Multi-walled carbon nanotube)/nylon 66 nanocomposites looks like "nano-hybrid shish kebab" in which carbon nanotube is the *shish* and nylon 66 crystals is the kebab [40–42]. Melting temperature also depends on the crystalline behavior of the polymers. So the change in crystallization due to nanofiller incorporation also leads to change in melting behavior of the polymers.

3 Thermal Properties of Polymer–Carbon Nanotube Composites

Carbon nanotubes possess very high aspect ratio (>1000). CNTs also possess high modulus and tensile strength. They show comparatively less mass density. Carbon nanotubes can be classified into different types depending on their structure and their representation is shown in Fig. 2.

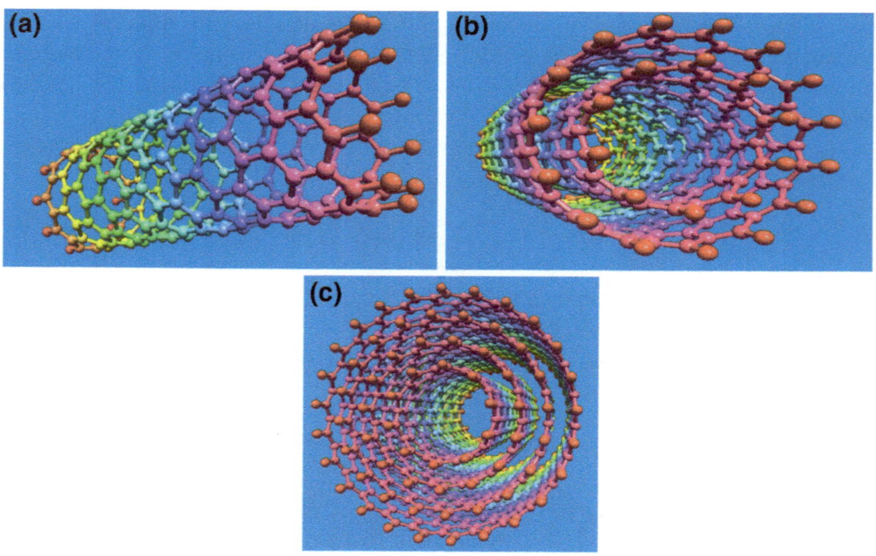

Fig. 2 Visualization of **a** SWCNT **b** DWCNT **c** MWCNT (The geometries were modeled using Nanotube Modeler, ©JCrystalSoft Trial version)

The morphology of SWCNTs can be visualized by wrapping a graphene layer into a cylinder. MWCNTs consist of number of concentric graphene layers which are bounded with weak van der Waals forces. MWCNTs have relatively less tensile strength and modulus than SWCNTs. Single-walled CNTS have diameters in the range of 20–100 nm which is generally lower compared to MWCNTs. MWCNTs are stiff because of their thick cross-section than SWCNTs. MWCNTs and SWCNTs can act as reinforcing fillers and they also increase the thermal properties of polymers [43]. The key methods to prepare carbon nanotubes are catalytic methods, arc-discharge method, and laser ablation techniques. Among these methods catalytic methods and laser ablation techniques yield high purity nanotubes with desired structure. Higher amount of CNTs can be prepared by chemical vapor deposition (CVD) which is a catalytic method. This process possesses reasonable potential for the industrial production of carbon nanotubes [44]. Few specific characteristics of CNTs are listed in Table 1.

3.1 Thermal Stability of Polymer/Carbon Nanotube Nanocomposites

CNTs show promising characteristics for various applications such as microelectronics applications, biosensors, and hydrogen storage, etc. So it is quite important to gain knowledge about the thermal stability of polymer–CNT composites [45–48]. Liu et al. [49] studied thermal stability of CNTs and showed that the SWNTs placed in air for 5 days at very high temperatures were changed into the amorphous carbon. It indicates that SWNTs were unstable in hot air and easily oxidized. Xu et al. [50] studied the heat capacity (C_p) of CNTs with dissimilar sizes. The results showed that C_p increases with decrease in the length of CNTs when the diameters of CNTs were in between 60 and 100 nm. But when the length of CNT is in between 1 and 2 μm, C_p is independent of CNTs diameters. Similarly when the diameter is below 10 nm, C_p of CNT is independent of its length. DSC analysis of CNTs with different dimensions showed that the release of heat during thermal scan

Table 1 Properties of carbon nanotubes [44]

Property	SWCNT	DWCNT	MWCNT
Tensile strength (GPa)	50–500	23–63	10–60
Elastic modulus (TPa)	~1	–	0.3–1
Elongation at break (%)	5.8	28	–
Density (g/cm^3)	1.3–1.5	1.5	1.8–2.0
Electrical conductivity (S/m)	~10^6	~10^6	~10^6
Thermal stability	>700 °C (in air)	>700 °C (in air)	>700 °C (in air)
Average diameter	1 nm	~5 nm	~20 nm
Specific surface area	10–20 m^2/g	–	–

increases with the increase in diameter of CNT, and independent of the lengths of CNTs.

Thermal stability of polymers can be enhanced by the incorporation of carbon nanotubes. The enhancement in the thermal resistance is attributed to different mechanisms such as physical absorption effect, absorption of free radicals by CNT, reduction of mobility of molecular chains, etc. This absorption causes a reduction in the volatilization of polymer without affecting the onset degradation temperature [8]. Free radicals produced during the degradation of polymer may be entrapped/ absorbed by CNT that also leads to the improvement in polymer stability [51]. The mobility of the polymer chains also decreases in the presence of CNTs. This stabilization effect in polymer due to carbon nanotube is similar to the effect of layered silicate in polymer even though there is no intercalation mechanism between CNT and polymer chains [51, 52]. This effect is clear in the degradation of polymer–CNT nanocomposites in inert condition and it is proved by displaying a change of onset degradation temperature towards high temperatures [53]. Marosfői research group studied the thermal stabilization effect of nanofillers in composites and proved that the improved interfacial connections between the nano-additive and polymer, direct to an enhancement in the activation energy of degradation which prevents the early degradation of polymer chains [54]. They observed that the presence of MWCNTs delayed volatilization under the heat without modifying the degradation mechanism. Thermal degradation of the polymer–CNT composites in the air was deferred by about 100 °C and it was independent on the amount of loading of CNT [55, 56]. The stabilization effect is due to the generation of a thin layer of char made up of MWCNT/polyaromatic carbon which acts as a protective layer in composites. This char is the association of isolated MWCNTS knotted to form a nonwoven configuration [52]. Chipara et al. [57] proved that the concentration of the nanofiller is also a major reason behind the enhanced thermal stability. The results showed the connections between polymer chains and CNTS lead to the generation of polymer–nanoparticle interface. This creation of polymer–filler interface reduces the thermal degradation and enhances the thermal stability. But in some studies, the incorporation of carbon nanoparticles accelerated the commencement of polymer disintegration under oxidative environment [58]. This is attributed to the presence of the trace amount of inorganic elements in the nanoparticles [59]. The thermal degradation nature of polymer/carbon nanotube composites also depends on the nature of surface modification of the CNTs [60, 61]. The reactive chemical modifications lead to enhanced distribution of nanofillers in the matrix. This can change degradation pathway of the nanocomposites and leads to a better thermal stability. For example, the thermal strength of polyimide can be enhanced with the incorporation of plasma-modified CNTs with maleic anhydride. This is achieved through better dispersion of CNTs in the polyimide as well as the chemical/physical binding among CNTs and polyimide molecular chains [62]. Surface modification of CNTs using amine groups significantly enhanced the thermal resistance of the epoxy matrix due to the covalent interaction between the surface modifier and macromolecular chains of epoxy

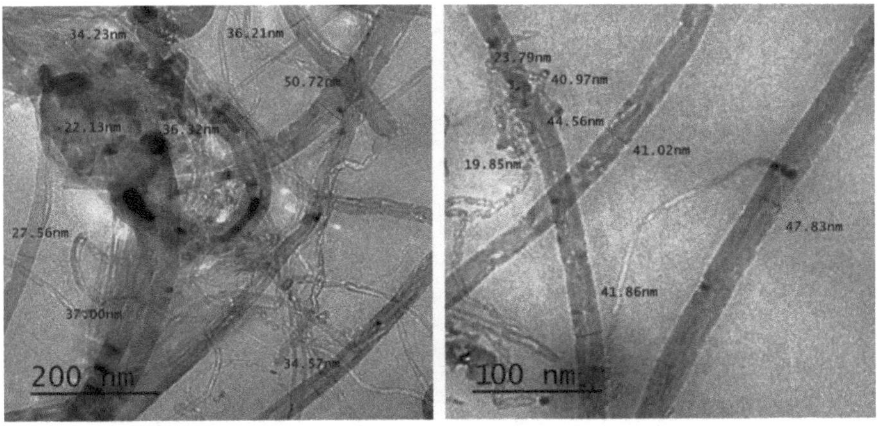

0.04 Wt.% SWCNTs 0.06 Wt.% SW CNTs

Fig. 3 Transmission electron micrographs of carboxylic acid functionalized SWCNT filled PU/PVC nanocomposite at different loadings of nanotube. *Source* [64]

polymer [63]. Transmission electron micrographs of carboxylic acid functionalized SWCNT-filled PU/PVC nanocomposite in Fig. 3 show that carbon nanotubes are properly dispersed in the polymer matrix.

Functionalization of the CNT is important to reduce the agglomeration and to increase the interfacial surface area connecting CNTs and the matrix. The modification of CNTs is classified into covalent or non-covalent based on the bonding to the nanotube surface to the polymer matrix. Non-covalent modification includes physical adsorption of molecules and/or polymer wrapping to the surface of the CNTs. For example, the graphitic sidewalls of CNTs allow the interaction with conjugated polymers such as poly (phenylene vinylene). Similarly the graphitic sidewalls also provide an interaction with polymers containing hetero-atoms with free electron pair such as poly(3-hexyl)selenophene [65]. Ma et al. [66] proved that 0.25% of silane-modified CNT improves the T_5, and T_d of epoxy polymer. Also modification of CNT shows better thermal stability than unmodified CNT-epoxy nanocomposites. The physical barrier effect is prominent in the silane-CNT composite due to the enhanced degree of cross-linking reactions by the epoxy polymer and the end-groups of silane-CNT than the untreated-CNT counterpart. Barick et al. [67] studied the temperature resistance of Thermoplastic Polyurethane (TPU)-carboxylic acid functionalized multi-walled carbon nanotube nanocomposites prepared by melt incorporation method. The results shown in Fig. 4 indicate that the thermal stability of the TPU is significantly enhanced by the inclusion of the MWNT. This increase is due to the outstanding thermal stability of CNTs as well as the connections between the TPU matrix and the carboxylic acid portion on the shell of MWCNT.

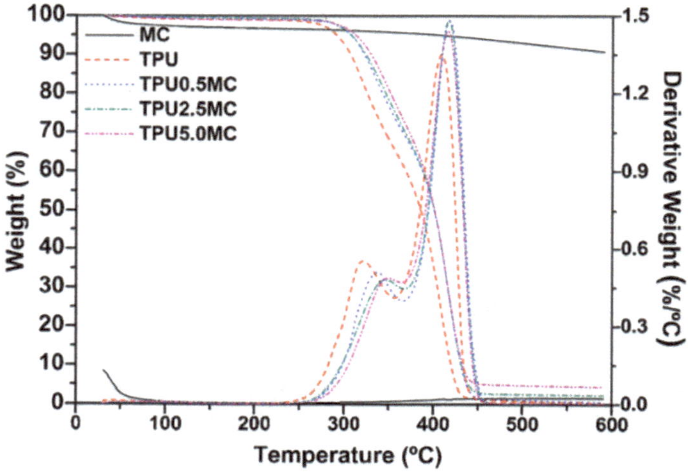

Fig. 4 Thermogravimetric analysis of neat TPU and its nanocomposites containing different loading of MWNT. *Source* [67]

3.2 Transition Temperatures of Polymer/CNT Nanocomposites

The choice of a given application of a polymer nanocomposite is also based on the glass transition temperature (T_g). The T_g dictates the potential application of a given polymer as a function of the service temperature range. Table 2 consists of the collection of data from few literatures on glass transition temperature of the CNT/polymer composites.

Pham et al. [70] studied the influence of the incorporation of both functionalized and pristine SWCNTs in the polystyrene matrix. The incorporation of fillers, irrespective to the functionalization increases the T_g of the polymer. In addition to that, the width of the glass transition, as measured using DSC, is broader for the SWNT nanocomposites than for the virgin polymer. Using molecular dynamics simulations, Wei et al. [74] showed that the enhancement in T_g due to the insertion of nanofillers into the matrix is associated with the higher density of the nanocomposites over the polymer. The increase in T_g can be also be due to the reduced dynamics of polymer chains in the composites. If the SWCNT fillers are very well dispersed in the matrix, polymer chains locally in contact with SWNTs would exhibit slower dynamics than chains in the neat polymers. The surface functionalization of carbon nanotube has an important role in deciding the relaxation of the polymer chains in the polymer composites. No/improper surface functionalization leads to the aggregation of nanotubes to form bundles and ropes throughout the composite. Mitchell et al. [75] prepared surface-functionalized SWCNTs using 4-(10-hydroxy) decyl benzoate moieties by covalent bonding. Even 1% SWCNT

Table 2 Glass transition temperatures of CNT/polymer composites from selected literatures

Reference	Polymer	CNT type	wt% of CNT (%)	CNT aspect ratio	T_g of the composite (°C)	$\Delta T = T_g$ of the composite-T_g of the polymer (°C)
Zhou et al. [68]	Epoxy	MWNT	0.4	150	147	22
Ganguli et al. [69]	Epoxy	MWNT	0.1	125	85	24
Pham et al. [70]	Polystyrene (PS)	SWNT	0.75	–	–	3
Wang et al. [71]	Epoxy	SWNT-NH_2	0.5	–	170	−11
Gojny et al. [72]	Epoxy	MWCNT	0.75	–	80	16
Sterzynski et al. [73]	Poly vinyl chloride (PVC)	MWCNT	0.05	–	93	8

loading was able to form a percolated network in the matrix and this gives the strength of the polymer–SWNT (functionalized) interactions.

Figure 5 shows that incorporation of small amount of CNTs increases the T_g of UHMWPE-based composites.

CNTs can interact with semi-crystalline polymers and can modify their crystalline structure and crystallization behavior [77]. CNTs can induce nucleation in

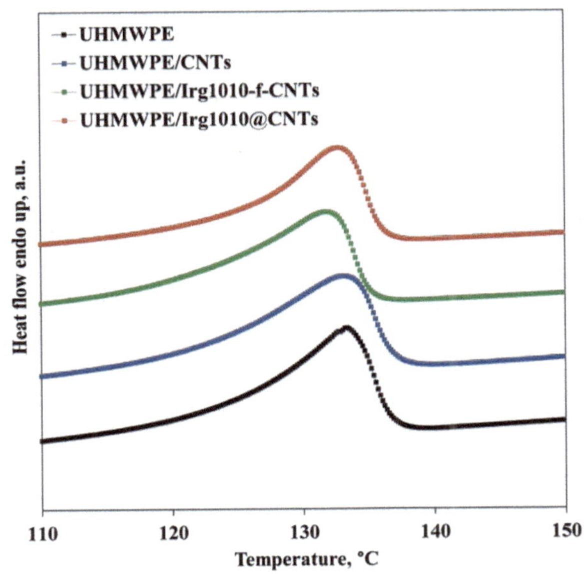

Fig. 5 Thermograms for neat UHMWPE and UHMWPE-based nanocomposites. *Source* [76]

polyethylene (PE) [78] and polypropylene (PP) [79]. Orientation of CNT also influences the crystallinity orientation in the polymers. Brosse et al. [80] shows that CNT induces trans-crystalline lamellae in the interphase of CNT surface and polyamide nanocomposites. Studies showed that incorporation of SWCNT induces an increase in isothermal crystallization rate, and also alters the spherulitic morphology of polypropylene [30, 79]. Valentini et al. [81] have reported that the crystallization kinetics was affected by the distance between the nanotube bundles. The presence of SWCNTs also decreased the size of the spherulites of PP to one-tenth-fold in PP/SWCNT composites. The microphotographs of neat PP and 5 wt% SWCNT filled PP composites is shown in Fig. 6.

Grady et al. [35] shows that the crystallization rate and the % crystallinity were higher in PP/SWCNT composites than virgin PP. Few studies show that introduction of SWCNT into semi-crystalline polymers increases the crystallization rate but not changing the crystallized form of the polymer [82, 83]. MWCNT can affect the kinetics of the tetragonal to hexagonal polymorphic transformation in Poly (1-butene). In addition to that, MWCNT can also increase the nucleating ability of the polymer matrix. This means that the activation energy required for nucleation have been decreased with the presence of MWCNTs [84]. Liu et al. [84] have shown that incorporation of MWCNT induces nucleation but do not favor the formation of γ crystalline form of nylon-6 at elevated temperatures. The nucleating effect of carbon nanotube in polyimide have studied by Yudin et al. [85] and they observed 15% degree of crystallinity in polyimide in the presence of CNTs. Lai et al. [86] have manufactured nanocomposites of a biodegradable thermoplastic polyester poly hydroxyl butarate co-hydroxyvalerate (PHBV) with MWCNTs by solution mixing and ultra-sonication. The crystallization study by using differential scanning calorimetry showed that the inclusion of MWCNTs enhanced the crystallization temperature. Also the melting and crystallization peaks were narrower for the nanocomposites than the pristine polymer. Arjmand et al. [87] explored the crystallization behavior of carbon nanotube filled PVDF. Figure 7 shows the effect of CNTs (prepared using different catalysts) on the crystallization temperature of

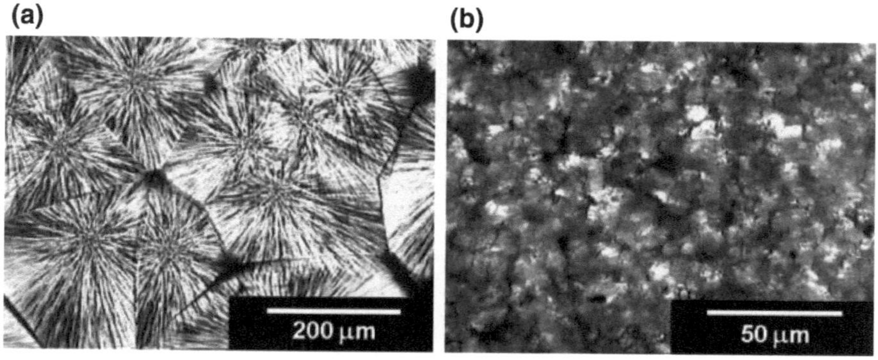

Fig. 6 Microphotographs of: **a** virgin PP and **b** 5% SWNT filled PP composite. *Source* [81]

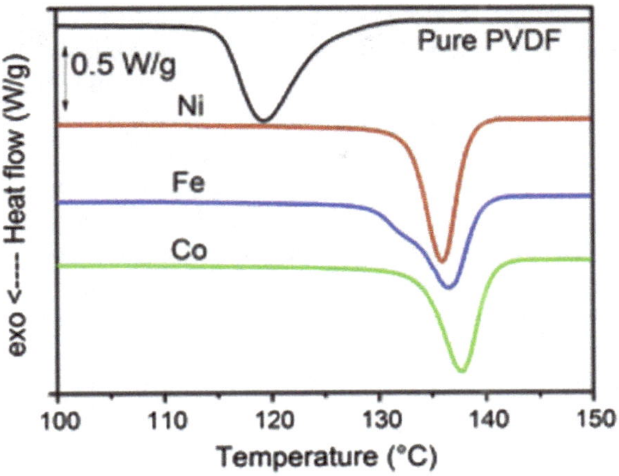

Fig. 7 DSC cooling runs of pure PVDF and PVDF filled with 2.0 wt% CNT synthesized using different catalysts. *Source* [87]

polyvinylidene fluoride (PVDF). The crystallization temperatures of the nanocomposites were shifted to higher values compared to pure PVDF.

4 Thermal Properties of Polymer–Graphene Nanocomposites

Graphene is 2D carbon nanofiller with sp^2hybridized carbon atoms that are compactly packed in a honeycomb crystal lattice. Graphene is considered to be the "thinnest and strongest material in the universe" with excellent physical and chemical properties [88, 89]. Graphene possesses high thermal conductivity, excellent electrical conductivity as well as excellent mechanical strength [90–92]. But these characteristics of graphene depend mainly on structural precision and this is governed by the method of synthesis. Graphene can be prepared by micromechanical exfoliation of graphite [93], epitaxial growth on silicon carbide or metals [94], chemical vapor deposition on substrates [95], chemical or thermal reduction of graphene oxide [96], and exfoliation of graphite intercalation compounds [97]. One of the most efficient methods for the synthesis of high-quality graphene is the micromechanical cleavage method. Graphene has certain advantages over other conventional nanofillers because of its very high surface area, aspect ratio, EMI shielding ability, tensile strength (TS) thermal, and electrical conductivity [92, 98].

Since 2010, large quantity of studies has been published in the field of synthesis and characterization of polymer/graphene composites. Also lots of studies reported about the optical properties, electrical properties, and permeability of the polymer/graphene nanocomposites. The category of graphene generally implies not only the pure graphene but also on the modified graphene such as graphene oxide (GO) [99–101]. Polymer/graphene nanocomposites possess better thermal, electrical, mechanical, and barrier properties than virgin polymers [102–107]. Even though carbon nanotubes possess similar mechanical strength compared to graphene, graphene is superior filler than carbon nanotube in terms of electrical and thermal conductivity. So if the application needs a polymer composites with higher thermal or/and electrical conductivity, graphene will be the right option as the reinforcing filler. The physicochemical and dynamic properties of the graphene-based nanocomposites depend on several factors such as the dispersion of graphene layers and its interfacial adhesion between the macromolecular chains in the matrix. Due to the incompatibility with organic polymers, pristine graphene cannot be considered as reinforcing filler. The surface modification of graphene to achieve better interfacial adhesion between polymer and graphene is a vital step to get proper dispersion of graphene layers. This morphology leads to better mechanical and dynamic mechanical properties. The surface of the GO is highly oxygenated and contains different functional groups such as hydroxyl, carboxyl, diols etc. These groups can be modified by chemical functionalization. These modifications on the graphene change the van der Waals interactions and leads to the compatibility between filler and the polymers [108, 109]. The carbonyl and carboxyl groups positioned at the periphery of the GO induce hydrophilic nature to the graphene oxide. This makes graphene to easily swell and uniformly distribute in water. This property of GO is fascinated by polymer technologist because of manufacturing point of view of the nanocomposites. However, sheets of graphene oxide can only be uniformly distributed in aqueous media, and the aqueous media is incompatible with most of the organic polymers. Being electrically insulating in nature, graphene oxide is not suitable for the production of conducting polymer nanocomposites [110, 111]. Few characteristic properties of graphene are listed in Table 3.

Pure/neat graphene tends to aggregate and agglomerate in the polymer matrix. Oxidation of graphene followed by chemical functionalization can solve the above issue. The surface modification leads to improved dispersion of graphene in polymer matrix. The surface modification can be done either by attaching molecules or polymer chains on the surface of the graphene [116]. Few examples on the surface modification of graphene are given in Table 4.

Mode of preparation of polymer–graphene nanocomposites plays an important part in deciding the nature of the composites. The degree of distribution and exfoliation of graphene is mainly influenced by the shear force exerted during processing, temperature, and the type of solvents. To obtain polymer–graphene

Table 3 Characteristics of graphene

Property	Value
Surface area [112, 113]	$\sim$2630 (m^2/g)
Tensile strength [114–116]	130 $\pm$ 10 GPa
Electrical conductivity [116, 117]	7200 S/m
Thermal conductivity at room temperature [118]	$(4.84 \pm 0.44) \times 10^3$ to $(5.30 \pm 0.48) \times 10^3$ W/mk

Table 4 Literatures on different types of surface modification of graphene

Sl No.	Surface modification	Literatures
1	Amination	Du et al. [119], McAllister et al. [120], Niyogi et al. [121]
2	Esterification	Cano et al. [122]
3	Isocyanate modification	Jiang et al. [123], Sadasivuni et al. [124]
4	Polymer wrapping	Samanta et al. [125], Ye et al. [126], Fujigaya et al. [127]
5	Ionic Liquids	Liu et al. [128], Kim et al. [129], Yang et al. [130], Nuvoli et al. [131], Zhang et al. [132]

nanocomposites with excellent properties, several criteria such as effective restacking of graphene layers, wrinkling, and aggregation of graphene should be fulfilled. Major preparation/fabrication methods for polymer–graphene composites are given in Fig. 8.

4.1 Thermal Stability of Graphene-Polymer Composites

Reinforcement of polymers using graphene imparts better thermal stability to the host polymers. The carbon framework of graphene oxide is stable up to certain high temperatures but functional groups of graphene oxide will be degraded before the degradation of carbon framework. Effective graphene incorporation can increase thermal degradation temperature (T_d) of the polymers, up to 10–100 °C [133, 134]. Interaction with graphene and polymer will restrict the chain movement of macromolecules near the surface of graphene and leads to better thermal stability [135].

Graphene oxide imparts thermal stability to polymer at even low loading. Onset temperature of degradation and maximum degradation temperature were increased to higher temperatures even at low loading (<0.7 wt%) of GO into Polyvinyl alcohol (PVA) [136]. The onset degradation temperature of polyethylene is increased by 30 °C with the incorporation of 5.4 wt% of graphene. Similarly 15.3 wt% of graphene increased the maximum degradation temperature of polyethylene by 30 °C. Being thermally stable and mechanically stiffer, graphene induces rigidity to the polymeric matrix. This rigidity also decreases the chain mobility and reducing the thermal degradation rate of the polymer. The thermal behavior of graphene-based composites depends on the degree of dispersion of nanoparticles and the extent of loading [137].

Different trends have been observed for the thermal stability behavior of polymer–graphene oxide nanocomposites. Both thermally and chemically reduced GO improves thermal stability but non-reduced GO is not affecting thermal stability of polymer matrix [138–141]. The superior thermal stability of the polymer/reduced

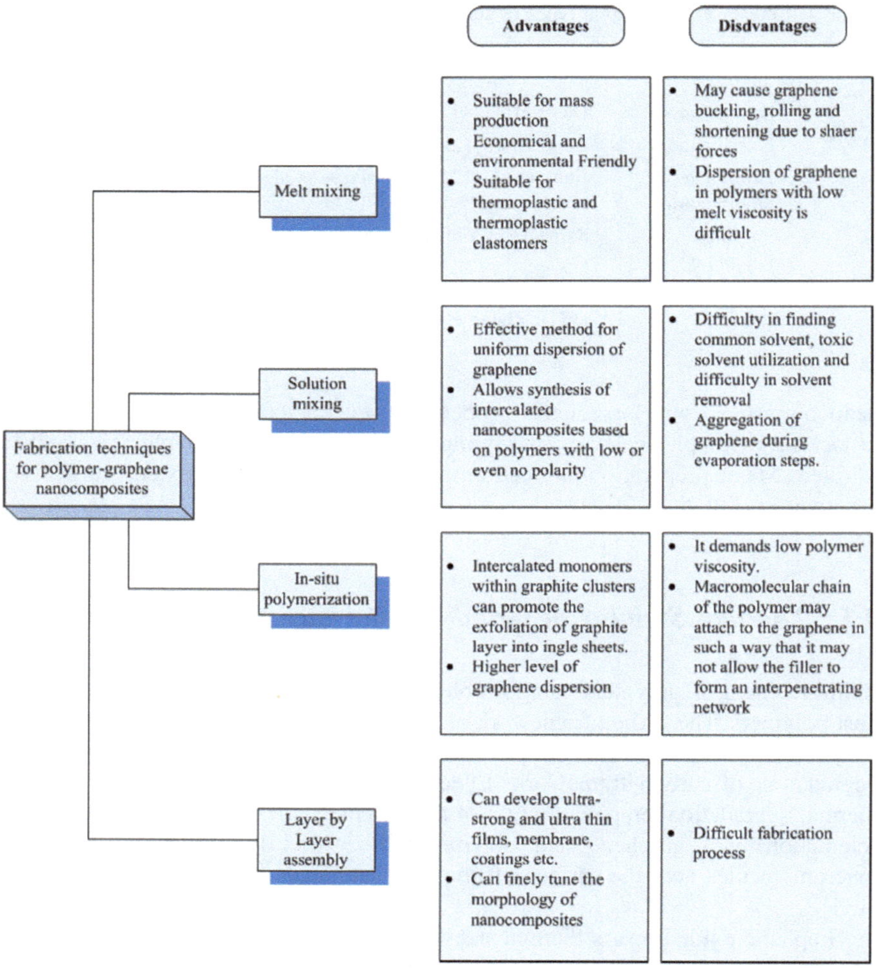

Fig. 8 Characteristics of fabrication techniques of polymer-graphene nanocomposites

graphene oxide nanocomposites was due to the high surface area and good distribution of the layers of filler in the matrix.

4.2 Transition Temperatures of Graphene-Polymer Composites

Several studies have reported a drastic change in the Tg in polymers occurs due to the small loading of graphene. Even 0.05% of graphene oxide can change the transition temperature of polymers to a significant extent. The change in glass transition temperatures is mainly due to reduced mobility via covalent bonds between the polymer and graphene, mechanical interlocking, and adhesion due to the wrinkled topology of the exfoliated GO, or H-bonding between the polymer chains and the oxygen functionalities on the graphene/graphene oxide layers [15]. Few literatures related to the influence of the inclusion of graphene in the T_g of the polymer matrix are compiled in Table 5.

Yoonessi et al. [144] proved that the incorporation of graphene in Polyimide resin does not change the Tg of the polyimide. However, the area under damping peak (tan δ) was decreased with increasing graphene content due to the weak attractive forces between the graphene and the resin. These weak forces are insufficient to hinder segmental motion of the polymer chains and this leads to no change in the T_g. The relaxation behavior of polymer chains grafted to the nanofiller surface is distinct from that of free or entangled chains.

1 wt% of oxygenated graphene can increase the T_g of the polyacrylonitrile by 40 °C due a strong particle–polymer interaction [138]. T_g of the nanocomposite also depends on the size of the grafting polymer and particles and the grafting density [89]. The influence of graphene oxide in the T_g of polar polymers seems to be higher than that in the nonpolar polymers. For example, 12 wt% of graphene oxide increases the T_g of PS only by 15 °C. The increased T_g of the graphene-polymer composite is ascribed to the sturdy confinement effect of graphene sheets as well as the covalent bonding with the polymer chains [143].

Polymer crystallization behavior mainly depends on the thermal history, the fabrication/preparation method and the interfacial adhesion between the fillers. Studies show that there is no particular trend in the crystallization behavior of polymer–graphene nanocomposites. Degree of crystallization was observed to increase in few polymer composites. Similarly no change in crystallization behavior was also observed in few studies [141]. Nanometer-sized graphene acts as

Table 5 Compilation of data from few literatures on glass transition temperature of graphene/polymer composites

Reference	Polymer	Type of graphene	wt% of graphene (%)	$\Delta T = T_g$ of the composite-T_g of the polymer (°C)
Ramanathan et al. [138]	PMMA	Thermally exfoliated graphene oxide	0.05	30
Salavagione et al. [142]	PVA	hydrazine reduced GO	10	20
Fang et al. [143]	PS	PS-grafted GO/	12	15
Yoonessi et al. [144]	Polyimide	Oxygenated graphene	0.1–4	No change
Yoonessi et al. [145]	Polycarbonate	Graphene platelets	1.1 2.2	2.2–2.1

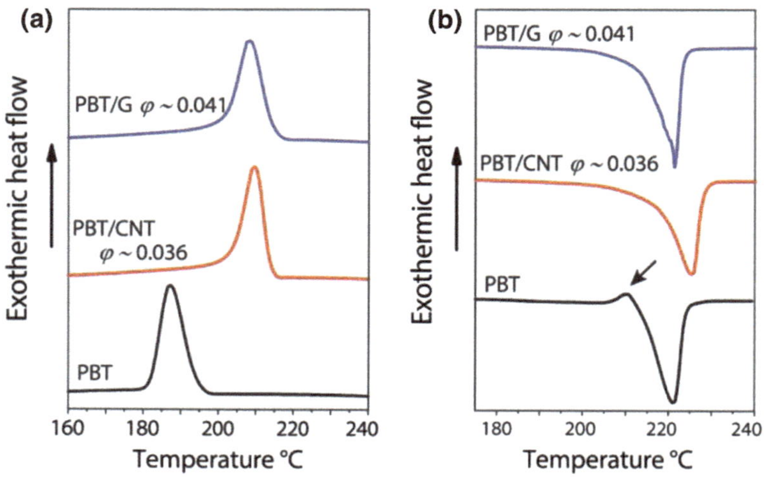

Fig. 9 Cooling (**a**) and heating (**b**) scans of DSC measurements of PBT/graphene nanocomposites and arrow indicates the recrystallization. *Source* [147]

nucleating agents which enhances the crystallization rate of the polymer in the composites. Graphenes can regulate/change the grain size of the crystals and enhance the rate of crystallization through heterogeneous nucleation even at a low loading of less than 1 wt% [146].

Liang et al. [114] investigated the effect of graphene oxide on glass transition temperatures and the melting point of polyvinyl acetate and found that introduction of GO in PVA increases both the T_g and melting point of PVA. Similarly the introduction of graphene into polybutylene terephthalate (PBT) increases the crystallization and melting temperatures of the base polymer and the DSC curves are shown in Fig. 9.

5 Thermal Properties of Polymer–Carbon Nanofiber Nanocomposites

Carbon nanofibers are discontinuous filaments with aspect ratio >100. CNF has linear structure made up of carbon atoms which are sp^2 hybridized [148]. The layers of graphitic planes of most of the CNFs are not oriented in the direction of the axis of the fiber. Based on the angle of the alignment of the graphene layers, carbon nanofibers have been classified as (a) Stacked and Herringbone/Cup-stacked [149]. Other classification of carbon nanofiber is vapor-grown carbon nanofibers (VGCNFs) and electrospun carbon nanofibers (ECNFs). The synthesis method of CNFs mainly decides its structure. VGCNFs comprise of a single or double layer with hollow core structure [65]. On comparison with carbon nanotube, VGCNFs have gained relatively less importance because of their comparatively lower mechanical properties and higher density compared to carbon nanotubes [150]. But VFCNFs can be synthesized very easily with cheaper manufacturing steps and these makes it as a suitable good reinforcing filler in many industrial applications. VGCNFs are synthesized by chemical vapor deposition of carbon monoxide or a hydrocarbon over a surface of a metal or metal alloy catalyst [151–153]. The important catalysts used in the preparation of VGCNFs are Fe, Ni, Cu, Ni–Cu, and Fe–Ni, etc. VGCNFs possess high aspect ratio and their diameters are higher than CNTs, but lesser than carbon fibers. Similarly, the length of VGCNFs is similar to that of carbon nanotubes, but shorter than that of carbon fibers [65, 154]. Good carbon fiber-polymer adhesion can be achieved through interactions taking place at the fiber–polymer interface. The interactions can be either chemical or physical linkages [155]. Surface treatment on VGCNF can be done by etching in air at high temperatures, soaking in sulfuric/nitric acid mixtures, or in peracetic acid, etc. These treatments can improve the fiber-polymer adhesion [156]. Oxidation of VGCNF with nitric acid [157], grafting an aromatic (ether–ketone) in the polymer backbone [158], addition of oxidizing gases such as CO_2, to VGCNF [159] are the few other methods to increase the adhesion between polymer and fiber matrix.

5.1 Thermal Stability of Polymer–Carbon Fiber Nanocomposites

Because of its good barrier property, CNF can enhance the thermal stability of polymer/CNF composites. The addition of VGNF into polymers increases their thermal stability. Few literatures on the improvement in the thermal stability of polymers are shown in Table 6.

Nanofibers impose restriction on the polymer chain movements and it increases the thermal resistance of the nanocomposites [155]. The higher thermal conductivity

Table 6 Compilation of data from few literatures on CNF/polymer composites T_g of the composites

Reference	Polymer	wt% of CNF (%)	Difference in degradation temperature of nanocomposite and virgin polymer (°C)
Yang et al. [160]	HDPE	20	~30
Choi et al. [161]	Epoxy	5	80
Lozano et al. [162]	PP	5	~35
Xu et al. [163]	PS	2	~35
Choi et al. [164]	PC	10	~15
Zeng et al. [165]	PMMA	5	~60
Kumar et al. [166]	PC	5	~10
Higgins et al. [167]	PC	9	~40

of the composite can also enhance the thermal stability, because thermally conducting materials can dissipate the heat within the polymer composite [37]. The onset degradation temperature of polymers will also be increased with the increase in nanofiber content [162, 163]. VGCNF/Polycarbonate (PC) composites show higher stability than virgin polycarbonate. The increase in the thermal stability is due to the adsorption of free radicals produced by the polymer degradation by VGCNFs [164, 167]. Chatterjee et al. [51] measured thermal stability of polypropylene/carbon nanofiber composite and found that CNF improves the thermal stability of polypropylene to a considerable amount. The reasons behind the increase in thermal stability are (a) increase in the crystallinity (b) constraints imparted by nanofibers on the chain mobility of polymers at the interphase of polymer–nanofibers, and (c) the char formation. The formed char acts as a physical constraint for the degradation of macromolecular chains. Careful oxidation of nanofiber creates moderate amount of oxygenated groups on the nanofiber surface. This oxidized CNF possesses better interaction between the polymers and it leads to better mechanical and thermal properties. The intermolecular hydrogen bonds formed between the oxidized CNF and the matrix also leads to better dispersion of the nanofillers and this also leads to betterment of mechanical and thermal properties [168].

Few studies show that the incorporation of CNF decreased the onset degradation temperature of polymers. The onset degradation temperature was decreased by 30 ° C with the incorporation of 3 phr CNF into polyetherimide composites. This may be due to the presence of impurities or catalyst present in the CNF. But surface modification of CNF using the treatment with acid increases both the initial and maximum degradation temperature of polyetherimide. This may be due to the elimination of impurities on the surface of CNF and enhanced interfacial interactions between CNFs and the polymer due to the acid treatment [169].

Barick et al. [170] prepared and characterized melt-mixed TPU-CNF nanocomposites. The high-resolution transmission electron microscopy (HRTEM) of the prepared nanocomposites is shown in Fig. 10 [171]. The images show that the nanofibers are properly embedded in the TPU matrix.

Thermogravimetric analysis of the above nanocomposites showed that the thermal stability of TPU is increased considerably with the introduction of CNF into the TPU matrix. The reasons for the improvement of thermal stability are (a) homogenous dispersion of CNF (b) CNF forms a good network structure and it restricts mobility of TPU chains and this will not allow the gaseous molecules produced during the early degradation stage to contact with the outer atmosphere in degradation process and (c) higher thermal conductivity of CNF avoids heat concentration in the composites.

5.2 *Transition Temperatures of Polymer–CNF Composites*

Addition of fillers can increase or decrease the crystallinity of the polymers depending on the characteristics of polymers and fillers and the amount of fillers in

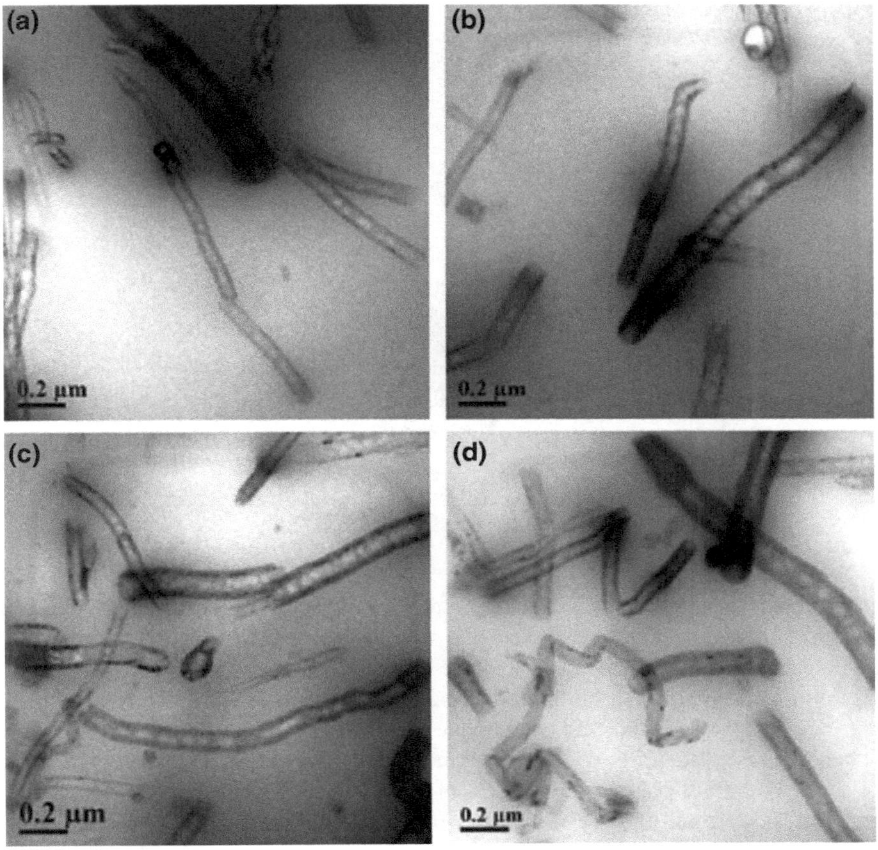

Fig. 10 High-resolution TEM of TPU/CNF nanocomposites with **a** 1 wt%, **b** 4 wt%, **c** 7 wt%, **d** 10 wt% loading of CNF. *Source* [170]

the composites [161]. Reduction of crystallinity in polyethylene due to the incorporation of VGCNF was reported by Yang et al. [160]. Contrary to this, an increase in crystallinity and the crystallization rate of PP was reported with the introduction of VGCNF. Takahashi et al. [172] reported the incorporation of graphitized VGCNF to PC increases the crystallization rate of PC. On contrary to this, the non-graphitized VGCNF decreases the crystallinity of the PC.

Few studies on PS [163], PEEK [173], and epoxy polymer [164] show that the incorporation of CNF does not affect the Tg of the composites. But in all these composites, the tan δ peak decreased with the increase in CNF content. There are a significant number of studies which shows the incorporation of CNF increases the T_g of the polymers [174] and few of them are listed in Table 7.

Barick et al. [171] investigated the effect of CNF in the thermal characteristics of thermoplastic polyurethane. The incorporation of CNF did not alter the T_g of the

Table 7 Compilation of data from the literature on CNF/polymer composites on glass transition temperature of the composites

Reference	Polymer	Type of CNF	wt% of CNF (%)	$\Delta T = T_g$ of the composite-T_g of the polymer (°C)
Zeng et al. [175]	PMMA	VGCNF	5	5
Choi et al. [161]	Epoxy	VGCF	20	26
Zhou et al. [176]	Epoxy	VGCNF	3	8
Green et al. [177]	Epoxy	Pyrolytically stripped carbon nanofiber	0.1 1	25 25
Kumar et al. [169]	Polyetherimide	Carboxylic acid functionalized CNF	3	11
Kumar et al. [169]	Polyetherimide	Unmodified CNF	3	8
Rasheed et al. [168]	Poly(styrene-co-vinyl phenol)	Oxidized CNF	5	9
Kumar et al. [166]	PC	Oxidized CNF	5	10

soft segment of thermoplastic polyurethane. The reasons behind the observation is (a) the change in heat capacity associated with incorporation of CNF was too low to be detected and (b) the restriction in the soft segment chains due to the incorporation of CNF is nullified by the reduction in the hard phase of TPU. Even though the incorporation of CNF did not change the hard domains of the TPU matrix, it decreases the crystallinity of its soft segment. The factors affecting the crystallinity of semi-crystalline polymers by the incorporation of CNF are the filler geometry, amount of filler, and the surface groups present on the fillers [178].

6 Polymer–Fullerene Nanocomposites

Harry Kroto, Richard Smalley, and Robert Curl won the nobel prize in 1996 for the discovery of soccer ball-shaped Buckminsterfullerene [179, 180]. Fullerenes show superconductivity at comparatively higher temperatures [181] and possess good mechanical stability and unique chemical reactions [182]. Eletskii et al. [183] and Millican et al. [184] reported that C_{60} molecule is stable in an inert (argon) atmosphere at temperatures up to ~1200 K. Kolodney et al. [185] showed that that C_{60} retains thermal stability up to 1700 K [186]. Diki et al. indicated that the decomposition of C_{60} and C_{70} molecules in the gas phase begins only at 2650 and 2440 K, respectively. Few studies show that C_{60} possesses a thermal degradation temperature even up to 3000 K [187, 188]. All these studies give us an impression that fullerene can be used as a reinforcing filler if there is an interaction between the polymer and fullerene. Olah et al. [189] reported the first polymer–fullerene

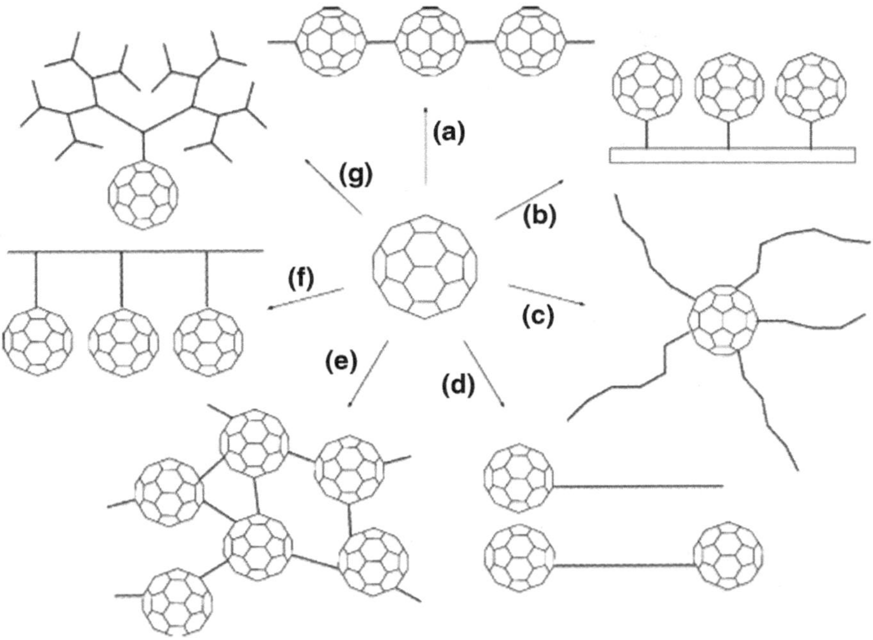

Fig. 11 Structural classifications of polymeric fullerene derivatives: **a** main-chain fullerene polymer; **b** fullerenes on solid surface; **c** star-shaped fullerene polymers; **d** fullerene-end-capped polymers; **e** crosslinked polymeric fullerene derivatives; **f** side-chain fullerene polymers; **g** fullerodendrimer. *Source* [189]

composite with polystyrene. With proper modification of C_{60}, fullerene polymers with different structures can be prepared and these structural classifications of polymeric fullerene derivatives are shown in Fig. 11.

6.1 Thermal Stability of Polymer–Fullerene Nanocomposites

Fullerene, especially C_{60} has the capacity for radical absorption which makes it as good anti-oxidizing agent for polymer matrix. Zeinalov et al. [190] showed that the incorporation of C_{60} improved the onset degradation temperature. Troitskii et al. [191] studied the thermal degradation behavior of the composites of fullerene with Poly (methyl methacrylate) (PMMA) and PS. They observed that fullerene reacts with the low molecular weight alkyl radicals produced from the early degradation of polymers and this mechanism prevents the further thermo-oxidative degradation of the above polymers. Zuev et al. [192] found that the addition of C_{60} fullerene into acrylic polymers shifted the thermal degradation mechanism of acrylic

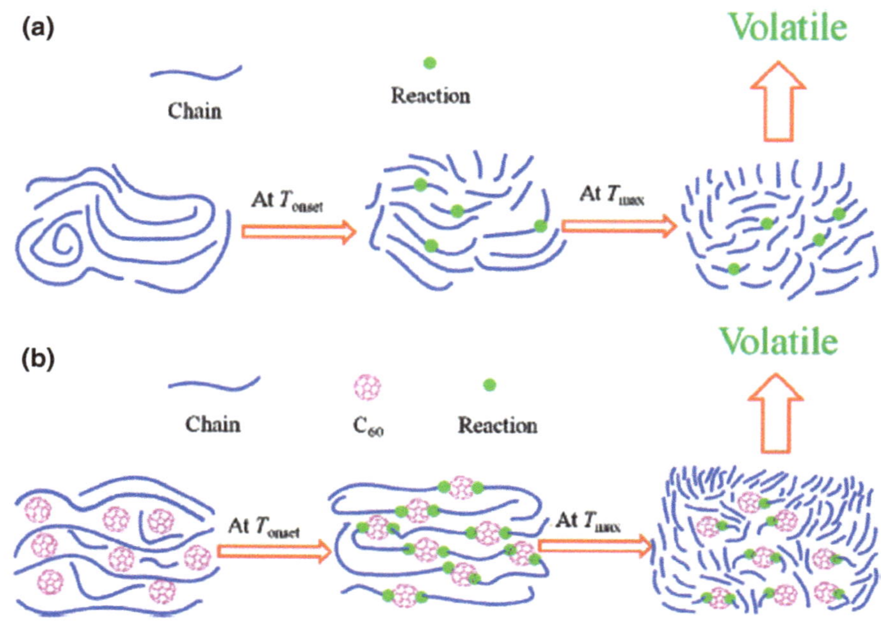

Fig. 12 Scheme of the mechanism for the thermal degradation of **a** HDPE and **b** composite. *Source* [193]

polymers from free radical mechanism to non-radical pathway. Song et al. [193] and Zhao et al. [193] investigated and thermal degradation behavior of fullerene-filled PP and HDPE, respectively, and found that the incorporation of C_{60} remarkably retards the thermal degradation of both the polymers. Zhao et al. [193] put forward a mechanism for the thermal degradation of HDPE/C_{60} composites and it is shown in Fig. 12. The incorporation of 2.5 wt% C_{60} increases the onset degradation temperature in air of HDPE by 91 °C.

The change in the degradation temperatures due to the incorporation of fullerene into the polymers are listed in Table 8.

6.2 Transition Temperatures in Polymer–Fullerene Composites

Literatures on the effect of fullerene on crystallinity and Tg of the polymers are limited. Wozniak-Braszak et al. [195] analyzed the effect of fullerene in the thermal properties of Poly (butylenes terephthalate) (PBT). Decylamine-C_{60}(DA-C_{60}) and tetracyanoethylene oxide-C_{60} (TCNEO-C_{60})were used as the fullerene derivative for this study. Glass transition temperature of the nanocomposites shows higher

Table 8 Compilation of data from the literature on degradation temperatures of fullerene/polymer composites

Reference	Polymer	Type of fullerene	wt% of fullerene (%)	Difference in onset degradation temperature of nanocomposite and virgin polymer (°C)	Difference in maximum degradation temperature of nanocomposite and virgin polymer (°C)
Song et al. [194]	PP	C_{60}	2	20	61
Troitskii et al. [191]	PMMA	C_{60}	0.13 mol %	34 (in oxygen atmosphere)	–
Zeinalov et al. [190]	Polystyrene	C_{60}	1	~20	30
Zeinalov et al. [190]	Polydimethylsiloxane (PDMS)	C_{60}	1	~30	20
Zuev et al. [192]	poly-n-alkyl acrylates (PAA-n),	C_{60}	1	–	1
Zuev et al. [192]	poly-n-alkyl methacrylates (PMA-n)	C_{60}	1	–	19

value compared to virgin PBT and the increase was prominent with high loading of fullerene. Nanocomposites showed wider melting range than the virgin PBT. The addition of fullerene also increased the temperature of the first melting of peak of PBT. The crystallization behavior showed that the incorporation of fullerenes causes a change in the nucleation process and accelerating the crystallization of the PBT. Sinitsin et al. [196] also found that the fullerene acts as a nucleating agent and increased the crystallinity in polyamide (PA6). Sanz et al. [197] studied the effect of C_{60} fullerene on the glass transition temperature of polystyrene. Up to 4 wt% of fullerene content, T_g of polystyrene increased and further addition decreased the T_g. On contrary to the above studies, Tayfun et al. [198] found that the addition of fullerene into TPU decreased the T_g of TPU. Similar observations were made by Alekseeva et al. [199] with fullerene-filled polystyrene. This change may be due to the plasticizing effect of fullerene in the composites.

7 Conclusions and Future Prospects

Polymer–carbon nanocomposites achieved great attention in the industry and academia due to their excellent mechanical, thermal, and electrical properties. The excellent thermal properties of polymer–carbon nanocomposites make them to be used as materials for sensors, high temperature withstanding seals and gaskets, electrodes, flame resistant bearings, etc. In industry point of view, it is very essential to establish a technique to properly disperse the carbon nanofillers in the polymer matrix using conventional composite manufacturing techniques. In the same time, the enhancement in the interaction between the fillers and the polymers is also a potential area and the surface modification techniques also should be extensively studied especially with respect to the nature of the chemistry of polymer matrix. Green approach for the surface functionalization of the fillers also will get considerable attention in the future. Hybrid composites with different fillers and their synergism in the composite are also expected to be an interesting area in the polymer–carbon composites in future. Defect-free carbon nanomaterials preparation at low cost will also be an important research area in the polymer nanocomposites.

Acknowledgements The author acknowledge the support of MS Ramaiah University of Applied Sciences, Bangalore

References

1. Manias E, Touny A, Wu L, Strawhecker K, Lu B, Chung T (2001) Polypropylene/montmorillonite nanocomposites. Review of the synthetic routes and materials properties. Chem Mater 13(10):3516–3523
2. Pavlidou S, Papaspyrides C (2008) A review on polymer–layered silicate nanocomposites. Prog Polym Sci 33(12):1119–1198

3. Jordan J, Jacob KI, Tannenbaum R, Sharaf MA, Jasiuk I (2005) Experimental trends in polymer nanocomposites—a review. Mater Sci Eng, A 393(1):1–11
4. Yang Y, Gupta MC, Dudley KL, Lawrence RW (2005) A comparative study of EMI shielding properties of carbon nanofiber and multi-walled carbon nanotube filled polymer composites. J Nanosci Nanotechnol 5(6):927–931
5. Sullivan J, Friedmann T, Hjort K (2001) Diamond and amorphous carbon MEMS. MRS Bull 26(04):309–311
6. Chiang K, Yang L, Wei R, Coulter K (2008) Development of diamond-like carbon-coated electrodes for corrosion sensor applications at high temperatures. Thin Solid Films 517(3): 1120–1124
7. Roy N, Sengupta R, Bhowmick AK (2012) Modifications of carbon for polymer composites and nanocomposites. Prog Polym Sci 37(6):781–819
8. Dasari A, Yu Z-Z, Mai Y-W (2016) Polymer nanocomposites: towards multi-functionality. Springer
9. Crompton TR (2006) Polymer reference book. Rapra Technology Limited, UK
10. Crompton TR (1993) Types of polymers and their uses. Practical polymer analysis. Springer, US, Boston, MA, pp 1–9. https://doi.org/10.1007/978-1-4615-2874-6_1
11. Bourbigot S, Gilman JW, Wilkie CA (2004) Kinetic analysis of the thermal degradation of polystyrene–montmorillonite nanocomposite. Polym Degrad Stab 84(3):483–492
12. Jang BN, Wilkie CA (2005) The thermal degradation of polystyrene nanocomposite. Polymer 46(9):2933–2942
13. Chrissafis K, Paraskevopoulos K, Jannakoudakis A, Beslikas T, Bikiaris D (2010) Oxidized multiwalled carbon nanotubes as effective reinforcement and thermal stability agents of poly (lactic acid) ligaments. J Appl Polym Sci 118(5):2712–2721
14. Liu H, Pa Song, Fang Z, Shen L, Peng M (2010) Thermal degradation and flammability properties of HDPE/EVA/C 60 nanocomposites. Thermochim Acta 506(1):98–101
15. Zhu J, Uhl FM, Morgan AB, Wilkie CA (2001) Studies on the mechanism by which the formation of nanocomposites enhances thermal stability. Chem Mater 13(12):4649–4654
16. Qin H, Zhang S, Zhao C, Feng M, Yang M, Shu Z, Yang S (2004) Thermal stability and flammability of polypropylene/montmorillonite composites. Polym Degrad Stab 85(2): 807–813
17. Zanetti M, Camino G, Reichert P, Mülhaupt R (2001) Thermal behaviour of poly (propylene) layered silicate nanocomposites. Macromol Rapid Commun 22(3):176–180
18. Kropka JM, Pryamitsyn V, Ganesan V (2008) Relation between glass transition temperatures in polymer nanocomposites and polymer thin films. Phys Rev Lett 101 (7):075702
19. Ayippadath Gopi J, Patel S, Chandra A, Tripathy D (2011) SBR-clay-carbon black hybrid nanocomposites for tire tread application. J Polym Res 18(6):1625–1634. https://doi.org/10.1007/s10965-011-9567-9
20. Begtrup GE, Ray KG, Kessler BM, Yuzvinsky TD, Garcia H, Zettl A (2007) Probing nanoscale solids at thermal extremes. Phys Rev Lett 99(15):155901
21. Rittigstein P, Priestley RD, Broadbelt LJ, Torkelson JM (2007) Model polymer nanocomposites provide an understanding of confinement effects in real nanocomposites. Nat Mater 6 (4):278–282
22. Ramanathan T, Stankovich S, Dikin D, Liu H, Shen H, Nguyen S, Brinson L (2007) Graphitic nanofillers in PMMA nanocomposites—an investigation of particle size and dispersion and their influence on nanocomposite properties. J Polym Sci, Part B: Polym Phys 45(15):2097–2112
23. Sun Y, Zhang Z, Moon KS, Wong C (2004) Glass transition and relaxation behavior of epoxy nanocomposites. J Polym Sci, Part B: Polym Phys 42(21):3849–3858
24. Ash B, Schadler L, Siegel R (2002) Glass transition behavior of alumina/polymethylmethacrylate nanocomposites. Mater Lett 55(1):83–87
25. Starr FW, Schroder T, Glotzer SC (2002) Molecular dynamics simulation of a polymer melt with a nanoscopic particle. Macromolecules 35:4481

26. van Zanten JH, Wallace WE, W-l Wu (1996) Effect of strongly favorable substrate interactions on the thermal properties of ultrathin polymer films. Phys Rev E 53(3):R2053
27. Rittigstein P, Torkelson JM (2006) Polymer–nanoparticle interfacial interactions in polymer nanocomposites: confinement effects on glass transition temperature and suppression of physical aging. J Polym Sci, Part B: Polym Phys 44(20):2935–2943. https://doi.org/10.1002/polb.20925
28. Qiao R, Deng H, Putz KW, Brinson LC (2011) Effect of particle agglomeration and interphase on the glass transition temperature of polymer nanocomposites. J Polym Sci, Part B: Polym Phys 49(10):740–748
29. Gopi JA, Patel SK, Tripathy DK, Chandra AK (2016) Development of cooler running PCR tire tread using SBR–ENR–nano clay composite. Int J Plast Technol 20(2):345–363
30. Jog J (2006) Crystallisation in polymer nanocomposites. Mater Sci Technol 22(7):797–806
31. Avrami M (1939) Kinetics of phase change. I general theory. J Chem Phys 7(12):1103–1112. https://doi.org/10.1063/1.1750380
32. Avrami M (1940) Kinetics of phase change. II transformation-time relations for random distribution of nuclei. J Chem Phys 8(2):212–224. https://doi.org/10.1063/1.1750631
33. Ozawa T (1971) Kinetics of non-isothermal crystallization. Polymer 12(3):150–158. https://doi.org/10.1016/0032-3861(71)90041-3
34. Kissinger HE (1956) Variation of peak temperature with heating rate in differential thermal analysis. J Res Natl Bur Stan 57(4):217–221
35. Grady BP, Pompeo F, Shambaugh RL, Resasco DE (2002) Nucleation of polypropylene crystallization by single-walled carbon nanotubes. J Phys Chem B 106(23):5852–5858
36. Manafi P, Ghasemi I, Karrabi M, Azizi H, Ehsaninamin P (2014) Effect of graphene nanoplatelets on crystallization kinetics of poly (lactic acid). Soft Mater 12(4):433–444
37. Moniruzzaman M, Winey KI (2006) Polymer nanocomposites containing carbon nanotubes. Macromolecules 39(16):5194–5205
38. Lu K, Grossiord N, Koning CE, Miltner HE, Bv Mele, Loos J (2008) Carbon nanotube/isotactic polypropylene composites prepared by latex technology: morphology analysis of CNT-induced nucleation. Macromolecules 41(21):8081–8085
39. Cheng S, Chen X, Hsuan YG, Li CY (2011) Reduced graphene oxide-induced polyethylene crystallization in solution and nanocomposites. Macromolecules 45(2):993–1000
40. Yang J, Wang C, Wang K, Zhang Q, Chen F, Du R, Fu Q (2009) Direct formation of nanohybrid shish-kebab in the injection molded bar of polyethylene/multiwalled carbon nanotubes composite. Macromolecules 42(18):7016–7023
41. Li CY, Li L, Cai W, Kodjie SL, Tenneti KK (2005) Nanohybrid shish-kebabs: periodically functionalized carbon nanotubes. Adv Mater 17(9):1198–1202
42. Li L, Li CY, Ni C, Rong L, Hsiao B (2007) Structure and crystallization behavior of Nylon 66/multi-walled carbon nanotube nanocomposites at low carbon nanotube contents. Polymer 48(12):3452–3460
43. Sahoo NG, Rana S, Cho JW, Li L, Chan SH (2010) Polymer nanocomposites based on functionalized carbon nanotubes. Prog Polym Sci 35(7):837–867
44. Chrissafis K, Bikiaris D (2011) Can nanoparticles really enhance thermal stability of polymers? Part I: an overview on thermal decomposition of addition polymers. Thermochim Acta 523(1):1–24
45. Dillon AC, Jones K, Bekkedahl T, Kiang C (1997) Storage of hydrogen in single-walled carbon nanotubes. Nature 386(6623):377
46. Liu C, Fan Y, Liu M, Cong H, Cheng H, Dresselhaus MS (1999) Hydrogen storage in single-walled carbon nanotubes at room temperature. Science 286(5442):1127–1129
47. Ly SY (2006) Detection of dopamine in the pharmacy with a carbon nanotube paste electrode using voltammetry. Bioelectrochemistry 68(2):227–231
48. Minot ED, Janssens AM, Heller I, Heering HA, Dekker C, Lemay SG (2007) Carbon nanotube biosensors: the critical role of the reference electrode. Appl Phys Lett 91(9):093507

49. Liu H, Liu Z, Liu J (2006) Investigations on stability of single-walled carbon nanotubes. J Synth Cryst 35(3):494
50. Xu F, Sun LX, Zhang J, Qi YN, Yang LN, Ru HY, Wang CY, Meng X, Lan XF, Jiao QZ, Huang FL (2010) Thermal stability of carbon nanotubes. J Therm Anal Calorim 102(2):785–791. https://doi.org/10.1007/s10973-010-0793-x
51. Chatterjee A, Deopura B (2006) Thermal stability of polypropylene/carbon nanofiber composite. J Appl Polym Sci 100(5):3574–3578
52. Kashiwagi T, Grulke E, Hilding J, Harris R, Awad W, Douglas J (2002) Thermal degradation and flammability properties of poly (propylene)/carbon nanotube composites. Macromol Rapid Commun 23(13):761–765
53. Bikiaris D, Vassiliou A, Chrissafis K, Paraskevopoulos K, Jannakoudakis A, Docoslis A (2008) Effect of acid treated multi-walled carbon nanotubes on the mechanical, permeability, thermal properties and thermo-oxidative stability of isotactic polypropylene. Polym Degrad Stab 93(5):952–967
54. Marosfői B, Marosi GJ, Szep A, Anna P, Keszei S, Nagy B, Martvonova H, Sajo I (2006) Complex activity of clay and CNT particles in flame retarded EVA copolymer. Polym Adv Technol 17(4):255–262
55. Bocchini S, Frache A, Camino G, Claes M (2007) Polyethylene thermal oxidative stabilisation in carbon nanotubes based nanocomposites. Eur Polymer J 43(8):3222–3235
56. Davis RD, Gilman JW, VanderHart DL (2003) Processing degradation of polyamide 6/montmorillonite clay nanocomposites and clay organic modifier. Polym Degrad Stab 79(1):111–121
57. Chipara M, Lozano K, Hernandez A, Chipara M (2008) TGA analysis of polypropylene–carbon nanofibers composites. Polym Degrad Stab 93(4):871–876
58. Sarno M, Gorrasi G, Sannino D, Sorrentino A, Ciambelli P, Vittoria V (2004) Polymorphism and thermal behaviour of syndiotactic poly (propylene)/carbon nanotube composites. Macromol Rapid Commun 25(23):1963–1967
59. Hirschler M (1984) Reduction of smoke formation from and flammability of thermoplastic polymers by metal oxides. Polymer 25(3):405–411
60. Mai F, Habibi Y, Raquez J-M, Dubois P, Feller J-F, Peijs T, Bilotti E (2013) Poly (lactic acid)/carbon nanotube nanocomposites with integrated degradation sensing. Polymer 54(25):6818–6823
61. Di Blasi C, Galgano A, Branca C (2013) Modeling the thermal degradation of poly (methyl methacrylate)/carbon nanotube nanocomposites. Polym Degrad Stab 98(1):266–275
62. Chou W-J, Wang C-C, Chen C-Y (2008) Thermal behaviors of polyimide with plasma-modified carbon nanotubes. Polym Degrad Stab 93(3):745–752
63. Chen X, Wang J, Lin M, Zhong W, Feng T, Chen X, Chen J, Xue F (2008) Mechanical and thermal properties of epoxy nanocomposites reinforced with amino-functionalized multi-walled carbon nanotubes. Mater Sci Eng, A 492(1):236–242
64. Hezma A, Elashmawi I, Abdelrazek E, Rajeh A, Kamal M (2017) Enhancement of the thermal and mechanical properties of polyurethane/polyvinyl chloride blend by loading single walled carbon nanotubes. Prog Nat Sci: Mater Int 27(3):338–343
65. Breuer O, Sundararaj U (2004) Big returns from small fibers: a review of polymer/carbon nanotube composites. Polym Compos 25(6):630–645. https://doi.org/10.1002/pc.20058
66. Ma PC, Kim J-K, Tang BZ (2007) Effects of silane functionalization on the properties of carbon nanotube/epoxy nanocomposites. Compos Sci Technol 67(14):2965–2972
67. Barick AK, Tripathy DK (2011) Preparation, characterization and properties of acid functionalized multi-walled carbon nanotube reinforced thermoplastic polyurethane nanocomposites. Mater Sci Eng, B 176(18):1435–1447. https://doi.org/10.1016/j.mseb.2011.08.001
68. Zhou Y, Pervin F, Lewis L, Jeelani S (2007) Experimental study on the thermal and mechanical properties of multi-walled carbon nanotube-reinforced epoxy. Mater Sci Eng, A 452:657–664

69. Ganguli S, Aglan H, Dennig P, Irvin G (2006) Effect of loading and surface modification of MWCNTs on the fracture behavior of epoxy nanocomposites. J Reinf Plast Compos 25 (2):175–188
70. Pham JQ, Mitchell CA, Bahr JL, Tour JM, Krishanamoorti R, Green PF (2003) Glass transition of polymer/single-walled carbon nanotube composite films. J Polym Sci, Part B: Polym Phys 41(24):3339–3345
71. Wang S, Liang Z, Liu T, Wang B, Zhang C (2006) Effective amino-functionalization of carbon nanotubes for reinforcing epoxy polymer composites. Nanotechnology 17(6):1551
72. Gojny FH, Schulte K (2004) Functionalisation effect on the thermo-mechanical behaviour of multi-wall carbon nanotube/epoxy-composites. Compos Sci Technol 64(15):2303–2308
73. Sterzyński T, Tomaszewska J, Piszczek K, Skórczewska K (2010) The influence of carbon nanotubes on the PVC glass transition temperature. Compos Sci Technol 70(6):966–969
74. Wei C, Srivastava D, Cho K (2004) Structural ordering in nanotube polymer composites. Nano Lett 4(10):1949–1952
75. Mitchell CA, Bahr JL, Arepalli S, Tour JM, Krishnamoorti R (2002) Macromolecules 35:8825
76. Arrigo R, Bellavia S, Gambarotti C, Dintcheva NT, Carroccio S (2017) Carbon nanotubes-based nanohybrids for multifunctional nanocomposites. J King Saud Univ-Sci 29(4):502–509
77. Barrau S, Vanmansart C, Moreau M, Addad A, Stoclet G, Lefebvre J-M, Seguela R (2011) Crystallization behavior of carbon nanotube—polylactide nanocomposites. Macromolecules 44(16):6496–6502
78. Haggenmueller R, Fischer JE, Winey KI (2006) Single wall carbon nanotube/polyethylene nanocomposites: nucleating and templating polyethylene crystallites. Macromolecules 39 (8):2964–2971
79. Bhattacharyya AR, Sreekumar T, Liu T, Kumar S, Ericson LM, Hauge RH, Smalley RE (2003) Crystallization and orientation studies in polypropylene/single wall carbon nanotube composite. Polymer 44(8):2373–2377
80. Brosse A-C, Tencé-Girault S, Piccione PM, Leibler L (2008) Effect of multi-walled carbon nanotubes on the lamellae morphology of polyamide-6. Polymer 49(21):4680–4686
81. Valentini L, Biagiotti J, Kenny JM, Santucci S (2003) Morphological characterization of single-walled carbon nanotubes-PP composites. Compos Sci Technol 63(8):1149–1153. https://doi.org/10.1016/S0266-3538(03)00036-8
82. Manchado MAL, Valentini L, Biagiotti J, Kenny JM (2005) Thermal and mechanical properties of single-walled carbon nanotubes–polypropylene composites prepared by melt processing. Carbon 43(7):1499–1505. https://doi.org/10.1016/j.carbon.2005.01.031
83. Sarno M, Gorrasi G, Sannino D, Sorrentino A, Ciambelli P, Vittoria V (2004) Polymorphism and thermal behaviour of syndiotactic poly(propylene)/carbon nanotube composites. Macromol Rapid Commun 25(23):1963–1967. https://doi.org/10.1002/marc.200400344
84. Liu Phang IY, Shen L, Chow SY, Zhang W-D (2004) Morphology and mechanical properties of multiwalled carbon nanotubes reinforced nylon-6 composites. Macromolecules 37(19):7214–7222. https://doi.org/10.1021/ma049132t
85. Yudin VE, Svetlichnyi VM, Shumakov AN, Letenko DG, Feldman AY, Marom G (2005) The nucleating effect of carbon nanotubes on crystallinity in R-BAPB-type thermoplastic polyimide. Macromol Rapid Commun 26(11):885–888. https://doi.org/10.1002/marc.200500085
86. Lai M, Li J, Yang J, Liu J, Tong X, Cheng H (2004) The morphology and thermal properties of multi-walled carbon nanotube and poly(hydroxybutyrate-co-hydroxyvalerate) composite. Polym Int 53(10):1479–1484. https://doi.org/10.1002/pi.1566
87. Arjmand M, Chizari K, Krause B, Pötschke P, Sundararaj U (2016) Effect of synthesis catalyst on structure of nitrogen-doped carbon nanotubes and electrical conductivity and electromagnetic interference shielding of their polymeric nanocomposites. Carbon 98:358–372

88. MacDonald AKGAH (2007) Graphene: exploring carbon flatland. Phys Today 60(8):35–41. https://doi.org/10.1063/1.2774096
89. Si Y, Samulski ET (2008) Synthesis of water soluble graphene. Nano Lett 8(6):1679–1682
90. Dreyer DR, Park S, Bielawski CW, Ruoff RS (2010) The chemistry of graphene oxide. Chem Soc Rev 39(1):228–240. https://doi.org/10.1039/B917103G
91. Wang G, Yang J, Park J, Gou X, Wang B, Liu H, Yao J (2008) Facile synthesis and characterization of graphene nanosheets. J Phys Chem C 112(22):8192–8195. https://doi.org/10.1021/jp710931h
92. Blake P, Brimicombe PD, Nair RR, Booth TJ, Jiang D, Schedin F, Ponomarenko LA, Morozov SV, Gleeson HF, Hill EW, Geim AK, Novoselov KS (2008) Graphene-based liquid crystal device. Nano Lett 8(6):1704–1708. https://doi.org/10.1021/nl080649i
93. Novoselov KS, Geim AK, Morozov SV, Jiang D, Zhang Y, Dubonos SV, Grigorieva IV, Firsov AA (2004) Electric field effect in atomically thin carbon films. Science 306 (5696):666–669. https://doi.org/10.1126/science.1102896
94. Berger C, Song Z, Li X, Wu X, Brown N, Naud C, Mayou D, Li T, Hass J, Marchenkov AN, Conrad EH, First PN, de Heer WA (2006) Electronic confinement and coherence in patterned epitaxial graphene. Science 312(5777):1191–1196. https://doi.org/10.1126/science.1125925
95. Reina A, Jia X, Ho J, Nezich D, Son H, Bulovic V, Dresselhaus MS, Kong J (2009) Large area, few-layer graphene films on arbitrary substrates by chemical vapor deposition. Nano Lett 9(1):30–35. https://doi.org/10.1021/nl801827v
96. Wu Z-S, Ren W, Gao L, Liu B, Jiang C, Cheng H-M (2009) Synthesis of high-quality graphene with a pre-determined number of layers. Carbon 47(2):493–499. https://doi.org/10.1016/j.carbon.2008.10.031
97. Fu W, Kiggans J, Overbury SH, Schwartz V, Liang C (2011) Low-temperature exfoliation of multilayer-graphene material from FeCl3 and CH3NO2 co-intercalated graphite compound. Chem Commun 47(18):5265–5267. https://doi.org/10.1039/C1CC10508F
98. Allen MJ, Tung VC, Kaner RB (2010) Honeycomb carbon: a review of graphene. Chem Rev 110(1):132–145. https://doi.org/10.1021/cr900070d
99. Shahil KMF, Balandin AA (2012) Thermal properties of graphene and multilayer graphene: applications in thermal interface materials. Solid State Commun 152(15):1331–1340. https://doi.org/10.1016/j.ssc.2012.04.034
100. Britnell L, Ribeiro RM, Eckmann A, Jalil R, Belle BD, Mishchenko A, Kim Y-J, Gorbachev RV, Georgiou T, Morozov SV, Grigorenko AN, Geim AK, Casiraghi C, Neto AHC, Novoselov KS (2013) Strong Light-matter interactions in heterostructures of atomically thin films. Science 340(6138):1311–1314. https://doi.org/10.1126/science.1235547
101. Kim KS, Zhao Y, Jang H, Lee SY, Kim JM, Kim KS, Ahn J-H, Kim P, Choi J-Y, Hong BH (2009) Large-scale pattern growth of graphene films for stretchable transparent electrodes. Nature 457 (7230):706–710. doi:http://www.nature.com/nature/journal/v457/n7230/suppinfo/nature07719_S1.html
102. Ansari S, Giannelis EP (2009) Functionalized graphene sheet—poly(vinylidene fluoride) conductive nanocomposites. J Polym Sci, Part B: Polym Phys 47(9):888–897. https://doi.org/10.1002/polb.21695
103. RamanathanT, Abdala AA, StankovichS, Dikin DA, Herrera Alonso M, Piner RD, Adamson DH, Schniepp HC, ChenX, Ruoff RS, Nguyen ST, Aksay IA, Prud'Homme RK, Brinson LC (2008) Functionalized graphene sheets for polymer nanocomposites. Nat Nano 3 (6):327–331. http://www.nature.com/nnano/journal/v3/n6/suppinfo/nnano.2008.96_S1.html
104. Xu Y, Wang Y, Liang J, Huang Y, Ma Y, Wan X, Chen Y (2009) A hybrid material of graphene and poly (3,4-ethyldioxythiophene) with high conductivity, flexibility, and transparency. Nano Res 2(4):343–348. https://doi.org/10.1007/s12274-009-9032-9
105. Quan H, B-q Zhang, Zhao Q, Yuen RKK, Li RKY (2009) Facile preparation and thermal degradation studies of graphite nanoplatelets (GNPs) filled thermoplastic polyurethane (TPU) nanocomposites. Compos A Appl Sci Manuf 40(9):1506–1513. https://doi.org/10.1016/j.compositesa.2009.06.012

106. Eda G, Chhowalla M (2009) Graphene-based composite thin films for electronics. Nano Lett 9(2):814–818. https://doi.org/10.1021/nl8035367
107. Liang J, Xu Y, Huang Y, Zhang L, Wang Y, Ma Y, Li F, Guo T, Chen Y (2009) Infrared-triggered actuators from graphene-based nanocomposites. J Phys Chem C 113 (22):9921–9927. https://doi.org/10.1021/jp901284d
108. Becerril HA, Mao J, Liu Z, Stoltenberg RM, Bao Z, Chen Y (2008) Evaluation of solution-processed reduced graphene oxide films as transparent conductors. ACS Nano 2 (3):463–470. https://doi.org/10.1021/nn700375n
109. Dikin DA, Stankovich S, Zimney EJ, Piner RD, Dommett GHB, Evmenenko G, Nguyen ST, Ruoff RS (2007) Preparation and characterization of graphene oxide paper. Nature 448 (7152):457–460. http://www.nature.com/nature/journal/v448/n7152/suppinfo/nature06016_S1.html
110. Szabó T, Szeri A, Dékány I (2005) Composite graphitic nanolayers prepared by self-assembly between finely dispersed graphite oxide and a cationic polymer. Carbon 43 (1):87–94. https://doi.org/10.1016/j.carbon.2004.08.025
111. Castro Neto AH, Guinea F, Peres NMR, Novoselov KS, Geim AK (2009) The electronic properties of graphene. Rev Mod Phys 81(1):109–162
112. Sadasivuni KK, Ponnamma D, Thomas S, Grohens Y (2014) Evolution from graphite to graphene elastomer composites. Prog Polym Sci 39(4):749–780. https://doi.org/10.1016/j.progpolymsci.2013.08.003
113. Peigney A, Laurent C, Flahaut E, Bacsa RR, Rousset A (2001) Specific surface area of carbon nanotubes and bundles of carbon nanotubes. Carbon 39(4):507–514. https://doi.org/10.1016/S0008-6223(00)00155-X
114. Liang J, Huang Y, Zhang L, Wang Y, Ma Y, Guo T, Chen Y (2009) Molecular-Level dispersion of graphene into poly(vinyl alcohol) and effective reinforcement of their nanocomposites. Adv Func Mater 19(14):2297–2302. https://doi.org/10.1002/adfm.200801776
115. Lee C, Wei X, Kysar JW, Hone J (2008) Measurement of the elastic properties and intrinsic strength of monolayer graphene. Science 321(5887):385–388. https://doi.org/10.1126/science.1157996
116. Kuilla T, Bhadra S, Yao D, Kim NH, Bose S, Lee JH (2010) Recent advances in graphene based polymer composites. Prog Polym Sci 35(11):1350–1375. https://doi.org/10.1016/j.progpolymsci.2010.07.005
117. Kholmanov IN, Magnuson CW, Aliev AE, Li H, Zhang B, Suk JW, Zhang LL, Peng E, Mousavi SH, Khanikaev AB, Piner R, Shvets G, Ruoff RS (2012) Improved electrical conductivity of graphene films integrated with metal nanowires. Nano Lett 12(11):5679–5683. https://doi.org/10.1021/nl302870x
118. Balandin AA, Ghosh S, Bao W, Calizo I, Teweldebrhan D, Miao F, Lau CN (2008) Superior thermal conductivity of single-layer graphene. Nano Lett 8(3):902–907. https://doi.org/10.1021/nl0731872
119. Du J, Cheng H-M (2012) The fabrication, properties, and uses of graphene/polymer composites. Macromol Chem Phys 213(10–11):1060–1077. https://doi.org/10.1002/macp.201200029
120. McAllister MJ, Li J-L, Adamson DH, Schniepp HC, Abdala AA, Liu J, Herrera-Alonso M, Milius DL, Car R, Prud'homme RK, Aksay IA (2007) Single sheet functionalized graphene by oxidation and thermal expansion of graphite. Chem Mater 19(18):4396–4404. https://doi.org/10.1021/cm0630800
121. Niyogi S, Bekyarova E, Itkis ME, McWilliams JL, Hamon MA, Haddon RC (2006) solution properties of graphite and graphene. J Am Chem Soc 128(24):7720–7721. https://doi.org/10.1021/ja060680r
122. Cano M, Khan U, Sainsbury T, O'Neill A, Wang Z, McGovern IT, Maser WK, Benito AM, Coleman JN (2013) Improving the mechanical properties of graphene oxide based materials by covalent attachment of polymer chains. Carbon 52:363–371. https://doi.org/10.1016/j.carbon.2012.09.046

123. Jiang T, Kuila T, Kim NH, Lee JH (2014) Effects of surface-modified silica nanoparticles attached graphene oxide using isocyanate-terminated flexible polymer chains on the mechanical properties of epoxy composites. J Mater Chem A 2(27):10557–10567. https://doi.org/10.1039/C4TA00584H
124. Sadasivuni KK, Ponnamma D, Kumar B, Strankowski M, Cardinaels R, Moldenaers P, Thomas S, Grohens Y (2014) Dielectric properties of modified graphene oxide filled polyurethane nanocomposites and its correlation with rheology. Compos Sci Technol 104:18–25. https://doi.org/10.1016/j.compscitech.2014.08.025
125. Samanta SK, Fritsch M, Scherf U, Gomulya W, Bisri SZ, Loi MA (2014) Conjugated polymer-assisted dispersion of single-wall carbon nanotubes: the power of polymer wrapping. Acc Chem Res 47(8):2446–2456. https://doi.org/10.1021/ar500141j
126. Ye W, Shi X, Su J, Chen Y, Fu J, Zhao X, Zhou F, Wang C, Xue D (2014) One-step reduction and functionalization protocol to synthesize polydopamine wrapping Ag/graphene hybrid for efficient oxidation of hydroquinone to benzoquinone. Appl Catal B 160–161:400–407. https://doi.org/10.1016/j.apcatb.2014.05.042
127. Fujigaya T, Nakashima N (2015) Non-covalent polymer wrapping of carbon nanotubes and the role of wrapped polymers as functional dispersants. Sci Technol Adv Mater 16 (2):024802. https://doi.org/10.1088/1468-6996/16/2/024802
128. Liu N, Luo F, Wu H, Liu Y, Zhang C, Chen J (2008) One-step ionic-liquid-assisted electrochemical synthesis of ionic-liquid-functionalized graphene sheets directly from graphite. Adv Func Mater 18(10):1518–1525. https://doi.org/10.1002/adfm.200700797
129. Kim TY, Lee HW, Stoller M, Dreyer DR, Bielawski CW, Ruoff RS, Suh KS (2011) High-performance supercapacitors based on poly(ionic liquid)-modified graphene electrodes. ACS Nano 5(1):436–442. https://doi.org/10.1021/nn101968p
130. Yang H, Shan C, Li F, Han D, Zhang Q, Niu L (2009) Covalent functionalization of polydisperse chemically-converted graphene sheets with amine-terminated ionic liquid. Chem Commun 26:3880–3882. https://doi.org/10.1039/B905085J
131. Nuvoli D, Valentini L, Alzari V, Scognamillo S, Bon SB, Piccinini M, Illescas J, Mariani A (2011) High concentration few-layer graphene sheets obtained by liquid phase exfoliation of graphite in ionic liquid. J Mater Chem 21(10):3428–3431. https://doi.org/10.1039/C0JM02461A
132. Zhang Q, Wu S, Zhang L, Lu J, Verproot F, Liu Y, Xing Z, Li J, Song X-M (2011) Fabrication of polymeric ionic liquid/graphene nanocomposite for glucose oxidase immobilization and direct electrochemistry. Biosens Bioelectron 26(5):2632–2637. https://doi.org/10.1016/j.bios.2010.11.024
133. Shante VK, Kirkpatrick S (1971) An introduction to percolation theory. Adv Phys 20 (85):325–357
134. Liu N, Luo F, Wu H, Liu Y, Zhang C, Chen J (2008) One-step ionic-liquid-assisted electrochemical synthesis of ionic-liquid-functionalized graphene sheets directly from graphite. Adv Func Mater 18(10):1518–1525
135. Mukhopadhyay P, Gupta RK (2012) Graphite, graphene, and their polymer nanocomposites. CRC Press
136. Liang J, Huang Y, Zhang L, Wang Y, Ma Y, Guo T, Chen Y (2009) Molecular-level dispersion of graphene into poly (vinyl alcohol) and effective reinforcement of their nanocomposites. Adv Func Mater 19(14):2297–2302
137. Fim FdC, Basso NR, Graebin AP, Azambuja DS, Galland GB (2013) Thermal, electrical, and mechanical properties of polyethylene–graphene nanocomposites obtained by in situ polymerization. J Appl Polym Sci 128(5):2630–2637
138. Ramanathan T, Abdala A, Stankovich S, Dikin D, Herrera-Alonso M, Piner R, Adamson D, Schniepp H, Chen X, Ruoff R (2008) Functionalized graphene sheets for polymer nanocomposites. Nat Nanotechnol 3(6):327–331
139. Verdejo R, Barroso-Bujans F, Rodriguez-Perez MA, de Saja JA, Lopez-Manchado MA (2008) Functionalized graphene sheet filled silicone foam nanocomposites. J Mater Chem 18 (19):2221–2226

140. Higginbotham AL, Lomeda JR, Morgan AB, Tour JM (2009) Graphite oxide flame-retardant polymer nanocomposites. ACS Appl Mater Interfaces 1(10):2256–2261
141. Verdejo R, Bernal MM, Romasanta LJ, Lopez-Manchado MA (2011) Graphene filled polymer nanocomposites. J Mater Chem 21(10):3301–3310
142. Salavagione HJ, Martínez G, Gómez MA (2009) Synthesis of poly (vinyl alcohol)/reduced graphite oxide nanocomposites with improved thermal and electrical properties. J Mater Chem 19(28):5027–5032
143. Fang M, Wang K, Lu H, Yang Y, Nutt S (2009) Covalent polymer functionalization of graphene nanosheets and mechanical properties of composites. J Mater Chem 19(38):7098–7105
144. Yoonessi M, Shi Y, Scheiman DA, Lebron-Colon M, Tigelaar DM, Weiss R, Meador MA (2012) Graphene polyimide nanocomposites; thermal, mechanical, and high-temperature shape memory effects. ACS Nano 6(9):7644–7655
145. Yoonessi M, Gaier JR (2010) Highly conductive multifunctional graphene polycarbonate nanocomposites. ACS Nano 4(12):7211–7220
146. Lee S, Hong JY, Jang J (2013) The effect of graphene nanofiller on the crystallization behavior and mechanical properties of poly (vinyl alcohol). Polym Int 62(6):901–908
147. Gnanasekaran K, Heijmans T, van Bennekom S, Woldhuis H, Wijnia S, Friedrich H (2017) 3D printing of CNT-and graphene-based conductive polymer nanocomposites by fused deposition modeling. Appl Mater Today 9:21–28
148. Bhushan B (2010) Springer handbook of nanotechnology. Springer Science & Business Media
149. Teo KB, Singh C, Chhowalla M, Milne WI (2003) Catalytic synthesis of carbon nanotubes and nanofibers. Encycl Nanosci Nanotechnol 10(1)
150. Homenick CM, Lawson G, Adronov A (2007) Polymer grafting of carbon nanotubes using living free-radical polymerization. Polym Rev 47(2):265–290
151. Coquay P, Peigney A, De Grave E, Flahaut E, Vandenberghe RE, Laurent C (2005) Fe/Co alloys for the catalytic chemical vapor deposition synthesis of single-and double-walled carbon nanotubes (CNTs). 1. The CNT−Fe/Co−MgO system. J Phys Chem B 109 (38):17813–17824
152. Hammel E, Tang X, Trampert M, Schmitt T, Mauthner K, Eder A, Pötschke P (2004) Carbon nanofibers for composite applications. Carbon 42(5):1153–1158
153. Kuzuya C, In-Hwang W, Hirako S, Hishikawa Y, Motojima S (2002) Preparation, morphology, and growth mechanism of carbon nanocoils. Chem Vap Deposition 8(2):57–62
154. Al-Saleh MH, Sundararaj U (2009) A review of vapor grown carbon nanofiber/polymer conductive composites. Carbon 47(1):2–22
155. Tibbetts GG, Lake ML, Strong KL, Rice BP (2007) A review of the fabrication and properties of vapor-grown carbon nanofiber/polymer composites. Compos Sci Technol 67 (7):1709–1718
156. Glasgow D, Tibbetts G, Matuszewski J, Walters K, Lake M (May 2004) Surface treatment of carbon nano fibers for improved composite mechanical properties. In: International society for advancement of materials and process engineering (SAMPE) symposium and exhibition, Long Beach, CA, pp 16–20
157. Lakshminarayanan PV, Toghiani H, Pittman CU (2004) Nitric acid oxidation of vapor grown carbon nanofibers. Carbon 42(12):2433–2442
158. Baek J-B, Lyons CB, Tan L-S (2004) Grafting of vapor-grown carbon nanofibers via in-situ polycondensation of 3-phenoxybenzoic acid in poly (phosphoric acid). Macromolecules 37 (22):8278–8285
159. Glasgow DG, Lake ML (2003) Production of high surface energy, high surface area vapor grown carbon fiber. Google Patents
160. Yang S, Taha-Tijerina J, Serrato-Diaz V, Hernandez K, Lozano K (2007) Dynamic mechanical and thermal analysis of aligned vapor grown carbon nanofiber reinforced polyethylene. Compos B Eng 38(2):228–235

161. Choi Y-K, K-i Sugimoto, Song S-M, Gotoh Y, Ohkoshi Y, Endo M (2005) Mechanical and physical properties of epoxy composites reinforced by vapor grown carbon nanofibers. Carbon 43(10):2199–2208
162. Lozano K, Barrera EV (2001) Nanofiber-reinforced thermoplastic composites. I. Thermoanalytical and mechanical analyses. J Appl Polym Sci 79(1):125–133. https://doi.org/10.1002/1097-4628(20010103)79:1%3c125::aid-app150%3e3.0.co;2-d
163. Xu Y, Higgins B, Brittain WJ (2005) Bottom-up synthesis of PS–CNF nanocomposites. Polymer 46(3):799–810
164. Choi YK, Sugimoto KI, Song SM, Endo M (2005) Mechanical and thermal properties of vapor-grown carbon nanofiber and polycarbonate composite sheets. Mater Lett 59 (27):3514–3520
165. Zeng J, Saltysiak B, Johnson W, Schiraldi DA, Kumar S (2004) Processing and properties of poly (methyl methacrylate)/carbon nano fiber composites. Compos B Eng 35(2):173–178
166. Kumar S, Lively B, Sun L, Li B, Zhong W (2010) Highly dispersed and electrically conductive polycarbonate/oxidized carbon nanofiber composites for electrostatic dissipation applications. Carbon 48(13):3846–3857
167. Higgins BA, Brittain WJ (2005) Polycarbonate carbon nanofiber composites. Eur Polym J 41(5):889–893
168. Rasheed A, Dadmun MD, Britt PF (2006) Polymer-nanofiber composites: Enhancing composite properties by nanofiber oxidation. J Polym Sci, Part B: Polym Phys 44(21):3053–3061
169. Kumar S, Rath T, Mahaling R, Reddy C, Das C, Pandey K, Srivastava R, Yadaw S (2007) Study on mechanical, morphological and electrical properties of carbon nanofiber/polyetherimide composites. Mater Sci Eng, B 141(1):61–70
170. Barick AK, Tripathy DK (2010) Effect of nanofiber on material properties of vapor-grown carbon nanofiber reinforced thermoplastic polyurethane (TPU/CNF) nanocomposites prepared by melt compounding. Compos A Appl Sci Manuf 41(10):1471–1482. https://doi.org/10.1016/j.compositesa.2010.06.009
171. Barick A, Tripathy D (2012) Preparation and characterization of carbon nanofiber reinforced thermoplastic polyurethane nanocomposites. J Appl Polym Sci 124(1):765–780
172. Takahashi T, Yonetake K, Koyama K, Kikuchi T (2003) Polycarbonate crystallization by vapor-grown carbon fiber with and without magnetic field. Macromol Rapid Commun 24 (13):763–767
173. Sandler J, Werner P, Shaffer MS, Demchuk V, Altstädt V, Windle AH (2002) Carbon-nanofibre-reinforced poly (ether ether ketone) composites. Compos A Appl Sci Manuf 33(8):1033–1039
174. Al-Saleh MH, Sundararaj U (2011) Review of the mechanical properties of carbon nanofiber/polymer composites. Compos A Appl Sci Manuf 42(12):2126–2142
175. Zeng J, Saltysiak B, Johnson WS, Schiraldi DA, Kumar S (2004) Processing and properties of poly(methyl methacrylate)/carbon nano fiber composites. Compos B Eng 35(2):173–178. https://doi.org/10.1016/S1359-8368(03)00051-9
176. Zhou Y, Pervin F, Jeelani S (2007) Effect vapor grown carbon nanofiber on thermal and mechanical properties of epoxy. J Mater Sci 42(17):7544–7553
177. Green KJ, Dean DR, Vaidya UK, Nyairo E (2009) Multiscale fiber reinforced composites based on a carbon nanofiber/epoxy nanophased polymer matrix: synthesis, mechanical, and thermomechanical behavior. Compos A Appl Sci Manuf 40(9):1470–1475
178. Gunes IS, Cao F, Jana SC (2008) Evaluation of nanoparticulate fillers for development of shape memory polyurethane nanocomposites. Polymer 49(9):2223–2234
179. Kroto HW, Heath JR, O'Brien SC, Curl RF, Smalley RE (1985) C 60: buckminsterfullerene. Nature 318(6042):162–163
180. Arndt M, Nairz O, Vos-Andreae J, Keller C, Van der Zouw G, Zeilinger A (1999) Wave–particle duality of C60 molecules. Nature 401(6754):680–682
181. Chakravarty S, Kivelson S (1991) Superconductivity of doped fullerenes. EPL (Europhys Lett) 16(8):751

182. Culotta E, Koshland DE Jr (1991) Buckyballs: wide open playing field for chemists. Science 254(5039):1706
183. Eletskiĭ A, Smirnov BM (1993) Fullerenes. Phys Usp 36(3):202–224
184. Milliken J, Keller TM, Baronavski A, McElvany SW, Callahan JH, Nelson H (1991) Thermal and oxidative analyses of buckminsterfullerene, C60. Chem Mater 3(3):386–387
185. Diky VV, Kabo GJ (2000) Thermodynamic properties of C 60 and C 70 fullerenes. Russ Chem Rev 69(2):95–104
186. Kolodney E, Tsipinyuk B, Budrevich A (1994) The thermal stability and fragmentation of C60 molecule up to 2000 K on the milliseconds time scale. J Chem Phys 100(11):8542–8545
187. Zhang B, Wang C, Chan CT, Ho K (1993) Thermal disintegration of carbon fullerenes. Phys Rev B 48(15):11381
188. Kim SG, Tománek D (1994) Melting the fullerenes: a molecular dynamics study. Phys Rev Lett 72(15):2418
189. Wang C, Guo Z-X, Fu S, Wu W, Zhu D (2004) Polymers containing fullerene or carbon nanotube structures. Prog Polym Sci 29(11):1079–1141
190. Zeinalov EB, Koßmehl G (2001) Fullerene C60 as an antioxidant for polymers. Polym Degrad Stab 71(2):197–202. https://doi.org/10.1016/S0141-3910(00)00109-9
191. Troitskii BB, Troitskaya LS, Dmitriev AA, Yakhnov AS (2000) Inhibition of thermo-oxidative degradation of poly(methyl methacrylate) and polystyrene by C60. Eur Polym J 36(5):1073–1084. https://doi.org/10.1016/S0014-3057(99)00156-1
192. Zuev VV, Bertini F, Audisio G (2005) Fullerene C60 as stabiliser for acrylic polymers. Polym Degrad Stab 90(1):28–33. https://doi.org/10.1016/j.polymdegradstab.2005.02.011
193. Zhao L, Guo Z, Cao Z, Zhang T, Fang Z, Peng M (2013) Thermal and thermo-oxidative degradation of high density polyethylene/fullerene composites. Polym Degrad Stab 98 (10):1953–1962
194. Pa Song, Zhu Y, Tong L, Fang Z (2008) C60 reduces the flammability of polypropylene nanocomposites by in situ forming a gelled-ball network. Nanotechnology 19(22):225707
195. Wozniak-Braszak A, Jurga K, Jurga J, Baranowski M, Grzesiak W, Brycki B, Holderna-Natkaniec K (2012) Effect of fullerene derivates on thermal and crystallization behavior of PBT/Decylamine-C60 and PBT/TCNEO-C60 nanocomposites. J Nanomater 2012:8. https://doi.org/10.1155/2012/932573
196. Sinitsin AN, Zuev VV (2016) Dielectric relaxation of fulleroid materials filled PA 6 composites and the study of its mechanical and tribological performance. Mater Chem Phys 176:152–160. https://doi.org/10.1016/j.matchemphys.2016.04.007
197. Sanz A, Wong HC, Nedoma AJ, Douglas JF, Cabral JT (2015) Influence of C60 fullerenes on the glass formation of polystyrene. Polymer 68:47–56. https://doi.org/10.1016/j.polymer.2015.05.001
198. Tayfun U, Kanbur Y, Abaci U, Guney HY, Bayramli E (2015) Mechanical, flow and electrical properties of thermoplastic polyurethane/fullerene composites: Effect of surface modification of fullerene. Compos B Eng 80:101–107. https://doi.org/10.1016/j.compositesb.2015.05.013
199. Alekseeva OV, Barannikov VP, Bagrovskaya NA, Noskov AV (2012) DSC investigation of the polystyrene films filled with fullerene. J Therm Anal Calorim 109(2):1033–1038. https://doi.org/10.1007/s10973-011-1936-4

Rheological Properties of Polymer–Carbon Composites

Sayan Ganguly and Narayan Ch Das

Abstract Polymer rheology is an enormously sensitive indicator of polymer long-chain branching, and consequently can be exploited as a tool to evaluate polymer structures. Carbonaceous fillers are most abundantly used filler due to its reinforcing nature and its low cost. Among the carbon filler family, the most widely uttered names are carbon black and carbon nanotubes (CNTs) because of their relatively low cost, ease of processing, surprising dispersibility and mechanical strength. Several researchers did work on the polymer solution or polymer melt-based composite processing methods in order to distribute the fillers at a greater extent. It has been noticed by the various researchers that carbon black is easier to distribute than CNTs due to low aspect ratio of CNTs. Several rheological models have been discussed for filler-polymer composite systems. The relation among the rheological parameters is discussed also in light of yield stress value, shear rates and steady-state shear character of composites. We also discussed how the polymer/filler ratio affects the rheological nature of nanocomposites. Basically, dilute domains (rheology dominated by polymer concentration) and semi-dilute domains (dominated by the filler particles, filler fractals/cluster, filler agglomerates, etc.) have been analysed by various hypothesizes as told by researchers. However, we tried to contextualize the rheological resultant effects of carbonaceous filler impregnated polymer composites through the underlying structure-dispersion relationship and cultivate the interplay of different filler-polymer forces in the nanocomposites.

Keywords Carbonaceous fillers · Rheological models · Carbon black
Carbon nanotubes · Shear rate

S. Ganguly · N. C. Das (✉)
Rubber Technology Centre, Indian Institute of Technology Kharagpur,
Kharagpur 721302, India
e-mail: ncdas@rtc.iitkgp.ernet.in

S. Ganguly
e-mail: sayanganguly2206@gmail.com

M. Rahaman et al. (eds.), *Carbon-Containing Polymer Composites*,
Springer Series on Polymer and Composite Materials,
https://doi.org/10.1007/978-981-13-2688-2_8

Abbreviations

CNTs	Carbon nanotubes
SPM	Scanning probe microscopy
T_g	Glass transition temperature
MC	Monte Carlo
$-CH_2$	Methylene
$-CH_3$	Methyl
PE	Polyethylene
G^*	Complex modulus
WLF	Williams–Landel–Ferry
PTT	Phan–Thein–Tanner
UCM	Upper-Convected Maxwell
K-BKZ	Kaye and Bernstein
EPDM	Ethylene propylene diene monomer
SWCNT	Single wall carbon nanotubes
MWCNT	Multi wall carbon nanotubes
GPa	Giga pascal
η^*	Complex viscosity
f_c	Percolation threshold
σ_{max}	Maximum stress

1 Introduction

Composite materials are imprinted by assembling two or more than two components that have quite dissimilar attributes. The dissimilar constituent materials function wholly as an integrated system to afford the composite specialty properties. Composites constitute a crucial family of engineering and technology material where mostly carbonaceous materials are vehemently heralded due to their enormous compatibility into polymer matrix, strength increment and other properties related to their service functionalities. New demand and new product development have headed to thrusting the frontiers of polymer–carbon composites for more and better improvement in their service life. Lightweight, durable, flexible and high-performance polymer–carbon composites had replaced metals in versatile applications because of the cheap and low specific gravity of the polymers [1]. The several applications of polymer composites range from the fabricating of advanced engineering constructions like war tanks, advanced armoury systems, aircraft interior furnishings, pipes, support beams to the building of leisure and sporting specifics like golf balls, racquets, skis, yachts and boats. Fundamentally, polymer composites may be assorted into the following families: [2] (Fig. 1).

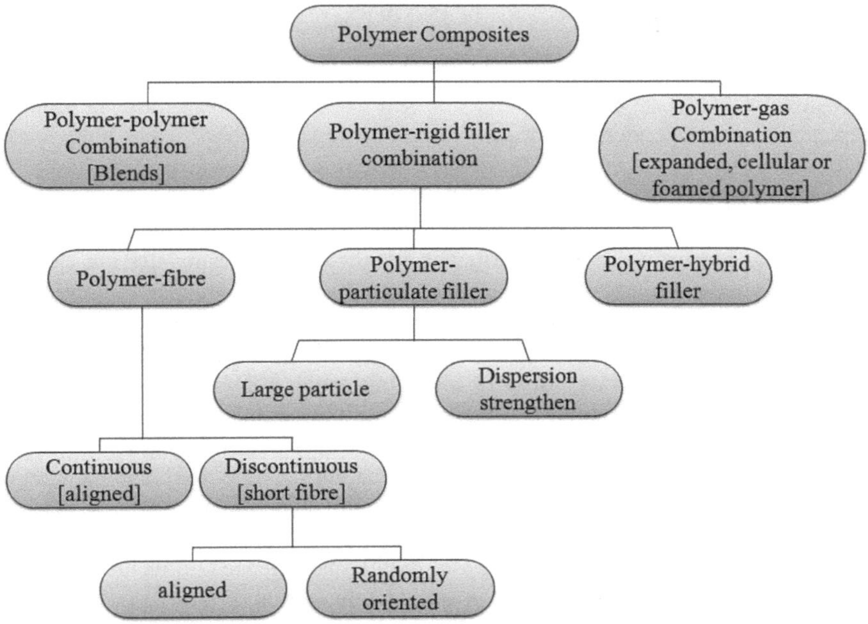

Fig. 1 Classification of polymer composites

1.1 Overview of Polymer Rheology

1.1.1 Why the Polymer Rheology Is Significant?

The term rheology was proposed by Professor Bingham of Lafayette College, Indiana, on the advice of a fellow, the professor of classics which intends: 'the study of the deformation and flow of matter'. The type of deformation that occurs is, however, dependent on the state of matter. For example, gases and liquids will flow when a force is applied, whilst solid materials generally defined by a fixed amount or up to a certain degree. The study of the rheology of a particular sample allows the experimentalist to gain a greater insight into the forces that operate in the sample. It is a useful tool that is employed to study the control and flow of complex Fluids such as polymers, pastes, suspensions and foods. But this tool is not suitable for describing simple fluids like water and air. There is a marked difference in the response of a solid and liquid when subjected to an applied force. When considering their rheology, it is convenient to think of a behaviour scale, where they are at opposite extremes. The solid response is explained by Robert Hook's 'true theory of elasticity' 1678, where is the behaviour of liquids was proposed in Isaac Newton's Principia published in 1687.

1.1.2 Basic Flow Characteristics of Polymers

Rheology is a subdivision of physics which contends on the deformation features and flow property of matters under stressed conditions where time is either a dependent or independent factor. It is especially, pertained with the holdings of matter that ascertain its trend of flow when an external mechanical force is applied thereon. Rheology is differentiated from fluid dynamics because it is implicated with all the common states of matters instead of only liquid and gaseous state. Rheological properties have authoritative significances in many diverse applications. Among these, carbonated filler occupies a vehement wide area of application in the polymer–carbon composites, organo-clay products and are also used as additives composites. Applications of rheology are crucial in lots of areas in commercial aspects such as industries involving metal, polymers and several other stuffs. Mathematically, rheology can provide the viscoelastic nature of any system either in pure state or in any compounded form. An apprehension of the rheology of a polymer/composite system is crucial in the fabrication and manufacture of composites. The job is planning an injection moulded component or fixing the total mould cycle for a pre-moldable material. Terminologically, rheology is a semi-quantitative common practicing tool in polymer science and technology area. The relationship between the structural architecture and rheology of a polymer is of common concern for rheological attributes which are structure sensitive less and complicated in practical applications. Besides this, the rheological characteristics that regularize the deformation pattern of polymers at the time of processing in the molten state. Deliberating the structures of polymers by nature of the shape and size of molecules and the distribution of these features amongst the molecules, structure organization and governed assembly are the obvious focus of various experiments assisted with mathematical modelling of the flow behaviours wither in form of Newtonian analogous features or non-Newtonians [3]. The primitive investigations of the dispersed polymer system in liquid form are majorly creditworthy for testifying that the high-molecular-weight polymers are not a colonial clustering of colloids, but these are definitely macromolecules adjudged together by the means of strong covalent bonds. Figure 2 shows the basic parameters related with the flow features and the processing of the polymers and filler impregnated polymer matrix. It is trustworthy to highlight that the flow property of polymer long-chain molecules and dispersions contemplates their size, shape and interaction behaviours in a flowing field. Approximately since the last decades, solution attributes of macromolecules, e.g. their viscosities, solubility, light scattering behaviour and colligative properties have been examined. Composites having completely novel property combination are accomplished by blending together, at least two polymers. In case of several specialty polymer-polymer alloys, immiscibility is a major concern which is observed by shearing astatic phase-separated blend. Generally, theoretical expectancy depicts that shear-induced mixing in low and high shear rates, with de-mixing vicinity at the medium shear rates. Incompatible blends adopt specific rheological model since they provide single-phase morphology; the flow behaviour of incompatible and phase-separated

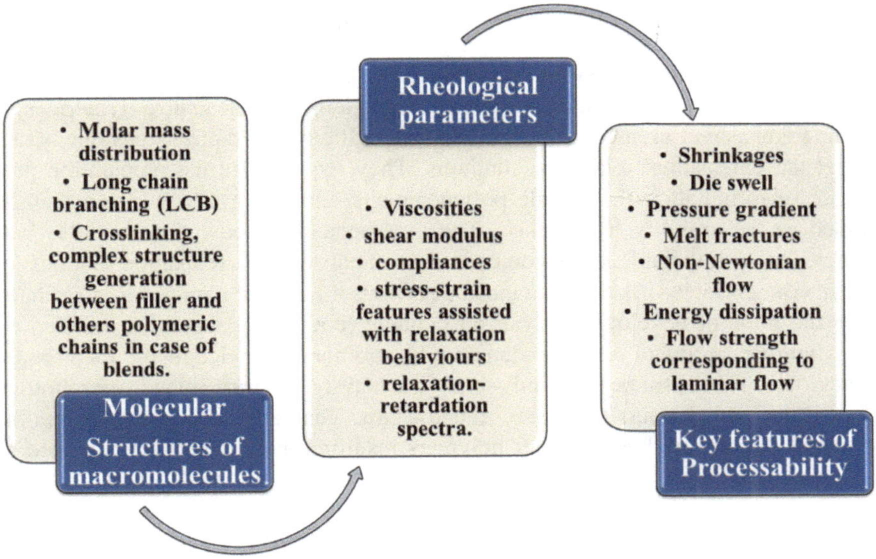

Fig. 2 Overview of intermittency among the rheological parameters

system is relatively complex than single-phase systems. The phase morphology which is a significant component, as it sternly impacts the alterations of dimension in shear flow, as discovered by viscosity. The continuous phase of the incompatible blend is decided by two substantial components, like viscosity and volume fraction. High volume fraction and lowering of viscosity privilege phase persistence in the mixture [4].

1.1.3 Polymer Melts and Blends

Appreciable advancement has been built in the rheology of polymer either in liquid dispersions or in molten conditions. The measurement of rheological dimensions of polymeric systems in melt system is essential in order to acquire primal realizing of the processability. This is due to the rheological features potently regulated by the polymer composites' construction and the interface-related features. The flow property of polymer in molten state is effectively requisite for two grounds; primarily, it has served to dissolve several polymer difficulties, like wide thickness variations in films, poor optical clarity of sheets, low production rates, dimensional imbalance and inadequate mechanical attributes and later, it also has been utilized for the analysis of processing parameters, like pendant groups, degree of grafting/branching, molecular structures, molecular chain entanglements and so on. Nichetti and Manas-Zloczower applied an uncomplicated superposition model relating to the inherent molecular chain physics [5]. The consequence of polydispersity on rheological flow curves was estimated where simple models are more feasible in numerical simulations. This can simplify also the extent of complex geometries in polymer design and processing segments. Their study also casts some supporting documentation over the entanglement-disentanglement theory in polymeric systems. Takahashi et al. assessed stress relaxation afterwards rapid extensional strain to get the extensional relaxation modulus. They hypothesize the preparation procedure of formulating the mobile polymer melts, through a new adopted method named as 'separability hypothesis'. The time-dependent relaxation modulus was detected to be invariable in tension or in shear, analysing the relaxation modulus of linear viscoelasticity [6]. The enhanced activities of polymer blends are interrelated with the hope of developing advanced eminent functioning materials founded on long-familiar products and the demand for primary knowledge on their phase behaviour. These studies extend some tractability to tune blend morphology throughout processing. Polymer blends are generally mixtures of unlike homopolymers, copolymers or ter-polymers resulting homogeneous designated as miscible blend or heterogeneous which is termed as multiphasic blend system. A mixture of polymers are generally chemically different and their miscibility depends on the thermodynamics of mixing rule. There is a significant component involving the blend thermodynamics equated to polymer dispersions consisting of high-molecular-weight of both polymers. Polymer processing technique has a prominent effect on the ensuing optical and mechanical belongings of the output. For example, dimensional integrity in injection moulding, elasticity and also tear strength of films are impacted by the viscoelasticity. The flow property is associated with particular molecular chain arrangements. In the common fabrication techniques of polymeric items, the majority of the shape forming is accomplished in the molten conditions due to easy deformable features of the molten dough. Melt processing affects the interaction between fluid mechanics and heat transfer in rheologically complex power law obeyed liquids, and taken wholly, it is a pertinent instance of the significance of 'coupled transport' techniques. The viscoelastic quality of molten polymer doughs ponders the chain entanglement and brings a crucial function in developing the product end features. Viscoelasticity has a trivial impression on the development of several processing flows, where the mechanics are prevailed by the temperature, shear and viscosity of the polymer mixture. This is especially true for several moulding techniques, but it evenly enforces few extensional flows. Lots of reports based on these have been published. Seemann and co-workers considered the constancy conditions of thin (3–300 nm) films on versatile substrates. The fundamental function was acted by the interfacial air, film or substrate which ascertained the de-wetting assumption for unstable films. As per their report, they accounted to distinguish spinodal de-wetting assumption by examining the issuing morphology of the film surface. They also admitted line tension experimentations of very small droplets, revealing that the long-range portion of effective interface potential may affect the drop profile. The active attributes of the films are assessed via several scientific methods, viz., the configuration of the de-wetting front which was judged by electron microscopy (SPM) which is a very strong tool to solve out the boundary specification. They further accounted experiments analysing the viscosity of the polymer system and the T_g of nanometer-thick films using ellipsometry. Ellipsometry is an essential optical imaging technique for inquiring the dielectric features and complex refractive index of thin coherent films. They showed that tiny molecular chain of polymers revealed lowering in T_g [7]. Daoulas and his co-workers delivered an atomistic modelling approach path for modelling the present interface between a molten state and crystalline domain of solid polymeric substrate. According to the hypothesis, atomistic simulations are carried on by tracing the PE and PP chains with the afore-mentioned model. The novel advance has been admitted by the authors to study the morphological and conformational features on the length scale of barely a few angstroms from both surfaces. Elaborated outcomes are delivered for the local mass density, morphology and conformation of PE at the two interfacial junctions, received from simulations by models which are strictly monodisperse polyethylene specimens of molecular chain length having repeating unit of around 400. Surplus structural characteristics of the physisorped contact, suchlike the statistical distribution of backbone carbon atoms in loop-tail and train conformations having their 'permutation-combination' statistics. Their atomistic simulation data manifested a firmer reliability of such signifiers of adhered layered architecture on main chain length than the estimated mesoscopic Scheutjens–Fleer lattice model based main chain length [8, 9].

Brief Development of Multi-component Polymer Blends

The reality to be a polymer miscible to another one in order to explicate a uniform and homogeneous system at segmental or molecular scale has been the utmost thrust for a huge amount of reports on multiphase polymeric systems. It has been experienced for a long time that the physical features could be evaluated by their morphology. Reactive polymers are widely applied for several pragmatic functions, viz., the alteration of surface tension between dissimilar phases via which internal microstructure of multiphase polymeric systems. These strategies are professionally termed as reactive compatibility [10]. Various types of polymer blend are already depicted in the Fig. 3. From the types of polymer, blends are suitably diverged to constitute of many diversified combinations of polymeric systems for cultivating both education and commercial pursuits. The basic distinction of blends demands their phase characteristics which is specifically the compatibility or miscibility versus phase-detachment. The elementary advantage of miscible versus immiscible/phase-separated blends is the blend quality visibility (as a function of mechanical properties of blends), which are typically intermediate between that of the unblended components. The engineering regards in polymer blends admit a development of new area of interest in polymer alloy where compositions letting in elastomeric blends, techno-commercial polymer blends, high resilient polymer blends and so on.

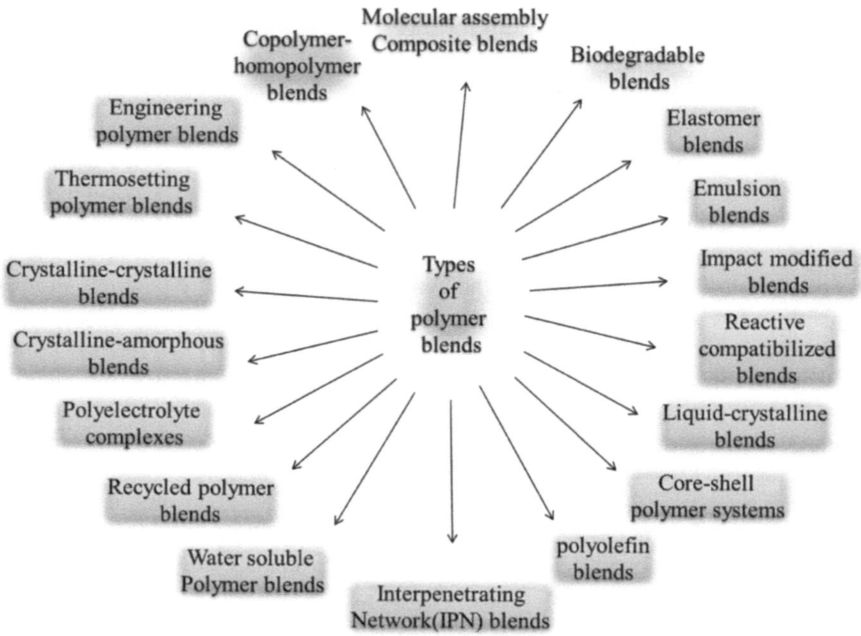

Fig. 3 Various types of polymer blend systems

Polymer Blends with Effective Miscibility

Miscible polymer blends are getting progressively significant for commercially viable because of the discriminative tuning in blend properties are possible by alteration on mixing strategy. An apprehension of the rheological characteristics of such blends is essential for anticipation of processing techniques. For multiphase polymeric blends, the empirical rule related to the time–temperature superposition is unable to deliver scientific logic [11]. In a hypothesis it is manifested that time–temperature superposition disables for a theoretically compatible blend. The deviation of superposition was demonstrated to be easy, unique and allows for substantial penetration into the temperature dependency of flow property for all single-phase polymer systems [12]. The empiric rule of time–temperature superposition expresses that the complex modulus G^* (ω, T) is a function of frequency which is

$$G^*(\omega, T) = b_T G^*(\omega, T_0) \tag{1}$$

This above principle is obeyed for both pristine component systems and also was applied to build master curves. a_T and b_T are the changeable parameters related to temperature only. The modulus scale changes b_T parameters were noticed to be very little. In case of pure components, the temperature dependence dominates more with respect to the frequency which shifts can be assigned via WLF equation [11].

$$\log a_{\mathrm{T}} = -\frac{C_1(T - T_g)}{C_2 + (T - T_g)} \tag{2}$$

The rheological attributes and the behaviour of extrudate profiles of molten doughs are of fundamental significance in the casting of polymeric goods. The melt-rheological features of fibres, viz., pineapple, sisal and coir fiber-reinforced polymer composites have been considered in several reports [13, 14]. The limitations in the research of self-governing features are the fibre length, matrix-fibre fundamental bonding and the length to diameter ratio of green-fibre-reinforced thermoplastic composites [15, 16]. The expression is very much attached with viscoelastic system having complicated comparable to its Newtonian similitudes. For an authentically viscoelastic system, this relationship is progressively sustained. This reality is the supposed to be the cause of the challenging nature of this subject. The leading equations not only discover the rheological characteristics of the molten polymer system but also establish the concluding fibre preference of the developed system. That is the reason it have utmost primal acceptability to assess the purpose of the composite's flow behaviour and the fibre–polymer bonding. It also has been detected that the evaluated stress of the developed system enhances as fibres are contributed; accordingly, an acceptable related equation could be accomplished by summing a surplus stress to factual element equations for the molten polymeric systems [17, 18]. Consequently, such equations established in the research literature that report the vehement use of molten polymeric system could be better explored in fabrication of bio-based composite systems. Especially, focus will inclined to nonlinear rheological features of viscoelastic materials. The cross WFL, power law, Casson, Hershel Bulkleym and Bird-Carreau models are vehemently heralded among the favoured rheological models because of their power to anticipate velocity and pressure statistical distribution in consistent flows in add-on to their elementary behaviour of shear thinning [18–22]. Even so, for the high shear stress of viscoelastic systems, the prognostic ability of these models is substantially foreshortened.

Viscoelasticity of Materials

Viscoelasticity is named as a special dimension of a material's property to establish both Newtonian (viscous) as well as Hookian (elastic) attributes under the identical considerations when it tends to some sort of flow. The shear stress of such materials is a function of strain when the strain is carried out and then completed; they counter to their previous positions. Some commonly uttered viscoelastic materials are polymer melt, paints, honey, blood, ketchup, mayonnaise, polymer suspension, polymer solution, cornstarch, shampoo, etc. At invariable temperature; common fluids are exemplified as ideal Newtonian fluids. This constant temperature is not altered with shear rate. Also, the plot communicates via the primitive point: that is, the shear rate is measured to be zero at the point of zero shear stress [23]. A fluid that does not act as Newtonian (ideal viscous material) fashion after experiencing the shear stress and shear rate, then it can be called as anomalous Newtonian system or non-Newtonian systems. This corroborates that in these non-Newtonian cases, the shear stress versus deformation cannot be a straight line, but it may be a curve. High-molecular-weight polymer melts and dispersions, i.e. in liquid state are commonly non-Newtonian. In this situation, the slope of the stress–strain curve will be variable with alteration of the shear strain rate. When viscosity drop-offs with rising of shear strain rate, the fluid is anticipated as shear thinning. Oppositely, where the viscosity enhances after application of the external applied stress, the fluid is defined as shear thickening. Shear thinning substances also are named as pseudoplastic systems [24–26]. A different type of non-Newtonian fluid is also known in this nomenclature which is called as viscoplastic or 'yield stress' fluid. The shear stress must surmount a vital value recognized as the 'yield stress' accounting on verge of the fluid flowing. If it oversteps the yield stress, the fluid flow of those viscoplastic fluids will assume to be an ordinary fluid. Sidewise, some divisions of fluids demonstrate time-dependent character. This class of material appoints thixotropic as well as rheopectic fluids, whose features are in contraries. The viscosities of such materials will drop-off with time under an unvarying employed shear stress. Nevertheless, in the absented of stress, the viscosity stepwise recovers as a function of time as well. As a contrast, the viscosity sometimes increases after application of invariable shear force for a prolonged time which is known as 'rheopectic fluid'.

1.1.4 Rheological Modelling on Viscoelasticity

Mathematical simulations of viscoelasticity are predominately established on a differential or integral theatrical performance. Nevertheless, the integration deductions are able to predict the time dependence character more efficiently. The specific feature of linear viscoelastic system is that they follow a linear tread in stress–strain plot with memorizing their deformation history. Linear viscoelasticity is generally relevant exclusively for little distortions, low rate of shear, low stress, and linearity of the molecular chain configuration. However, actually, nearly 90% of mobile systems are nonlinear showing huge distortions/flow properties and with nonlinear reaction in the front of such strains. Nonlinear viscoelastic feature be is generally demonstrated when the strain is prominent [27–32]. Accordingly, non-linear viscoelastic frameworks are necessitated. Surviving nonlinear numerical rheological patterns are frequently built via adjustments and denotations to higher order stress or strain conditions of the linear theory [31]. As mentioned earlier, from a calculative point of view, the integration constitutes viscoelastic equation which is harder to execute than the derivative function. Thus, various examples/models are conditioned after combination of elastic, viscous, and inertial nonlinear components with several ramifications. However, in these exemplified models, regarding the mathematical complexities, only the elastic or viscous nonlinearity is frequently allowed and the inertial part is neglected. Furthermore, there are merely a couple of theoretical prototypes developed with fixed-value rheological parameters.

On account of this situation, nonlinear models with invariable rheological coefficients are mandatory. The combination of simultaneous elastic and viscous phenomena is generally considered by Maxwell fluid model [33, 34]. The modified versions of the Maxwell model were elucidated by a group of researchers which are named as Phan–Thein–Tanner (PTT), Upper-Convected Maxwell (UCM) and K-BKZ model [35–37]. The models are suitable for polymer solutions and melt polymer systems. The features related to the non-Newtonian flow behaviours are given as follows:

The governing equations for the polymer melt or solution are deducted by many researchers as follows:

The equation for continuity of incompressible fluids is

$$\nabla . \upsilon = 0$$

where υ is the velocity vector.

The equation for momentum will be

$$\rho\left(\frac{\delta V}{\delta t} + V.\nabla V\right) = -\nabla p + \nabla.\tau + \rho g$$

Here, ρ, p and τ are the density, pressure and stress tensor, respectively.

Upper-Convected Maxwell (UCM) Model

UCM model is modified generalizable class of Maxwell model by differential interpretations based on large deformation. The model can be interpreted as follows:

$$\tau + \lambda_1 \tau^0 = 2\eta D$$

$$D = \frac{1}{2}[(\nabla u) + (\nabla u)^\tau]$$

D corresponds to the strain tensor.

Now the flow equation will be

$$\tau + \lambda_1 \tau^0 = \eta \ddot{\Upsilon}$$

Here, λ, η and τ^0 are the relaxation time, zero shear viscosity and upper-convected time derivative of the stress tensor.

The τ^0 can be written as

$$\tau^0 = \frac{\delta}{\delta t} + V.\nabla\tau - (\nabla V)^{\mathrm{T}}\tau - \tau(\nabla V)$$

The UCM model can also interpret memory effects of materials with invariable viscosities at several shear rates.

White-Metzner Model
According to White and Metzner network, theory associated with polymers at 1963, viscosity and relaxation parameters both are functions of shear rate. This model demonstrates shear thinning behaviours, not because of non-affine movement, but due to the accelerated relaxation at high shear strain rates, where the rate of relaxation is faster than any kind of deformation [38]. The viscoelastic differential component model assumes the pattern

$$\tau_1 + \lambda(\gamma)\tau^0 = 2\eta(\gamma)D$$

Here, $\eta(\gamma)$ can be derived from the experimental shear viscosity curve and the function $\lambda(\gamma)$ can also be found from observational stress deviation plot.

Phan–Thien–Tanner (PTT) Model
The PTT model concerns to a quasi-linear viscoelastic constitutional simulated relationship which is significantly exploited in designing of polymer solutions and melts in flows. The original PTT equation was published utilizing the adopting adjustments concurrently. It uses particular categories for the construction and destruction rates of the network adjunctions reported elsewhere [39]. In the PTT model, the surplus stress tensor is studied as the algebraic sum of the viscoelastic factor τ_1 and the solely Newtonian factor τ_2.

$$\tau = \tau_1 + \tau_1$$

where τ_2 is given by

$$\tau_2 = 2\eta_2 D$$

where D stands for the rate of strain tensor. The overall form of the PTT equation for the viscoelastic substances, τ_1 is as below

$$f(\mathrm{tr}_\tau)\tau + \lambda\tau^0 + \frac{\varepsilon}{2}\lambda(\gamma\tau + \tau\gamma) = \eta\ddot{\Upsilon}$$

or,

$$\tau^0 + \xi\left(D.\tau - \tau.D^{\mathrm{T}}\right) + \frac{Y}{\lambda}.\tau = 2GD$$

where the Oldroyd's upper convective derivative τ^0 is defined by

$$\tau^0 = \frac{DV}{Dt} - \tau.\nabla V - \nabla V^{\mathrm{T}}.\tau$$

where V is the velocity matrix, V^{T} corresponds to the transpose of the velocity matrix and $D(V)/Dt$ is the material derivative of the velocity matrix.

After studying the above formula, the beginning term corresponds to the stress tensor transport and the transient component of the respective deformations. The second term corroborates the dislocation between the mobile macromolecular chains is computed. The third term admits the elasticity. Eventually, the term on the right side of the equality represents the diffusive effects. τ is the stress tensor, D the deformation rate tensor, G the relaxation module and λ the fluid relaxation time. The parameter ξ controls the amount of movement between the fluid polymeric chains [40]. For $\xi = 0$, the model is called 'PTT Affine' and the sliding among the polymeric chains is unheeded. The function Y depends on the rate of building and destruction of the links between the chains. Furthermore, the PTT model comprises the retentivity effect of polymeric systems and its viscosity can change with the change of the shear rate. When ξ is zero, PTT constitutional equation is deduced to its modified form (SPTT):

$$f(\mathrm{tr}(\tau))\tau + \lambda\tau^0 = \eta\ddot{\Upsilon}$$

According to PTT model, specific classes for the construction and destruction rates of the network joints and deduced a constitutive equation holding in two free parameters, ε and ξ [41]. The exponential constitutive model accepts the following form:

$$\exp\left[\varepsilon\frac{\lambda}{\eta}\mathrm{tr}(\tau_1)\right]\tau_1 + \lambda\left[\left(1 - \frac{\xi}{2}\right)\tau_1^0 + \frac{\xi}{2}\tau_1^0\right] = 2\eta D$$

η and λ are the viscosity and relaxation time, respectively, evaluated from the equilibrium relaxation spectrum of the fluid. The PTT model can be figured out utilizing a single or multiple relaxation times, as per the Giesekus model. The linear trend of the PTT model anticipates shear thickening at high elongation shear rates followed by a plateau [39–41].

Giesekus-Leonov Model

Giesekus projected a constitutive model established on a concept of configuration-dependent molecular mobility. On account of the highly nonlinear nature of the model, equations are as follows [42–46]:

$$\tau_{\mathrm{p}} + \lambda_1\tau_{\mathrm{p}} - \frac{\alpha\lambda_1}{\eta}\tau_{\mathrm{p}}^2 + \eta\ddot{\Upsilon}$$

The α is the dimensionless Giesekus model mobility factor and ascertains the extensional viscosity and the ratio of 2nd normal stress deviation to the 1st one. The α is applied to assess the anisotropic drag which has been estimated as $0\ 1 \ll \alpha$. For $\alpha = 0$, the model turns into the isotropic UCM model. But in the case of for $\alpha = 1$, the model is just an anisotropic drag, and for $\alpha > 0$, the model constitutes shear thinning feature. The Giesekus model anticipates the 'tension-thickening' domain for longitudinal flow followed by a plateau region [47–50].

1.2 Polymer–Carbon Black Composites

Elastomer carbon black composites have long been a very authoritative category of commercial materials and it has been accomplished via vehement utilization of rheological knowledge. It dates back to 1931, when Scott and co-workers, on the groundwork of compression plastometer experimentations, aimed that rubber filled with fillers resulted a definite yield value [51]. This decision was also attained in 1933 paper by Dillon and Johnston using capillary rheometer experiments. Both Scott and Dillon and Johnston suggested constituting the shear stress throughout shear flow by the algebraic sum of a yield value *Y*, and a power law dependency on shear rate [52]. Later on, Mullins and Whorlow drew attention to thixotropic features applied to elastomer by high loading levels of carbon black as filler [53, 54]. The recent tenure of this field commences on the work of Zakharenko et al. and afterwards Vinogradov and co-workers [55, 56]. They manifested from deliberate low shear rate evaluations that elastomeric materials do display yield values as already calculated by Chapman and Lee on talc-filled polypropylene [57]. In consequent years, one of the contemporary authors and his co-workers have worked on extensive filler loading in polymer matrix to investigate the particulate characteristics of carbon black in elastomer matrix [58]. Besides this, calcium carbonate [59, 60] and titanium dioxide [61] were also studied in polymer matrix which revealed gradual increment in yield values of the composites. Various rheological experimentations are accounted by G. J. Osanaiye et al. also. Sandwich rheometer was utilized to study the creep and strain recovery at low shear stresses [62]. They trialled over several carbon black-filled rubber composites to measure rheological features. In the case of carbon black-rubber composites creep and strain sweep are immensely necessary because of their viscoelastic nature. Primarily to judge the viscoelasticity capillary rheometers were used. The data were combined and allowed to represent as function of shear rate or shear stress over a very wide domain of shear rates or stresses. Most extensively studied rheological property is creep. It is basically deformation of material in constant stress. Strain recovery is another rheological character of viscoelastic material which has also been studied widely. The creep experiments were carried out as strain versus time (Fig. 4a, b).

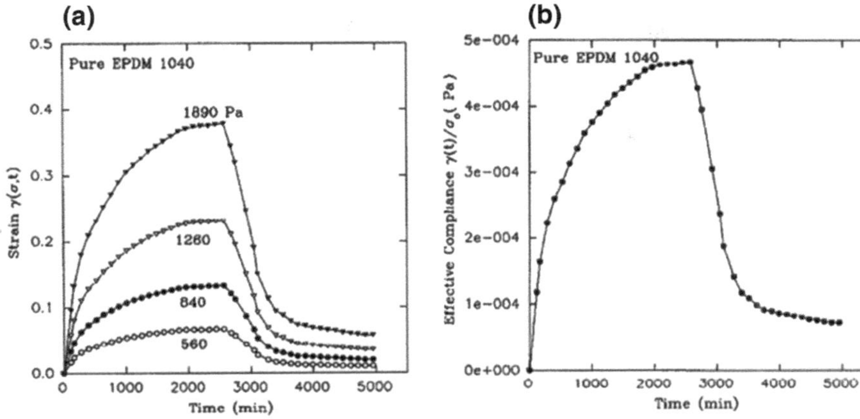

Fig. 4 Creep (**a**) and stress recovery (**b**) plot for unloaded EPDM with variation of stresses at [62]

Generally for a material which is under a specific load, the strain/deformation increases monotonically. The upper positions of the plot are relatively flattened with respect to the steep rise in the plot. Such flat portions are due to the viscous steady flow behaviours of the rubber-carbon black composites. The multiple plots in Fig. 4a showed the effect of stress in creep behaviours. It is noticed that with increasing the constant load (stress), the strain is almost proportional to the corresponding applied stress. It can give idea of linear viscoelasticity.

Shear viscosity can also be applied to evaluate the carbon black-rubber composites' rheological behaviours. Researchers report the shear stress versus time plot as a shear rate as shown in Fig. 5a. Rubber normally exhibits a decreasing order of zero shear viscosity with increase in shear rate. But in the case of filled rubber, this trend not always persists. In the case of carbon black-rubber composites, no such zero shear viscosity is found but it can be assumed that with increasing the stress up to infinite limit the shear viscosity decreases monotonically. That means at infinite shear stress/load, the zero shear viscosity may appear. Figure 5b is the pictorial proof of such features exhibited by the rubber-carbon black composites.

The yield stress also played an important role in filled rubber processing where rheological assumptions cannot be ruled out. As shown in Fig. 6, the upper graph plot exhibited yield stress value after extrapolation. It shows a common trend of filler loaded rubber composites. The fillers dispersed into the rubber matrix and after homogeneous distribution the yield stress value enhances. The rubber polymer chains adsorb on to the carbon black surface which results increasing bound rubber content. This bound rubber concept is an effective way to interpret rubber-carbon black interactions. Increasing the amount of bound rubber enhances the tensile strength of the rubber-carbon black composites. This can effectively interpret the tensile strength increment data. But after a certain order of filler loading, the yield stress value decreases due to lowering of rubber-filler interfacial adhesion and

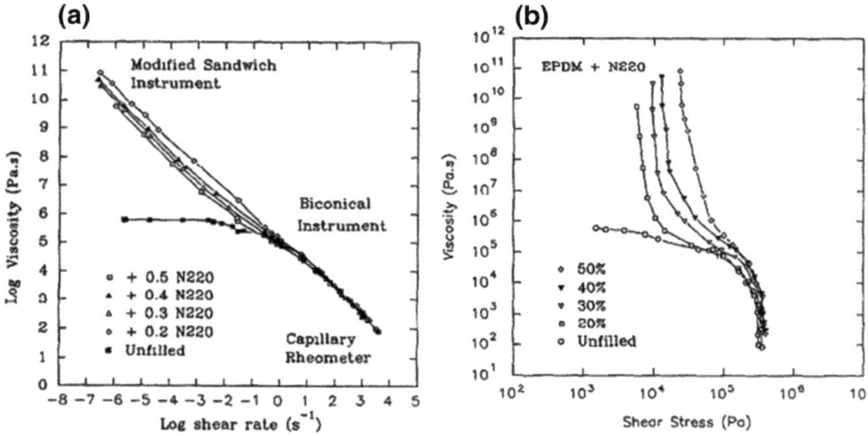

Fig. 5 **a** Shear viscosity versus shear rate plot for filled EPDM composites at **b** shear viscosity versus shear stress [62]

increase in filler-filler aggregate formation. When a material is loaded with high amount of filler with nominal dispersion and distribution of filler due to lack of rubber/filler weight ratio, the stress dissipation becomes inhomogeneous. This inhomogeneity in stress or the localized effect/ stress concentration become the cause of rubber composite failure during service. The crack propagation through the interfaces of filler particles become so prominent that the rubber-carbon black composites normally fail easily.

Fig. 6 Yield stress as a function of filler loading; 1 corresponds to the sandwich instrument ($3 < 10\ m^3\ s^{-1}$ and 2, implies the flow behaviour in rotational instrument ($y > 10^{-1}\ s^{-1}$). [62]

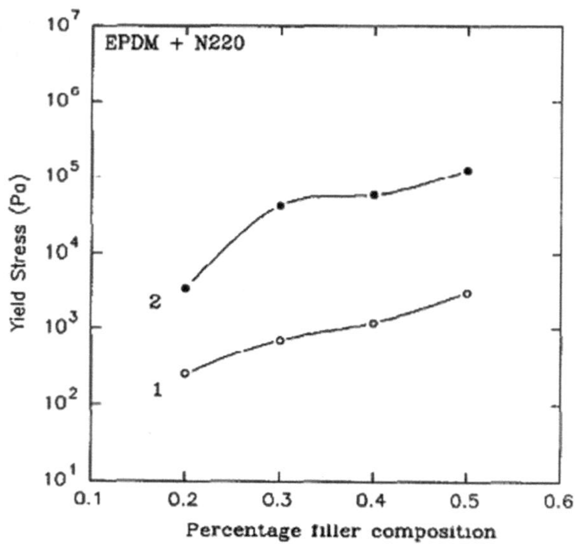

1.3 Polymer–Carbon Nanotube (CNT) Composites

CNTs-filled polymeric composites have appealed appreciable attention in the research and techno-commercial area because of their unparalleled electrical and mechanical attributes. CNT–polymer nanocomposites own immense stiffness, high tensile strength and reasonable electrical conductivity even at comparatively low concentrations [63–67]. These uprising eventually stem from the distinguishable features of the CNTs themselves. Several experiments affecting single-walled carbon nanotubes (SWCNTs) and multi-walled carbon nanotubes (MWCNTs) have established that CNT can have moduli and strength levels in the range 200–1000 GPa and 200–900 MPa, respectively [68–70]. Additionally, CNTs have very large aspect ratios (length-to-diameter ratio) in the range of 100–1000 [63]. Such exceptional properties construct CNTs splendid compound for high strength and electrically conductive polymer composite fabrication. Several commercially viable methods adapted by industries are schematically summed up in Fig. 7.

Currently, two major classifications are present regarding the in which CNTs are being employed, electronics and automotive. In electronic area, especially in the semiconductor applications, CNTs are utilized to dissipate undesirable static charge build-up. This dissipative consequence is attained by good dispersing CNTs in the polymeric composites such that an interlinking morphology of CNT is organized. This interconnecting network thus provides a conductive pathway. In the automotive manufacturing area, CNTs are used as a conducting agent to assist in electrostatic painting. These are produced via melt processing

Generically, the phase morphology and characteristics of macroscopically unoriented CNTs dispersed in a matrix can be sorted into three classes according to the concentration of nanotubes and the internal morphology correlations between CNTs. First one is at low concentration portion; where dispersed CNTs stay and act as separately differentiated tubes or secondarily as little diffused bundles. In this

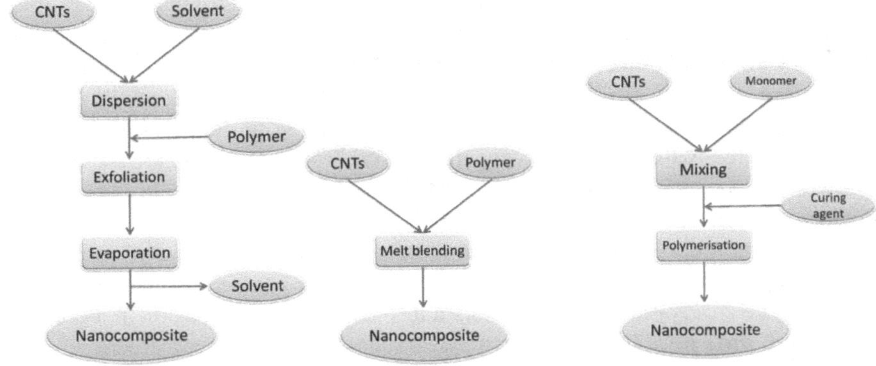

Fig. 7 Various conventional methods/techniques used to develop polymer–CNT composites

subclass, there are some short-range inter-tube interactions but no such long-range interactions are attributed. Therefore, the morphological strengthening in such dispersions can be considered to take place due to the influence of individual nanotubes and their aggregation with the features of the polymer matrix. Changeover from dilute to semi-dilute region concurs with a percolation phenomenon (i.e. the establishment of a matrix crossing backbone path/network made of nanotubes).

Masterbatch dilution technique is another effective method to disperse CNTs into polymers matrix. In case of plastics, twin screw extruder is used for mixing purposes followed by compression moulding. Polycarbonate, a commonly uttered name in engineering plastic region, was used for polycarbonate CNT nanocomposite matrix. The increasing viscosity after the addition of CNT is much higher for Carbon nanofibres due to their higher aspect ratio. High loading of nanotubes normally exhibits non-Newtonian behaviour at low frequencies. A sudden jump in viscosity for high loading of nanotubes may observe which is called step increase. Such step increment in the viscosity is a specific composition implies rheological threshold [71].

The type of dispersion can be measured through the percolation threshold (f_c, in vol%). For example, f_c is assessed via the concentration of CNTs where linear viscoelastic attributes of the CNT-based nanocomposite convert from liquid-like to solid-like state. At this particular concentration, the primarily developed network traversing path (or backbone property) is acquired. Likely, in the case of a molten polymer or a polymer solution in liquid phase under its application tenure, elastic modulus and viscous modulus (G' and G'', respectively) displays final behaviour with $G' \sim \omega^2$ and $G'' \sim \omega$, respectively. Loading of CNTs in the polymer matrix gradually interprets the liquid-like state to solid-like system (i.e. both $G0$ and G'' are independent of ω as $\omega \to 0$ (Fig. 8); in this particular case, CNTs are diffused in polycarbonate, an engineering polymer). Similarly, the low-frequency complex viscosity deviates with $|\eta^*| \sim 1/\omega$ [72].

Elastic character of CNT–polymer composites can also be analysed by steady-state shear properties of the materials. Elastic networks formed by CNT dispersions in a polymer matrix are anticipated to reveal high tensile stress value and better manifested through their steady-state shear behaviour (Fig. 8) [74]. For high CNT concentrations, the percolation threshold ($f \gg f_c$), application of steady shear (from rest) typically results in a stress wave-off (independent of the polymer matrix) which balances to a steady-state value ($f < f_c$) at prolonged times. In contrary, for dilute mixture of CNTs (specifically SWNTs) in superacid (102% H_2SO_4), at low shear rates (Peclet number $\ll 1$), Brownian motion manipulates the relaxation process and the consequence gyration.

CNT suspensions do not follow Cox-Merz rule in solvent system or in polymer matrix. The dynamic viscosities represent the dormant or near-dormant state morphology. The steady-state viscosity constitutes steady-state dimension at a constant shear rate during flow (Fig. 9). It is supposed that the network construction, under steady shear, primarily self-assembled to adapt the deformation. The enhancement in the stress is a direct reflection of structural changes which is due to the

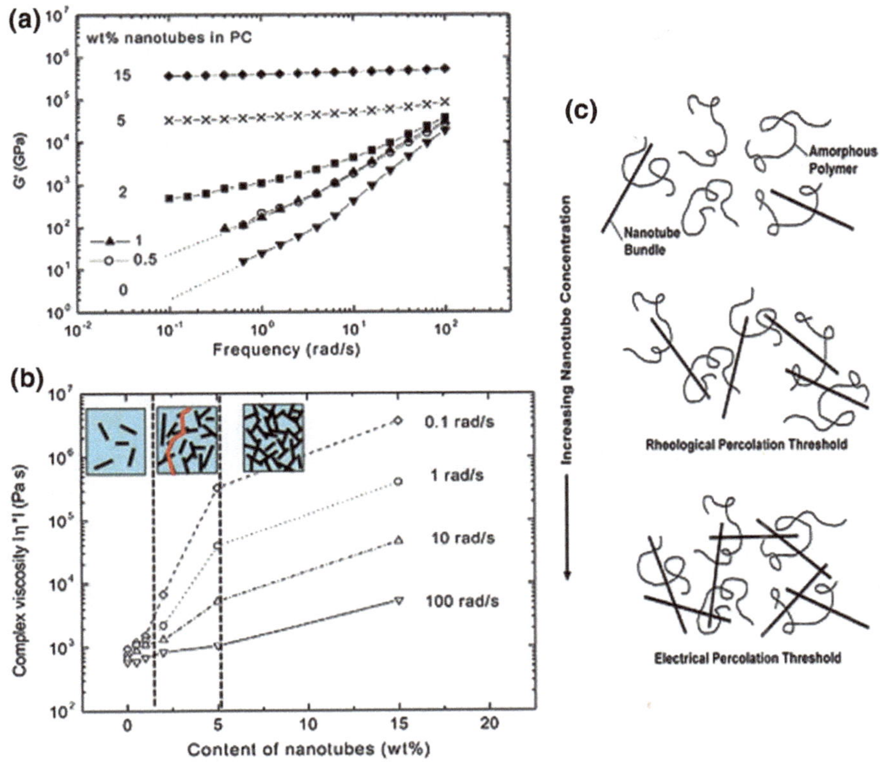

Fig. 8 **a** Frequency sweep experiment for polycarbonate loaded with nanotubes at 260 °C [71]. **b** Change in complex viscosity with respect to nanotube concentration in polymer matrix at various frequencies [71] (Inset) graphical illustration of (left) distinct nanotubes in matrix below the percolation threshold, (centre) onset of percolation, **c** CNTs—polymer nanocomposites with highly agglomerated bundles having definite array. (Top) at low nanotube concentrations, (Middle). The onset of solid-like viscoelastic behaviour after adsorption of polymer chain onto nanotube surface, (Bottom). The onset of electrical conductivity or formation of 3D spatial percolating network [73]

agglomeration or cluster formation. For times bigger than the time comparable to the maximal in the stress, the fractal/clustered network collapses and the effectual stress affirmed by the network minimizes. Thus, the targeted stress, or the maximum stress (σ_{max}) is correspondent to a yield stress limit beyond which the network begins to flow. This idea is further defended by optical microscopic evaluation of CNTs agglomeration under flow [75]. For strains just above the lower limit in shear stress, the fractional domain of the formed network/three-dimensional network rupturing (bond breaking) and the system deforms until a steady shear viscosity is attained. The steady-state flow character is controlled by an organization of balance between bond making and breaking procedures [76].

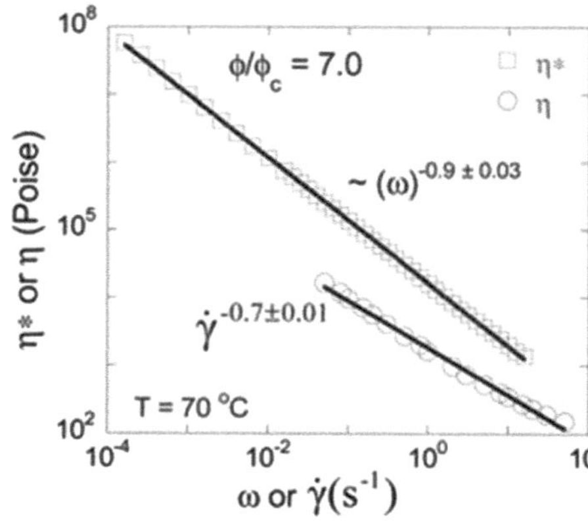

Fig. 9 Comparison of the complex and steady shear viscosities in the light of the Cox-Merz rule; the nanocomposites fail to conform the Cox-Merz rule presumably due to an adjustment in the meso-scale microstructure [74]

In the case of dense colloidal sols, Silbert and co-workers established that under uninterrupted strain solely, a fraction of the initial (motionless state) amount of CNT clusters fundamentally accepts the stress and assures the reaction [77]. Contrarily, non-equilibrium molecular dynamics modelling studies of polymer nanocomposites have discovered the elastic uniaxial elongation of the nanoparticle–polymer network as the principal effect for the stress dissipation [78].

2 Summary

In this chapter, we a brief discussion on rheological significances of carbonaceous fillers on polymer matrix has been documented and attempted to explicate those results in terms of morphology and structure related flow behaviours. Carbon black, very abundant filler used in polymer matrix (specifically in tyre industries) has been nurtured by various researchers in light of flow and time-dependent flow behaviours. The relation between filler dispersion in polymer matrix and their rheological features also have been briefly discussed. After summing up the chapter, it can be said that filler dispersion in a polymer matrix can restrict the mobility of the polymer chains which results in higher stiffness, modulus and low elongation at break of the composites. Besides carbon black, carbon nanotube-filled polymer composites also have been studied. CNTs behaviour in dilute region, semi-dilute region in polymer composites affects its rheology. The aggregation of CNTs in polymer matrix was also supported by the small angle scattering phenomena. To enhance the physicomechanical properties (tensile strength, modulus, hardness, serviceability, etc.), dispersion of CNTs in polymer matrix in a high degree is highly demanded. Meanwhile, in a well-dispersed state, the nanotubes can act as stress transfer domains under shear force. To achieve better dispersion, high shear force is required. In the case of low shear, CNTs can form cylindrical fractals which were discussed in the chapter. Finally, we conclude that significant amount of results were published in such core polymer-nanofiller rheology. Hence, there are plenty of scopes to investigate rheology of filler-polymer composites. Carbon black and CNTs-based products are normally marketed, where carbon black-filled composites are very common but CNTs based composites have enough scopes. Detail understanding in polymer–carbon filler composites is crucial for industrial and academic success for application and research.

References

1. Sheldon RP (1982) Composite polymeric materials: applied science. Elsevier Science distributor
2. McCabe KA, Rassenti SJ, Smith VL (1990) Auction design for composite goods: the natural gas industry. J Econ Behav Organ 14(1):127–149
3. Almomen A, Cho S, Yang C-H, Li Z, Jarboe EA, Peterson CM et al (2015) Thermosensitive progesterone hydrogel: a safe and effective new formulation for vaginal application. Pharm Res 32(7):2266–2279
4. Robeson LM (2007) Polymer blends. A comprehensive review. Leseprobe
5. Nichetti D, Manas-Zloczower I (1998) Viscosity model for polydisperse polymer melts. J Rheol (1978-present) 42(4):951–969
6. Takahashi M, Isaki T, Takigawa T, Masuda T (1993) Measurement of biaxial and uniaxial extensional flow behavior of polymer melts at constant strain rates. J Rheol (1978-present) 37 (5):827–846
7. Seemann R, Herminghaus S, Neto C, Schlagowski S, Podzimek D, Konrad R et al (2005) Dynamics and structure formation in thin polymer melt films. J Phys: Condens Matter 17(9): S267
8. Daoulas KC, Harmandaris VA, Mavrantzas VG (2005) Detailed atomistic simulation of a polymer melt/solid interface: structure, density, and conformation of a thin film of polyethylene melt adsorbed on graphite. Macromolecules 38(13):5780–5795
9. Bojan MJ, Steele WA (1987) Interactions of diatomic molecules with graphite. Langmuir 3 (6):1123–1127
10. Xanthos M, Dagli S (1991) Compatibilization of polymer blends by reactive processing. Polym Eng Sci 31(13):929–935
11. Ferry JD (1980) Viscoelastic properties of polymers. Wiley
12. Prest W, Porter RS (1972) Rheological properties of poly (2, 6-dimethylphenylene oxide)—polystyrene blends. J Polym Sci Part A-2: Polym Phys 10(9):1639–1655
13. Ogah AO, Afiukwa JN, Nduji A (2014) Characterization and comparison of rheological properties of agro fiber filled high-density polyethylene bio-composites. Open J Polym Chem
14. Moutee M, Fortin Y, Fafard M (2007) A global rheological model of wood cantilever as applied to wood drying. Wood Sci Technol 41(3):209–234
15. Xie F, Halley PJ, Avérous L (2012) Rheology to understand and optimize processibility, structures and properties of starch polymeric materials. Prog Polym Sci 37(4):595–623
16. Qiao X, Li W, Watanabe H, Sun K, Chen X (2009) Rheological behavior of biocomposites of silk fibroin fiber and poly (ε-caprolactone): effect of fiber network. J Polym Sci, Part B: Polym Phys 47(20):1957–1970
17. Zhang Y, Lim CT, Ramakrishna S, Huang Z-M (2005) Recent development of polymer nanofibers for biomedical and biotechnological applications. J Mater Sci—Mater Med 16 (10):933–946
18. Marynowski K (2006) Two-dimensional rheological element in modelling of axially moving viscoelastic web. Eur J Mech-A/Solids 25(5):729–744

19. Dealy JM, Wissbrun KF (2012) Melt rheology and its role in plastics processing: theory and applications. Springer Science & Business Media
20. Ansari M, Hatzikiriakos SG, Mitsoulis E (2012) Slip effects in HDPE flows. J Nonnewton Fluid Mech 167:18–29
21. Ansari M, Alabbas A, Hatzikiriakos S, Mitsoulis E (2010) Entry flow of polyethylene melts in tapered dies. Int Polym Proc 25(4):287–296
22. Phillips T, Owens R (2002) Computational rheology. Imperial College Press, London, UK
23. Willett J, Jasberg B, Swanson C (1995) Rheology of thermoplastic starch: effects of temperature, moisture content, and additives on melt viscosity. Polym Eng Sci 35(2):202–210
24. James DF (2009) Boger fluids. Annu Rev Fluid Mech 41:129–142
25. Jarecki L, Lewandowski Z (2009) Mathematical modelling of pneumatic melt spinning of isotactic polypropylene. Part III. Computations of the process dynamics. Fibres Text Eastern Eur 17(1):75–80
26. Higashitani K, Pritchard W (1972) A kinematic calculation of intrinsic errors in pressure measurements made with holes. Trans Soc Rheol (1957–1977) 16(4):687–696
27. Cherizol R, Sain M, Tjong J (2015) Review of non-Newtonian mathematical models for rheological characteristics of viscoelastic composites. Green Sustain Chem 5(01):6
28. Crochet M, Legat V (1992) The consistent streamline-upwind/Petrov-Galerkin method for viscoelastic flow revisited. J Nonnewton Fluid Mech 42(3):283–299
29. Bird RB, Armstrong R, Hassager O (1987) Dynamics of polymeric liquids, vol 1. Fluid mechanics
30. Tucker C, Dessenberger R (1994) Flow and rheology in polymer composites manufacturing, chapter 8. Elsevier Science
31. Grafe T, Graham K (2003) Polymeric nanofibers and nanofiber webs: a new class of nonwovens. Nonwoven Technol Rev 12:51–55
32. Zhou C, Kumar S (2010) Thermal instabilities in melt spinning of viscoelastic fibers. J Nonnewton Fluid Mech 165(15):879–891
33. da Silva LJ, Panzera TH, Velloso VR, Christoforo AL, Scarpa F (2012) Hybrid polymeric composites reinforced with sisal fibres and silica microparticles. Compos B Eng 43(8):3436–3444
34. Devereux BM, Denn MM (1994) Frequency response analysis of polymer melt spinning. Ind Eng Chem Res 33(10):2384–2390
35. Ellison CJ, Phatak A, Giles DW, Macosko CW, Bates FS (2007) Melt blown nanofibers: fiber diameter distributions and onset of fiber breakup. Polymer 48(11):3306–3316
36. Ellyin F, Vaziri R, Bigot L (2007) Predictions of two nonlinear viscoelastic constitutive relations for polymers under multiaxial loadings. Polym Eng Sci 47(5):593–607
37. Burghardt W, Fuller G (1989) Note: end effects in flow birefringence measurements. J Rheol (1978-present) 33(5):771–779
38. Maders H, Vergnes B, Demay Y, Agassant J (1992) Steady flow of a White-Metzner fluid in a 2-D abrupt contraction: computation and experiments. J Nonnewton Fluid Mech 45(1):63–80
39. Thien NP, Tanner RI (1977) A new constitutive equation derived from network theory. J Nonnewton Fluid Mech 2(4):353–365
40. Brondani WM, Coradin HT, Franco AT, Morales RE, Martins AL (2007) Numerical study of a PTT visco elastic fluid flow through a concentric annular. In: Proceedings of COBEM (Brasillia)
41. Phan-Thien N (1978) A nonlinear network viscoelastic model. J Rheol (1978-present) 22 (3):259–283
42. Giesekus H (1982) A simple constitutive equation for polymer fluids based on the concept of deformation-dependent tensorial mobility. J Nonnewton Fluid Mech 11(1–2):69–109
43. Giesekus H (1985) Constitutive equations for polymer fluids based on the concept of configuration-dependent molecular mobility: a generalized mean-configuration model. J Nonnewton Fluid Mech 17(3):349–372
44. Mostafaiyan M, Khodabandehlou K, Sharif F (2004) Analysis of a viscoelastic fluid in an annulus using Giesekus model. J Nonnewton Fluid Mech 118(1):49–55

45. Isaki T, Takahashi M, Takigawa T, Masuda T (1991) Comparison between uniaxial and biaxial elongational flow behavior of viscoelastic fluids as predicted by differential constitutive equations. Rheol Acta 30(6):530–539
46. Yarin AL, Sinha-Ray S, Pourdeyhimi B (2010) Meltblowing: II-linear and nonlinear waves on viscoelastic polymer jets. J Appl Phys 108(3):034913
47. Luo X-L, Tanner R (1987) A pseudo-time integral method for non-isothermal viscoelastic flows and its application to extrusion simulation. Rheol Acta 26(6):499–507
48. Chauvière C, Owens RG (2002) A robust spectral element method for simulations of time-dependent viscoelastic flows, derived from the Brownian configuration field method. J Sci Comput 17(1–4):191–199
49. Cirulis JT, Keeley FW, James DF (2009) Viscoelastic properties and gelation of an elastin-like polypeptide. J Rheol (1978-present) 53(5):1215–1228
50. Oldroyd J (1950) On the formulation of rheological equations of state. Proc Roy Soc Lond A 200:523–541
51. Scott J (1935) Theory and application of the parallel-plate plastimeter. Part 2. Rubber Chem Technol 8(4):587–596
52. Dillon J, Johnston N (1933) The plastic properties of several types of unvulcanized rubber stocks at high rates of shear. J Appl Phys 4(6):225–235
53. Mullins L (1969) Softening of rubber by deformation. Rubber Chem Technol 42(1):339–362
54. White JL, Soos I (1993) The development of elastomer rheological processability quality control instruments. Rubber Chem Technol 66(3):435–454
55. Zakharenko N, Tolstukhina F, Bartenev G (1962) Flow of rubberlike polymers with and without carbon black. Rubber Chem Technol 35(2):326–334
56. Bershtein V, Yegorov V, Marikhin V, Myasnikova L (1985) Specific features of molecular motion in lamellar polyethylene between 100 and 400 K. Polym Sci USSR 27(4):864–874
57. Toki S, White JL (1982) Rheological and solid wall boundary condition characterization of unvulcanized elastomers and their compounds. J Appl Polym Sci 27(8):3171–3184
58. Lobe VM, White JL (1979) An experimental study of the influence of carbon black on the rheological properties of a polystyrene melt. Polym Eng Sci 19(9):617–624
59. Tanaka H, White JL (1980) Experimental investigations of shear and elongational flow properties of polystyrene melts reinforced with calcium carbonate, titanium dioxide, and carbon black. Polym Eng Sci 20(14):949–956
60. Suetsugu Y, White JL (1983) The influence of particle size and surface coating of calcium carbonate on the rheological properties of its suspensions in molten polystyrene. J Appl Polym Sci 28(4):1481–1501
61. Suh CH, White JL (1996) Talc-thermoplastic compounds: particle orientation in flow and rheological properties. J Nonnewton Fluid Mech 62(2):175–206
62. Osanaiye GJ, Leonov AI, White JL (1993) Investigations of the rheological behavior of rubber-carbon black compounds over a wide range of stresses including very low stresses. J Nonnewton Fluid Mech 49(1):87–101
63. Subramoney S (1998) Novel nanocarbons—structure, properties, and potential applications. Adv Mater 10(15):1157–1171
64. Haggenmueller R, Gommans H, Rinzler A, Fischer JE, Winey K (2000) Aligned single-wall carbon nanotubes in composites by melt processing methods. Chem Phys Lett 330(3):219–225
65. Lozano K, Barrera E (2001) Nanofiber-reinforced thermoplastic composites. I. Thermoanalytical and mechanical analyses. J Appl Polym Sci 79(1):125–133
66. Lozano K, Bonilla-Rios J, Barrera E (2001) A study on nanofiber-reinforced thermoplastic composites (II): investigation of the mixing rheology and conduction properties. J Appl Polym Sci 80(8):1162–1172
67. Qian D, Dickey EC, Andrews R, Rantell T (2000) Load transfer and deformation mechanisms in carbon nanotube-polystyrene composites. Appl Phys Lett 76(20):2868–2870
68. Salvetat J-P, Briggs GAD, Bonard J-M, Bacsa RR, Kulik AJ, Stöckli T et al (1999) Elastic and shear moduli of single-walled carbon nanotube ropes. Phys Rev Lett 82(5):944

69. Walters D, Ericson L, Casavant M, Liu J, Colbert D, Smith K et al (1999) Elastic strain of freely suspended single-wall carbon nanotube ropes. Appl Phys Lett 74(25):3803–3805
70. Li F, Cheng H, Bai S, Su G, Dresselhaus M (2000) Tensile strength of single-walled carbon nanotubes directly measured from their macroscopic ropes. Appl Phys Lett 77(20):3161–3163
71. Pötschke P, Fornes T, Paul D (2002) Rheological behavior of multiwalled carbon nanotube/polycarbonate composites. Polymer 43(11):3247–3255
72. Chatterjee T, Krishnamoorti R (2013) Rheology of polymer carbon nanotubes composites. Soft Matter 9(40):9515–9529
73. Du F, Scogna RC, Zhou W, Brand S, Fischer JE, Winey KI (2004) Nanotube networks in polymer nanocomposites: rheology and electrical conductivity. Macromolecules 37 (24):9048–9055
74. Chatterjee T, Krishnamoorti R (2008) Steady shear response of carbon nanotube networks dispersed in poly (ethylene oxide). Macromolecules 41(14):5333–5338
75. Moreira L, Fulchiron R, Seytre G, Dubois P, Cassagnau P (2010) Aggregation of carbon nanotubes in semidilute suspension. Macromolecules 43(3):1467–1472
76. Whittle M, Dickinson E (1997) Stress overshoot in a model particle gel. J Chem Phys 107 (23):10191–10200
77. Silbert L, Farr R, Melrose JR, Ball R (1999) Stress distributions in flowing aggregated colloidal suspensions. J Chem Phys 111(10):4780–4789
78. Thomin JD, Keblinski P, Kumar SK (2008) Network effects on the nonlinear rheology of polymer nanocomposites. Macromolecules 16(41):5988–5991

Morphology and Spectroscopy of Polymer–Carbon Composites

Purabi Bhagabati, Mostafizur Rahaman and Dipak Khastgir

Abstract Over several decades, polymer/carbon filler composites have been serving the human society through technological prospects to the greater dimension. While carbon-based fillers like low-cost carbon black have emerged as the most functional and effective additive in elastomers; similarly, carbon fillers in nanometric range (e.g., carbon nanotube, graphene, etc.) are capable of escalating electrical and thermal conductivity of insulating polymer matrices. However, in all such cases, the morphology of these carbon fillers in polymer matrix and existence of any chemical or physical interaction between these fillers and polymer matrix dictates the overall composite properties. This book chapter discusses the importance and role of morphology and spectroscopic analysis of carbon fillers on the properties of polymer composites. Also, a detailed understanding on different types of carbon fillers, its morphological aspects, and its nature of interaction with various polymer matrices are thoroughly covered.

Keywords Carbon filler · Morphology · Spectroscopy · Polymer composites
Polymer–filler interaction

P. Bhagabati (✉)
Chemical Engineering Department, Indian Institute of Technology Guwahati,
Guwahati 781039, India
e-mail: purabi.bhagabati08@gmail.com

M. Rahaman
Department of Chemistry, College of Science, King Saud University,
Riyadh 11451, Saudi Arabia
e-mail: mrahaman@ksu.edu.sa

D. Khastgir
Rubber Technology Centre, Indian Institute of Technology Kharagpur,
Kharagpur 721302, India
e-mail: khasdi@rtc.iitkgp.ernet.in

M. Rahaman et al. (eds.), *Carbon-Containing Polymer Composites*,
Springer Series on Polymer and Composite Materials,
https://doi.org/10.1007/978-981-13-2688-2_9

1 Introduction

Over the last two centuries, massive advancements in science and technology have led the human race to the next level of understanding various materials such as metals, ceramics, and polymers. Impact of materials on the society is mainly dictated by its usability in various applications. Both the natural and synthetic polymers have attracted extensive interests of researchers, scientists, and of course the industrialists due to its enormous beneficial properties. Besides all the application-oriented properties, polymers have low specific gravity that contributes the maximum to its widespread popularity. However, most of the neat polymers have restricted mechanical and thermal performances. Composite materials are the mixture of two or more materials with good degree of homogeneity with superior properties over its constituents. Compounding of polymers with different fillers leads to the development of hybrid material of their parent constituents called "composites," which have usually synergistic properties. By inclusion of fillers into a polymer matrix, one can attain a combination of lightweight, toughness, and transparency of polymers along with the excellent mechanical, physical, and other properties of rigid fillers. The success in exploiting polymers in load-bearing applications or applications with the condition of high thermal stability is provoked by the concept of composite preparation of the polymer. Polymer composite shows new physicochemical properties by maintaining the individual traits of each phase.

Carbon "C" is a chemical element with atomic number 6, and exists in three different isotopes; ^{12}C, ^{13}C, and ^{14}C. Among these, the ^{13}C isotope is most stable and abundant in several forms. The most available form of carbon known as carbon black was used as pigments back in the third century B.C. There are several other allotropes of carbon are available, best known as graphite, diamond, fullerene, carbon nanotubes, carbon fiber, etc. Polymer composites containing carbonaceous fillers like carbon black (CB), natural graphite (NG), graphenes, carbon nanotubes (CNTs), carbon fiber and nanofibers (CF and CNF), fullerenes, quantum dots, diamonds, etc., were been studied for years due to its ability to retain low specific gravity and impart high electrical and thermal conductivity, and good corrosion and ultraviolet (UV) resistance to the commodity polymers. All these carbon fillers differ from each other in many aspects and subsequently imparts several different properties onto polymer matrices. Carbon black was the first filler to be used as reinforcing filler in polymer matrices. Also till the present time, carbon black covers the maximum share as both reinforcing and non-reinforcing filler of polymer matrices, especially in case of rubber industries in several fields of application. Electrically conducting polymers are the important group of materials with several engineering applications including electrically conducting adhesive, electrical sensors and actuators, antistatic coatings, electromagnetic interference (EMI) shielding layer in electrical components, etc. Polymers are insulating in nature and conventionally micron-size metal powders are incorporated into the polymers to enhance the overall electrical conductivity. But the prime disadvantage in such polymer/metal composite systems are inhomogeneity and lack of interfacial adhesion between the phases that retards mechanical properties of the polymer composite and increase in specific gravity of the composite, which is frequently an undesirable characteristic in several applications. Carbons in both sp^2 and sp^3 hybridized state are capable of conducting electrons, and hence carbon-based particles like carbon black (CB), graphites and graphenes, carbon nanotubes (CNTs), carbon nanofibers (CNFs), diamonds, fullerenes, etc., have emerged as most effective alternative to the metal powders. These carbonaceous particles are available both in micro and nano size, and they are proven to reduce the filler content in attaining adequate conductivity [1].

Different Types of Carbon Fillers

Carbon-based filler possesses diversified geometry: globular or spherical structure, hollow cylindrical tube-like structure, rod-like structure, and flake or platelet-like structure. Such variation in geometry of these carbonaceous fillers has a profound effect in improving several properties of the respective polymer composites. All carbon fillers are more or less electrically conducting in nature due to the unique state of hybridization of its carbon atoms [2].

In this chapter of the book, we provide a brief description of the morphology and spectroscopy of carbon filler-based polymer composites. Special emphasis is made on understanding the effect of shape, structure, geometry, chemical functionality, and physical interaction of various carbon fillers on the polymer composites. Besides, the type and effect of different polymer matrices on the composite property were also explained for further clarity. Recent and up-to-date advances on carbon filler-based polymer composites were also described along with special case studies to serve our motto for updating the readers with current and most trending research scopes in the related field.

2 Morphology and Spectroscopic Characterization

2.1 Morphology

Morphology of polymer composites often creates enigma among researchers due to the overlapping of several features like length or other dimensions with scales that are a formidable task for accurate interpretation. However, material scientists show tremendous interest in understanding the microstructural information in order to estimate its physical and chemical behavior. The morphology study of polymer composites under different conditions is highly informative in understanding the complete behavior of filler distribution and interaction with polymer matrices. In this case, cryo-fracture, tensile fracture, solvent etching, drop casting, etc., are some of the commonly used methods to follow sample preparation and analysis. While SEM, TEM are versatile techniques to analyze fillers of nanometer range, the AFM and OM are helpful to study polymer composites with micron size. Moreover, crystallization behavior of base polymer matrices in the presence of fillers are also essential to analyze as it provides important information about the physical and chemical behavior of polymer composites. In morphology through POM study, we can understand the physical and material characteristics of polymer composites. A brief discussion on some of the commonly used techniques to study the morphology of polymer composites is mentioned below.

Optical Microscopy (OM)
Optical microscopy of polymer composites renders information about the type of micro-filler distribution within the polymer matrix. They are not effective in analyzing the morphology of nanofillers within the matrix. It is the early and most basic form of microscopic technique that is mostly used to study the surface roughness and presence of microparticles in samples. Biological samples are mostly analyzed with this microscopic technique.

Optical Microscopy (POM)
Polarized optical microscopy (POM) is the microscopic technique to visualize spherulites and its formation process through recrystallization of polymer matrices both in absence and or presence of fillers. The rate of crystallization of polymer during cooling and the effect of fillers on the crystalline behavior of polymer can well be understood through detailed analysis via POM. While nanofillers like CNT accelerates nucleation in the crystallization process of polymer, the micron-size fillers like CB hinders early formation of spherulites, which directly affects the overall physical properties of polymer composites.

Scanning Electron Microscopy (SEM)
Scanning electron microscopy (SEM) is a commonly used microscopic technique by the material scientists to evaluate the morphology of polymer composites. It cannot only provide information regarding filler dispersion and distribution but also indicates polymer–filler interaction in the composites. Besides, SEM is also utilized in showing up the elemental composition of a particular component. Cryo-fractured and tensile-fractured polymer composite samples are proven to be highly informative in terms of adhesion of fillers with polymer matrix. However, chemical or physical modification of fillers cannot be analyzed through SEM as the technique involves two-dimensional reflection photomicrographs.

Transmission Electron Microscopy (TEM)
Transmission electron microscopy (TEM) is another sophisticated instrumentation technique used for the observation of minute change in the morphology of polymer composites containing nanofillers. It can also be utilized for checking any morphological change in nanofillers during and after modification. The transmittance mode of TEM allows us to have an insight into negligibly small change in the nanoparticles.

Atomic Force Microscopy (AFM)
Atomic force microscopy (AFM) is the technique to have an overall look of any microcomposites of polymers in three-dimensional mode as well. However, it is

important to understand the working principles of AFM that may help us in determining the modulus of any filler and polymer composites. AFM provides three-dimensional view of filler distribution in polymer matrix.

2.2 *Spectroscopy*

Spectroscopy is a technique to understand various chemical modifications of fillers or polymers taking place during the composite formation. It helps us understand the different physical and chemical interactions between polymers and fillers. Hence, the spectroscopic study of a polymer composite comes in the first place while carrying out a systematic study.

UV Spectroscopy
Carbon blacks are effective as UV stabilizer to polymer matrices as they are capable of absorbing large amount of UV light and thereby, quench the excited induced by the UV light in polymer matrices. Hence, it can be understood that UV spectrometer helps us understand the type of UV light being absorbed the filler in polymer composites. Besides, reports are also available utilizing UV spectroscopy in understanding the polymer–filler interaction, where the system contains any UV active functional groups.

FTIR Spectroscopy
FTIR offers both qualitative as well as quantitative analysis of various functional groups in chemical compounds. It helps scientists or researchers to estimate the chemical behavior of materials. Nowadays, it is becoming almost a mandate in any research activities in the material science field to carry out FTIR study to find its reactivity in different chemical environments. FTIR is an essential tool to find any interaction between polymer and filler in polymer composites. Besides, it also ensures any chemical or physical modification in fillers that may act as an important aspect in improving polymer/filler adhesion.

NMR Spectroscopy
NMR spectra help us understand the chemical composition of polymers in composites and also quantify mixture components. It can be used for characterization of polymers by structural analysis, monomer or comonomer ratios, end group analysis, branching of polymer chains, and calculation of average molecular weights.

Raman spectroscopy
Raman spectra of polymer composites provide information about the state of filler dispersion. The spectra of polymer composites with fillers embedded within the polymer matrix undergo shifting of the characteristic peaks of fillers if only good chemical interaction between polymer and filler exists. Thus, polymer composites are characterized by Raman spectroscopy that leads us to enhance the knowledge about filler dispersion and interaction with polymer matrices.

X-Ray Diffraction (XRD)
XRD is the most common and highly facilitating technique to measure the degree of crystallinity of polymers and its composites. It can directly reflect the percent of crystalline and amorphous regions through its spectra. It is the most important tool in investigating the type of crystallites, its orientation, its size, and extent of crystallinity in materials. In polymer composites, the presence of interacting fillers may act as nucleating agent and hence can improve the degree of crystallinity in the resulting polymer composites. However, a large number of instances support a decrease in percent crystallinity on the addition of fillers into the polymer matrix due to hindrance exerted by the fillers during crystallization. All these information can easily be acquired through XRD analysis.

X-Ray Photoelectron Spectroscopy (XPS)
XPS analysis of solid surfaces provides information about chemical composition and chemical bonds existing in the samples. During analysis, the samples are illuminated with specified monochromatic X-rays that produced energy due to emitting photoelectrons from the surface of the samples. A spectrum of this emitted photoelectron against binding energy is recorded, which provide chemical compositional information of surfaces of samples. Chemical modification of polymer and polymer composites can well be analyzed with XPS. Further, the chemical composition of carbon and carbon-based fillers may also be analyzed and surface modification of these carbon-based fillers can be estimated with the help of XPS.

3 Morphology and Spectroscopy of Polymer/Carbon Filler Composites

Carbon filler possesses a wide variety of morphology as mentioned in the introduction part, which has an enormous impact on the overall physical as well as mechanical properties of its polymer composites. While spherical anisotropic morphology of carbon fillers imparts improvement in the properties of polymer composites in all directions, the isotropic 2D carbon fillers are capable of shooting up various properties of the respective polymer composites along the direction of filler alignment. Further, nano-dimensional carbon fillers are far more effective in synergizing the properties of polymer composites due to higher surface area. Thus, it is much of the interest of researchers and scientists to understand the morphology of fillers before incorporating it into the polymer matrix. Similarly, spectroscopic analysis of the carbon fillers is also necessary as it has tremendous influence on the chemical as well as other physico-mechanical properties of polymer composites. The critical agglomerated structure of carbon filler forms mechanical interlocking with polymer chains during processing of composites. Besides, the functional groups present on the surface of carbon black also takes part an essential role in improving the characteristic properties of polymer composites through the formation of chemical or physical bonds with polymers chains. Therefore, the process of functionalization of other carbon fillers like carbon nanotube (CNT), carbon nanofiber (CNF), graphene platelets, fullerenes, nanodiamonds, etc., are adopted to improve the polymer/fillers interaction in respective composites. Spectroscopic analysis of these modified fillers is necessary that enhance our understanding on the type of interaction with various polymer matrices. The presence of conducting carbon fillers at a concentration exceeding its critical value, known as "percolation threshold" in polymer matrices are capable of inculcating electrical conductivity in the insulating polymers. The percolation threshold of the conducting fillers in the polymer composites can be characterized by the abrupt rise in the conductivity of the composites by many degrees. The percolation threshold is due to the formation of a three-dimensional conductive network structure of the conductive fillers within the polymer matrix.

3.1 Polymer/Carbon Black Composites

Carbon Blacks (CBs)

As soon as carbon black was discovered, it was started to be recognized as "active" filler in rubbers. Presently, carbon black is known for its unique ability to reinforce weak polymer matrices like rubbers. It is the prime reinforcing filler used in rubber industries like tire applications. Carbon black is particulate filler, which is commonly used in the rubber industry. CB has globular nanometric particle size which mostly remains in agglomerated form. The main parameters of carbon black for its reinforcing characteristic are bestowed to its size and shape of the aggregate structure as shown in the TEM image of various CB in Fig. 1.

An aggregate of carbon black consists of primary particles, which are of spherical shape in nanometer dimension. These primary particles are chemically bonded with each other in a much random fashion. Generally, carbon blacks are characterized by two important properties: its surface area and structure. Surface area is the measure of the area available for physical interaction with polymeric materials. Surface area is inversely dependent upon the primary particle size of carbon black. Smaller the size of primary particles of carbon black, higher is the surface area. However, the structure of carbon black is directly dependent upon the size of the aggregates and the type of functional groups present on the surface of the primary particles. Carbon blacks having aggregates with larger number of primary particle implies higher structure, which is due to the more complex shape and physical voids. Surface activity of carbon black indicates the strength of interaction with rubber either from nonbonded physical interactions or through chemisorption. Presence of small quantity of functional groups like sulfur (S), hydrogen (H), oxygen (O), carbonyl (C=O), etc., onto the surface of primary particles causes increase in surface activity. All these parameters play a significant role in reinforcing polymer matrices through different mechanisms.

Carbon blacks are the first ever fillers used in polymers including elastomers to enhance the several physical properties of polymers. Due to its ease of availability

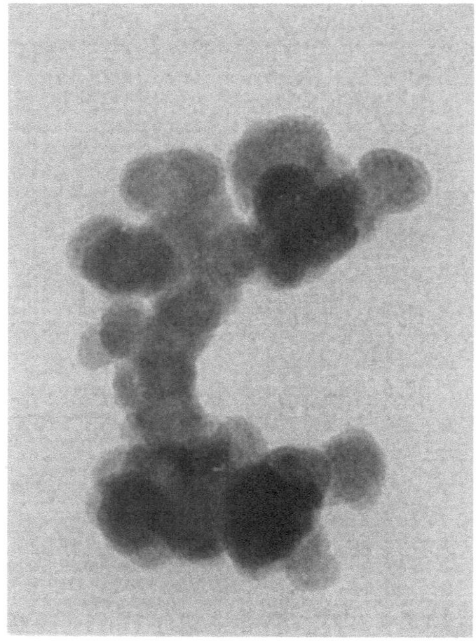

Fig. 1 Transmission electron micrographs of carbon [1]

and in production and variation in properties, it has attained tremendous interest as additive of polymer. It is the first ever known "Active Filler" for elastomers from the beginning of the nineteenth century. Carbon blacks in the polymer are capable of imparting several unique properties into polymer matrices: excellent UV resistance, reinforcement, thermal stability, and chemical resistance in certain cases as well. The reinforcement capability of carbon black has a profound impact on today's tire industries. The whole credit for such unique degree of reinforcement of carbon black goes to two important factors:

1. The secondary CB structure that contains an aggregate of large number of primary nano CB particles resulting in complex geometries that form physical interlocking with the amorphous entangled elastomer chains through occlusion,
2. The surface activity of the CB particles arising from the functional groups attached to the surface of CB that forms chemical links with elastomer chains.

Hence, the interfacial interaction between CB and elastomer chains is reduced by the abovementioned factors that further enhance reinforcement through reduced chain mobility. Hence, the addition of CB in elastomers increases the number of immobilized elastomer chains, which is also known as "bound rubber." Higher the number of bound rubber, higher is the degree of reinforcement. In order to increase bound rubber content, it is recommended to modify the surface of CB fillers or of the polymers through chemical or physical methods. The common technique of improving the surface activity of CB fillers are oxidation in air or acid treatment, which leads to the attachment of various oxygen-containing functional groups like

–COOH, –COOR, –OH, etc. Similarly, the polymeric chains are also modified via copolymerization or adding various chemical compatibilizers like maleic anhydride or dicumyl peroxide (DCP). Grafting of polymer chains onto CB surface further improves the overall polymer–filler interaction and thereby enhances the final properties. Based on the environmental condition, plasma treatment of CB also alters the surface chemistry of CB. The addition of such surface-activated CB into the polymer matrix increases the compound viscosity due to immobilization of the entangled polymer chains. In other words, elastomeric chains are physically crosslinked, which causes enhancement of mechanical and thermal properties of the resultant polymer composites. In the subsequent section, detailed information on various aspects of CB towards polymer composites is discussed in order to induce an easy understanding on the subject matter.

Carbon Black Composites

Carbon blacks are intimately related to elastomers in the form of reinforcing fillers. Addition of CB fillers not only improves elastomer mechanical properties but also converts it from insulating material to a semiconducting or conducting material. The electrical conductivity of various polymer components is highly important for manufacturers in several applications like in electrical components. From the electrical point of view, CB is very much homogeneous with amorphous rubber matrices. While, rubber materials are insulating by nature ($\sigma < 10^{-13}\ \Omega^{-1}\ m^{-1}$), the CB particles are semiconducting ($\sigma \sim 10^{5}\ \Omega^{-1}\ m^{-1}$). CB imparts conductivity in composites essentially in the dielectric range, and no physical contact would be there between individual particles. However, at higher CB loading, three-dimensional conducting network structure gets introduced that leads to abrupt rise in conductivity of the composite, noted as critical volume fraction of the conducting material also known as percolation threshold. Interestingly, above this critical concentration of the CB filler, nominal change in conductivity is observed. Conductivity of the polymer composite depends upon the number of contacts between CB particles within the polymer matrix. Again, the critical volume of conducting filler to attain percolation threshold depends upon the degree of uniformity in the dispersion and extent of polymer–filler interaction. Enhanced polymer–filler interaction along with improved dispersion lowers the critical volume fraction. Further, the interaction between polymer and filler is well dictated by the chemical nature of the surface of CB particles and its primary and secondary structure, while the chemical structure of polymer matrix also influences to a greater extent. Hence, the properties of polymer/CB composites are mainly dependent upon three broad factors: the degree and state of CB filler dispersion in polymer matrix, the inherent principal properties of the CB fillers like particle size and shape, surface area, geometry of aggregates, and surface activity, and third the polymer–filler interaction [3]. The state of dispersion of these CB fillers within polymer matrix is one of the deciding factors of physical properties of polymers composites. Sumita and coworkers have made an extensive study on understanding the mode of dispersion of these fillers into polymer matrices [4–6]. They reported increasing properties of composites on the addition of surface modified CB over untreated CB.

Park and Kim also have made significant study in understanding the effect of surface treatment of CB with polar or nonpolar functional groups on polymer–rubber interaction. They reported to have obtained large improvement in polymer–filler interfacial interaction parameter on organic modification of CB, which further caused significant increment in mechanical properties of the rubber composites [7]. CB can enhance properties of polymers even up to the loading as high as >60 wt%; however, further addition may lead to reduction in properties. While the incremental aspect of properties in amorphous polymers like rubbers is much prominent, the effect is comparatively insignificant in case of semicrystalline plastic materials. The electrical property of CB/epoxy resin composites were studied by Ma et al., in the year 2009, where they have made a detailed analysis of the CB/CNT morphology onto polymer electrical properties and the mechanical properties as well [8]. According to the report, there is a significant increase in conductivity of epoxy composites on the addition of CB along with CNT at low loading, e.g., 1–2 wt%. Such hybrid composites of carbon material are more efficient in enhancing electrical conductivity through developing percolation even at lower loading. Both CB and CNT form some conducting channeled networks for electrons to pass through the insulating polymer matrix, and thereby result in significant increment in the overall conductivity. TEM images of prepared composites with 0.2 wt% of CNT, 0.2 wt% of CB and both the fillers are shown in Fig. 2a–c, respectively. While, the CNT fillers are more of agglomerated form in its composite, the spherical shape of CB in the composite itself is unable to form any network structure at such low loading. However, in the case of hybrid composite of CB/CNT, well-defined channeled structures were observed, in which the CB particles are attached with and around the end of CNTs. In this case, the gap between the CNT particles was filled by CB particles, thereby formation of conducting networks. For better understanding, the authors have made an illustrative schematic of the images in Fig. 2d–f, respectively. Hence, a low percolation threshold was observed with only 0.2 wt% CB and CNT in the composites.

Besides improving the electrical property of composites, the mechanical properties were also synergized by the addition of CB fillers in the composites. While, the epoxy composite with only CNT has insignificant improvement in mechanical properties, the addition of CB in 0.2 wt% was capable of greatly enhance both the ductility as well as fracture toughness of the hybrid composites. Thus, CB acts as efficient low-cost multifunctional filler in polymer matrices.

CB is also known as a stabilizer to polymer matrices in its composites and it can be well understood that the polymers are getting protection from UV light. Hence saturated polymers like polyethylene (PE), polypropylene (PP), polyesters, etc., are some of the polymers that are benefited with CB fillers in specific to outdoor applications like water tank, covering sheets, etc. [9]. Further, the addition of CB into polyester matrices can improve its optical properties along with UV light stability. However, it is important to mention that the nature of optical properties is strongly dependent upon the type of polyester used. For polyesters with different aromatic rings in the main or side chain can show varying fluorescence band at longer wavelength [10]. The peak observed in the UV spectra is due to absorption

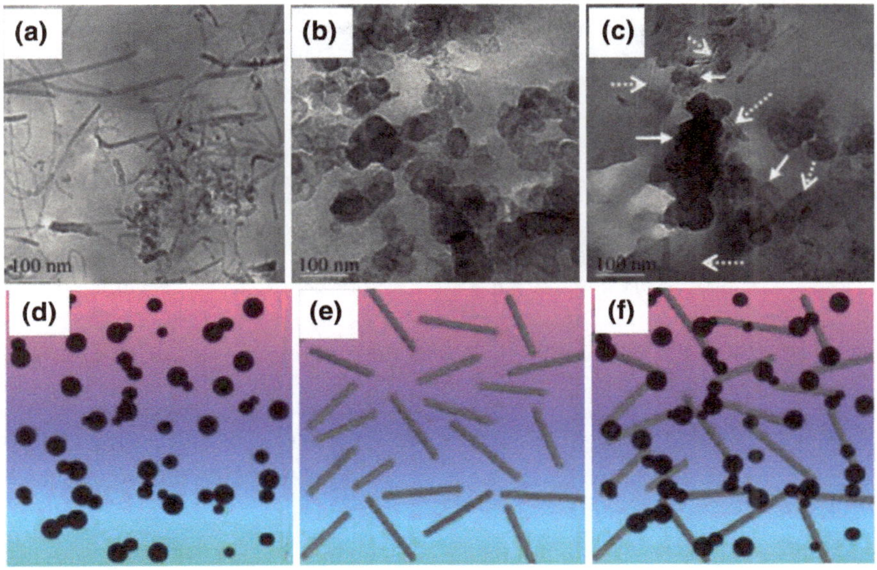

Fig. 2 TEM images of **a** CNT/epoxy composites with 0.2 wt% loading, **b** CB/epoxy composites with 0.2 wt% loading, and **c** CB/epoxy composites with 0.2 wt% CNT and 0.2 wt% CB loading [8]

of light energy by the chromophoric functional groups in UV and visible region, which involves excitation of electrons from n to π^* orbital. This is only possible because such transition falls under the region of spectrum 200–700, which is within the convenient region of the instrument. The increased absorbance value of composites with increased CB content may be interpreted with the increased proximity of interchain interaction between polymer chains and increased local concentration of functional groups of CB. The UV–visible–NIR absorption spectra of carbon black are shown in Fig. 3, where the carbon blacks are seen to show great absorption in the whole UV region [11].

This indicates great UV blocking effect of carbon black, which has an obvious positive influence on its rubber composites. The UV shielding effect of the CB is helpful in protecting unsaturated rubbers from being degraded by ozone attack.

3.2 Polymer/Carbon (Nano) Fiber Composites

Carbon (Nano) Fiber

The history of carbon fiber was dated more than a century back in the year 1889. In the years 1970 or 1980, a new research trend has taken over the polymer composite field, with the addition of carbon nanofiber (CNF) into various polymeric matrices,

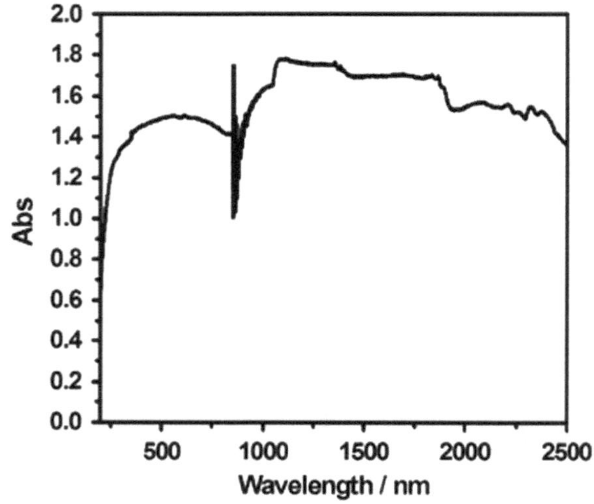

Fig. 3 UV–visible–NIR absorption spectroscopy of carbon black [11]

especially plastics. The reason behind the growing interests in CNFs is because they are capable of providing many solutions to certain challenging problems in existing in glass fiber-reinforced composites. CNFs provide a complete package of superior mechanical, thermal, structural performance, along with the cost advantages over other carbon-based nanofillers like fullerene, graphene, and CNTs. CNFs can be produced in bulk and in that case, its production cost is very much comparable to E-glass fibers, while the properties like mechanical, thermal and electrical are far better than its glass fiber counterparts. CNFs is produced by catalytic chemical vapor deposition (CVD) method which involves an effective catalytic degradation of hydrocarbon gases or carbon monoxide (CO) at temperatures in the range of 500–700 °C, in presence of several metals like iron (Fe), cobalt (Co), Nickel (Ni), or some alloys. Based on the production technique and condition, a wide variety of morphology can be observed in the case of CNF. The vapor grown CNF (VGCNF) are another type of CNF with diameter ranging from 60 nm to 200 nm and length of ~7–10 μm. VGCNF has hollow cylindrical morphology with single or double ordered graphitic layers. Besides, bamboo-like morphology, graphitized cup stacked-like morphology, and nested conical shell structure-like morphology are also available [12, 13]. The addition of two-dimensional fibers into polymer matrices improves various properties along a particular direction more effectively than anisotropic fillers. Carbon fibers are the most common type of 2D fillers used in polymer matrices. However, carbon nanofibers are found to be better in terms of improving its properties that are not only due to its higher aspect ratio but also due to better adhesion with polymers.

Carbon (Nano) Fiber Composites

Inclusion of CNF into polymer matrices is to serve several application-oriented purposes, in specific to improving mechanical properties, electrical and thermal conductivity and sometimes to improve thermal properties. Hence, CNF/polymer composites are used in many hi-technology applications like in spacecraft, airplanes, artificial satellites, etc. Tensile strength and modulus of CNF is intrinsically at least two to threefold higher than neat polymer matrices. Other than its excellent mechanical properties, good polymer–filler adhesion, and uniform dispersion and distribution of individual nanofibers within the polymer nanocomposites are important that determines the overall mechanical properties of nanocomposites. Several techniques are been followed to improve the dispersion of CNF into various polymeric matrices. CNFs exhibit strong van der Waals force of attraction that is responsible for the enhanced tendency to agglomerate. Formation of agglomeration of CNFs causes an improper state of filler dispersion in the polymer matrix, which in turn affects the mechanical, thermal and other physical properties of polymer composites. However, the defects present due to side wall and open end H atoms of sp^2 and sp^3 defects can also take part in the electrophilic substitution reaction with some polymer matrices like via in situ polymerization of 3-phenoxybenzoic acid in the medium of poly(phosphoric acid)/phosphorus pentoxide (PPA/P_2O_5; 1:4 w/w) [14]. The nanocomposites have shown improved thermal properties like increased T_g and oxidative degradation stability. Further, CNFs were modified by 4-(3-aminophenoxy) benzoic acid through Friedel–Crafts acylation reaction, which was later grafted with polyimide chains via in situ polymerization technique [15]. Heat treatment of CNF at an optimum temperature of 1500 °C graphitizes the outer surface of the nanofibers along the fiber axis and improves physico-mechanical and electrical properties of nanocomposites. Attaching oxygen onto the surface of the CNF leads to polymer–filler adhesion in the composites by increasing the overall surface free energy of the nanofibers. Grafting of oxygen onto the CNF surface is mostly carried out by following three different procedures: etching of the nanofibers in oxygen environment at temperature of around 400–600 °C, soaking and subsequent rigorous treatment with strong inorganic acids like sulphuric acid, or sulphuric acid/nitric acid, etc., and by oxygen plasma treatment [16, 17]. Such surface modification of CNF can be characterized by a decreased water contact angle value. Plasma treatment of CNF in oxygenated environment leads to attachment of four different types of functional groups: carboxylic, carbonyl, phenolic, and hydroxylic groups. Though the polymer–filler interaction rises in case of treated CNF over neat CNF, but the increase in mechanical properties of plasma-treated VGCNF/PP composites are marginal, which can be ascribed to poor dispersion of these filers [18]. However, at higher filler loading (15 vol%), air-etched VGCNF has shown far better mechanical property and polymer–filler adhesion over graphitized VGCNF in PP matrix. In the latter case of graphitized VGCNF/PP nanocomposite, the fractured surface contained longer pulled out fiber than the air-etched VGCNF/PP nanocomposite as shown in Fig. 4, indicating better adhesion between CNF and PP polymer matrix [19].

Further, aligned CNF within polymer matrix can introduce a certain degree of improvement in properties. Gordeyev et al. successfully inculcated CNF into semicrystalline polymer matrices and stretched into fibrillar form to produce self-reinforced structure [20]. VGCNF/PP composite was spun and drawn from

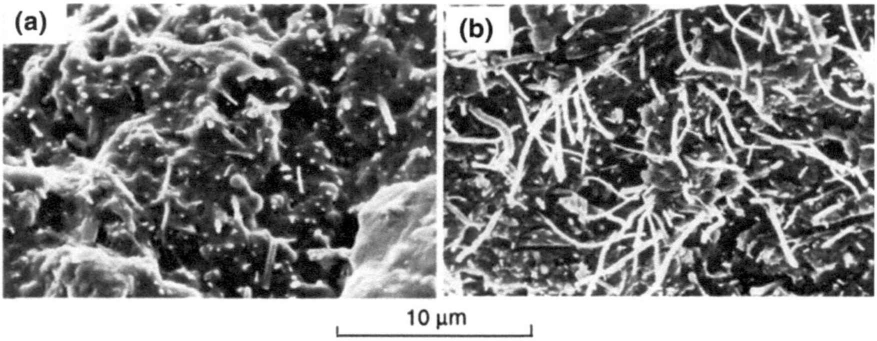

Fig. 4 SEM micrographs of **a** air-etched VGCNF/PP composite; **b** graphitized VGCNF/PP composite [19]

melt extruder to produce high strength filaments with improved properties. In such case, the drawing ratio and filler volume percentage affect the final property of the composite filament. Thus, the nanofibers can be aligned within polymer matrix during the processing stage through melt spinning or extruding, where the degree of fiber alignment can further be increased by increasing the overall residence time in the die. In case of resins-like epoxy, the composites were prepared by casting solution of epoxy onto CNF mats under vacuum followed by curing and it leads to improved flexural strength and modulus by 30 and 5%, respectively. Alignment of CNF in polymer composites always leads to improvement in several properties including mechanical, thermal and electrical conductivity. Application of AC electric field has been established as one of the most promising methods for developing alignment of CNF in the polymer matrix. Lim et al. have made an extensive study on the effect of AC electric field induced alignment of carboxylic acid and amine functionalized CNF onto the mechanical and electrical conductivity of epoxy resins at low filler concentration (maximum 4.5 wt%) [21]. Placing the epoxy/functionalized CNF suspension in high voltage AC current leads to alignment of the fibers within epoxy matrix. From Fig. 5, the aligned structure of both carboxylic acid and amine functionalized CNF can be estimated through the optical micrographs.

From Fig. 5, it can be realized that the functionalized CNF are aligned along the direction of the applied electric field. Besides, it was reported that the extent of alignment of the nanofibers is concentration dependent in both the cases. At higher filler concentration, the alignment structures of the functionalized CNF are more of dense web-like tiny bundles. It was also observed that the compressive strength, as well as modulus of both functionalized CNF composites, was much higher than the neat epoxy resins. Hence, the alignment of nanofibers has within the polymer matrix improved the mechanical properties of epoxy resin composites. Similarly, the electrical resistivity of the composites was observed to be fourfold lower than the neat epoxy at a concentration of 4.5 wt% of functionalized CNF.

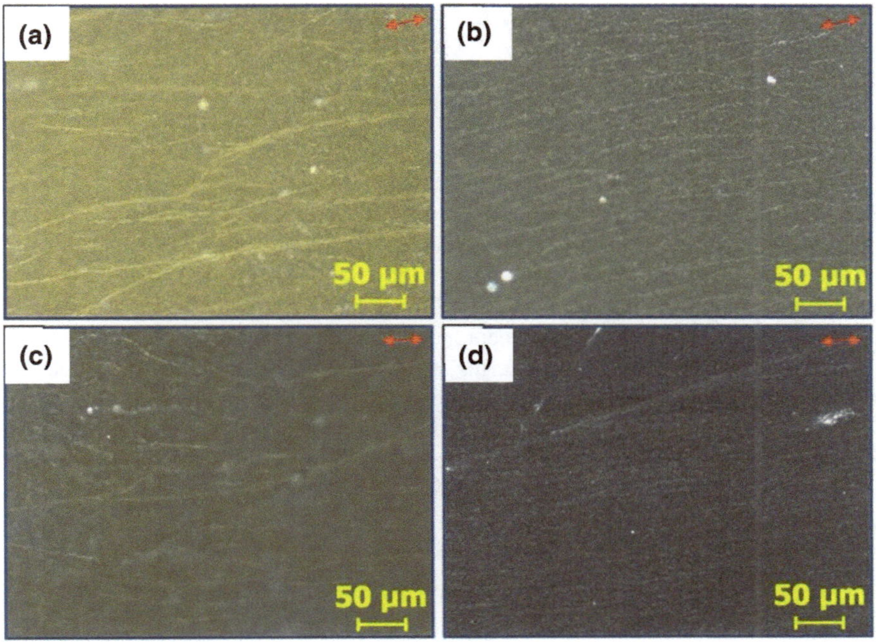

Fig. 5 Optical images of alignment structure of composites **a** Epoxy/carboxylic acid functionalized CNF (0.5 wt%), **b** Epoxy/carboxylic acid functionalized CNF (1.5 wt%), **c** Epoxy/amine functionalized CNF (0.5 wt%), **d** Epoxy/amine functionalized CNF (1.5 wt%). The red color arrow indicates the direction of aligned CNF [21]

Functionalization of CNF with 3-glycidoxypropyltrimethoxysilane was reported to improve mechanical properties of epoxy composites due to uniform state of filler dispersion [22]. While CNF contributes in improving the tribological properties like wear resistance of poly(ether ether ketones) (PEEK) [23], it also enhances the tensile strength and stiffness of the nanocomposites [24]. Also, the thermal stability of polycarbonates (PCs) was increased by 40 °C via following in situ polymerization of cyclic oligomeric carbonates [25]. Besides, silver doped CNF (AgCNF) and silver-modified natural zeolite-CNF (AgZCNF) were incorporated into epoxy resin and was successfully demonstrated to use as a detection tool for Ibuprofen (IBP) in aqueous solution [26]. Hence, a multidirectional attributes of CNF can be observed from a wide literature available to the scientific community towards the benefit of mankind in every aspect. While, the chemical and physical modification of these nanofibers has resulted in improved morphology of respective nancomposites, the enhanced adhesion with the base polymer matrix has got direct influence onto its mechanical, thermal and electrical properties as well. Thus, the cost advantage of CNF can commercially be explored by carrying out different economically viable chemical and physical modification of the CNF.

3.3 Polymer/Carbon Nanotube (CNT) Composites

Carbon Nanotube (CNT)

After the discovery of fullerene, another important allotrope of carbon was synthesized by Iijima and coworkers in the early 1990s. Through the extraordinarily extended π-electron system, carbon nanotubes (CNTs) exhibit different properties that are much relevant to be considered as smart material. It shows excellent electrical properties of conductivity ranges 1000–200,000 S cm^{-1}. Also, it is capable of exchanging electrons with electron donors or acceptors and thereby forms negative or positive charged ions [27, 28]. Though diamond has higher thermal conductivity over CNTs, but the recent technique of synthesizing CNTs are capable to overshadow the diamond in terms of thermal conductivity. A considerable number of polymers are employed to prepare CNT-based nanocomposites, where improvement in various properties like mechanical, thermal, and electrical were reported when compared with the base polymer matrix. However, researchers continually are an effort to modify the nonreactive surface of the CNTs to make it reactive towards organic polymers. In the coming section, a broader discussion on the various methods followed to prepare polymer/CNT nanocomposites through spectroscopic illustration and effect of the nanotube morphology on the general properties of the nanocomposites will be discussed.

Carbon Nanotube (CNT) Composites

From the time of the discovery of CNTs, a number of approaches have been reported to prepare CNT/polymer nanocomposites with the purpose to improve certain application specific properties like electrical conductivity, thermal conductivity while retaining the mechanical properties and thermal degradation stability of the polymer uncompromised. Like graphene nanocomposites, the final properties of polymer/CNT nanocomposites are also very much dependent upon several factors. The major hurdle in preparing the nanocomposites of CNT is its nonreactive surface property towards polymeric chains. The defect-free sp^2 hybridized network structure of carbon atoms makes it chemically inert to polymer matrices. However, it is a necessary requirement for nanofillers to have good adhesion or interaction with polymer matrices so that there would be effective stress transfer across the polymer–filler interfaces during mechanical testing. Surface modification of CNTs via different physical and chemical route enhances the interaction between polymer matrices and the CNT fillers. In this case, the nature or type of modification is dependent upon the chemical nature of the polymer matrices. More detailed discussion of various modification techniques is already discussed in the previous chapter of the book. Also, it is important to mention that good polymer–filler interfacial adhesion leads to uniform dispersion of the nanofillers within the polymer matrix, which is important in dictating the material characteristics.

The surface modification of CNTs can broadly be classified into two categories: in the first case, carboxylic acids will be directly bound to the end of the CNT openings in the process of chemical oxidation through acid treatment and amine treatment [29]. In the process, the CNTs are mostly treated with strong acids such as

HNO_3 and H_2SO_4, which creates carboxylic acid moieties onto the nanotube surfaces. The disadvantage of this process is that the strong acids are capable of breaking down the CNTs into smaller pieces [30]. Besides, silanization of the functionalized CNTs using organo-silane as coupling agent has gained popularity. A commonly used silane coupling agent is 3-aminopropyltriethoxysilane (APTES) and Fig. 6a, b shows the FTIR spectra of raw CNT and silanized CNT, respectively.

Whereas, Fig. 6b indicates chemical reaction between APTES and functionalized CNTs [31]. As shown in the FESEM images of unmodified-, oxidized-, and silane-modified CNTs in Fig. 7, it was reported to have improved the de-agglomeration of the CNT bundles as shown in FESEM images of the CNTs in Fig. 7 on silane functionalization of the CNTs. Hence, silane modification of CNTs

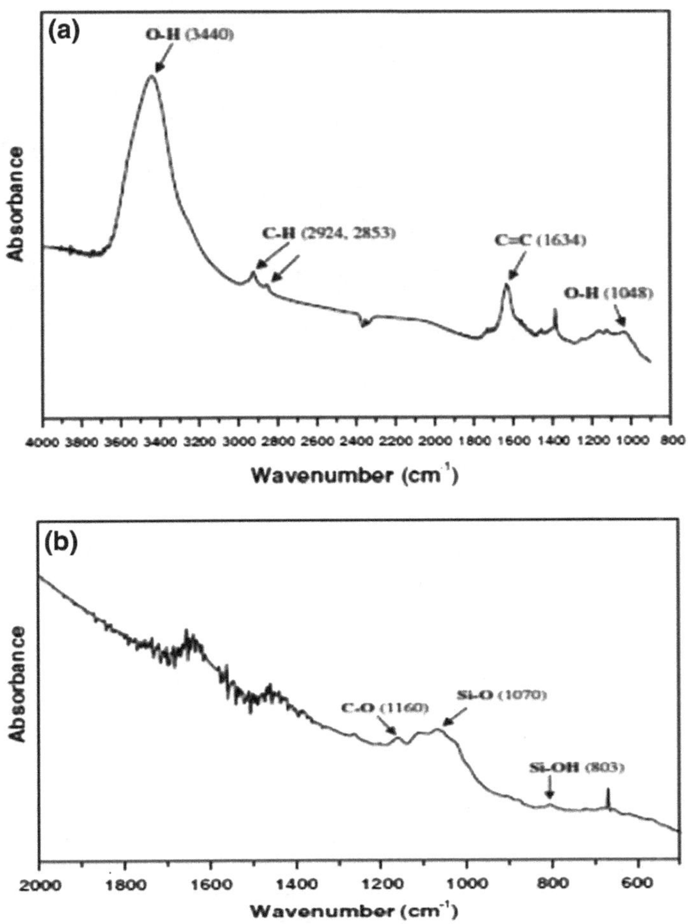

Fig. 6 **a** FTIR spectrum of raw multi-walled carbon nanotubes; **b** FTIR spectrum of silanized multi-walled carbon nanotubes [31]

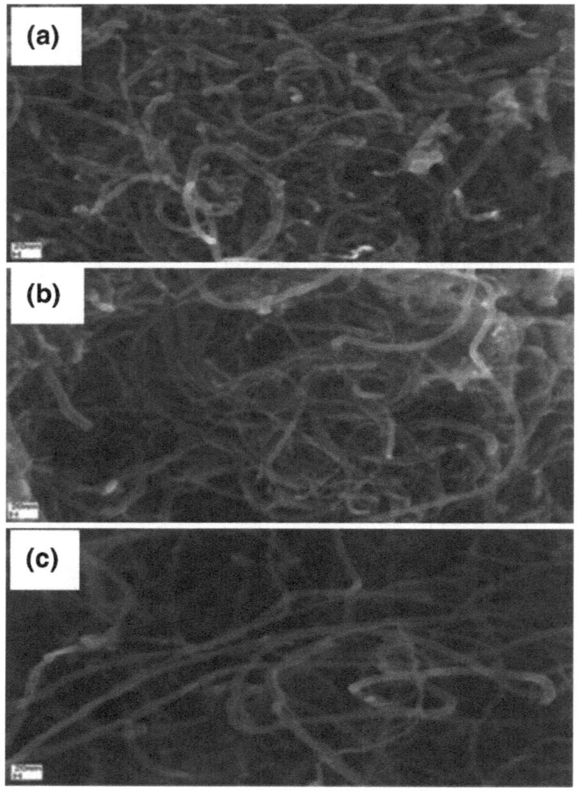

Fig. 7 FESEM images of multi-walled nanotubes **a** as received **b** oxidized, and **c** silanized with 3-aminopropyltriethoxysilane (APTES) [31]

is capable of improving the dispersion of CNTs within polymer matrices without causing any significant damage to its surface. Hence, silanization of CNTs are is of the promising techniques of improving its adhesion with organic polymer matrices.

There is several other organo-silanes available in the market that differs in their activity with respect to the process of modification followed. The second category includes the plasma treatment that permits fixation of various functional groups onto the surface of the CNTs. Plasma treatment of the CNTs is advantageous over the acid functionalization for its noncorrosive nature onto the surfaces. Thus, this kind of treatment almost retains the overall characteristics of the CNTs [30]. The XPS C 1 s spectra of raw CNT, acid functionalized CNT, and plasma-treated CNT are shown in Fig. 7a–c, respectively.

While the O 1s spectra of the all the mentioned three samples are shown in Fig. 8d–f, respectively. The typical peak in the raw CNT in Fig. 8a can be observed in 284.6 eV for the sp^2 hybridized C atom, while in case of acid functionalized CNTs in Fig. 8b, the peak at 286.2 eV corresponds to the C atoms that are bound to oxygen atoms of carboxylic acid of CNTs. Finally, the peak at 287.2 eV in plasma-treated CNTs as shown in Fig. 8c is due to the C atom that is bound to an oxygen atom by

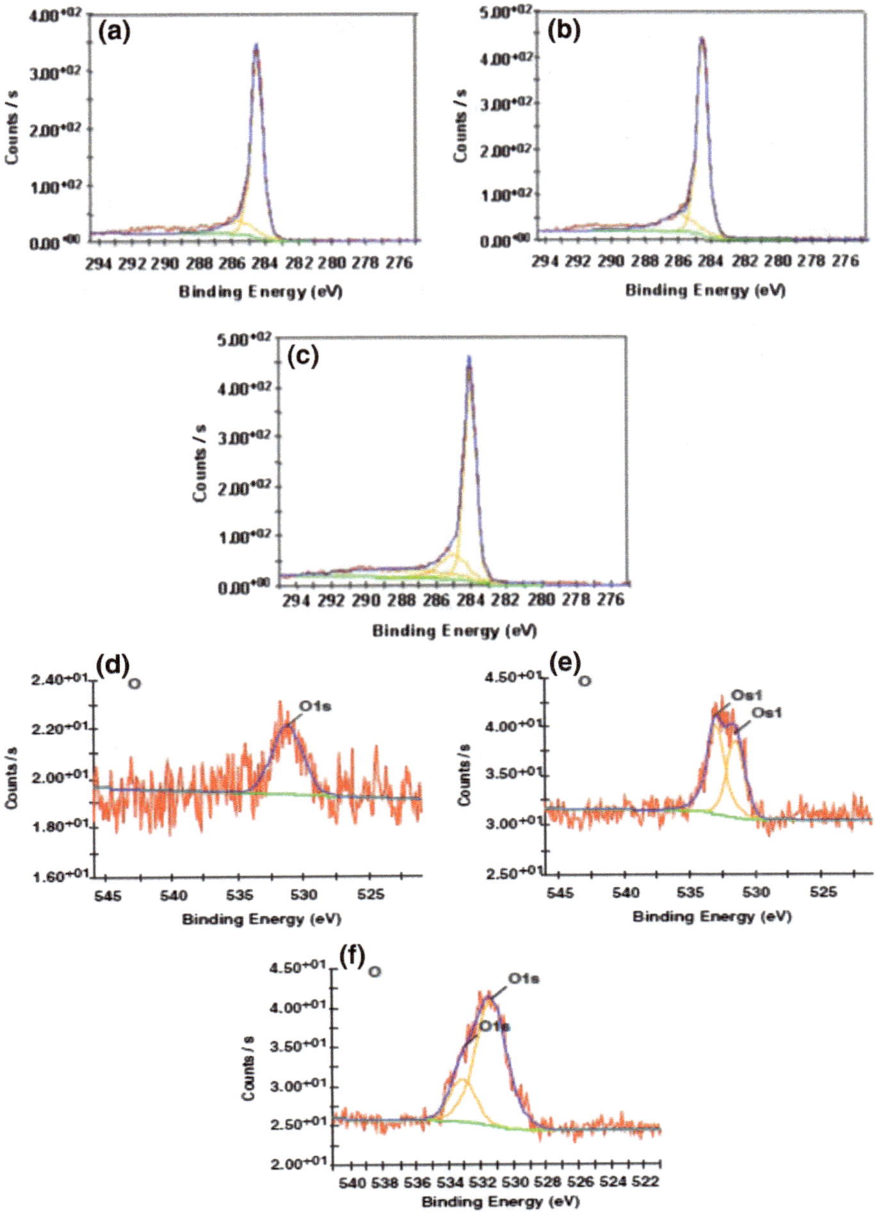

Fig. 8 C ls XPS spectra of **a** untreated CNTs, **b** acid-treated CNTs, **c** plasma-treated CNTs; and O 1s XPS spectra of **d** untreated CNTs, **e** acid-treated CNTs, **f** plasma-treated CNTs [32]

double bond due to the influence of plasma. Similarly, in Fig. 8d, there is no O 1s peak for raw CNT, while the two peaks at 531.7 and 533.0 eV in Fig. 8e of acid-treated CNTs are due to C–OH and C–OOH bonds developed in the nanotube surface. The peak for plasma-treated CNTs in Fig. 8f is at 287.5 eV that indicates the characteristic O 1s peak of carbonyl group (C=O) [32]. From the XPS spectra of all CNT samples, the introduction of oxygenated functional groups was established. Hence, the spectroscopic study of the raw and modified CNTs is important characterization technique to understand the degree and nature of modification in the surface of nanotubes. Also, analysis of FTIR spectra of the treated CNTs can provide detailed information about the type of functional groups attached during the modification process. As for examples, oxidation of CNTs by ultraviolet or ozone treatment (UV/O_3), reduction with lithium aluminum hydride ($LiAlH_4$), or silanization with the organo-silane coupling agent named 3-glycidoxypropyltrimethoxy silane (GPTMS) can be qualitatively characterized by FTIR analysis. For better understanding, Fig. 9 shown here is from the work reported by Ma et al., where the FTIR spectra of all the raw and modified CNTs are shown in a sequence [33].

The presence of two peaks at 1715 and 1160 cm^{-1} are attributed to the C=O and C–O stretching vibration of carboxylic acid. Whereas, the disappearance of the peak at 1715 cm^{-1} indicates reduction of the –COOH group into –OH group. Finally, in the case of silane functionalized, CNTs have shown broader band at 3410 cm^{-1} and also the band at 1384 cm^{-1} was disappeared, while the appearance of two new peaks at 2914 and 2848 cm^{-1} refers to the methylene groups from the GPTMS molecules. Similarly, both ^1H and ^{13}C NMR study also provides useful information regarding functionalization of CNTs. Hence, spectroscopic studies are the important characterization techniques to understand the

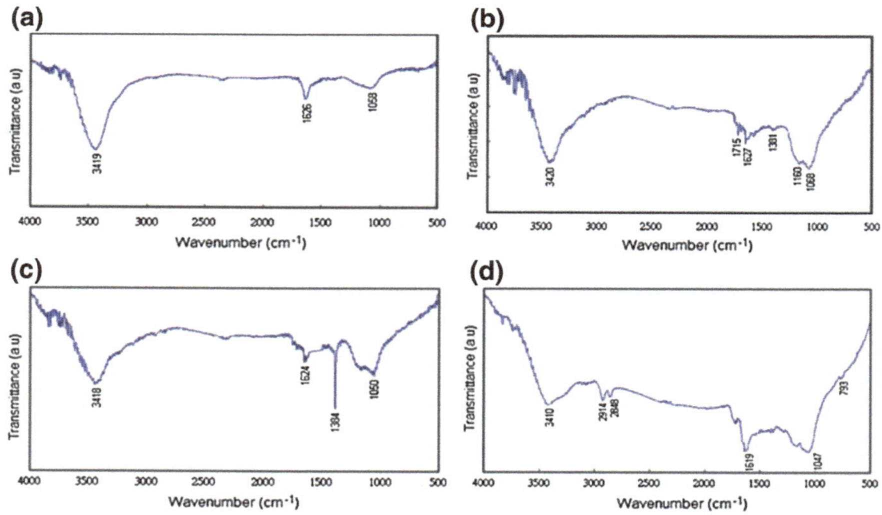

Fig. 9 FTIR spectra of the MWCNTs **a** Pristine, **b** UV/O3-treated, **c** Reduced, **d** Silanized [33]

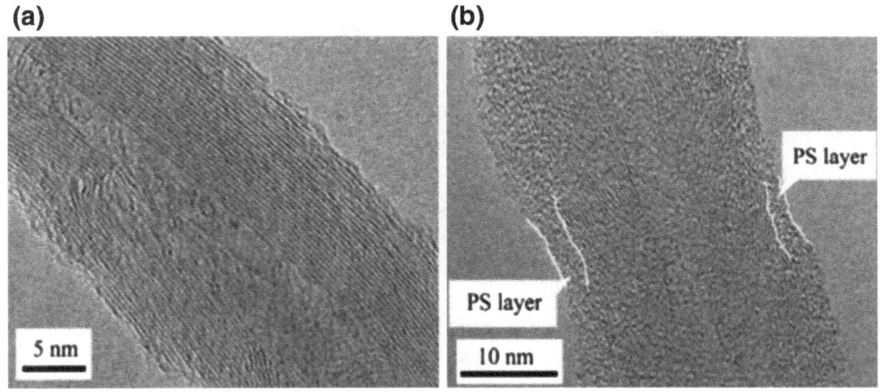

Fig. 10 TEM images of **a** purified and **b** PS-modified MWNTs [34]

modification of various nanofillers including CNTs. Furthermore, these techniques are also widely used for the characterization of polymer nanocomposites of CNTs, which includes some of the common morphological studies through FESEM, TEM and XRD analysis. Currently, the most popular technique to modify the surface of the CNTs by means of grafting of various polymers is through in situ polymerization technique. A large variety of polymers are been physically or chemically grafted or coated onto the surface of CNTs. Figure 10a, b refers to TEM images of neat CNT and PS modified CNT, wherein Fig. 10b, a clear layer of polystyrene (PS) over the surface of CNT can be noticed.

In the work, the CNTs are covalently modified with PS through polymerization of styrene monomer under microwave irradiation. It is universally known that the morphology of nanofillers within the polymer matrix affect overall properties of polymer nanocomposites. Functionalization or modification of CNTs can alter the surface and structural morphology of the nanotubes. Generally, organic polymers barely interact with the nonreactive CNTs, while functionalization of its surface can improve polymer–filler interfacial adhesion. Besides, surface modification of CNTs can also improve the overall degree of dispersion of the nanotubes within the polymer matrix, which further depends upon the type of polymer used. It is obvious that the morphology of the CNTs within polymer matrices dictate the final properties of the nanocomposites like electrical, thermal, mechanical, and other physical properties. In the subsequent paragraph, a more detailed and scientific discussion will be made on the effect of morphology on various properties of nanocomposites.

Electrical properties of CNT-based polymer nanocomposites is excellent and the conductive mechanism is mainly by the electron transport through tunneling or electron hopping that occurs along the interconnected CNT networks [34–36]. The high aspect ratio of the CNTs is an important factor contributing to the electrical properties of the polymer nanocomposites. CNTs in polymer nanocomposites are capable of exhibiting very low percolation threshold for electrical conductivity due to its nano-dimension and high aspect ratio. Presence of well-aligned CNTs in polymer nanocomposites shows much higher electrical conductivity along the direction of the alignment rather its cross section. Furthermore, proper alignment of the CNTs in the polymer matrix enhances the electrical conductivity of the polymer nanocomposites to the maximum extent. Nevertheless, the process of aligning CNTs in the viscous polymer matrix is a much tough and tedious job to accomplish due to its high aspect ratio and flexibility. The incorporation of tiny quantity of CNT nanofillers into polymer matrices can improve the electrical properties to several times higher, which is due to the formation of an extra electrically conductive network structure [37]. However, higher concentration of CNTs in polymer matrices has detrimental effect on the electrical properties of nanocomposites which is due to the formation of aggregation [38]. In other words, the electrical percolation threshold of CNTs in polymer matrices is dependent upon aspect ratio, dispersion and alignment of the nanotubes within the polymer matrix. Uniform and well dispersion of nanotubes generally have high aspect ratio, thereby resulting in lower percolation threshold in the nanocomposites [39]. Liang and Tjong developed nanocomposites of LDPE/CNT with only 1.1 vol% of CNT, which they reported to have high electrical properties, dielectric constant in special with minimum agglomeration and excellent state of nanofillers dispersion [40]. Figure 11 shows the SEM images of the fractured surface of uniformly dispersed CNTs which are the white dots well distribute within the LDPE matrix.

A large number of researches have already been published with the focus on improving the electrical properties of flexible insulating polymeric material for special purpose applications like light-emitting diode (LED) applications, electromagnetic interference (EMI) shielding in electronic equipment and sensors, etc. Jung et al. have fabricated an extremely flexible hybrid composite structure of multiwall carbon nanotube (MWCNT) arrays in poly(dimethylsiloxane) (PDMS) matrix using lithographic patterning technique [41]. These flexible nanotube composites are sufficiently robust against high stress retaining its conducting behavior and exhibit unique electromechanical properties. The PDMS/MWCNT composite has the potential to be utilized directly in the form of flexible field emission devices and as gas sensors.

Fig. 11 SEM micrographs of LDPE/1. 9 vol% CNT nanocomposites [40]

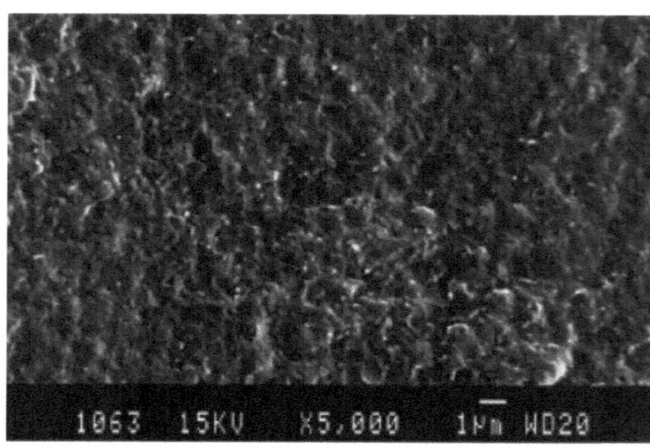

Though the mechanical property of CNTs is appreciably higher than polymeric materials, but many times, it becomes a tough process to transfer its high mechanical properties into polymer matrices. Poor polymer–filler interfacial interaction is the main factor to act on this problem, and it is extremely necessary to overcome this difficulty. Nevertheless, the researchers are capable to improve the interfacial bonding between polymer and CNTs through following various techniques like covalent or non-covalent surface modification. Modification of CNTs surfaces with amine treatment followed by plasma oxidation to improve the surface functionality helps improving the interfacial bonding with epoxy resins, thereby improves the overall mechanical properties of the resulting epoxy/CNT nanocomposites [42]. Such modification in the surface of CNTs is turned out to be highly effective in improving mechanical properties of the polymer matrix. In this case of epoxy matrix, the plasma treatment has outperformed the acid and amine treatment and the subsequent nanocomposite showed highest increment in ultimate tensile strength from 26 to 58 MPa [32]. The excellent state of dispersion of the CNTs within the epoxy polymer matrix along with strong polymer–filler interaction are the main reasons behind the stupendous property improvement in epoxy matrix, which can be understood from the FESEM images of the fracture surface of epoxy composites containing 1 wt% of plasma-treated CNTs in Fig. 12.

Similarly, acid treatment of MWCNT leads to the formation of large number of functional groups by creating defects onto its graphitic surface, and mostly produce carboxylic groups which act as the most reactive location with polyamide 6 (PA 6) [43]. In such case, simple melt compounding can even result in uniform dispersion of the CNTs within the PA 6 matrix. A direct influence of such morphology can be observed from its mechanical properties, where inclusion of 2 wt% of MWCNT improved the elastic modulus of PA 6 by 214% and yield strength by 162% [44]. Similarly, acid-treated CNTs bearing large number of functional groups are able to improve the mechanical properties of high-density polyethylene (HDPE) nanocomposites. The CNT/HDPE nanocomposites with increasing concentration (vol%) of nanotubes from 0.11, 0.22, 0.33 to 0.44% lead to improvement in Young's modulus from 6.7 to 100%; and strain at fracture from 863 to 1069% [45]. Hence, researchers are continuously trying to elevate and transfer the beneficial

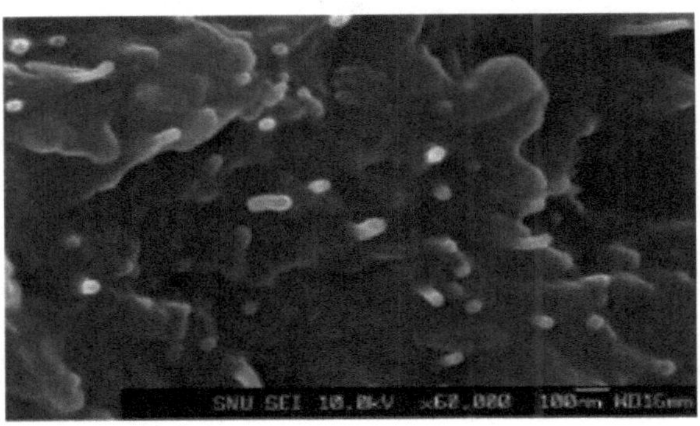

Fig. 12 FESEM images of the fracture surface in composites containing 1 wt% of plasma-treated CNTs [32]

properties of CNTs into polymer matrices. In nanocomposites good adhesion between CNTs and polymer matrix are key to the effective transfer of applied load or stress from the weak matrix phase to strong CNT fillers. Besides, aggregation or agglomeration of the CNTs in the polymer matrix can lead to serious deterioration in mechanical properties, as these are the prime points of stress concentration during the testing. Besides, the well-separated CNTs have maximum available surface area that can be utilized in the formation of bonds with polymer matrices.

Like mechanical properties, the thermal properties of polymer matrices are also influenced by the presence of CNTs in different forms and in this case morphology plays a vital role in dictating the property. The thermal conductivity of polymers is also affected by the incorporation of CNTs. The thermal conductivity of polymeric material is considerably low; and incorporation of conducting CNTs is efficient enough to improve the property. Furthermore, a well-aligned morphology of the CNTs in the polymer matrix further improves the thermal conductivity to a much greater extent along the direction of alignment. Figure 13a is the SEM images of CNT/silicon rubber nanocomposites, while Fig. 13b shows increasing thermal conductivity of silicon rubber nanocomposites along the direction of the aligned CNTs [46].

Besides, CNTs are also reported to improve thermal degradation stability of certain polymers, which is considered to have similar flame retardancy like clay particles [47]. Similarly, the poly(ethylene vinyl acetate) copolymer (EVA)/MWCNT also showed high flame resistance property that surpasses the EVA/clay nanocomposites [48]. Also, the thermal stability of polyurethane (PU) was improved by the addition of minor concentration of MWCNTs that is expected to arise not only from the higher thermal stability of nanotubes, but also from the strong polymer–filler interaction in the nanocomposites system [49]. The mechanism of improving thermal degradation stability of the base polymers can be attributed the MWCNTs, where the nanotubes are capable of reducing the heat release rate of the base polymer matrix significantly. The MWCNT network enhances the mechanical integrity of a thermally insulating layer that also acts as a barrier to the degraded products to the gaseous phase [50]. Besides, the presence of minor amount of iron in the PP/MWCNT nanocomposites has direct influences on the flame retardancy efficiency. During the combustion process, the iron particles convert into iron oxides, it can act as flame retardant additive in certain polymeric systems.

3.4 Polymer/Fullerene Composites

Fullerenes

In the year 1985, Kroto et al. discovered Buckminsterfullerene C_{60} in Rice University [51]. Structurally, a single C_{60} fullerene molecule consists of 12 pentagons and 20 hexagons of sp^2 hybridized carbon atoms. There are several other forms of fullerenes that consist of more number of carbon atoms like C_{70} and its

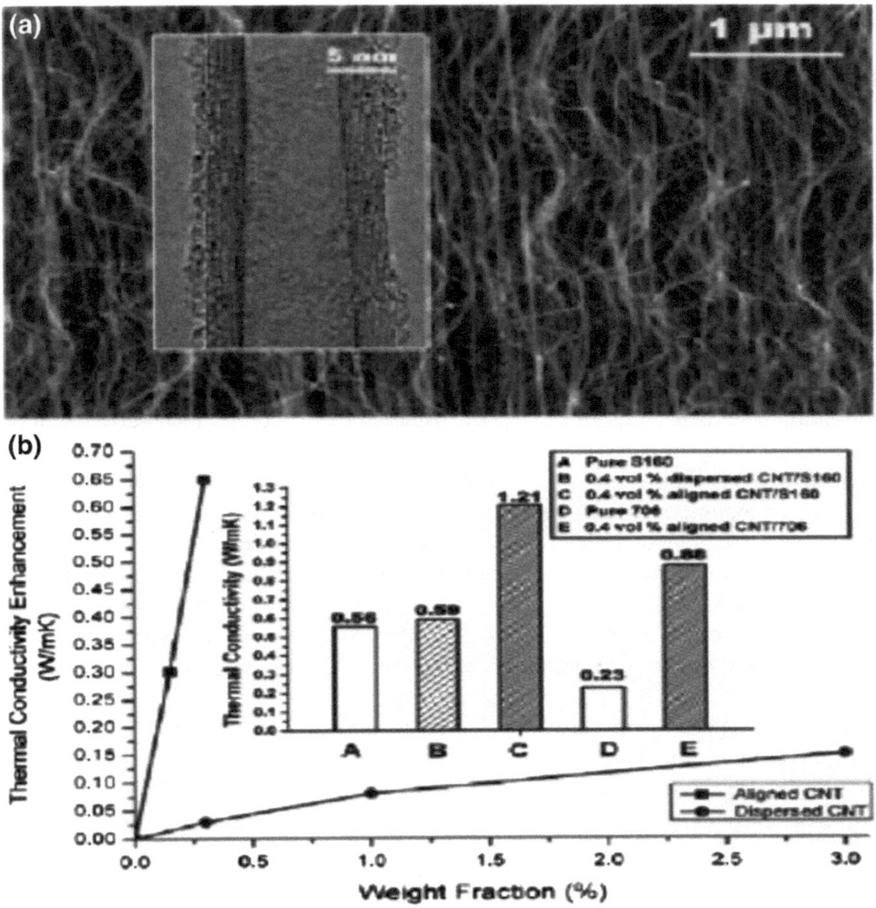

Fig. 13 **a** SEM image of the side view of aligned MWCNT in silicon rubber matrix (inset: a high-resolution TEM image of an isolated MWNT showing eight nested nanotubes) **b** Thermal conductivity enhancement (composite—matrix) versus weight fraction of MWNT in both aligned and unaligned nanocomposites (inset: thermal conductivity for selected MWNT/silicone elastomer composites) [46]

other homologs like C_{76}, C_{78}, C_{80}, C_{82}, C_{84}, etc. The stability of the fullerene homologs increases with decreasing number of carbon atoms.

Fullerene Composites

Introduction of nanoparticles into polymer matrices generally leads to many changes on material properties, which is difficult to comprehend by extension of macroscopic fillers. Well-dispersed nanoparticles can dramatically enhance the material stiffness, optical, gas barrier properties, electrical, thermal, and fire-retardant properties of composites. Such a unique range of properties of polymer nanocomposites is usually rationalized in terms of the large surface area-to-volume ratio of the nanoparticles. The unusual hollow ball-like structure of fullerene molecules has tremendous influence on optical and transport properties of its polymer composites.

Fullerene and its derivatives show a wide range of applications in thin film and photovoltaic applications [51]. The unique structure of fullerenes in its composites with polymer develops unique electronic structure and subsequent properties. Furthermore, its easy accessibility to doping and charge transfer results customized control on the electronic properties of composites. Therefore, polymer/fullerene composites have attracted more attention for electronic applications. Alejandro Sanz et al. reported in-depth study in structure–property relationship of polymer composites of the glass forming polymer polystyrene (PS) with fullerene [52]. Extensive work was carried out to understand the impact of fullerene on the glass formation of PS in terms of degree of dispersion of fullerene nanoparticles, morphology and spectroscopy. Morphology of the PS-C_{60} nanocomposites was investigated to find the type of nanoparticle dispersion in PS matrix. The spun-cast films of nanocomposites with a C_{60} concentration of around 4–5% exhibited uniform and homogeneous state of dispersion of nanoparticles within the matrix. However, within minutes after annealing at 180 °C, the fullerene nanoparticles at concentration >1% C_{60} started to cluster and develop nucleation. At sufficiently high fullerene concentration (4–5%), spinodal-like patterns were emerged, which was speculated due to the correlated nucleation of C_{60} as represented in the optical micrographs in Fig. 14.

Dielectric spectroscopy provides further insight into the influence of fullerenes (C_{60}) onto the segmental dynamic properties of PS as in Fig. 15. Due to very small β-relaxation of PS, more focus was made in the α-relaxation. The maxima observed in the dielectric loss data as a function of frequency corresponds to the relaxation. Addition of fullerene to PS resulted shifting of the maximum peak towards lower frequencies, indicating slowing down of the PS segmental motions.

The effect of fullerene agglomeration on the dielectric response, the in situ thermal annealing of the nanocomposites was carried out at 175 °C for 60 min that is shown in Figs. 15 and 16, respectively. A shifting of the peak maxima toward higher frequencies after annealing is the indication of reduced restriction of polymer chain, which is due to lowering of the overall surface-to-volume ratio of fillers in the nanocomposites.

The application spectrum of fullerene in polymer matrices is quite broad. While the water-insoluble fullerene can be made soluble in water via developing nanocomposites of certain water-soluble polymers like polyvinyl alcohol (PVOH), PVP, PEO, PEG, etc. Incorporation of fullerene C_{60} into a complex like PVP changes its electronic structure, as polymer affects its p-electron system, which can be understood from the UV spectroscopy [53]. Such complexation is also accompanied by local changes in the chain conformation, i.e., the formation of a loop around a C_{60} molecule. Though the loop formation does not lead to any conformational changes in a polymer chain as a whole but the authors have reported that the overall electron density transfer onto a fullerene molecule to be higher than 8%

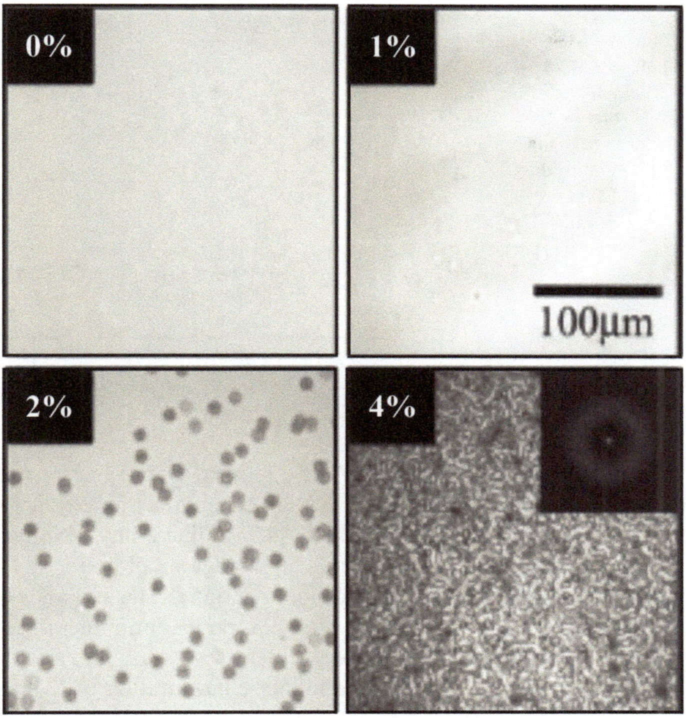

Fig. 14 Optical micrographs of PS and PS-C_{60} thin films after annealing at 180 °C. The inset shows the FFT of the demixed morphology, which occurs between 1 and 2 wt% C_{60} [52]

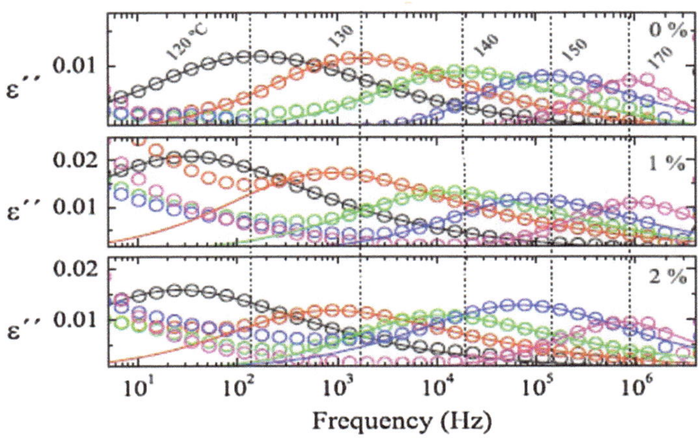

Fig. 15 Dielectric loss spectra of neat PS270k, PS-C_{60} 1 and 2% as a function of temperature showing the relaxation [52]

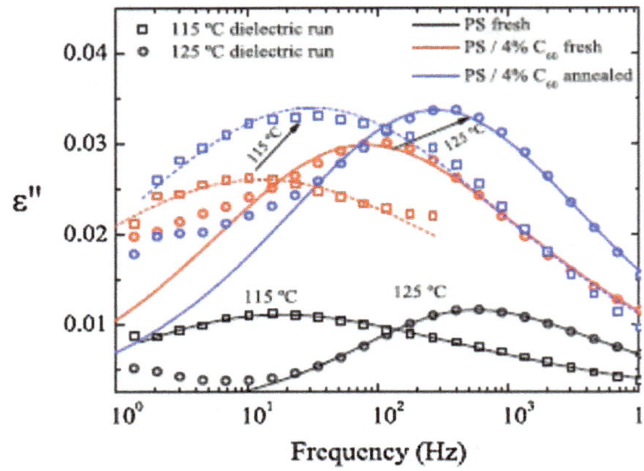

Fig. 16 Selected dielectric loss spectra obtained via isothermal frequency sweep (T 1/4 115 °C (,) and 125 °C (B)) for PS270k (black) and PS-C_{60} 4% before (red) and after (blue) thermal annealing (175 °C, 60 min) [52]

which is expected due to the charge transfer complex. Doping raw natural rubber with 0.065–0.75 wt% fullerene C_{60} is characterized by a higher elastic modulus at different tensile strains and a higher hardness [54]. Adding a tiny quantity of 2 wt% of C_{60} in an epoxy matrix has led to a significant hardening of polymer–filler bonds in the system of the carbon fiber–epoxy polymer [55]. The polymer PMMA is known for its valuable properties such as strength, optical brightness, transparency and stability to photoaging. Incorporation of even 1 wt% of fullerene has higher effect in improving its properties than the higher concentration of C_{60} [56].

From the spectroscopic studies of SAXS and WAXS in Figs. 17 and 18, it was understood that at low fullerene concentration up to 5 wt%, there is no agglomeration; rather a homogeneous dispersion was found in the composite. However, at 10 wt% of C_{60}, some inhomogeneity started appearing that can be understood from the increased intensity of the primary beam in SAXS result in Fig. 17. Evidence for similar observation was also found from the appearance of two sharp halos (or diffuse reflections), which is indicative of the nucleation of a crystalline fullerene phase. Similar improvement in mechanical properties of low-density polyethylene (LDPE) was observed with the addition of very small concentration of fullerene

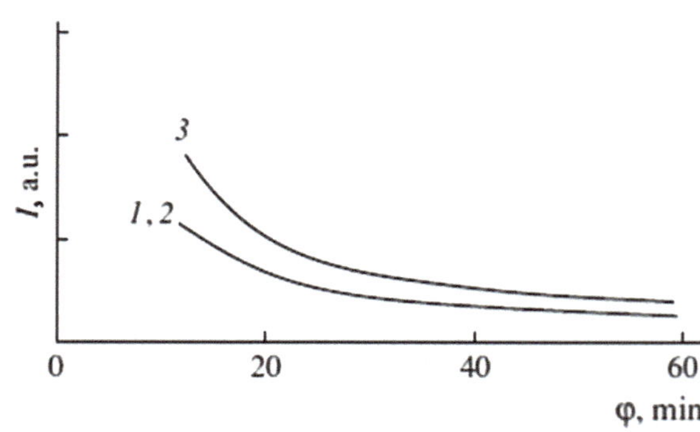

Fig. 17 SAXS patterns of PMMA films with various concentrations of C_{60} fullerene: (*1*) 0 (pure PMMA); (2) 1 and 5% C_{60}; (*3*) 10% C_{60} [56]

Fig. 18 WAXS patterns of PMMA films with various concentrations of C_{60} fullerene: (*1*) 0 (pure PMMA); (*2*) 1 and 5% C60; (*3*) 10% C_{60} [56]

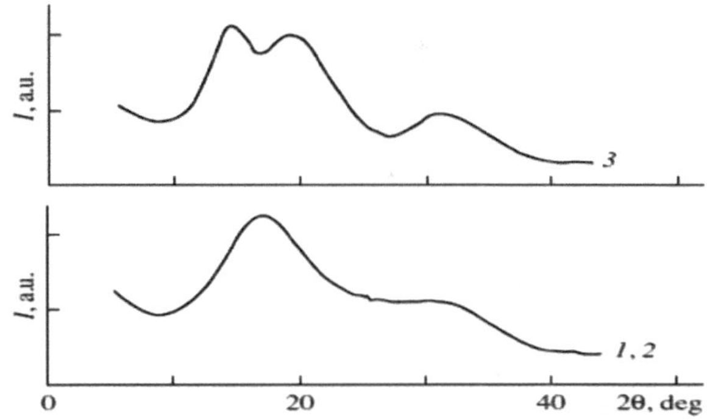

(C_{60}) [57]. Solution-processed polymer: fullerene thin films were developed for use as photovoltaic cells. The efficiency of the fullerene nanocomposites in photovoltaic cell is largely determined by the nanoscopic and mesoscopic morphology formed during solvent evaporation of these blends [58].

Fullerenes are ubiquitous electron acceptor due to its high electron affinity and its ability for charge transport effectively. Chemical modification of fullerenes can alter their properties and device performance, which is generally helpful in developing new electronically active components. Prato reaction is a simple pathway to modify fullerenes; and various fullerene derivatives were already been synthesized and characterized for organic transistors and solar cells [57]. Pyrrolidinofullerenes synthesized with triphenylamine or cyanobenzene via Prato reaction for a photosensitizer in a photoconductive polymer composite was reported [59, 60].

Fullerene/polymer composites are bi-continuous systems effective for solar cell applications. The low-bandgap polymers discussed are the polymers that can lead to materials with better absorption solar radiation, while retaining high voltages and charge carrier mobility [60]. Over the last 12 years since the discovery of the polymer–fullerene bulk heterojunction (BHJ) solar cells, dramatic improvements in the fundamental understanding, device construction, and processing of the conducting active layer been achieved by several research groups using blends of poly (3-hexylthiophene) (P3HT) and the fullerene derivative [6,6]-phenyl-C_{61}-butyric acid methyl ester (PCBM). The BHJ solar cell made of the MDMO-PPV/PCBM couple, it was observed that changing the solvent to chlorobenzene led to improvement in the efficiency by 2.5% [60, 61]. A detailed spectroscopic study proved that the difference was mainly due to morphological changes. The micrometer-sized PCBM cluster morphology of toluene-cast films was embedded in a polymer "skin." Such arrangement of blend was non-bi-continuous, in which fullerene photoluminescence is not fully quenched by the polymer. This clearly indicates the macroscale phase segregation in the system. However, chlorobenzene-cast films, PCBM clusters with sizes less than 100 nm are observed in a significantly more homogenous and bicontinuous composite. The above illustration clearly shows the critical role of choice of solvent for generating homogenous solutions. Inherent miscibility between the two components is another decisive factor in determining the film morphology that shows composition dependent behavior [62].

In other words, several work evidences the ability of fullerene nanoparticles to reinforce polymer matrices, and also improve thermal and electrical conductivity useful for several applications at the lowest concentration. Also, the fullerenes have significant influence in improving thermal properties of polymers like thermal stability, glass transition temperature (T_g), etc.

3.5 Polymer/(Nano) Diamond Composites

Diamond or Nanodiamonds

Diamonds are naturally available toughest allotropes of carbon that also can synthetically be prepared. There are two forms of such synthetically prepared diamond powders: nanodiamonds and microdiamonds, which can be used as filler in polymer matrices. Detonation nanodiamonds (DNDs) are synthetically prepared another form of low-cost nanodiamonds that are mostly used in the preparation of thermally conductive polymer composites [63]. Microdiamonds prepared by the application of high pressure and temperature are also capable of improving the thermal conductivity of its polymer nanocomposites [64, 65]. Hence, the use of nano and microdiamonds into polymer matrices has attained a massive interest in the recent times to improve thermal conductivity. The inclusion of diamond powders at high volume percentage into epoxy resins, polyethylene (PE) and polypropylene (PP) leads to significant improvement in the thermal conductivity of the resulting composites [66–68]. All these mentioned composites are prepared by conventional solvent casting technique, which is of much time-consuming, tedious, and are not amenable for fabrication of structures with complex geometry. However, the 3D printing is another frequently used technique to fabricate intricate geometrical structures that would be essential for applications for rapid prototyping and microelectronics.

Polymer/Diamond or Nanodiamond Composites

Diamond is another important allotrope of carbon that is one of the most inert forms in the Earth available. Its importance also comes into focus with the virtue of being in the list of the strongest material with superior hardness and wear resistance [69]. Due to its high inert nature, it is very difficult to carry out chemical modification of diamonds under common conditions. Hence, unlike other carbon-based fillers, diamonds need vigorous conditions to functionalize. Nevertheless, several research groups in the past were capable of successfully functionalizing diamond surfaces through rigorous oxidation, bromination, fluorination, UV irradiation, and Diels–Alder reaction [70–74]. Transfer of the advanced properties of diamond to lightweight polymer matrices lies on the efficacy of the polymer–filler adhesion in the composites. The initial work on polymer/diamond nanocomposites was pioneered by Dolmatov, where substantial improvement in tensile strength and modulus values of elastomers was reported [75]. Recently, Zhang et al. and Shenderova et al. in 2007 reported to have developed nanocomposites of poly(amic acid) (PAA)/ nanodiamond and polyimide/nanodiamond with improved thermal stability and mechanical properties [76, 77]. Commercially, nanodiamonds can be developed by high temperature and high-pressure detonation followed by oxidation process that leads to attachment of carboxylic (–COOH) functional groups onto its surface. Refluxing these nanodiamonds with thionyl chloride leads to the formation of acyl chloride functional group onto its surface, which helps improving polymer–filler interfacial interaction.

The functionalization of nanodiamonds with acyl chloride can be affirmed by EDAX analysis as referred to the Cl Kα peak observed at 2.65 keV as shown in Fig. 19a and the FTIR analysis in Fig. 19b. A better state of dispersion of acyl functionalized nanodiamond was observed in the polyimide matrix over the non-functionalized nanodiamonds. The inclusion of functionalized nanodiamonds even at loading of 1 wt% to polyimide polymer matrix results in tremendous increment in the mechanical properties like hardness that was found to be much less in case of non-functionalized nanodiamonds. These results infer that functionalization of

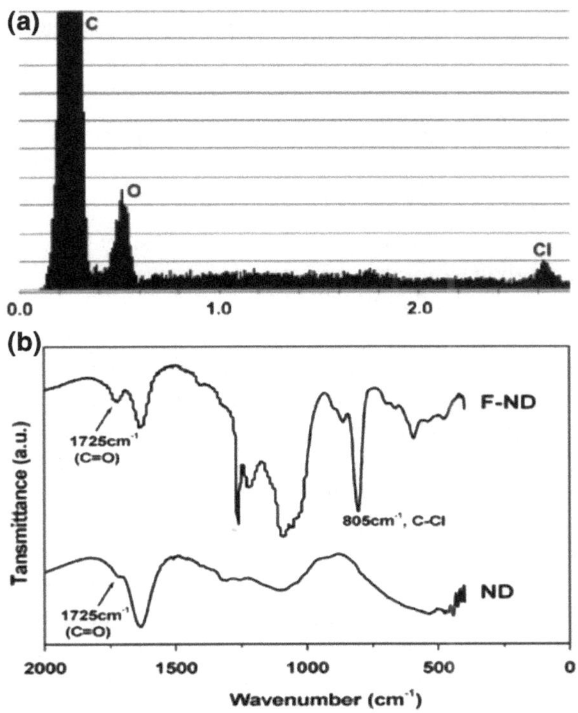

Fig. 19 **a** EDAX of acyl chloride functionalized nanodiamonds and **b** FTIR spectra of acyl chloride functionalized nanodiamonds and untreated nanodiamonds [78]

nanodiamonds with acyl chloride is effective enough to improve polymer/filler interfacial interaction in polyimide matrix [78]. Similarly, the inclusion of nanodiamonds in 2 wt% of loading along with onion-like carbon (OLC) are reported to be quite effective in improving the thermal conductivity of polydimethylsiloxane (PDMS) up to 15% and loss tangent of both PDMS and polyurethane composites. Further, the thermal conductivity, thermal stability and adhesion with polyimide and poly(methyl methacrylate) (PMMA) composites was found to be improved on incorporation of nanodiamonds into these polymer matrices [77]. Recently, Kalsoom et al. have developed thermally conducting 3D printed prototype of diamond microparticle/acrylate polymer composites with a variation of filler loading in 10, 20, 25 and 30% (w/v) using a low-cost stereolithographic 3D printer [79]. A significant improvement in heat transfer of nanocomposites with 30% (w/v) diamond microparticles was observed, which is mainly be attributed to the interconnected diamond aggregates networks at higher loading of diamond microparticles.

This can clearly be observed in the scanning electron microscope (SEM) images in Fig. 20 of all the composites with different diamond microparticle loadings [79]. In Fig. 20, low loading of diamond microparticles are well separated from each other (Fig. 20a–c). On the other hand, at higher loading of filler, the number of contact points between diamond microparticles that are interconnected with each other are much higher in numbers (Fig. 20d–f). Such highly connected network of diamond microparticles leads to substantial improvement in the heat transfer efficiency of the composite with 30% w/v of the diamond microparticles.

3.6 *Polymer/Graphite or Graphene Composites*

Graphite or Graphene

Graphene, the most versatile two-dimensional allotrope of carbon has a single layer of carbon atoms that are arranged in a honeycomb lattice. Graphene is mentioned as the mother of different carbon allotropes like the 0D fullerene, the 1D carbon nanotube, and the 3D graphite and carbon blacks, etc. Though graphene has been known to the world much back from the time of 1947, but for the first time, Geim and Novoselov have demonstrated for successful isolation of graphene from graphite [80]. Later, this discovery has led them to win the 2010 Nobel Prize in the field of Physics. Graphene is mostly regarded as the "thinnest material in the universe" that has enormous potential in several applications. Due to some of its inherent remarkable properties like high aspect ratio, high surface area, high mechanical property, excellent electrical and thermal conductivity, electromagnetic interference (EMI) shielding ability, excellent flexibility and transparency, and low coefficient of thermal expansion (CTE), it has generated high potential in the field of polymer nanocomposites. As a nanofiller, graphene has proven to be the preferred choice over several other conventional white and black fillers like montmorillonites (MMTs), layered double hydroxide (LDH), carbon nanotubes (CNTs), carbon nanofiber (CNF), etc. [81, 82].

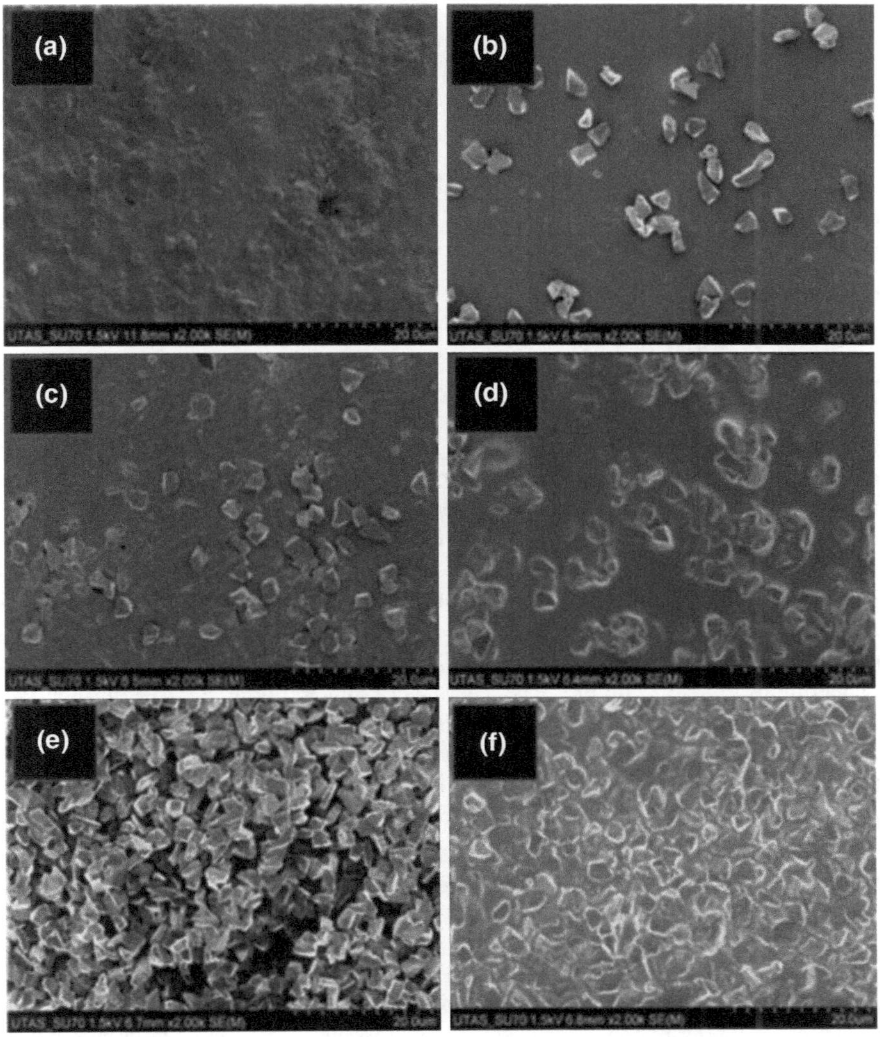

Fig. 20 SEM images of **a** Acrylate polymer. **b** Acrylate polymer diamond-10% (w/v). **c** Acrylate polymer diamond-20%. **d** Acrylate polymer diamond-25%. and **e–f** Acrylate polymer diamond-30% composite materials [79]

Graphene, as a nanofiller, may be preferred over other conventional nanofillers (Na-MMT, LDH, CNT, CNF, EG, etc.) owing to high surface area, aspect ratio, tensile strength (TS), thermal conductivity and electrical conductivity, EMI shielding ability, flexibility, transparency, and low CTE. Table 1 shows various properties of graphene and different types of polymers (plastics and rubbers), which can provide a clearer picture about their properties.

Table 1 Mechanical, thermal, and electrical properties of various materials along with graphene

Materials	Tensile strength	Thermal conductivity at room temperature (W/mk)	Electrical conductivity (S/m)	References
Graphene	130 ± 10 GPa	$(4.84 \pm 0.44) \times 10^3$ to $(5.30 \pm 0.48) \times 10^3$	7200	[89, 104, 105]
High-density polyethylene (HDPE)	18–20 MPa	0.46–0.52	Insulator	[106, 107]
Kevlar®	3620 MPa	0.04	Insulator	[108]
Natural rubber (NR)	20–30 MPa	0.13–0.142	Insulator	[109]

From Table 1, it can be well understood that graphene possesses much higher mechanical, thermal and electrical properties over the rest of the counterparts mentioned in the table. Subsequently, the superior property of graphene also gets reflected in its polymer nanocomposites. Hence, graphene/polymer nanocomposites show superior mechanical, thermal, and electrical properties as compared to neat polymer [83–85]. Moreover, it is also observed that graphene/polymer nanocomposites show higher improvement in mechanical, thermal and electrical properties over other nanofiller like clay, CNT, carbon black, and CNF-based polymer nanocomposites [86]. In this case, it is very much important to consider two important factors that can dictate the behavior of final nanocomposites. First, the degree of dispersion of the graphene nanosheets in the polymer matrix, and second, the polymer–filler interfacial interaction between the graphene nanosheets and the polymer matrix. While, uniform dispersion of graphene nanosheet in polymer matrix increases the mechanical as well as thermal and electrical properties; a good interfacial bonding between the graphene nanosheet and the host polymer finalizes the optimal property of the graphene reinforced polymer nanocomposites [87]. A defect-free graphene nanosheet has much lesser compatibility with organic polymeric matrices, which is the main reason behind less number of reported defect-free graphene/polymer nanocomposites in the literature. However, heavy oxidation of the graphene nanosheets can lead to the formation of large number of defects that bears large number of hydroxyl, ketone, alcohol, epoxide and carboxyl functional groups. Such oxygenated graphenes are commonly known as graphene oxides (GOs), which can significantly alter the intermolecular interaction with hydrophilic polymer matrix, and result in more compatibility with it [88]. However, most of the polymers being hydrophobic are incompatible with GO. In that case, such GO are organically modified by covalent grafting technique. However, the surface modification of graphene with different organic molecules is most trending because of its effective adherence and interaction with hydrophobic polymers, while maintaining defect-free structure. In that case, the surface modified graphene nanosheets retain its excellent thermal and electrical conductivity that is very specific to certain applications.

Graphite or Graphene Composites

In the recent years, the development of effective nano level dispersion of graphene nanofillers have opened up an interesting facet in the field of material science. Graphite in different forms like, expanded graphite, exfoliated graphite, etc., are incorporated in various polymer matrices with the purpose of improving various physico-mechanical properties of the composites. It is well known to the whole scientific community that converting graphite into defect-free graphene is a tedious job to accomplish. This is the reason why more and more researchers are coming up to develop new ideas of inculcating the costly graphene nanosheets into polymers matrices in a more uniform manner along with high degree of interaction with the matrix. In this section of the chapter, the author will provide detailed up-to-date discussion on the recent advances in graphene/polymer nanocomposites, which by default also includes information on graphite/polymer nanocomposites as well.

Incorporation of modified graphite-like expanded graphite (ExG) or exfoliated graphite (EG) into polymer matrices many times improves the degree of dispersion of the graphene nanosheets within the polymer matrix. Graphene/polymer nanocomposites can be prepared by in situ polymerization of monomers into polymers. In such process, the monomers are first allowed to get inside the inter-spacing gallery between two consecutive graphene nanosheets, which is followed by its polymerization. The polymerization leads to separation of the nanosheets apart from each other, thereby results in the formation of graphene-based polymer nanocomposites. Some of the examples of such nanocomposites are poly(vinyl alcohol) (PVA), poly(methyl methacrylate) (PMMA), epoxy resin, silicon foam, and polyurethane (PU) nanocomposites [84, 89–93]. It is important to mention that all the cases, the polymerization was carried out in solvent state due to the high viscosity of the graphenes. Furthermore, the graphene-based polymer nanocomposites were also been fabricated by grafting the graphenes nanosheets onto the reactive functional groups of the base polymer matrix. In that case, there must be some reactivity between the functional group of graphene sheets and the functional group of polymers. Examples of such polymer nanocomposites are PVA and PU nanocomposites [94, 95].

Another effective method of preparation of polymer/graphene nanocomposites is melt and solution mixing technique of thermally reduced graphene (TRG) with the polymer matrix, out of which the melt technique is commercially most viable. The only difficulty in this technique is the poor thermal stability of chemically modified TRG, which limits its popularity in the research. Raghu et al. in the year of 2008 [96] reported preparation of nanocomposites with finely dispersed functionalized graphene sheets (FGSs) in waterborne PU (WPU) matrix by emulsion polymerization technique. Here, they have prepared the FGS by following Staudenmaier method [97]. The visible wrinkled structures of individual graphene nanosheets within the polymer matrix in the TEM images (Fig. 21) of the WPU/FGS nanocomposites indicate good state of dispersion of these graphene nanosheets. By following a similar procedure, Nguyen et al. in the same year reported to have prepared polycaprolactone (PCL) diol-based thermoplastic polyurethane/FGS

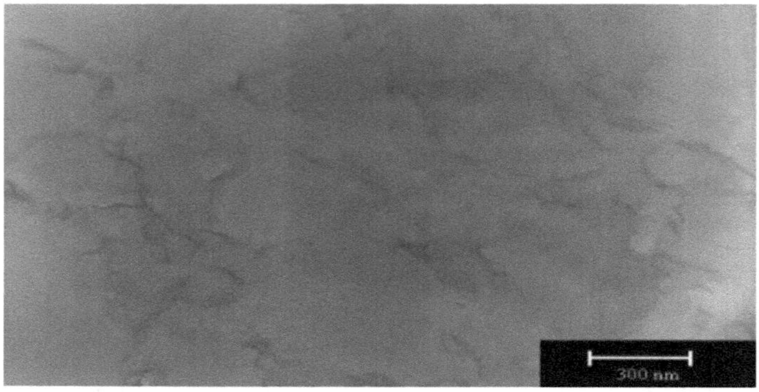

Fig. 21 TEM micrographs of waterborne polyurethane/functionalized graphene sheets (FGSs) nanocomposites with 3 wt% of FGS per 100 parts of the polymer [96]

nanocomposites by solution polymerization technique in methyl ethyl ketone (MEK) under inert atmospheric condition [98].

The wide-angle XRD patterns of the prepared nanocomposites with varying FGS concentration are shown in (Fig. 22). Two peaks found at $2\theta = 21.1°$ and 23.3° are due to the (110) and (200) plane of the PCL crystals, while the absence of any peak in the region above 2° is the clear indication of excellent dispersion of the graphene nanosheets within the polymer matrix. In this regards, it is essential to mention that neat graphite normally shows a diffraction peak for the (002) plan in the region of 26° that corresponds to ordered arrangement of the graphene nanosheets at a particular interlayer spacing [99]. This result of morphology study of the

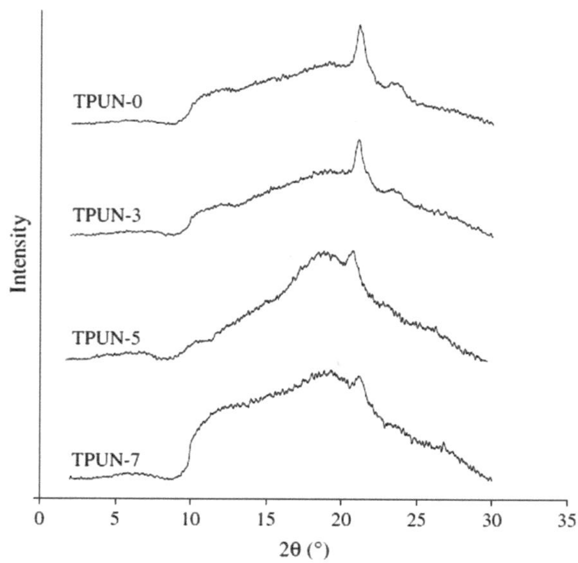

Fig. 22 XRD patterns of thermoplastic polyurethane TPU/FGS nanocomposites with varying FGS concentration (FGS concentration of 0, 3, 5, and 7 wt% in TPUN-0, TPUN-3, TPUN-5, TPUN-7, respectively) [98]

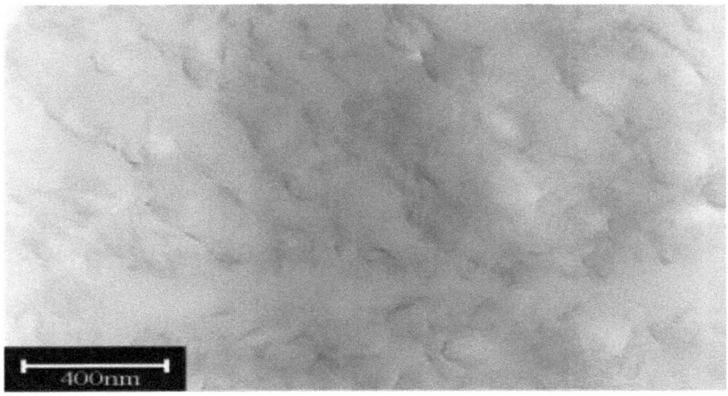

Fig. 23 TEM images of thermoplastic polyurethane (TPU)/functionalized graphene sheets (FGSs) nanocomposites with 4 wt% of FGS concentration (TPUN-4) [98]

nanocomposites with wide-angle XRD is further supported by TEM images of the nanocomposites in Fig. 23.

Here also, nano-sized wrinkled, hair-like structures of FGSs visible in the TPU polymer matrix refers to fine and even dispersion of separate and individual graphene nanosheets. This is the indication of good polymer–filler compatibility in the prepared nanocomposites, which is also evident from the evaluation of thermal and mechanical properties of the nanocomposites. It is obvious that such morphology of the FGSs within polymer matrices clearly refers to the reinforcing activity of the nanofillers with the matrix, which can be understood from the increased ultimate tensile property and modulus values of the nanocomposites [84]. Hu et al. have reported the preparation of graphene nanosheets–polystyrene nanocomposites via emulsion polymerization technique, where they had reduced the graphene oxides into graphene using a strong reducing agent hydrazine [100]. Figure 24a shows the Raman spectra and Fig. 24b shows the FTIR spectra of pure graphene nanosheets and its PS nanocomposites. Two intense peaks of graphene nanosheets in Fig. 24a corresponds to the *D* and *G* band in the region of 1331 and 1591 cm^{-1}, respectively, and the ratio of the intensities of these peaks, i.e., $I(D)/I(G)$ for the nanocomposites was found to be of 1.156. This refers to the presence of some localized sp^3 defects in the sp^2 structure of graphene that could cause chemical grafting of the PS chains onto the surface of graphene. Thus, creating defects can help improving grafting of polymer chains onto the graphene nanosheets at the cost of reduced electrical, thermal, and barrier properties.

Furthermore, the thermal degradation stability of the prepared FGS-based nanocomposites was also improved by manifold, which can be attributed to the barrier effect of the sheet-like graphene and high its thermal stability up to 700 °C. Besides, the well-dispersed morphology of individual graphene nanosheets is capable of enhancing the electrical conductivity of the polymer matrix through developing a proper channel for electrical conduction. Uniform dispersion of graphene nanosheets within the polymer matrices generally provide a percolated route

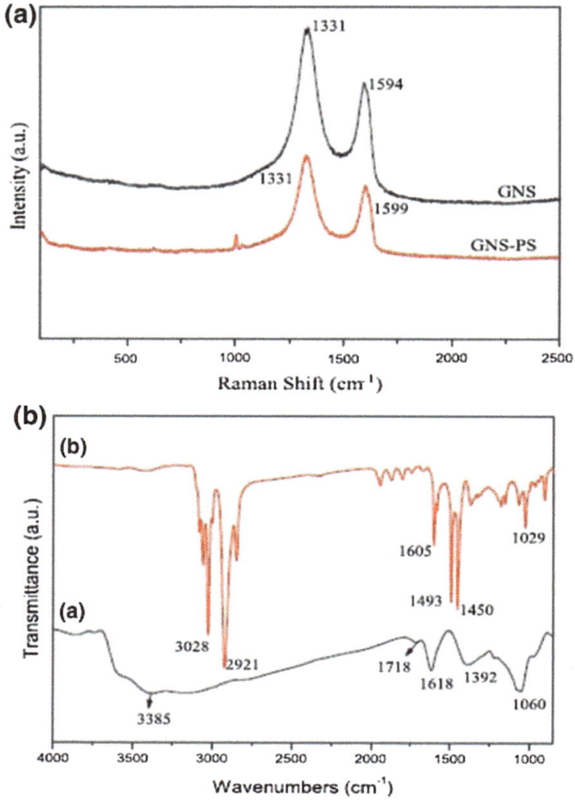

Fig. 24 **a** Raman spectra and **b** FTIR spectra of graphene nanosheets and its polystyrene-based nanocomposite [100]

for the transfer of electrons and thereby making the nanocomposites electrically and thermally conductive. Furthermore, an aligned direction of the sheets can provide the electrical percolation threshold at lower concentration. Another extraordinary advantage of graphene nanocomposites over other carbon-based nanocomposites is its barrier characteristics towards oxygen and other gases including moisture. It is generally observed that low nanofiller concentration results in better dispersion with minimum aggregation, which helps in improving other physical and chemical properties of the nanocomposites. Hence, the morphology of the resulting polymer nanocomposites can dictate nanocomposites properties to a greater extent. The high barrier property of defect-free graphene nanosheets can alter the barrier properties of polymers through its incorporation in a much careful manner without harming the integrity structure of each graphene nanosheets. Thus, the polymers acting as a host material for the graphene sheets is possible to develop large-scale barrier membranes. It is also found that graphene/polymer nanocomposites show better barrier properties over other graphitic structure-based carbon filler/polymer nanocomposites, which is attributed to its sheet-like structure [101–103].

4 Conclusion

There are several applications in the engineering field where carbon fillers play important role in improving the application specific properties of its polymer composites. The morphological differences in various carbon fillers affect the properties of resultant polymer composites. While carbon black is capable of forming mechanical interlocking with elastomer chains along with chemical interaction through the dangling organic functional groups, some other carbon fillers like carbon nanotube (CNT), carbon nanofiber (CNF), graphene nanosheets, etc., need modification either through chemical treatment like acid/base or organic treatment or through physical impulse-like plasma treatment. Such modification of carbon fillers is necessary to improve polymer–filler interfacial interaction in polymer composites. Introduction of active functional groups into carbon filler not only modifies the morphology of polymer composites but also enhances overall physico-mechanical properties of respective composites. Thus, both morphological and spectroscopic study is an integral part of study in polymer/carbon filler composites.

References

1. Medalia AI, Heckman FA (1969) Morphology of aggregates—II. Size and shape factors of carbon black aggregates from electron microscopy. Carbon 7:567–582
2. Ma PC, Kim JK (2011) Carbon nanotubes for polymer reinforcement. Taylor & Francis
3. Boonstra BB, Medalia AI (1963) Effect of carbon black dispersion on the mechanical properties of rubber vulcanizates. Rubber Chem Technol 36:115–142
4. Zhang C, Yi XS, Yui H, Asai S, Sumita M (1998) Selective location and double percolation of short carbon fiber filled polymer blends: high-density polyethylene/isotactic polypropylene. Mater Lett 36:186–190
5. Sumita M, Sakata K, Hayakawa Y, Asai S, Miyasaka K, Tanemura M (1992) Double percolation effect on the electrical conductivity of conductive particles filled polymer blends. Colloid Polym Sci 270:134–139
6. Sumita M, Takenaka K, Asai S (1995) Characterization of dispersion and percolation of filled polymers: molding time and temperature dependence of percolation time in carbon black filled low density polyethylene. Compos Interfaces 3:253–262
7. Park SJ, Kim JS (2000) Role of chemically modified carbon black surfaces in enhancing interfacial adhesion between carbon black and rubber in a composite system. J Colloid Interface Sci 232:311–316
8. Ma PC, Liu MY, Zhang H, Wang SQ, Wang R, Wang K, Wong YK, Tang BZ, Hong SH, Paik KW, Kim JK (2009) Enhanced electrical conductivity of nanocomposites containing hybrid fillers of carbon nanotubes and carbon black. ACS Appl Mater Interfaces 1:1090–1096
9. Sumita M, Sakata K, Asai S, Miyasaka K, Nakagawa H (1991) Dispersion of fillers and the electrical conductivity of polymer blends filled with carbon black. Polym Bull 25:265–271
10. Al-Mosawi Ali I, Al-Maamori Mohammad H, Al-Mayalee Khalidah H (2013) Spectroscopic studies of polyester-carbon black composites. Res J Mater Sci 2320:6055
11. Han D, Meng Z, Wu D, Zhang C, Zhu H (2011) Thermal properties of carbon black aqueous nanofluids for solar absorption. Nanoscale Res Lett 6:457

12. Endo M, Kim YA, Hayashi T, Fukai Y, Oshida K, Terrones M, Yanagisawa T, Higaki S, Dresselhaus MS (2002) Structural characterization of cup-stacked-type nanofibers with an entirely hollow core. Appl Phys Lett 80:1267–1269
13. Endo M, Kim YA, Ezaka M, Osada K, Yanagisawa T, Hayashi T, Terrones M, Dresselhaus MS (2003) Selective and efficient impregnation of metal nanoparticles on cup-stacked-type carbon nanofibers. Nano Lett 3:723–726
14. Baek JB, Lyons CB, Tan LS (2004) Grafting of vapor-grown carbon nanofibers via in-situ polycondensation of 3-phenoxybenzoic acid in poly (phosphoric acid). Macromolecules 37:8278–8285
15. Arlen MJ, Wang D, Jacobs JD, Justice R, Trionfi A, Hsu JW, Schaffer D, Tan LS, Vaia RA (2008) Thermal–electrical character of in situ synthesized polyimide-grafted carbon nanofiber composites. Macromolecules 41(21):8053–8062
16. Chung DD, Chung D (2012) Carbon fiber composites. Butterworth-Heinemann
17. Chirila V, Marginean G, Iclanzan T, Merino C, Brandl W (2007) Method for modifying mechanical properties of carbon nano-fiber polymeric composites. J Thermoplast Compos Mater 20:277–289
18. Brandl W, Marginean G, Chirila V, Warschewski W (2004) Production and characterisation of vapour grown carbon fiber/polypropylene composites. Carbon 42:5–9
19. Finegan IC, Tibbetts GG, Glasgow DG, Ting JM, Lake ML (2003) Surface treatments for improving the mechanical properties of carbon nanofiber/thermoplastic composites. J Mater Sci 38:3485–3490
20. Gordeyev SA, Ferreira JA, Bernardo CA, Ward IM (2001) A promising conductive material: highly oriented polypropylene filled with short vapour-grown carbon fibres. Mater Lett 51:32–36
21. Lim CS, Rodriguez AJ, Guzman ME, Schaefer JD, Minaie B (2011) Processing and properties of polymer composites containing aligned functionalized carbon nanofibers. Carbon 49:1873–1883
22. Nie Y, Hübert T (2012) Surface modification of carbon nanofibers by glycidoxysilane for altering the conductive and mechanical properties of epoxy composites. Compos A 43:1357–1364
23. Werner P, Altstädt V, Jaskulka R, Jacobs O, Sandler JK, Shaffer MS, Windle AH (2004) Tribological behaviour of carbon-nanofibre-reinforced poly (ether ether ketone). Wear 257:1006–1014
24. Sandler J, Werner P, Shaffer MS, Demchuk V, Altstädt V, Windle AH (2002) Carbon-nanofibre-reinforced poly (ether ether ketone) composites. Compos A 33:1033–1039
25. Higgins BA, Brittain WJ (2005) Polycarbonate carbon nanofiber composites. Eur Polym J 41:889–893
26. Manea F, Motoc S, Pop A, Remes A, Schoonman J (2012) Silver-functionalized carbon nanofiber composite electrodes for ibuprofen detection. Nanoscale Res Lett 7:1–4
27. Min C, Shen X, Shi Z, Chen L, Xu Z (2010) The electrical properties and conducting mechanisms of carbon nanotube/polymer nanocomposites: a review. Polym Plast Technol Eng 49:1172–1181
28. Maity A, Ray SS, Pillai SK (2007) Morphology and electrical conductivity of poly(*N*-vinylcarbazole)/carbon nanotubes nanocomposite synthesized by solid state polymerization. Macromol Rapid Commun 28:2224–2229
29. Sun YP, Fu K, Lin Y, Huang W (2002) Functionalized carbon nanotubes: properties and applications. Acc Chem Res 35:1096–1104
30. Tasis D, Tagmatarchis N, Bianco A, Prato M (2006) Chemistry of carbon nanotubes. Chem Rev 106:1105–1136
31. Kathi J, Rhee KY (2008) Surface modification of multi-walled carbon nanotubes using 3-aminopropyltriethoxysilane. J Mater Sci 43:33–37
32. Kim JA, Seong DG, Kang TJ, Youn JR (2006) Effects of surface modification on rheological and mechanical properties of CNT/epoxy composites. Carbon 44:1898–1905

33. Ma PC, Kim JK, Tang BZ (2006) Functionalization of carbon nanotubes using a silane coupling agent. Carbon 44:3232–3238
34. Yuan JM, Fan ZF, Chen XH, Chen XH, Wu ZJ, He LP (2009) Preparation of polystyrene–multiwalled carbon nanotube composites with individual-dispersed nanotubes and strong interfacial adhesion. Polymer 50:3285–3291
35. Zainal NFA, Azira AA, Nik SF, Rusop M (2009) The electrical and optical properties of PMMA/MWCNTs nanocomposite thin films. In: Rusop M, Soga T (eds) AIP conference proceedings, vol 1136, pp 750–754
36. Kymakis E, Alexandou I, Amaratunga GAJ (2002) Single-walled carbon nanotube–polymer composites: electrical, optical and structural investigation. Synth Met 127:59–62
37. Liao SH, Hung CH, Ma CCM, Yen CY, Lin YF, Weng CC (2008) Preparation and properties of carbon nanotube-reinforced vinyl ester/nanocomposite bipolar plates for polymer electrolyte membrane fuel cells. J Power Sources 176:175–182
38. Liao SH, Yen CY, Weng CC, Lin YF, Ma CCM, Yang CH, Tsai MC, Yen MY, Hsiao MC, Lee SJ, Xie XF (2008) Preparation and properties of carbon nanotube/polypropylene nanocomposite bipolar plates for polymer electrolyte membrane fuel cells. J Power Sources 185:1225–1232
39. Bai JB, Allaoui A (2003) Effect of the length and the aggregate size of MWNTs on the improvement efficiency of the mechanical and electrical properties of nanocomposites—experimental investigation. Compos A 34:689–694
40. Liang GD, Tjong SC (2006) Electrical properties of low-density polyethylene/multiwalled carbon nanotube nanocomposites. Mater Chem Phys 100:132–137
41. Jung YJ, Kar S, Talapatra S, Soldano C, Viswanathan G, Li X, Yao Z, Ou FS, Avadhanula A, Vajtai R, Curran S (2006) Aligned carbon nanotube–polymer hybrid architectures for diverse flexible electronic applications. Nano Lett 6:413–418
42. Gojny FH, Nastalczyk J, Roslaniec Z, Schulte K (2003) Surface modified multi-walled carbon nanotubes in CNT/epoxy-composites. Chem Phys Lett 370:820–824
43. Ago H, Kugler T, Cacialli F, Salaneck WR, Shaffer MS, Windle AH, Friend RH (1999) Work functions and surface functional groups of multiwall carbon nanotubes. J Phys Chem B 103:8116–8121
44. Liu T, Phang IY, Shen L, Chow SY, Zhang WD (2004) Morphology and mechanical properties of multiwalled carbon nanotubes reinforced nylon-6 composites. Macromolecules 37:7214–7222
45. Kanagaraj S, Varanda FR, Zhil'tsova TV, Oliveira MS, Simões JA (2007) Mechanical properties of high density polyethylene/carbon nanotube composites. Compos Sci Technol 67:3071–3077
46. Huang H, Liu CH, Wu Y, Fan SH (2005) Aligned carbon nanotube composite films for thermal management. Adv Mater 17:1652–1656
47. Kashiwagi T, Grulke E, Hilding J, Harris R, Awad W, Douglas J (2002) Thermal degradation and flammability properties of poly (propylene)/carbon nanotube composites. Macromol Rapid Commun 23:761–765
48. Beyer G (2002) Improvements of the fire performance of nanocomposites. In: Thirteenth annual BCC conference on flame retardancy. Stamford, CT
49. Xiong J, Zheng Z, Qin X, Li M, Li H, Wang X (2006) The thermal and mechanical properties of a polyurethane/multi-walled carbon nanotube composite. Carbon 44:2701–2707
50. Kashiwagi T, Grulke E, Hilding J, Groth K, Harris R, Butler K, Shields J, Kharchenko S, Douglas J (2004) Thermal and flammability properties of polypropylene/carbon nanotube nanocomposites. Polymer 45:4227–4239
51. Krato H, Heath J, O'Brien SC, Curl RF, Smalley RE (1985) Nature 318:162–163
52. Sanz A, Wong HC, Nedoma AJ, Douglas JF, Cabral JT (2015) Influence of C_{60} fullerenes on the glass formation of polystyrene. Polymer 68:47–56
53. Sariciftci NS, Smilowitz L, Heeger AJ, Wudl F (1992) Photoinduced electron transfer from a conducting polymer to buckminsterfullerene. Science 258:1474–1476

54. Vinogradova LV, Melenevskaya EY, Khachaturov AS, Kever EE, Litvinova LS, Novokreshchenova AV, Sushko MA, Klenin SI, Zgonnik VN (1998) Water-soluble complexes of C_{60} fulleren with poly(*N*-vinylpyrrolidone). Polym Sci Ser B Polym Chem 40:1152–1159
55. Jurkowska B, Jurkowski B, Kamrowski P, Pesetskii SS, Koval VN, Pinchuk LS, Olkhov YA (2006) Properties of fullerene-containing natural rubber. J Appl Polym Sci 100:390–398
56. Ginzburg BM, Tabarov SK, Tuichiev S, Shepelevskii AA (2007) Effect of C_{60} fullerene additives on the structure and mechanical properties of thin organic glass films. Tech Phys Lett 33:1007–1010
57. Tuichiev S, Tabarov SK, Rashidov D, Shoimov U, Ginzburg BM (2008) Effect of C_{60} fullerene additives on the mechanical properties of low-density polyethylene films. Tech Phys Lett 34:56–57
58. Chirvase D, Parisi J, Hummelen JC, Dyakonov V (2004) Influence of nanomorphology on the photovoltaic action of polymer–fullerene composites. Nanotechnology 15:1317
59. Zeng HP, Wang T, Sandanayaka AS, Araki Y, Ito O (2005) Photoinduced charge separation and charge recombination in [60]fullerene–ethylcarbazole and [60]fullerene–triphenylamines in polar solvents. J Phys Chem A 109:4713–4720
60. Hoppe H, Niggemann M, Winder C, Kraut J, Hiesgen R, Hinsch A, Meissner D, Sariciftci NS (2004) Nanoscale morphology of conjugated polymer/fullerene-based bulk-heterojunction solar cells. Adv Funct Mater 14:1005–1011
61. Shaheen SE, Brabec CJ, Sariciftci NS, Padinger F, Fromherz T, Hummelen JC (2001) 2.5% efficient organic plastic solar cells. Appl Phys Lett 78:841–843
62. Thompson BC, Fréchet JM (2008) Polymer–fullerene composite solar cells. Angew Chem Int Ed 47:58–77
63. Şen F, Kahraman MV (2014) Thermal conductivity and properties of cyanate Ester/nanodiamond composites. Polym Adv Technol 25:1020–1026
64. Kidalov SV, Shakhov FM, Vul AY (2008) Thermal conductivity of sintered nanodiamonds and microdiamonds. Diam Relat Mater 17:844–847
65. Mochalin VN, Gogotsi Y (2015) Nanodiamond–polymer composites. Diam Relat Mater 58:161–171
66. Zhang Y, Hu X, Zhao JH, Sheng K, Cannon WR, Wang X, Fursin L (2009) Rheology and thermal conductivity of diamond powder-filled liquid epoxy encapsulants for electronic packaging. IEEE Trans Compon Packag Technol 32:716–723
67. Jee AY, Lee M (2011) Thermal and mechanical properties of alkyl-functionalized nanodiamond composites. Curr Appl Phys 11:1183–1187
68. Zubrowska A, Masirek R, Piorkowska E, Pietrzak L (2015) Structure, thermal and mechanical properties of polypropylene composites with nano-and micro-diamonds. Polimery 60:331–336
69. Dubrovinskaia N, Dub S, Dubrovinsky L (2006) Superior wear resistance of aggregated diamond nanorods. Nano Lett 6:824–826
70. Li L, Davidson JL, Lukehart CM (2006) Surface functionalization of nanodiamond particles via atom transfer radical polymerization. Carbon 44:2308–2315
71. Liu Y, Gu Z, Margrave JL, Khabashesku VN (2004) Functionalization of nanoscale diamond powder: fluoro-, alkyl-, amino-, and amino acid-nanodiamond derivatives. Chem Mater 16:3924–3930
72. Schreiner PR, Fokina NA, Tkachenko BA, Hausmann H, Serafin M, Dahl JE, Liu S, Carlson RM, Fokin AA (2006) Functionalized nanodiamonds: triamantane and tetramantane. J Org Chem 71:6709–6720
73. Liu Y, Khabashesku VN, Halas NJ (2005) Fluorinated nanodiamond as a wet chemistry precursor for diamond coatings covalently bonded to glass surface. J Am Chem Soc 127:3712–3713
74. Yang JH, Song KS, Zhang GJ, Degawa M, Sasaki Y, Ohdomari I, Kawarada H (2006) Characterization of DNA hybridization on partially aminated diamond by aromatic compounds. Langmuir 22:11245–11250

75. Dolmatov VY (2001) Detonation synthesis ultradispersed diamonds: properties and applications. Russ Chem Rev 70:607–626
76. Zhang Q, Naito K, Tanaka Y, Kagawa Y (2007) Polyimide/diamond nanocomposites: microstructure and indentation behavior. Macromol Rapid Commun 28:2069–2073
77. Shenderova O, Tyler T, Cunningham G, Ray M, Walsh J, Casulli M, Hens S, McGuire G, Kuznetsov V, Lipa S (2007) Nanodiamond and onion-like carbon polymer nanocomposites. Diam Relat Mater 16:1213–1217
78. Zhang Q, Naito K, Tanaka Y, Kagawa Y (2008) Grafting polyimides from nanodiamonds. Macromolecules 41:536–538
79. Kalsoom U, Peristyy A, Nesterenko PN, Paull B (2016) A 3D printable diamond polymer composite: a novel material for fabrication of low cost thermally conducting devices. RSC Adv 6:38140–38147
80. Geim AK, Novoselov KS (2007) The rise of graphene. Nat Mater 6:183–191
81. Blake P, Brimicombe PD, Nair RR, Booth TJ, Jiang D, Schedin F, Ponomarenko LA, Morozov SV, Gleeson HF, Hill EW, Geim AK (2008) Graphene-based liquid crystal device. Nano Lett 8:1704–1708
82. Miranda R, Vázquez de Parga AL (2009) Graphene: surfing ripples towards new devices. Nat Nanotechnol 4:549–550
83. Ramanathan T, Abdala AA, Stankovich S, Dikin DA, Herrera-Alonso M, Piner RD, Adamson DH, Schniepp HC, Chen XRRS, Ruoff RS, Nguyen ST (2008) Functionalized graphene sheets for polymer nanocomposites. Nat Nanotechnol 3:327–331
84. Lee YR, Raghu AV, Jeong HM, Kim BK (2009) Properties of waterborne polyurethane/functionalized graphene sheet nanocomposites prepared by an in situ method. Macromol Chem Phys 210:1247–1254
85. Yamaguchi H, Eda G, Mattevi C, Kim H, Chhowalla M (2010) Highly uniform 300 mm wafer-scale deposition of single and multilayered chemically derived graphene thin films. ACS Nano 4:524–528
86. Quan H, Zhang BQ, Zhao Q, Yuen RK, Li RK (2009) Facile preparation and thermal degradation studies of graphite nanoplatelets (GNPs) filled thermoplastic polyurethane (TPU) nanocomposites. Compos A 40:1506–1513
87. Yuen SM, Ma CCM, Chiang CL, Chang JA, Huang SW, Chen SC, Chuang CY, Yang CC, Wei MH (2007) Silane-modified MWCNT/PMMA composites–preparation, electrical resistivity, thermal conductivity and thermal stability. Compos A 38:2527–2535
88. Becerril HA, Mao J, Liu Z, Stoltenberg RM, Bao Z, Chen Y (2008) Evaluation of solution-processed reduced graphene oxide films as transparent conductors. ACS Nano 2:463–470
89. Liang J, Huang Y, Zhang L, Wang Y, Ma Y, Guo T, Chen Y (2009) Molecular-level dispersion of graphene into poly (vinyl alcohol) and effective reinforcement of their nanocomposites. Adv Funct Mater 19:2297–2302
90. Chen G, Weng W, Wu D, Wu C (2003) PMMA/graphite nanosheets composite and its conducting properties. Eur Polym J 39:2329–2335
91. Jović N, Dudić D, Montone A, Antisari MV, Mitrić M, Djoković V (2008) Temperature dependence of the electrical conductivity of epoxy/expanded graphite nanosheet composites. Scr Mater 58:846–849
92. Mu Q, Feng S (2007) Thermal conductivity of graphite/silicone rubber prepared by solution intercalation. Thermochim Acta 462:70–75
93. Hirata M, Gotou T, Horiuchi S, Fujiwara M, Ohba M (2004) Thin-film particles of graphite oxide 1: high-yield synthesis and flexibility of the particles. Carbon 42:2929–2937
94. Yang X, Li L, Shang S, Tao XM (2010) Synthesis and characterization of layer-aligned poly (vinyl alcohol)/graphene nanocomposites. Polymer 51:3431–3435
95. Kim H, Miura Y, Macosko CW (2010) Graphene/polyurethane nanocomposites for improved gas barrier and electrical conductivity. Chem Mater 22(11):3441–3450

96. Raghu AV, Lee YR, Jeong HM, Shin CM (2008) Preparation and physical properties of waterborne polyurethane/functionalized graphene sheet nanocomposites. Macromol Chem Phys 209:2487–2493
97. Hontoria-Lucas C, Lopez-Peinado AJ, López-González JDD, Rojas-Cervantes ML, Martin-Aranda RM (1995) Study of oxygen-containing groups in a series of graphite oxides: physical and chemical characterization. Carbon 33:1585–1592
98. Nguyen DA, Lee YR, Raghu AV, Jeong HM, Shin CM, Kim BK (2009) Morphological and physical properties of a thermoplastic polyurethane reinforced with functionalized graphene sheet. Polym Int 58:412–417
99. Shin HJ, Kim KK, Benayad A, Yoon SM, Park HK, Jung IS, Jin MH, Jeong HK, Kim JM, Choi JY, Lee YH (2009) Efficient reduction of graphite oxide by sodium borohydride and its effect on electrical conductance. Adv Funct Mater 19:1987–1992
100. Hu H, Wang X, Wang J, Wan L, Liu F, Zheng H, Chen R, Xu C (2010) Preparation and properties of graphene nanosheets–polystyrene nanocomposites via in situ emulsion polymerization. Chem Phys Lett 484:247–253
101. Kim H, Miura Y, Macosko CW (2010) Graphene/polyurethane nanocomposites for improved gas barrier and electrical conductivity. Chem Mater 22:3441–3450
102. Kim H, Macosko CW (2008) Morphology and properties of polyester/exfoliated graphite nanocomposites. Macromolecules 41:3317–3327
103. Kim H, Abdala AA, Macosko CW (2010) Graphene/polymer nanocomposites. Macromolecules 43:6515–6530
104. Zhao H, Min K, Aluru NR (2009) Size and chirality dependent elastic properties of graphene nanoribbons under uniaxial tension. Nano Lett 9:3012–3015
105. Scarpa F, Adhikari S, Phani AS (2009) Effective elastic mechanical properties of single layer graphene sheets. Nanotechnol 20:065709
106. Chrissafis K, Paraskevopoulos KM, Pavlidou E, Bikiaris D (2009) Thermal degradation mechanism of HDPE nanocomposites containing fumed silica nanoparticles. Thermochim Acta 485:65–71
107. Woo MW, Wong P, Tang Y, Triacca V, Gloor PE, Hrymak AN, Hamielec AE (1995) Melting behavior and thermal properties of high density polythylene. Polym Eng Sci 35:151–156
108. Ventura G, Martelli V (2009) Thermal conductivity of Kevlar 49 between 7 and 290K. Cryogenics 49:735–737
109. Sun Y, Luo Y, Jia D (2008) Preparation and properties of natural rubber nanocomposites with solid-state organomodified montmorillonite. J Appl Polym Sci 107:2786–2792

Electromagnetic Interference (EMI) Shielding Effectiveness (SE) of Polymer-Carbon Composites

Ranvijai Ram, Mostafizur Rahaman and Dipak Khastgir

Abstract In this chapter, the electromagnetic interference shielding effectiveness (EMISE) of carbon based polymer composites is discussed in details. The basic principle of EMI, EMI shielding, and its theory are mentioned herein. The basic requirement of EMI SE of a material is its electrical conductivity. It has been mentioned that electrical conductivity of 0.5 S/cm is required to produce at least 30 dB attenuation. As non-conducting materials exhibit negligible EMI SE, hence EMI SE of polymer-carbon composites based on only conducting carbons like carbon black, carbon fiber, carbon nanotubes, and graphene are reported within this chapter. EMI SE depends on many factors like nature of filler, filler concentration, nature of polymer, filler geometry, polymer blending, sample thickness, frequency of radiation, etc. These governing factors of EMI SE are discussed is details at the end of this chapter.

Keywords Polymer Composites · Carbons · Electrical conductivity
EMI SE · Dependent phenomena

R. Ram · D. Khastgir (✉)
Rubber Technology Centre, Indian Institute of Technology Kharagpur,
Kharagpur 721302, India
e-mail: khasdi@rtc.iitkgp.ernet.in

R. Ram
e-mail: ranvijai@rtc.iitkgp.ernet.in

R. Ram
Department of Chemistry, Maharaja Agrasen College, University of Delhi,
Vasundhara Enclave, New Delhi 110096, India

M. Rahaman (✉)
Department of Chemistry, College of Science, King Saud University,
Riyadh 11451, Saudi Arabia
e-mail: mrahaman@ksu.edu.sa

M. Rahaman et al. (eds.), *Carbon-Containing Polymer Composites*,
Springer Series on Polymer and Composite Materials,
https://doi.org/10.1007/978-981-13-2688-2_10

1 Introduction

The modern age is considered to be electronic age. There are numerous sources of generation of electromagnetic radiation. The radio frequency range (0.3–1000 MHz) are used for communication signals like television, radio, and cellular telephones. The generation of the electromagnetic field by operation of television, radio, and cellular telephones may interfere the operation of other electronic devices which is known as EMI. The connectivity problems of mobile phones, data corruption of computer hard drives, and interrupted television signals are caused by EMI. The EMI has also some biological harmful effect. For example, overexposure to EMI radiation may cause Leukemia in the human body [1]. EM radiation may also interfere the function or operation of artificial biological devices like Pacemaker [2].

Conventional plastics are electrically insulating and transparent to EM radiation. Electrical conductivity plays a major role to prevent EMI radiations [1–4]. Hence, it is very interesting to develop flexible polymer compounds which are electrically conductive [4, 5]. Some electrically conductive fillers like carbon black, carbon fibers, carbon nanotube, graphene, and metal fibers are used to develop polymer-based EMI shielding materials [6–9]. Appropriate doping of insulation polymer is also another method to produce transparent moldable EMI shielding materials.

The composition of molded compounds is different to that of coated compounds. Coating of a material needs lesser amount of conducting fillers than molded one. The higher amount of conducting filler leads to increase the cost of conducting molded compounds [10, 11]. The amount of conducting fillers can be reduced by using higher aspect ratio of filler. The aspect ratio of filler is defined as the ratio of length to diameter. The geometry of conducting filler is another factor governing the conductivity of composite [8]. High-structure carbon fillers has low percolation threshold than low structure filler which leads to reduction of cost of the final product [8, 10, 11]. Actually, high structure carbon fillers form the continuous conducting network at lower concentration. Hence, high EMI shielding can be achieved by using high structure and large aspect ratio carbon fillers at low concentration.

2 EMI

EMI is defined as radiated or conducted electromagnetic signals emitted by electrical circuits during operation cause damage to biological or living species or disturb proper operation of electrical equipment [12]. EMI is an environmental pollutant. With the national emphasis today on the elimination or reduction of environmental pollution, most people readily recognize and understand water, air, noise and other forms of pollution. Most people probably have not heard or know much about spectrum pollution. It cannot be directly seen, tasted, smelled, or felt. Therefore, how can it be a problem?

It is readily known that certain types of electronic devices will jam household radios. The resulting buzzing or crackling noise results in the inability to listen to the radio while the device is in use. Conducted or radiated electrical noise jams radios picking up broadcast signals or cellular phones trying to make calls. Spectrum (EMI) pollution is very harmful for the patient using pacemaker. The improper operation of instrument like pacemaker can cause of death. Another example is an unsuppressed automobile idling outside of a house causing interference to the television picture in the house by blotching, developing intermittent dash lines or even total loss of the picture. Electrical noises from radiated automobile ignition systems cause these problems.

Interference caused by high electricity or noise background can adversely affect or cross the communication lines. This results adverse effect on electronic transaction in the business community. The EMI/Radio frequency interference (RFI) can be a serious threat in modern age of information technology.

The spectrum pollution problem can affect human life and the global economy [13, 14]. Thus, high EMI/RFI is causing spectrum pollution and it is a national concern. Because of that, the government and some industries have certain regulations and specifications on electromechanical, electrical, electrochemical, and electronic equipment. However, enforcing these regulations is difficult. Many commercial goods may have to meet a regulation when they are purchased, but many consumers modify these devices to their own needs or desires.

Like air, water and another form of pollution, EMI is also a kind of environmental pollution. EMI pollution may also be called as spectrum pollution. Certain types of electronic devices (like a mobile phone) jam the household radio resulting cracking or buzzing noise of radio when the devices are in use. Similarly, the unsuppressed sound of an automobile causes interference to picture of the television by rising intermittent dash, blotching, or even total loss of picture. EMI is not good during medical treatment in which patient is using sensitive devices like a pacemaker. During the electronic transaction, EMI or RFI (Radio Frequency Interference) can cause serious threat [15–17].

3 EMI Shielding

Prevention of the propagation of magnetic or electric waves (radiation) from one region to another by means of magnetic or conducting materials is called EMI shielding. The capacity of prevention of EMI of material is called as EMI SE [12]. The better EMI SE can be achieved by minimizing the radiation passing through the material by absorption, reflection, or dissipation of radiation inside the material. EMI SE of metal's sheet is caused by reflection but for extrinsically or intrinsically polymer composites, the EMI SE is caused by reflection, absorption, and dissipation of electromagnetic radiations.

The designing and packaging of electronic materials free from electromagnetic radiation (EMI shielding) is a critical issue. Previously, many designers took the attitude of the device first and figure how to prepare it compliant later. But nowadays it is non-viable strategy. In the present market, the pressure is rising to prepare EMI shielding device in a shorter period and product life will be longer [18]. Now strategy of manufacturers has been changed, they do not afford the added cost, design complexity of over shielding, weight. They also don't want to tolerate the system quality failure that causes under shielding. Now manufacturers are also started to realize the importance of flexible design geometry, low cost to manufacture, and parts integration of EMI shielding material over electron equipment [18].

Plastics are the good choice because of transparency and flexibility of their thin films. But they are also transparent to electromagnetic radiation interferences. This problem of poor EMI shielding behavior of plastics can be resolved by incorporating conducting fillers in their matrices in the desired quantity. Traditional shielding methods are conductive plating, painting, and applying metal liners provides sufficient EMI SE. The higher microprocessor frequency, higher currents, and lower voltages create new problem [18].

The EMI SE caused by the metallic sheet is called a Faraday cage effect. Metal sheet has poor mechanical flexibility because of high weight density, high stiffness, limited tuning of the SE, and propensity to corrosion [2]. The polymeric EMI shielding materials have the advantage over metallic one because of their low cost, lightness, easy shaping, etc. There is three main scheme has been developed to process EM radiation absorbing polymer-based composites. The 1st one is a dispersion of fibers, metallic fillers, and nanoparticles in the polymer matrix to increase the interaction with EM radiation [12, 19–23]. Carbon Nanotube–Polystyrene Foam, [23]. The 2nd scheme is blending of conventional polymers with conducting polymers to adequate mechanical properties and interaction with EM radiation [23]. Polyaniline and Polypyrrole are the examples of blending of intrinsically conducting (2nd scheme) polymers [12, 24–30]. The EMI SE of composites rises with increasing electrical conductivity which is subjected to addition of conducting filler in the polymer matrix. The 3rd scheme is a dispersion of conducting carbon fillers in the insulating polymer matrix. This type of conducting composites are called extrinsically conducting composites. Conducting carbon fillers are carbon fiber, carbon nanotube, conducting carbon black, graphene, etc. [31–34].

It is observed that electrical conductivity increases rapidly before percolation threshold whereas there is a sudden improvement in electrical conductivity at percolation threshold and after percolation threshold increment in electrical conductivity is marginal [35–37]. EMI SE exhibits different path to that of conduction mechanism path. It is reverse to conducting mechanism, i.e., EMI SE increases gradually before percolation whereas it increases rapidly after percolation threshold [30, 38]. For good EMI SE extrinsic composite material, it is necessary that all sieves of insulating polymer (transparent to EMI SE) to be filled by conducting fillers so that no EM radiation can pass through the EMI SE material (polymer/filler conducting composite) [24, 39–41].

4 EMI SE Theory

The ratio of transmitted power to incident power is known as EMI SE [42]. The total EMI shielding is due to reflection of electromagnetic radiation, multiple reflections of electromagnetic radiation, and absorption electromagnetic radiation. Mathematically total EMI SE (SE_{Total}) can be represented as follows [1, 42–46].

$$SE_{Total} = 10\log\frac{P_i}{P_0} = SE_A + SE_R + SE_M, \quad (1)$$

where P_i is the power of transmitted electromagnetic radiation, P_0 is the power of incident electromagnetic radiation, SE_A is EMI SE due to absorption, SE_R is EMI SE due to reflection, and SE_M is EMI SE due to multiple internal reflections of electromagnetic radiation. The pictorial representation of EMI SE is shown in Fig. 1.

The skin depth (*d*) of conducting material is defined as a distance within the material in which the intensity (*I*) of the electromagnetic (EM) wave decreases to *1*/*e* of its original strength (I_0) [42, 47]. Mathematically it can be expressed as;

$$d = \sqrt{\frac{2}{\mu_r \omega \sigma_s}}, \quad (2)$$

where $\sigma_s; (\sigma_s = \omega\varepsilon_0\varepsilon'')$ is frequency-dependent electrical conductivity, $\omega; (\omega = 2\pi f)$ is angular frequency [48]. EMI shielding due to absorption (SE_A) can be expressed as follows:

$$SE_A = -RL, \quad (3)$$

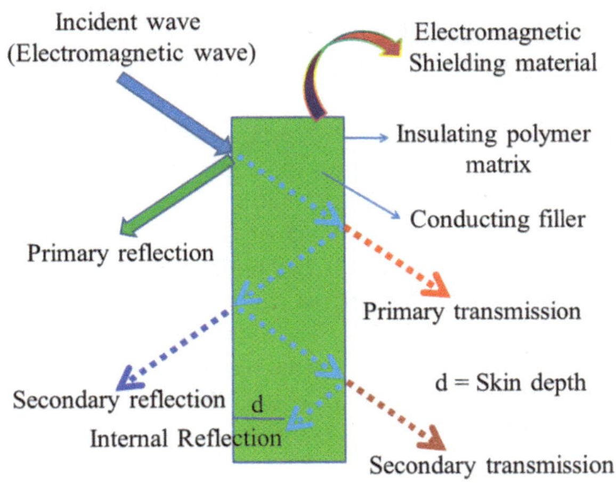

Fig. 1 Schematic representation of EMI shielding

where RL is reflection loss and can be expressed as

$$\mathrm{RL(dB)} = 20\log\left|\frac{Z_{\mathrm{in}} - Z_0}{Z_{\mathrm{in}} + Z_0}\right|, \tag{4}$$

where Z_{in} is the input impedance on the interface, it can be expressed as follows [49].

$$Z_{\mathrm{in}} = Z_0\sqrt{\frac{\mu_r}{\varepsilon_r}}\tan h\left(j\frac{2\pi fD}{c}\sqrt{\varepsilon_r\mu_r}\right), \tag{5}$$

where μ_r is complex permeability ($\mu_r = \mu' - j\mu''$), Z_0 is the impedance (free space), c is the velocity of light, ε_r is complex permittivity ($\varepsilon_r = \varepsilon' - j\varepsilon''$), and D is the thickness of the material. For electrically thick sample ($D > d$), the EMI shielding due to reflection can be expressed as follows [42, 49].

$$\mathrm{SE_R(dB)} = 10\log\left(\frac{\sigma_s}{16\omega\mu'\varepsilon_0}\right) \tag{6}$$

When total EMI SE ($\mathrm{SE_{Total}}$) is greater or equal to 10, then EMI SE due to multiple internal reflections of EM radiation ($\mathrm{SE_M}$) can be neglected [50].

5 Requirements for Shielding

The EMI shielding materials have high demand in recent years. It is used in military and other commercial purposes. EMI shielding materials are also used in highly sensitive electronic devices for communication and for packaging purposes. It is also the very interesting topic of research at the present time [51, 52]. Now a day it is the demand of customers to build lighter, faster, and cost-effective vehicles (space crafts, aircraft, etc.) which have forced engineers to integrate discrete avionics boxes into a firm electronic device [53]. Therefore, there is a requirement of preparing EMI shielding material with lightweight, somewhat transparent, low cost, and durable for protecting electromagnetic equipment that is vulnerable to EMI phenomenon.

In past, EMI shielding materials are made up of metals but, now a day conducting polymer composites are used because of low cost, low weight, restlessness, easier handling, easy processing, and moldable shape. Generally, carbon fillers like carbon fibers, graphene, carbon nanoparticles, carbon blacks, etc. are used as conducting filler in extrinsically prepared polymer composites used for EMI shielding material [54]. The composite materials have lower conductivity than metal but it is sufficient for the requirement of EMI shielding application. There are many desired shape of enclosures (Fig. 2) [55] made up of EMI shielding materials are used for protecting electronic equipment. Previously, these enclosures are made

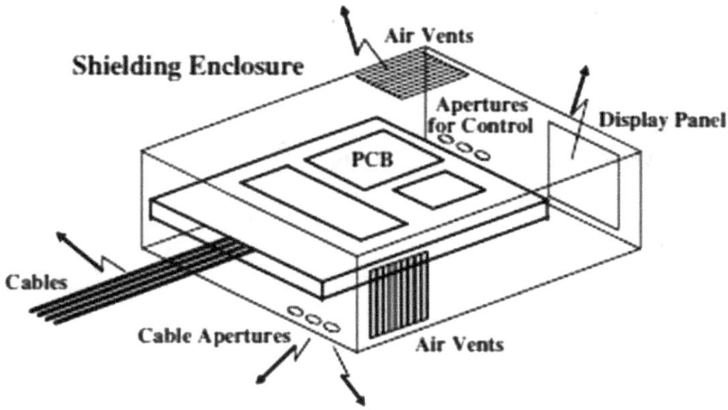

Fig. 2 Schematic representation of a typical shielding enclosure. Reproduced from [53]

up of metals in rectangular parallelepiped shape (Fig. 2) but now conducting polymer composites are used in various shapes.

The typical shielding enclosure (Fig. 2) is required for cable passages. The apertures are necessary for the cable passages, air vent for heat dissipation and other components [51, 55]. There are many parameters are taken under consideration during designing an EMI shielding apparatus like materials thickness, size, and geometry, number and position of aperture, the requirement of EMI shielding to inner equipment, and other physical properties of cashing material like solvent resistant, heat resistant, flame and fire resistant, etc.

Electrical conductivity of a material is the basic requirement for EMI SE. Some other factors includes applied electric field, packaging, and thickness of materials. The EMI SE refers to protection from direct charge, electrostatic field, and static charges. According to Bigg a material must possess a volume conductivity of 0.5 S/cm or more to produce at least 30 dB attenuation [56] which prevents 99.9% of an converging signal. It is observed that 20–30 dB of attenuation is sufficient for consumer and most industrial applications. The relation between EMI SE (dB) and attenuation (%) is presented in Table 1. Bigg [56] states that electrical conductivity at percolation threshold (transition from insulator to conductor), there is continuous conducting network is formed in the insulating polymer matrix. Electrical conductivity beyond percolation threshold leads to marginal increment of electrical conductivity but improves EMI SE remarkably. Actually, bigger conducting particles achieves percolation at very low concentration and exhibits good EMI SE. [56].

Table 1 EMI SE and % attenuation

EMI SE (dB)	20	30	40	50	60	70
Attenuation (%)	99	99.9	99.99	99.999	99.9999	99.99999

Reproduced from [56]

6 Compounding Considerations for Shielding

There are many factors governing EMI SE like polymer matrix, filler (concentration, conductivity, and aspect ratio), and processing. Compounding of elastomer undergo through mixing. During mixing there are various parameters responsible for compounded product like:

(i) Viscosity
(ii) Molecular weight distribution (MWD)
(iii) Plasticity
(iv) Filler
(v) Duration of mixing
(vi) Plasticizer and
(vii) Incorporation.

Mixing of elastomers requires various instruments viz. mills, internal mixers, and extruders. The compounded elastomers are characterize by measuring plastic stress, viscous stress, die swell, and dynamic shear stress.

The electrical conductivity of compounded elastomer governs the EMI SE of composite. Electrical conductivity is affected by viscosity of polymer matrix [57]. EMI SE of composite rises with increasing conductivity. Hence, EMI SE of composite is highly influenced by compounding.

The effect of filler type and its concentration on EMI SE has been shown in Fig. 3. The physical bonding (interaction) between polymer and PAN carbon fibers better than nickel-coated carbon fibers and stainless steel where EMI SE of polymer/stainless steel fibers is highest among PAN carbon fibers and nickel-coated carbon fibers. The reason is stainless steel is highly conducting material. Stainless steel based polymer composites have high abrasion resistance. Carbon-fiber-based composites suffers reduction in aspect ratio during processing which causes reduction in electrical conductivity as well as EMI SE.

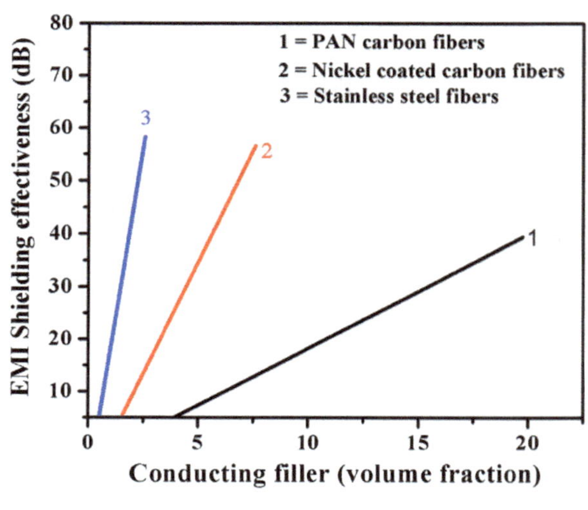

Fig. 3 Shielding effectiveness (dB) versus volume % of filler. Reproduced from [57]

Carbon fibers offer the best mechanical reinforcement while having a moderate conductivity. Nickel-coated graphite fibers offer the most conductivity and potential for shielding but lack easy processing. Stainless steel fibers are efficient in producing adequate conductivity. Loading levels of stainless steel from 5 to 7% wt are currently being used in industry.

Stainless steel offers minimal mold shrinkage, but difficult processing. Carbon fibers were chosen for this project because of easy processing. Using metal-filled surface coatings is still prevalent in large volume shielding applications. Cathode ray tubes (CRT), keyboards, computer housings, and cellular phone casings are industries that were once dominated by metal enclosures. Thermoplastics are overtaking metals in these applications and also creating new markets.

Polymer-based conductive composites are cost-effective, light in weight, ease in molding, and resistant to corrosion and chemicals. The products of plastic based composites are prepared by injection molding process. These products are widely used and replace the use of metal-based products which are heavy, costly, prone to corrosion, and highly reactive in polar solvents.

Both near field and far field EMI shielding are important to enhance the efficiency as well as the lifetime of electrical equipment and devices. Near field EMI shielding is defined as the product of wavenumber (k) and the distance between EM radiation sources to the detector (r) is less than 1, i.e., ($kr < 1$) where as far field is $kr > 1$. Actually, polymers are synthesized easily and are flexible in spite of that mass production of polymers is easy and cost-effective than metal. The performance of polymer-based conducting composites is better than metal for EMI shielding application [58–60].

Metals like aluminum, which are mostly used for EMI/EMC applications till date, enjoy the inherent advantage of possessing very high electrical conductivity ($\sigma \approx 10^7$ S/m) and hence possess intrinsic Electromagnetic (EM) shielding capabilities. However, polymer composites in general, are very poor conductors of electricity ($\sigma \approx 10^{-14}$ S/m). Thus, for them to be capable of shielding EM wave, one must devise innovative means of tailoring them to meet the required level of electrical conductivity. In fact, if we place the entire gamut of materials in the resistivity scale, we will have the family of polymers in one end ($\rho \approx 10^{14}$ to 10^{18} Ω m) and the metals in the other ($\rho \approx 10^{-7}$ to 10^{-5} Ω m). The EMI shielding materials come somewhere in between ($\rho \approx 10^{0}$ to 10^{4} Ω m). Hence, the basic idea of designing such a composite is to select two or more materials from this scale such that the combination, on a whole, attains a resistivity in the desired range.

By the addition of certain conductive fillers in critical quantity (depending upon the percolation threshold of the particular filler in the selected matrix), the conductivity of these composites had been reportedly increased to the order of 10^1 Ω/m. Similar technologies are extensively used in consumer electronics like cellular phone covers, telecom equipment, microwave ovens, etc. where the operating frequency band is narrow and the desired level of shielding efficiency (SE, measured in dB) is relatively nominal. But, for critical applications in defense and aerospace sectors where a high SE is recommended over a wide band of frequency, the criticalities of the material design is much more demanding.

Lightweight conductive composites that have been developed and studied use highly conductive graphite ($\sigma \approx 10^5$ Ω/m), bromine intercalated graphite (σ 10^6 Ω/m) fibers, nickel-coated graphite (NCG), carbon fibers, carbon nanotubes (CNT) and other special reinforcements/fillers [61–64] Once the desired electrical conductivity is attained, one can get significant SE from them as well [58, 65–67]. The EMI SE of a polymer-filler composite depends on the electrical conductivity, permittivity, permeability, impedance, and aspect ratio of fillers [6, 39]. The CNT is excellent EMI shielding material because of its high aspect ratio, mechanical strength, and electrical conductivity.

Plastics are an integral part of the electronics/electrical industry. They are used in packaging, handling, and interconnections. Their increased growth in this industry is proof of their acceptance as a material of preference for design engineers. The inherent property of electrical insulation is a great asset. However, EMI/RFI protection and electrostatic discharge (ESD) protection are also important uses.

Therefore, this creates a need for electrically conductive composites. The most widely accepted solution which allows plastics to meet the necessary requirements for EMI/RFI shielding and ESD is to apply a conductive metal coating through a secondary post-molding operation. Conductive metal coatings have proven well in shielding tests, but they have flaws that only plastics can solve.

Zinc spray, nickel paint, and electroless nickel painting all provide excellent protection for shielding. The disadvantages of conductive coatings are centered on secondary handling and long-term adhesion. The latter is often cause for concern in terms of reliability. In small intricate parts, the application costs can be extremely high. If the part is very complex, the application can be very difficult. Any additional handling caused by secondary molding or handling adds in extra costs [57]. By having to coat a surface with many layers of shielded conductive resins becomes very expensive and adds more operations to the manufacturing process.

One advantage of highly conductive plastics as an EMI shield is the elimination of expensive and time-consuming secondary coatings, which were necessary at one time [57]. A new way of thinking is to incorporate conductive constituents or fillers into the plastic matrix through means of compounding (i.e., extrusion and molding).

Conductive composites are easily injection molded into parts and conductive coatings are not needed. Only normal assembly is required. The smaller and more complex the part is, the more favorable the economics for conductive resins. Parts weighing as little as a gram have been EMI-shielded with conductive composites. Conductive composites have a few disadvantages. In some applications, a concern with conductive resins can be the raw material cost and appearance. This can prevent their use in large electronic plastic enclosures such as cathode ray tube housing [3, 57]. A goal of this project is to manufacture a cost-effective conductive resin.

The growth of the electronics industry has led to EMI which is causing environmental electronic pollution. In general, elastomers and plastics are electrically insulating and transparent to electromagnetic radiation; i.e., their EMI SE is zero. However, both conductivity and EMI SE is enhanced by using electrically conductive fillers like carbon black, graphite fibers, metal fibers, and metal particulates

[68]. The EMI SE of exfoliated polyester/graphite composites has been studied [69]. Recently, the EMI SE of carbon fiber-reinforced cement has been reported [70]. The polyethylene/carbon-fiber and polyethylene/metal-fiber composites exhibit poor EMI SE [71]. Abdelazeez and co-workers have recently observed that polypropylene/metal-fiber and polypropylene/aluminum-coated glass-fiber composites are not effective for EMI shielding purposes [71, 72].

Electromagnetic EM compatibility of conductive plastics and rubber gaskets having volume conductivity 10^2 to 10^3 S/m enables an application for EMI shielding and hermetic sealing simultaneously. These types of materials are made by using silicon rubber filled with tiny silver balls or silvered brass balls [71, 73, 74]. Recently, it has been investigated that short carbon fiber (SCF) filled rubber composites shows composition and aspect ratio dependent electrical conductivity and EMI SE [75, 76].

7 Polymer/Carbon Filler Based Composites as EMI Shielding Materials

There are several Polymer/carbon filler based composites used as EMI shielding material. Few of them are discussed as follows.

7.1 Polymer/Carbon Fiber (CF) Based Composites

Carbon fibers (CFs) are prepared by graphitization of Polyacrylonitrile (PAN) or petroleum pitch [77, 78]. Ethylene-vinyl acetate (EVA) based conducting carbon composites gives better performance to that of EPDM and natural rubber [12]. For instance, 8 wt% of CNF is required to 20 dB EMI SE in EVA matrix whereas 17.5 wt% CNF is required to get the same degree of EMI SE in case of natural rubber matrix (frequency = 12 GHz, thickness = 3.5 mm). The EPDM/CNF composites EMI SE is closer to that of EVA/CNF but EVA/CNF composites exhibit better EMI SE because of its polarity [12, 79]. When EVA and EPDM matrices are blended together in 50:50 then it exhibits better EMI SE than an individual. The 9 wt% CNF loaded EVA-EPDM blend (50:50) exhibits 40 dB EMI SE for composite thickness 3.5 mm from frequency 8–12 GHz [12, 79]. The effect of mixing speed, mixing time, and mixing temperature for HDPE/VGCNF composites was performed by Al-Saleh et al. [80, 81]. It was observed that increase in mixing speed, time, and temperature causes decrease in EMI SE of HDPE/VGCNF composites which is shown in Fig. 4a, b. Actually, in melt mixing process, the aspect ratio of CNF depends on the speed of mixing, mixing time, and temperature. The aspect ratio of CNF decreases with increase in speed of mixing because of rising in shear stress. Increase in temperature during mixing causes preventing the reduction of the

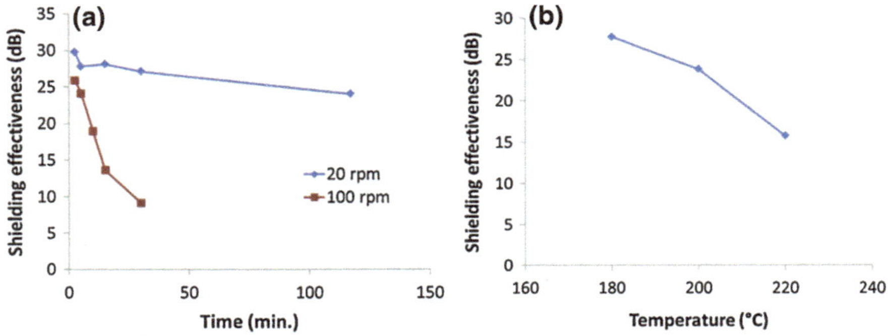

Fig. 4 **a** EMI SE of 7.5 vol.% VGCNF/HDPE composites at 1 GHz and at different processing condition (20 rpm and 100 rpm) as a function of time; **b** EMI SE of 7.5 vol.% VGCNF/HDPE composites processed at 50 rpm as a function of the processing temperature. Reproduced from [12]

breakdown of CNF which results in retention of aspect ratio. Reduction in aspect ratio causes a lower quality of dispersion of CNF in the polymer matrix results lowering EMI SE.

The orientation of CNF plays a major role in determining EMI SE in the polymer matrix. Jou et al. observed that oriented CNF (using a liquid crystalline polymer based on aromatic polyester in injection molding) in polymer matric increases EMI SE 30–50 dB by addition of 20 wt% CNF having composite thickness 3 mm measured 1 GHz frequency.

It is observed from solution casting method made PVDF/CNF and PVOH/CNF composites that the EMI SE of polymer/filler based composite is not directly proportional to electrical conductivity like specific surface area of the conducting filler which is responsible for optimum contact among the carbon particle resulting high magnitude of EMI SE performance of composites [12, 82, 83]. Hence EMI SE effectiveness of CNF based composites can be improved by modification of CNF surface in such a way that it can increase the specific surface area of CNF by the formation of pores by chemical or heat treatment [84, 85].

The variation of EMI SE due to the incorporation of CNF in the blend of EVA/NBR reveals that EMI SE of EVA/NBR/CNF composite increases with increase in EVA and CNF content presented in Fig. 5 [86]. The EMI SE is measured at 10 GHz having sample thickness 5 mm. It can be concluded that EVA/CNF composite has better interaction responsible for EMI SE than NBR/CNF composites.

7.2 *Polymer/Carbon Black (CB) Based Composites*

Different types of Carbon blacks(CBs) are prepared by thermal decomposition of hydrocarbons in the gas phase [87, 88]. CB is always formed in an aggregated form,

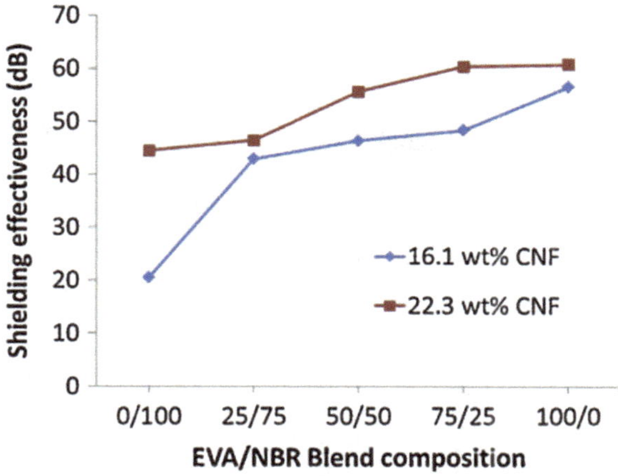

Fig. 5 Variation of the EMI SE of EVA/NBR/CNF blends as a function of the blend composition at 10 GHz (t = 5 mm). Reproduced from [12]

not in the single particle. It is widely used in rubber industry for reinforcement of rubber compound. It is also used as a pigment. The degree of aggregation and shape of CB is referred as its primary structure [88]. Generally, hardness, modulus, electrical conductivity, dispersibility, and viscosity of polymer/CB composites increase with an increase in the primary structure of CB [89, 90]. The CB based ethylene acrylic, EVA/NBR, BR/LDPE, EPDM, SBR, siloxane, and NR/BR composites are used as EMI shielding material [12, 91–94]. These types of EMI shielding materials are generally prepared by melt mixing followed by a vulcanization process. It is observed that SBR/CB composites exhibit poor EMI SE. The 20 dB of EMI SE is obtained by 35 wt% SBR/CB, thickness = 0.65 cm or a large thickness (7 cm) to that of 15.3 wt% SBR/CB composite from 8 to 12 GHz frequency [93]. The vulcanized EPDM/CB composites exhibit better EMI SE that of SBR/CB composites. The 20 dB of EMI SE was observed by 35 wt% of EPDM/CB composite having sample thickness of 0.55 cm measured frequency from 8 to 12 GHz. The composite of CB with a blend of EVA and NBR shows better performance than NBR/CB and EPDM/CB composites.

The EMI SE of CB based composites of chlorinated polyethylene (Cl-PE), linear low-density polyethylene (LLDPE), polypropylene (PP), epoxy resins, and blend of polystyrene (PS) and PP was studied [12, 95–97]. It was found that electrical conductivity and EMI SE of PP/CB composite is higher to that of PS/CB composite. Actually, the filler prefers crystalline part in comparison to that of amorphous part of the polymer. When conducting filler enters in amorphous part, it undergoes proper dispersion than the crystalline part [97]. In crystalline part of the polymer, the fillers tend to agglomerate which causes enhanced electrical and EMI SE property. In fact, electrical conductivity and EMI SE of Polymer/filler composites depend on aggregation behavior of conducting filler in the insulating polymer matrix. The filler-filler and polymer-filler interaction play a vital role in the magnitude of physical properties (mechanical, electrical, thermal, dielectric, etc.).

7.3 Polymer/Carbon Nanotube (CNT) Based Composites

Carbon nanotubes (CNTs) are regarded as the conducting filler used for EMI shielding application because of its large surface area and high aspect ratio [98–100]. Multiwall carbon nanotubes (MWCNTs) is nothing but vapor-grown carbon nanofibers (VGCFs) having a typical diameter of 50–200 nm. Vapor-grown carbon nanofibers (VGCFs) have good electrical and thermal conductivity. The aspect ratio is around 1000 and its cost is low [99–101].

Carbon nanotubes are of two types, namely, single-walled carbon nanotube (SWCNT) and multiwalled carbon nanotube (MWCNT) [102]. The SWCNT can be considered as graphene sheet is rolled up to hallow cylinder [102–104]. The concentric cylinders of graphene are considered to be MWCNT. The CNTs are mainly synthesized by three methods like laser ablation, chemical vapor deposition, and electric arc discharge. The aspect ratio of CNT is very high as compared to other carbon fillers which cause substantial advantages than conventional carbon fillers. It is observed that percolation threshold of polymer-CNT-based composites is very low (<2 wt%) [104–106].

The melt mixing technique is used for the preparation of polymer-CNT composite for commercial viability. There are various polymer-based composites of CNT are prepared by melt mixing technique like- poly(ethylene terephthalate) (PET), poly(ethylene) (PE), poly(trimethylene terephthalate) (PMTT), polycarbonate (PC), poly(ethylene-vinyl acetate) (EVA), poly(propylene) (PP), poly(ether ether ketone) (PEEK), poly(ethylene methyl acrylate) (EMA), acrylonitrile-butadiene-styrene copolymer (ABS), poly(L-lactide) (PLLA), andpoly (caprolactone) (PCL) [107–109]. It was observed that CNTs shows high EMI shielding in PMTT (20 dB at 5 wt%, 12–18 GHz, $t = 2$ mm) due to absorption of EMI radiation [12, 110, 111]. The method of dispersion of CNT in polymer matrix plays an important role to achieve desired electrical conductivity and EMI shielding effectiveness. If the polymer-CNT composite is prepared by melt mixing process, then there is a chance of decrease in an aspect ratio (length/diameter) of CNT. The extent of decrease in an aspect ratio of CNT increases with increase in shear rate during mixing. The reduction in an aspect ratio of CNT in polymer matrix leads to decrease in electrical conductivity which in turns reduction of EMI shielding effectiveness (SE). If the polymer-CNT composite is prepared by co-precipitation method the aspect ratio of CNT-based composites retains which signifies the highest electrical conductivity and EMI SE.

Solution casting method is also a very effective method of preparation of polymer-CNT composite. The aspect ratio of CNT remains as such which results in the best electrical conductivity and EMI SE. There is many CNT-based polymer (PUR, PC, PS, PMMA, PVDF, PVP, cellulose triacetate, LDPE) composites are prepared. It was observed that the EMI SE and electric properties of PVDF/PVP blends were much higher compared to homopolymer composites (20 dB at 0.4 wt%, 0.1–1.5 GHz). The functionalization of CNT by carboxylic group reduces the electrical conductivity by disrupting the electronic network but it improves the dispersion of CNT in the polymer matrix which in turn increases the permittivity of composite resulting higher EMI SE to that of non-functionalized CNT-based composite. The EMI shielding of composite not merely dependent on electrical conductivity, but it also depends on permittivity, permeability, impedance, and angular frequency. The electrical conductivity of polymer composite is dependent on its percolation threshold. The improvement of the aspect ratio of conducting filler results reduction in percolation threshold but the increment in EMI SE and electrical conductivity. The heat treatment of CNT at 1000 °C before dispersion in polymer matrix improves electrical conductivity but it adversely affects the EMI SE.

There are several studied have been done cured resin/CNT composite. Three different aspect ratio/or wall integrity of CNT was compared with respect to EMI SE. It was observed that the highest aspect ratio and wall integrity of CNT imparts most efficient EMI SE. The EMI SE of polymer composite increases with a better dispersion of CNT in the polymer matrix. The EMI SE of resin with and without functionalized CNT is presented in Fig. 6.

In situ polymerization is another interesting method of preparation of conducting composites used for EMI shielding application. In this method, the monomer is polymerized in presence of conducting nano-fillers. By using this technique, better electrical conductivity and EMI SE achieved by doping polyaniline with CNT. The absorption of electromagnetic radiation is significantly improved by the addition of CNT in polyaniline which is shown in Fig. 7. The dispersion of CNT in polyaniline can be improved by oxyfluorination of CNT to achieve better EMI SE. The EMI SE of PANI/CNT composites can be improved by treatment with the O_2/F_2 ratio which is presented in Fig. 8.

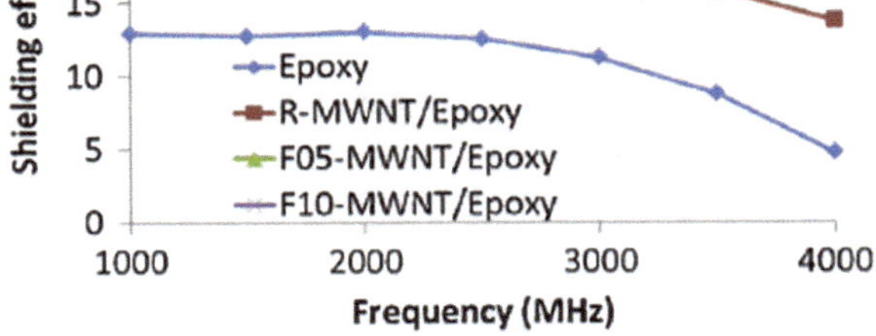

Fig. 6 EMI SE as a function of the frequency of epoxy resin filled with raw MWCNT (R-MWCNT) and MWCNTs treated with fluorine gas at a pressure of 0.5 bar (F05-MWCNT) and 1 bar (F01-MWCNT). Reproduced from [12]

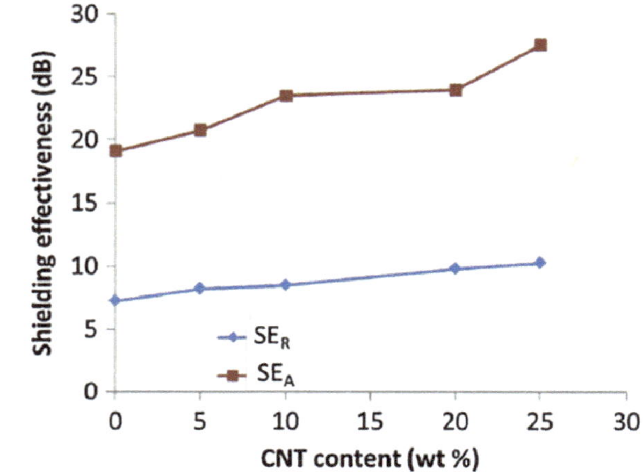

Fig. 7 Contribution of absorption (SE_A) and reflection (SE_R) to the EMI SE of PANI/MWCNTs nanocomposites prepared by in situ polymerization (thickness 2 mm). Reproduced from [12]

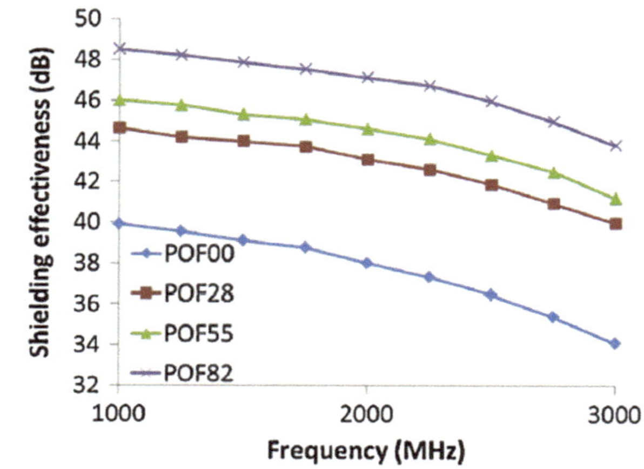

Fig. 8 EMI SE of polyaniline filled with 1 wt% of raw MWNTs (POF00) and MWNTs treated with a different mixture of O2/F2: 8/2 (POF82), 5/5 (POF55) and 2/8 (POF28). Reproduced from [12]

7.4 Polymer/Graphene-Based Composites

Monolayer Graphene sheet has attracted the scientific community because of its excellent low density, large aspect ratios, high surface area, versatile processing, ultrahigh electrical conductivity, thermal, and excellent mechanical stiffness [112, 113]. The percolation threshold of graphene (anatomically thick two-dimensional structure) is low. This attributes adequate EMI shielding at a low content of graphene in the polymer matrix. There are many kinds of polymer/graphene composite like foam, sandwich, porous, and segregated structured materials have been developed for EMI shielding application. The polymer/graphene composites are costlier material, hence, the recent research progress is to develop cheaper and viable manufacturing of composites by making the composite more conducting by functionalization of either polymer or carbon fillers or both. It is reported in the literature [114] that the annealing (at 250 °C) sample of epoxy resin—graphene composite enhances electrical conductivity and it shows EMI SE of 20 dB (15 wt%, 8–12 GHz). When graphene (10 wt%) is dispersed in polyurethane, then the composite exhibits higher EMI SE than the composites obtained from a combination of polyurethane-carbon nanotube composite [115]. The better EMI SE can be achieved by graphene with electrically conductive polymers [116]. Graphene sheet (GS) were dispersed in conducting polymer like polyaniline [117–119] and (Poly (3-hexylthiophene) (P3HT) [116] in order to improve the shielding properties. The aggregation of P3HT particles on the surface of graphene sheet was reduced by adding drops of methanol in a chloroform solution of GS/P3HT. The combination of equal quantity of GS and P3HT showed improved SE which was 20 dB (8–12 GHz, $t = 2$ mm). The polyaniline/GS composites exhibit superior SE than P3HT/GS composite. The synergistic behavior of SWCNT and GS was also observed. The SWCNT acts as conducting wire which connects GS.

It is reported that 20 wt% lightweight composites of reduced graphene oxide (rGO) exhibit EMI shielding of 38 dB [120]. The ultrathin composites of graphene show absorption value of 15 dB and reflection of 30 dB [121]. The magnetic nanoparticles ($NiFe_2O_4$) decorated r-GO composites show remarkable EMI shielding attenuation capacity [122]. The recent progress of the research is to develop flexible EMI shielding material having high electrical conductivity and low-density carbon-based polymer composite. There are various approaches like vacuum-assisted, layer-by-layer assembly, and solution casting, etc., have been developed to fabricate graphene-reinforced thin film composites [123–125]. However, orientation and distribution of graphene are difficult to control in the polymer matrix. It is also very difficult to prepare thin films of polymer/graphene composite having a high electrical conductivity which is due to poor dispersion, interconnectivity, and insufficient reduction of rGO in the insulating polymer matrix. It is the fact that uniform dispersion of graphene is very difficult, as it agglomerates and forms multilayer without any preferential arrangement instead of a single layer in the polymer matrix. Hence, the preferred orientation of nanofillers in the polymer matrix is essential to achieve the highest electrical conductivity and EMI shielding. The internal morphology of polymer/graphene composite depends on polymer-filler interaction and content of filler. Low filler content in the polymer matrix exhibits poor connectivity and random filler distribution whereas high filler content above percolation threshold confirms the inter-filler connectivity layer.

The EMI SE of flexible asymmetric PVDF-HFP/rGO composite exhibits totals EMI SE (SE_T); $\sim$30 dB, EMI SE due to reflection (SE_R); $\sim$10 dB and EMI SE due absorption (SE_A) $\sim$20 dB [126]. The graphene is aligned in PVDF-HFP matrix which is confirmed by electrical conductivity measurement. The SE_T, $\sim$30 dB of PVDF-HFP/rGO composite is sufficient EMI SE for practical application.

Effect of Different Factors on EMI Shielding Effectiveness of Polymer Composites

Nature and Types of Fillers

EMI Shielding effectiveness of polymer composites various according to the nature and types of fillers used in polymer matrices. The used fillers in insulating polymer matrices may be metallic, carbonaceous, conductive polymers. etc. We investigated the EMI SE of EVA and NBR polymers filled with carbon compared to carbon black filled ones. Actually, the EMI SE polymer composite depends on the conductivity of fillers and their connectivity in the polymer matrix. Both the conductivity and connectivity of fillers improve the EMI SE of the polymer composites by absorbing EM radiation. The conductivity and connectivity of filler vary according to their nature and types. Through the conductivity of some fillers are higher compared to other fillers, but whether it will show higher EMI SE or not, that will depend on the connectivity in the polymer matrix, which will be discussed later on the section "geometry of filler".

Concentration/Loading of Additives

Conductive additive concentration greatly affects the EMI SE polymer composites. Insulating polymer matrices are transparent to EM radiation. These polymers cannot interfere with the radiation, hence their SE is very low (0–3 dB) depending on the nature of polymer that is either polar or nonpolar. The addition of conductive additive in the polymer matrices increases the EMI SE of polymer composites. It was reported that, initially, the increment in EMI SE is marginal with the addition of fillers because the conductive particle cannot from connectivity at low loading a number of absorption site is also less. On raising the concentration of fillers, the number of the conductive site for absorption of variation is increased, and hence the EMI SE is also increased. After a certain loading, there is the increase in the number of the more conductive site as well as the formation of connective conductive filler network. This is called the percolation threshold of EMI SE. At this stage is the drastic increase is EMI SE is observed. Further addition of conductive filler into the polymer matrices increases the number of conductive site and connectivity by the formation of the cluster. However, the increment in EMI SE is less compared to their percolation stage. Actually, after percolation threshold, the increment in electrical conductivity is marginal and hence the EMI SE mostly governed by the connectivity of the conductive additives, that means the increase in a number of absorption site. Thus the increment in the electrical conductivity and the increment in EMI SE after percolation threshold do not follow the same path. It was observed that the increment is EMI SE is always higher compared to the electrical conductivity after percolation threshold at their similar loading of filler [127, 128]. The conductive composites filled with different fillers showed the different value of EMI SE at their similar filler loading.

Geometry of Fillers

The connectivity and black conductivity of fillers within the polymer matrices are greatly affected by the geometry of filler particles. Actually conductive fillers have different geometry like spherical, rod-like, tube-like, flat surface, granular, etc. with different shape and size. Spherical, particles forms aggregates with different size known as the structure of fillers. Rod-like/tube-like fillers differ in diameter and length known as the aspect ratio of fillers (length/diameter). Graphene has monoatomic thick planer geometry with the flat surface area. The connectivity and bulk conductivity increase with the increase in structure/aspect ratio of fillers and hence we can get higher EMI SE of the composites at a relatively lower loading of filers. We tested the effected of the geometry of low carbon blacks with different structure and short carbon filler with high aspect ratio. It was revealed that high structure carbon exhibited a higher value of EMI SE compared to low structure one at their similar loading of fillers. Moreover, the carbon fiber with high aspect ratio exhibited the highest EMI SE due to its better connectivity compared to the others at their similar filler loading. The effect of carbon black structure on the EMI SE of polymer composites was also investigated by other authors and reported the same phenomenon [1, 39].

Nature of Polymers

The characteristics of polymers play an important role in the dispersion of filler within the polymer matrices. The nature of dispersion of fillers will be different in polar and nonpolar polymers. The interface energy between the filler and polymer varies according to their polarity. Polar polymers are having higher interface energy compared to nonpolar one and hence produce a better dispersion of filler within the polymer matrices. Thus, the EMI SE will be different for polar and nonpolar polymer filled with the same type of filler article. However, which polymer composite will exhibit higher EMI SE that not only governed by their polarity but also by other factors of polymer like its viscosity, crystallinity, molecular weight, etc., also play the important roles. It has been reported that the filler mostly distributes in the amorphous region of polymer and less to the crystalline region. Some parts of the filler distribute at the interface between crystalline and amorphous region. Hence for crystalline polymers, the electrical percolation threshold of conductivity is observed at a relatively lower loading of fillers compared to amorphous polymers. However, the results of EMI SE may be contradictory. The SE of the crystalline polymer may be higher or lower compared to amorphous polymer depending on the level of crystallinity of polymer and connectivity of filler within the polymer matrix. Higher crystallinity results in low electrical percolation threshold and higher electrical conductivity to the polymer composites at the same level of filler loading but on the contrary, it gives better connectivity at the amorphous region and worse connectivity in the crystalline region. Thus, the EM radiation, though absorbed and dissipated in the amorphous region due to better connectivity in the filler networks, but some part of it transmitted through the crystalline region due to less quantity of filler and connectivity. We reported the EMI SE of low crystalline EVA and amorphous NBR filled with car blacks and short carbon fiber. It was shown that the EMI SE of NBR composites was higher compared semi-crystalline EVA at their similar filler loading. This was attributed to the better dispersion and distribution of filler in the amorphous polymer NBR. The role of polymer viscosity is also a crucial factor. The dispersion of fillers becomes the case with low the viscous polymer. High viscous polymer polymers cause the breakdown of conductive particles during the processing of composite due to high shearing force. Thus, the particle size (structure/aspect ratio) is processed with high viscous polymers, which adversely affect the EMI SE, because it negatively affects the filler connectively within the polymer matrices. It is reported that the polymer with low viscosity exhibited higher EMI SE compared to high viscosity one at the similar loading of filler. Polymer molecular weight (Mw) also plays on an imported role in controlling the EMI SE of polymer composites. Actually, a polymer with high Mw will exhibit high viscosity compared to low Mw polymer.

Blending of Polymers

The blending of polymers is done by mixing two or more polymers together. When two polymers are mixed together, then it is known as a binary blend and when their polymers are mixed together then it is known as a ternary blend and likewise. Blending may be polar–polar, nonpolar–nonpolar and polar–nonpolar. Polar–polar and nonpolar–nonpolar blending are expected to be miscible, whereas, polar–nonpolar blending is immiscible. For miscible/compatible/blends, there is no any distinguishable interface between the polymers, and hence the distribution of filler particles expected uniformly within the polymer matrices depending upon the level of crystallinity in the polymers. But for immiscible blend/incompatible blend, the filler particle may distribute within both phases as well as at the interface of both polymers. In literature [129], it was reported that the blending of two immiscible polymers resulted in low electrical percolation threshold and high and electrical conductivity compared to miscible blend because of the formation of a continuous network of filler at the interface of two polymers. We investigated the electrical conductivity and EMI SE of EVA/NBR blend composites of their different proportions filled with carbon blacks and carbon fiber as mentioned earlier [86]. We found that NBR-rich composites exhibited higher conductivity EMI SE compared to EVA rich composite at their similar loading of carbon filler. Through the blend composites were compatible, it was shown that the fillers particle was distributed more to the amorphous region of the polymer. As NBR was almost completely amorphous in nature, hence NBR-rich blend exhibited a higher value of EMI SE and electrical conductivity compared to EVA rich to EVA rich blend composite. A. Saleh et al. reported the EMI SE of PC/ABS blend composite filled with CNTs nanofiber [130]. The increase in the amount of PC in ABS resulted in the decrease in EMI SE of the blend nanocomposites. It was reported that PC adversely affected the processability of composites by increasing the torque inside the mixing chamber, resulting in the higher shearing force on the filler particles and consequently more break down of particles and lowering of the average aspect ratio.

There is also the phase separation between PC and ABS, and this phase separation increase with the increases with the increase in PC composition in ABS matrix.

Thickness of Sample

The increase in thickness of a test sample/composite may be considered as a combination of different layer players placed alternatively one after another. If any EM radiation is incident on the first layer of the sample, then some part of it will be absorbed and the rest part will be transmitted through it. The emitted radiation from the first layer will then pass through the next layer and some part gets absorbed by it while the remaining part if any will pass through the subsequent layer and so on. Thus, with the increase in thickness, the absorption of EM radiation increases within the polymer composites and will result in the overall increase in EMI SE. The increment in EMI SE with the increase in sample thickness follows a linear relationship as shown in our previous investigation with the formula

$$Y = mX + C,$$

where '*Y*' is the EMI SE of the composite at the particular thickness "*X*" "*m*" is the slope of the straight line and '*C*' is the intercept. It was reported that the slope value increases with the filler loading in the composites and the increase in the structure of filler. Actually, with the increase in filler loading, the concentration of fillers per unit volume increase, and hence the mess size decrease, leading to the less passage of EM radiation and higher value of the slope.

Functionalization/Modification

Mostly conductive additives are used as fillers in polymer matrices for making EMI SE materials. These conductive additives are associated with either conjugated double-bond or the delocalized lone-pair of electrons. The modification of these additives is done by the reaction of a modifier with the double-bond or lone-pair of electrons of the additives. Thus, due to modification, the conjugation of the additives is lost, and hence the conductivity and connectivity are reduced. Thus, theoretically, may adversely affect the EMI SE of the polymer and polymer composite. The conductive polymer polyaniline, polypyrrole, etc., are also modified with different acids to improve their electrical conductivity and/or dispersion within the matrices. There were some contradictory reports about the effect of the modification on EMI SE. In some cases, there is the improvement in EMI SE after modification, whereas, in other cases, there is the decrement in EMI SE.

Method of Preparation/Synthesis

The EMI SE of polymer composites is greatly affected by the technique in which the composites are being prepared. The dispersion of filler and the breakage of filler particle vary according to the preparation methods of the polymer composites. The breakage of filler particles in the polymer matrix during mixing process depends on the exerted shearing force on it. Higher the shearing force during the mixing process, higher is the breakage of filler particles in the polymer composites. A polymer composite prepared by melt mixing technique and solvent casting technique widely differ in their mechanical, electrical, thermal and EMI SE properties. When a composite is prepared by melt mixing technique, the exerted shearing force on the filler particle is comparatively very high compared to solvent casting technique. Thus, a composite prepared through solvent casting technique would exhibit higher EMI SE compared to the composite preparer through melt mixing technique.

Effect of Frequency

The EMI SE of polymer composites is greatly affected by the frequency of electric/electromagnetic field Sau et al. [14] had reported the EMI SE of carbon black and fiber-filled polymer composites at S-band (100–2000 MHz) and X-band (8–12 GHz) regions. The EMI SE at X-band region were reported much higher compared to S-band region. Actually, the conductivity is increased with the increase in frequency and EMI SE almost proportionally increases with the increase in conductivity. This is why the EMI SE is increased with the increase in frequency.

Electrical Conductivity

The EMI SE of the polymer-filler composite is highly influenced by electrical conductivity. According to the literature [42, 51], Eq. 6, it is clear that EMI SE is directly proportional to the electrical conductivity of the composite. In extrinsically conducting composite, the electrical conductivity of the composite depends on percolation threshold. Before percolation threshold electrical conductivity gradually increases but at percolation threshold, electrical conductivity increases instantly and after percolation marginal improvement in electrical conductivity is observed. Whereas EMI SE is independent of percolation threshold. There is a tremendous increment in EMI SE after percolation, unlike electrical conductivity.

Processing Parameters/Synthesis Parameters

It has been mentioned earlier that the polymer and polymer composites are prepared by different techniques. However, the processing/synthesis of polymer composites and polymers are dependent on different parameters. As for example, the processing of polymer composites by melt mixing technique using twin-screw extruder is dependent on mixing time, mixing temperature, rotor speed, the sequence of addition of ingredients, volumetric capacity inside the barrel, molding pressure, etc. In dry mixing, through two the roll-mill, the mixing time, temperature, rpm of the roller and their speed ratio play important role in the mixing of ingredients. The properties of the composite prepared through solution mixing process depend on the quantity and types of taken solution in a beaker, rpm of the mechanical/magnetic stirrer, mixing time, temperature, etc. In the ultra-sonication technique (either bath type or probe type), the mixing time, temperature, frequency, pulse rate, types and quantity of solvent, etc., play a major role in controlling the properties of the composites.

In situ *Process*

In this method, the synthesis of one component is carried out in the presence of other ingredients. For the preparation of polymer composite, either the polymer or the additive is synthesized in presence of the other component. Mostly, the insulating polymer is synthesized from its monomer in the presence of conductive additives. In many cases, the conducting polymers are synthesized from its monomer in the presence of insulating polymer to prepare insulator–conductor polymer composite. In this case, the synthesized conducting polymer is used as a conductive additive. This method is also used to prepare a functional composite. The advantage associated with better homogeneity and grafting between polymer and additive can be carried out. Hence the composite with the high volume fraction of filler can be prepared. The disadvantage of this method is that the removal of final traces of reaction by-product produce in this method is very difficult and hence these acts as an impurity which reduce the electrical conductivity of the composite.

8 Conclusion

The adsorption and/or reflection of EM radiation by a material are referred to as EMI shielding. Actually, EMI shielding material shields the penetration of EMI radiation through the material. There are different methods have been developed to prepare polymer/carbon filler based EMI shielding materials. Solution casting method is preferred during mixing of SCF in the polymer matrix to avoid reduction of aspect ratio of SCF. Moderately high electrical conductivity (~ 1 S/m) and/or high permittivity is the basic requirements for making an efficient screen. The EMI SE of composites depends on permittivity, permeability, electrical conductivity, and thickness of material. Polymer/carbon filler based composites are better than metallic EMI shielding materials because of its low cost of production, non-rusting, inertness to polar solvents, light weight, durability, and flexibility. Compounding and processing parameters like mixing time, temperature, and rotor speed are another factors governing EMI SE of polymer/filler composites. The intrinsic electrical conductivity, geometry (structure), and aspect ratio of fillers affects the overall conductivity and EMI SE of composites. The order of intrinsic conductivity of carbon fillers are GS > CNT > CF > CB.

The 20–30 dB attenuation is enough to prevent 99.9% EMI radiations which can be achieved by the electrical conductivity of 1–2 S/cm of composites. This range of attenuation is sufficient for most of the industrial and consumer application. The investigation is going on to prepare cost-effective, transparent, and flexible thin film of polymer composites used for EMI shielding application. To achieve this target, functionalization of the insulating polymer matrix, as well as a coating of conducting fillers, are required. The EMI shielding materials are used in the form of foam, woven structures, and multilayered casing. Nowadays EMI shielding materials have high demands in the military and other commercial purposes. It is used in highly sensitive electronic devices and as a packaging material.

Acknowledgements Authors are thankful to Rubber Technology Centre, Indian Institute of Technology Kharagpur and Department of Chemistry, College of Science, King Saud University, Riyadh, 11451, Saudi Arabia, for financial support to write this book chapter.

References

1. Yang Y, Gupta MC, Dudley KL, Lawrence RW (2005) Novel carbon nanotube–polystyrene foam composites for electromagnetic interference shielding. Nano Lett 5(11):2131–2134
2. Geetha S, Satheesh KKK, Rao CR, Vijayan M, Trivedi DC (2009) EMI shielding: methods and materials—a review. J Appl Polym Sci 112(4):2073–2086
3. Mottahed BD, Manoochehri S (1995) A review of research in materials, modeling and simulation, design factors, testing, and measurements related to electromagnetic interference shielding. Polym Plast Technol Eng 34(2):271–346
4. Maiti S, Shrivastava NK, Suin S, Khatua B (2013) Polystyrene/MWCNT/graphite nanoplate nanocomposites: efficient electromagnetic interference shielding material through graphite nanoplate–MWCNT–graphite nanoplate networking. ACS Appl Mater Inter 5(11):4712–4724
5. Mazurkiewicz PH, Hewlett-Packard Development Company LP (2005) Board-level conformal EMI shield having an electrically-conductive polymer coating over a thermally-conductive dielectric coating. US Patent 6,849,800
6. Chung D (2001) Electromagnetic interference shielding effectiveness of carbon materials. Carbon 39(2):279–285
7. Huang JC (1995) EMI shielding plastics: a review. Adv Polym Technol 14(2):137–150
8. Gamble J, Yats LD (1995) The Dow Chemical Company. EMI shielding composites. US Patent 5,399,295
9. Luch D (2004) Electromagnetic interference shields and methods of manufacture. US Patent 6,697,248
10. Bigg D (1984) The effect of compounding on the conductive properties of EMI shielding compounds. Adv Polym Tech 4(3–4):255–256
11. Pukánszk B, Maurer FH (1995) Composition dependence of the fracture toughness of heterogeneous polymer systems. Polymer 36(8):1617–1625
12. Thomassin J-M, Jérôme C, Pardoen T, Bailly C, Huynen I, Detrembleur C (2013) Polymer/carbon based composites as electromagnetic interference (EMI) shielding materials. Mater Sci Eng, R 74(7):211–232
13. Prasad V (2012) Low temperature charge transport and microwave absorption of carbon coated iron nanoparticles–polymer composite films. Mater Res Bull 47(6):1529–1532
14. Chen SC, Chien RD, Lee PH, Huang JS (2005) Effects of molding conditions on the electromagnetic interference performance of conductive ABS parts. J Appl Polym Sci 98(3):1072–1080
15. Huynen I, Quiévy N, Bailly C, Bollen P, Detrembleur C, Eggermont S, Molenberg I, Thomassin JM, Urbanczyk L, Pardoen T (2011) Multifunctional hybrids for electromagnetic absorption. Acta Mater 59(8):3255–3266
16. Saini P, Choudhary V, Singh B, Mathur R, Dhawan S (2011) Enhanced microwave absorption behavior of polyaniline-CNT/polystyrene blend in 12.4–18.0 GHz range. Synth Met 161(15):1522–1526

17. Saini P, Choudhary V (2013) Structural details, electrical properties, and electromagnetic interference shielding response of processable copolymers of aniline. J Mater Sci 48(2):797–804
18. Markham D (1999) Shielding: quantifying the shielding requirements for portable electronic design and providing new solutions by using a combination of materials and design. Mater Des 21(1):45–50
19. Bagwell RM, McManaman JM, Wetherhold RC (2006) Short shaped copper fibers in an epoxy matrix: their role in a multifunctional composite. Compos Sci Technol 66(3):522–530
20. Biswas KK, Somiya S (2001) Effect of isothermal physical aging on creep behavior of stainless-fiber/PPE composites. J Phys Soc Jpn 50(9Appendix):172–177
21. Chen JH, Hsu KC, Hsieh MY (2016) Effects of preparation parameters of a one-pot approach on the conductivity, structure, and chemical composition of silver/reduced-graphene oxide composite. Ind Eng Chem Res 55(16):4390–4402
22. Huang CJ, Chang TC (2004) Studies on the electromagnetic interference shielding effectiveness of metallized PVAc-AgNO3/PET conductive films. J Appl Polym Sci 91(1):270–273
23. Yang Y, Gupta M, Dudley K (2007) Studies on electromagnetic interference shielding characteristics of metal nanoparticle-and carbon nanostructure-filled polymer composites in the Ku-band frequency. Micro Nano Lett 2(4):85–89
24. Kim M, Kim H, Byun S, Jeong S, Hong Y, Joo J, Song KT, Kim JK, Lee CJ, Lee JY (2002) PET fabric/polypyrrole composite with high electrical conductivity for EMI shielding. Synth Met 126(2):233–239
25. Bhadra S, Singha NK, Khastgir D (2008) Semiconductive composites from ethylene 1-octene copolymer and polyaniline coated nylon 6: studies on mechanical, thermal, processability, electrical, and EMI shielding properties. Polym Eng Sci 48(5):995–1006
26. Phang SW, Tadokoro M, Watanabe J, Kuramoto N (2009) Effect of Fe_3O_4 and TiO_2 addition on the microwave absorption property of polyaniline micro/nanocomposites. Polym Adv Technol 20(6):550–557
27. Geetha S, Kumar KKS, Trivedi DC (2005) Conducting fabric-reinforced polyaniline film using p-chlorophenol as secondary dopant for the control of electromagnetic radiations. J Compos Mater 39(7):647–658
28. Lakshmi K, John H, Mathew K, Joseph R, George K (2009) Microwave absorption, reflection and EMI shielding of PU–PANI composite. Acta Mater 57(2):371–375
29. Niu Y (2006) Preparation of polyaniline/polyacrylate composites and their application for electromagnetic interference shielding. Polym Compos 27(6):627–632
30. Vulpe S, Nastase F, Nastase C, Stamatin I (2006) PAN–PAni nanocomposites obtained in thermocentrifugal fields. Thin Solid Films 495(1):113–117
31. Ma PC, Siddiqui NA, Marom G, Kim JK (2010) Dispersion and functionalization of carbon nanotubes for polymer-based nanocomposites: a review. Compos A 41(10):1345–1367
32. Baughman RH, Zakhidov AA, De Heer WA (2002) Carbon nanotubes-the route toward applications. Science 297(5582):787–792
33. Lau AKT, Hui D (2002) The revolutionary creation of new advanced materials-carbon nanotube composites. Compos B 33(4):263–277
34. Chou TW, Gao L, Thostenson ET, Zhang Z, Byun JH (2010) An assessment of the science and technology of carbon nanotube-based fibers and composites. Compos Sci Technol 70(1):1–19
35. Sandler J, Kirk J, Kinloch I, Shaffer M, Windle A (2003) Ultra-low electrical percolation threshold in carbon-nanotube-epoxy composites. Polymer 44(19):5893–5899
36. Lorenz CD, Ziff RM (2001) Precise determination of the critical percolation threshold for the three-dimensional "Swiss cheese" model using a growth algorithm. J Chem Phys 114(8):3659–3661
37. Stanley HE (1977) Cluster shapes at the percolation threshold: and effective cluster dimensionality and its connection with critical-point exponents. J Phys A: Math Gen 10(11):L211

38. Duan Y, Liu S, Guan H (2006) Investigation of electromagnetic characteristics of polyaniline composite. J Compos Mater 40(12):1093–1104
39. Li N, Huang Y, Du F, He X, Lin X, Gao H, Ma Y, Li F, Chen Y, Eklund PC (2006) Electromagnetic interference (EMI) shielding of single-walled carbon nanotube epoxy composites. Nano Lett 6(6):1141–1145
40. Saini P, Choudhary V, Singh B, Mathur R, Dhawan S (2009) Polyaniline–MWCNT nanocomposites for microwave absorption and EMI shielding. Mater Chem Phys 113 (2):919–926
41. Yang Y, Gupta MC, Dudley KL, Lawrence RW (2005) A comparative study of EMI shielding properties of carbon nanofiber and multi-walled carbon nanotube filled polymer composites. J Nanosci Nanotechnol 5(6):927–931
42. Colaneri NF, Schacklette L (1992) EMI shielding measurements of conductive polymer blends. IEEE Trans Instrum Meas 41(2):291–297
43. Lee C, Song H, Jang K, Oh E, Epstein A, Joo J (1999) Electromagnetic interference shielding efficiency of polyaniline mixtures and multilayer films. Synth Met 102(1):1346–1349
44. Kim B, Lee H, Park S, Kim H (2011) Electromagnetic interference shielding characteristics and shielding effectiveness of polyaniline-coated films. Thin Solid Films 519(11):3492–3496
45. Li Y, Pei X, Shen B, Zhai W, Zhang L, Zheng W (2015) Polyimide/graphene composite foam sheets with ultrahigh thermostability for electromagnetic interference shielding. RSC Adv 5(31):24342–24351
46. Cao M-S, Wang X-X, Cao W-Q, Yuan J (2015) Ultrathin graphene: electrical properties and highly efficient electromagnetic interference shielding. J Mater Chem C 3(26):6589–6599
47. Singh AP, Garg P, Alam F, Singh K, Mathur R, Tandon RP, Chandra A, Dhawan SK (2012) Phenolic resin-based composite sheets filled with mixtures of reduced graphene oxide, γ-Fe_2O_3 and carbon fibers for excellent electromagnetic interference shielding in the X-band. Carbon 50(10):3868–3875
48. Du L, Du Y, Li Y, Wang J, Wang C, Wang X, Xu P, Han X (2010) Surfactant-Assisted Solvothermal Synthesis of Ba (CoTi) x Fe12–2xO19 Nanoparticles and Enhancement in Microwave Absorption Properties of Polyaniline. J Phys Chem C 114(46):19600–19606
49. Naito Y, Suetake K (1971) Application of ferrite to electromagnetic wave absorber and its characteristics. IEEE Trans Microw Theory Techn 19(1):65–72
50. Wu F, Xu Z, Wang Y, Wang M (2014) Electromagnetic interference shielding properties of solid-state polymerization conducting polymer. RSC Adv 4(73):38797–38803
51. Mottahed BD (2000) Enhanced shielding effectiveness of polymer matrix composite enclosures utilizing constraint-based optimization. Polym Eng Sci 40(1):61–69
52. Nam I, Lee H, Jang J (2011) Electromagnetic interference shielding/absorbing characteristics of CNT-embedded epoxy composites. Compos A 42(9):1110–1118
53. Bahadorzadeh M, Lotfi-Neyestanak AA (2012) A novel and efficient technique for improving shielding effectiveness of a rectangular enclosure using optimized aperture load. Elektronika ir Elektrotechnika 18(10):89–92
54. Luo X, Chung D (1999) Electromagnetic interference shielding using continuous carbon-fiber carbon-matrix and polymer-matrix composites. Compos B 30(3):227–231
55. Li M, Drewniak JL, Radu S, Nuebel J, Hubing TH, DuBroff RE, Doren TPV (2001) An EMI estimate for shielding-enclosure evaluation. IEEE Trans Electromagn Compat 43(3):295–304
56. Bigg D (1987) The effect of chemical exposure on the EMI shielding of conductive plastics. Polym Compos 8(1):1–7
57. Gerteisen S, Nangrani K (1988) Plastics that shield against EMI/RFI. Society of Plastic Engineers, Inc., Chicago Section Electric and Electronics Division, Rosemont IL
58. Kim H, Kim K, Lee C, Joo J, Cho S, Yoon HS, Pejaković DA, Yoo JW, Epstein AJ (2004) Electrical conductivity and electromagnetic interference shielding of multiwalled carbon nanotube composites containing Fe catalyst. Appl Phys Lett 84(4):589–591

59. Joo J, Lee C (2000) High frequency electromagnetic interference shielding response of mixtures and multilayer films based on conducting polymers. J Appl Phys 88(1):513–518
60. Lee C, Lee D, Jeong C, Hong Y, Shim J, Joo J, Kim MS, Lee JY, Jeong SH, Byun SW, Zang DS, Yang HG (2002) Electromagnetic interference shielding by using conductive polypyrrole and metal compound coated on fabrics. Polym Adv Technol 13(8):577–583
61. Sankaran S, Dasgupta S, Sekhar KR, Kumar MJ (2006) Thermosetting polymer composites for EMI shielding applications. In: Proceedings of the 9th international conference on electromagnetic interference and compatibility (INCEMIC), IEEE, pp 1–6
62. Dixon D, Masi J (1989) Development of a composite material with long-term EMI shielding properties. SAMPE J 25:31–37
63. Bai J, Allaoui A (2003) Effect of the length and the aggregate size of MWNTs on the improvement efficiency of the mechanical and electrical properties of nanocomposites—experimental investigation. Compos A 34(8):689–694
64. Tsotra P, Friedrich K (2003) Electrical and mechanical properties of functionally graded epoxy-resin/carbon fibre composites. Compos A 34(1):75–82
65. Olivero DA, Radford DW (1997) Integrating EMI shielding into composite structure. SAMPE J 33(1):51–57
66. Sarto MS, Michele S, Leerkamp P, Thuis H (2001) An innovative shielding concept for EMI reduction. IEEE EMC Soc Newsl Summer 22–28
67. Guire T, Varadan VV, Varadan V (1990) Influence of chirality on the reflection of EM waves by planar dielectric slabs. IEEE Trans Electromagn Compat 32(4):300–303
68. Bigg DM (1979) Mechanical, thermal, and electrical properties of metal fiber-filled polymer composites. Polym Eng Sci 19(16):1188–1192
69. Foy JV, Lindt JT (1987) Electrical properties of exfoliated-graphite filled polyester based composites. Polym Compos 8(6):419–426
70. Chiou J-M, Zheng Q, Chung D (1989) Electromagnetic interference shielding by carbon fibre reinforced cement. Composites 20(4):379–381
71. Jana P, Mallick A, De S (1991) Electromagnetic interference shielding by carbon fibre-filled polychloroprene rubber composites. Composites 22(6):451–455
72. Ahmad M, Abdelazeez M, Zihlif A (1989) Microwave properties of the talc filled polypropylene. J Mater Sci 24(5):1795–1800
73. Awan FG, Sheikh NM, Qureshi SA Ali A (2008) A generic model for the classification of radiation emission data in electromagnetic compatibility measurement. In: Radio and wireless symposium, IEEE pp 315–318
74. Arai K, Xu G, Sugimoto O (1989) Micro-gap discharge waveshapes and radiated electromagnetic waves in atmospheric air and sulphur hexafluoride gas. Electrical Engineering in Japan 109(4):7–16
75. Jana PB, Chaudhuri S, Pal A, De S (1992) Electrical conductivity of short carbon fiber-reinforced polychloroprene rubber and mechanism of conduction. Polym Engin Sci 32 (6):448–456
76. Běhal M, Ducháček V (1988) Thermovulcanization of polychloroprene rubber and its blends with poly (vinyl chloride). J Appl Polym Sci 35(2):507–515
77. Tibbetts GG, Lake ML, Strong KL, Rice BP (2007) A review of the fabrication and properties of vapor-grown carbon nanofiber/polymer composites. Compos Sci Technol 67 (7):1709–1718
78. Al-Saleh MH, Sundararaj U (2009) A review of vapor grown carbon nanofiber/polymer conductive composites. Carbon 47(1):2–22
79. Wang L-L, Tay B-K, See K-Y, Sun Z, Tan L-K, Lua D (2009) Electromagnetic interference shielding effectiveness of carbon-based materials prepared by screen printing. Carbon 47 (8):1905–1910
80. Al-Saleh MH, Sundararaj U (2010) Processing-microstructure-property relationship in conductive polymer nanocomposites. Polymer 51(12):2740–2747
81. Al-Saleh MH, Sundararaj U (2011) Electrically conductive carbon nanofiber/polyethylene composite: effect of melt mixing conditions. Polym Adv Technol 22(2):246–253

82. Lee BO, Woo WJ, Kim MS (2001) EMI shielding effectiveness of carbon nanofiber filled poly (vinyl alcohol) coating materials. Macromol Mater Eng 286(2):114–118
83. Lee B, Woo W, Park H, Hahm H, Wu J, Kim M (2002) Influence of aspect ratio and skin effect on EMI shielding of coating materials fabricated with carbon nanofiber/PVDF. J Mater Sci 37(9):1839–1843
84. Wu J, Chung D (2002) Increasing the electromagnetic interference shielding effectiveness of carbon fiber polymer–matrix composite by using activated carbon fibers. Carbon 40(3):445–447
85. Zou T, Zhao N, Shi C, Li J (2011) Microwave absorbing properties of activated carbon fibre polymer composites. Bull Mater Sci 34(1):75–79
86. Rahaman M, Chaki TK, Khastgir D (2011) High-performance EMI shielding materials based on short carbon fiber-filled ethylene vinyl acetate copolymer, acrylonitrile-butadiene copolymer, and their blends. Polym Compos 32(11):1790–1805
87. Antolini E (2009) Carbon supports for low-temperature fuel cell catalysts. Appl Cat B: Environ 88(1):1–24
88. Ahmed S, Aitani A, Rahman F, Al-Dawood A, Al-Muhaish F (2009) Decomposition of hydrocarbons to hydrogen and carbon. Appl Cat A: Gen 359(1):1–24
89. Huang JC (2002) Carbon black filled conducting polymers and polymer blends. Adv Polym Technol 21(4):299–313
90. Sánchez-González J, Macías-García A, Alexandre-Franco M, Gómez-Serrano V (2005) Electrical conductivity of carbon blacks under compression. Carbon 43(4):741–747
91. Sahoo BP, Naskar K, Tripathy DK (2012) Conductive carbon black-filled ethylene acrylic elastomer vulcanizates: physico-mechanical, thermal, and electrical properties. J Mater Sci 47(5):2421–2433
92. Ghosh P, Chakrabarti A (2000) Conducting carbon black filled EPDM vulcanizates: assessment of dependence of physical and mechanical properties and conducting character on variation of filler loading. Eur Polym J 36(5):1043–1054
93. Mohanraj G, Chaki T, Chakraborty A, Khastgir D (2006) AC impedance analysis and EMI shielding effectiveness of conductive SBR composites. Polym Eng Sci 46(10):1342–1349
94. Madani M (2010) Conducting carbon black filled NR/IIR blend vulcanizates: Assessment of the dependence of physical and mechanical properties and electromagnetic interference shielding on variation of filler loading. J Polym Res 17(1):53
95. Ling Q, Sun J, Zhao Q, Zhou Q (2011) Effects of carbon black content on microwave absorbing and mechanical properties of linear low density polyethylene/ethylene-octene copolymer/calcium carbonate composites. Polym-Plast Technol Eng 50(1):89–94
96. Oh J-H, Oh K-S, Kim C-G, Hong C-S (2004) Design of radar absorbing structures using glass/epoxy composite containing carbon black in X-band frequency ranges. Compos B 35 (1):49–56
97. Al-Saleh MH, Sundararaj U (2008) Electromagnetic interference (EMI) shielding effectiveness of PP/PS polymer blends containing high structure carbon black. Macromol Mater Eng 293(7):621–630
98. Ma CCM, Huang YL, Kuan HC, Chiu YS (2005) Preparation and electromagnetic interference shielding characteristics of novel carbon-nanotube/siloxane/poly-(urea urethane) nanocomposites. J Polym Sci, Part B: Polym Phys 43(4):345–358
99. Heremans J, Beetz C Jr (1985) Thermal conductivity and thermopower of vapor-grown graphite fibers. Phys Rev B 32(4):1981
100. Heremans J (1985) Electrical conductivity of vapor-grown carbon fibers. Carbon 23(4):431–436
101. Endo M, Kim Y, Hayashi T, Nishimura K, Matusita T, Miyashita K, Dresselhaus MS (2001) Vapor-grown carbon fibers (VGCFs): basic properties and their battery applications. Carbon 39(9):1287–1297
102. Gamaly EG, Ebbesen TW (1995) Mechanism of carbon nanotube formation in the arc discharge. Phys Rev B 52(3):2083

103. Hirahara K, Suenaga K, Bandow S, Kato H, Okazaki T, Shinohara H, Iijima S (2000) One-dimensional metallofullerene crystal generated inside single-walled carbon nanotubes. Phys Rev Lett 85(25):5384
104. Scott CD, Arepalli S, Nikolaev P, Smalley RE (2001) Growth mechanisms for single-wall carbon nanotubes in a laser-ablation process. Appl Phys A Mater Sci Process 72(5):573–580
105. Ren ZF, Huang ZP, Wang DZ, Wen JG, Xu JW, Wang JH, Calvet LE, Chen J, Klemic JF, Reed MA (1999) Growth of a single freestanding multiwall carbon nanotube on each nanonickel dot. Appl Phys Lett 75(8):1086–1088
106. Kong J, Cassell AM, Dai H (1998) Chemical vapor deposition of methane for single-walled carbon nanotubes. Chem Phys Lett 292(4):567–574
107. Fan Z, Luo G, Zhang Z, Zhou L, Wei F (2006) Electromagnetic and microwave absorbing properties of multi-walled carbon nanotubes/polymer composites. Mater Sci Eng, B 132 (1):85–89
108. Arjmand M, Mahmoodi M, Gelves GA, Park S, Sundararaj U (2011) Electrical and electromagnetic interference shielding properties of flow-induced oriented carbon nanotubes in polycarbonate. Carbon 49(11):3430–3440
109. Thomassin J-M, Lou X, Pagnoulle C, Saib A, Bednarz L, Huynen I, Jérôme R, Detrembleur C (2007) Multiwalled carbon nanotube/poly (ε-caprolactone) nanocomposites with exceptional electromagnetic interference shielding properties. J Phys Chem C 111 (30):11186–11192
110. Gupta A, Choudhary V (2011) Electromagnetic interference shielding behavior of poly (trimethylene terephthalate)/multi-walled carbon nanotube composites. Compos Sci Technol 71(13):1563–1568
111. Song K, Zhang Y, Meng J, Green EC, Tajaddod N, Li H, Minus ML (2013) Structural polymer-based carbon nanotube composite fibers: understanding the processing–structure–performance relationship. Mater 6(6):2543–2577
112. Novoselov KS, Geim AK, Morozov SV, Jiang D, Zhang Y, Dubonos SV, Grigorieva IV (2004) Electric field effect in atomically thin carbon films. Firsov, Science (Washington, DC, U. S.) 306:666–669
113. Moon JS, Gaskill DK (2011) Graphene: its fundamentals to future applications. IEEE Trans Microwave Theory Tech 59:2702–2708
114. Liang J, Wang Y, Huang Y, Ma Y, Liu Z, Cai J, Zhang C, Gao H, Chen Y (2009) Electromagnetic interference shielding of graphene/epoxy composites. Carbon 47:922–925
115. Bhattacharya P, Das CK, Kalra SS (2012) Graphene and MWCNT: potential candidate for microwave absorbing materials. J Mater Sci Res 1:126–132
116. Fang M, Tang Z, Lu H, Nutt SJ (2012) Strategies for chemical modification of graphene and applications of chemically modified graphene. Mater Chem 22:109–114
117. Basavaraja C, Kim WJ, Kim YD, Huh DS (2011) Synthesis of polyaniline-gold/graphene oxide composite and microwave absorption characteristics of the composite films. Mater Lett 65:3120–3123
118. Chen YJ, Nguyen DD, Li YA, Yip MC, Hsu WK, Tai NH (2011) Investigation of the electric conductivity and the electromagnetic interference shielding efficiency of SWCNTs/GNS/PAni nanocomposites. Diamond Relat Mater 20:1183–1187
119. Yu H, Wang T, Wen B, Lu M, Xu Z, Zhu C, Chen Y, Xue X, Sun C, Cao MJ (2012) Graphene/polyaniline nanorod arrays: synthesis and excellent electromagnetic absorption properties. Mater Chem 22:21679–21685
120. Wen B, Cao M, Lu M, Cao W, Shi H, Liu J, Wang X, Jin H, Fang X, Wang W, Yuan J (2014) Reduced graphene oxides: light-weight and high-efficiency electromagnetic interference shielding at elevated temperatures. Adv Mater 26:3484
121. Cao WQ, Wang XX, Yuan J, Wang WZ, Cao MS (2015) Temperature dependent microwave absorption of ultrathin graphene composites. J Mater Chem C 3:10017
122. He JZ, Wang XX, Zhang YL, Cao MS (2016) Small magnetic nanoparticles decorating reduced graphene oxides to tune the electromagnetic attenuation capacity. J Mater Chem C 4:7130

123. Yousefi N, Sun X, Lin X, Shen X, Jia J, Zhang B, Tang B, Chan M, Kim JK (2014) Highly aligned graphene/polymer nanocomposites with excellent dielectric properties for high-performance electromagnetic interference shielding. Adv Mater 26:5480
124. Putz KW, Compton OC, Palmeri MJ, Nguyen ST, Brinson LC (2010) High Nanofiller-Content Nanoocomposites via Vacuum-Assisted Self-Assembly. Adv Funct Mater 20:3322–3329
125. Potts JR, Dreyer DR, Bielawski CW, Ruoff RS (2011) Graphene-based polymer nanocomposites. Polymer 52:5–25
126. Kumar P, Kumar A, Cho KY, Das TK, Sudarsan V (2017) An asymmetric electrically conducting self-aligned graphene/polymer composite thin film for efficient electromagnetic interference shielding, American Institute of Physics. AIP Adv 7:015103. https://doi.org/10.1063/1.4973535
127. Wong K, Pickering S, Rudd C (2010) Recycled carbon fibre reinforced polymer composite for electromagnetic interference shielding. Compos A 41(6):693–702
128. Rahaman M, Chaki T, Khastgir D (2012) Modeling of DC conductivity for ethylene vinyl acetate (EVA)/polyaniline conductive composites prepared through insitu polymerization of aniline in EVA matrix. Compos Sci Technol 72(13):1575–1580
129. Su C, Xu L, Zhang C, Zhu J (2011) Selective location and conductive network formation of multiwalled carbon nanotubes in polycarbonate/poly (vinylidene fluoride) blends. Compos Sci Technol 71(7):1016–1021
130. Al-Saleh MH, Sundararaj U (2009) Electromagnetic interference shielding mechanisms of CNT/polymer composites. Carbon 47(7):1738–1746

Thermal Conductivity of Polymer–Carbon Composites

Soumya Mondal and Dipak Khastgir

Abstract Thermally conductive polymer–carbon composites whirled the motion and created a new era in polymer composites field by replacing the metal parts and inorganic filler-polymer composites in several application areas like heat exchanger, electric motors, power exchanger and generators. The advantages that create the possibilities of polymer–carbon composites over others lie in the several advantages such as lightweight, non-corrosive and most importantly ease of processing. Carbonaceous fillers generally have very high thermal conductivity, but their polymer composites suffer from achieving such high level of thermal conductivity though it attains the requirement in field application or even more in some particular filler-polymer combinations. The challenge mainly comes from the high interfacial resistance between filler and polymer matrix which alter the phonon transfer path, resulting in low thermal conductivity of polymer composites. This chapter reviews the thermal conductivity of various polymer–carbon filler composites with special references to the role of crystallinity developed in polymer due to the presence of filler, size and shape of filler, surface modification to improves interfacial adhesion and their dispersion ability in the polymer matrix in controlling the thermal conductivity of composites.

Keywords Polymer · Composites · Thermal Conductivity · Filler

1 Introduction

The thermal conductivity of material has always been one of the most important properties, since the ancient age, when people started looking for something like animal wool for use as cloth, which has kept them warm in winter for very long

---

S. Mondal (✉) · D. Khastgir
Rubber Technology Centre, Indian Institute of Technology Kharagpur, Kharagpur 721302, India
e-mail: soumya971@gmail.com

D. Khastgir
e-mail: dkhastgir2011@gmail.com

M. Rahaman et al. (eds.), *Carbon-Containing Polymer Composites*, Springer Series on Polymer and Composite Materials,
https://doi.org/10.1007/978-981-13-2688-2_11

time. As animal wool are the materials with very low thermal conductivity, which restricts the passes of human body heat quickly to the environment and kept them warming for a long time. This concept of heat transfer is later on used by the scientist at the time of industrial revolution, where they used metal as very high thermal conductivity materials. In the past two decades, various metal like copper, silver, gold and aluminium have been utilized for the manufacturing of conductors. However, certain demerits like heavyweight, high cost and some area of applications where some insulating material is needed to perform the heat transfer away from the conductor, polymers are one of the peerless substituents for metal. As polymers are the most preferred materials for the past few decades in every direction of the materialistic world due to its outstanding combination of light weight, ease of production, ease of handling, low cost and most interestingly can be given any shape by using simplest method. High thermal conductivity is highly required in electronic circuits for transferring heat quickly in order to resist the dielectric failure due to overheating. Generally, polymers exhibit low thermal conductivity due to its lower atomic density, feeble atomic interactions, very complicated crystal structure and finally higher level of anharmonicity in the molecular vibrations restricts the phonon transport [1]. Introduction of fillers in polymers is one of the common practices that have been adopted by the researchers to improve its thermal conductivity. Metallic fillers have the capability to improve the thermal conductivity of polymers manifold [2–5]. Due to incorporation of high amount of metallic filler the density of material increases consequently it alters the mechanical properties, hence it find limited application where light weight is the matter of concern [6]. Other than metallic fillers, a lot of carbonaceous filler like carbon black, graphite, carbon nanotube, carbon fibres and most recently graphene have been introduced as thermally conductive fillers for polymers [7–9]. This chapter contains theoretical explanations of thermal conductivity, measurement of thermal conductivity of only carbon-based polymer composites. Details explanations of lower thermal conductivity of polymers and the process of improvement are given. The uses of various carbonaceous fillers for thermally conductive composites are explained elaborately. Effort has been made to discuss about different thermal conductivity model of polymer-carbon composites and its explanations to fit thermal conductivity results for various filler-matrix combinations. The role of size and shape, surface modifications, concentrations of various fillers on thermal conductivity of polymer composites is discussed. How the interface of polymer and filler acts in controlling the thermal conductivity is also explained.

1.1 Definition of Thermal Conductivity

The thermal conductivity (denoted by 'λ' or often by 'k') is the property of a material that describes its ability to conduct heat. It is measured on the basis of Fourier's Law (1822) for conduction of heat. Heat conduction is the phenomenon of transfer heat (i.e. internal energy) that evolves through microscopic collisions of particles and movement of electrons within a particular material, which may be in the form of solids, liquids, gases or plasmas. A microscopic collision generally takes place between molecules, atoms and most importantly between electrons and it transfers the disorganized microscopic kinetic and potential energy. In SI units, thermal conductivity is measured in watts per metre per Kelvin [W/(m.k)].

1.2 Thermal Conductivity Measurement

Depending upon various materials, consequently their thermal properties and the medium temperature where the experiment will be carried out, thermal conductivity can be measured in several ways. But the most common practice of measurement is the guarded hot plate (GHP) technique, using the steady-state phenomenon, in a temperature range of 20–70 °C according to ASTM C177 [10, 11]. This GHP method has shown schematically in Fig. 1. In the steady-state heat flow (q), a temperature gradient (dT/dx) is created inside the material along the direction of heat flow. Thermal conductivity of a material based on this GHP method can be calculated using the following equation:

$$\lambda = -\frac{Q}{A}\frac{d}{\Delta T} \tag{1}$$

where λ is the proportionality constant and is known as thermal conductivity of a material [12]. Q is the total heat flow supplied by the main heater and $\Delta T = T_H - T_c$. The guarded heater and ring play the crucial role of thermal barriers to ensure that heat can be flow in one direction, i.e. from the main heater to cold

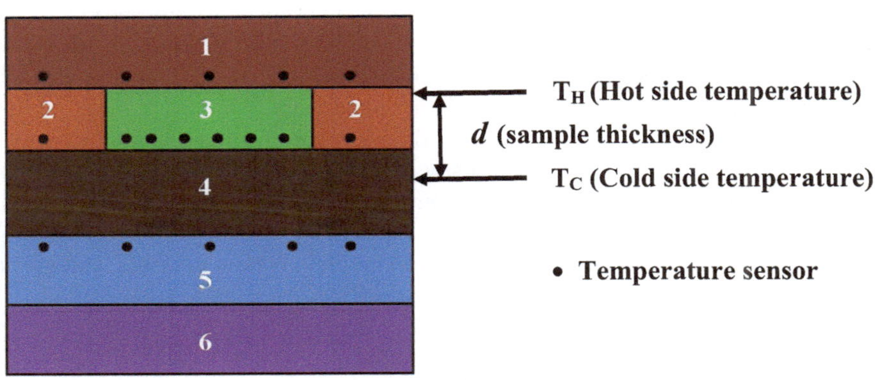

Fig. 1 A schematic representation of measurement process of thermal conductivity; (1) guarded heater, (2) guarded ring, (3) main heater, (4) sample specimen, (5) cold plate and (6) heat sink

plate. This test method is very useful in measuring the thermal conductivity of flexible material with high accuracy level.

The proportionality constant λ in Eq. (1), in other form may be defined as the quantity of heat that passes through a unit cube of material in a unit time when the temperature difference between the opposite faces of the cube is 1 k and Eq. (1) is modified as follows:

$$q = -\lambda \frac{\mathrm{d}T}{\mathrm{d}x} \quad (2)$$

1.3 Thermal Conductivity Behaviour of Low Dielectrics Polymer

The process of heat transfer depicts the phenomena of transfer of energy from one place to other having sufficient level of energy difference by certain energy carriers. Unlike gas or liquids where the heat energy transferred through molecules by random molecular motion (diffusion) or by an overall drift of the molecules in a certain direction (advection) process, in dielectric solids (polymer), the heat energy is transferred through phonons. Phonons, the quantized modes of vibrational energy, arise from the oscillating atoms within a crystal. This vibrational energy is the primary mechanism of heat transfer in polymeric materials as the free movements of electrons are impossible [13]. This state of the process of thermal conductivity in dielectric solids was successfully explained by the following Debye kinetic model:

$$\lambda = \frac{1}{3} C \cdot v \cdot l \quad (3)$$

where C is the specific heat per unit volume, v is the average phonon speed and l is the mean free path of phonon. Depending upon the operating temperature, the mean free path (l) varies from few nanometres to few micrometres while the velocity of phonon remain unchanged with variation in temperature.

The specific heat of a material can be written in other form as $C = C_P \cdot \rho$, where C_P and ρ are the heat capacity and density of that particular material. The velocity (v) and the mean free path (l) of a phonon is related to each other in the form of thermal diffusivity (D) by $D = v \cdot l$.

Hence, thermal conductivity of the dielectric solids is now reformed and given by the following equation [14]:

$$\lambda = \frac{1}{3} C_P \cdot \rho \cdot D \quad (4)$$

Generally, polymeric materials show very low thermal conductivity and the reason lies in their molecular structure. In polymers, the mean free path (l) is very small nearly few angstroms due to scattering of phonon from numerous defects, mainly from the interfaces of the amorphous and crystalline region, hence its' contribution to thermal conductivity is very low [15]. Due to the chain line structure of it, the contribution of heat capacity comes from in the two forms: (a) lattice vibrations or acoustics vibration and (b) characteristics vibrations from the internal motion of the repeating unit since the free movement of electrons is not possible here [16]. The lattice vibrations have major contribution in thermal conductivity of polymer composites at low temperatures. The contribution of the side groups to the net thermal conductivity comes only at the temperature above 100 K [17] and this is the reason behind the net rise in thermal conductivity with rise in temperature. Table 1 displays the thermal conductivity of some polymers that are commonly used in various applications.

Table 1 Thermal conductivity of some polymers

Material	Thermal conductivity at 25 °C (W/mK)
Low density polyethylene (LDPE)	0.30
High-density polyethylene (HDPE)	0.44
Polymethylmethacrylate (PMMA)	0.21
Epoxy resin	0.19
Polypropylene (PP)	0.11
Polystyrene (PS)	0.14
Poly(ethylene terephthalate) (PET)	0.15
Poly(butylene terephthalate) (PBT)	0.29
Poly(acrylonitrile-butadiene-styrene) copolymer (ABS)	0.33
Polyetheretherketone (PEEK)	0.25
Polyphenylene sulphide (PPS)	0.30
Polysulfone (PSU)	0.22
Polycarbonate (PC)	0.20
Poly(dimethylsiloxane) (PDMS)	0.25
Polyvinylidene difluoride (PVDF)	0.19
Polytetrafluoroethylene (PTFE)	0.27
Poly(ethylene vinyl acetate) (EVA)	0.34
Polyimide, thermoplastic (PI)	0.11
Polyphenylsulfone (PPSU)	0.35
Polyvinyl chloride (PVC)	0.19
Nylon-6 (PA6)	0.25
Nylon-6.6 (PA66)	0.26

Reproduce after permission from [21]

Table 2 Effect of cooling rate on crystallinity and thermal conductivity of PTFE

Cooling rate (°C min^{-1})	ΔH_f (J g^{-1})	Crystallinity (%)	λ (mW m^{-1} °C^{-1}) at 232 °C
Pristine samples	37.4	75 ± 2	300 ± 12
1	37.9	76 ± 2	298 ± 12
2.5	35.6	71 ± 2	292 ± 12
5	34.0	68 ± 2	287 ± 11
10	32.8	66 ± 2	286 ± 11
20	31.5	63 ± 2	283 ± 11
40	30.2	61 ± 2	279 ± 11

Reproduce after permission from [6]

1.4 Effect of Crystallinity and Temperature on the Thermal Conductivity of Polymers

The crystallinity of polymers highly influences its thermal conductivity behaviour over a wide range of temperatures. The conductivity of highly crystalline polymers namely high-density polyethylene (HDPE) is nearly 0.44 W/m K and it is always higher than an amorphous polymer like polystyrene (PS) with thermal conductivity of 0.14 W/m K. Higher the percent of crystallinity higher will be the thermal conductivity. This observation was studied by the Duncan M Price et al. by preparing different level of crystalline samples using the DSC (differential scanning calorimeter) instrument with varying the cooling rate. The values of thermal conductivity at different crystalline levels are given in Table 2 [18–21].

Kline [22] cited in his article about the variation of thermal conductivity of both the semicrystalline and amorphous polymers with temperature varied from 0 to 100 °C. He nicely represented the work done by Cherkasova [23] in his article. It has been seen that the thermal conductivity of all the amorphous polymers increases with temperatures but the same for semicrystalline polymers decreases with temperatures though the value of thermal conductivities of crystalline polymers is always higher than amorphous one. The increase in conductivity of amorphous polymers with temperatures lies in their internal molecular mobility caused by the increase in temperature and also the behaviour of the amorphous polymers are tend to that of glasses, in which also the conductivity varies in the same fashion. Cherkasova did the study with several polymers in which the thermal conductivity of amorphous polymers namely polystyrene (PS) started to increase very slowly with temperatures as it is closed to 100 °C. The conductivity behaviour of semicrystalline low-density polyethylene remains unchanged near about 40 °C but after that, it started to decrease due to melting of the crystalline zone, though its conductivity value is quite higher than the amorphous polymers. The thermal conductivity behaviours of highly crystalline polymers like poly (hexamethylene adipamide) and poly (tetrafluroethylene) remain unchanged in this entire temperature zone. As both these polymers are highly crystalline, its crystallite does not melt at this temperature and thus its conductivity remains unchanged and hence they possess very high thermal conductivity value. But the thermal conductivity value of polymers at low temperatures is reviewed with minute attention to distinguish the thermal conductivity behaviours of semicrystalline and amorphous polymers. At very low temperatures (1 °K), the thermal conductivity behaviours of semicrystalline polymers are very similar to that of highly imperfect crystals having a maximum thermal conductivity near 100 °K and moves to lower temperatures and higher conductivities as the crystallinity started to increase. On the other hand, the amorphous polymers exhibit temperature dependence similar to that of glass as stated earlier with no maximum peak but with significant plateau region in the range between 5 and 15 °K [24]. In continuation with that, the thermal conductivity of an amorphous polymers increases up to the glass transition temperature (T_g) but it decreases above T_g [25, 26]. This may be explained as with the increase of temperature up to T_g, the polymer chain straighten out and thus increase in path length of phonon transport leads to the increasing in thermal conductivity. In the vicinage of glass transition temperature, the density of the polymers becomes minimum and thus the mean free path becomes maximum resulting in highest value of thermal conductivity at T_g. Above T_g, the polymeric material undergoes transition from rubbery phase to molten state and increase in thermal motion of small chain create some micro-voids resulting in scattering of phonons. This scattering phenomenon in turn slightly decreases the thermal conductivity of polymer.

From the all the information described in the anteceding, it is quite clear that very low thermal conductivity is the characteristics of polymers. On the other hand, there are numerous applications where the high thermal conductivity of material is required and polymers being of very lightweight and having a lot of interesting features, it can be a very good material for many applications if their thermal behaviour can be improved. In several industrial application domains like electronic packaging and encapsulations, circuit board in power electronics, heat exchangers, materials for power equipment, computer chips, satellite devices and a lot of others crucial sectors where good dissipation of heat is one of the significant factors for the sustainability of the material are very much needed. This versatile need of modern era has forced the researchers to think about developing high thermally conductive polymer composites [27].

2 Various Carbon Fillers for Thermally Conductive Polymer Composites

Introduction of carbon fillers into the insulating polymer matrix is the common procedure to improve their electrical, mechanical and most importantly their thermal properties. The thermal conductivity of the composites has improved a lot with the addition of fillers like carbon black, carbon fibre, carbon nanotubes, graphite

Table 3 Thermal conductivities of some thermally conductive carbon fillers

Material	Thermal conductivity at 25 °C (W/mK)
Carbon black	6–174
Graphite (on plane)	100–400
Carbon fibre	21–180
Vapour grown carbon fibre (VGCF)	
longitudinal direction	~2000
transverse direction	10–100
Carbon nanotubes	2000–6000
Graphene	3000–5500

Reproduce after permission from [6]

and most recently the graphene [28–30]. The thermal conductivity of some carbon fillers is given in Table 3 [6, 31]. As we have discussed earlier that there are a lot of disadvantages of using the metallic fillers and thus obtaining a polymer composites with thermal conductivity greater than 4 W/mK is becoming a great challenge to the researchers. The outcome of the detail investigation shows that carbon-based fillers are the most promising filler for achieving higher thermal conductivity composites material with varying lightweight. High thermal conductivity, lightweight, low cost and easy dispersibility in polymer matrix make graphite a very interesting material for making polymer composites. Haoming et al. [32] have shown that with the introduction of 34 vol.% colloidal graphite, the thermal conductivity of polystyrene (PS) improved to 1.95 W/mK from 0.14 W/mk. Higher dispersion characteristics of graphite into the polymer matrix improve the interface of PS/graphite and higher degree of exfoliation with very good conducting network helps in improving the thermal conductivity of the composites. Haddon et al. [33] have shown that few layer graphene (GNP), where $n \sim 4$ (n = number of layer), with a thickness of ~2 nm is a very efficient filler for epoxy composites for enhancing thermal conductivity up to 6.44 W/mK over the conductivity of neat epoxy (0.26 W/mK) at the loading of ~25 vol.%. Thus, this use of such low volume fraction is very much helpful to produce flexible composites, where conventional filler requires a loading of ~70 vol.% to achieve that much amount of conductivity. Lafdi et al. [34] stated that a sufficient amount of increase in thermal conductivity 2.8 W/mK was achieved by using vapour grown carbon fibre (VGCF) at a level of 20 wt% in epoxy matrix with very good level of mechanical properties. Kuriger et al. [35] have prepared VGCF/polypropylene (PP) composites through melt mixing and have shown that the thermal conductivity value in different directions of carbon fibre is different. At the loading of 23 vol.% of fibre, the thermal conductivity value of the composites is 5.38 W/mK in longitudinal direction and 2.49 W/mK in the transverse direction. Abdel-Aal et al. [36] have shown that the thermal conductivity of epoxy-carbon black (CB) composites increases with increase in CB loading and reached a moderate value of 1.5 W/mK at the loading of 30 wt% of CB. With the increase in CB concentration, particles come close to each other and favour the phonon transfer process as a very thin layer of epoxy is present between the CB particles. On the other case, the increase in crystallinity may be another reason for the increase in thermal conductivity as discussed earlier. In spite of being a low-cost filler, the use of CB black for enhancing thermal conductivity is limited due to very low increase in thermal conductivity with high amount of loading; it is mainly used for an increase in thermal conductivity. As the investigation of M. Bououdina et al. shows that a very limited increase in thermal conductivity is observed after incorporation of 30 wt% of CB. Hence, the main carbon-based filler for polymer is shifted towards the carbon fibre, carbon nanotube, graphite and graphite nanoplatelet due to high increase in thermal conductivity with minimum loading and at reasonable cost. Starting from the use of CB in polymer for enhancing thermal conductivity and passing through graphite, it landed in the era of carbon nanotube (CNT). From Table 3, it can be seen that the pristine CNT beers high amount of thermal conductivity in between 2000 and 6000 W/mK and use of it in polymer matrix people experienced composites with high thermal conductivity compared to that of graphite and CB. A. T. Jhonson and the co-workers introduced single-wall carbon nanotube (SWCNT) into the epoxy matrix and measured the thermal conductivity behaviour [37]. Composites having only 1 wt% SWCNT give 70% rise in thermal conductivity value at 40 K and it raises up to 125% as the temperature reached to room temperature. They have compared the enhancement of thermal conductivity value with the loading of VGCF in the same matrix and have shown that at equal amount of loading, i.e. 1 wt %, the rise in conductivity value for WCNT is 125% at room temperature where it is only 40% for VGCF as given in Fig. 2. This information proves the superior acceptability of SWCNT over VGCF as filler for polymer. Apart from room temperature, a study throughout temperature range starting from 40 K to 300 K has shown (give in Fig. 3) that SWCNT-epoxy composite shows a much higher conductivity value at all temperature compared to VGCF-epoxy composites. Effect of alignment of carbon nanotube has a significant role in increasing the thermal

Fig. 2 Enhancement in thermal conductivity relative to pristine epoxy as a function of SWNT and VGCF loading. Reproduce after permission from [37]

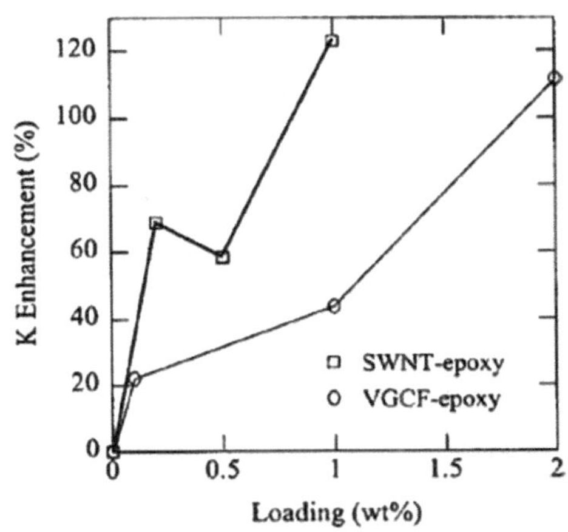

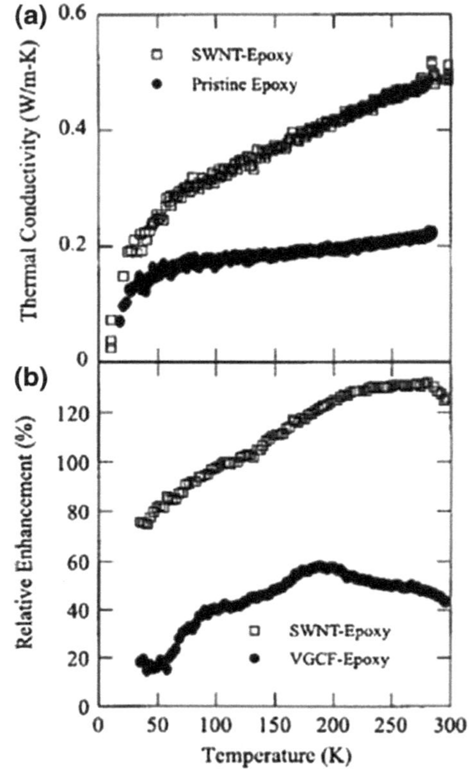

Fig. 3 **a** Thermal conductivity vs temperature for the pristine epoxy and epoxy with 1 wt% SWNT loading. **b** Relative enhancement in the thermal conductivity vs temperature for composites loaded with 1 wt% SWNT and VGCF. Reproduce after permission from [37]

conductivity value compare to normally prepared composites. D. L. Eaton et al. reported that thermal conductivity value rises up to 300% with 3 wt% loading of SWCNT and it is further increased by 10% with the application of 25T magnetic field [38]. This may be explained as follows with the application of magnetic field SWCNT aligning themselves in a particular direction, which in turn helps in improving the thermal conductivity value.

After the era of CNT, a new emerging field in polymer nanocomposites has introduced with the synthesis of graphene. Although sufficient advancement in various properties of polymer has been made by the CNT, there are many such unresolved issues like tendency of agglomeration of nanotubes during the processing along with the availability of high quality nanotubes because of high production cost which made the researchers to think about the alternative. Graphene is one of the most superior nanofillers with all-around properties for diversified application of polymer–graphene nanocomposites and most interestingly the precursor of graphene is naturally available graphite [30]. One atom thick two-dimensional layers of sp^2 bonded carbon is known as graphene and it is predicted to have unusual range properties having high thermal conductivity value. Yu et al. first introduced

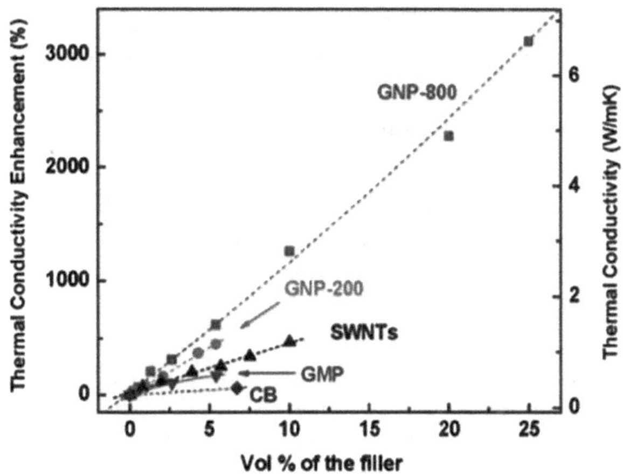

Fig. 4 Improvement of thermal conductivity of epoxy-based composites at 30 °C using CB graphite microplatelates (GMP), purified SWCNT, TRGO (at 200 and 800 °C). Reproduce after permission from [39]

thermally reduced few layer graphene (TRGO also known as GNP_n, where n = number of layer stacked) in epoxy matrix and investigated the thermal conductivity behaviours of the composites [28]. Thermal conductivity measurement has shown that few layer graphene (where, $n \sim 4$) is a very efficient filler for epoxy matrix. At the loading of ~25 vol.% of GNP_4, the increase in thermal conductivity is nearly ~3000% and conductivity value is 6.44 W/mK at room temperature as shown in Fig. 4. This value of thermal conductivity is very much higher than conventional filler at the same loading. The enhancement in thermal conductivity value for GNP is very much higher than SWCNT at the same loading. The epoxy composites prepared by 10 wt% SWCNT give the thermal conductivity value k = 0.85 W/mK whereas, K = 1.49 W/mK for GNP.

Fukushima et al. [39] investigated the effect of the addition of GNP on different thermoplastic to enhance thermal conductivity behaviour. They have shown that incorporation of only 20 vol.% of GNP in Nylon 66 enhances in thermal conductivity up to ~3.6 W/mK from ~0.4 W/mK and incorporation of the same percentage of GNP increases the thermal conductivity up to ~4.1 W/mK from ~0.35 W/mK. Chemically reduced graphene oxide (CRGO) improves the thermal conductivity of polystyrene (PS) matrix up to 260% with the incorporation of only 2 wt% of it [40]. Hence, it can be seen that among the carbon filler for enhancing, the thermal conductivity characteristics CNT and graphene are the most two promising filler for next-generation applicability of thermal interference material (TIM).

3 Role of Different Parameter of Filler on Thermal Conductivity Behaviour of Composites

3.1 Role of Size and Shape of Filler

As we have already seen that phonon transfer is the main mechanism of thermal conductivity and for this to happen, mean free path of phonon transfer is the primary requirement. A particular filler having larger length is thought to be possessed with higher phonon mean free path than that of with lower length. The phonon mean free path is found to be 500 nm for multiwall carbon nanotube (MWCNT) and it is even larger for single-wall carbon nanotube (SWCNT) [41–43]. Thus, the shape and size of the filler are key parameters for controlling the thermal conductivity. Carbon fillers like CNT, carbon fibres, or carbon nanofibres are characterized by their aspect ratio (i.e. ratio of length to diameter) and surface area but for carbon black, it is very difficult to calculate their length and aspect ratio. Hence, it is expected that the dependency diameter and the length of carbon black on thermal conductivity are much less affected and thus it is less investigated. The layered structure filler, i.e. graphite is also characterized by the length and aspect ratio. The dependency of these two key parameters on thermal conductivity of polymer composites is discussed in the following paragraph.

3.2 Role of Diameter

It is very interesting to note that the thermal conductivity (λ) of CNT decreases with the increase in diameter. As explained by Cao et al. [44], the thermal conductivity of SWCNT at 300 K is inversely proportional to its diameter as shown in Fig. 5. It

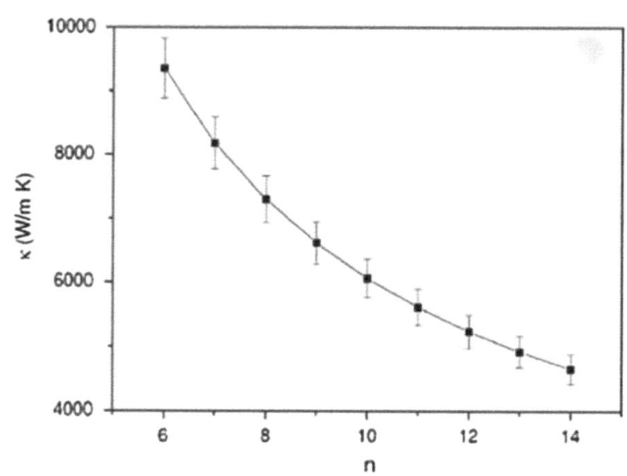

Fig. 5 Thermal conductivity of (n, 0) SWCNTs (n = 6–14) at 300 K. Reproduce after permission from [44]

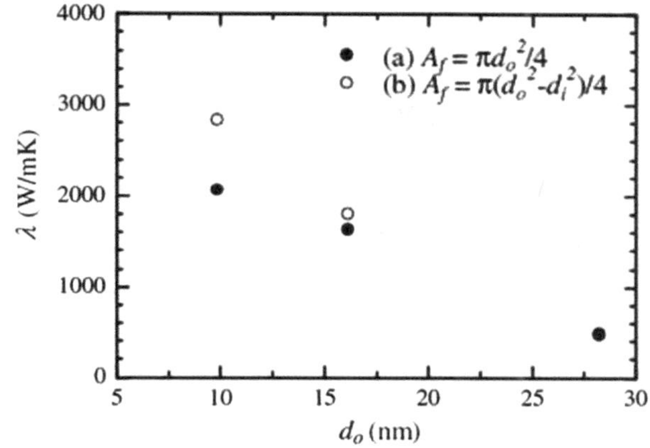

Fig. 6 Thermal conductivity of a single carbon nanotube of different diameters. Reproduce after permission from [45]

is seen that CNT with diameter $n = 14$ (this value of n is denoted by the number unit vector, i.e. number of hexagonal unit at a particular direction and it is known as chirality index) shows lower thermal conductivity value than with CNT having diameter of $n = 6$. It may be explained by the decrease in group velocity of phonon and most importantly due to umklapp scattering process. Umklapp scattering is the unusual scattering phenomenon where due to the increase in phonon momentum, the normal scattering process is hampered and as a result, the translation wave vector to another Brillouin zone occurs.

At any given temperature, the chances of umklapp process are higher in SWCNT's with larger diameter than compared that one with smaller diameter. As the umklapp scattering process and group velocity reduce the thermal conductivity (λ) in SWCNT with the increase in its diameter. Fujii et al. [45] nicely explained the effect of diameter on the thermal conductivity behaviour in both SWCNT and MWCNT. The thermal conductivity (λ) of the CNTs having different diameters is shown in Fig. 6.

The cross sectional is (A_f) of two different diameter is calculated by using the following equation (1) $A_f = \pi d_0^2/4$ and (2) $A_f = \pi(d_0^2 - d_i^2)/4$, where d_0^2 is the outer diameter of CNT and d_i^2 is the diameter of inner lumen. It is clearly understandable that with the decrease of the outer diameter, the thermal conductivity of CNTs increases at room temperature. This variation in thermal conductivity with variation in diameter indicates the interactions of phonon and electrons between multiwalled layers affect the thermal conductivity of the CNTs. Hence, it can be said that as the number of walls decreases, the thermal conductivity value increases. Thus, it can be said an SWCNT always gives better thermal conductivity behaviour than MWCNT.

3.3 Role of Length

At a given temperature, here 300 K, the thermal conductance of filler say CNT with length between 50 and 350 nm obeys the power law relation given below λ.

$$\sigma = \beta L^{-\gamma}$$

where $\gamma = (1 - \alpha) < 1$. Here the exponent α is more or less equal to 0.54 ($\alpha \approx 0.54$) for nanotubes of different diameters and the exponent γ also is almost constant with a value $\gamma \approx 0.46$. The prefactor β has been found to be dependent linearly with the chirality indices n along the circumference. The dependency of β on n is stated by the following equation:

$$\beta = 2(n - 1)10^{-9}$$

Now with utilizations of both the above equations, the thermal conductance of SWCNT can be measured.

Alaghemandi et al. [46] have investigated vastly the effect of tube length of CNT on thermal conductivity having different chirality index (n). They have shown that all CNTs with different chirality index behave almost equally that is for all nanotubes, the thermal conductivity value increases with an increase in tube length. The values of thermal conductivity of different chirality index (5, 5; 7, 7; 10, 10; 20, 20; 10, 0; 17, 0) are given in Table 4.

Hence, it can be said that with the increase in tube length the thermal conductivity of the CNT increases at particular temperature. Shigeo Maruyama [47] has investigated the effect of tube length (from 6 to 404 nm) in different SWCNTs with chirality index (5, 5 and 10, 10). The variation of thermal conductivity on tube length is shown in Fig. 7.

It is quite interesting to note that the thermal conductivity of (10, 10) nanotube was found to be almost constant with increase in tube length while the same for (5, 5) nanotube was found to be diverging with tube length at higher diameter value. Between tube lengths 50–350 nm, it follows the power law equations as given earlier but at higher length above 350 nm, the thermal conductivity is expected to constant as it is getting higher than the mean free path of phonon. Gharagozloo-Hubmann et al. [48] have compared the variation of thermal conductivity of MWCNTs and Pyrograf carbon fibres with the variation of their length. It can be easily said from Fig. 8 that throughout the entire length window MWCNT gives better thermal conductivity value than carbon fibre. Most interestingly, for the same volume fraction of carbon fibres, fibres with length 12 μm give 30% higher thermal conductivity than with length of 2 μm. This result is true for MWCNTs also and the slope of increment in thermal conductivity is always high for MWCNTs than fibres at all tube length value.

Table 4 Thermal conductivities λ of different single-walled nanotubes as a function of the tube length at 300 K

Length (nm)	Chirality index					
	5, 5	7, 7	10, 10	20, 20	10, 0	17, 0
	Thermal conductivity, λ (W/m K)					
5	37.59 ± 0.46	49.95 ± 1.39	35.56 ± 1.19	51.15 ± 1.21	42.28 ± 1.10	33.30 ± 0.72
10	68.99 ± 0.44	77.27 ± 1.71	78.68 ± 1.55	83.77 ± 1.61	65.01 ± 0.82	79.14 ± 1.63
25	141.57 ± 1.46	142.02 ± 0.82	164.83 ± 3.25	149.05 ± 1.71	149.53 ± 0.56	169.18 ± 4.41
50	232.00 ± 1.31	218.32 ± 1.13	223.93 ± 2.72	223.08 ± 2.77	234.38 ± 0.83	221.84 ± 0.86
75	290.64 ± 1.01	276.77 ± 1.18	279.24 ± 2.85	271.95 ± 1.43	279.25 ± 1.32	278.81 ± 1.07
100	345.14 ± 1.01	302.79 ± 1.12	305.87 ± 3.07	326.77 ± 1.51	338.84 ± 1.81	304.66 ± 2.49
250	477.65 ± 0.92	518.04 ± 0.78	549.70 ± 5.51	511.63 ± 6.96	542.78 ± 1.65	532.15 ± 3.16
350	546.07 ± 0.57	621.53 ± 1.29	611.97 ± 3.24	635.35 ± 7.58	602.82 ± 0.82	620.13 ± 1.83

Reproduce after permission from [46]

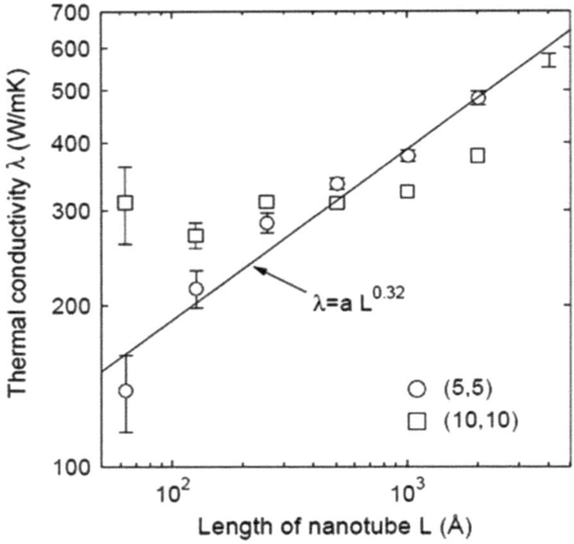

Fig. 7 Dependence of thermal conductivity on length of nanotubes for 300 K. Reproduce after permission from [47]

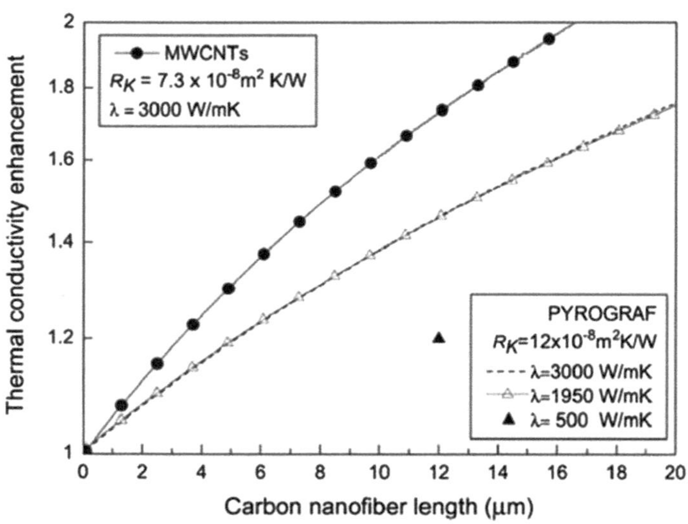

Fig. 8 Effect of fibre length on thermal conductivity enhancement of MWCNTs and Pyrograf fibres. Reproduce after permission from [48]

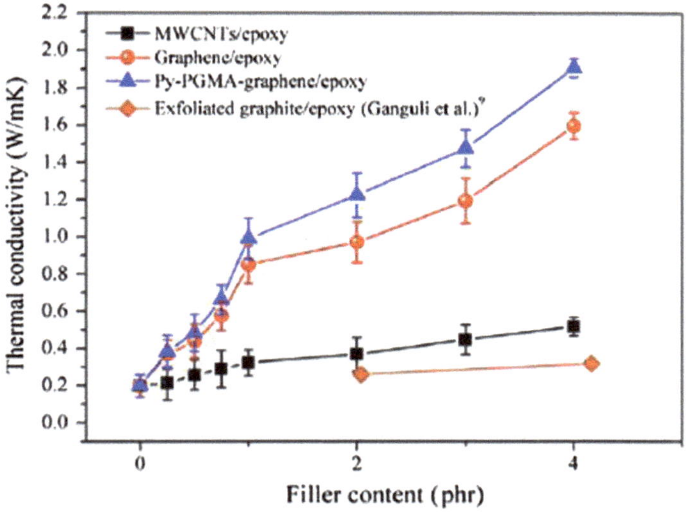

Fig. 9 Thermal conductivity with various filler contents of MWCNTs/epoxy, graphene/epoxy, and Py-PGMA–graphene. Reproduce after permission from [50]

3.4 Role of Surface Modifications

Surface modifications or functionalizations of nanofillers (e.g. CNT, graphite, graphene) is an interesting approach to enhance their dispersion behaviour in polymer matrix and as well as their interaction with it for successful preparation of polymer nanocomposites [49]. This process helps in improving several physical-mechanical properties with concurrent rise in electrical and thermal conductivity behaviours. Among all the functionalizations, process non-covalent process is the most promising method as it improves the thermal conductivity behaviours remarkably. Teng et al. [50] investigated the role of non-covalent functionalizations of graphene nanosheets (GNSs) through functional segmented poly(glycidyl methacrylate) containing localized pyrene groups (Py-PGMA) to prepare graphene-epoxy composites for high thermal conductivity composites. The π–π stacking of pyrene molecules with functional segmented polymer chain on GNS surface facilitates both the dispersion of GNS in epoxy matrix as well as improves the polymer–filler interactions. This interaction helps in improving GNS-epoxy contact area which in turn effectively improved the thermal conductivity behaviours. Figure 9 shows the gradual increase in thermal conductivity GNS-epoxy composites with an increase in GNS content and the rate of increase of thermal conductivity is even greater than that of MWCNT-epoxy composites. This can be explained in terms of surface area as GNS possess higher surface area than MWCNT which effectively increase the epoxy-filler contact area, the thinner

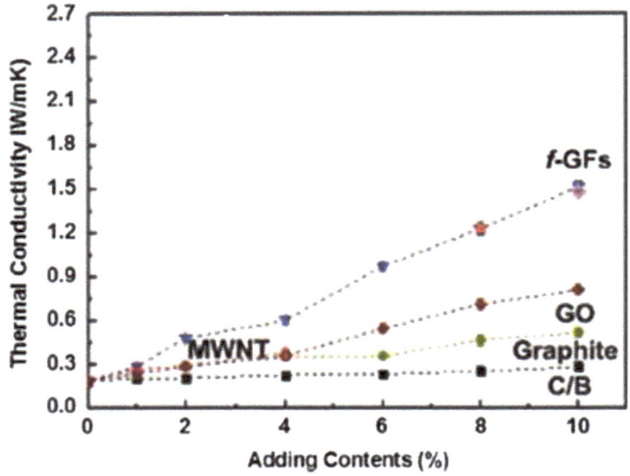

Fig. 10 Thermal conductivity of epoxy nanocomposites with C/B, graphite, MWNT, GO and *f*-GFs. The highest value of thermal conductivity is 1.53 W/mK at a 10 wt% loading of *f*-GFs in epoxy. Reproduce after permission from [51]

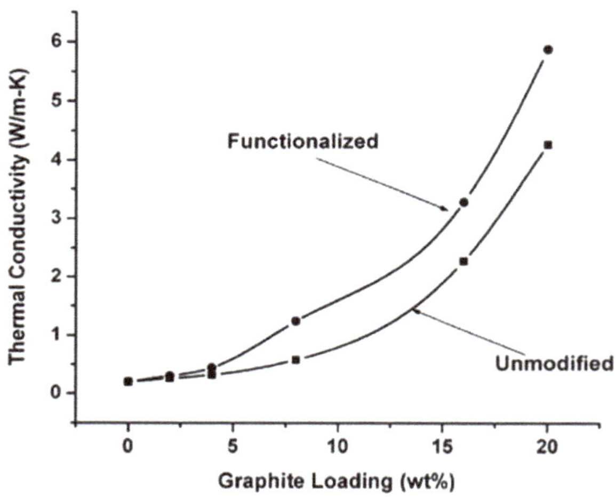

Fig. 11 Thermal conductivity plot for the graphite–epoxy composite. Reproduce after permission from [52]

polymer layer on the GNS surface can improve the phonon transport process. But the Py-PGMA/epoxy composites show the best thermal conductivity behaviour which confirms the better polymer–filler interactions by the π–π stacking. The thermal conductivity of Py-PGMA/epoxy composites is greater than GNS/epoxy composites nearly by 16.4% at the same amount of filler loading of 1 phr and it is much higher than MWCNT/epoxy composites. Song et al. [51] also prepared the non-covalent modified graphene flakes (GFs)/epoxy composites and investigated the thermal conductivity behaviour. They have prepared the functionalized GFs (f-GFs) using 1-pyrenebutyric acid (PBA) that are highly soluble in various solvents. Figure 10 depicts the thermal conductivity behaviour of various nano fillers-epoxy composites with variation of filler content. The thermal conductivity of

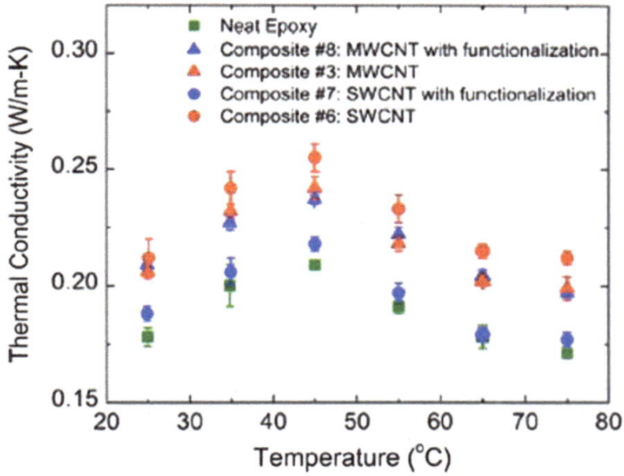

Fig. 12 Thermal conductivity of the epoxy-based nanocomposites with functionalized and unfunctionalized SWCNT and MWCNT at 3 wt% loading fraction. Reproduce after permission from [53]

f-GFs/epoxy composites at 10 wt% loading is nearly 1.53 W/mK and it much higher than other nanocomposites. Here also, the functionalizations improve the polymer–filler interaction, dispersion behaviour of GFs in epoxy matrix which in reality reduces the surface defects for better transport of phonon, which is the main reason of the superior enhancement in thermal conductivity.

Ganguli et al. [52] have investigated the effect of chemical functionalizations of graphite (silylated modified) on thermal conductivity behaviour of exfoliated-epoxy composites. Figure 11 shows the comparison of thermal conductivity behaviour between the pristine-graphite/epoxy and functionalized graphite/epoxy composites. It is observed that at 20 wt% loading level, the thermal conductivity of functionalized graphite/epoxy composites is 5.8 W/mK and it is nearly 35% greater than pristine-graphite/epoxy composites. Basically, the functionalizations process improves the polymer–filler interaction and improves the overall phonon transport phenomenon by minimizing the interfacial phonon scattering compared to normal graphite.

Formation of covalent bonds between the epoxy matrix and the graphite platelets improve the compatibility due to chemical functionalizations which helps in reducing the phonon scattering and ensure a steady phonon transport process which is impossible in normal graphite-epoxy composites. Besides the so much improvement, Gulotty et al. [53] have investigated the effect of carboxyilic (–COOH) functionalizations of MWCNT and SWCNT on thermal conductivity behaviour of epoxy nanocomposites with variation in temperature. Figure 12 shows the thermal conductivity behaviour of neat epoxy and epoxy-based composites prepared by functionalized and unfunctionalized SWCNT and MWCNT with variation of temperature at a particular filler loading of 3 wt%. It can be seen that the addition of SWCNT enhances the thermal conductivity of epoxy by several magnitude but functionalizations of it decrease this behaviour throughout all temperature window and this decreased value is almost equal to that of neat epoxy. Addition of the same amount of MWCNT also increases the thermal conductivity value but functionalization of it does not bring out any changes in thermal conductivity value. The reason behind this contradictory behaviour of these two types of CNT lies between the presence of number of layers in MWCNT. Functionalization of CNT creates some sidewall defects when it gets coupled with the matrix and its strongly hinders the natural phonon transfer process along the length of SWCNT and as a result, it affects the thermal conductivity behaviour to a great extent. But in case of MWCNT due to the presence of multiple layer in spite of the infected outer layer by the functionalization process, the inner layers take part (as it remains unaffected in the functionalization process) in the phonon conduction process. Thus, in functionalized MWCNT/epoxy nanocomposites, the thermal conductivity value does not change so much compared to unfunctionalized MWCNT/epoxy composites. Ma et al. [54] investigated the effect of different functionalization (amide or amine functional groups) of CNT on their thermal conductivity behaviour of their polymer composites. He has shown that functionalization does not increase the thermal conductivity value rather it decreases with modifications. Hence in summary of this part, we can conclude that the increase in thermal conductivity behaviour due to functionalization is very much dependent on the type of filler and type of functional groups that are being incorporated and mostly the interaction with polymer matrix after their functionalization.

3.5 Role of Interface Resistance and Dispersion of Filler in Matrix Polymer

Apart from all the previous factor which controls the thermal conductivity of carbon-polymer composites, the dispersion of filler particles into the matrix and their interfacial interaction also make significant contribution in controlling the thermal conductivity of carbon-polymer composites. The interfacial resistance seemed to be the bottleneck for preparing highly thermally conductive polymer composites. Due to the low thermal conductivity of polymer, the heat transfer from filler to polymer does not come out to an effective one for overall heat transfer. According to the theory of thermal conduction, phonon transfer occurs through preferably from filler particles. Hence, any alteration of this conduction path may affect the thermal conductivity of the polymer composites. The phonon conduction in material is controlled by scattering defects, i.e. scattering generated from the interfacial boundary and it is proportional to the effective interface of filler towards the matrix polymer. The higher will be the defect density lower will be its efficiency as thermal conductor. Thus, interface of all the nanoparticles–polymer composites plays an important in controlling the overall thermal conductivities. Gojny et al. [55] have investigated the effect of this interfacial resistance by using different CNT. As SWCNT possess highest diameter dependent specific surface area (SAA) and thus it creates more scattering defects hence it acts as the lowest enhancement in thermal conductivity of the polymer composites followed by DWCNT (intermediate SSA) and MWCNT (lowest SSA). This can be explained as follows being a low SSA filler generally MWCNT creates low interfacial defects and also being combined of multilayer, the interlayer of its responsible for conduction of phonon and thus it effectively enhance the thermal conductivity compare to the other two CNTs i.e. SWCNT & DWCNT.

It is very difficult to predict any general rule that can explain the dependency of thermal conductivity of polymer composites on filler dispersion. Because the dispersion behaviour of filler fully depends on the method of preparation of composites. Sometimes, thermal conductivity of well disperse filler systems may be lower than the agglomerated systems as the agglomerated regions may help in better transport of phonon compare to the well-dispersed systems.

4 Overview of Different Model to Explain the Thermal Conductivity Behaviour of Polymer Composites

As we have already discussed that thermal conductivity of carbon-polymer composites finds a lot of interesting application and its theoretical model for calculation of exact value is very much needed for better understanding of the fundamentals. A lot of theoretical models has been proposed to find out the effective thermal conductivity of polymer composites. For two-component composites systems depending upon the direction of heat flow, it has been categorized in parallel and series model composites. Parallel and series model give the upper and lower limit of the effective thermal conductivity of the composites with the arrangement of filler according to the heat flow.

For the simplest rule of mixture, i.e. parallel model or arithmetic mean model are as follows:

$$\lambda_C = \phi\lambda_f + (1 - \phi)\lambda_m \tag{4.1}$$

For the inverse rule of mixture, i.e. series model or harmonic mean model is as follows:

$$\frac{1}{\lambda_c} = \frac{\phi}{\lambda_f} + \frac{1 - \phi}{\lambda_m} \tag{4.2}$$

Apart from these, two another model namely geometric mean equations are widely used and given as follows:

$$\lambda_c = \lambda_f^{\phi} \cdot \lambda_m^{(1-\phi)} \tag{4.3}$$

where λ_c, λ_f, λ_m are the thermal conductivities of composites, filler material and polymer matrix respectively and ϕ is the volume fraction of the filler [56].

Agri and Uno [57] have converted Eqs. (4.1) and (4.2) to Eq. (4.4) as such that when constant $n = 1$ or -1, it corresponds to Eqs. (4.1) or (4.2). The value of n is smaller than +1 and larger than −1.

$$\lambda_c^n = \phi.(\lambda_f)^n - (1 - \phi).(\lambda_m)^n \tag{4.4}$$

Later they have modified the above equation on the contribution of filler particles to the effective thermal conductivity and the state of dispersion as it seemed that the filler particles contribute less compare to the matrix to bring significant changes in thermal conductivity. Thus, they have replaced the constant (n) by constant (C_2), where C_2 is the factor that takes care of the ease of forming conductive chain and its value is smaller than 1 and greater than 0.

The more the tendency of the filler particles towards forming conductive network more it will contribute to the effective thermal conductivity and the value of C_2 will be close to 1. It is also established that the filler particles affect the crystallinity of the polymer matrix when it is inserted into it, hence they have introduced another constant C_1 and replace the λ_1 by $C_1 \cdot \lambda_1$. Now if the dispersion state of filler is uniform throughout the matrix and n is assumed to be close to 0, and λ^n can be transformed into $(1+n) \cdot \log \lambda$.

Consequently Eq. (4.4) can be converted into the following equation:

$$\log \lambda_c = \phi \cdot C_2 \cdot \log \lambda_f + (1 + \phi) \cdot \log(C_1 \cdot \lambda_m) \tag{4.5}$$

Hence, Eq. (4.5) can be very useful to determine the thermal conductivity of the composites while the dispersion state of filler is different for the same composite.

But above these models find limitations and do not give a well prediction of thermal conductivity results as soon as the filler particles are randomly distributed in the polymer matrix, i.e. they are not oriented either parallel or in series method with respect to the heat flow.

Maxwell [56] first investigated to find out the effective thermal conductivities of the randomly oriented fillers in a continuous polymer matrix. By using potential theory and solving the Laplace equation, he proposed a model that can give the more or less exact value of thermal conductivity of spherical fillers with non-interacting behaviour in a continuous polymer matrix. Thus, the effective thermal conductivity of the system is expressed as follows:

$$\lambda_c = \lambda_m \frac{2\lambda_m + \lambda_f + 2\phi(\lambda_m - \lambda_f)}{2\lambda_m + \lambda_f + \phi(\lambda_m - \lambda_f)} \tag{4.6}$$

Bruggeman [58, 59] modified Maxwell's model by using different assumptions related to permeability and field strength and proposed a new model applicable for a dilute suspension, i.e. lower amount of filler (<10 vol.%) in matrix of non-interacting fillers dispersed in homogeneous matrix. Thus, new Bruggeman model for derivation of the effective thermal conductivity is given as follows:

$$1-\phi=\frac{\lambda_f-\lambda_c}{\lambda_f-\lambda_m}\left(\frac{\lambda_m}{\lambda_c}\right)^{1/3} \tag{4.7}$$

Cheng and Vachon [60] have proposed another equation related to thermal conductivity of polymer composites where $\lambda_f \rangle \lambda_m$ is given as follows by the equation (4.8).

$$\frac{1}{\lambda_c}=\frac{1-B}{\lambda_m}+\frac{1}{\sqrt{K\lambda_d(\lambda_f+B\lambda_d)}}\cdot\ln\left[\frac{\sqrt{\lambda_m+B\lambda_d}+B/2\cdot\sqrt{K\cdot\lambda_d}}{\sqrt{\lambda_m+B\lambda_d}-B/2\cdot\sqrt{K\cdot\lambda_d}}\right] \tag{4.8}$$

where $B=\sqrt{3\phi/2}$, $K=-4\sqrt{2/3\phi}$ and $\lambda_d=\lambda_f-\lambda_m$.

Nielsen [61] modified the Halpin-Tsai [62] equation, which was based on the determining of thermal conductivity of composites formed by the general rule of mixtures, for a two-phase systems having spherical shaped dispersed particles.

$$\lambda=\frac{1+AB\phi_f}{1-B\psi\phi_f} \tag{4.9}$$

$$A=\kappa_E-1 \quad B=\frac{\lambda_2/\lambda_1-1}{\lambda_2/\lambda_1+A} \quad \psi=1+\frac{\phi_f(1-\phi_m)}{\phi_m^2}$$

where λ_1 and λ_2 are the thermal conductivity of continuous and dispersed phase, respectively. The constant 'A' related to the shape of the dispersed particles and how they are oriented with respect to the heat flow and it is linked with the Einstein coefficient κ_E. The factor ψ is related to the maximum packing fraction ϕ_m of the dispersed particles. This model finds very good applicability on an average low and medium loading of filler (<20 vol.%) and covers wide range of particles shapes and patterns but find limitations at higher loading of filler. In modifications of these models, another empirical model was introduced by Agari and Uno [63] for composites containing higher amount of filler. They have considered the effect of conductive chain network formed by the filler particles on the higher thermal conductivity of the composites and introduced new parameters given as follows:

$$\lambda_c=\lambda_m\frac{2\lambda_m+\lambda_f+2\phi_{af}(\lambda_m-\lambda_f)}{2\lambda_m+\lambda_f-2\phi_{af}(\lambda_m-\lambda_f)}+\phi_{bf}\phi_{bf}^{\phi_{bf}^{-2/3}}C^2\phi_f \tag{4.10}$$

where ϕ_{af} is the volume fraction of fillers that are not contributing to the formation of conductive network and ϕ_{bf} is the volume fraction of fillers that are contributing to the formation of conductive network and C_2 is the geometrical factor connected to the observable conductivity with the random assembling of conductive chains. As it is already discussed that this model is only applicable for higher loading of fillers as the calculation of these tricky parameters is very difficult.

Till now, the empirical equation which we have described is mainly applicable for spherical or oblong spherical shape of filler particles. But the fillers having an aspect ratio (i.e. carbon nano tube, CNT and short carbon fibre, SCF) need to be explored, i.e. the dependency of this type of filler particles on the effective thermal conductivity of polymer composites. In SCF, composites prepared by injection moulding there are always the possibilities of having two type of orientation, i.e. one is parallel and another is transverse to the direction of heat flow. It has been investigated that [64–66] the Halpin-Tsai empirical equation can be used to explain the thermal conductivities of the SCF reinforced polymer composites having unidirectional orientation of fibre. For unidirectional orientation of filler, the thermal conductivities in parallel direction (λ_1) and perpendicular direction (λ_2) are given by the following equations:

$$\lambda_1 = \frac{1 + 2\alpha\mu_1\phi_\mathrm{f}}{1 - \mu_1\phi_\mathrm{f}}\lambda_\mathrm{m} \tag{4.11}$$

$$\lambda_2 = \frac{1 + 2\mu_2\phi_\mathrm{f}}{1 - \mu_2\phi_\mathrm{f}}\lambda_\mathrm{m} \tag{4.12}$$

where $\alpha = L/d_\mathrm{f}$, and d_f is the fibre diameter. ϕ_f is the volume fraction of fibre and λ_m is the matrix thermal conductivity given in the equation of μ_1 and μ_2 as follows:

$$\mu_1 = \frac{\lambda_{\mathrm{f}1}/\lambda_\mathrm{m} - 1}{\lambda_{\mathrm{f}1}/\lambda_\mathrm{m} + 2\alpha} \tag{4.13}$$

$$\mu_2 = \frac{\lambda_{\mathrm{f}2}/\lambda_\mathrm{m} - 1}{\lambda_{\mathrm{f}2}/\lambda_\mathrm{m} + 2} \tag{4.14}$$

where $\lambda_{\mathrm{f}1}$ and $\lambda_{\mathrm{f}2}$ are the thermal conductivity of the fibre in parallel and transverse direction with respect to the fibre axis, respectively.

For randomly distributed CNT-polymer composites, the measurement of thermal conductivity is a very difficult one for both the high as well as low loading of filler. Deng et al. [67] have introduced new empirical method for obtaining thermal conductivity of composites even at high loading of CNT so that interaction between the particulates can be considered. For a CNT-polymer composites with randomly oriented distribution and having length L and diameter, the effective thermal conductivities of the composites can be represented as follows:

$$\frac{\lambda}{\lambda_\mathrm{m}} = 1 + \frac{\phi}{3}\left[\frac{1}{\left(\lambda_{22}^{c}/\lambda_\mathrm{m} - 1\right)^{-1} + H} + \frac{2}{\left(\lambda_{11}^{c}/\lambda_\mathrm{m} - 1\right)^{-1} + (1 - H)/2}\right] \tag{4.15}$$

where ϕ represents the volume fraction of CNT, and λ_{11} and λ_{22} represent the transverse and axial thermal conductivities of the CNTs. λ_m represents the thermal conductivity of isotropic matrix and H denotes the influence of aspect ratio given in the below, where $p = L/d$.

$$H(p) = \frac{1}{p^2 - 1}\left[\frac{p}{\sqrt{p^2 - 1}}\ln(p + \sqrt{(p^2 - 1)} - 1\right] \quad (4.16)$$

The beauty of Eq. (4.15) lies in its capability of taking account of both the thermal conductivity of the system in spite of the presence of anisotropy behaviour of CNT through the two dimensionless parameter λ_{22}^c/λ_m and λ_{11}^c/λ_m and by introducing the dimensionless factor H, which considers the aspect ratio part of CNT, assumed to be all straight in length.

5 Summary

The diversified application of polymeric materials originated from its outstanding all-round properties including superior processability, ease of handling, lightweight and most importantly its low manufacturing cost in addition easy to mix with filler or other polymers. But it finds some limitations due to its low electrical and thermal conductivity. Various electrically and thermally conductive fillers are introduced into polymers to make polymer–filler composites to overcome these limitations. It is found that all the carbonaceous filler more or less improve the thermal conductivity behaviours of polymer depending on their structure and phonon transfer capabilities. It is observed that filler size, aspect ratio, its concentration, dispersion capability, the presence of functional group in its surface and polymer–filler interface affect the conductivity of the composites with great extent. The more important parameter relating to phonon transfer (i.e. the mechanism of thermal conductivity) is the length of their moving path. As the length of their moving path increases, the thermal conductivity value increases. It is observed that for carbon nanotube, those having higher length exhibiting higher thermal conductivity than those nanotube having lower length. Different models based on the different filler have been discussed elaborately to explain the thermal conductivity behaviour of different composites having different filler geometry as one model is insufficient to explain the behaviour of all composites. Based on the overview given here, the thermal conductivity of spherical fillers in both high and low loading can be determined in addition composites prepared by fillers having aspect ratio can also be calculated by using particular equation discussed here.

References

1. Shindé SL, Goela J (2006) High thermal conductivity materials. Springer
2. Tanaka T, Kozako M, Okamoto K (2012) Toward high thermal conductivity nano micro epoxy composites with sufficient endurance voltage. J Int Couns Electr Eng 2(1):90–98
3. Jordan J, Jacob KI, Tannenbaum R, Sharaf MA, Jasiuk I (2005) Experimental trends in polymer nanocomposites—a review. Mater Sci Eng, A 393(1):1–11

4. Raman C, Meneghetti P (2008) Boron nitride finds new applications in thermoplastic compounds. Plast Additvs Comp 10(3):26–31
5. Ishida H, Rimdusit S (1998) Very high thermal conductivity obtained by boron nitride-filled polybenzoxazine. Thermochim Acta 320(1):177–186
6. Han Z, Fina A (2011) Thermal conductivity of carbon nanotubes and their polymer nanocomposites: a review. Prog Polym Sci 36(7):914–944
7. Chen Y-M, Ting J-M (2002) Ultra high thermal conductivity polymer composites. Carbon 40 (3):359–362
8. Yang S-Y, Lin W-N, Huang Y-L, Tien H-W, Wang J-Y, Ma C-CM et al (2011) Synergetic effects of graphene platelets and carbon nanotubes on the mechanical and thermal properties of epoxy composites. Carbon 49(3):793–803
9. Yu A, Itkis ME, Bekyarova E, Haddon RC (2006) Effect of single-walled carbon nanotube purity on the thermal conductivity of carbon nanotube-based composites. Appl Phys Lett 89 (13):133102
10. Hammerschmidt U (2002) Guarded hot-plate (GHP) method: uncertainty assessment. Int J Thermphys 23(6):1551–1570
11. Kwon SY, Kim Y-G, Lee S, Kim JC (2011) Evaluation system for figure of merit of thermoelectric devices. Jpn J Appl Phys 50(11S):11RE02
12. Chung D (1995) Materials for electronic packaging. Butterworth-Heinemann
13. Jakubinek MB (2013) Thermal conductivity of nanotube assemblies and superfiber materials. Elsevier Inc. Nanotube Superfiber Materials: Chapter 16
14. Araki T, Shibayama M, Tran-Cong Q (1998) Structure and properties of multiphase polymeric materials. CRC Press
15. Agari Y, Ueda A, Omura Y, Nagai S (1997) Thermal diffusivity and conductivity of PMMA/ PC blends. Polymer 38(4):801–807
16. Rohsenow WM, Hartnett JP, Cho YI (1998) Handbook of heat transfer. McGraw-Hill, New York
17. Godovsky YK (2012) Thermophysical properties of polymers. Springer Science & Business Media
18. T'Joen C, Park Y, Wang Q, Sommers A, Han X, Jacobi A (2009) A review on polymer heat exchangers for HVAC&R applications. Int J Refrig 32(5):763–779
19. Hu M, Yu D, Wei J (2007) Thermal conductivity determination of small polymer samples by differential scanning calorimetry. Polym Test 26(3):333–337
20. Wang C-F, Huang P-K, Hung P-R, Hsu C-C, Jian S-R, Yang P-F et al (2015) Synthesis and characterization of polydopamine modified carbon nanotube (CNT)/polydimethylsiloxane (PDMS) composites. In: 2015 International Conference on: IEEE, Electronics Packaging and iMAPS All Asia Conference (ICEP-IACC), pp 826–829
21. Price DM, Jarratt M (2002) Thermal conductivity of PTFE and PTFE composites. Thermochim Acta 392:231–236
22. Kline DE (1961) Thermal conductivity studies of polymers. J Polym Sci 50(154):441–450
23. Cherkasova L (1959) Effect of structure on the thermal conductivity of polymers. Zh Fiz Khim 33(9):1928–1932
24. Reese W (1969) Thermal properties of polymers at low temperatures. J Macromol Sci Chem 3 (7):1257–1295
25. Zhong C, Yang Q, Wang W (2001) Corrigendum to "Correlation and prediction of thermal conductivity of amorphous polymers" [Fluid Phase Equilibria 181 (2001) 195–202]. Fluid Phase Equilib 192(1):209
26. Mathur V, Sharma K (2016) Thermal response of polystyrene/poly methyl methacrylate (PS/ PMMA) polymeric blends. Heat and Mass Transfer 1–11
27. King JA, Tucker KW, Vogt BD, Weber EH, Quan C (1999) Electrically and thermally conductive nylon 6, 6. Polym Compos 20(5):643–654
28. Yu A, Ramesh P, Itkis ME, Bekyarova E, Haddon RC (2007) Graphite nanoplatelet-epoxy composite thermal interface materials. J Phys Chem C 111(21):7565–7569

29. Kalaitzidou K, Fukushima H, Drzal LT (2007) Multifunctional polypropylene composites produced by incorporation of exfoliated graphite nanoplatelets. Carbon 5(7):1446–1452
30. Verdejo R, Bernal MM, Romasanta LJ, Lopez-Manchado MA (2011) Graphene filled polymer nanocomposites. J Mat Chem 21(10):3301–3310
31. Pierson HO (1993) Handbook of carbon, graphite, diamonds and fullerenes: processing, properties and applications (materials science and process technology). William Andrew Inc., Norwich, NY
32. Tu H, Ye L (2009) Thermal conductive PS/graphite composites. Polym Advan Technol 20 (1):21–27
33. Yu A, Ramesh P, Sun X, Bekyarova E, Itkis ME, Haddon RC (2008) Enhanced thermal conductivity in a hybrid graphite nanoplatelet–carbon nanotube filler for epoxy composites. Advan Mater 20(24):4740–4744
34. Tibbetts GG, Lake ML, Strong KL, Rice BP (2007) A review of the fabrication and properties of vapor-grown carbon nanofiber/polymer composites. Compos Sci Technol 67(7):1709–1718
35. Kuriger RJ, Alam MK, Anderson DP, Jacobsen RL (2002) Processing and characterization of aligned vapor grown carbon fiber reinforced polypropylene. Compos Part A: Appl Sci 33 (1):53–62
36. Abdel-Aal N, El-Tantawy F, Al-Hajry A, Bououdina M (2008) Epoxy resin/plasticized carbon black composites. Part I. Electrical and thermal properties and their applications. Polym Compos 29(5):511–517
37. Biercuk M, Llaguno MC, Radosavljevic M, Hyun J, Johnson AT, Fischer JE (2002) Carbon nanotube composites for thermal management. Appl Phys Lett 80(15):2767–2769
38. Choi E, Brooks J, Eaton D, Al-Haik M, Hussaini M, Garmestani H et al (2003) Enhancement of thermal and electrical properties of carbon nanotube polymer composites by magnetic field processing. J Appl Phys 94(9):6034–6039
39. Fukushima H, Drzal L, Rook B, Rich M (2006) Thermal conductivity of exfoliated graphite nanocomposites. J Therm Anal Calorim 85(1):235–238
40. Fang M, Wang K, Lu H, Yang Y, Nutt S (2010) Single-layer graphene nanosheets with controlled grafting of polymer chains. J Mater Chem 20(10):1982–1992
41. Yu C, Shi L, Yao Z, Li D, Majumdar A (2005) Thermal conductance and thermopower of an individual single-wall carbon nanotube. Nano Lett 5(9):1842–1846
42. Chang C-W, Okawa D, Garcia H, Majumdar A, Zettl A (2008) Breakdown of Fourier's law in nanotube thermal conductors. Phys Rev Lett 101(7):075903
43. Donadio D, Galli G (2007) Thermal conductivity of isolated and interacting carbon nanotubes: comparing results from molecular dynamics and the Boltzmann transport equation. Phys Rev Lett 99(25):255502
44. Cao J, Yan X, Xiao Y, Ding J (2004) Thermal conductivity of zigzag single-walled carbon nanotubes: role of the umklapp process. Phys Rev B 69(7):073407
45. Fujii M, Zhang X, Xie H, Ago H, Takahashi K, Ikuta T et al (2005) Measuring the thermal conductivity of a single carbon nanotube. Phys Rev Lett 95(6):065502
46. Alaghemandi M, Algaer E, Böhm MC, Müller-Plathe F (2009) The thermal conductivity and thermal rectification of carbon nanotubes studied using reverse non-equilibrium molecular dynamics simulations. Nanotechnology 20(11):115704
47. Maruyama S (2002) A molecular dynamics simulation of heat conduction in finite length SWNTs. Phys Rev B 323(1):193–195
48. Gharagozloo-Hubmann K, Boden A, Czempiel GJ, Firkowska I, Reich S (2013) Filler geometry and interface resistance of carbon nanofibres: key parameters in thermally conductive polymer composites. Appl Phys Lett 102(21):213103
49. Sahoo NG, Rana S, Cho JW, Li L, Chan SH (2010) Polymer nanocomposites based on functionalized carbon nanotubes. Prog Polym Sci 35(7):837–867
50. Teng C-C, Ma C-CM, Lu C-H, Yang S-Y, Lee S-H, Hsiao M-C et al (2011) Thermal conductivity and structure of non-covalent functionalized graphene/epoxy composites. Carbon 49(15):5107–5116

51. Song SH, Park KH, Kim BH, Choi YW, Jun GH, Lee DJ et al (2013) Enhanced thermal conductivity of epoxy-graphene composites by using non-oxidized graphene flakes with non-covalent functionalization. Advan Mater 25(5):732–737
52. Ganguli S, Roy AK, Anderson DP (2008) Improved thermal conductivity for chemically functionalized exfoliated graphite/epoxy composites. Carbon 46(5):806–817
53. Gulotty R, Castellino M, Jagdale P, Tagliaferro A, Balandin AA (2013) Effects of functionalization on thermal properties of single-wall and multi-wall carbon nanotube–polymer nanocomposites. ACS Nano 7(6):5114–5121
54. Ma P-C, Siddiqui NA, Marom G, Kim J-K (2010) Dispersion and functionalization of carbon nanotubes for polymer-based nanocomposites: a review. Compos Part A: Appl Sci 41 (10):1345–1367
55. Gojny FH, Wichmann MH, Fiedler B, Kinloch IA, Bauhofer W, Windle AH et al (2006) Evaluation and identification of electrical and thermal conduction mechanisms in carbon nanotube/epoxy composites. Polymer 47(6):2036–2045
56. Kochetov R (2012) Thermal and electrical properties of nanocomposites, including material properties: TU Delft, Delft University of Technology
57. Agari Y, Uno T (1986) Estimation on thermal conductivities of filled polymers. J Appl Polym Sci 32(7):5705–5712
58. Chai YH, Yusup S, Chok VS, Arpin MT, Irawan S (2016) Investigation of thermal conductivity of multi walled carbon nanotube dispersed in hydrogenated oil based drilling fluids. Appl Therm Eng 107:1019–1025
59. Bruggeman VD (1935) Berechnung verschiedener physikalischer Konstanten von heterogenen Substanzen. I. Dielektrizitätskonstanten und Leitfähigkeiten der Mischkörper aus isotropen Substanzen. Ann Phys 416(7):636–664
60. Cheng S, Vachon R (1969) The prediction of the thermal conductivity of two and three phase solid heterogeneous mixtures. Int J Heat Mass Tran 12(3):249–264
61. Nielsen LE (1974) The thermal and electrical conductivity of two-phase systems. Ind Eng Chem Fund 13(1):17–20
62. Ashton J, Halpin J, Petit P (1969) Primer on composite analysis. Technomic, Stamford, CN
63. Agari Y, Uno T (1985) Thermal conductivity of polymer filled with carbon materials: effect of conductive particle chains on thermal conductivity. J Appl Polym Sci 30(5):2225–2235
64. Choy C, Kwok K, Leung W, Lau FP (1994) Thermal conductivity of poly (ether ether ketone) and its short-fiber composites. J Polym Sci Polym Phys 32(8):1389–1397
65. Progelhof R, Throne J, Ruetsch R (1976) Methods for predicting the thermal conductivity of composite systems: a review. Polym Eng Sci 16(9):615–625
66. Fu S-Y, Yue C-Y, Hu X, Mai Y-W (2001) Characterization of fiber length distribution of short-fiber reinforced thermoplastics. J Mater Sci Lett 20(1):31–33
67. Deng F, Zheng Q-S, Wang L-F, Nan C-W (2007) Effects of anisotropy, aspect ratio, and nonstraightness of carbon nanotubes on thermal conductivity of carbon nanotube composites. Appl Phys Lett 90(2):021914

Electrical and Electronic Application of Polymer–Carbon Composites

Sambhu Bhadra, Mostafizur Rahaman and P. Noorunnisa Khanam

Abstract Carbon, the most important element in the periodic table, has various structures, such as carbon black, graphite, graphene, fullerenes, carbon nanotubes, etc. These possess an excellent physical and chemical properties. As a result, they can be used in numerous applications directly or using as a filler in the polymer composite. In this chapter, the application of different carbon materials as specialized fillers in the polymer composites has been discussed. This chapter includes the discussion on various types of carbon fillers, their basic features, their composites with polymers, percolation phenomena for electrically conductive composites and finally electrical and electronic applications of polymer/carbon composites. Applications of polymer/carbon composites in microelectronics, transparent conductive coating and flexible conductors, displays, organic light-emitting diode (OLED), electroluminescent device, photovoltaic device, sensor, actuator, electrode, battery, capacitor, supercapacitor or ultra-capacitor, ESD and EMI shielding, memory devices, field-effect transistor are discussed in details.

Keywords Polymer composites · Carbons · Electrical conductivity
Electrical percolation · Electrical applications · Electronic applications

S. Bhadra
Steer Engineering Private Limited, Bengaluru 560058, India
e-mail: sambhu2008@gmail.com

M. Rahaman (✉)
Department of Chemistry, College of Science, King Saud University,
Riyadh 11451, Saudi Arabia
e-mail: mrahaman@ksu.edu.sa; mrahaman1997@gmail.com

P. Noorunnisa Khanam
Department of Mechanical and Industrial Engineering, Qatar University,
2713 Doha, Qatar
e-mail: pnkhanam_phd@yahoo.com

M. Rahaman et al. (eds.), *Carbon-Containing Polymer Composites*,
Springer Series on Polymer and Composite Materials,
https://doi.org/10.1007/978-981-13-2688-2_12

1 Introduction

The most universal element in the periodic table is carbon. Carbon is bonded with other elements/chemical compounds with sp, sp^2, and sp^3 hybridization to form single (sigma bond) or multiple covalent bonds (pi bonds). Using the different hybridization states, it forms multidimensional structures of substances, which is responsible for its several physical and chemical properties [1]. The strongest mechanical strength of C–C σ bond and the creation of C–C π bond in carbon allows this element to form many types of materials like natural and synthetic with their different superior properties like electrical, mechanical, biological, etc. It can be used as filler in different matrix to improve basic material properties [1].

1.1 Different Carbon-Based Materials Used in Polymer Composites

Following are the main carbon materials used as filler for making carbon-reinforced polymer composites

- Graphite
- Graphene
- Fullerenes
- Carbon nanotubes
- Carbon fibers
- Carbon black.

1.1.1 Graphite

Graphite is one of the widely used carbon materials. Crystal structure graphite is flat equilateral (top right of Fig. 1) [2, 3]. In graphite, C–C bonds are sigma bonds with sp^2 hybridization. Sharing of electrons among the carbon atoms of graphite layer is responsible for electrical properties like in metals.

Graphite is used as a precursor for the synthesis of diamond. It is used to manufacture bearings, electrodes, seals, etc. It is also applied as filler in different polymers-based composites. This is because it has excellent electrical, thermal, and lubricating properties and hence is widely used in many industries [2, 3].

1.1.2 Graphene

Graphene (top left of Fig. 1), a lattice of carbon atoms with honeycomb structure is nearly transparent, zero gap semi-conductor. It forms the basic structure of other

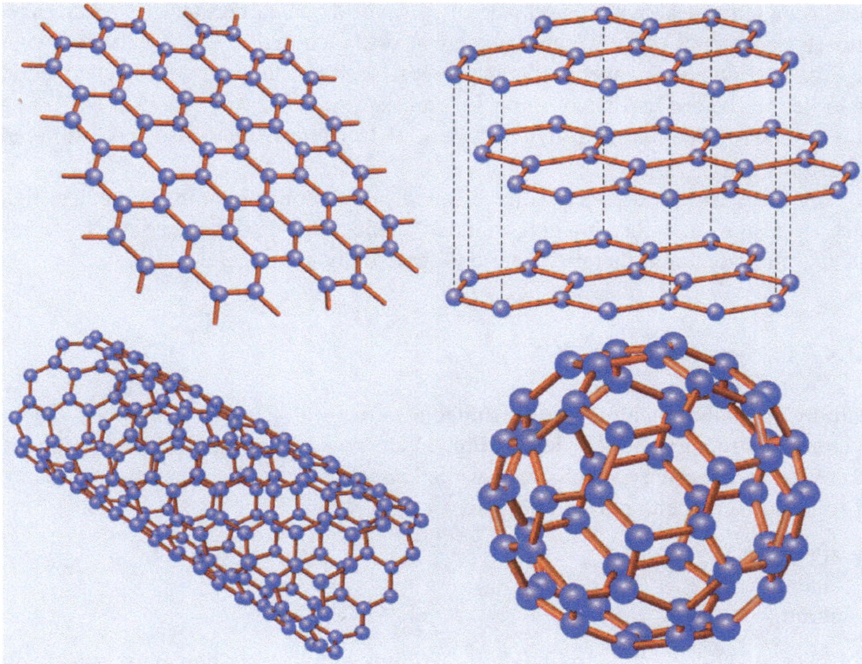

Fig. 1 Graphene (top left) is a honeycomb, single-layer lattice of carbon atoms. Graphite (top right) is a stack of graphene layers. Carbon nanotubes are rolled-up cylinders of graphene (bottom left). Fullerenes (C60) are molecules consisting of wrapped graphene with pentagons on the hexagonal lattice. Adopted from [4]

carbon allotropes and is having many outstanding properties such as conduction of electricity and heat, and if compared is about 100 times stronger than steel. It can be manufactured either by top down or bottom up approach [4].

Top down approach: In this method micromechanical exfoliation of graphite is possible either by 'scotch tape' or 'peel-off' method and this type of graphene oxide reduction is solution-based reduction.

Bottom up approach: Chemical vapor deposition, epitaxial growth, decomposition of ethylene on nickel surfaces.

Graphene can be used in different fields such as electrical, electronic, strong or lightweight composites, energy storage, bio-engineering, filtration and photovoltaic, etc. [5].

1.1.3 Fullerenes

Fullerenes are the molecules of a carbon atom family with contains of different shapes like hollow sphere shapes, tube, or ellipsoid shapes. These are described as "multi-walled cages" which are made of sp^2 hybridized carbon atoms and arranged into 12 pentagonal ($n/2-10$) and hexagonal walls wherein $n \geq 20$" (bottom right of Fig. 1). Fullerenes can be of many types, such as standard, giant, egzohedral, endo hedral, hetero and nano onion fullerenes.

The above types are the main models of fullerenes, they also have different subgroups. [6].

Fullerenes and its derivatives have found great applications in biomedical field such as design of X-Ray imaging contrast agents, high-performance MRI contrast agents, photodynamic therapy, drug and gene delivery, etc.

1.1.4 Carbon Nanotubes (CNTs)

Carbon nanotubes belong to the fullerene structural family. Nanotubes having cylindrical structure (bottom left of Fig. 1) are made by rolling a plane graphene. There are three basic types of single-walled nanotubes classified based on direction of rolling of graphene plane. They are

- zig zag
- chair
- chiral.

CNTs with diameters from 0.3 to 2.6 nm with larger diameters, which are thermodynamically unstable. CNTs can be either single-walled (SWCNT) or multi-walled Carbon nanotubes (MWCNT). These both CNTs possess good electrical and heat conductivity properties. Based on the structure and purity nanotubes may have semi-conductive properties to metallic conductivity [7].

Carbon nanotubes are given the application in the area of nanotechnology, especially in electronics, optics and other fields of science and technology.

1.1.5 Carbon Fibers and Nanofibers

Carbon nanofibers are linear, non-continuous filaments that are different from continuous, several micrometer diameter carbon fibers (Fig. 2). Generally, carbon nanofibers can be prepared by various methods such as catalytic chemical vapor deposition (CVD), electro-spinning, thermal treatment, etc. Carbon nanofibers have high specific surface area, great strength, and flexibility that makes them to be used as electrode materials for filler in carbon fiber reinforced plastic composites, energy-storage devices, and bone tissue scaffold.

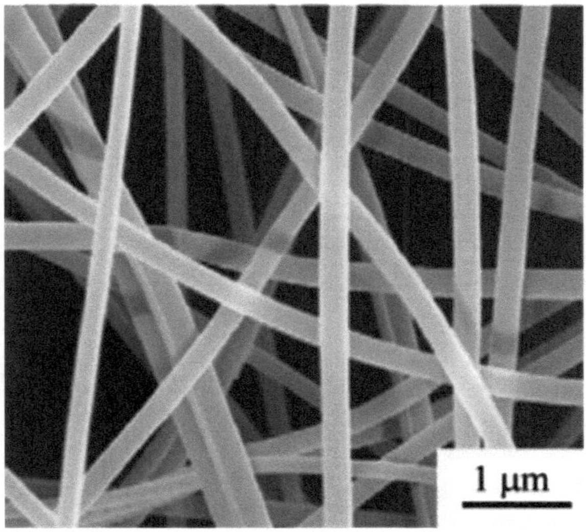

Fig. 2 Carbon nanofiber. Adopted from [247]

1.1.6 Carbon Black

Carbon black materials are very important materials with black colored reinforcing filler especially for thermosetting elastomers. These materials are in amorphous form of coal, which normally contains of spherical particles of elemental carbon. Size of these molecules is less than few dozen nanometers. Surface of the particles may contain different chemical groups, such as hydroxyl, carbonyl, carboxylic, etc. Particles form agglomerates with different shape and configuration. Several agglomerates combined together to form aggregated structure. The structure and surface characteristics of carbon black is shown in Fig. 3. Particle structure and configurations are effect on properties of carbon black. This filler is mainly obtained from incomplete combustion of carbonaceous materials. Carbon block is obtained from coal, wood, natural gas, and hydrocarbons. Furnace method, lamp methods are used for production and now plasma method is also widely used to produce the carbon black. The main production methods for carbon black are furnace, lamp methods and now more widely used plasma method. Carbon black is mainly used as filler in composites, rubber components during rubber processing, in black master batch, paints, and varnishes [8, 9].

1.1.7 Conductive Carbon Black

Carbon black is available in many types depending upon structure and properties. Some carbon blacks are semi-conducting, some are highly conducting. The factors which affect the electrical performance of carbon black and its compound are the particle size, structure, and porosity of the carbon black.

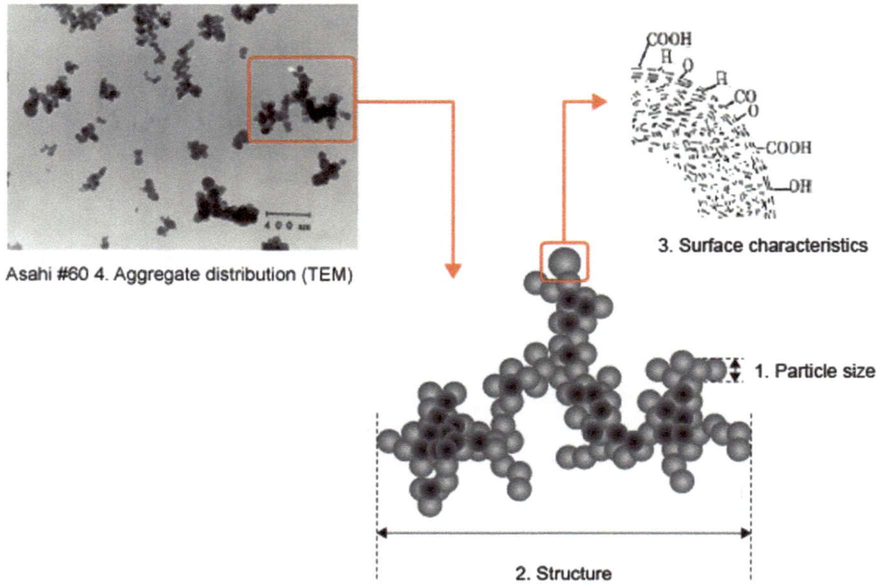

Fig. 3 Structure of carbon black. Adopted from [248, 249]

High structure: High structure means formation of agglomerates with long and branched chains. Carbon blacks with high structure exhibits high electrical conductivity.

Small particle size: Carbon black with smaller particle exhibits higher electrical conductivity. If the particles of carbon black are small and uniform in shape, their surface area is larger which gives the higher electrical conductivity.

High porosity: Higher particle porosity gives the good electrical conductivity.

2 Polymer–Carbon Filler Composites

Composites are combination of different materials, which forms heterogeneous substance structure, given the material with specific design, function, and provide required properties. Composites mainly consists of matrix and reinforcement. Polymer-based composites can be classified in different ways based on type of polymers and fillers used. Classification of polymers is shown in Fig. 4. Polymer can be either thermoplastic or thermoset. After curing, thermosets have well-bonded three-dimensional molecular structure and they also decompose on heating instead of melting unlike thermoplastics and as a result they cannot be recycled. Whereas, thermoplastics are used as it is without crosslinking. They melt when heated. That is why they can be reprocessed and reused.

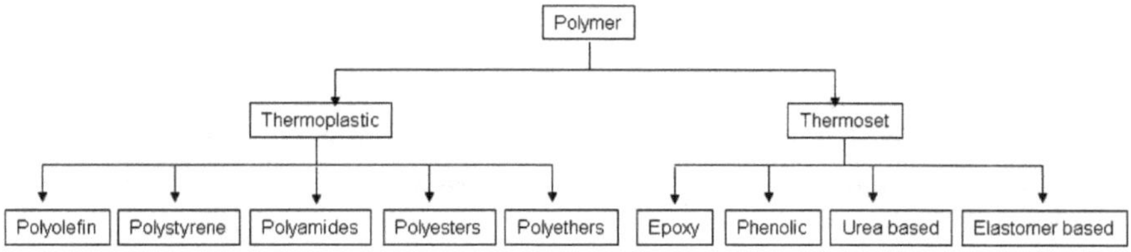

Fig. 4 Classification of main polymers

Classification of filler based on physical structure is shown in Fig. 5. Fillers can be of fiber, whisker, flake, and particulate type. Fibers are very long, either often circular or near circular with very high L/D. Particles have no particular shape and orientation. But whiskers have specific shapes, which are small both in length and diameter as compared to the fibers.

3 Electrically Conductive Polymer/Composites

Electrically conductive polymer systems can be divided in two types: (i) intrinsically conductive polymers which contain the delocalized pi electron system which is responsible for the electrical conduction. These types of polymers are chemically synthesized from appropriate monomer; and (ii) extrinsically conductive composites which are prepared by incorporation of conducting fillers into insulating polymer matrix. Generally, extrinsically conductive composite systems are cheaper than intrinsically conductive polymer systems. Besides this, the mechanical properties, electrical conductivity, and processability of extrinsically conductive composites can be easily adjusted through the suitable selection of filler, polymer matrix, and their relative proportion [10].

3.1 Percolation

Electrical conductivity of carbon filler based polymer composites depends on the arrangement of conductive constituent in the polymer matrix which is insulating material. Main objective of development of good conductive composite is to get an

Fig. 5 Different type of fillers

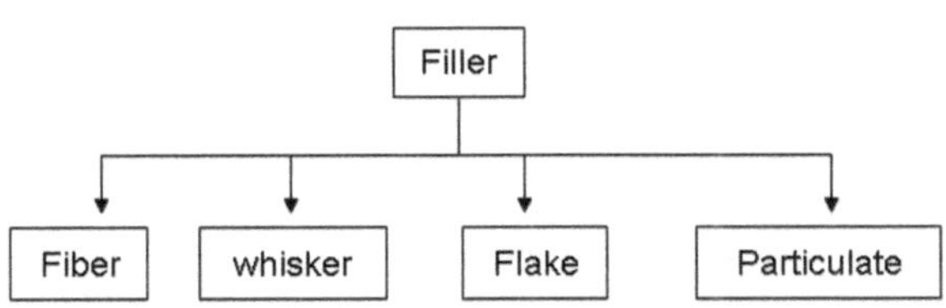

excellent conductivity with the minimum concentration of conductive filler in the polymer matrix. In carbon filler filled polymer composite, the conductive filler should be well dispersed which generates the interconnected network throughout the polymer matrix. The electrical conductivity of such composite system depends on the concentration, type, quality of contact, shape and size of the conductive particles in the composite.

Percolation threshold is that minimum filler content in the polymer matrix after which there is no significant change in the electrical properties of the composites. Before percolation in polymer composites, conductive particles in insulating polymer matrix remain isolated from each other and continuous conductive network formed at the percolation. Therefore, due to transition state from insulating to conducting state, there is an abrupt change in the conductivity of the composite system. This effects on the conductivity which sharply increases at the percolation threshold. The electrical conductivity of the composite significantly not only increases with further increment of the concentration of the conducting filler beyond the percolation limit. The possible situation which may arise with increasing filler concentration can be explained as follows. Conductive fillers can be considered as small metal wires. By addition of enough conductive filler which is called percolated quantity in an insulating polymer matrix as if all small wires are connected and form continuous wires. Before the percolation in composites, these metal wires are not continuous and as a result net conductivity is very less. There is only increase in the number of conductive chains only after the percolation, just like increase in the diameter of metal wire, which relatively increase (small increment)

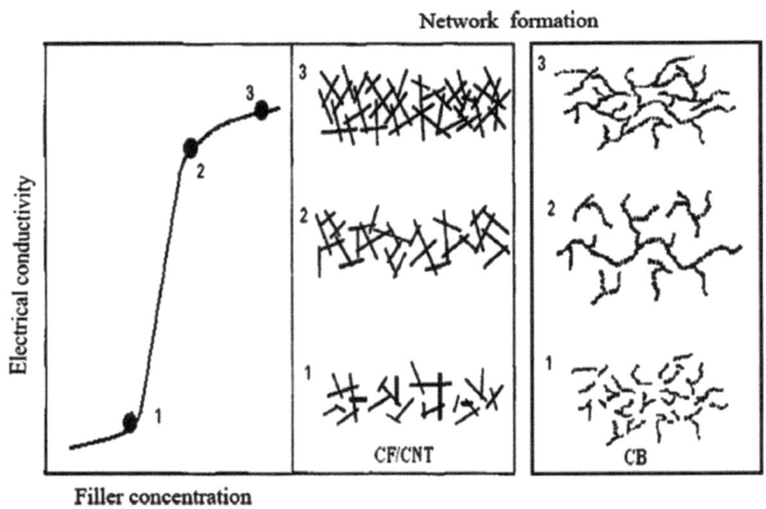

Fig. 6 Different types of network structure formation with increasing concentration of the conducting fillers, carbon black (CB), carbon fiber (CF), carbon nanotube (CNT); (1) before percolation, (2) at percolation, (3) above percolation and variation of conductivity of the composites because of the formation of network structure. Adopted from [10]

the electrical conductivity of the composites (Fig. 6). Hence, a sharp increase in electrical conductivity is observed in polymer composite only at the percolation threshold [10].

4 Electrical and Electronic Applications of Carbon Filler Filled Polymer Composites

In polymer matrix composite (PMC), nano- and micro-sized carbon fillers, including carbon fibers (CFs), graphene nanoplatelets (GNPs), and carbon nanotubes (especially MWCNTs) have been incorporated into the polymers to develop the properties of the polymer composites which can be used as multifunctional materials [11]. The most-used carbon-based reinforcing fillers are MWCNTs, GNPs, and CFs due to their superior thermal, electrical, and mechanical properties. Carbon filler based reinforced polymer composites have different superior properties, such as low density and are able to be compression molded or injection molded into complex shapes, lightweight, high thermal conductivity, high strength, good manufacturability, and increased reliability, etc. Because of good mechanical properties and thermal conductivity, these carbon-based composites are gaining popularity in different fields, such as aerospace, microelectronics industries, and automotive for electronic packaging and thermal management applications [11–13].

There are numerous electrical and electronic applications of carbon-filled polymer composites. Few specific applications are in microelectronics, transparent conductive coating, flexible conductors, displays, organic light-emitting diode (OLED), electroluminescent device, photovoltaic device, sensor, actuator, electrode, battery, capacitor, supercapacitor or ultra-capacitor, ESD, EMI shielding, memory devices, field-effect transistor, etc. Application of carbon-filled polymer composites in those fields is detailed in the following sections.

4.1 Microelectronics

Microelectronics is very small electronic designs and components, typically made from semi-conductor materials. Example of such devices includes transistors, capacitors, inductors, resistors, diodes, etc. A schematic of a microelectronic device is shown in Fig. 7. The applications of carbon filler filled polymer composites in microelectronics include integrated circuit, printed circuit board, interconnections, die attach, solder, lids/enclosure, heat sinks, connectors, electrical contacts, and thermal interface materials, etc.

Materials with high thermal conductivity (to improve heat dissipation) and low thermal expansion are required for their application in heat sink, lids, and die attach of microelectronic devices. For this reason, polymer and metal matrix based

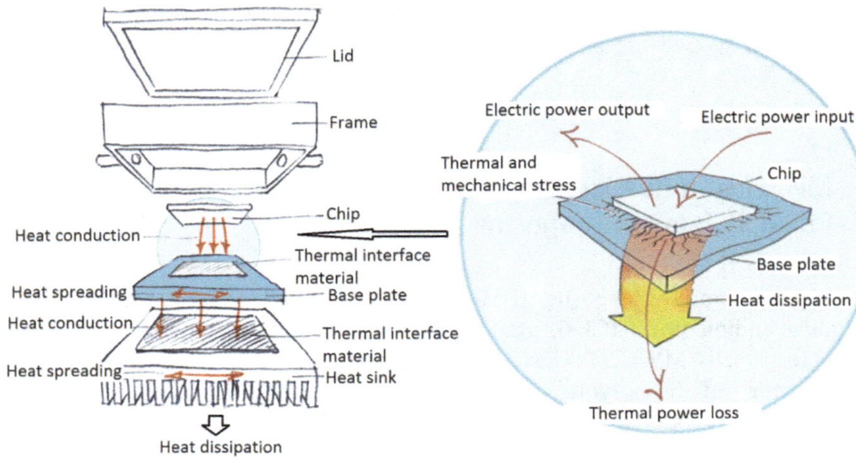

Fig. 7 A typical design of a microelectronic device. Adopted from [250]

composites have been developed. Low density of carbon filler filled polymer composites is helpful to reduce the weight of the product and easier processability enables them attractive for end application. Carbon-filled composites are also cheaper than metals such as silver when used for interconnection. Moreover, metals are more prone to oxidation resulting decrease in electrical conductivity over a period of time. That is why, carbon fillers filled polymer composite are used for thermal management and electronic packaging, such as heat sink, enclosures, die attach, conducting adhesive, electrical interconnections and, thermal interface materials. Polymer can be either thermoplastic or thermoset. Thermoplastic matrices have the advantages of reworking or repairing by heating. On the other hand, thermosetting matrices are not allowed to reworking but are thermally more stable than thermoplastics.

4.2 *Integrated Circuit*

These integrated circuit chips are attached to the substrate or printed circuit board on where the interconnection lines are written on multilayer substrate (each layer) or board. Another involving multilayer of thinner layer conductors and interlayer dielectrics are applied to the substrate before the chip attachment to increase the interconnection density. Soldered joints connect among electrical contract pads on the chip, board, or substrate. The board or substrate is attached to the heat sink. Thermal interface material is placed between the substrate or board to improve the quality of the thermal contact of the heat sink. The whole group of materials and other things are placed in a thermally conducting housing.

Nano-filler including carbon nano-filler-based smart polymer composites were prepared for their application in integrated circuit [14, 15].

Nanostructured composites of polyvinylidine fluoride (PVDF) matrix based composites which contains of aluminum nitride (AlN) and exfoliated graphene nanoplatelets (xGNP) nanoparticles and these composites were prepared for integrated circuit or 3D electronic packaging applications. Addition of GNP-exhibited significance improvement in electrical and thermal conductivity [16].

4.3 *Printed Circuit Board*

Printed circuit board is a sheet which contains of chips and these chips are attached in the board. Generally, this board is a polymer or ceramics matrix composites which is electrically insulating materials with four conductor lines. Polymers have lower thermal conductivity than ceramics, whereas ceramics have higher dielectric constant than polymers. High thermal conductivity is required for heat dissipation and lower dielectric constant is important for a smaller capacitive effect resulting smaller signal delay.

A base film which is prepared for flexible printed circuit boards is made with polymer matrix composites which are embedded with carbon nanotube bundles. Nanotube bindles are spaced apart from each other and first and second surface contains of flexible polymer. Because of higher thermal conductivity of carbon nanotubes, heat can be performed to the printed circuit board through the first surface to the second surface of base film [17].

4.4 *Interconnection*

An interconnection is a conductor line thick film with thickness of >1 μm for power, signal transmission, or ground. This interconnection can be on a chip, a substrate, or a printed circuit boards. Thick film conductor pastes containing silver/copper particle as a conductor and a glass frit as a binder are extensively used to make thick film conductor lines or for interconnections on substrate by screen printing and subsequent firing. These films have lower electrical conductivity because of glass present as binding agent and development of porosity in the film due to firing. Metals used as conductor get oxidized over a period of time. Huge mismatch of coefficient of thermal expansion (CTE) with silicon causes the thermo-mechanical stress which finally decreases the reliability of the devices. Best composite materials are required for future downscaling, where they are overcome these present limitations and also required for compatibility. Carbon nano-fillers are favorable materials with superior thermal, mechanical, and properties and these are suitable for electronic packaging applications either as fillers in polymer composites or as in pure forms.

4.5 Die Attach

A die attach is used to join a die (a chip) to a substrate, which is normally a material. A metal alloy (a solder paste), a glass or a polymeric adhesive can be used as die attach materials. B-stage polymer films are one of the most widely used adhesive films in electronic packaging applications such as anisotropic conductive films, die attach films, thermal interface materials for electronic packaging [18].

4.6 Solder Joints

A solder enhances heat dissipation. It has to have high thermal conductivity to enable heat dissipation. A polymer–metal paste can be used as solder material, where polymer acts as a binder and metal power, such as silver acts as a thermal conductor. However, due to high cost of silver and its tendency for oxidation, it can be replaced with carbon fillers.

Nai et al. [from the National University of Singapore] investigated the effect of nanotube concentration on electrical resistivity and shear strength. There was not significant improvement in electrical conductivity but noticeable increase in shear strength with increasing nanotube concentration. This work proved that improvement in mechanical and thermal properties can be achieved without effects on electrical properties for polymer-nanotube composites [19].

A CNT-solder composite paste comprising solder powder and 30–40% (volume fraction) CNTs was developed to provide high electrical conductivity and applied to at least one of a die or a package substrate [20].

Epoxy-based conductive adhesives were developed by filled with MWCNTs [21]. CNPs were dispersed in epoxy by using Ultrasonic mixing process for disperse them more uniformly and also to make them contact better, resulting in higher electrical conductivity. Increment of CNT loading decreased the volume resistivity and contact resistance of the conductive adhesive. The percolation threshold was observed at 3 wt% loading of MWCNTs. The average contact resistance was similar with solder joints even at high frequency. Moreover, a high mechanical strength was constant even by replacing the metal particle containing fillers with CNTS in the conductive adhesive. For example, 0.8 wt% of CNT content in polymer matrix retained the 80% of shear strength, while less than 28% of the shear strength was retained with common metal based filled conductive adhesive.

Effect of addition of various loading of CNTs in silver-filled conductive adhesive on electrical conductivity was reported [22]. CNT loading greatly increases the electrical conductivity of conductive adhesive with the silver loading below percolation threshold. Resistivity of 66.5 wt% silver-filled conductive adhesive without CNT loading had 10^4 Ω cm whereas 10^4 Ω cm of resistivity was shown for adding of 0.27 wt% CNT. Therefore, it s concludes that equal level of electrical conductivity is achieved with the addition of small quantity of CNT to replacing the conventional silver metal filler.

Polymer composites containing a polymer matrix, a graphene, and metal particles were developed to use it as a lead-free solder material. Metal particles and the graphene together in the polymer matrix reduced the electrical percolation threshold. [23].

Epoxy-based solderable conductive adhesive (ICA) was developed with functionalized graphene. Properties of this ICA were compared with three kinds of conventional ICAs, which were filled with Ag particles. Great metallurgical network interconnection formed between upper and lower electrode with the functionalized graphene-filled solderable ICA. The functionalized graphene-filled ICA exhibited higher mechanical strength and electrical conductivity compared with the conventional ICAs. Moreover, addition of functionalized graphene caused the improvement of thermal conductivity by 20% than solderable ICA without graphene [24].

4.7 Heat Sink

Slow dissipation of heat from microelectronics causes the reduction of its performance and early damage because of thermal stress, joint failure, and warpage of the chip. For these reasons, thermal management is very important in the field of microelectronics. Heat sink is a thermal conductor which is bonded to printed circuit boards, that serves as conductor and radiator which makes the heat away from the circuit. Carbon fibers are mostly used in heat sink and lid/enclosure due to good conductivity and reinforcement and low thermal expansion [25–29].

For heat sunk applications, thermal conductivity of the polymer composites needed around 1–30 $W\ K^{-1}\ m^{-1}$. Pure polymer thermal conductivity can be improved by the addition of thermally conductive fillers such as graphite, graphene, carbon nanotubes, carbon fibers, carbon black, ceramic, or metal particles. Typical thermal conductivity of these fillers is presented in Table 1 [30].

Boron nitride (BN) was added in polyamide 6 and graphite was added in polyphenylene sulfide (PPS) to develop composites for the application of heat sink.

Table 1 Thermal conductivities of some conductive fillers

Material	Thermal conductivity at 25 °C ($W\ K^{-1}\ m^{-1}$)
Graphite	100–400 (on plane)
Carbon black	6–174
Carbon nanotube	2000–6000
Carbon fiber	8–100
Copper	483
Silver	450

The influence of composition and size of the filler on mechanical and thermal properties of these polymer composites were studied. The addition of thermally conductive fillers, such as graphite and BN rise to increase the thermal conductivity about two times of the respective polymer composites. In PPS/graphite composite system, smaller sized graphite particles are more effective to elevate the conductivity of the composites. This thermally conductive composite was developed for their application as heat sink [31].

MWCNTs, graphite fibers, and BN platelets were used as fillers in epoxy matrix to develop conductive nanocomposites. Composite containing MWCNTs produced by chemical vapor deposition were exhibited better thermal diffusivity than that of MWCNTs composites which were produced by the combustion method. Nanocomposites based on boron nitride filled epoxy exhibited the highest thermal conductivity among all the composites mentioned above [32]. These composites can be used as heat sink.

4.8 Lid/Enclosure

Generally, a lid is a cover for a chip which protests physically and normally chip is mounted in a well in a substrates and lid covers the well. A metal is preferred for making the lid because of heat dissipation which is joined with the substrate by soldering.

For avionic electronic enclosure, lightweight is essential to save aircraft fuel. Carbon fiber reinforced epoxy composite has been developed because of its good mechanical, environmental, and electrical properties with reduced weight, cost, and ease of manufacturing [33]. Carbon fiber based polymer composites having high strength, modulus, high thermal conductivity, and low density which are promising for the thermal management of satellites [34].

4.9 Thermal Interface Material

Graphene is one of the most efficient fillers for thermal interface materials [35]. Exfoliated graphene-filled (10–15 vol.%) polymer composites were prepared, which exhibited very high thermal conductivity. As a thermal interface material, it outperformed those with carbon nanotubes or metal nanoparticles. This is because of higher aspect ratio and lower resistance at the graphene–polymer interface [36]. Generally, epoxy resins have poor thermal conductivity and electrically conductive or insulate materials should be introduced into the epoxy to enhance the thermal conductivity. Significant improvement in thermal conductivity of epoxy can be achieved with the addition of graphene into epoxy. Four times of thermal conductivity increased with the addition of 5 wt% of GO in GO-filled epoxy composites [37, 38]. This is the reason to say that graphene material is suitable as thermal interface materials for heat dissipation.

Aligned graphene sheet-polymer composite was prepared for thermal interface material. A mixture was prepared with the dispersed graphene sheets in the polymer fluid. The graphene filament bundles substantially paralleled to each other are formed by a sequence of aligned graphene sheets in the polymer fluids when a field was applied. Finally, the mixture was solidified. An anisotropic index in a range of 1.00–2.00 was obtained in an aligned graphene sheet-polymer composite by calculating the ratio of the coefficient of thermal conductivity in a parallel direction and also the one in perpendicular direction [39].

Thermal interface materials comprising graphene fillers as conducting filler with or without other metallic fillers in a polymer matrix (epoxy) were prepared. At least 150% of the thermal conductivity increased for the polymer matrix in polymer nanocomposites [40].

School of Chemical and Process Engineering, University of Leeds developed several carbon filler filled polymer composites for the application of thermal interface materials [41].

Marion et al. made thermal interface material from metal (Cu, Sn) decorated MWCNTs and Vinnapas® BP 600 polymer. The sample was functionalized by nitric acid (HNO3) treatment (mid oxidation) or, with N-Methly-2-Pyrrolidone (NMP). There was significant improvement in thermal conductivity while retaining mechanical properties of the polymer [42].

Thermal interface materials were prepared comprising polystyrene (PS) and polybutene as polymer matrix and graphite, graphene, and carbon nanotubes as conducting filler [43].

4.10 Electrically Conductive Adhesives

These types of conductive adhesives gave applications in electronic packaging such as solder-less interconnections, die attachment, display interconnections, component repair, and heat dissipation.

Heimann et al. [44] developed the epoxy adhesives filled with CNTs polymer composites and these composites were tested for possible electronics packaging applications. Thermal conductivity of these composites was increased by nearly 4.4 percentage with CNTs in the matrix than the composite without CNTs. CNTs also improved the mechanical properties of the composites. In the same study, researchers modified the nanotubes by chemically modification for getting better dispersion, which gives the additional advantage to improve the mechanical performance [44]. Same group researchers develop the CNT-filled- (3 wt%) and Ag-filled (82 wt%) adhesive composites. There was an increase in stiffness by 75% for the CNT-filled adhesive compared to that of Ag-filled adhesive with a significant improvement in thermal characteristics [45]. Isotropically conductive adhesives (ICAs) were prepared by using CNTs as filler in epoxy matrix and different concentrations of MWCNTs were used in these composites. Thermal, impact and conductivity properties were studied for these epoxy composites and found improve properties with very lower filler loading of ICA [46].

Adhesive resins were filled with carbon fibers of variable lengths and widths. The fiber-filled adhesive exhibited high thermal conductivity. Electronic systems having components bonded by a layer of these adhesives resins have high through-the-thickness thermal conductivity [47].

4.11 Transparent Conductive Coatings/Flexible Conductors

Generally transparent conducting films (TCFs) are electrically conductive and optically transparent in nature. TCFs have applications in electrically conductive coating, photovoltaic, smart windows, etc. To give the TCFs in photovoltaic applications, these must be fabricated mainly from ITO and zinc oxide. These are brittle and may break easily. Organic films using carbon nanotube networks and graphene have been developed, which are highly conductive and transparent [48].

Carbon nanotubes (CNTs) have attracted to researchers because of their excellent properties such as high tensile strength ($\sim$13–53 GPa), high elastic modulus ($\sim$1–2 TPa), and a high electrical conductivity (theoretically metallic tubes can carry an electrical current density of 4×10^9 Acm^{-2}, which is $\sim$1000 times higher than for other metals such as copper) [49]. Because of good electrical properties, thin films of CNT have been used in transparent electrodes in TCFs.

Transparent, flexible and conducting composites film was prepared with poly (methyl methacrylate) (PMMA) by using $SOCl_2$ which is doped with single-walled carbon nanotubes (SWCNTs). The electrical conductivity of the $SOCl_2$-doped SWCNT-filled composite was five times higher compared to that of un-doped SWCNT-filled composite. SWCNT loaded composite with 0.1 wt% showed the optical transmittance of visible light at 500 nm with 92% and at 0.5 wt% of SWCNT loading showed [50].

SWCNTs were deposited on a transparent substrate. It was used to grow a thin film of polypyrrole (PPy) or polyaniline (PANI) by electrochemical polymerization. The resulting hetero-structure was sufficiently transparent and electrically conductive to be used in practical applications [51].

Polyimide nanocomposites with SWCNT were synthesized by using in situ polymerization method. The resultant well-dispersed SWCNT–polyimide nanocomposite films were optically transparent with good electrically conductivity [52].

SWCNTs reinforced Poly(3,4-ethylenedioxythiophene) (PEDOT), poly(styrenesulfonate) (PSS) composite thin films were prepared which have transparent in nature, flexible, and high conductivity.

The films exhibited high optical uniformity and conductivity. DC conductivity was >105 S cm^{-1} for >50 wt% of SWCNT [53].

PEDOT/sulphonated graphene composite were prepared by in situ polymerization method and these composites showed excellent electrical conductivity, high thermal stability, transparency, good flexibility. These composites were easily processable in both organic and aqueous and solvents [54].

4.12 Display

An electronic visual display performs as a video display for presentation of images transmitted electronically. Common examples of electronic visual displays are television, computer monitors, mobile, etc. Display system can be of mainly two types; (i) light-emitting diodes (LED) and (ii) liquid crystal display (LCD). LED is a two-lead semi-conductor light source, which emits light because of electroluminescent effect when activated by the application of suitable voltage. The color of the light is determined by the energy band gap of the semi-conductor. A LCD utilizes properties of liquid crystals such as the light modulating properties. The LCD has different advantages which has low power consumption than the LED [55, 56].

A liquid crystal cell contains of about 10 μm thin layer of a liquid crystal which is kept between two sheets of glass. Two electrodes which were transparent are deposited on the inside faces of the glass sheets and a reflective mirror is setup in the back side. Indium tin oxide (ITO) contains electrode plane kept on top and at polarizing film with a glass were kept at the bottom side. The whole part of the LCD is covered with common electrode and above area is placed with liquid crystal substance. An additional rectangle part of glass with an electrode is kept on the bottom. Another polarizing film is placed on top. In the absence of current light passes via sfront of the LCD which is reflected by the mirror and it bounced back. An electrode is connected to a temporary battery and the current causes the liquid crystals between the common-plane electrode and the rectangle-shaped electrode to untwist. Hence the light is blocked by passing through causing the rectangular area blank [57]. The construction of typical LCD is shown in Fig. 8.

High transparency (T) and Low sheet resistance R_s are required for materials which are given applications in optoelectronic devices such as, light-emitting diodes, touch-screens, displays, and solar cells. Mostly ITO can be used as transparent conductive (TC) material for flexible transparent optoelectronic devices. But this ITO has some disadvantages like high cost, brittle, difficult to pattern and sensitive to both acidic and basic environments. Therefore, development of new low cost, flexible, transparent, easily processable TC materials with improved performance is required. SWCNTs and graphene also show great promise as a transparent TC material.

San et al. dispersed fullerene, graphene, single- and multi-walled carbon nanotube in liquid crystal to enhance its properties. Fullerene is found to be the best compatible material for optical aims [58]. Qi et al. highlighted the effect of carbon nanotubes on current and future generation's liquid crystal displays [59].

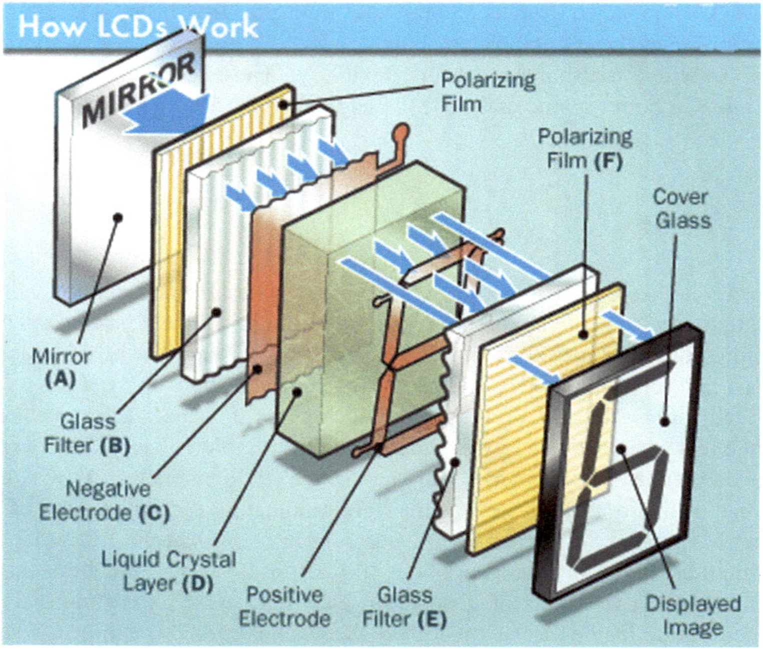

Fig. 8 Structure of a typical LCD. Adopted from [57]

4.13 Organic Light-Emitting Diode (OLED)

An organic light-emitting diode (OLED) is a light-emitting diode (LED), in which an organic emissive electroluminescent layer is applied, which emits light when electric current is applied. A typical OLED consists of an organic materials layer which is placed between two electrodes (anode and cathode). All the materials are deposited on an electrically conductive transparent substrate, such as ITO. OLEDs have attracted attention in recent years as next-generation lighting and displays because of their excellent performance, good flexibility, ease of fabrication, etc. Further scope of improvement of OLED devices may include increasing flexibility, reducing cost, improving performance. This can be done by replacing the conventional ITO with better alternative transparent conducting electrodes (TCEs) [60].

Typical construction of an OLED is shown in Fig. 9. It consists of transparent anode, glass substrate, hole-injection layer, electron transport layer, organic emitters, and metal cathode. The charges are starting to move in the device by applying voltage to the electrodes under the influence of the electric field. Electrons leave the cathode and holes move in opposite direction from the anode. Creation of a photon is possible with the recombination of this charges with a frequency given by the energy gap ($E = h\nu$) between the LUMO and HOMO levels of the emitting

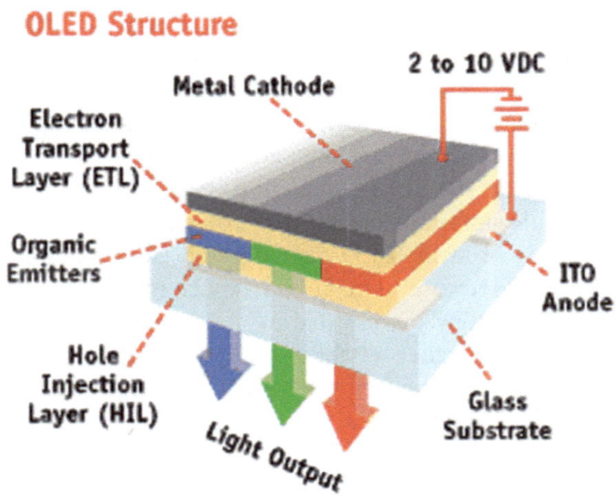

Fig. 9 Typical construction of an OLED. Adopted from [61]

molecules. Thus, the electrical power applied to the electrodes is transformed into light [61].

Graphene has of 4.5 eV work function, which is similar to ITO with flexibility and low cost material making it an ideal candidate as OLED anode. Graphene-based polymer composites have been used in OLED and field emission devices because of high electrical conductivity, high carrier mobility and moderately high optical transmittance in the visible range of spectrum. [62, 63].

Polymer composite which is made with poly(m-phenylene vinylene-co-2,5-dioctoxy-p-phenylene) (PmPV) and SWCNTs were used in Organic light-emitting diodes and in this composite SWCNTs were well dispersed in the polymer. This device is fabricated with up to 0.1 wt% SWCNT, results showed that oscillator strength increases of the green emission with a dominant emission peak near 500 nm. The shift in the emission indicates that the SWCNTs in the PmPV matrix act as a hole-blocking material [64].

SWCNT film was deposited on poly(ethyleneterephthalate) substrate and used as flexible, transparent anode for OLED. The operational lifetime of the device was comparable with ITO/PET anode [65].

Electrodes with SWCNT-polymer composite as both the electron and hole-injection electrodes are used to prepare Polymer light-emitting devices. The prepared devices were metal-free and linearly stretchable up to 45%. This is a good example of highly stretchable semi-conductor light-emitting device [66].

PS was coated on top of a CNT/PEDOT:PSS composite film, soaked in HNO_3 acid and then encapsulated with polymer. This transparent conductive composite was used as anode for OLED. It exhibited maximum lamination of $\sim$9000 Cd m^{-2}, and efficiency of $\sim$10 Cd A-1, which is very close to ITO-based OLED device performance. 30–450 times better performance is achieved with devices in which CNT anodes were using if we compare with devices [67].

Poly 9,9-dioctyl fluorene-alt-bithiophene and reduced graphene oxide were used as an emissive layer to prepared Polymer light-emitting diode. Reduced graphene oxide (rGO2) was synthesized by repeating two times Hummers oxidation followed by hydrazine hydrate reduction method. The rGO2 thus produced had more uniform distribution in size and thickness with increased crystallite size. This increased in in-plane crystallite size and high localize aromatic confinement of rGO2 causes favorable shift in balance of electron and hole recombination zone toward the center of emissive layer. LED device comprising rGO2-based nanocomposite exhibited five times increase in maximum device efficiency (Cd A^{-1}) and three times increase in maximum brightness (Cd m^{-2}) compared to that of neat polymer [68].

To improve the wettability on PEDOT:PSS composite electrode, modification of hydrophobic graphene on PET substrate is possible by using interface engineering. It exhibited high optical transmittance, low sheet resistance, and high work function. The resulting device exhibited excellent flexibility and good electroluminescence emission [69].

Graphene oxides was synthesized and then different concentrations of graphene oxide was mixed with PEDOT:PSS polymer solution. Reduction of graphene was done in the PEDOT:PSS composite through heat treatment. This composite with particular doping concentration was used as a hole injection for the fabrication of PLED devices. Device performance was studied by using different device properties such as influence of doping concentration. It was found that the resistance in the hole-injection layer and the turn-on voltage were effectively decreased by the doping of graphene which gives the higher overall efficiency. Increasing concentration of doping increases the conductivity of the hole injection. The optimal doping concentration was found to be 0.03 wt% [70].

4.14 Electroluminescent Device

Electroluminescence (EL) is an opto-electrical property in which a material emits light in response to a strong electric field or an electric current. It is the result of recombination and radiative of holes and electrons in a material, generally this is a semi-conductor. Energy is released in photon form from excited electrons. Prior to recombination and also to form a p-n junction (in semi-conductor electroluminescent devices such as LED), doping can be separated by the electrons and holes or through excitation with high-energy electrons which is accelerated by a strong electric field (as with the phosphors in electroluminescent displays) [71].

In recent years, photo-induced electron transfer between semi-conducting conjugated polymers and SWCNTs filler has attracted to researchers. Electroluminescent properties (EL) of SWCNT-polymer composites have also been studied. EL and photovoltaic (PV) response were improved with doping of SWCNTs into poly [2-methoxy-5-(2′-ethylhexyloxy)-1, 4-phenylenevinylene] (MEH-PPV) type conjugated polymer based on single-layer light-emitting diodes [72].

CNT/PEDOT:PSS textile-based conducting composites were developed with transparent PET mesh fabrics which were dip and coated with CNT followed by inkjet printed with PEDOT:PSS. These coated fabrics exhibited good transparency, conductivity, and flexibility which is suitable for flexible displays applications. EL devices exhibit superior flexibility which were prepared with textile-based CNT/PEDOT:PSS composite conductors as the front electrodes under stretching and bending compared to that of ITO-coated PET film-based electrodes [73].

4.15 Photovoltaic Device

Photovoltaic (PV) device converts the solar energy into electricity by using semi-conducting materials which shows the photovoltaic effect.

A photovoltaic system, which is solar panels are made with number of solar cells to create electricity from solar energy [74]. P-type and N-type semi-conductors with different electrical properties and joined together were used in the making of photovoltaic cell. P-N junction refers as the joints between these two semi-conductors. Sunlight which consists of some particles of solar energy is called photons and these hotons are reflected, absorbed, or transmitted by the PV when a PV cell is exposed to the sunlight. Electrons are getting excited and freed from the negative semi-conductor material when the enough photons are absorbed by the negative layer of the photovoltaic cell. Electrons which were freed were migrated to the positive layer thus creating the voltage differential. The electrons (−) is collected in the N-type semi-conductor, and the holes (+) in the P-type semi-conductor. Electricity flows in the cell when electric bulb or an electric motor types of an external load is connected between the front and back electrodes [75]. Basic working principle of a photovoltaic device is shown in Fig. 10.

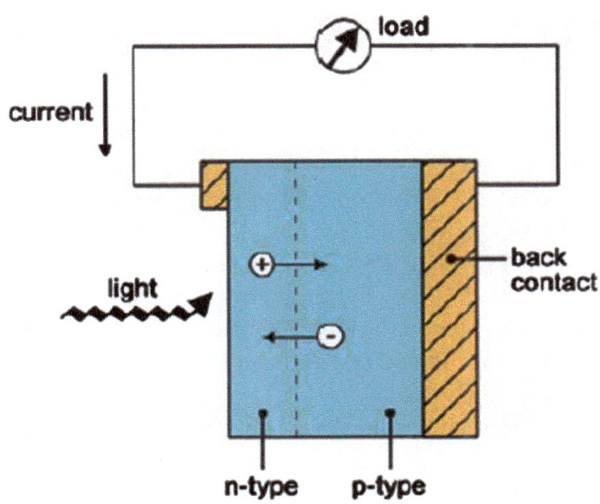

Fig. 10 Working principle of PV. Adopted from [75]

Silicon based solar cells are the first-generation solar cells having very high efficiency (~25%) of converting solar energy to electrical energy. However, this type of solar cells has high weight and is very expensive. The second-generation solar cells comprise of amorphous Si, cadmium-telluride (CdTe), gallium-arsenide (GaAs), copper–indium–gallium–selenide (CIGS), copper–indium–selenide (CIS), etc. These are having lighter weight and lower cost, but the conversion efficiency is also lower (<15%) with lower stability and increased toxicity. Dye-sensitized hybrid solar cells and organic photovoltaic (OPV) are the third-generation solar cells with lightweight, flexible, low fabrication cost, ease of fabrication but the low conversion efficiency (<6%). Reason for low efficiency of OPV is less absorption of light (more reflection, scattering, and transmission and high band energy of polymer), recombination, and junction loss. OPV consists of a bi-layer or multilayer or bulk organic semi-conducting (p-type and n-type) sandwiched polymers which are sandwiched between two metal contacts. OPV can be (i) bi-layer, where N-type and P-type semi-conductors are placed layer-by-layer or (ii) bulk, where N-type and P-type semi-conductors are mixed together. Fullerene makes them the best acceptor component for OPV due to its high electron affinity and superior ability to transport charge [74].

Casting of semi-conducting polymer layer which is containing an inorganic material is other procedure to make a heterojunction OPV. In literature, many types of inorganic material reinforced polymer combinations as heterojunction OPVs have been investigated. Until now, MEH-PPV and soluble fullerene derivatives given the most efficient such bulk heterojunction OPVs. [76].

N-(1-pyrenyl)maleimide (PM) absorbs strongly in the peak green region of the solar spectrum. SWCNTs were functionalized with PM and PM-functionalized SWCNTs are blended into the semi-conducting polymer poly(3-octylthiophene) (P3OT) to form organic photovoltaic cells. Because of PM functionalization, the interaction between SWCNT and P3OT was improved resulting 30 times improvement in short circuit current under AM 1.5 lights [77].

OPV based on SWCNTs-polymer composites with PPV [78], MEH-PPV [79], P3HT [80, 81], P3OT [82] have also been developed.

Bulk heterojunctions of β-carotene/C60 and MEH-PPV/C60 through Photo-induced electron transfer was reported in literature before 20 years ago [83], where they used the fullerenes as acceptors. By combination of these both ideas, C60 is used as electron acceptor and the high mobility of SWCNTs, C60-SWCNT-P3HT composites have been reported [84].

Graphene which is chemically modified were also used as electron-acceptors with P3HT and P3OT as donors in OPV, achieving conversion efficiency, $\eta \sim 1.4\%$ [85]. An hybrid composite of PEDOT:PSS/GO was used as a counter electrode, where $\eta = 4.5\%$, comparable to 6.3%, for other counter electrode which is made with Pt and tested under the same conditions, but achieved with a cheaper material [86].

Polymer composites with graphene filler have high carrier mobility, high electrical conductivity, and visible range of spectrum with quite high optical transmittance, which are essential requirements PV material. That is why graphene filler based polymer composites have been widely used in dye-sensitized solar cells [87–89] and in organic solar cells as electrodes [90–94].

Conducting polymer such as PEDOT:PSS, which are doped with graphene have shown good efficiency conversion in dye-sensitized solar cells than that of un-doped PEDOT:PSS polymers [95].

Wang et al. [96] reported that the effect of polymer/P3HT-fullerene-graphene structure on OPV. Device which is made with solution-processable functionalized graphene (SPFGraphene) with low concentration of graphene also shows good performance than device which has no SPE graphene. The improved performance of OPV is because of the formation of charge-transport pathways by SPFGraphene. However, there is a decrease in efficiency of OPV at higher concentration of SPFGraphene due to the recombination of free carriers [96].

Graphene/tetracyanoquinodimethane (TCNQ)/graphene sandwiched stacked films were prepared by Hsu et al. [97] and they used the layer-by-layer molecular doping process on graphene. They reported that these polymer composite films can be used in solar cell anodes. Heterojunction polymer solar cells were prepared by using P3HT/Phenyl-C61-butyric acid methyl ester (PCBM) and multilayered grapheme/TCNQ anodes were fabricated, which showed high conductivity and transparency [97].

Functionalized graphene (SPFGraphene) doped with P3HT hybrid thin films were prepared by Wang and his co-workers [98] by using solution processing for using in organic photovoltaic cells. The effect of annealing temperature on different properties and solar cell performance was studied. The maximum efficiency obtained with 10 wt% SPFGraphene content, and then decreased. After annealing at 210 °C, charge carrier-transport mobility increased greatly [98].

Polymer composites with poly(2-methoxy-5-(3′,7′-dimethyloctyloxy)-1,4-phenylene vinylene), MDMO-PPV and PCBM (1:3 weight ratio) were developed for using in solar cells [99–101].

4.16 Sensor

A sensor is a device which senses or detects different characteristics, such as motion, light, temperature, gravity, magnetic fields, gravity, electrical fields, moisture, humidity, pressure, vibration, sound, stretch, different chemicals, biological entities and other physical aspects of the external environment of its environment.

Sensor detects events or changes and provides a corresponding output, generally as an optical or electrical signal. For example, a thermocouple converts temperature to an output voltage. A good example is mercury-in-glass thermometer; where temperature causes expansion or contraction of mercury and it reaches to a particular mark providing the indication for the temperature [102]. There are various types of sensors. Those are

- Temperature sensor
- pH sensor
- Chemical sensor
- Biosensor
- Pressure sensor
- Ultrasonic sensor
- The acceleration sensor
- Displacement sensor.

Carbon fillers such as CNTs, fullerene, graphene have proved to be suitable materials to detect different types of molecules such as gases, to pH, pressure, temperature, bio-molecules, etc. Because of such properties, they are very good candidate for the application in various types of sensors.

Generally, conductive polymer nanocomposites showed a positive temperature coefficient but opposite behavior (negative temperature coefficient) shows in the thermally reduced graphene-filled PVDF composites. This behavior makes them use as use as temperature sensor [103].

Shan et al. [104] developed the glucose biosensor by using graphene- and PVP-protected Graphene/polyethylenimine (PEI)-functionalised ionic liquid/GOD (glucose oxidase)-modified electrode [104].

Xue et al. developed the organo-phosphorus pesticides (OPs) determined sensor with graphene-nafion matrix modified glassy carbon electrode (Graphene-Nafion/GCE). This sensor has been successfully used to identify the methyl parathion present in vegetable samples [105].

Zhu et al. [106] developed the strain sensors with a percolation threshold of 1 wt % by using CNF/elastomer (VM2) composites which exhibited large mechanical deformation. The electrical conductivity inversely changed 10^2 to 10^3 orders of magnitude upon stretching to 120% strain and recovery to 40% strain [106]. Gas sensor prepared with CNF/poly(acrylate) composites [107] and this sensor could detect vapor through the change in electrical conductivity (five orders of magnitude) [107].

CNF/PPy composites were fabricated by using vapor deposition polymerization method and these composites were used as coaxial nano-cable toxic gas sensor to detect the irritant gases such as HCl and NH_3. The structure of the sensor comprise of an ultra-thin PPy layer on the CNF surface. The oxidation level of the PPy is changed when it reacts with the NH_3 or HCl. Because of the change in oxidation level of PPy, its electrical conductivity is also changed which is used as an indicator for the detection of NH_3 or HCl [108].

Carbon black polymer composite based electronic nose was developed that could mimic biological human nose, which can detect different types of odorant. This electronic nose is able to detect gases and gives respond just like the actual gas sensor. Carbon black polymer composite based sensor was developed, current source circuit was designed, and change in resistance in presence of different gases was observed via LabVIEW [109].

An array of 32 polymer–carbon black composite sensor was developed. The functionalities of the polymers used could be varied to identify variety of compounds and even to distinguish isomers and enantiomers [110].

CNT was perpendicularly aligned and coated with polymer given a novel and advanced idea for invent of new sensors which have high stability, high sensitivity, and have high selectivity which detects variety of chemicals. The working principle is change in resistivity in the presence of those chemicals [111].

Tetrahydrofuran (THF) based sensor comprises of cellulose-based polymer and carbon black composite was fabricated and reproducibility, durability, desorption time, and the sensitivity percent as a function of the amount of solvent were studied [112].

Coordination polymer–fullerene (C_{60}) composite sensor was developed for the detection of 2,4,6-trinitrophenol (TNP) with high sensitivity even at low concentration. This sensor is highly sensitive and selective towards TNP even in presence of nitro aromatics such as 1,4-dinitro benzene (1,4-DNB), 1,3-dinitro benzene (1,3-DNB), 2,4-dinitro toluene (DNT), nitrobenzene (NB) and 4-nitrotoluene (NT)) and also high sensate for normal metal ions such as K^+, Na^+, Mg^{2+}, Ca^{2+}, Fe^{3+} [113].

Sensor based on chip-mounted source–drain electrode coated with a fullerene composite polymer layer containing an immobilized immunoglobulin molecule was developed which can detect antibody–antigen reactions in normal saline solutions. Addition of fullerene causes the improvement in accuracy and reproducibility of detection [114].

Shih et al. [115] prepared the temperature sensor with graphite filler reinforced polydimethylsiloxane composites on polyimide films which is flexible and this composite can be used as artificial skin for sensation humanoid robots system. The sensor array comprises of 64 sensing cells in a 4×4 cm^2 area with copper electrodes to determine the composites resistivity changes with change in temperature [115].

The platelet graphite nanofibers (PGNF)/CNF/graphite micro-particles filled polysulfone (PSf) nanocomposites sensors for glucose were prepared by facile phase-inversion method and their electrochemical performance was compared. It was found that PGNF/PSf composites given excellent amperometric and voltammetric performance for sensing and bio-sensing compared to that of other two composites [116].

Metallic strain sensors cannot sustain more than 5% strain. However, for multifunctional applications, sensor with higher strain is required. High-strain sensors which were stretchable and flexible were fabricated by dispersing MWCNTs or graphite flakes in natural rubber. The working principle is the formation of gaps among conducting particles with stretching, results in the increase of the electrical resistance of the sensor. Graphite sensor with strains of up to ~250% and ~600% of carbon nanotube sensor increased the electrical resistance by ~50 and ~120 times, respectively [117].

Eswaraiah et al. prepared graphite oxide-PVDF nanocomposite by in situ reduction method of graphite oxide, electromagnetic radiation was used to polymer melting and embedding of reduced graphite oxide nanoflakes in polymer. They proposed that these graphite oxide-PVDF nanocomposites will be given the applications in an electromechanical applications [118].

4.17 Actuator

An actuator is a device which is responsible for controlling or moving a system or mechanism. Actuators transform an input signal, mainly electrical signal into motion. Actuator is operated by using energy, typically electrical energy, pneumatic pressure or hydraulic pressure and converts that energy into the motion. These actuators are used to introduce a motion, or to clamp an object so as to prevent motion [119].

Mohamadi et al. prepared graphene-reinforced PMMA nanocomposite by using in situ polymerization [120]. Polydiacetylene (PDA)/graphene nanocomposite were fabricated by Liang et al. by solution processing method [121]. The resulting nanocomposite showed the excellent actuation character with controllable motion, high-frequency resonance and fast response rate.

Mechanical actuator based on MWCNTs-Silicone elastomer composites was developed by Ahir et al. Photomechanical response of this actuator was very fast and possible models and explanations for this was given by them [122].

Bending actuator was fabricated by embedding aligned CNT into PDMS as shown in Fig. 11. At very low driving DC voltages ($<700\ V\ m^{-1}$), the actuator could generate large bending actuation with controllable motion. The actuator could also be operated without electrolytes in the air. This kind of actuator can be used in microrobotics, artificial muscles, micro-sensors, micromanipulation, microtransducers, and micro-cantilever for medical applications [123].

Sulfonated poly(styrene-b-ethylene-co-butylene-b-styrene) (SSEBS) ionomer and fullerene-C60 composite-based ionic polymer–metal composite (IPMC) actuators were developed. Actuation performance or bending response of IPMC under both direct current (DC) and alternating current (AC) excitations was much better compared to that of SSEBS ionomer [124].

Traditional ionic polymer–metal composite actuators have many drawbacks; to overcome some of those fullerene-reinforced ionic polymers membrane was developed. Fullerene-reinforced ionic polymers membrane actuators exhibited higher tensile strength, Young's modulus, proton conductivity and water-uptake and hence better response [125].

Exfoliated graphite oxide (A-aMEGO)/polymer (poly (vinylidene fluoride/ chlorotrifluoroethylene) [P(VDF-CTFE)]) nanocomposites were prepared by using aligned activated microwave to develop the electromechanical actuators properties. It was found that the elastic energy density and elastic modulus of the ionic actuators could be adjusted over a large range by different polymer concentration in the nanocomposite [126].

Fig. 11 Bending actuator based on aligned CNT/PDMS. Adopted from [123]

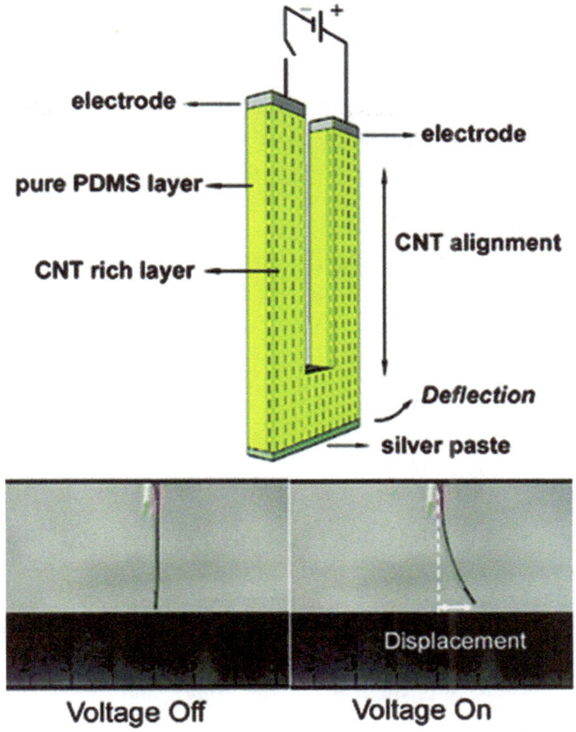

Muralidharan et al. [127] developed a photochemical actuator using thermally reduced graphene oxide (TRGO)/thermoplastic polyurethane (TPU) composite with tunable photomechanical response by controlling pre-strain and the filler loading. A very high photomechanical strain of 50.2% with an excellent stress of 1680 kPa at a pre-strain of 220% was obtained with a low filler loading of 2 wt% TRGO. This is a promising material for light triggered actuators for many potential applications, such as robotics and biomedical devices [127].

Lian et al. [128] fabricated GO (0.5–10 wt%)-Nafion actuators. CNT was applied on both side of the membrane by dip coating method, which was used as electrode. The electrochemical behavior of this actuator was four times higher and displacement was nearly double than those of with virgin Nafion [128].

Graphene nanoplatelets with different concentration (0.10, 0.25, and 0.50 wt%) were incorporated into cellulose matrix to develop cellulose-based composite actuators and studied their performances, such as the minimum tip displacement, maximum tip displacement and time constant [129].

MWCNT–graphene-based IPMC actuator was developed to replace typical platinum or gold electrode. MWCNT and graphene, used as electrode, were coated on both sides of the IPMC by spray coating and baking. The IPMC with a MWCNT electrode exhibited higher displacement compared to that of platinum electrode.

When the ratio of graphene to MWCNT was increased in electrode material, displacement was reduced but the fundamental natural frequency of the polymer actuator was improved [130].

PDMS composites reinforced with 0.1–5 wt% of GNP were prepared for photomechanical actuators. Four orders of magnitude changes were found in stress upon IR illumination when compared with pure PDMS polymer. The mode of actuation is dependent on the applied pre-strains. The actuators exhibited reversible expansion at low % of pre-strain (at 3–9%) and actuators exhibited reversible contraction at high levels (15–40%). GNP-reinforced PDMS composites showed higher actuation stresses than the other forms of PDMS nanostructured composites which were reinforced with carbon materials such as carbon nanotubes with same fabrication method [131].

4.18 Electrode

An electrode is an electrochemical cell which is either an anode or a cathode. The oxidation occurs at anode through the release of electron from it and reduction occurs at cathode through the acceptance of electron by it. Depends on the direction of the current through the cell, each electrode may become either the anode or the cathode. Generally, electrodes are used to supply the current through non-metal objects to alter them in various routes and to measure the conductivity for various purposes. Examples of some applications of electrodes include are [132]:

- Fuel cells
- solar cells
- Medical devices, such as EEG, ECG, ECT, defibrillator
- Electrophysiology techniques in biomedical research
- Electroplating
- welding
- Electrochemical reactions
- Chemical analysis using electrochemical methods.

PANI/graphene composite electrode which is high-performance electrode was prepared by Wang et al. [133] by using a spin coating method. In this work they disperse the graphene oxide film in aqueous solution and this film was deposited on the quartz substrate with deep coating method and also thermal reduction is applied to get a graphene film. Then spin coating was used to deposit the PANI in n-methylpyrrolidone (NMP) dark blue solution on graphene films. This prepared PANI/graphene electrode can be used to make electrochromic devices [133].

Graphene-modified electrodes were prepared from Tris (2,2′-bipyridyl) ruthenium (II) $[Ru(bpy)_3]^{+2}$/nafion/graphene by using the solution mixing of graphene and nafion. This electrode was immersed in 1 M $[Ru(bpy)_3]^{+2}$ solution for getting modified electrode which shows the good stability, sensitivity, and selectivity [134].

Hole-only and double-layer-type light-emitting devices were fabricated based on SWCNT–polyvinylcarbazole (PVK) composite. Mobility and device performance were significantly improved with the increment of weight fraction in the range of 0–1.0 wt%. This increment is because of the nanotubes which are highly conductive and this conducting mobility is passed to the whole system. Devices with higher external quantum efficiencies were achieved with the dye-doped composites [135].

Carbon-polymer composite electrodes were given the applications in HPLC as electrochemical detector and these carbon electrodes were good than glass-carbon electrode. The carbon-polymer composite electrode was found to be better in terms of machinability, cost, stabilization time, accessible potential range, and noise levels and inferior in terms of the potentials which are required for the oxidation of epinephrine and norepinephrine model compounds as compared to those of glassy carbon electrode [136].

PPy-coated carbon nanocoils (CNCs) and PANI were used in supercapacitors which were efficient binder-free electrode materials. PANI/CNCs and PPy/CNCs exhibited specific capacitance up to 360 and 202 F g^{-1}, respectively. The PANI/CNC-based supercapacitors exhibited the maximum storage energy of 44.61 W h kg^{-1}. These devices lost capacitance initially due to the instability of the polymer, however they become stable after 500 charge–discharge cycles [137].

SWCNT/PEDOT:PSS composite electrode of various patterns such as circle, line, and square were prepared by solution transfer method by micro-contact printing using PDMS stamp. High electrical conductivity and stability showed the device which pattern is single parallel line due to the formation and alignment of SWCNTs. It is a good method to fabricate patterned, transparent, conductive SWCNT/polymer composite electrode for optoelectronic applications [138].

Graphite powder was used as filler and mixed with Araldite® (epoxy), polyurethane, and silicon rubber matrices to prepare the composite materials which can be used in the preparation of electrodes. In all the cases, voltammetric response obtained when at least 50% of graphite (wt./wt.) is present in the composites [139].

Graphite-filled poly(methyl methacrylate) based plastic chip electrode (PCE) was fabricated by a simple solution casting method. This electrode is economically inexpensive, mechanically stable, high surface area, multipurpose, disposable electrode for various applications. This type of electrodes is useful in cyclic voltammetry, electro-polymerization, stripping voltammetry, amperometric sensing, and electro-winning [140].

Polymer/graphite fiber composite electrodes were fabricated by using single-step electro-polymerization PPy onto a porous graphite fiber matrix and these electrode materials are high charge density conducting materials. The graphite substrate provides a lightweight structure with high surface area. The available charge capacity of the composite electrodes was proportional to the polymerization time and the mass of electro-active polymer [141].

An electrode for supercapacitors was fabricated by using nanocomposite of graphene (G)–PANI material and these materials were synthesized by using chemical precipitation method. This supercapacitor exhibited a 300–500 F g^{-1} specific capacitance at 0.1 A g^{-1} current density [142].

4.19 Battery

A lithium-ion battery (Li-ion battery or LIB) is one of the rechargeable batteries. Ions move from the negative electrode to the positive electrode in LIB, during discharge and back when charging. In LIB, intercalated lithium compound is used as one the electrode material and the metallic lithium is used in non-rechargeable lithium batteries. Because of low atomic weight of M = 6.94 g mol^{-1} and absolute high potential against the standard hydrogen cell (−3.04 V), LIB considered as the one of the most important batteries in storage system battery which has the large theoretical energy density up to ~400 W h Kg^{-1}. LIBs are common in consumer electronics, military, battery electric vehicle, aerospace applications, etc. LIBs are popular for the replacement of lead acid batteries because of its lightweight [143].

Lithium battery comprises a positive electrode, negative electrode, a separator, and an electrolyte which allows for ionic movement. The separator is a micro-perforated plastic which is a very thin sheet and this separates the both positive and negative electrodes while allowing ions to pass through. For making the positive electrode, lithium salt is used and carbon or other material is used for making the negative electrode.

During charging, lithium ions move through the electrolyte from the positive electrode to the negative electrode and while discharge time lithium ion moves back to the negative electrode as shown in Fig. 12 [144].

Green cathode was developed by Z. Song and his co-workers with combination of graphene and two polymer promising cathode materials, polyimide and poly-(anthraquinonyl sulfide). They improve this composite to use as cathode in rechargeable lithium batteries (RLB). In situ polymerization was used with graphene sheets to make these composites. Because of better dispersion, electronic conductivity and electrochemical activity improved significantly (more than 100 mA h g^1 witihin few seconds) [145].

Flexible energy-storage devices have growingly attracted with the demand of flexible electronic systems. However, obtaining electrode materials with high mechanical strength, high flexibility and good electron and lithium-ion conductivity is a big challenge. Lee et al. developed stretchable, highly conductive and porous polymer–CNT nanocomposite electrode material. First, blend of PDMS, PMMA, and CNT was prepared and then PMMA was removed to generate well-controlled pore networks. The porous PDMS nanocomposites which were embedded with CNT-exhibited good mechanical flexibility with good electrochemical performance and found suitable for anode material in flexible lithium-ion batteries [146].

A carbon (amorphous)/polymer (polypropylene oxide) composite electrode was fabricated for its application as an anode in an electrochemical cell, or battery [147].

Sivakkumar et al. [148] synthesized PANI-MWCNT nanocomposite by using the in situ chemical polymerization of aniline in well-dispersed CNT solution. Composite of PANI/CNT was used as an active cathode material and porous poly (vinylidenefluoride-*co*-hexafluoropropylene) membrane was used as an electrolyte material in rechargeable polymer lithium cells. The cell exhibited a discharge

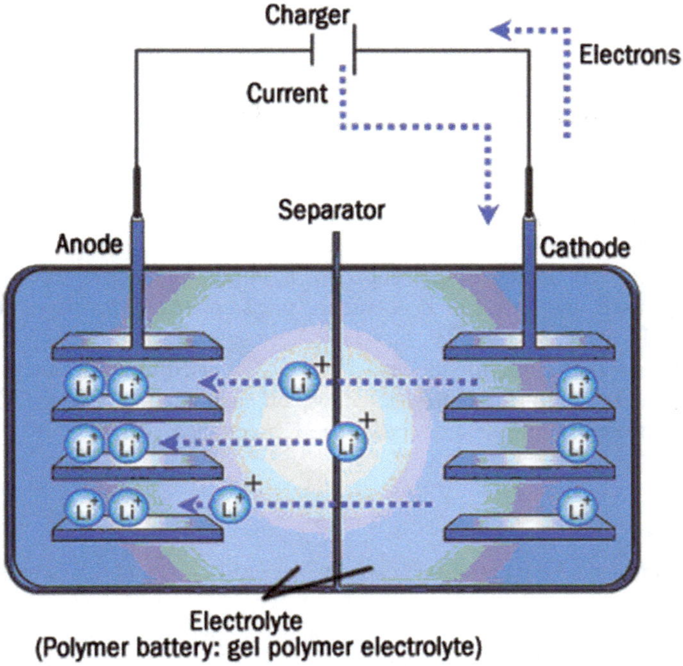

Fig. 12 Working principle of Li-ion battery. Adopted from [144]

capacity maximum of 86 mA h g^{-1} at the 80th cycle with an average columbic efficiency of 98% [148].

Polypyrrole graphite composite was developed by Veeraraghavan et al. [149] to use as anode material in secondary lithium-ion batteries. They polymerize the Pyrrole and commercial SFG10 graphite by using in situ polymerization method. This graphite composites exhibited higher columbic efficiency, good reversibility, good rate capability, and better cycle life compared to those of pristine SFG10 graphite [149].

Li et al. [150] developed the PPy/graphene composite to use in biocompatible zinc/polymer battery as cathode material. This composite has the high specific surface area of 561 m^2 g^{-1} with high conductivity of 141 S cm^{-1}. Efficiency of this composite as cathode material was higher than pure PPy. The battery which is prepared with PPy/graphene as cathode and Zn as anode exhibited with energy density of 264 mW h g^{-1} in 0.1 M of phosphate buffer saline [150].

Recent advances and applications of graphene-based composites in cathode materials in lithium-ion batteries were reviewed by Chen et al. [151]. Their focus of discussion was mainly on the processing methods and electrochemical performance of graphene-based composites as cathode materials for lithium-ion batteries [151].

4.20 Capacitor

A capacitor is electrical component which passes two-terminal electrical to store energy in an electric field. In capacitor, energy is stored in electrostatic field formed between the plates. These capacitors are mostly used in electrical devices as parts of electrical circuits. Two electrical conductors (plates) needed to make capacitors which are separated by a dielectric (i.e., insulator) as shown in Fig. 13. Thin films, sintered beads of metal, or conductive electrolyte and foils, etc., are used to make the capacitors. To increase the capacitor's charge capacity, non-conducting dielectric is used and this can be ceramic, glass, plastic film, paper, air, mica, vacuums, oxide layer, etc. [152].

A single constant value for its capacitance; *C*. is used to characterize the ideal capacitor. *Q* is the ratio of the electric charge on each conductor to the *V* potential difference between them is expressed as the Capacitance.

$$C = \frac{Q}{V} \tag{1}$$

Farad (F) is used for measure the capacitance in SI units and this is equals to one Coulomb per volt (1 C/V).

Zhang et al. [153] developed graphene and polypyrrole (PPy/GNS) composite for its application as electrode material for supercapacitor. This composite shows better rate capability, higher specific capacitance, and cycling stability than of pure polypyrrole. PPy/GNS composite has high specific capacitance of 482 F g^{-1} at current density of 0.5 A g^{-1} with good cycling performance (specific capacitance is lower than 5% at after 1000 cycles,) [153].

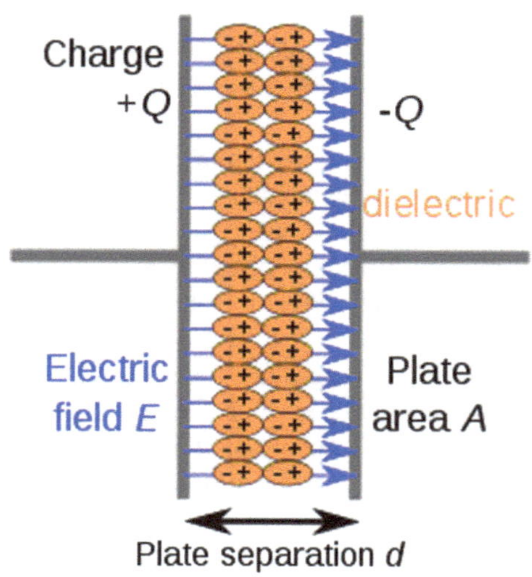

Fig. 13 Schematic of a typical parallel-plate capacitor. Adopted from [152]

Liao et al. [154] prepared mesophase-activated carbon (MAC) from coal tar pitch and then activated with potassium hydroxide (KOH). The activated carbon electrode made from MAC showed an excellent electrostatic capacitance with 196 F g^{-1} and 72 F cm^{-3}.

A capacitor comprising carbon fiber reinforced epoxy composite with a paper interlayer was prepared, which exhibited a capacitance of 1.2 μF m^{-2} at 2 MHz. The high capacitance is due to the large area of the surface of a fiber layer (twice the flat area) sandwiching the paper interlayer [155].

Li et al. [156] developed the structural carbon fiber (surface modified by oxidation) and epoxy-based gel polymer composites to be used as electrodes and electrolyte in capacitor.

The electrochemical stability of the electrolyte reached to 2.75 V and the ionic conductivity reached to 10^{-5} S cm^{-1} at room temperature. Energy-storage stability increases with the enlargement of surface area of carbon fiber and this can be possible by surface modification of the fiber by oxidation reaction. This carbon fiber reinforced epoxy composite has specific capacitance of 3.0 F g^{-1} with good cycling performance [156].

Graphite oxide was incorporated in poly(ethylene oxide)-based polymer used as electric double-layer capacitors. This graphite oxide composite showed a double-layer capacitance of 130 F g^{-1} [157].

Electrochemical capacitors were developed with composite of Reduced graphene oxide (rGO) and poly(vinyl pyrrolidone) (PVP) which contains the 55 and 71 wt% rGO. Thi rGO composite can be directly used as the electrode material without conductive additives or binders.

The ECs exhibited excellent flexibility, high volumetric, and gravimetric specific capacitances and satisfactory electrochemical stability. The EC constructed with rGO/PVP composite containing 71 wt% of rGO showed a higher power density (40 W cm^{-3}) and high energy density of 2.5 mW h cm^{-3} at a current density of 1 A g^{-1} [158].

Graphene-epoxy flexible transparent capacitor was developed by Sangermano. SU8 resin was used as flexible and transparent dielectric layer on the surface of graphene, which acts as conductive layer. Two layers of graphene/SU8 were bonded together with the help of photo-curable epoxy resin. This capacitor exhibited a clear and stable capacitive behavior [159].

4.21 Supercapacitor/Ultracapacitor

One of important device for electrochemical energy storage is Supercapacitor or Ultracapacitor which provides the >10 kW kg^{-1} of high power density and short charge/discharge time as compared with other battery devices.

Graphene derivatives and conducting polymers are widely used in supercapacitor. Conducting polymers like PANI is mostly incorporated with rGO and GO sheets and these composites were used in supercapacitors.

Wu et al. [160] fabricated the graphene-/PANI-based supercapacitor with very high mechanical stability and flexibility. They prepared the chemically converted graphene composite films (CCG) with polyaniline nanofibers (PANI-NFs) by using vacuum filtration method. Electrical conductivity of the composite film was 5.5×10^2 S m^{-1} at 44% of CCG and had the 210 F g^{-1} electrochemical capacitance at a 0.3 A g^{-1} discharge rate [160].

Fibrillary composites of PANI doped with graphene oxide sheets were prepared by using in situ polymerization and in this reaction they polymerize the monomer in the presence of graphene oxide. The nanocomposite exhibited a conductivity of 10 S cm^{-1} at 22 °C at 100:1 weight ratio of aniline and graphite oxide. At potential range of 0–0.45 V, Pristine PANI exhibited specific capacitance of 216 F g^{-1} whereas it is 531 F g^{-1} for PANI/graphene oxide composite. The degree of doping and the concentration of graphene oxide have been effects on the performance and electrochemical capacitance of the nanocomposites [161].

Graphene nanosheet/carbon nanotube/polyaniline composite was fabricated by Yan et al. [162] by using in situ polymerization. They want to use this composite material as electrode material for supercapacitors applications. This composite showed the 1035 F g^{-1} (1 mV s^{-1}) of specific capacitance in 6 M of KOH and this capacitance was lower than the GNS/PANI composite (1046 F g^{-1}), but capacitance higher than the CNT/PANI composite (780 F g^{-1}) and also higher than pure PANI (115 F g^{-1}). With the addition of 1wt% CNT, cycle stability increases more for GNS/CNT/PANI composite. This increment is because of the keeping the highly mechanical strength as well as high conductive path of the electrode during the doping and dedoping processes. The loss of capacitance was less for GNS/CNT/PANI composite (only 6%) 1000 cycles as compared with of GNS/PANI (52%) and CNT/PANI (67%) composites [162]. Multi-walled CNT reinforced PANI, PPy, and PEDOT composites were prepared by in situ chemical polymerization for using these composites supercapacitor applications [163]. These composites were optimized by using different varying ratios of ECPs to MWCNT. All the materials exhibited good energy-storage performance. PPy/MWCNT composites showed superior supercapacitor characteristics than other composites.

Composite of exfoliated graphite and electrochemically active material was fabricated for the application as a composite electrode in supercapacitor. A supercapacitor made from such a composite electrode exhibited an exceptionally high capacitance value and low equivalent series resistance [164].

GO/PPy composites were prepared for giving the applications in Supercapacitor. Charge storage improvement shows cause exfoliation of graphite oxide sheets through intercalation of PPy, a specific capacitance of $\sim$181 F g^{-1} in 1 M Na_2SO_4 aqueous electrolyte achieved with a $\sim$56.5 W h kg^{-1} of corresponding specific energy density [165].

Liu et al. [166] incorporated various types of nanomaterials, such as silicon nanoparticles, carbon nanotubes, titanium oxide particles and graphene flakes into the conducting polymer PANI to form nanocomposite materials for making flexible supercapacitor sheets. Incorporation of nanomaterials in PANI has significantly improved the energy and power capabilities of the capacitor.

4.22 ESD and EMI Shielding

Electrostatic discharge (ESD) is the flow of electricity between two electrically charged objects caused by contact. Static electricity can be possible by electrostatic or tri bocharging induction. ESB occurs when the opposite charged objects come closely together or when the dielectric between them breaks down which often creates the visible spark. There is possibility of some harmful effect with ESD such as explosions of fuel vapor, gas, and coal dust and also possibility of failure of some integrated circuits in solid state electronics components. This can be avoided by preventing static charge formation through the formation of conducting path, providing antistatic devices, and controlling humidity [167].

Electromagnetic shielding is the practice of reducing the electromagnetic field in a space by blocking the field barriers which are made with magnetic or conductive materials. To isolate electrical devices from the 'outside world', this Shielding can be used in devices [168].

Electromagnetic interference (EMI) is the disturbance which is generated by the performance of an electronic device electric field which is from another nearly placed device. For any material, electromagnetic interference shielding efficiency (EMI SE) refers as the efficiency of absorb and/or reflect electromagnetic radiation and this is needed for low observable aircraft and radomes [169].

The EMI SE is presented in percentage as

$$\mathrm{EMI\,SE}(\%) = \frac{P_{\mathrm{A}} + P_{\mathrm{B}} + P_{\mathrm{R}}}{P_{\mathrm{I}}} \times 100$$

In the above equation, P_{I} refers as material body incident power, P_{R} is reflected power, P_{A} shows the absorbed power, internally reflected power from the material body is shows with P_{B} and transmitted power from the material body is P_{T} which is schematically shown in Fig. 14 [170, 171].

The percolation thresholds of graphene, nanotubes, nanofibers can be achieved at very low concentration because of their high intrinsic conductivity and aspect ratio. The surface resistivity of ESD materials ranges between 10^{12} and 10^{5} Ω cm^{-2}, whereas in EMI shielding material it should be lower than 10^{5} Ω cm^{-2}. ESD and EMI materials given the applications is many fields including mobile phone parts, carpeting floor mats, electronics packing, telecommunication antenna, and also can be used in aircraft as shielding coating material.

Lee et al. [172] developed PP/MWCNTs nanocomposites for ESD applications. MWCNTs were surface modified using octadecylamine (ODA) via carbon tetrafluoride (CF_4) plasma-assisted fluorination and subsequent alkylamination, which was then incorporated in PP. PP/MWCNTs composite with 2 wt% of MWCNTs loading showed the high electrical conductivity along with good mechanical properties and also found to be suitable for the use as ESD material [172].

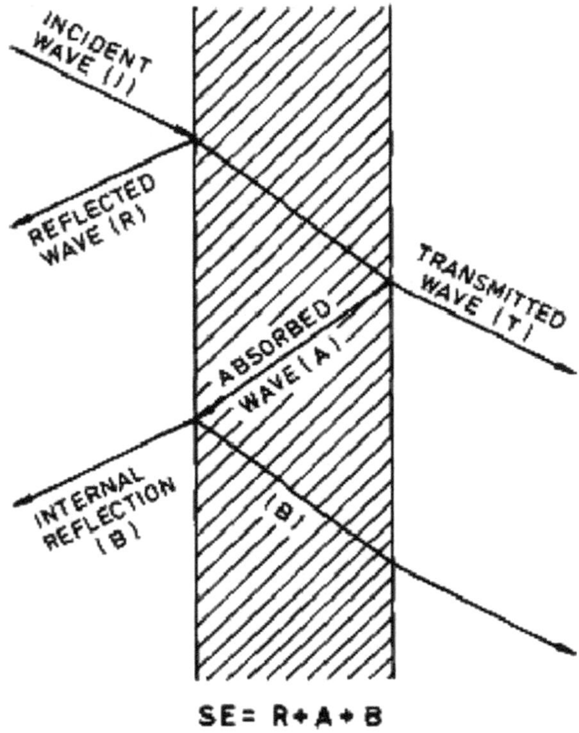

Fig. 14 Schematic diagram of shielding efficiency (SE) in a material body. Adopted from [171]

CNT (single-, double-, and multi-walled) was used as filler in different types of polymer for preparing electrically conductive composites for their application of ESD material [173].

ESD adhesives with resistance in a range from 10 kΩ to about 100 MΩ were prepared using a mixture of a polymeric thin film and electrically conductive fillers. Polymers include polyurethane, polyester, polyacrylate, polymethacrylate and epoxy, and electrically conductive fillers include carbon black, carbon nanotubes, silver nanoparticles, gold nanoparticles, and fullerene compounds. The adhesive can be applied on the item to be protected against electrostatic charging dispersed, then solvent to be evaporated to form continuous and smooth hard coating of ESD material [174].

Poosala et al. [175] incorporated oxygen-plasma treated graphene nanoplatelets (OGNP) (0.0 to 5.0 wt%) and MWCNT (2.0 wt%) in polycarbonate (PC) by melt processing through twin-screw extruder. OGNP/MWCNT/PC nanocomposites found to be suitable for ESD applications [175].

Liang et al. [176] developed the CMG/epoxy nanocomposites with 15 wt% loading weight fraction which shows the efficiency EMI shielding to achieve commercially suitable level (around 20 dB in the X-band). These composites were prepared by reduction of GO by suing hydrazine.

Graphite flake (GF) filled PPS composites were prepared by Goyal et al. [177] by using hot press molding for given applications in EMI shielding. They obtained percolation threshold of electrical conductivity with 5 wt% of GF. These composites were found to be useful for EMI shielding application because of the significant improvement in electrical properties [177].

Yang et al. [178] developed PS-carbon nanofiber composites for EMI shielding application. They observed that the EMI shielding effectiveness is independent on the applied frequency but dependent on the concentration of conducting filler. EMI shielding efficiency increased with increasing carbon nanofiber loading within the applied frequency range 12.4–18 GHz (Ku-band). EMI shielding effectiveness of 36 dB for the polymer composite obtained at 20 wt% carbon nanofibers [178].

EMI shielding was examined for Epoxy/graphene composites and these composites were prepared by using in situ polymerization. The EMI shielding effectiveness increased with increasing graphene loading over the entire frequency range. These epoxy/graphene composites found to be suitable in electromagnetic radiation as effective lightweight shielding material [179, 180].

Luo et al. [181] developed carbon fiber/polymer composite, which exhibited excellent EMI shielding efficiency of 124 dB in 0.3 MHz to 1.5 GHz frequency range.

SWCNT composites have been fabricated and EMI SE was evaluated for these composites in the frequency range of 10 MHz to 1.5 GHz. 15 wt% SWCNT composites given the highest

EMI SE of 49 dB at 10 MHz frequency. 15–20 dB of EMI SE was obtained in the 500 MHz to 1.5 GHz frequency range. EMI SE found to be proportional to DC conductivity [182].

Al-Saleh et al. [183] studied the electrical resistivity (ER), morphological and EMI SE properties of PS, PP, PP/PS, and PP/PS/styrene–butadiene–styrene (SBS) blends which were filled with 10 vol.% of carbon black (CB).

Morari et al. [184] developed the composites from silicone rubber with carbon powder and ferrite powder and measured their electrical conductivity and EMI SE in microwave frequency. They observed that there is negligible variation of shielding performance with frequency for samples with higher electrical conductivity; whereas it decreases significantly with increasing frequency for samples with lower.

Maiti et al. [185] prepared electrically conducting nanocomposites of PS/MWCNT/GNP by using in situ polymerization. Polymerization reaction of styrene/MWCNTs was occurred in the presence of PS/GNP microbeads which are suspension polymerized. This composite exhibited very high electrical conductivity and EMI SE ($\sim$20.2 dB) at low loading of MWCNTs filler ($\sim$2 wt%) and GNP filler ($\sim$1.5 wt%) because of the random distribution of both fillers in the PS matrix.

PVDF-based functionalized graphene (f-G) foam composites were prepared by Eswaraiah et al. [186] for using these composites in shielding applications. Concentration of f-G effects on the electrical conductivity of the composites and electrical conductivity increases with increment of filler loading in matrix. 5 wt%

f-G loading composites showed the EMI SE of ~20 dB in X-band (8–12 GHz) region 18 dB in broadband (1–8 GHz) region.

4.23 Memory Devices

Memory can be defined as like a brain of humans. This memory generally used to store data and followed the instruction. In computers, there is a memory storage place where data is processed and stored and this memory can be divided into more numbers of small parts. Each of these part is called cell and every cell or location has different address which is varies from zero to memory size minus one [187].

For example, if computer has 64 k words, then this memory unit has 64 * 1024 = 65536 memory location. These address locations varies from 0 to 65,535.

Mamo et al. [188] developed the composites of poly(vinyl phenol) (PVP) and carbon nanotubes by sandwiched between Al electrodes and by using these composites they prepared the write-once-read-many times (WORM) memory devices. They used three kinds of nanotubes, such as un-doped, nitrogen-doped and boron-doped MWCNTs. It was found that the OFF to ON state switching threshold depends on types of nanotube whereas the ON/OFF current ratio is dependent on both concentration and type of nanotubes [188].

Similarly, Sustaita et al. [189]. also developed WORM memory devices from functionalized carbon nanoshells and carbon sphere based PVP composites, which showed ON/OFF current ratio of 10^5 [189].

Machado et al. [190] also developed WORM memory devices by using cross-linked PVP and carbon spheres composites. They used three kinds of nanotubes, such as un-doped, N-doped and B-doped MWCNTs and their effects on memory characteristics were studied. These memory devices exhibited quick OFF to ON (high resistance to low resistance) transition (less than 1 μs) and it can happen at very low voltages (2 V) [190].

Electrical bi-stability and large conductance switching at room temperature was observed in functionalized carbon nanotube (CNT)-conjugated polymer composites. The bi-stability was dependent on the CNTs concentration in the composite [191].

Jo et al. [192] developed organic nonvolatile resistive switching memory by spin casting method using PCBM filled poly(styrene-b-methyl methacrylate (PS(10)-b-PMMA(130))) self-assembled di-block copolymer [192]. The device exhibited better bipolar-switching behavior with stable and reproducible properties at low operating voltages (RESET at 1.3 V and SET at −1.5 V) and under ambient conditions, [192].

Khan et al. [193] developed the memory devices with ferroelectric poly (vinylidenefluoride–trifluoroethylene (P(VDF-TrFE)) and n-type PCBM blends with conducting PEDOT:PSS polymer electrodes by using spin casting method. These blends can be used as nonvolatile bistable memory devices. The devices showed high I_{on}/I_{off} ratios ($\approx 3 \times 10^3$), low read voltages (≈ 5 V), superb retention characteristics up to 10,000 s and better dielectric properties at high frequencies ($\epsilon_r \approx 8.3$ at 1 MHz). [193].

Kanwal et al. [194] developed memory devices based on C60 fullerene, PS and PVP polymers with mild thermal annealing to improve the stability of the device [194].

Flexible organic memory (BOM) bi-stable devices were prepared by using ultra-thin graphite sheets (UGS) which were sandwiched between PMMA insulating polymer layers. Efficiency of BOB devices even after bending was similar and was stable during repeated bending [195].

Mamo et al. [196] developed WORM memory devices using rGO-PVP composite as the active layer. These devices exhibited a high ON/OFF current ratio of 10^5 at 1 V, with good retention. Very low energy is required for this device to perform write and read operations at 5 V.

Kafy et al. [197] modified GO with hexamethylene diisocyanate and then incorporated in cellulose to develop nanocomposite for given applications memory storage devices and flexible energy devices. The composite was environmentally stable and found to be suitable for electronic devices and energy storage [197].

Zhuang et al. [198] developed rewritable memory devices with ITO/ triphenylamine-based polyazomethine (TPAPAM)-GO/Al and these are non-volatile devices. These devices showed an ON/OFF-state with more than 10^3 of current ratio and a memory effect with a turn-on voltage of −1.0 V. These devices worked up to 10^8 read cycles at a −1.0 V of read voltage and both ON and OFF states were stable under a constant voltage.

4.24 Fuel Cells

A device which converts the chemical energy from a fuel into electricity by chemical reaction with oxygen or other type of oxidizing agent is called fuel cell.

If inputs are supplied, electricity can produce continuously by fuel cell. Fuel cells can be used as primary and backup power in commercial, industrial, commercial, residential buildings, and in remote areas. Fuel cells can also be used in power fuel cell vehicles such as automobiles, forklifts, motorcycles, buses, sub-marines, and boats.

Different kind of fuel cells are divided into different categories based on the type of electrolyte used.

Different varieties of fuel cells are available which includes in the following:

- Proton exchange membrane fuel cell (PEMFC)
- Direct methanol fuel cell (DMFC)
- Direct formic acid fuel cell (DFAFC)
- Phosphoric acid fuel cell (PAFC)
- Metal hydride fuel cell (MHFC)

- Alkaline fuel cell (AFC)
- Solid acid fuel cell (SAFC)
- Solid oxide fuel cell (SAFC)
- Molten carbonate fuel cell (MCFC).

However, working principle of all fuel cells is similar. To produce the direct current electricity, electrons are drawn from the anode to the cathode through an external circuit as shown in Fig. 15 [199, 200].

Generally, a single fuel cell contains the electrolyte and two catalyst-coated electrodes (anode and cathode). Different components which are available in fuel cell are electrodes, bipolar plates, catalyst, membrane, and necessary hardware.

Bipolar plate is mainly prepared from metals. However, metals have problem with corrosion. Because of this corrosion, metals form the conductive oxide layer which is effects on the electrical conductivity of the component. Moreover, catalyst layer in the membrane electrode may be poisoned due to the ions which are formed from the metallic surface. Small amount of ions also reduce the total power density of the fuel cell. Graphite–polymer composites have more advantages than metallic materials because those are corrosion resistant, low weight easy and economical production (injection or compression, transfer molding processes). Many types of graphite polymer composites are commercially available such as PVDF, PPS, PP, and phenolic resin [201–204]. However, metallic plates have higher bulk electrical conductivity than these polymer composites. Therefore, researchers try to develop the composites with higher electrical conductivity to given the applications in different fields. This can be achieved by incorporating other carbon-based conductive fillers, except conventional graphite filler, in the polymer matrix. The other filler which can be using in polymer matrices are SWCNT, MWCNT, carbon black (CB), expanded graphite, graphene, carbon fibers (CF) and combinations thereof [205–209]. Electrical conductivity and mechanical strength of these types of filler reinforced polymer composites depends on the filler content, size of particles,

Fig. 15 A block diagram of a fuel cell. Adopted from [199]

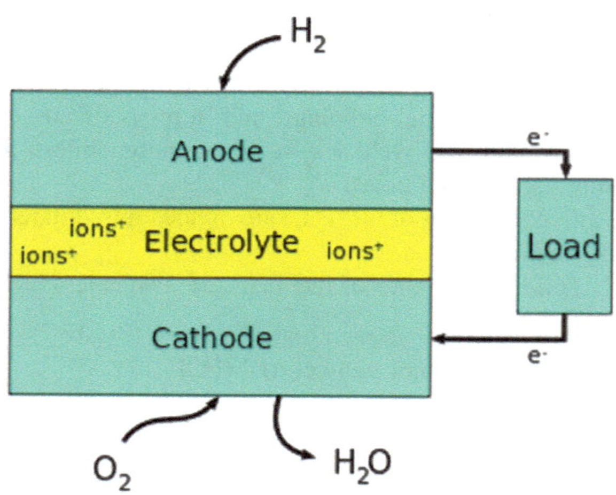

morphology, etc. Mechanical and electrical properties of graphite-epoxy composite bipolar plate which is reinforced with CF, MWCNT, and CB were investigated by Lee et al. [205]. They reported that electrical conductivity of composites was increased with the addition of CB up to 5 vol.% of concentration and the addition of MWCNTs up to a concentration of 2 vol.%. However, at higher filler loading electrical conductivity decreased to some extent may be due to aggregation of conducting filler or improper wetting of the filler by the polymer matrix [205].

Allaoui et al. [210] reported that the small filler content between 0.5 and 1 wt% of filler concentration in MWCNT-reinforced epoxy polymer composites reaches the percolation threshold in epoxy polymer composites [210]. Same type of behavior was confirmed by Munson-McGee [211] and Celzard et al. [211]. They reported that filler with high aspect ratio with rod-like particles showed the percolation threshold at 0.24–1.35 vol.% of filler and also lower values are also often reported in literature [212, 213]. Celzard et al. [211] and Munson-McGee [214] published the deviations from the results which were described by Gojny et al. [215] and these people develop the model in the absence of molecular level considerations. Literature reports shows that the dispersion of CNTs in polymer matrices can be improved by surface modification or functionalization of conducting nanoparticles with different chemical groups [216], [217] and [218]. NH_2-functionalized MWCNTs was prepared by ball milling process with nanotubes with and ammonia. Due to this ball milling process, the aspect ratio is lower for MWCNT which were functionalized with NH_2 than that of pristine nanoparticles and these nanotubes were break during ball milling. Results showed that functionalized MWCNT composites has higher percolation threshold than that of untreated nano-tubes composites. Functionalization of carbon nanotubes improves the bond between filler and polymer matrix where may be formation of insulating surface layer on the nanotubes. Functionalization is frequently used to enhance dispersion of CNTs in polymer matrices [215–217]. Zhu et al. [219] also reported that aspect ratio of CNTs also reduce after the functionalization of carbon nanotubes with some chemicals for polyimide composites. In this work, they used the mixture of sulfuric acid and nitric acid to treat the CNTs. At 7.0 wt%, treated CNT composite showed percolation threshold which is higher than often reported. Lee et al. [220] successfully functionalized the CNTS with treatment of mixture of nitric acid and sulfuric acid and then incorporated in PP using twin-screw extruder. Composite with treated CNT exhibited higher electrical conductivity compared to that of untreated CNT. In this work they used the Maleic anhydride grafted polypropylene (MA-g-PP) as coupling agent and maleic anhydride grafted styrene–ethylene–butylene–styrene (MA-g-SEBS) as compatibilizers. Coupling agent helped in better dispersion of CNT and also enhanced the conductivity of the resultant composite.

Aligned particles of MWCNTs in an epoxy matrix may also be reduced the percolation threshold instead of entangled nanotubes [Sandler et al. [212]. This alignment can be possible during the injection chemical vapor deposition manufacturing process of the nanotubes [212].

Carbon fillers are mainly used in bipolar plate. Some researchers have also worked on the application of carbon filler in polymeric electrolyte membrane

(PEM). Mechanical and/or dimensional stability of Nafion membranes, especially at high temperature, is moderate. It may cause problems during the fabrication of membrane and during operation. Contraction and swelling of Nafion membrane is possible because of the differences in hydration level during operation of the fuel cell which may change the dimension of membrane and also effects on the fuel cell performance and working life. To overcome this problem CNT is incorporated in Nafion to improve the mechanical strength of the membrane [221].

Liu et al. [222] prepared the composite membranes with Nafion and CNTs by solution casting method for using applications in H_2O_2 fuel cell. Addition of 1 wt% CNTs improved the mechanical strength of the membrane significantly but decrease the dimensional change of membrane, whilst maintain the proton conductivity. CNTs being electrically conducting in nature may cause short circuiting between two electrodes. By keeping the carbon nanotubescontent lower than the percolation threshold, we can minimize the short-circuits risks [222]. They proposed an electrolyte membrane with sandwich structured, prepared with a platinum (Pt)/CNT/Nafion layer which is inserted between two plain Nafion layers.

Mechanical property this membrane was improved. Moreover, the Pt/CNTs present in the membrane helped to react H_2 and O_2 to form water. The water directly humidifies the membrane and allows to the operation of PEMFCs without external humidification system. CNTs being electrically conducting in nature may cause short circuiting between two electrodes. This was overcome by providing pristine Nafion layers on both sides of Pt/CNT/Nafion layer [223].

Thomassion et al. [224] also incorporated functionalized MWCNTs in Nafion by melt extrusion process. Grafted of MWCNT was done by carboxylic acid containing alkyl radicals. They want to decrease the methanol permeability without affecting ionic conductivity. CNTs concentration was less then percolation level to avoid short circuit. Methanol permeability of the functionalized Nafion/MWCNTs membrane was decreased by approximately 60%, while ionic conductivity remained unchanged; 140 and 160% increment was observed in Young's modulus and this increment was achieved with the addition of 1 and 2 wt% of functionalized MWCNTs, respectively [224].

Wang et al. [225] oxidized MWCNTs using H_2O_2 and then incorporated in Nafion by solution casting method. Tensile strength of the Nafion membrane was increased by 54% and the elongation at break was increased by 27% because of the addition of oxidized MWCNTs.

While using CNTs to improve mechanical strength of a fuel cell membrane, one has to be careful about the increased electrical conductivity of the membrane. CNTs being electrically conducting in nature develops conduction pathway across the membrane, which may cause short circuiting between two electrodes. To avoid short circuiting either concentration of CNT should be below percolation level [222] or there should be an insulating coating on both side of the membrane [223]. 2–3 wt% can be given the threshold value [224, 225], for any specific system, real values can only be established by trials and errors, because aspect ratio and intrinsic electrical conductivity of CNTs effects on the development of electron conductive network in polymer matrix. Electrical conductivity also depends on filler dispersion and orientation in polymer matrix, interactions with the matrix at the interfaces, phase separation, etc. [226].

One of the most important factors towards the efficiency of a fuel cell is the proton conductivity of the electrolyte membrane. Many researchers tried to improve the proton conductivity of Nafion by introducing functionalized or pristine CNT, fullerene, graphene, etc.

Chen et al. [227] prepared polysiloxane-functionalized MWCNTs by covalently grafted hydrophilic layers of poly(oxyalkylene) diamines and tetraethyl orthosilicate-reinforced polysiloxane. To form a proton-conducting membrane, functionalized MWCNTs were blended with Nafion. The functionalization of MWCNTs was performed to improve the dispersion and also to reduce the electrical conductivity of MWCNTs. The composite membrane exhibited good proton conduction of $\sigma = 2.8 \times 10^{-2}$ S cm^{-1} at 30 °C and also maintains the proton conduction at high temperatures ($\sigma = 6.3 \times 10^{-2}$ S cm^{-1} at 130 °C) [227].

Asgari et al. [228] developed a Nafion®/histidine composite membrane, where they modified the carbon nanotube with imidazole groups (Im-CNT) to use in methanol fuel cell (DMFC) applications. Nafion®/CNT composites with 0.5% of CNT loading membranes showed high proton conductivities and lower methanol crossover than those of commercial Nafion® membranes at elevated temperatures. 61 mW cm^{-2} of power density have these nanocomposite membranes, whereas 42 mW cm^{-2} for Nafion® 117 (at 0.5 V and 5 M methanol concentration).

In Nafion sulfonic acid groups form ion channels, through which proton or ion conduction takes place under humid condition. Hence, by increasing the number of sulfonic acid groups in the Nafion, its proton conductivity could be improved. Keeping this principle in mind, Kannan et al. [229] incorporated sulfonic acid functionalized SWCNTs in Nafion, which increase the proton conductivity by one order of magnitude.

Polybenzimidazole (PBI) membrane is generally doped with phosphoric acid to achieve desired level of proton conductivity. Keeping this principle in mind, Kannan et al.) [230] incorporated MWCNTs into polybenzimidazole membrane and MWCNTs were functionalized with 2-aminoethylphosphonic acid. They reported that proton conductivity is improved by 50% from 0.07 to 0.11 S.cm^{-1} [230].

Although most of the researcher found the improvement of ionic conductivity by incorporating functionalized CNTs in Nafion membrane, Cele et al. found a small decrease in ionic conductivity when oxidized MWCNTs or hexadecylamine functionalized MWCNTs were incorporated in Nafion membrane. Their justification for the decrease in ionic conductivity by using functionalized MWCNTs was because of the poor dispersion of CNTs in Nafion or the presence of amine surfactant chains on the outer layer of tubes [231].

Tasaki et al. [232] prepared the conducting composite membranes with doped Nafion® 117 and fullerenes [C_{60} and polyhydroxy fullerene, $C_{60}(OH)n$ ($n \sim 12$)] to be used as proton-conducting composite membranes. These Nafion®–fullerene–composite membranes passes higher proton conductivities than Nafion® 117 at both low (20 °C) and high (80 °C) temperatures under low relative humidity (<50% RH).

DeSousa et al. [233] prepared composite membrane of Nafion, fullerene, and poly-hydroxyfullerene (PHF) by solution casting method. Some of these composites exhibited higher ion conductivity at low humidity conductivity compared to that of Nafion [233].

GO/Nafion composite membranes were developed by Kumar et al. [234] for given the applications in PEMFC. The proton conductivity at 4 wt% of GO/Nafion composite was 0.078 S cm^{-1} which was higher than the Nafion 212 conductivity (0.068 S cm^{-1}) and Nafion recast membrane conductivity (0.043 S cm^{-1}) at 30 °C and 100% humidity. The fuel cell which is prepared with GO (4 wt%)/Nafion composite membrane performance was good than that of Nafion recast and Nafion 212 under 25% relative humidity [234].

Xu et al. [235] developed the composites of GO/PBI and sulfonated GO (SGO)/ PBI composite membranes for their application in high temperature PEMFC. The ionic conductivities of phosphoric acid doped PBI/GO and PBI/SGO membranes were 0.027–0.052 S cm^{-1} at 175 °C and 0% RH. The highest power density of fuel cell with SGO/PBI membrane was 600 mW cm^{-2} at 175 °C.

Lee et al. [236] synthesized the GO from graphite by using Hummer's method. They also deposited Pt nanoparticles onto GO by a microwave method, finally producing Pt-G. Then they blended GO and Pt-G separately with Nafion to produce GO/Nafion and Pt-G/Nafion composite membranes. Fuel cell was prepared with the Nafion/GO composite membrane exhibited significant improvement in performance. However, fuel cell which is prepared with the Pt-G/Nafion/composite membrane did not exhibit significant performance.

Zarrin et al. [237] developed functionalized GO–Nafion (F-GO/Nafion) composite membrane for using in high temperature PEMFC applications. They produced GO from graphite by using modified Hummer's method and 3-mercaptopropyl trimethoxysilane (MPTMS) was used to functionalized as the sulfonic acid functional group precursor. Solution casting method was used to make the F-GO/Nafion composite membranes. There was a significant improvement in proton conductivity and single-cell performance (4 times) for F-GO/Nafion membranes as compared with those of Nafion at 120 °C and 25% RH [237].

4.25 Field-Effect Transistor

The field-effect transistor (FET) is a transistor made of a semi-conductor material, in where electric filed is control the one type of channel conductivity and shape charge carrier. FETs involve in single-carrier-type operation with unipolar transistors [238]. A scheme of FET is shown in Fig. 16. In FET, load path contains the uniformly doped p- or n-doped piece of silicon which is called *CHANNEL*. This channel has to very thin for functioning of the FET. "*SOURCE*" and "*DRAIN*" are the terminals of the channel.

The source could be compared with the emitter and the drain with the collector of bipolar transistors. Current can flow in any direction as the channel contains no

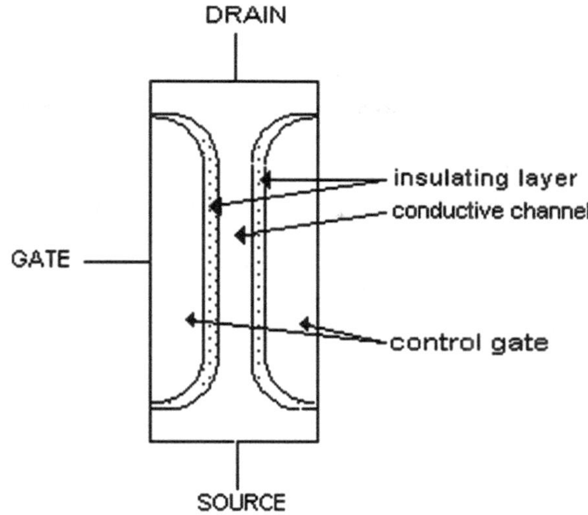

Fig. 16 The schematic diagram showing the principle of the FET. Adopted from [239]

junction. The control terminal of the FET is called the "*GATE*" which is located at the sides of the channel. This gate is insulated from the channel to avoid flow of current between gate and channel. A voltage is applied to the gate. If it repels the majority of carriers of the doped channel, its resistance will increase if the conductive area of the channel will decrease. So we can control the resistance of the channel by control the electric filed of the gate [239].

Schießl et al. [240] developed the carbon nanotube polymer-sorted semi-conducting networks for high-performance ambipolar FETs. They used atactic poly (9-dodecyl-9-methyl-fluorene) (a-PF-1-12), a polyfluorene derivative with asymmetric side-chains, for the selective dispersion of semi-conducting SWCNTs. These FETs exhibited gains up to 61, can be operated at <3 V low voltages and not shown any current hysteresis.

Chua et al. [241] developed a hybrid carbon nanotube/polymer hetero-interface FET by drop-casting carbon nanotubes on top of the semi-conducting polymer. The charge mobility in this FET is 10-fold higher than that of control device with improved ON/OFF ratio.

Derenskyi et al. developed carbon nanotube-based ambipolar FET with 10^8 ON/OFF ratio. They improved the dispersion of CNT by wrapping it with polymer and semi-aligned networks were obtained by blade coating. The performance of FET could be tuned by the choice of wrapping polymer, and its concentration [242].

Organic field-effect transistor was developed by Yasin et al. [243] by using the blend of P3HT and PCBM as active layer [243]. Strain sensing properties of this device was studied and the results found were satisfactory.

Marjanović et al. [244] developed the photoresponsive organic FET-based fullerene/conjugated polymer solid state mixtures as active semi-conductor layer, PVA or divinyl tetra methyl di siloxanebis (benzocyclobutene) (BCB) as the gate dielectrics and LiF/Al as the top source–drain contacts. These devices showed great n-type transistor behavior [244].

Organic field-effect transistors (OFETs) prepared by solution casting usually exhibit low mobilities with high ON/OFF ratios, where graphene-based transistors usually exhibits very high mobilities with low ON/OFF ratios. This problem was solved by developing the hybrid FETs with incorporation of both graphene and organic semi-conductors. Huang et al. [245] developed polymeric semi-conductor/graphene hybrid FET. Hybrid FETs exhibited up to 20 times high effective mobilities, and thus comparable ON/OFF ratios compared to those of OFETs. The results indicate that graphene can be incorporated in organic semi-conductor to improve the performance of OFETs [246]. Similar approach was taken by Gemayel et al. They blended exfoliated graphene with semi-conducting *n*-type polymer, P (NDI2OD-T2), in liquid phase and fabricated hybrid FET. There was improvement in device performance due to the addition of exfoliated graphene [245].

5 Conclusions

Carbon filler filled polymer composites are typically designed for given the applications in different types of microelectronics with having high thermal conductivity, low dielectric constant, low CTE, high electrical conductivity, low density, ease of processability and low cost. Application of such composites in microelectronics includes integrated circuit, printed circuit board, interconnections, die attach, solder, lids/enclosure, heat sinks, electrical contacts, connectors, thermal interface materials, etc.

Both carbon nanotubes and graphene are more suitable for a variety of optoelectronic applications. Over the past few years, research interest is going on SWCNT-based optoelectronics, in particular ultrafast pulse generation. This is because of their excellent optoelectronic properties and easy device fabrication. Graphene is also a strong competitor of CNTs in the photonics field and also in the optoelectronics.

Nanotechnology is entirely involved in the design of advanced devices for optoelectronic and electronic applications. Electronic devices dimensions entered in the nanorange.

The use of carbon filler filled polymer nanocomposites in electronics and electrical areas are quite diverse involving many developed applications and have been proposed to use in different applications.

Carbon filler filled polymer nanocomposites are used in chemical sensors, electro-catalysis, electroluminescent devices, batteries, memory devices and smart windows. Another potential application includes supercapacitors, photovoltaic (PV) cells and photodiodes, light-emitting diodes (LEDs) printable conductors and field-effect transistors.

The high electrical conductivity of the carbon materials converts the insulator polymers to an electrical conductor and these conducting composites are attracted to researchers due to its exceptional properties. The possible applications include electromagnetic–reflective materials, shielding, static charge–dissipative materials, conductive coatings, electronic dissipation, semi-conductor layers in high voltage cables, super capacitors, electromechanical actuators and also applications in electrodes.

Conjugated polymers with different types of nano-carbon fillers given the applications in sensor field, including biosensors, gas sensors, and chemical sensors. Nano-fillers which can be used in sensor applications are carbon nanotubes, fullerene, graphene, etc.

The interest of graphene materials in electronics application is parallel with carbon nanotubes. Sheet form of graphene may offer promising materials in replacement to silicon. Polymer semi-conductor nanocomposites have many possibilities of applications in gas sensors, photovoltaic, catalysis, optical displays, electrical devices, photoconductors, mechanics, and superconductor devices.

Although challenges are there with the processing of polymer nanocomposites such as incorporation and alignment of nanomaterials inside a polymer matrices and also dispersion of carbon-based nanomaterials. Despite these issues, literature reported that their promising electrical and thermal properties make them an ideal candidate for incorporation into a different types of functional materials and these nanocomposites already been explored in a wide range of applications in electrical and electronic sectors.

Researchers focused on their research to find out the methods which improve the electrical and thermal properties of carbon materials and also try to achieve a good dispersion of a carbon which enhances mechanical properties of the resultant polymer nanocomposites. However, we believe that carbon nano-fillers filled polymer nanocomposites might merge amazing nanomaterials for electronic devices.

References

1. Scharff P (1998) New carbon materials for research and technology. Carbon 36(5–6): 481–486
2. Dobrzaski LA (2002) Fundamentals of materials science and physical metallurgy. In: Engineering materials with elements of materials design, WNT, Warsaw
3. Skoczkowski K (1995) The production technology of carbon-graphite elements. Slask, Katowice, pp 20–177
4. Kuilla T, Bhadra S, Yao D, Kim NH, Bose S, Lee JH (2010) Recent advances in graphene based polymer composites. Prog Polym Sci 35:1350–1375
5. Fuente J, Graphenea. http://www.graphenea.com/pages/graphene–usesapplications#.VTnmIiGqqko
6. Lalwani G, Sitharaman B (2013) Multifunctional fullerene and metallofullerene based nanobiomaterials. Nano LIFE 3(3):1342003 (22 pages)
7. Przygocki W, Wlochowicz A (2001) Fullerenes and nanotubes: properties and applications. WNT, Warsaw

8. Huczko A (2004) Carbon nanotubes. Black diamonds of the twenty-first century, BEL Studio, Warsaw
9. Zielinski T, Kijenski J (2004) Technical-grade plasma carbon black used as an active modifier of plastics. Chem Ind 83(10):517–521
10. Sohi NJS, Bhadra S, Khastgir D (2011) The effect of different carbon fillers on the electrical conductivity of ethylene vinyl acetate copolymer-based composites and the applicability of different conductivity models. Carbon 49:1349–1361
11. Endo M, Strano MS, Ajayan PM (2008) Potential applications of carbon nanotubes. In: Carbon nanotubes. Advanced topics in the synthesis, structure, properties and applications. Springer, Berlin, Germany, pp 12–61
12. Mighri F, Huneault MA, Champagne MG (2004) Electrically conductive thermoplastic blends for injection and compression molding of bipolar plates in fuel cell application. Polym Eng Sci 44(9):1755–1765
13. Njuguma J, Pielichowski K (2003) Polymer nanocomposites for aerospace applications: properties. Adv Eng Mat 5(11):769–778
14. Chakrapani N, Chris J, Matayabas JR, Wakharkar V (2010) Applications of smart polymer composites to integrated circuit packaging. US 20100237513 A1
15. Chakrapani N, Chris J, Matayabas JR, Wakharkar V (2011) Applications of smart polymer composites to integrated circuit packaging, US 7952212 B2
16. Lingamneni S, Marconnet AM, Goodson KE (2013) 3D Packaging materials based on graphite nanoplatelet and aluminum nitride nanocomposites. In: Proceedings of the ASME 2013 international mechanical engineering congress & exposition IMECE 2013 13–21 Nov 2013, San Diego, California, USA, Final Paper IMECE 2013-66419
17. Yeh TH, Chang HY, Liou ST (2012) Flexible printed circuit boards including carbon nanotubes bundle. US 8,164,000 B2
18. Paik KW (2014) Aligned graphene-epoxy composite B-stage films. http://www.mae.ust.hk/news_n_events/event_n_seminars_details.html?uid=69ca957c-7415-11e3-8838-001cc47a7474
19. Kreupl F, Graham AP, Liebau M, Duesberg GS, Seidel R, Unger E (2004) Carbon nanotubes for interconnect applications. In: Proceedings of the IEEE International Electron Devices Meeting (IEDM'04), pp. 683–686, December 2004
20. Suh DW (2012) Carbon nanotubes solder composite for high performance interconnect. US 8,100,314 B2
21. Li J, Lumpp JK (2006) Electrical and mechanical characterization of carbon nanotube filled conductive adhesive. In: Proceedings of aerospace conference. IEEE, NJ, 2006, pp 1–6
22. Lin XC, Lin F (2004) Improvement on the properties of silver-containing conductive adhesives by the addition of carbon nanotube. In: Proceedings of high density microsystem design and packaging. IEEE, NJ, 2004, pp 382–384
23. Bullock S, Vanderwlel RW (2012) Electrically conductive polymer compositions containing metal particles and a graphene and methods for production and use thereof. US 8167190 B1
24. Yim BS, Oh SH, Kim J, Kim J, Kim JM (2012) Characteristics of graphene-filled solderable isotropically conductive adhesive (ICA). Mater Trans 53(3):578–581
25. Bertram A, Beasley K, De La Torre W (1992) An overview of navy composite developments for thermal management. Naval Eng J 104:276
26. Fleming TF, Rwey WC, Proc. SPIE The international society for optical engineering, 1997 (1993) 136–147
27. Fleming TF, Levan CD, Riley WC (1995) Proceedings technical conference, international electronics packaging conference pp 493–503
28. Ibrahim AM (1992) SAMPEE electronics conference, pp 556–567
29. Spicer JWM, Wilson DW, Osinader R, Thomas J, Oni BO (1999) Proc. SPIE—the internal society for optical engineering, 3700: 40
30. Ebadi-Dehaghani H, Nazempour M (2012) Thermal conductivity of nanoparticles filled polymers. http://cdn.intechopen.com/pdfs-wm/35438.pdf

31. Yoon YS, Oh MH, Kim AY, Kim N (2012) The development of thermal conductive polymer composites for heat sink. J Chem Chem Eng 6:515–519
32. Chiguma J, Johnson E, Shah P, Gornopolskaya N, Jones WE Jr (2013) Thermal diffusivity and thermal conductivity of epoxy-based nanocomposites by the laser flash and differential scanning calorimetry techniques. Open J Compos Mater 3:51–62
33. Smaldone PL (1995) 27th international SAMPE technical conference, pp 819–829
34. Glatz JJ, Vrable DL, Schmedake T, Johnson C (1992) 6th international SAMPE electronics conference, pp 334–346
35. Berger M (2012) Graphene sets new record as the most efficient filler for thermal interface materials. http://www.nanowerk.com/spotlight/spotid=24109.php
36. Khan MFS, Alexander AB (2011) Graphene—based nanocomposites as highly efficient thermal interface materials. https://arxiv.org/ftp/arxiv/papers/1201/1201.0796.pdf
37. Yu A, Ramesh P, Sun X, Bekyarova E, Itkis ME, Haddon RC (2008) Enhanced thermal conductivity in a hybrid graphite nanoplatelet-carbon nanotubes filler for epoxy composites. Adv Mater 20:4740–4744
38. Yu A, Ramesh P, Itkis ME, Elena B, Haddon RC (2007) Graphite nanoplatelet-epoxy composite thermal interface materials. J Phys Chem C 111:7565–7569
39. Wu TY, Lin JC et al (2014) Aligned graphene sheets-polymer compositeand method for manufacturing the same. US 20140097380 A1
40. Balandin AA (2013) Graphene based thermal interface materials and methods of manufacturing the same. US 20140120399 A1
41. Nanocarbons and nanocarbon-filled polymer composites for electronic thermal management materials. School of Chemical and Process Engineering, Institute for Materials Research. http://www.engineering.leeds.ac.uk/imr/research/carbon/nanocarbons.shtml
42. Okoth MO (2010) Synthesis of thermal interface materials made of metal decorated carbon nanotubes and polymers. Dissertation, Texas A&M University
43. Arora H, Matayabas Jr JC (2014) Thermal interface material composition including polymeric matrix and carbon filler. US 8920919 B2
44. Heimann M, Wirts-Ruetters M, Boehme B, Wolter KJ (2008) Investigations of carbon nanotubes epoxy composites for electronics packaging. In: Proceedings of the 58th electronic components and technology conference (ECTC'08), May 2008, pp 1731–1736
45. Matthias H, Boehme B, Sebastian S, Wirts-Ruetters M, Wolter KJ (2009) CNTs—a comparable study of CNT-filled adhesives with common materials. In: Proceedings of the 59th electronic components and technology conference (ECTC'09), San Diego, California, USA, May 2009, pp 1871–1878
46. Mir IA, Kumar D (2012) Carbon nanotube-filled conductive adhesives for electronic applications. Nanosci Meth 1:183–193
47. Adams JT, Yost BA (1991) Matrix filled with three-dimensional arrangement of carbon fibers, thermoplastic, thermosetting or elastomeric resins; bonding electronic components US 5026748 A
48. Transparent conducting film From Wikipedia. http://en.wikipedia.org/wiki/Transparent_conducting_film
49. Hong S, Myung S (2007) Nanotube electronics: a flexible approach to mobility. Nat Nanotech 2(4):207–208
50. Dettlaff-Weglikowska U, Kaempgen M, Hornbostel B, Skakalova V, Wang J, Liang J, Roth S (2006) Conducting and transparent SWNT/polymer composites. Phys Stat Sol (b) 243:3440–3444
51. Ferrer-Anglada N, Kaempgen M, Skakalova V, Dettlaf-Weglikowska U, Roth S (2004) Synthesis and characterization of carbon nanotube-conducting polymer thin films. Diamond Relat Mater 13:256–260
52. Park C, Ounaies Z, Watson KA, Crooks RE, Smith J, Lowther SE, Connell JW, Siochi EJ, Harrison JS, Clair TLS (2002) Dispersion of single wall carbon nanotubes by in situ polymerization under sonication. Chem Phys Lett 364:303–308

53. De S, Lyons PE, Sorel S, Doherty EM, King PJ, Blau WJ, Nirmalraj PN, Boland JJ, Scardaci V, Joimel J, Coleman JN (2009) Transparent, flexible, and highly conductive thin films based on polymer-nanotube composites. ACS Nano 3:714–720
54. Xu Y, Wang Y, Jiajie L, Huang Y, Ma Y, Wan X et al (2009) Ahybrid material of graphene and poly (3,4-ethyldioxythiophene) with high conductivity, flexibility, and transparency. Nano Res 2:343–348
55. Liuid crystal display. From Wikipedia. http://en.wikipedia.org/wiki/Liquid-crystal_display
56. LED display. From Wikipedia. http://en.wikipedia.org/wiki/LED_display
57. John (2010) Nano C Inc., Liquid crystal display (LCD)-working. http://www.circuitstoday.com/liquid-crystal-displays-lcd-working
58. Eren San S, Okutan M, Köysal O, Yerli Y (2008) Carbon nanoparticles in nematic liquid crystals. Chin Phys Lett 25(1):212
59. Qi H, Hegmann T (2008) Impact of nanoscale particles and carbon nanotubes on current and future generations of liquid crystal displays. J Mater Chem 18:3288–3294
60. OLED. From Wikipedia. http://en.wikipedia.org/wiki/OLED
61. Moni-X Ltd. (2005) Organic light emitting diode (OLED). http://qxwujoey.tripod.com/oled.htm
62. Eda G, Unalan HE, Rupesinghe NL, Amaratunga GAJ, Chhowalla M (2008) Field emission from graphene based composite thin films. Appl Phys Lett 93:233502–233503
63. Verma VP, Das S, Lahiri I, Choi W (2010) Large-area graphene on polymer film for flexible and transparent anode in field emission device. Appl Phys Lett 96:203108 / 1–3
64. Woo HS, Czerw R, Webster S, Carroll DL, Ballato J, Strevens AE, O'Brien D, Blau WJ (2000) Hole blocking in carbon nanotube-polymer composite organic light-emitting diodes based on poly (m-phenylene vinylene-co-2, 5-dioctoxy-p-phenylene vinylene). Appl Phys Lett 77(9):1393–1395
65. Li J, Hu L, Wang L, Zhou Y, Gruner G, Marks TJ (2006) Organic light-emitting diodes having carbon nanotube anodes. Nano Lett 6:2472–2477
66. Yu Z, Niu X, Liu Z, Pei J (2011) Intrinsically stretchable polymer light-emitting devices using carbon nanotube-polymer composite electrodes. Adv Mater 23:3867–3994
67. Ou ECW, Hu L, Raymond GCR, Soo OK, Pan J, Zheng Z, Park Y, Hecht D, Irvin G, Drzaic P et al (2009) Surface-modified nanotube anodes for high performance organic light-emitting diode. ACS Nano 3:2258–2264
68. Singh JP, Saha U, Jaiswal R, Anand RS, Srivastava A, Goswami TH (2014) Enhanced polymer light-emitting diode property using fluorescent conducting polymer-reduced graphene oxide nanocomposite as active emissive layer. J Nanopart Res 16:1–20
69. Luo W, Chen W, Leng C, Huang D, Zhang Y, Yang J, Li Z, Shi H, Du C (2014) Graphene composite anode for flexible polymer light emitting diode. In: Proceedings SPIE 9272, optical design and testing VI, 927206, November 5, 2014
70. Lin CH, Chen KT, Ho JR, Cheng JWJ, Tsiang RCC (2012) PEDOT:PSS/graphene nanocomposite hole-injection layer in polymer light-emitting diodes. J Nanotech 2012:1–7
71. Electroluminescence. From Wikipedia. http://en.wikipedia.org/wiki/Electroluminescence
72. Xu Z, Wu Y, Hu B, Ivanov IN, Geohegan DB (2005) Carbon nanotube effects on electroluminescence and photovoltaic response in conjugated polymers. Appl Phys Lett 87:263118
73. Hu B, Li D, Manandharam P, Fan Q, Kasilingam D, Calvert P (2012) CNT/conducting polymer composite conductors impart high flexibility to textile electroluminescent devices. J Mater Chem 22:1598–1605
74. Photovoltaics. From Wikipedia. http://en.wikipedia.org/wiki/Photovoltaics
75. Edward LO (2008) PV cell—working principle and applications. http://cd1.edb.hkedcity.net/cd/science/physics/NSS/Energy01_Dec08/PhotoVoltaicsCells.pdf
76. O'Connell MJ, Boul P, Ericson LM, Huffman C, Wang Y, Haroz E, Kuper C, Tour J, Ausman KD, Smalley RE (2001) Reversible water-solubilization of single-walled carbon nanotubes by polymer wrapping. Chem Phys Lett 342(3–4):265–271

77. Bhattacharyya S, Kymakis E, Amaratunga GAJ (2004) Photovoltaic properties of dye functionalized single-wall carbon nanotube/conjugated polymer devices. Chem Mater 16:4819–4823
78. Ago H, Petritsch K, Shaffer MSP, Windle AH, Friend RH (1999) Composites of carbon nanotubes and conjugated polymers for photovoltaic devices. Adv Mater 11:1281–1285
79. Kazaoui S, Minami N, Nalini B, Kim Y, Hara K (2005) Near-infrared photoconductive and photovoltaic devices using single-wall carbon nanotubes in conductive polymer films. J Appl Phys 98:084314
80. Li C, Chen Y, Wang YIZ, Chhowalla M, Mitra S (2007) A fullerene-single wall carbon nanotube complex for polymer bulk heterojunction photovoltaic cells. J Mater Chem 17:2406–2411
81. Pradhan B, Batabyal SK, Pal AJ (2006) Functionalized carbon nanotubes in donor/acceptortype photovoltaic devices. Appl Phys Lett 88:093106
82. Kymakis E, Alexandrou I, Amaratunga GAJ (2003) High open-circuit voltage photovoltaic devices from carbon-nanotube-polymer composites. J Appl Phys 93:1764–1768
83. Sariciftci NS, Smilowitz L, Heeger AJ, Wudl F (1992) Photoinduced electron transfer from a conducting polymer to buckminsterfullerene. Science 258:1474–1476
84. Alley NJ, Liao KS, Andreoli E, Dias S, Dillon EP, Orbaek AW, Barron AR, Byrne HJ, Curran SA (2012) Effect of carbon nanotube-fullerene hybrid additive on P3HT:PCBM bulk-heterojunction organic photovoltaics. Synth Met 162(1–2):95–101
85. Liu Z, Liu Q, Huang Y, Ma Y, Yin S, Zhang X, Sun W, Chen Y (2008) Organic photovoltaic devices based on a novel acceptor material: graphene. Adv Mater 20:3924–3930
86. Hong W, Xu Y, Lu G, Li C, Shi G (2008) Transparent graphene/PEDOT-PSS composite films as counter electrodes of dye-sensitized solar cells. Electrochem Commun 10:1555–1558
87. Eda G, Lin YY, Miller S, Chen CW, Su WF, Chhowalla M (2008) Transparent and conducting electrodes for organic electronics from reduced graphene oxide. Appl Phys Lett 92:233305 / 1–3
88. Wu J, Becerril HA, Bao Z, Liu Z, Chen Y, Peumans P (2008) Organic solar cells with solution processed graphene transparent electrodes. Appl Phys Lett 92:263302 / 1–3
89. Lim SP, Pandikumar A, Lim YS, Huang NM, Lim HN (2014) In-situ electrochemically deposited polypyrrole nanoparticles incorporated reduced graphene oxide as an efficient counter electrode for platinum-free dye-sensitized solar cells. Sci Rep 4. https://doi.org/10.1038/srep05305
90. Gomez De Arco L, Zhang Y, Schlenker CW, Ryu K, Thompson ME, Zhou C (2010) Continuous, highly flexible, and transparent graphene films by chemical vapor deposition for organic photovoltaics. ACS Nano 4(5):2865–2873
91. Li SS, Tu KH, Lin CC, Chen CW, Chhowalla M (2010) Solution processable grapheme oxide as an efficient hole transport layer in polymer solar cells. ACS Nano 4:3169–3174
92. Valentini L, Cardinali M, Bon SB, Bagnis D, Verdejo R, Lopez Manchado MA, Kenny JM (2010) Use of butylamine modified graphene sheets in polymer solar cells. J Mater Chem 20:995–1000
93. Wang X, Zhi L, Müllen K (2008) Transparent, conductive grapheNe electrodes for dye-sensitized solar cells. Nano Lett 8:323–327
94. Su Q (2012) Graphene based electrode materials for solar cell and electrochemical oxygen reduction. Ph.D. Dissertation, Max-Planck Institute for Polymer Research
95. Saranya K, Rameez Md, Subramania A (2015) Developments in conducting polymer based counter electrodes for dye-sensitized solar cells—an overview. Eur Polym J 66:207–227
96. Wang J, Wang Y, He D, Wu H, Wang H, Zhou P, Fu M (2012) Influence of polymer/fullerene-graphene structure on organic polymer solar devices. Integr Ferroelect 137(1):1–9
97. Hsu CL, Lin CT, Huang JH, Chu CW, Wei KH, Li LJ (2012) Layer-by-layer grapheme/TCNQ stacked films as conducting anodes for organic solar cells. ACS Nano 6(6):5031–5039

98. Wang J, Wang Y, He D, Wu H, Wang H, Zhou P, Fu M, Jiang K, Chen W (2011) Organic photovoltaic devices based on an acceptor of solution-processable functionalized graphene. J Nanosci Nanotechnol 11(11):9432–9438
99. Brabec CJ, Padinger F, Hummelen JC, Janssen RAJ, Sariciftci NS (1999) Realization of large area flexible fullerene-conjugated polymer photocells: a route to plastic solar cells. Synth Met 102(1–3):861–864
100. Fromherz T, Padinger F, Gebeyehu D, Brabec C, Hummelen JC, Sariciftci NS (2000) Comparison of photovoltaic devices containing various blends of polymer and fullerene derivatives. Sol En Mat 63(1):61–68
101. Gebeyehu D, Brabec CJ, Padinger F, Fromherz T, Hummelen JC, Badt D, Schindler H, Sariciftci NS (2001) The interplay of efficiency and morphology in photovoltaic devices based on interpenetrating networks of conjugated polymers with fullerenes. Synth Met 118(1–3):1–9
102. Sensor. From Wikipedia. http://en.wikipedia.org/wiki/Sensor
103. Ansari S, Giannelis EP (2009) Functionalized graphene sheet—Poly(vinylidene fluoride) conductive nanocomposites. J Polym Sci Pt B Polym Phys 47:888–889
104. Shan C, Yang H, Song J, Han D, Ivaska A, Niu L (2009) Direct electrochemistry of glucose oxidase and biosensing for glucose based on graphene. Anal Chem 81(6):2378–2382
105. Xue R, Kang TF, Lu LP, Cheng SY (2013) Electrochemical sensor based on the graphene-nafion matrix for sensitive determination of organophosphorus pesticides. Anal Lett 46(1):131–141
106. Zhu J, Wei S, Ryu J, Guo Z (2011) Strain-sensing elastomer/carbon nanofiber "metacomposites". J Phys Chem C 115:13215–13222
107. Li L, Li J, Lukehart CM (2008) Graphitic carbon nanofiber-poly (acrylate) polymer brushes as gas sensors. Sens Actuators B Chem 130:783–788
108. Jang J, Bae J (2007) Carbon nanofiber/polypyrrole nanocable as toxic gas sensor. Sens Actuators B Chem 122:7–13
109. Harun FKC, Jumadi AM, Mahmood NH (2011) Carbon black polymer composite gas sensor for electronic nose. Int J Sci Eng Res 2(11):1–7
110. Ryan MA, Shevade AV, Zhou H, Homer ML (2004) Polymer–carbon black composite sensors in an electronic nose for air-quality monitoring. MRS Bull 29(10):714–719
111. Wei C, Dai L, Roy A, Tolle TB (2006) Multifunctional chemical vapor sensors of aligned carbon nanotube and polymer composites. J Am Chem Soc 128(5):1412–1413
112. Hernández-López S, Vigueras-Santiago E, Mora MM, Mancilla JRF, Contreras EAZ (2013) Cellulose-based polymer composite with carbon black for tetrahydrofuran sensing. Int J Polym ScI 2013:1–7
113. Singha DK, Mahata P (2015) Luminescent coordination polymer–fullerene composite as a highly sensitive and selective optical detector for 2,4,6-trinitrophenol (TNP). RSC Adv 5:28092–28097
114. Isoda T, Sato H et al (2011) Evalution of immunoglobulne sensing function using a fullerene- composite-polymer-coated sensor electrode. Sens Mater 23(4):237–249
115. Shih WP, Tsao LC, Lee CW, Cheng MY, Chang C, Yang YJ, Fan KC (2010) Flexible temperature sensor array based on a graphite-polydimethylsiloxane composite. Sens 10(4): 3597–3610
116. Seah TH, Pumera M (2011) Platelet graphite nanofibers/soft polymer composites for electrochemical sensing and biosensing. Sens Actuators B: Chemical 156(1):79–83
117. Tadakaluru S, Thongsuwan W, Singjai P (2014) Stretchable and flexible high-strain sensors made using carbon nanotubes and graphite films on natural rubber. Sens 14:868–876
118. Eswaraiah V, Balasubramaniam K, Ramaprabhu S (2012) One-pot synthesis of conducting graphene-polymer composites and their strain sensing application. Nanoscale 4(4):1258–1262
119. Actuator. From Wikipedia. http://en.wikipedia.org/wiki/Actuator

120. Mohamadi S, Sanjani NS, Mahdavi H (2011) Functionalization of graphene sheets via chemically grafting of PMMA chains through in situ polymerization. J Macromol Sci Pt A 48(8):577–582
121. Liang J, Huang L, Li N, Huang Y, Wu Y, Fang S, Oh J, Kozlov M, Ma Y, Li F, Baughman R, Chen Y (2012) Electromechanical actuator with controllable motion, fast response rate, and highfrequency resonance based on graphene and polydiacetylene. ACS Nano 6(5):4508–4519
122. Ahir SV, Terentjev EM (2006) Fast relaxation of carbon nanotubes in polymer composite actuators. Phys Rev Lett 96(13):133902
123. Chen L, Liu C, Liu K, Meng C, Hu C, Wang J, Fan S (2011) High-performance, low-voltage, and easy-operable bending actuator based on aligned carbon nanotube/polymer composites. ACS Nano 5(3):1588–1593
124. Wang XL, Oh IK (2010) Sulfonated poly(styrene-b-ethylene-co-butylene-b-styrene) and fullerene composites for ionic polymer actuators. J Nanosci Nanotechnol 10(5):3203–3206
125. Jung JH, Vadahanambi S, Oh IK (2010) Electro-active nano-composite actuator based on fullerene-reinforced Nafion. Compos Sci Technol 70(4):584–592
126. Ghaffari Zhou MY, Lin M, Koo CM, Zhang QM (2014) High electromechanical reponses of ultra-high-density aligned nano-porous microwave exfoliated graphite oxide/polymer nano-composites ionic actuators. Int J Smart Nano Mater 5(2):114–122
127. Muralidharan MN, Ansari S (2013) Thermally reduced graphene oxide/thermoplastic polyurethane nanocomposites as photomechanical actuators. Adv Mat Lett 4(12):927–932
128. Lian Y, Liu Y, Jiang T, Shu J, Lian H, Cao M (2010) Enhanced electromechanical performance of graphite oxide-nafion nanocomposite actuator. J Phys Chem 114(21):9659–9663
129. Sen I, Seki Y, Sarikanat M, Cetin L, Gurses BQ, Ozdemir O, Yilmaz OC, Sever K, Akar E, Mermer O (2015) Electroactive behavior of graphene nanoplatelets loaded cellulose composite actuators. Compos Part B Eng 69:369–377
130. Yang W, Choi H, Choi S, Jeon M, Lee SY (2012) Carbon nanotube–graphene composite for ionic polymer actuators. Smart Mater Struct 21(5):055012
131. Loomis J, King B, Burkhead T, Xu P, Bessler N, Terentjev E, Panchapakesan B (2012) Graphene-nanoplatelet-based photomechanical actuators. Nanotechnol 23(4):045501
132. Electrode. From Wikipedia. http://en.wikipedia.org/wiki/Electrode
133. Wang DW, Li F, Zhao J, Ren W, Chen ZG, Tan J et al (2009) Fabrication of graphene / polyaniline composite paper via in situ anodic electropolymerization for high-performance flexible electrode. ACS Nano 7:1745–1752
134. Li H, Chen J, Han S, Niu W, Liu X, Xu G (2009) Electrochemiluminescence from tris (2,2-bipyridyl)ruthenium(II)-graphene-nafion modified electrode. Talanta 79:165–170
135. Kim JY, Kim M, Choi JH (2003) Characterization of light emitting devices based on a single-walled carbon nanotube–polymer composite. Synth Met 139(3):565–568
136. Kauffmann JM, Linders CR, Patriarche GJ, Smyth MR (1988) A comparison of glassy-carbon and carbon-polymer composite electrodes incorporated into electrochemical detection systems for high-performance liquid chromatography. Talanta 35(3):179–182
137. Rakhi RB, Chen W, Alshareef HN (2012) Conducting polymer/carbon nanocoil composite electrodes for efficient supercapacitors. J Mater Chem 22:5177–5183
138. Chang J, Najeeb CK, Lee JH, Kim JH (2011) Single-walled carbon nanotubes/polymer composite electrodes patterned directly from solution. Langmuir 27(11):7330–7336
139. Calixto CMF, Mendes RK, Oliveira AC, Ramos LA, Cervini P, Cavalheiro ETG (2007) Development of graphite-polymer composites as electrode materials. Mater Res 10(2):1439–1516
140. Perween M, Parmar DB, Bhadu GR, Srivastava DN (2014) Polymer–graphite composite: a versatile use and throw plastic chip electrode. Analyst 139:5919–5926
141. Coffey B, Madsen PV, Poehler TO, Searson PC (1995) High charge density conducting polymer/graphite fiber composite electrodes for battery applications. J Electrochem Soc 142(2):321–325

142. Gómez H, Ram MK, Alvi F, Villalba P, Stefanakos E, Kumar A (2011) Graphene-conducting polymer nanocomposite as novel electrode for supercapacitors. J Power Sources 196(8):4102–4108
143. Lithium ion battery. From Wikipedia. http://en.wikipedia.org/wiki/Lithium-ion_battery
144. Brain M, How Lithium-ion Batteries Work. http://electronics.howstuffworks.com/everyday-tech/lithium-ion-battery1.htm
145. Song Z, Xu T, Gordin ML, Jiang YB, Bae IT, Xiao Q, Zhan H, Liu J, Wang D (2012) Polymer—graphene nanocomposites as ultrafast-charge and discharge cathodes for rechargeable Lithium batteries. Nano Lett 12:22205–22211
146. Lee H, Yoo JK, Park JH, Kim JH, Kang K, Jung YS (2012) A stretchable polymer–carbon nanotube composite electrode for flexible lithium-ion batteries: porosity engineering by controlled phase separation. Adv Energ Mater 2(8):976–982
147. Fauteux D (1993) Carbon/polymer composite electrode for use in a lithium battery. EP0528557A1
148. Sivakkumar SR, Kim DW (2007) Polyaniline/carbon nanotube composite cathode for rechargeable lithium polymer batteries assembled with gel polymer electrolyte. J Electrochem Soc 154(2):A134–A139
149. Veeraraghavan B, Paul J, Haran B, Popov B (2002) Study of polypyrrole graphite composite as anode material for secondary lithium-ion batteries. J Power Sour 109:377–387
150. Li S, Shu K, Zhao C, Wang C, Guo Z, Wallace G, Liu HK (2014) One-step synthesis of graphene/polypyrrole nanofiber composites as cathode material for a biocompatible zinc/polymer battery. ACS Appl Mater Interfaces 6(19):16679–16686
151. Chen L, Zhang M, Wei W (2013) Graphene-based composites as cathode materials for lithium ion batteries. J Nanomat 2013: Article ID 940389, 8 pages
152. Capacitor. From Wikipedia. http://en.wikipedia.org/wiki/Capacitor
153. Zhang D, Zhang X, Chen Y, Yu P, Wang C, Ma Y (2011) Enhanced capacitance and rate capability of graphene/polypyrrole composite as electrode material for supercapacitors. J Power Sour 196:5990–5996
154. Liao WC, Liao FS, Tsai CT, Yang YP (2012) Preparation of activated carbon for electric double layer capacitors. China Steel Tech Rep 25:36–41
155. Luo X, Chang DDL (2001) Carbon fiber/ polymer matrix composites as capacitor. Compos Sci Technol 61:885–888
156. Li S, Zhao Y, Zhang Z, Tang H (2014) Preparation and characterization of epoxy/carbon fiber composite capacitors. Polym Compos. https://doi.org/10.1002/pc.23050
157. Tien CP, Teng H (2010) Polymer/graphite oxide composites as high-performance materials for electric double layer capacitors. J Power Sour 195(8):2414–2418
158. Huang L, Li C, Shi G (2014) High-performance and flexible electrochemical capacitors based on graphene/polymer composite films. J Mater Chem A 2:968–974
159. Sangermano M (2014) Graphene-epoxy flexible transparent capacitor obtained by graphene-polymer transfer and uv–induced bonding. http://www.radtechreport.com/2014_1ssue4_sangermano.html
160. Wu Q, Xu YX, Yao ZY, Liu AR, Shi GQ (2010) Supercapacitors based on flexible grapheme/ polyaniline nanofibre composite film. ACS Nano 4:1963–1970
161. Wang HL, Hao QL, Yang XJ, Lu LD, Wang X (2009) Graphene oxide doped polyaniline for super capacitors. Electrochem Commun 11:1158–1161
162. Yan J, Wei T, Fan ZJ, Qian WZ, Zhang ML, Shen XD, Wei F (2010) Preparation of graphene nanosheets/ carbon nanotubes/ polyaniline composite as electrode material for supercapaitors. J Power Sour 195:3041–3045
163. Lee KYT, Naguib H, Lian K (2014) Flexible multiwall carbon nano-tubes/conductive polymer composite electrode for supercapacitor applications. ASME 2014 Conference on Smart Materials, Adaptive Structures and Intelligent Systems, Paper No. SMASIS2014-7735, pp. V001T01A033; 7 pages. https://doi.org/10.1115/smasis2014-7735

164. Zhamu A, Jang BZ (2013) Method of producing graphite-carbon composite electrodes for supercapacitors. US 8,497,225 B2
165. Singh A, Chandra A (2013) Graphite oxide/polypyrrole composite electrodes for achieving high energy density supercapacitors. J Appl Electrochem 43:773–782
166. Liu Q, Nayfeh O, Nayfeh MH, Yau ST (2013) Flexible supercapacitor sheets based on hybrid nanocomposite materials. Nano Energ 2:133–137
167. Electrostatic discharge. From Wikipedia. http://en.wikipedia.org/wiki/Electrostatic_discharge
168. Electromagnetic Shielding From Wikipedia. http://en.wikipedia.org/wiki/Electromagnetic_shielding
169. Electromagnetic Interference. From Wikipedia. http://en.wikipedia.org/wiki/Electromagnetic_interference
170. Bhadra S, Singha NK, Khastgir D (2008) Semi-conductive composites from ethylene 1-octene copolymer and polyaniline coated nylon 6: studies on mechanical, thermal, processability, electrical and EMI shielding properties. Polym Eng Sci 48:995–1006
171. Bhadra S, Singha NK, Khastgir D (2009) Dielectric properties and EMI shielding efficiency of polyaniline and ethylene 1-octene based semi-conducting composites. Curr Appl Phys 9:396–403
172. Lee J, Yang SB, Jung HT (2009) Carbon nanotubes–polypropylene nanocomposites for electrostatic discharge applications. Macromol 42(21):8328–8334
173. Kim S, Kim S, Lee C (2012) Electrostatic discharge polymer filler containing carbon nanotube enclosed with thermoplatic resin layer and manufacturing method thereof. US 20120298925 A1
174. Boday DJ, Gentrupa MH, Iben IET (2014) Low viscosity electrostatic discharge (ESD) dissipating adhesive substantially free of agglomerates. US 8673462 B2
175. Poosal A, Kittipong Hrimchum K, Aussawasathien D, Pentrakoon D, The Effect of oxygen-plasma treated graphene nanoplatelets upon the properties of multiwalled carbon nanotube and polycarbonate hybrid nanocomposites used for electrostatic dissipative applications. J Nanomater 2015: 1–9, Article ID 470297
176. Liang JJ, Wang Y, Huang Y, Ma YF, Liu ZF, Cai FM, Zhang CD, Gao HJ, Chen YS (2009) Electromagnetic interference shielding of graphene/epoxy composites. Carbon 47(3): 922–925
177. Goyal RK, Kadam A (2010) Polyphenylene sulphide/graphite composites for EMI shielding applications. Adv Mat Lett 1(2):143–147
178. Yang Y, Guptal MC, Dudley KL, Lawrence RW (2007) Electromagnetic interference shielding characteristics of carbon nanofiber-polymer composites. J Nanosci Nanotechnol 7(2):549–554
179. Wang S, Tambraparni M, Qiu J, Tipton J, Dean D (2009) Thermal expansion of graphene composites. Macromol 42:5251–5255
180. Yu J, Lu K, Sourty E, Grossiord N, Koning CE, Loos J (2007) Characterization of conductive multiwall carbon nanotube/polystyrene composites prepared by latex technology. Carbon 45:2897–2903
181. Luo X, Chung DDL (1999) Electromagnetic interference shielding using continuous carbon-fiber carbon-matrix and polymer-matrix composites. Compos: Part B 30:227–231
182. Li N, Huang Y, Du F, He X, Lin X, Gao H, Ma Y, Li F, Chen Y, Eklund PC (2006) Electromagnetic interference (EMI) shielding of single-walled carbon nanotube epoxy composites. Nano Lett 6(6):1141–1145
183. Al-Saleh MH, Sundararaj U (2008) Electromagnetic interference (EMI) shielding effectiveness of PP/PS polymer blends containing high structure carbon black. Macromol Mater Eng 293(7):621–630
184. Morari C, Balan I, Pintea J, Chitanu E, Iordache I (2011) Electrical conductivity and electromagnetic shielding effectiveness of silicone rubber filled with ferrite and graphite powders. Prog Electromagnet Res M 21:93–104

185. Maiti S, Shrivastava NK, Suin S, Khatua BB (2013) Polystyrene/MWCNT/graphite nanoplate nanocomposites: efficient electromagnetic interference shielding material through graphite nanoplate–MWCNT–graphite nanoplate networking. ACS Appl Mater Interface 5(11):4712–4724
186. Eswaraiah V, Sankaranarayanan V, Ramaprabhu S (2011) Functionalized graphene–PVDF foam composites for EMI shielding. Macromol Mater Eng 296(10):894–898
187. Memory Devices. http://www.tutorialspoint.com/computer_logical_organization/memory_devices.htm
188. Mamo MA, Sustaita AO, Tetana ZN, Coville NJ, Hümmelgen IA (2013) Nitrogen-doped, boron-doped and undoped multiwalled carbon nanotube/polymer composites in WORM memory devices. Nanotechnol 24(12):125203
189. Sustaita AO, Mamo MA, Segura-Cardenas E, Reyes-Reyes M, López-Sandova R, Coville NJ, Hümmelgen IA (2013) Functionalized spherical carbon nanostructure/poly (vinylphenol) composites for application in low power consumption write-once-read-many times memories. J Nanosci Nanotechnol 13:1–7
190. Machado WS, Mamo MA, Coville NJ, Hümmelgen IA (2012) The OFF to ON switching time and ON state consolidation in write-once-read-many-times memory devices based on doped and undoped carbon-sphere/polymer composites. Thin Solid Films 520(13):4427–4431
191. Pradhan B, Batabyal SK, Pal AJ (2006) Electrical bistability and memory phenomenon in carbon nanotube-conjugated polymer matrixes. J Phys Chem B 110(16):8274–8277
192. Jo H, Ko J, Lim JA, Chang HJ, Kim YS (2013) Organic nonvolatile resistive switching memory based on molecularly entrapped fullerene derivative within a diblock copolymer nanostructure. Macromolecular Rapid Commun 34(4):355–361
193. Khan MA, Bhansali US, Cha D, Alshareef HN (2013) All-polymer bistable resistive memory device based on nanoscale phase-separated PCBM-ferroelectric blends. Adv Funct Mater 23:2145–2152
194. Kanwal A, Chhowalla M (2006) Stable, three layered organic memory devices from C60 molecules and insulating polymers. Appl Phys Lett 89:203103
195. Son DI, Shim JH, Park DH, Jung JH, Lee JM, Park WI, Kim TW, Choi WK (2011) Polymer-ultrathin graphite sheet-polymer composite structured flexible nonvolatile bistable organic memory devices. Nanotechnol 22(29):295203
196. Mamo MA, Sustaita AO, Coville NJ, Hümmelgen IA (2013) Polymer composite of poly (vinyl phenol)-reduced graphene oxide reduced by vitamin C in low energy consuming write-once–read-many times memory devices. Org Electron 14(1):175–181
197. Kafy A, Sadasivuni KK, Kim HC, Akther A, Kim J (2015) Designing flexible energy and memory storage materials using cellulose modified graphene oxide nanocomposites. Phys Chem Chem Phys 17:5923–5931
198. Zhuang XD, Chen Y, Liu G, Li PP, Zhu CX, Kang ET, Noeh KG, Zhang B, Zhu JH, Li YX (2010) Conjugated-polymer-functionalized graphene oxide: synthesis and nonvolatile rewritable memory effect. Appl Mater 22(15):1731–1735
199. Fuel Cell. From Wikipedia. http://en.wikipedia.org/wiki/Fuel_cell
200. Fuel Cell Principle. http://www.nedstack.com/technology/fuel-cell-principle
201. Dweiri R, Sahari J (2007) Electrical properties of carbon-based polypropylene composites for bipolar plates in polymer electrolyte membrane fuel cell (PEMFC). J Power Sour 171(2):424–432
202. Xia LG, Li AJ, Wang WQ, Yin Q, Lin H, Zhao YB (2008) Effects of resin content and preparing conditions on the properties of polyphenylene sulfide resin/graphite composite for bipolar plate. J Power Sour 178(1):363–367
203. Cunningham BD, Baird DG (2007) Development of bipolar plates for fuel cells from graphite filled wet-lay material and a compatible thermoplastic laminate skin layer. J Power Sour 168(2):418–425

204. Kakati BK, Deka D (2007) Differences in physico-mechanical behaviors of resol (e) and novolac type phenolic resin based composite bipolar plate for proton exchange membrane (PEM) fuel cell. Electrochim Acta 52:7330–7336
205. Lee JH, Jang YK, Hong CE, Kim NH, Li P, Lee HK (2009) Effect of carbon fillers on properties of polymer composite bipolar plates of fuel cells. J Power Sour 193(2):523–529
206. Liao SH, Yen CY, Weng CC, Lin YF, Ma CCM, Yang CH, Tsai MC, Yen MY, Hsiao MC, Lee SH, Xie XF, Hsiao YH (2008) Preparation and properties of carbon nanotube/polypropylene nanocomposite bipolar plates for polymer electrolyte membrane fuel cells. J Power Sour 185(2):1225–1232
207. Mathur RB, Dhakate SR, Gupta DK, Dhami TL, Aggarwal RK (2008) Effect of different carbon fillers on the properties of graphite composite bipolar plate. J Mater Process Technol 203(1–3):184–192
208. Song LN, Xiao M, Meng YZ (2006) Electrically conductive nanocomposites of aromatic polydisulfide/expanded graphite. Compos Sci Technol 66(13):2156–2162
209. Du C, Ming P, Hou M, Fu J, Shen Q, Liang D, Fu Y, Luo X, Shao Z, Yi B (2010) Preparation and properties of thin epoxy/compressed expanded graphite composite bipolar plates for proton exchange membrane fuel cells. J Power Sour 195(3):794–800
210. Allaoui A, Bai S, Cheng HM, Bai JB (2002) Mechanical and electrical properties of a MWNT/epoxy composite. Compos Sci Technol 62(15):1993–1998
211. Celzard A, McRae E, Deleuze C, Dufort M, Furdin G, Mareche JF (1996) Critical concentration in percolating systems containing a high-aspect-ratio filler. Phys Rev B 53:6209–6214
212. Sandler JKW, Kirk JE, Kinloch IA, Shaffer MSP, Windle AH (2003) Ultra-low electrical percolation threshold in carbon-nanotube-epoxy composites. Polymer 44(19):5893–5899
213. Martin CA, Sandler JKW, Shaffer MSP, Schwarz MK, Bauhofer W, Schulte K, Windle AH (2004) Formation of percolating networks in multi-wall carbon-nanotube–epoxy composites. Compos Sci Technol 64(15):2309–2316
214. Munson-McGee SH (1991) Estimation of the critical concentration in an anisotropic percolation network. Phys Rev B 43:3331–3336
215. Gojny FH, Wichmann MHG, Fiedler B, Kinloch IA, Bauhofer W, Windle AH, Schulte K (2006) Evaluation and identification of electrical and thermal conduction mechanisms in carbon nanotube/epoxy composites. Polymer 47(6):2036–2045
216. Shaffer MSP, Fan X, Windle AH (1998) Dispersion and packing of carbon nanotubes. Carbon 36(11):1603–1612
217. Sun J, Gao L (2001) Development of a dispersion process for carbon nanotubes in ceramic matrix by heterocoagulation. Carbon 41(5):1063–1068
218. Liu Y, Gao L (2005) A study of the electrical properties of carbon nanotube-NiFe2O4 composites: effect of the surface treatment of the carbon nanotubes. Carbon 43(1):47–52
219. Zhu BK, Xie SH, Xu ZK, Xu YY (2006) Preparation and properties of the polyimide/multi-walled carbon nanotubes (MWNTs) nanocomposites. Compos Sci Technol 66(3–4):548–554
220. Lee SH, Cho E, Jeon SH, Youn JR (2007) Rheological and electrical properties of polypropylene composites containing functionalized multi-walled carbon nanotubes and compatibilizers. Carbon 45(14):2810–2822
221. Cele NP, Ray SS (2009) Recent progress on nafion-based nanocomposite membranes for fuel cell applications. Macromol Mater Eng 294(11):719–738
222. Liu YH, Yi B, Shao ZG, Xing D, Zhang H (2006) Carbon nanotubes reinforced nafion composite membrane for fuel cell applications. Electrochem Solid-State Lett 9(7):A356–A359
223. Liu YH, Yi B, Shao ZG, Wang L, Xing D, Zhang H (2007) Pt/CNTs-Nafion reinforced and self-humidifying composite membrane for PEMFC applications. J Power Sour 163(2):807–813

224. Thomassin JM, Kollar J, Caldarella G, Germain A, Jerome R, Detrembleur C (2007) Beneficial effect of carbon nanotubes on the performances of Nafion membranes in fuel cell applications. J Membrane Sci 303(1–2):252–257
225. Wang L, Xing DM, Zhang HM, Yu HM, Liu YH, Yi BL (2008) MWCNTs reinforced Nafion® membrane prepared by a novel solution-cast method for PEMFC. J Power Sour 176(1):270–275
226. Zhang W, Dehghani-Sanij AA, Blackburn RS (2007) Carbon based conductive polymer composite. J Mater Sci 42(10):3408–3418
227. Chen WF, Wu JS, Kuo PL (2008) Poly(oxyalkylene)diamine-functionalized carbon nanotube/perfluorosulfonated polymer composites: synthesis, water state, and conductivity. Chem Mater 20(18):5756–5757
228. Asgari MS, Nikazar M, Molla-abbasi P, Hasani-Sadrabadi MM (2013) Nafion®/histidine functionalized carbon nanotube: high-performance fuel cell membranes. Int J Hydrogen Energy 38(14):5894–5902
229. Kannan R, Kakade BA, Pillai VK (2008) Polymer electrolyte fuel cells using Nafion-based composite membranes with functionalied carbon nanotubes. Angew Chem Int Ed 47(14): 2653–2656
230. Kannan R, Aher PP, Palaniselvam T, Kurungot S, Kharul UK, Pillai VK (2010) Artificially designed membranes using phosphonated multiwall carbon nanotube–polybenzimidazole composites for polymer electrolyte fuel cells. J Phys Chem Lett 1(14):2109–2113
231. Cele NP, Ray SS, Pillai SK, Ndwandwe Nonjola MS, Sikhwivhilu L (2009) Carbon nanotubes based nafion composite membranes for fuel cell applications. Fuel Cells 10(1): 64–71
232. Tasaki K, DeSousa R, Wang H, Gasa J, Venkatesan A, Pugazhendhi P, Loutfy RO (2006) Fullerene composite proton conducting membranes for polymer electrolyte fuel cells operating under low humidity conditions. J Membrane Sci 281(1–2):570–580
233. DeSousa R, Venkatesan A, Tasaki K, Wang H, Gasa J (2006) Fullerenes and their composites for proton conducting membranes in polymer electrolyte fuel cells. ECS Trans 1(6):175–181
234. Kumar R, Xu C, Scott K (2012) Graphite oxide/Nafion composite membranes for polymer electrolyte fuelcells. RSC Adv 2:8777–8782
235. Xu C, Cao Y, Kumar R, Wu X, Wang X, Scott K (2011) A polybenzimidazole/sulfonated graphite oxide composite membrane for high temperature polymer electrolyte membrane fuel cells. J Mater Chem 21:11359–11364
236. Lee DC, Yang HN, Park SH, Kim WJ (2014) Nafion/graphene oxide composite membranes for low humidifying polymer electrolyte membrane fuel cell. J Membrane Sci 452(15): 20–28
237. Zarrin H, Higgins D, Jun Y, Chen Z, Fowler M (2011) Functionalized graphene oxide nanocomposite membrane for low humidity and high temperature proton exchange membrane fuel cells. J Phys Chem C 115(42):20774–20781
238. Field effect transistor. From wikipedia. http://en.wikipedia.org/wiki/Field-effect_transistor
239. Field Effect Transistors (FET). http://www9.dw–world.de/rtc/infotheque/semiconamps/semiconductor_amps4.html
240. Schie SP, Fröhlich N, Held M, Gannott F, Schweiger M, Forster M, Scherf U, Zaumseil J (2015) Polymer-sorted semiconducting carbon nanotube networks for high-performance ambipolar field-effect transistors. ACS Appl Mater Interfaces 7(1):682–689
241. Chua CL, Yeoh KH, Woon KL (2014) Hybrid carbon nanotube/polymer heterointerface organic field effect transistor. Thin Solid Films 556(1):495–498
242. Derenskyi V, Gomulya W, Rios JMS, Fritsch M, Fröhlich N, Jung S, Allard S, Bisri SZ, Gordiichuk P, Herrmann A, Scherf U, Loi MA (2014) Carbon nanotube network ambipolar field-effect transistors with 10^8 on/off ratio. Adv Mater 26:5969–5975
243. Yasin M, Tauqeer T, Rahman HU, Karimov KS, San SE, Tunc AV (2015) Polymer-fullerene bulk heterojunction-based strain-sensitive flexible organic field-effect transistor. Arabian J Sci Eng 40(1):257–262

244. Marjanović N, Singh TB, Dennler G, Günes S, Neugebauer H, Sariciftci NS, Schwödiauer R, Bauer S (2006) Photoresponse of organic field-effect transistors based on conjugated polymer/fullerene blends. Org Electron 7(4):188–194
245. Gemayel ME, Haar S, Liscio F, Schlierf A, Melinte G, Milita S, Ersen O, Ciesielski A, Palermo V, Samorì P (2014) Leveraging the ambipolar transport in polymeric field-effect transistors via blending with liquid-phase exfoliated graphene. Adv Mater 26:4814–4819
246. Huang J, Hines DR, Jung BJ, Bronsgeest MS, Tunnell A, Ballarotto V, Katz HE, Fuhrer MS, Williams ED, Cumings J (2011) Polymeric semiconductor/graphene hybrid field-effect transistors. Org Electron 12:1471–1476
247. Inagaki M, Yang Y, Kang F (2012) Carbon nano fibres prepared via electrospinning. Adv Mat 24(19):2547–2566
248. Carbon black properties (2010). http://www.asahicarbon.co.jp/global_site/product/cb/characteristic.html
249. Conductive carbon black. Turning electrically conductive plastics into products. http://www.premixgroup.com/conductive-compounds/conductive-carbon-black
250. Prabhu L (2014) More effective cooling with PLANSEE's heat spreaders. http://www.plansee.com/en/Products-Heat-sinks-Heat-spreaders-495.htm

Structural/Load-Bearing Characteristics of Polymer–Carbon Composites

Madhab Bera, Pragya Gupta and Pradip K. Maji

Abstract In the last few years carbon-containing polymer composites have drawn significant attention due to their light weight, high thermal stability, excellent mechanical, and electrical property. Important characteristics of carbon-based materials (CBMs) like high specific surface area and high strength have made them as very good reinforcing filler for a wide range of polymers. The foremost aspiration of this chapter is to establish a relationship between the structures and load-bearing performance of carbon-containing polymer composites. Structural diversities of CBMs such as carbon black, carbon fiber, carbon nanofiber, diamond, nanodiamond, graphite, carbon nanotubes (CNTs), and graphene are reflected in the differential load-bearing characteristics of their polymer composites. The chapter also provides state-of-the-art information regarding the potential applications of polymer/carbon composites.

Keywords Graphene · Carbon nanotubes (CNTs) · Fullerene · Graphite
Carbon fiber · Load-bearing characteristics

1 Introduction

Carbon is one of the most important and versatile elements of the periodic table. It served the most to the mankind compared to any other elements. Sometimes, it spellbound the mankind as sparkling diamond and sometime it serves as fuel. It is

M. Bera · P. Gupta · P. K. Maji (✉)
Advanced Materials Research Laboratory (AMRL), Department of Polymer and Process Engineering, Indian Institute of Technology Roorkee, Saharanpur Campus, Saharanpur 247001 Uttar Pradesh, India
e-mail: pradip.fpt@iitr.ac.in

M. Bera
e-mail: madhabpst89@gmail.com

P. Gupta
e-mail: pragyagupta.iitr@gmail.com

M. Rahaman et al. (eds.), *Carbon-Containing Polymer Composites*,
Springer Series on Polymer and Composite Materials,
https://doi.org/10.1007/978-981-13-2688-2_13

almost impossible to imagine life in this planet without carbon, because most of the living organisms are made out of it. The most interesting feature of this material is its ability to assume variety of forms by simple carbon–carbon linking. It can exist in the form of diamond which is the strongest material in the world. It can also exist in the form of graphite which is soft in nature. Other forms of carbon like fullerene, carbon nanotubes (CNTs), graphene, carbon black, carbon fibers, and carbon nanofibers are unveiling day by day. Although, the carbon-based materials (CBMs) are structurally and property wise different from one another but one similarity exists among them and that is they are all made out of carbon only. Hybridization state of all these CBMs changes from sp^2 to sp^3 as shown in Fig. 1.

They have extraordinary physical and chemical properties like high strength, excellent thermal and electrical conductivity, good corrosion resistance and extraordinary thermal and mechanical stability. This uniqueness has drawn a great attention to the researchers and scientists working all over the world. They are working hard to make innovative products by utilizing these CBMs. Nowadays, the use of CBMs is ubiquitous. They are being used in biological field, medicines and energy storage, structural composite parts, etc.

Polymers are an indispensible part of carbon-containing polymer composites and are usually of high molecular weight materials. They have low modulus and strength. The deformability of complex structures and sliding of polymer chains over one another makes them weak. Various reinforcing fillers like clay, silica, $CaCO_3$ can be used to improve the load-bearing characteristics (ability to carry load) of polymers. One of the most important additions in the filler category is carbon-based materials. Although, research on carbon-based polymer composites is going from many decades but a major breakthrough in this field occurred after the auspicious discovery of CNTs in 1991 by Japanese physicist Iijima [1] and graphene by A. K. Geim and K. S. Novoselov from Manchester University in 2004 [2]. Thousands of research papers in SCI/Scopus database (between 2013 and 2018) with the keywords "carbon nanotubes (CNTs)" and "graphene" proves it. Surface chemistry of CBMs plays a significant role in determining the physical and chemical properties of carbon-based materials and is greatly influenced by the presence of heteroatoms like oxygen, nitrogen, sulfur, phosphorus, boron, etc. For example, the electrical conductivity of graphene and carbon nanotubes increases many fold with the incorporation of nitrogen, sulfur, or phosphorus heteroatoms in the structures. Incorporation of oxygen makes CBMs polar and become compatible with polar organic polymers. Unlike other traditional fillers like glass fiber, talc,

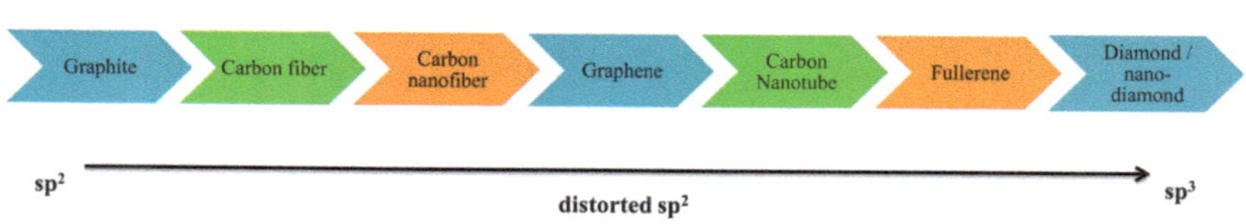

Fig. 1 Hybridization states of different carbon-based materials

calcium carbonate, etc., most of the CBMs are of light weight and has high specific surface area; hence, they can be used as efficient reinforcing material for a variety of polymers, starting from thermoplastic to thermoset, rubber to elastomer. A small amount of filler loading provides maximum reinforcement in the polymer matrix, resulting light weight composite materials. In this chapter, we have elucidated the effect of structure of CBMs and polymers on the load-bearing characteristics of the nanocomposites and also their important applications in solar cells [3–7], battery [8], supercapacitors [9], water purification membrane, automotive body parts [10, 11], packaging, infrastructure [12, 13], biomedical applications [14–22], electronics, etc. Factors related to load-bearing characteristics of a composite material are also important part of this chapter.

Overall, it gives state-of-the-art information regarding the load-bearing characteristics of carbon-containing polymer composites and their potential applications in various emerging areas.

2 Carbon-Based Materials

2.1 Classification of Carbon-Based Materials

Carbon atoms can link together in a variety of ways to produce structures having completely different properties and thus forms the allotropes. Based on the structure and properties, carbon-based materials can be classified into the following categories (Scheme 1).

Out of these eight categories carbon nanotube (CNT) and graphene are the two most emerging CBMs having maximum scientific curiosity in the recent time.

2.2 Structure of Carbon-Based Materials

Structure of different carbon allotrope is very important in the load-bearing characteristic of the composites. Concise information regarding every CBM has been discussed as follows.

2.2.1 Graphene

Graphene is one of the most important carbon-based nanomaterials which is recently at the center of scientific curiosity. The two-dimensional (2D) nanomaterial has hexagonal close pack structure where every carbon atom is sp^2 hybridized and arranged in a honeycomb crystal lattice [23]. Graphene is nothing but the single layer of multilayer material, graphite. Thickness of graphene layer is around

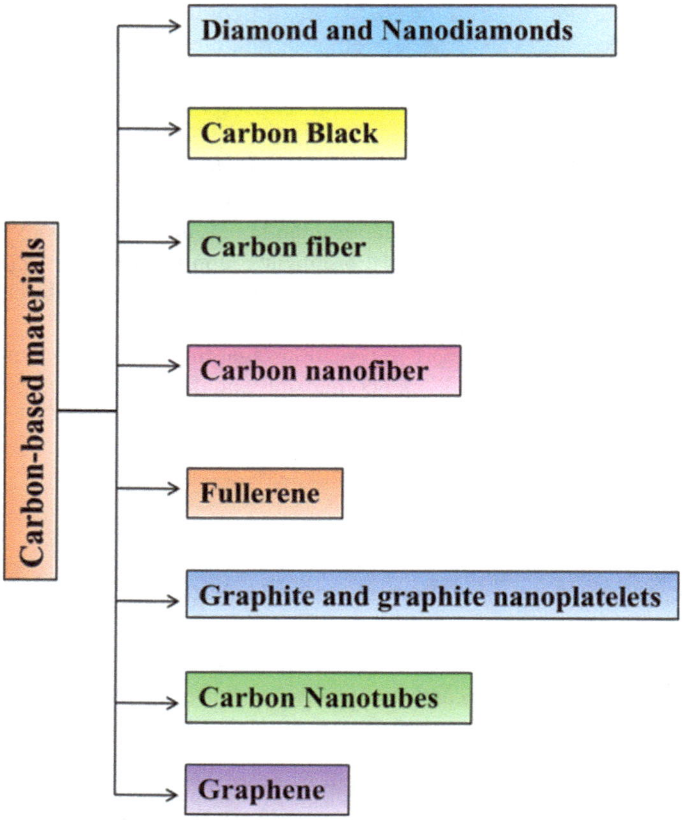

Scheme 1 Classification of carbon-based materials

0.30 nm. Due to the presence of long range π-conjugation, it has outstanding physical properties. It is one of the strongest materials in the world with Young's modulus of 1 TPa and tensile strength of 130 GPa [24]. It has exceptionally high thermal and electrical conductivity also. Thermal conductivity of graphene is as high as 5000 W m^{-1} K^{-1} [25] and electrical conductivity can be more than 200,000 cm^2 V^{-1} S^{-1} in suspended graphene [25]. Due to very high specific surface area ($\sim$2630 m^2/g) and high aspect ratio the material has an inherent tendency to agglomeration. Hence, dispersion of this material to a variety of polymers is a big challenge for the researchers. This limitation can be overcome by functionalization of the material with some polar groups so that the agglomeration tendency is reduced and the material become compatible with polymer matrix. In this context, reduced graphene oxide (RGO) comes into the field which is almost similar to pristine graphene in terms of structure and properties and has better compatibility with a range of polymers due to presence of very less number of oxygen containing functional groups.

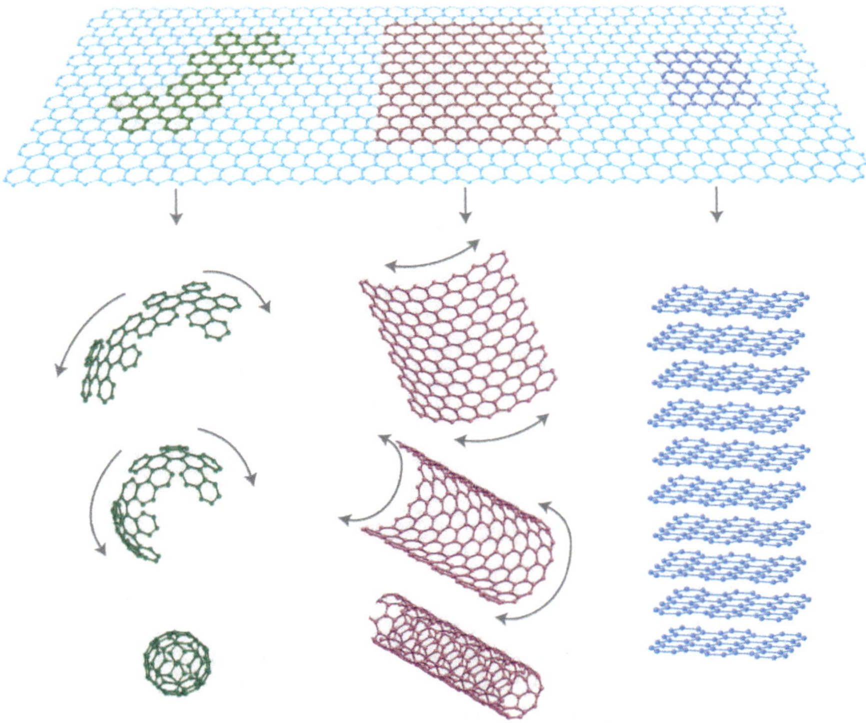

Fig. 2 Graphene (top) is the 2D building block of other carbon-based materials of all other dimensions. It can be wrapped up into 0D buckyballs (bottom left), rolled into 1D nanotubes (bottom middle) or stacked into 3D graphite (bottom right). Reproduced with permission from Ref. [23]. Copyright 2007, Springer Nature

Graphene is the building block of other carbon nanomaterials in other dimensions. It can be rolled-up to make 1D CNTs, wrapped to make 0D buckyballs and stacked together to make 3D graphite (Fig. 2).

2.2.2 Graphite and Graphite Nanoplatelets

Graphite is 3D soft and crystalline allotrope of carbon. It is formed by stacking more than 100 number of atomically thick carbon layers. In graphite structure, every carbon atom is sp^2 hybridized and covalently bonded with three other carbon atoms to forms hexagonal close pack structure. Although, the covalent bonding among the carbon atoms of individual layers are very strong but the bonding among the layers are very weak (Van der Waals forces). The layers can slide over one another and makes it a softer material compared to diamond [26]. The interlayer distance in graphite is 0.34 nm (Fig. 3). Inherent electrical conductivity is one of

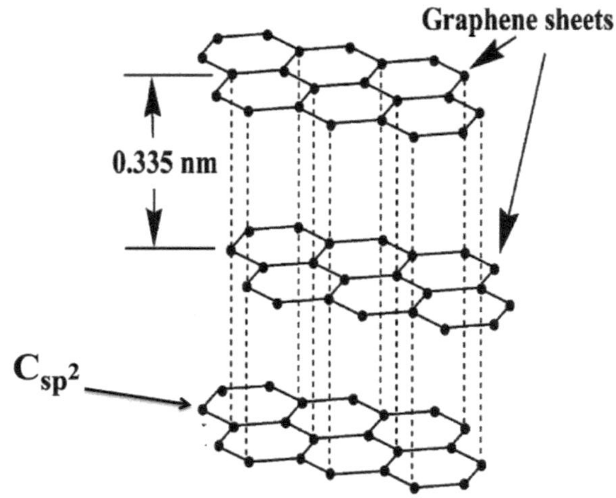

Fig. 3 Schematic crystal structure of graphite showing sp^2 hybridization state [26]

the most important properties of graphite that has been used in different speciality applications.

Graphite nanoplatelets have similar structure to that of graphite, only the inter-planer distance is greater than graphite. Its thickness is ~10 nm and size ranging from sub-micron to 15 μm or even greater. From the economic point of view, this material is very less expensive than most other carbon-based materials. It is 100 times cheaper than CNTs and has very good mechanical and electrical properties [27]. This material can be used to make polymer nanocomposites that are economically viable.

2.2.3 Carbon Nanotubes (CNTs)

Carbon nanotubes (CNTs) are one-dimensional (1D) crystalline allotropic form of carbon having diameter in nanometer scale and length of several micrometers. They have hollow cylindrical structure with wall made up with atomically thick 2D (graphene) material [28]. Based on the number of layers in the wall, CNTs can be classified into single walled carbon nanotube (SWNTs) and multiwalled carbon nanotubes (MWNTs) [29] (Fig. 4). The structure of SWNTs can be imagined to form by rolling up a single sheet of graphene into a cylindrical shape. Chirality of nanotubes is dependent on the way graphene sheet is rolled and represented by a pair of miller indices (n, m). The roll up vector is represented as $C_h = n\bar{a}_1 + m\bar{a}_2$. Here, the integer "$n$" and "$m$" denotes the number of unit vectors (a_1, a_2) along two directions of graphene crystal lattice. Depending on the values of "n" and "m" there exist three different configurations of CNTs. If $m = 0$, the nanotubes are called **zigzag** nanotubes. If, $n = m$ the nanotubes are called **armchair** nanotubes. In all other cases $(n \neq m)$ the nanotubes are called **chiral** nanotubes (Fig. 5) and its chiral angle (θ) is within 0°–30° [30]. Single-walled carbon nanotubes are an important

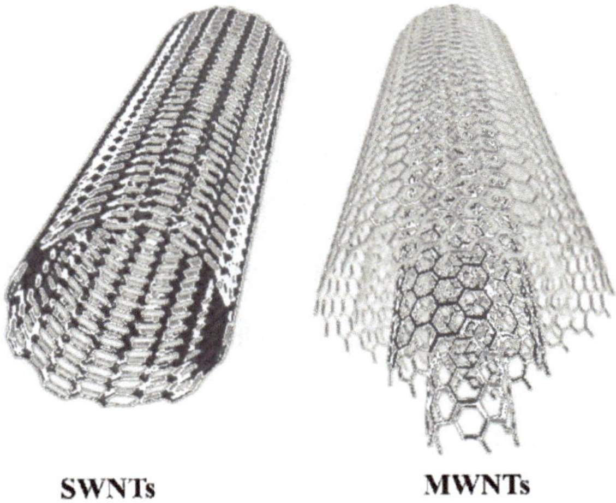

Fig. 4 Schematic diagram of single wall carbon nanotube (SWNTs) and multi wall carbon nanotubes (MWNTs) [29]

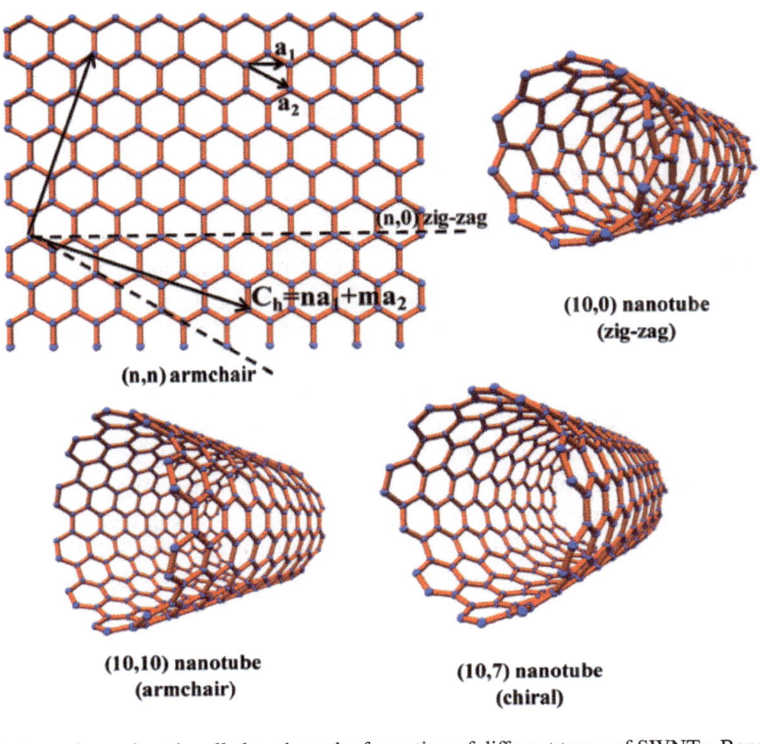

Fig. 5 A graphene sheet is rolled to show the formation of different types of SWNTs. Reproduced with permission from Ref. [30]. Copyright 2009, Royal Society of Chemistry

member of carbon filler family because their properties changes significantly with (*n*, *m*). Diameter (*d*) of the nanotubes can be calculated from the Eq. (1) by using the n and m value and chiral angle can be calculated from Eq. (2)

$$d = \frac{a}{\pi}\sqrt{(n^2 + nm + m^2)} \tag{1}$$

$$\cos\theta = \frac{(2n + m)}{2\sqrt{(n^2 + nm + m^2)}} \tag{2}$$

The diameter of SWNTs is ~1 nm where the same for MWNTs are 5–50 nm. SWNTs provide better reinforcing effect compared to MWNTs at the similar loading due to high specific surface area [30].

In case of multiwalled carbon nanotubes, the number of graphene layer in the wall is more than two and the interlayer distance between the layers is same as it found in graphite (0.34 nm). In MWNTs individual layers are held together by van der Waal forces or more specifically π–π stacking. Due to extended conjugated structure, CNTs have extraordinary thermal and electrical conductivity and very good mechanical properties.

2.2.4 Fullerene

Fullerenes are the third allotropic form of carbon after graphite and diamond and have a variety of shapes [31]. The zero dimensional (0D) nanomaterials may exist in the form of hollow sphere, ellipsoid and many other shapes. Spherical fullerene, similar to the shape of a football, containing 60 carbon atoms or buckminster fullerene (C_{60})is the most popular among the scientists and researchers. It is so popular that when we are talking about fullerene we mean to say about C_{60}. Although, other forms of fullerenes like C_{20}, C_{50}, C_{24}, C_{70}, etc., are also exist (Fig. 6). Structure of fullerene is quite similar to that of graphite, i.e., stacking of several layers of hexagonal graphene sheets. Fullerenes may have pentagonal or heptagonal rings in their structure. Every carbon atom in fullerene is sp^2 hybridized. The hexagonal graphene sheet is bent to form a closed sphere. This kind of bending introduces angular strain within the structure making fullerene highly reactive and susceptible for organic modification. Solubility of fullerenes is an important criterion for making polymer/fullerene composite by solution casting method [31]. Most of the fullerenes are insoluble or sparingly soluble in most of the solvents. However, only C_{60} is easily soluble in toluene and carbon disulfide at room temperature.

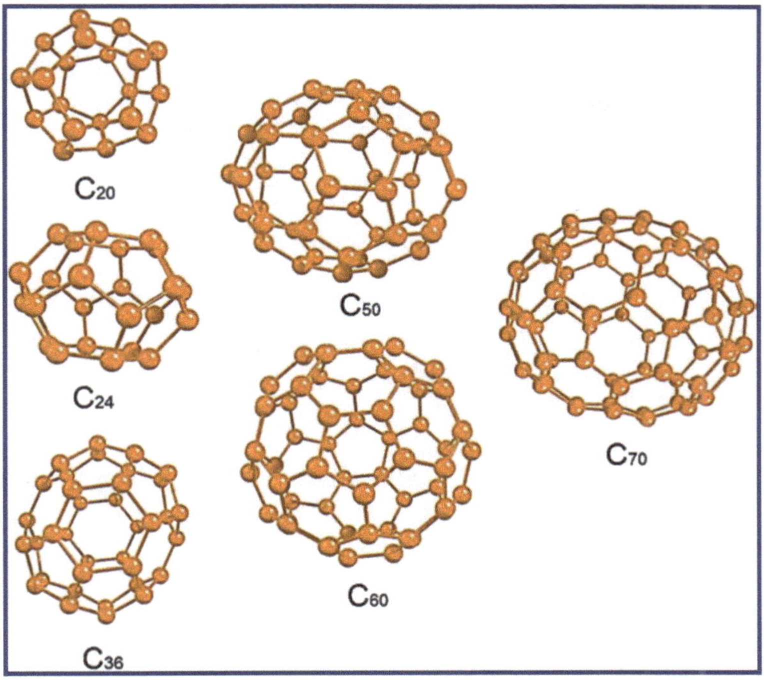

Fig. 6 Schematic structure of fullerene molecules [31]

2.2.5 Diamond

Diamond is extremely hard, transparent, and crystalline allotrope of carbon having tetrahedral structure. All the carbon atoms in diamond are sp^3 hybridized and strongly bonded (σ bonding) with neighboring carbon atoms forming a giant-like tetrahedral structure (Fig. 7). This kind of giant like structure makes it the toughest and strongest inert solid with high thermal conductivity and optical transparency. Since, the entire valence electros are used up for forming covalent bonds, there is no free electrons to conduct electricity. Diamond is electrically insulator. Silicon, Germanium and Gallium has similar kind of crystal structure.

2.2.6 Nanodiamond

Structure of nanodiamond is little bit complicated consisting of three layers; (i) the diamond core is made up with sp^3 hybridized carbon atoms; (ii) fullerene like shell of sp^2 hybridized carbon atoms partially covers the core and (iii) the outer layer consists of carbon atoms which forms functional groups by terminating with hydrogen and oxygen atoms (Fig. 8a). Various functional groups are identified on

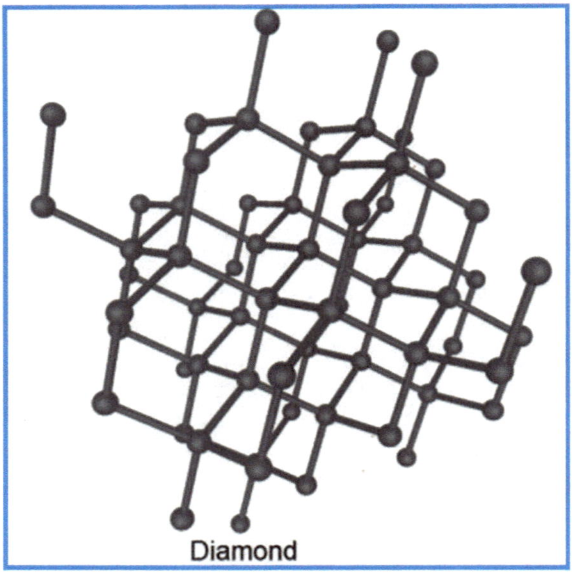

Fig. 7 Schematic crystal structure of diamond [32]

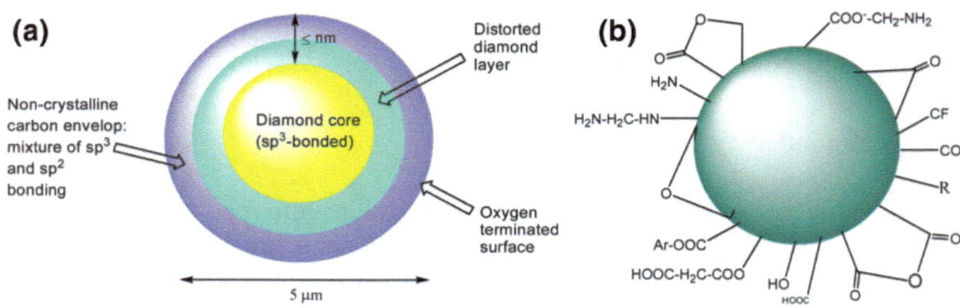

Fig. 8 **a** Model of nanodiamond structure; **b** schematic representation of various functional groups present on nanodiamond surface [33]

ND surface. They may be hydrocarbons like $-CH_3$, $-CH_2$ and –CH in different configurations or oxygen containing functional groups like ether (–C–O–C–), carboxylic acid (–COOH), carbonyl (–C=O) and lactone (–O–C=O), etc. (Fig. 8b). On the basis of particle size, nanodiamond particles are classified into three categories, (i) nanocrystalline particles (ii) ultra-nanocrystalline particles and (iii) diamondiods. Size of nanocrystalline particles are in the range of tens of micrometer while the same for ultra-nanocrystalline particles are within several nanometers and for diamondiods particles it normally lies within 1–2 nm. Average size of ND particles is 2–8 nm. Although, the size may be varied up to 20–25 nm depending on the process variables like weight of the explosive charge [32]. Nanodiamonds are non toxic. They have excellent mechanical properties, high surface area,

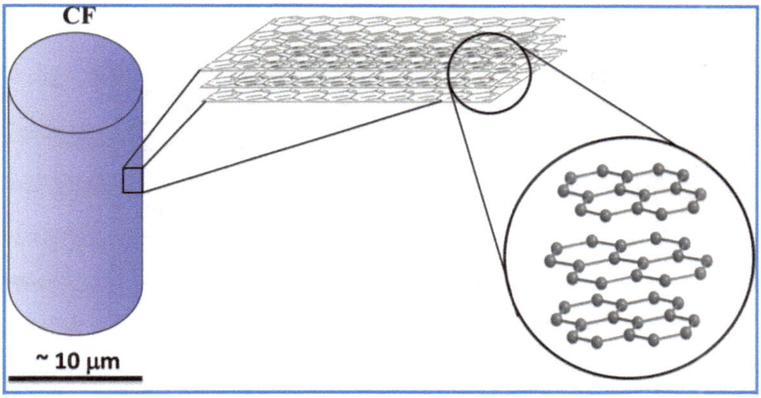

Fig. 9 Cross section of carbon fiber showing the microstructure

hardness, chemical stability and tunable surface chemistry and hence they can be used in microelectronic devices, super hard coating, antifriction coating, biomedical applications, etc.

2.2.7 Carbon Fiber (CFs)

Carbon fiber is an important member of carbon family. It is a fiber like non-crystalline material with diameter 5–10 μm and consists of mostly carbon atoms (Fig. 9). The material contains around 92% carbon where every carbon is sp^2 hybridized and strongly bonded with one another. This kind of strong inter-carbon bonding provides carbon fiber excellent mechanical property and chemical stability. It has very high tensile strength (2–7 GPa), tensile modulus (200–900 GPa), and low density (1.75–2.20 g/cm^3). Carbon fiber has also excellent thermal and electrical conductivity but has low thermal expansion coefficient. One interesting observation about carbon fiber is that although it is four times lighter than steel but it is many times stronger. This is the prime advantage of carbon fiber to develop light weight composite materials for structural applications [34, 35].

2.2.8 Carbon Nanofiber (CNFs)

Carbon nanofiber (CNFs) or vapor-grown carbon nanofiber (VGCNFs) are one-dimensional nanomaterials with cylindrical structure. In CNFs structure graphene layers are stacked together in different arrangements like cone, plate, or cup. Typical diameter for VGCNF is 50–200 nm and length can be up to 100 μm. Because of very high aspect ratio (250–2000) CNFs can act as efficient reinforcing filler for a wide range of polymers [35]. CNFs are usually synthesized by catalytic chemical vapor deposition (CCVD) or simply chemical vapor deposition

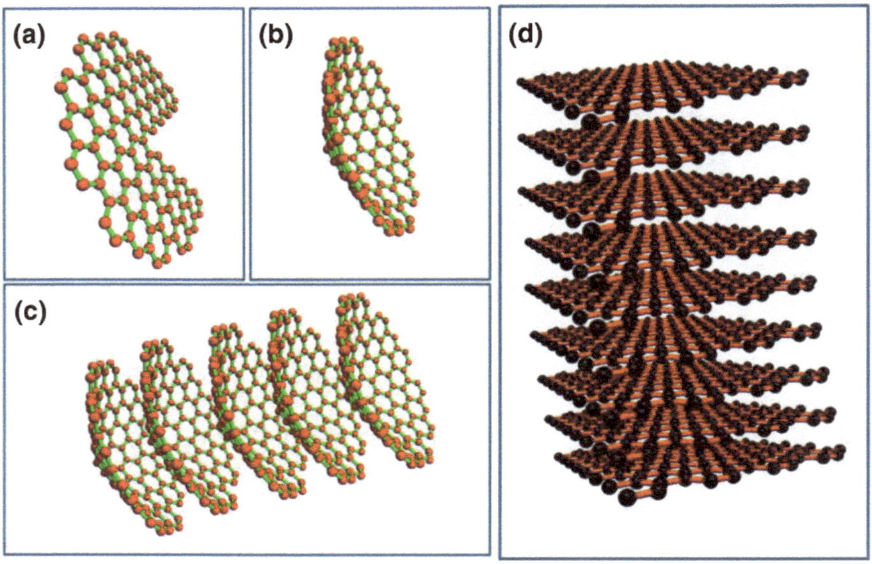

Fig. 10 Schematic representation of cup-stacked CNF structure **a–c**; and **d** platelet CNF structure. [36]

(CVD) and electro spinning method. Two types of CNFs are synthesized by CCVD method: cup-stacked CNF and the platelet CNF (Fig. 10). CNFs produced by electro-spinning method have a tendency to form web or mat like structure. This type of materials is specially recommended for batteries and/or supercapacitors applications [36].

There are certain basic differences between carbon fibers (CFs) and carbon nanofibers (CNFs): one important difference is obviously the diameter. CFs have diameter of several micrometer whereas the diameters of CNFs are 50–200 nm only. So, CNFs acts as better reinforcing filler compared to CFs due to its high surface area. The structure and preparation method of the two materials are quite different. These two materials are widely used in energy storage device and reinforced composites.

2.2.9 Carbon Black

Carbon black is a colloidal form of elemental carbon, synthesized by incomplete combustion of oils or natural gases. Carbon black particles are normally spherical in shape with particle size ranging from 10 nm to 500 nm. Particle size of carbon black is determined by Iodine (I_2) adsorption test [37] and surface area is determined by Brunauer–Emmett–Teller (BET) nitrogen adsorption test. Due to very high specific surface area and surface energy, carbon black particles have a natural

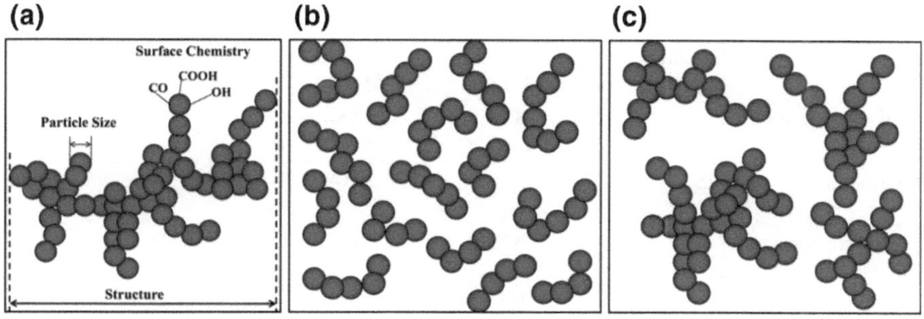

Fig. 11 **a** Schematic representation of particle size, structure and surface chemistry of carbon black, **b** low structure carbon black **c** high structure carbon black

tendency to fuse together and form aggregates. These are called the structure of carbon black (Fig. 11a). There are two types of structures for carbon black, low structure (Fig. 11b) and high structure (Fig. 11c). When large number of individual carbon black particles are fused together to form a long and branched chain (grape-like structure), it is called high structure. This kind of carbon black is perfect for reinforcement of compounds, where desired conductivity can be achieved at very lower loading. In case of low structure, small number of individual carbon black particles are fused together to form short and linear structures [38, 39]. Structure of carbon black is determined by dibutyl phthalate (DBP) absorption test and expressed as number of $cm^3$of DBP absorbed by 100 g of carbon black. Normally, increase in structure size leads to an increase in dispersibility of the particles but decreases the blackness. Other properties related to carbon black structure are shown in Table 1. The blackness is also related to particle size of the carbon black. Smaller the particle size, greater is the darkness/tinting strength of carbon black but poor dispersibility. This is due to strong cohesive forces among the carbon black particles. Based on the abrasion resistance, extrusion rate, manufacturing process, structure and surface area, ASTM D1765 [40] has classified carbon blacks into certain categories as shown in Table 2.

Table 1 Effect of structure on compound properties [39]

Property	High structure	Low structure
Dispersibility	Easier	More difficult
Adsorption	Low	High
Wetting	Slower	Faster
Gloss	Lower	Higher
Conductivity	Higher	Lower
Viscosity	Higher	Lower
Loading capacity	Lower	Higher
Tint strength	Lower	Higher

Table 2 Typical rubber grade carbon blacks with their particle size and surface area [40]

Name	Standard abbreviation	ASTM designation	Particle size (nm)	Avg. N_2 surface area (m^2/g)
Super abrasion furnace	SAF	N110	15–18	124–130
Intermediate SAF	ISAF	N220	20–25	112–115
High abrasion furnace	HAF	N330	28–36	76–80
Fast extrusion furnace	FEF	N550	39–55	39–41
General purpose furnace	GPF	N660	56–70	34–36
Semi reinforcing furnace	SRF	N770	71–96	31–32
Fine thermal	FT	N880	180–200	17–20
Medium thermal	MT	N990	250–350	7–9

In Table 2, the letter "*N*" indicates the normal rate of curing. In some cases the alphabet "*S*" is used which indicates slow curing rate. The first numeral, immediately after the alphabet indicates the particle size range which increases with the increase in number. The remaining two numerals are selected arbitrarily. In case of carbon black with standard level of structure, the second numeral is always the repeat of first numeral and the third numeral is kept as zero [39].

2.3 Comparison of Physical and Mechanical Properties of CBMs

The versatile material, carbon can exists in different allotropic forms having different crystal structures. Due to the structural diversity of CBMs, their physical and mechanical properties become distinctly different. The physical and mechanical properties of all the CBMs are summarized in Table 3 for better clarity and understanding.

3 Properties Related to Load-Bearing Characteristics of Carbon-Containing Polymer Composites

In order to understand the load-bearing characteristics of carbon-containing polymer composites, it is almost essential to have an elementary insight regarding the basic properties that are related to load-bearing characteristics. Strength, modulus of

Table 3 Important mechanical and physical properties of CBMs

Carbon-based materials	Morphology	Diameter[d]/ thickness[t] (nm)	Density (g/cm^3)	Aspect ratio (L/D)	Specific surface area (m^2/g)	Tensile strength	Modulus (TPa)	Reference No.
Nanodiamond	Particles: 3D	2–8[d]	…….	……..	……….	………	……..	[33]
Carbon black	Particles: 3D	10–500[d]	……..	…….	10–250	8–25.2 MPa	………	[39, 40]
Carbon fiber	Rod: 2D	5000–10,000[d]	1.75–2.2	440	……….	2–7 GPa	0.2–0.90	[74, 75]
Carbon nanofiber	Cylindrical: 1D	50–200	2	250–2000	………..	2.92 GPa	0.24	[75]
Fullerene	Dot: 0D	0.4–1.20[d]	……..	…….	………	……….	………	[59]
Graphite	Ellipsoid: 3D	2000–200,000	2.26	……	…….	130 GPa	1	[29]
Graphite nanoplatelets	Ellipsoid: 2D	10[t]	………	……..	………	……..	……	[27]
SWNTs	Needles: 1D	0.5–2[d]	0.80	100–10,000	40	50–500 GPa	~1	[29, 75]
MWNTs	Needles: 1D	5–50[d]	1.80	100–10000	……….	10–60 GPa	~0.3–1.0	[29, 75]
Graphene	Flakes: 2D	0.30[t]	…….	……..	2630	130 GPa	1	[2]

elasticity, stiffness, hardness, toughness, ductility, fatigue, creep, temperature resistance and corrosion resistance are the most important properties for load-bearing capacity of the composite.

3.1 Strength

Strength is the ability of a material to withstand load without being failure or plastic deformation. Strength may be of three types: **tensile strength**, **compressive strength** or compression strength and **flexural strength**. Amount of stress required for complete failure of a material is called its ultimate tensile strength (UTS) or tensile strength. It is remarkably different from the modulus of a material. Flexural strength is defined as the stress at failure in bending test. It is normally measured by three point bending test. Compressive strength or compression strength is the ability of a material or structure to withstand the load against compression. So, compression strength and tensile strength are just opposite to each other. The former withstand load which tends to reduce the size whereas the later withstands loads which tends to elongate the material. Strength of a carbon-containing composite material primarily dependent on the strength of the CBM used in composite and the strength of the polymer matrix. For polymer composites the term tensile strength has the maximum use.

3.2 Modulus of Elasticity

Modulus of elasticity or elastic modulus is defined as a measure of resistance of a material or an object against elastic deformation. It is normally expressed as stress/strain in the elastic region of stress versus strain curve (Eq. 3).

$$\text{Elastic modulus}\,(E) = \frac{\text{Stress}}{\text{Strain}} = \frac{(\text{Force}/\text{Area})}{(\text{Increased length}/\text{Initial length})} = \frac{(F/A)}{(l/L)}, \quad (3)$$

where, F is the applied force, A is the cross-sectional area of the test specimen; L is the gauge length of the specimen; l is the increased length.

3.3 Stiffness

Stiffness is the ability of a material to resists deformation when external force is applied. Stiffness is related to the structure of the material (Eq. 4).

$$\text{Stiffness } (k) = \frac{F}{l}, \tag{4}$$

where, F is the applied force and l is the change in length

So, a material having high elastic modulus is a stiff material.

3.4 *Hardness*

Hardness is an important mechanical property of a material used in coating, adhesives, hose pipes, etc. It is a measure of resistance of a material to plastic deformation. However, it can also be defined as the resistance to scratching, abrasion, or corrosion of a material. It is normally measured from the resistance of indention of an indenter. Durometer shore hardness is commonly used for polymers and its composites. It has two different scale **Shore A** and **Shore D**. Shore A is used for soft materials like rubber, elastomers, etc., and shore D scale is used for hard materials like cross-linked rubber and hard plastics. In both the scales, higher value indicates higher hardness of the material. The most well-known hard material is diamond and examples of soft materials are rubber, plastics, and their composites. Hardness is related to elastic modulus or stiffness of a material. Materials having high elastic modulus and stiffness will have high hardness.

3.5 *Toughness*

Toughness is the ability of a material to absorb energy and plastically deformed without being failure. So, in another term it is the absorbed energy per unit volume before failure. It is calculated by integrating a stress–strain curve (Eq. 5).

$$\text{Toughness} = \frac{\text{Energy}}{\text{Volume}} = \int_0^{\epsilon_f} \sigma d\epsilon \tag{5}$$

where,

ϵ is strain

ϵ_f is strain up to failure

σ is stress

Normally, the total area under the stress–strain curve is the measure of toughness. Toughness of a composite material is important to make a balance between ductility and brittleness.

3.6 Ductility

Ductility is the ability of a solid material to be deformed under tensile loading. There is a direct correlation between ductility and toughness. Because ductile materials are normally tough and non-ductile materials are brittle. When a material is stretched up to the plastic deformation limit and then the stress is withdrawn suddenly. The stress–strain curve is not ended up with zero strain although stress value is zero. That means some strain still remains within the material and that is the origin of the material's ductility. If the stress–strain curve of a material is completely linear up to the breaking point (like hard steel) then the material has very little ductility. Materials having stress–strain curve with very long region after the yield point (like rubber, plastics, and polymer composites) are ductile.

3.7 Fatigue

It is the weakening of a material caused by cyclic loading. If a material is subjected to alternating stresses then the material will fail at much lower stress than the tensile strength of the original material. The reason behind such behavior may be due to the internal structural breakdown or molecular chain scission by cyclic loading which creates microcracks within the materials and hence fails at much lower stress.

3.8 Creep

Creep is defined as the time-dependent permanent deformation or plastic deformation of a material when the applied load remains constant. Creep does not occur immediately after the application of stress, it is a time dependent phenomenon. Creep value depends on the amount of stress applied, time duration and temperature. It is one of the major limitations of polymeric composite materials used in structural application. Creep value normally increases with increase in temperature.

3.9 Temperature Resistance

It is defined as the ability of a material to withstand higher temperature without getting decomposed or changed. Normally, temperature effect on a composite material by three ways: (i) Individual components present inside the composite material may react chemically in presence of high temperature. Although this effect is not so prominent for in situ synthesized polymer composites, (ii) the difference in thermal expansion coefficient between polymer and filler generates internal stress and cracks, (iii) most materials exhibit creep behavior at high temperature. So, for higher load-bearing capacity, the composite material must have high temperature resistance.

3.10 Corrosion Resistance

Corrosion resistance of a material is its ability to withstand damage caused by temperature, chemicals or surrounding environment. Various cracks and scratches are on the composite surface because of corrosion which ultimately reduces the load-bearing capacity of the composite material. Hence, good corrosion resistance material has higher load-bearing capacity.

4 Load-Bearing Mechanisms of Carbon-Containing Polymer Composites

Mechanical properties like modulus, tensile strength, hardness, toughness, etc., of a composite material is related to its load-bearing efficiency. Normally, reinforcing fillers have greater modulus and tensile strength than polymer matrix [13]. Incorporation of reinforcing filler (CBMs) into the matrix leads to an increase in tensile strength and modulus. The question is how the load is being transferred. To get an explanation let us consider a simple fact that at a particular strain; filler carries more load than polymer. So, the load or stress is being transferred from matrix to the fillers via shear stress at filler/matrix interface and there is no stress at the end of the composite. The rate of load transfer is directly related to this interfacial shear stress [41]. Here, the structure of filler and the polymer plays an important role. Carbon-based materials have significantly different structures. Some are spherical or rod like, others are cylindrical- or flake-like structure. Their aspect ratio and mechanical properties are also different. They have different load-bearing capacity; but, the basic load-carrying mechanism is same for all cases. Thakur et al. [42] proposed a simple mechanism for the huge improvement in mechanical properties of hyper branched polyurethane (HPU)/functionalized reduced graphene oxide (f-RGO). According to their explanation toughness originates from the secondary bonds present between the polymer chains. At higher loading the secondary bonds ruptured and this allows to extend the long hidden polymer chains length, maintaining the structural integrity. The weak interlayer interaction between f-RGO sheets is also overcome by the applied load and the f-RGO sheets slide pass over one another, without damaging the original sheet like structure (Fig. 12).

During this process significant amount of energy is released. This increases the toughness of the material and the combine influence of polymer chain decoiling and

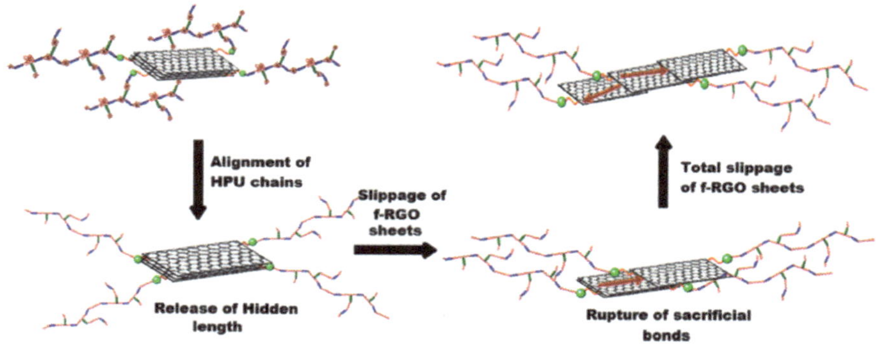

Fig. 12 Plausible Mechanism of High Elongation and Toughness of HPU/f-RGO Nanocomposites. Reproduced with permission from Ref. [42]. Copyright 2014, American Chemical Society

f-RGO sheets slipping increases the elongation at break of the composite materials. Due to the strong H-bonding interaction between f-RGO and polyurethane chains, the modulus and tensile strength of the composite materials increases.

5 Factors Influencing the Load-Bearing Characteristics of Carbon-Containing Polymer Composites

In any composite materials, load-bearing characteristics depends on a number of parameters: (i) nature of polymer and fillers (ii) dispersion of CBMs within the polymer matrix, (iii) loading of CBMs, (iv)interfacial interaction between polymer and CBMs, (v) notches and cracks on the materials and (vi) ambient conditions.

5.1 Nature of Polymer and Filler

Polymer composite consists of a continuous phase called polymer and a discontinuous phase or filler. Normally, fillers have better mechanical properties compared to that of polymers. In polymer composite most of the load is carried out by the fillers. For carbon-containing polymer composites CBMs acts as reinforcing filler and increases the load-bearing characteristics of the material. This can be proved from the basic equations of composites (Eqs. 6 and 7).

Modulus of polymer composite in longitudinal direction

$$(E_c) = (E_m V_m + E_f V_f) \quad (6)$$

$$\text{The same in transverse direction} \frac{1}{E'_c} = \left(\frac{E_m}{V_m} + \frac{E_f}{V_f}\right), \quad (7)$$

where, E_m and E_f are the modulus of polymer and filler respectively; V_m and V_f are the volume fraction of polymer and filler respectively; E_c and E'_c are the modulus of carbon-containing polymer nanocomposites in longitudinal and transverse direction respectively.

Table 3 illustrates the mechanical properties of different CBMs. From the basic equation of composite (Eqs. 6 and 7), it is clear that for different combinations of polymer and CBM, load-bearing characteristics will be different. For example, with the incorporation of 1 wt% of MWNTs to the polypropylene matrix, tensile strength increases by 20% and Young's modulus increases by 15%. In case of Nylon-6 under identical condition tensile strength increases by 120% and Young's modulus increases by 110% [43]. This is due to better mechanical properties of Nylon-6 over polypropylene.

5.2 *Dispersion of CBMs*

Homogeneous dispersion of CBMs into the polymer matrix is one of the most important criterions for designing high-performance composite materials. But, it is very difficult to disperse carbon-based nanofillers like graphene, CNTs, fullerene, carbon nanofibers, etc., within the polymer matrix. Because of very high specific surface area and high aspect ratio, the nanofillers have a natural tendency of aggregation. So, uniform dispersion of fillers is a great challenge to the material scientists and engineers. Non-uniform dispersion of filler leads to inferior load-bearing characteristics and hence poor mechanical properties of the composites [44]. Various studies suggest that three different kinds of morphological states (phase separated, intercalated, and exfoliated) exist in graphene, CNT, graphite nanoplatelets based polymer composites [45] based on the degree of filler dispersion (Fig. 13). Among these three morphological states exfoliated structure indicates the best filler dispersion, best load-bearing characteristics and hence best mechanical properties to the composites. The morphological differences can be observed through transmission electron microscopy (TEM), field emission scanning electron microscopy (FESEM) and atomic force microscopy (AFM) instruments and reflected in the mechanical and other properties.

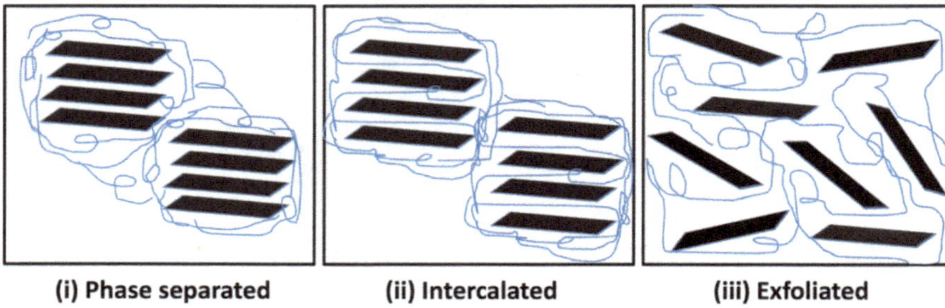

Fig. 13 Schematic representation of three different morphological states of graphene-polymer nanocomposites; **a** phase separated, **b** intercalated, **c** exfoliated [45]

5.3 Loading of CBMs

The concept of polymer composite was first come aiming to decrease the price of finished products by incorporating various kinds of low-cost fillers into the polymer matrix. But, with time it is found that filler not only reduce the price of final products but also increases the strength and load-bearing characteristics of the finished materials. It is quite obvious that increased filler loading leads to an increase in load-bearing characteristics of polymer composites but up to a certain limit; after which aggregation takes place causing a deterioration of the mechanical properties. For example, in case of PVDF/NDs composite tensile strength and Young's modulus increases with increase in NDs loading from 0 to 5% and at 5% loading, tensile strength increases by 105% and Young's modulus increases by

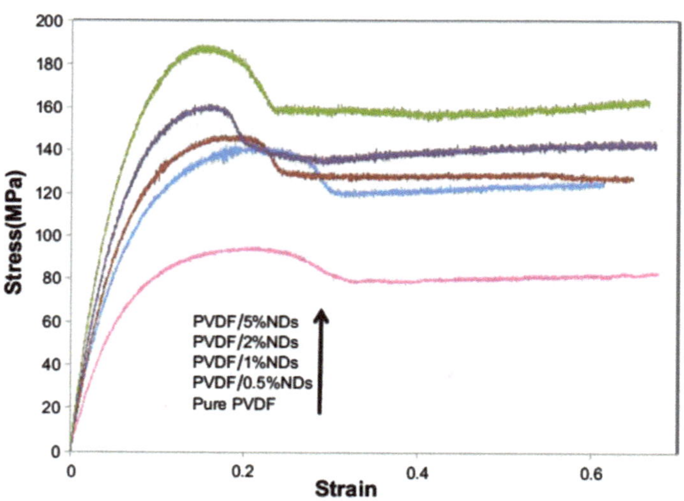

Fig. 14 stress–strain curve of PVDF/NDs composites at different loading of NDs [46]

110% [46]. The corresponding stress–strain curve is shown in Fig. 14. Similar type of improvement in mechanical properties is found with many other combinations of polymer/CBMs [47–49]. So, it is quite clear that the filler loading has an important role in the load-bearing capacity of carbon-containing polymer composites.

5.4 *Interfacial Interaction Between Polymer and CBMs*

Load-bearing characteristics of carbon-containing polymer composites depend to a great extent on the interfacial interaction between filler and polymer. For example, HIT and J-1 sizing agents are used to modify carbon fiber surface in order to improve the interfacial interaction between CFs and polymer as well as load-bearing capacity of the composite [50, 51]. Sizing agents are normally polymeric materials and are coated onto the carbon fiber surface during their manufacturing. Zhang et al. [52] studied the effect of sizing agent concentration on the mechanical properties of CFs and its composites at three different concentrations (1, 1.5, and 2 wt%) and found best results with 1.5 wt% loading of sizing agent. Ding et al. [53] also reported that interfacial interaction between polypropylene and carbon black increases with increase in the loading of silane coupling agent which was reflected from the improved mechanical properties of polypropylene/CB composites. Silane coupling agent was also used in HDPE/carbon black [54] and NR/carbon black composites [55]. Interfacial interaction between carbon black and SBR rubber can also be improved by physical activation of carbon black with CO_2.

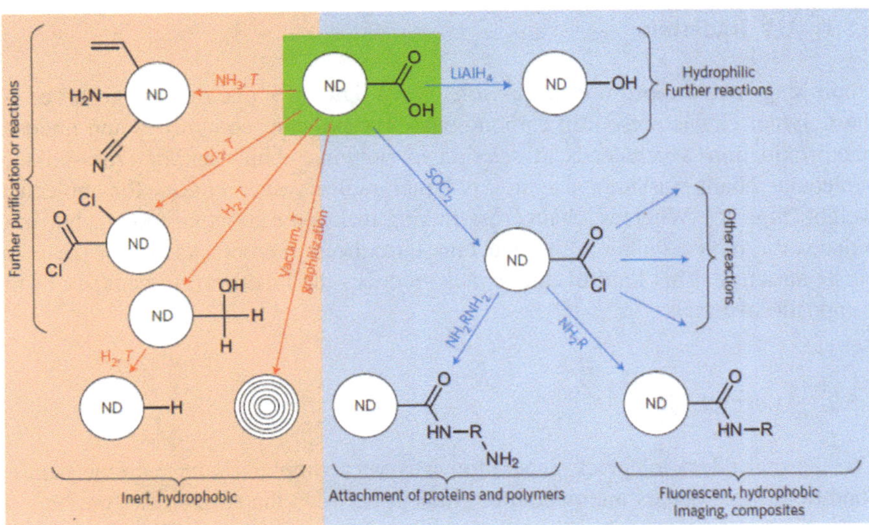

Fig. 15 Different types of surface modifications of NDs. Reproduced with permission from Ref. [57] Copyright 2011, Springer Nature

It develops surface heterogeneity of carbon black and hence increases the degree of adhesion between carbon black and SBR rubber [56]. In order to increase interfacial interactions, graphite is sometimes oxidized to synthesize graphite oxide which increases the interaction between polymer and filler. Kausar et al. [31] reported that the interfacial interaction between carbon fiber and epoxy resin increases with the addition of 2 wt% fullerene. Different type of surface modifications is also done to improve the interaction between NDs and polymers (Fig. 15). Similar to NDs, the surface of CNTs are also modified by attaching –NH_2, –COOH and –C=O functional groups to improve the interaction between CNTs and polymers [28].

5.5 *Notches and Cracks*

Microscopic cracks and notches generated in composite materials during processing, handling, transport, and service life also affect the load-bearing capacity of the composite materials. Cracks generated during service are mainly due to mechanical loading and long-term cyclic stress. Although, temperature, humidity, chemicals, UV radiation, and the presence of structural flaws also have some influence in crack generation. A detailed study about the formation, propagation, detection, and prevention of these notches and cracks are very important to monitor the load-bearing capacity. When energy required for crack growth is equal or larger than the energy required for the creation of a new surface, crack will propagate [58]. The reasons for crack generation are elaborated as follows.

5.5.1 UV Radiation

Upon long-term exposure of ultra-violet (UV) light, UV photons are absorbed by the polymer chains present in carbon-containing polymer composites and undergo photo-oxidation which leads to scission of polymer chains and/or cross-linking. Molecular chain scission generates polymer radicals and reduces the molecular weight (Fig. 16). Whereas, chain cross-linking makes the structure too much brittle, reduces the molecular mobility, and thus introduces microcracks within the composite structure. This kind of microcracks reduces the load-bearing capacity of the composite materials.

5.5.2 Thermal Effect

The thermal effect for crack generation it is very simple. Under extreme heating condition, there occurs a temperature gradient between the surface and the bulk of the composite material which introduces thermal stress to the material. This thermal

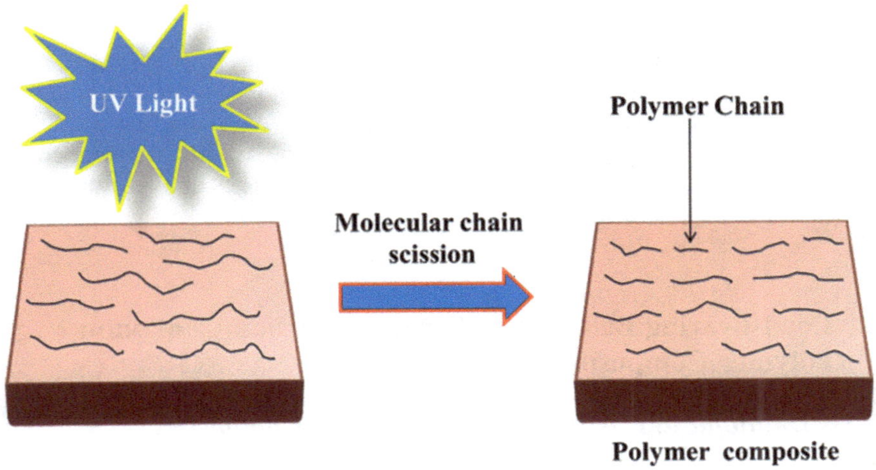

Fig. 16 Effect of UV light on molecular chain scission on long-term exposure

stress is responsible for the initiation and growth of cracks. This type of cracking observed in carbon-epoxy composite [43].

5.5.3 Hydrothermal Aging

Carbon-based polymer composites are used in many structural applications like aerospace, automotive structure, building materials, etc. which are directly exposed to hydrothermal environments. Hydrothermal environment is an environment where combined effect of heat and moisture or humidity is present. Since many polymers absorb water or moisture they are getting swelled out of it. Also, by the capillary action of CBMs, it is easily transported to the fiber–matrix interface and reduces the interfacial bonding interaction. Under the combined influence of absorbed water and high temperature, some microcracks are generated at the interface which propagates with time. Thus, the load-bearing capacity of the composite will decrease.

5.5.4 Chemical Environment

Sometimes polymer nanocomposites are used in some harsh chemical environment. For example, the material used in marine structure always remains in highly salty ambient. In some other cases harsh acid or basic ambient also facilitate the generation of microcracks. Vinyl ester, epoxy and polyester resin based composites are normally used under such chemical environments.

5.6 Surrounding Environmental Condition

Environmental conditions also play a key role in the load-bearing capacity of carbon-containing polymer composites. Parameters related to ambient conditions are temperature, humidity, chemicals, and UV radiation. All these ambient conditions are also related to the generation of cracks and notches and have been discussed in the respective section elaborately.

6 Load-Bearing Behavior of Different Carbon-Containing Polymer Composites

6.1 Diamond and Nanodiamond/Polymer Composites

Since nanodiamond particles are very small (2–8 nm) they have very high surface area and surface energy and have a natural tendency of aggregation. Various modifications of NDs surface has been done to increase the compatibility between polymer and NDs. Jee et al. [59] reported that Young's modulus and hardness of LDPE/NDs composites enhance with increase in NDs loading and length of alkyl groups (Fig. 17). According to their observation, hardness increases by 4.5 times and Young's modulus increases by 2.5 times with the incorporation of 11 wt% ethylhexyl modified NDs. The structure and mechanical properties of NDs reinforced polymer composites have been investigated by other researchers also. Zhang et al. [60] have investigated the hardness and Young's modulus of ND/polyimide composites by nanoindentation method and found 15% increase in hardness and 30% increase in Young's modulus at 5 wt% loading of NDs. Behler et al. [61] have also measured the mechanical properties of ND/polyamide composites and

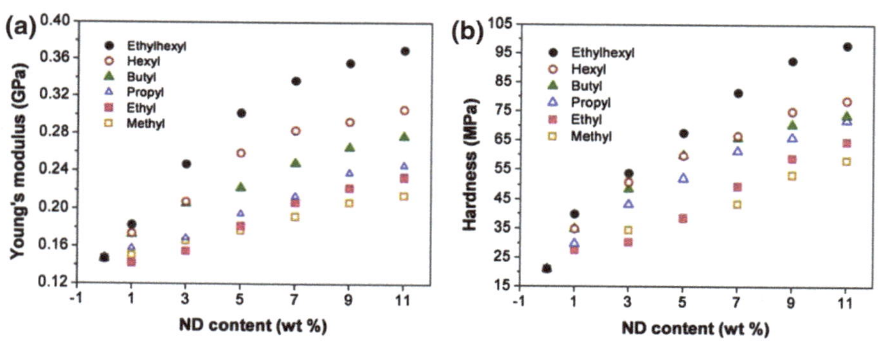

Fig. 17 Variation of **a** Young's modulus and **b** hardness of different alkyl chain functionalized NDs/LDPE composites as a function of NDs content. Reproduced with permission from Ref. [59] Copyright 2011, Elsevier

observed about two times increase in hardness and four times increase in Young's modulus with 20 wt% NDs. Neitzel et al. [62] reported 470% increase in Young's modulus and 300% increase in hardness of epoxy resin by adding 25 vol.% nanodiamond. Morimune et al. [63] synthesized poly- (vinyl alcohol) (PVA)/ND nanocomposites by solvent casting method and found 250% increase in Young's modulus and 40% increase in tensile strength of PVA with the addition of 1 wt% NDs.

6.2 *Carbon Black/Polymer Composites*

Carbon black is mostly used as a reinforcing filler to increase the mechanical, electrical and abrasion resistance property of polymers. About 90% of total carbon black manufactured worldwide is used in automotive tire industry to improve tear strength, modulus, and wear characteristics of the tires. The reinforcing nature of carbon black strongly depends on its surface area (type of carbon black) and their effect is reflected in the mechanical properties of their composite. Li et al. [64] observed that for ethylene-propylene-diene rubber(EPDM)/carbon black composite hardness, tensile strength, and modulus increases with increase in surface area of carbon black at 30 phr loading (Table 4). For SBR/carbon black (surface area = 80 m^2/g) composite tensile strength, tear strength and hardness increases by 895, 256, and 27% respectively with the addition of 20 phr carbon black after that the same decreases [65]. Natural rubber (NR)/carbon black composites are also used in the tire tread, sidewall to improve fuel efficiency (10%) and decreases rolling resistance (20%). Liang et al. [54] reported HDPE/silane coupling agent treated carbon black composites where mechanical properties like tensile strength, flexural strength, flexural modulus, and elongation at break increases up to 5 wt% loading of CB. Ao et al. [56] reported that mechanical property of N330 carbon black filled SBR rubber composite can be improved by using activated carbon black. According to the observation tensile strength and modulus increases up to 30 phr loading of carbon black. Shooli et al. [66] reported that tensile strength of 50:50 mixtures of SBR and epoxidized NR is increased by 158% with the addition of 35 phr carbon

Table 4 Mechanical properties of EPDM/carbon black composites. Reproduced with permission from Ref. [64] Copyright 2008, BME-PT and GTE

Properties	N770	N550	N330	N472
BET surface area (m^2/g)	15.80	37.80	81.10	1039.50
Hardness (Shore A)	52	60	59	79
300% modulus (MPa)	2.91	5.17	5.23	15.12
Tensile strength (MPa)	12.12	15.04	18.38	17.76
Elongation at break (%)	640	526	571	345

black (N330) and the improvement become more significant when mixed filler is used.

At the loading of 10 phr nanoclay and 20 phr carbon black, tensile strength, elastic modulus, loss factor, and abrasion resistance of the composites increases more compared to 35 phr carbon black filled composite. Also, the thermal stability, thermal conductivity and tensile strength of silicon rubber (SR) increase with increase carbon black (N990) loading. Thus the overall load-bearing capacity of SR/carbon black composite increases [67].

6.3 Graphite and Graphite Nanoplatelets/Polymer Composites

To improve the mechanical property of graphite/polymer composite, graphite is first treated with alkali metals (sodium, potassium, etc.) or halogens and halide ions or with concentrated nitric acid, sulfuric acid, perchloric acid to synthesize graphite intercalated compound (GIC) [68]. Now, this GIC interact better with polymer chains and forms composites having improved load-bearing property. Baptista et al. [69] reported an increase in mechanical properties of epoxy resin with the incorporation of graphite flake (up to 10 wt% loading). Due to greater interplanar distance, graphite nanoplatelets have very high aspect ratio and can be dispersed within the polymer matrix 100 times better than pristine graphite. Liu et al. [27] reported that with the addition of only 7 vol.% exfoliated graphite nanoplatelets (15 μm) to polyamide 6, flexural modulus increases by 140% and flexural strength increases by 22%. Kim et al. [70] reported that functionalized graphite sheet (FGS) exhibits better improvement in mechanical properties compared to graphite at the similar loading. Poly (ethylene-2, 6-naphthalate) (PEN)/graphite composite shows 17.50% increase in tensile modulus with the addition of 3 wt% graphite whereas for PEN/FGS composite the increase in tensile modulus is 31.50%. It can be concluded that FGS/PEN has better load-bearing characteristics than graphite/PEN.

6.4 Graphene/Polymer Composites

The exceptional mechanical property of graphene can be utilized to make high load-bearing polymer composites. Wang et al. [71] reported improved load-bearing capacity of functionalized graphene (FGs)/polystyrene (PS) composites where tensile strength increases by 103.7% and tensile modulus increases by 407% with the addition of 0.5 wt% FG (Fig. 18). The resultant mechanical properties are quite higher than that obtained from PS/MWNT or PS/thermally reduced graphene oxide

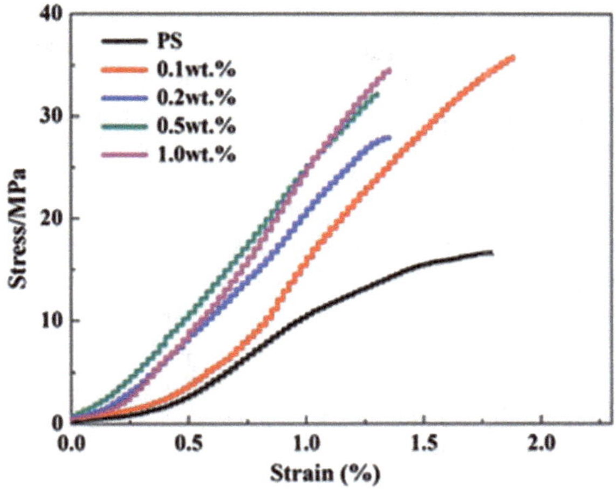

Fig. 18 Stress–strain curves for PS/FGs composites. Reproduced with permission from Ref. [71] Copyright 2016, Royal Society of Chemistry

(TRGO). This result is an indication of effective load transfer between FGs and PS. Fang et al. [72] has shown 70% increase in tensile strength and 57% increase in Young's modulus of 0.90 wt% graphene nanosheets filled polystyrene composite. Achaby et al. [73] reported that with the addition of 3 wt% graphene nanosheets (GNs) modulus and tensile strength of PP/GNs composite increases by 100 and 81% respectively. Also, the thermal stability of the composite increases by 11%. Chen et al. [74] investigated significant improvement in mechanical properties of graphene/polyurethane nanocomposite. According to their observation, tensile strength, modulus, and toughness increases by 51, 105, 48%, respectively. Bera et al. [45] investigated 245% increase in tensile strength, 180% increase in tensile modulus, 5% increase in hardness and 200% increase in toughness of thermoplastic polyurethane (TPU) by incorporating just 0.10 wt% of RGO. There are many other examples where load-bearing capacity of polymers increases with graphene loading.

6.5 *Carbon Nanotube/Polymer Composites*

Carbon nanotubes are one of the most important CBMs use to increase the load-bearing capacity of polymer composites. Sometimes it is used in unmodified form and sometimes need to modify the surface to increase the interaction with polymers. Surface modification can be done either by functionalization of the surface with hydroxyl, carbonyl, and carboxyl groups or by treatment with strong acids, plasma treatment, etc. [28]. Li et al. [48] reported 228% increase in tensile strength and 1040% increase in Young's modulus of poly[styrene-b-(ethylene-co-butylene)-b-styrene] triblock copolymer (SEBS), with the incorporation of 15 wt% pristine MWNTs. Pitchan et al. [75] investigated that mechanical property of polyetherimide (PEI)/MWNTs at 2 wt% loading increases 15% more in cases of acid and plasma treated MWNTs compared to untreated MWNTs. So, surface treatment provides better load-bearing property. Fatigue is one of the major issues for structural failure of composite materials. Zhang et al. [76] reported the suppression of fatigue crack growth rate of epoxy resin by incorporating CNTs. They also observe that the suppression of crack growth rate is dependent on the CNT loading and type of CNT used. For epoxy/MWNTs composite, the reduction of crack growth rate is ~1000% at 0.5% loading of MWNTs. Similar amount of crack growth reduction is observed at just 0.1–0.25% loading of SWNTs. So, SWNTs are more effective than MWNTs. With the help of simulation study, Gavrilov et al. [77] reported that mechanical property of elastomer/CNTs nanocomposites is dependent on the size of CNT used. Another important problem usually encounters with carbon-based polymer composites in the long-term structural application is creep. Zhang et al. [12] reported that creep behavior of thermosetting epoxy resin can be slowed down by incorporating 0.1–0.25 wt% of SWNTs. The epoxy/SWNTs composite can retain very good load-bearing capacity even at higher temperature.

6.6 Carbon Fiber/Polymer Composites

Carbon fiber/polymer composites are mainly used in structural applications for many decades. Since CFs is continuous fiber they can provide very high load-bearing property. One common problem encountered with carbon fiber is the lack of adhesion with polymer matrix. To improve the adhesion between CFs and polymer various sizing materials are coated on CFs surface. Jiang et al. [51] reported improved load-bearing capacity of epoxy resin/carbon fiber composite treated with GO and sizing agent. According to their observation, interfacial shear strength (IFSS) of carbon fiber/epoxy is increased by 37.2% for 2 wt% sizing agent and 1 wt% GO-treated CFs compared to CFs without GO. At this concentration of GO, the interfacial interaction between polymer/modified CFs are maximum and hence, mechanical property is better. Dhakate et al. [49] investigated excellent mechanical properties of carbon fiber semi-aligned electrospun carbon nanofiber hybrid polymer composites. With the addition of 1.1 wt% CNF bending strength increased by 175% and the modulus increased by 200%. Fu et al. [78] reported that at any loading CFs provide better tensile strength and tensile modulus to PP matrix compared to glass fiber (GF). This is probably due to superior mechanical properties of CFs compared to GFs. Creep behavior is very important for a material used in structural applications. Goertzen et al. [79] reported that for CF/epoxy composite, no creep failure occurs within 1600 h when the applied load is 77% of the ultimate tensile strength of the composite. But, with increase in temperature the failure time and load decreases, i.e., the material fails at lower load and within a shorter time. For example, stress requires inducing failure of CF/epoxy composite within 50 years, decreases from 84% of UTS at 30 °C to 42% at 50 °C. Load-bearing

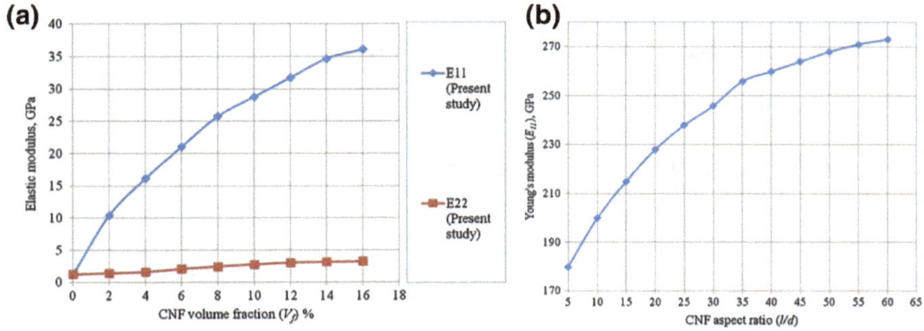

Fig. 19 **a** Variation of Young's modulus (E_{11} and E_{22}) with percentage of CNF in PP for fixed aspect ratio (*l*/*d*) = 10; **b** variation of Young's modulus (E_{11}) with aspect ratio (*l*/*d*) of CNF. Reproduced with permission from Ref. [81]. Copyright 2016, Springer Nature

capacity decreases with increase in temperature. Kawai et al. [80] also investigated the effects of stress ratio and temperature on fatigue life of a carbon fiber-reinforced polyamide-6 composite. According to their observation tensile and compressive strength decreases with increase in temperature from room temperature to 70 °C temperature. Also, the static stress level for a constant value of life under fatigue loading will deteriorate with increase in temperature.

6.7 *Carbon Nanofiber/Polymer Composites*

Carbon nanofiber (CNF) is one of the most important members of carbon-based filler. It has drawn tremendous scientific attention due to its diverse application opportunities. It can be used in electrical devices, battery, supercapacitors, sensors, structural applications, etc. The one-dimensional nanomaterial has very high aspect ratio (~250–2000) [36]. Hence, it can be used in high load-bearing polymer composites. With the help of molecular dynamic simulation, Sharma et al. [81] reported the effect of CNF volume fraction and aspect ratio (l/d) on the mechanical properties of CNF/PP nanocomposites. According to the simulation result, longitudinal modulus (E_{11}) increases with increase in CNF vol.% and at 2 vol.% loading E_{11} of PP increases by 748%. Young's modulus (E_{11}) also increases with increase in aspect ratio of CNF. With increasing the aspect ratio from 5 to 60, E_{11} increases by 48% (Fig. 19).

Although, there is very little effect of CNF volume fraction and aspect ratio on the transverse modulus (E_{22}). In all cases CNF/polymer nanocomposites have lower mechanical properties compared to CNT/polymer nanocomposites. This may be due to lower intrinsic mechanical properties of CNF over CNT [34]. Kumar et al. [82] reported that compressive strength and modulus of PP/nano carbon fiber increases by 100 and 50% respectively at 5% loading of nanocarbon fiber. Similar type of improvement in mechanical property was observed by Zeng et al. [83] for

PMMA/CNF composite. At 5% loading of CNF, axial tensile modulus increases by 50% and compressive strength increases by 170%.

6.8 Fullerene/Polymer Composites

Aggregation of fullerenes is one of the major drawbacks of polymer/fullerene composite. Various physical/chemical modifications of fullerene have been done to improve the thermal, mechanical, electrical, optical, and photovoltaic property of polymer/fullerene composite [31]. Doping of polymers with fullerene (mainly C_{60}) is another emerging area of intense research which improves the load-bearing characteristics of polymer/fullerene composites by efficient load transfer between polymer and fullerene. Rafieeet al. [84] reported improved load-bearing characteristics of epoxy/fullerene nanocomposite. According to their observation, toughness of fullerene/epoxy composite increase by 89%, ductility increases by 35%, Young's modulus increases by 18.5%, tensile strength increases by 20% with the addition of 1 wt% fullerene. The typical stress–strain curve of epoxy/fullerene composites at different loading of filler is shown in Fig. 20a. Crack propagation response of fullerene/epoxy composite at different stress intensity is shown in Fig. 20b. A significant lowering of crack growth rate was observed in fullerene/epoxy composites. Fracture toughness increases by ~50% with the addition of 1 wt% of fullerene to the epoxy resin which is much higher than that observed in 1 wt % amine functionalized MWNTs filled epoxy resin composite (26% only). So, the fullerene/epoxy composites improve the load-bearing capacity of epoxy resin and become useful in many structural applications like infrastructure, aerospace structure, automotive body parts, etc.

Kim et al. [85] reported the tensile properties and fracture mechanism of a bulk heterojunction (BHJ) films based on PTB7:PC_{71}BM. According to their observation, mixing ratio and amount of additive (1, 8-diiodoctane) play an important role in the tensile property. At higher concentration of fullerene, the composite becomes

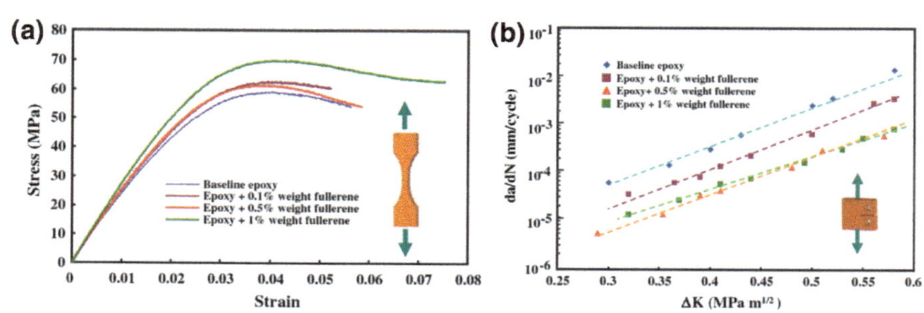

Fig. 20 **a** Stress–strain curve; **b** crack growth rate (da/dN) versus stress intensity factor amplitude (ΔK) graph for fullerene/epoxy resin composite at different fullerene loading. Reproduced with permission from Ref. [84]. Copyright 2010, Springer Nature

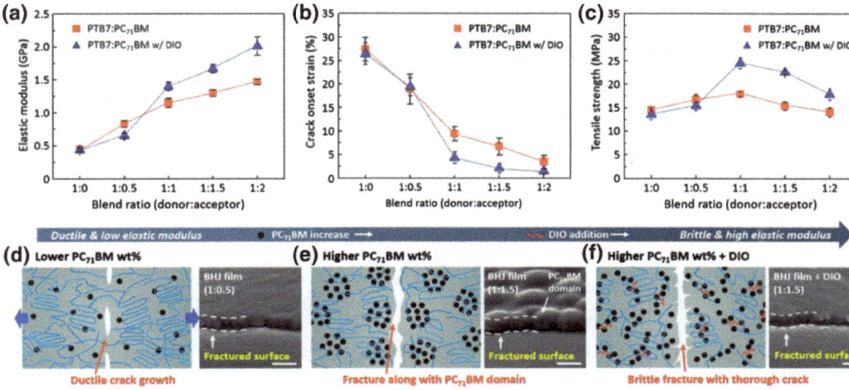

Fig. 21 Tensile properties and fracture mechanism of PTB7:PC_{71}BM blend films as a function of the PC71BM content and DIO additive. **a** Elastic modulus, **b** crack onset strain, and **c** tensile strength of PTB7: PC_{71}BM blend films as a function of blend ratio and DIO additive. Fracture mechanism and SEM images of fractured surface of PTB7:PC_{71}BM bulk heterojunction films with respect to PC71BM content and DIO additive: **d** 1:0.5 blend films, **e** 1:1.5 blend films without DIO additive, and **f** 1:1.5 blend films with DIO additive (scale bar 200 nm). Reproduced with permission from Ref. [84]. Copyright 2010, Springer Nature

stiffer and more brittle and fracture occurs along the aggregated fullerene domains. 1:1 mixture of fullerene and polymer provides the best mechanical properties. The presence of 1, 8-diiodoctane additive increases the stiffness and strength and reduces the ductility further. Tensile properties and fracture mechanism are shown in Fig. 21. Mechanical properties of polymer/fullerene composites as a function of molecular structure are elaborated by Savagatrupet al. [86]. They explored the presence of various pendent groups on the mechanical properties of their composites.

7 Applications

Metal, alloys, wood, and concrete are being replaced by polymeric composite materials in several advanced applications like aerospace, automotive, marine, infrastructure, etc., because of their light weight and superior mechanical properties. Nowadays, the applications of carbon-containing polymer composites are ubiquitous. The diversity in structure and properties of CBMs are reflected in their polymer composites which endow their applications in infrastructure, aerospace structure, automotive body parts, biomedical applications, sports equipment, energy storage, marine structures and others. However, every CBM cannot be used for all these applications. Each CBM has some special structural features which are utilized in some specific applications only. Microcracking and damages are common problems for these composite materials. Hence, the concept of self-healing become promising for the applications like aerospace, marine, automotive, pressure vessel, and infrastructure where the tendency of cracking and damage are the maximum. Self-healing composite materials are capable to heal or repair the cracks and damages by it and thus extend their shelf life. It also reduces the repairing and maintenance cost of the materials.

7.1 Infrastructure

Infrastructure refers to the fundamental facilities and services offered by a country, city or area for its economic growth as well as to improve the standard of living of people. Infrastructure includes bridges, roads, tunnels, water supply, electrical grids, etc. Reinforced cement concrete (RCC) has the maximum share in this area. Recently, carbon-containing polymer composites stepped into the field because of their high load-bearing capacity and corrosion resistance. Lighter weights of the polymer composites facilitate their transportation quite easily to the working sites. Carbon fiber/epoxy and carbon fiber/vinyl ester resin composites are commonly used in bridge and tunnels. Although, carbon-containing polymer composite has not any problem of rusting like steel cords, used in concrete, economically it is costlier but the bridge made with carbon/polymer composite vibrate too much during the traffic movement which affects in the stability of bridge [13]. Another important problem usually encounters with carbon-based polymer composites in long-term structural application is creep. Zhang et al. [12] reported that creep behavior of thermosetting epoxy resin can be slow down by incorporating 0.1–0.25 wt% of SWNTs. The epoxy/SWNTs composite can retain very good load-bearing capacity even at higher temperature.

7.2 Aerospace Structures

The primary requirement of aerospace structural materials is that they should have low density at the same time should be very strong and stiff. Carbon-based materials have relatively low density and outstanding all-round properties. Hence, these materials can be exploited to make lightweight and high-performance polymer composites which find their applications in aerospace structures. The most important criterion for a polymer to be used in any part of aeroplane structure is that it will not produce any toxic fumes during fire. Carbon-phenolic composites are used in floor and many other inner structures like inner walls, toilet, cubicles, etc., are also made of polymer composite materials. The structure of seats is made up of metals and alloys. But, some aeroplane companies like Boeing, Airbus, etc., have already started making the whole aeroplane structure by carbon/epoxy composite material [87, 88]. For example, in Boeing 787 (Fig. 22), polymer composite

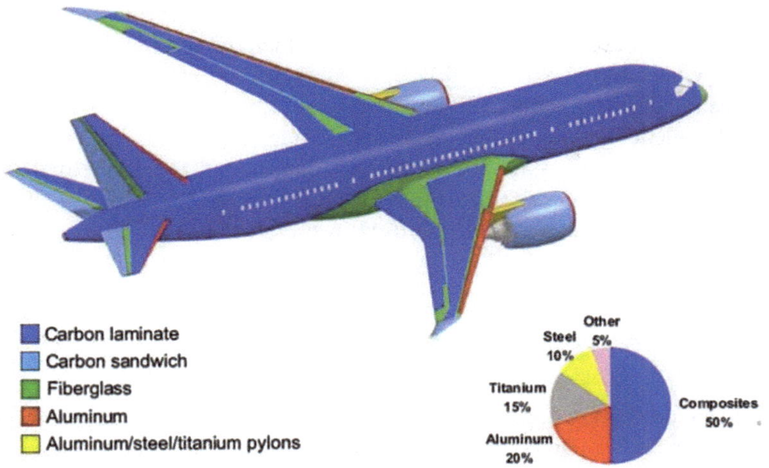

Fig. 22 Application of carbon fiber/polymer composite in Boeing 787 [87]

accounts 50% share of the total materials used which is relatively higher than Boeing 777 where only 12% polymer composite was used. Because of increased percentage of polymer composites, the aeroplane becomes 20% lighter than aluminum designs and also reduces the chances of fatigue failure [88]. Helicopters are mostly made with carbon-containing polymer composite materials. It not only reduces the weight of the whole structure but also increases the performance of the helicopters. Thus, carbon-containing polymer composites slowly replaced the conventional aerospace structural materials like aluminum and alloys. In space applications also composite materials are becoming very important. Rockets structures are made by filament-wound carbon fiber-epoxy composite material. Monetta et al. [89] reported graphene/epoxy anticorrosive composite coating for aircraft structure where only 1 wt% graphene is used.

7.3 Automotive Body Parts

Polymer composites have been a part of automotive industry since 1953 when it was first used in the corvette, small warship. The use of carbon-based polymer composites replaces the conventional materials like steel, aluminum, etc. Polymer composites are relatively lightweight materials with improved strength and corrosion resistance property. But, the cost of carbon fiber based polymer composite is too high (around 10 times higher than glass fibers) that is why they are used only in special cases. Now the challenge is to develop high-strength polymer composite which will be economic as well mechanically strong [11]. Wang et al. [10] reported shape memory induced self-healing property of crosslinked polyethylene/CB composites. According to their observation, this material can heal the scratches instantly which is an important requirement of a material used in automotive body parts. Bhowmick and coworkers [90] reported significant improvement in thermal, mechanical, and gas impermeability of bromo butyl rubber (BIIR)with the addition of para-phenylenediamine modified graphene oxide (GO-PPD). According to their observation, tensile strength (200%), storage modulus (189%), thermal stability (17 °C), and gas impermeability (44%) increases with the addition of 4 wt% GO-PPD. Hence, this material can be used in tire inner linear and tubes.

7.4 Biomedical Applications

Bone and tissue engineering, drug delivery, and sensor are the three major areas of biomedical emerging. Biocompatibility is the primary criteria of a material to be used in biomedical applications. Among the CBMs, ND, graphene and related materials, CNFs are mostly used in biomedical applications. There is a conflict regarding the biocompatibility of nanodiamond. According to some researchers, detonation nanodiamond has intrinsic biocompatibility while some are saying NDs induce toxicity under certain condition. Schrand et al. [14] reported that NDs has better biocompatibility than carbon black, SWNTs and MWNTs. Sun et al. [15] reported that chitosan (CS)/NDs composites scaffold can be used in bone tissue engineering applications. They measured the important mechanical properties of CS/NDs composites by nanoindentation method and found two important observations: (i) With the addition of 5% NDs, Young's modulus is increased by 340% and the hardness is increased by 120%, (ii) the improvement in mechanical property is less effective for NDs compared to functionalized NDs (F-NDs). Ochiai et al. [16] reported that boron-doped diamond powder (BDDP)-nafion composite made by flexible electrolysis can be used in dental treatment. Liu et al. [17] reported self-healing composite hydrogels based on polyacrylamide and graphene oxide (GO) used in biomedical applications. It can heal the cut surface very quickly and has very high degree of recovery (88%). Shuai et al. [19] reported that GO-reinforced poly (vinyl alcohol) composite scaffold can be used in bone tissue engineering application. 2.5 wt% loading of GO increases the compressive strength (60%), Young's modulus (152%) and tensile strength (69%) of poly (vinyl alcohol)/GO porous composite scaffold (Fig. 23) and make it a potential candidate for bone tissue engineering application. Graphene/polymer nanocomposites are also used as sensor to detect various gases, temperature [20], pressure, pH [21], etc. Li et al. [18] reported that CNF/poly(acrylate) composite can be used as gas sensor. Graphene-based polymer composites are also used in drug delivery and cancer therapy [22].

Fig. 23 **a**, **b** Photographs of the GO/PVA nanocomposite scaffold with 2.5 wt% GO loading fabricated via SLS, **c** the struts of the scaffold, **d** the pores of the strut. Reproduced with permission from Ref. [19]. Copyright 2015, Royal Society of Chemistry

7.5 Sports Equipment

Sporting equipment is an indispensable part of any sport. At the highest professional level of sport, people are ready to pay any amount of money for getting highly advanced equipment. Development of carbon-containing composite materials brings new hope to the athletes. Recently, CBMs/polymer composites are being used to make lightweight, durable and high-performance sporting equipments which reduces the chances of injury and at the same time enhances the performance of the athletes making the sports more enjoyable. Carbon-based materials like graphene, carbon nanotubes, fullerene, nanodiamond, carbon nanofibers, carbon black, etc., are used in sporting equipments like badminton and tennis rackets because of their high strength, stiffness, durability, lightweight, abrasion resistance, etc. Different CBMs have different specialties which are exploited in different sporting equipments as per the requirements. Table 5 shows the specific benefits of using a particular CBM for specific sporting equipments [91, 92].

Table 5 Advantages of using of CBMs for making various sporting equipments [92]

CBMs	Sports	Benefits
Carbon nanotubes	Tennis/ badminton	Increases strength, stiffness, durability, resiliency, impact, repulsion power and vibration control of rackets
	Golf	Reduce weight, lower torque/spin of the clubs
	Archery	Better vibration control in arrows
Graphene	Tennis/ badminton	Increases strength, stiffness and performance of rackets and decreases the weight. Overall, better control of rackets
	Ice-skidding	Increases strength and durability, decreases weight of skis
	Bicycle race	Increases the strength, durability, decreases the weight of bicycle wheels and safety helmets
	Car racing	Increases the strength, decreases the weight, decreases abrasion and improves the high-speed performance of the car
Fullerene	Tennis/ badminton	Reduces the weight and twisting of racket frames
	Golf	Facilitate flexible club whipping
	Bowling	Reduce chipping and cracking of balls
Carbon nanofibers	Cycling	Reduces the weight and increases the strength and stiffness of the cycle
Carbon black	Road racing	Decreases rolling resistance, increases grip and mileage of tires

7.6 Energy Storage

Ever-increasing demand for energy and environmental concern has led the scientists to think about the renewable energy technology. It requires a very large number of energy storage devices. Among the currently available energy storage devices, battery, and supercapacitors are most important. They reversibly convert electrical energy into chemical energy. Although, green energy technology like solar energy is also a rapidly growing area where solar cells are being used to convert solar energy into electrical energy. With the advancement of civilization the requirement for smart and portable storage devices like Li-Ion battery [8], high capacity supercapacitor becomes very important. Currently, scientists are working on Li-Ion batteries and supercapacitors to reduce their size and to enhance their energy storage capacity by utilizing graphene. Graphene-based microcapacitor is also a new area of research which may be utilized in smartphones and other small electronic devices. Calixtoet al. [69] reported that graphite powder can be used to make polymer composite with polyurethane, SR and epoxy resin and the graphite/ polymer composites can be used as electrode material. According to their observation best result was obtained at 60 wt% loading of graphite. Fullerene/polymer composites are mostly used in renewable solar energy application. Erb et al. [3] reported that poly(3-hexylthiophene-2,5-diyl)/[6,6]-phenyl C_{61} butyric acid methyl ester (P3HP/PCBM) films could be used as active layer of plastic solar cells. Mohajeri et al. [4] proposed four fullerene derivatives (Fig. 24) which can act as

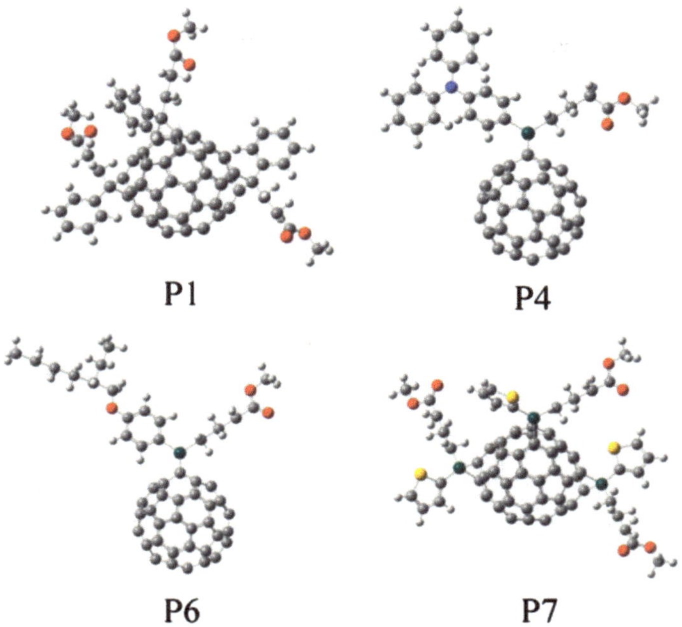

Fig. 24 Structures of fullerene derivatives. C, H, S, N, Si and O atoms are shown in gray, white, yellow, blue, green and red, respectively. Reproduced with permission from Ref. [4]. Copyright 2015, Royal Society of Chemistry

better acceptor than $PC_{60}BM$ in high-performance polymeric solar cells with conjugated polymers, P3HT. Graphene-based polymer nanocomposites are also increasingly used in battery [8] and supercapacitors [9]. Wang et al. [5] reported that polyaniline/graphene hybrids can be used as a counter electrode for dye-sensitized solar cells. The conversion efficiency of the solar cell with this counter electrode reached up to 6.09%. Lin et al. [8] investigated that composite made with 5-sulfoisophthalic acid (SPA) coated silicon nanoparticles doped poly aniline (core/shell SiNPs@PANi/SPA) can be used as anode material of Li-ion battery. It slows down the lithiation/delithiation process and increases the capacity up to 925 mAh g^{-1} and efficiency of 99.60%.

Savagatrup et al. [6] reviewed about the stability of polymer/fullerene bulk heterojunction solar cells and showed that the stability and performance of the solar cell are dependent on the size and purity of fullerene used. Also, the interaction between the fullerene molecules with the polymer chains plays an important role. Beal et al. [7] reported the morphology of P3HT: PCBM on the performance of BHJ solar cells.

7.7 Marine Structures

Polymer composite materials have been used in marine structure for many decades. They are being used in boats, ships, submersibles, offshore structure and other marine structures like hull, shells, shaft, ducts, etc. The first polymer composite used in marine structure was glass fiber (GF) reinforced unsaturated polyester resin (UPR) composite in early 1950s and continues to be an important material for marine structure. But, there are some disadvantages of using UPR like emission of toxic materials during the curing process can have some serious health and safety issues. Substantial improvement in this field has occurred to tailor the properties and to improve the performance of the composite materials as per the demand of modern technologies. Recently, there is an increasing demand for high-speed marine vehicles (e.g., high-speed boat) where strength and weight of the vehicles are the two most important criteria. Filament wound carbon fibers are mainly used in these cases replacing the glass fibers. Because, carbon fibers provide superior mechanical strength to the boat compared to glass fiber and also lighter in weight. But, carbon fibers are costly. Hence, some balance in cost and properties are required. [93]. Deshmukh et al. [94] reported that GO reinforced poly (3,4-ethylene dioxythiophene)-block-poly(ethylene glycol)(PEDOT-block-PEG)/polyvinylidenefluoride (PVDF)can be used in marine structure because the material has high strength and high corrosion resistance property which fits the marine conditions.

7.8 Pipelines and Chemical Plants

Fiber-reinforced polymer composites are widely used in pipelines, chemical storage tanks, and pressure vessel because of their good mechanical strength, corrosion resistance and chemical resistance property. They can be used to transport or store gaseous chlorine, bromine, carbon monoxide, dilute or concentrated acid and alkalis, etc. Their performance is better than steel. They are lighter in weight, so can be transported quite easily to the site where they need to install. They do not require schedule check-up for corrosion and thus reduces the maintenance cost substantially. Also, the low thermal conductivity reduces the cost of thermal insulation. Excellent load-bearing property is specifically important for chemical storage tank and pressure vessel where high pressure is created. Carbon fibers/epoxy resin is a good combination for this purpose. The electrical conductivity of carbon fiber can be utilized to heat the material by passing electricity through it as and when required. But there is one issue of using carbon fiber which is costly enough [13]. Kotal et al. [90] reported a polymer nanocomposite based on para-phenylenediamine modified graphene oxide (GO-PPD) and bromobutyl rubber (BIIR) which has very high thermal, mechanical and gas impermeability. This material can be used in the inner layer of pipes and pressure vessel.

7.9 Others

Other applications include body armor and in weapons for soldiers, electronic noses, antimicrobial composites, etc. Traditionally metal and alloys are being used in body armor, helmet, arms and transport equipment, etc. Because of the heavy weight of the armor, it reduces the efficiency of the soldiers. Further, the metal parts have an inherent tendency of rusting and corrosion. Hence, carbon–polymer composites have partially replaced the traditional armors. Carbon fiber or carbon nanofiber based epoxy resin composites are mainly used in body armor and in weapons [79]. Polymer composites based of carbon black, PVPD (Poly (Vinyl Pyrrolidone)) and PEG (Poly (Ethylene glycol)) is used as electronic noses (E-nose) which are used to detect various odorant and gases [95]. Santos et al. [96] reported for the first time that with the addition of 3 wt% of GO antimicrobial activity of poly-*N*-vinyl carbazole (PVK)/GO composite increases by 90%. So, the film can be used in antimicrobial composites.

8 Conclusions

Load-bearing capacity is one of the most import properties of polymer composites because it is the basic criteria of a composite material. Together with other parameters, structure of the CBMs and polymers plays an important role in it. The chapter provides state-of-the-art information regarding the load-bearing capacity of different carbon-based polymer composites, load-bearing mechanism, factors influencing the load-bearing property and important applications of load-bearing polymer composites.

Acknowledgements The authors gratefully acknowledge the unconditional support of every member of **Advanced Materials Research Laboratory (AMRL)**, IIT Roorkee to give proper shape of this book chapter.

References

1. Iijima S (1991) Helical microtubules of graphitic carbon. Nature 354(6348):56
2. Novoselov KS, Geim AK, Morozov SV, Jiang D, Zhang Y, Dubonos SV, Firsov AA (2004) Electric field effect in atomically thin carbon films. Science 306(5696):666–669
3. Erb T, Zhokhavets U, Gobsch G, Raleva S, Stühn B, Schilinsky P, Brabec CJ (2005) Correlation between structural and optical properties of composite polymer/fullerene films for organic solar cells. Adv Funct Mater 15(7):1193–1196
4. Mohajeri A, Omidvar A (2015) Fullerene-based materials for solar cell applications: design of novel acceptors for efficient polymer solar cells–a DFT study. Phys Chem Chem Phys 17 (34):22367–22376

5. Wang G, Xing W, Zhuo S (2012) The production of polyaniline/graphene hybrids for use as a counter electrode in dye-sensitized solar cells. Electrochim Acta 66:151–157
6. Savagatrup S, Printz AD, O'Connor TF, Zaretski AV, Rodriquez D, Sawyer EJ, Lipomi DJ (2015) Mechanical degradation and stability of organic solar cells: molecular and microstructural determinants. Energ Environ Sci 8(1):55–80
7. Beal RM, Stavrinadis A, Warner JH, Smith JM, Assender HE, Watt AA (2010) The molecular structure of polymer–fullerene composite solar cells and its influence on device performance. Macromolecules 43(5):2343–2348
8. Lin HY, Li CH, Wang DY, Chen CC (2016) Chemical doping of a core–shell silicon nanoparticles@ polyaniline nanocomposite for the performance enhancement of a lithium ion battery anode. Nanoscale 8(3):1280–1287
9. Zhang X, Samorì P (2017) Graphene/polymer nanocomposites for supercapacitors. ChemNanoMat 3(6):362–372
10. Wang X, Zhao J, Chen M, Ma L, Zhao X, Dang ZM, Wang Z (2013) Improved self-healing of polyethylene/carbon black nanocomposites by their shape memory effect. J Phys Chem B 117 (5):1467–1474
11. Ghassemieh E (2011) Materials in automotive application, state of the art and prospects. In: New trends and developments in automotive industry. InTech
12. Zhang W, Joshi A, Wang Z, Kane RS, Koratkar N (2007) Creep mitigation in composites using carbon nanotube additives. Nanotechnology 18(18):185703
13. Piggott M (2002) Load bearing fibre composites. Springer Science & Business Media, Berlin
14. Schrand AM, Huang H, Carlson C, Schlager JJ, Ōsawa E, Hussain SM, Dai L (2007) Are diamond nanoparticles cytotoxic? J Phys Chem B 111(1):2–7
15. Sun Y, Yang Q, Wang H (2016) Synthesis and characterization of nanodiamond reinforced chitosan for bone tissue engineering. J Funct Biomater 7(3):27
16. Ochiai T, Tago S, Hayashi M, Hirota K, Kondo T, Satomura K, Fujishima A (2016) Boron-doped diamond powder (BDDP)-based polymer composites for dental treatment using flexible pinpoint electrolysis unit. Electrochem Commun 68:49–53
17. Liu J, Song G, He C, Wang H (2013) Self-healing in tough graphene oxide composite hydrogels. Macromol Rapid Commun 34(12):1002–1007
18. Li L, Li J, Lukehart CM (2008) Graphitic carbon nanofiber-poly (acrylate) polymer brushes as gas sensors. Sens Actuators B Chem 130(2):783–788
19. Shuai C, Feng P, Gao C, Shuai X, Xiao T, Peng S (2015) Graphene oxide reinforced poly (vinyl alcohol): nanocomposite scaffolds for tissue engineering applications. RSC Adv 5 (32):25416–25423
20. Ansari S, Giannelis EP (2009) Functionalized graphene sheet—poly (vinylidene fluoride) conductive nanocomposites. J Polym Sci Part B Polym Phys 47(9):888–897
21. Liu J, Tao L, Yang W, Li D, Boyer C, Wuhrer R et al (2010). Synthesis, characterization, and multilayer assembly of pH sensitive graphene–polymer nanocomposites. Langmuir 26 (12):10068–10075
22. Das TK, Prusty S (2013) Graphene-based polymer composites and their applications. Polym Plast Technol Eng 52(4):319–331
23. Geim AK, Novoselov KS (2007) The rise of graphene. Nat Mater 6(3):183–191
24. Lee C, Wei X, Kysar JW, Hone J (2008) Measurement of the elastic properties and intrinsic strength of monolayer graphene. Science 321(5887):385–388
25. Zhu Y, Murali S, Cai W, Li X, Suk JW, Potts JR, Ruoff RS (2010) Graphene and graphene oxide: synthesis, properties, and applications. Adv Mater 22(35):3906–3924
26. Sengupta R, Bhattacharya M, Bandyopadhyay S, Bhowmick AK (2011) A review on the mechanical and electrical properties of graphite and modified graphite reinforced polymer composites. Prog Polym Sci 36(5):638–670
27. Liu W, Fukushima H, Drzal LT (2010) Influence of processing on morphology, electrical conductivity and flexural properties of exfoliated graphite nanoplatelets-polyamide nanocomposites. Carbon lett 11(4):279–284

28. Andrews R, Weisenberger MC (2004) Carbon nanotube polymer composites. Curr Opin Solid State Mater Sci 8(1):31–37
29. Choudhary V, Gupta A (2011) Polymer/carbon nanotube nanocomposites. In: Carbon nanotubes-Polymer nanocomposites. Intech
30. Rao CNR, Voggu R, Govindaraj A (2009) Selective generation of single-walled carbon nanotubes with metallic, semiconducting and other unique electronic properties. Nanoscale 1 (1):96–105
31. Kausar A (2017) Advances in polymer/fullerene nanocomposite: a review on essential features and applications. Polym Plast Technol Eng 56(6):594–605
32. Eight Allotropes of Carbon Michael Ströck—Created by Michael Ströck. Licensed under CC BY-SA 3.0 via Wikimedia Commons
33. Jabeen S, Kausar A, Muhammad B, Gul S, Farooq M (2015) A review on polymeric nanocomposites of nanodiamond, carbon nanotube, and nanobifiller: structure, preparation and properties. Polym Plast Technol Eng 54(13):1379–1409
34. Fu SY, Lauke B, Mäder E, Hu X, Yue CY (1999) Fracture resistance of short-glass-fiber-reinforced and short-carbon-fiber-reinforced polypropylene under Charpy impact load and its dependence on processing. J Mater Process Technol 89:501–507
35. Al-Saleh MH, Sundararaj U (2011) Review of the mechanical properties of carbon nanofiber/ polymer composites. Compos Part A Appl Sci Manuf 42(12):2126–2142
36. Feng L, Xie N, Zhong J (2014) Carbon nanofibers and their composites: a review of synthesizing, properties and applications. Materials 7(5):3919–3945
37. ASTM D1510-99, Standard test method for carbon black-iodine adsorption number
38. Mather PJ, Thomas KM (1997) Carbon black/high density polyethylene conducting composite materials: part I Structural modification of a carbon black by gasification in carbon dioxide and the effect on the electrical and mechanical properties of the composite. J Mater Sci 32(2):401–407
39. Donnet JB (ed) (1993) Carbon black: science and technology. CRC Press, Boca Raton
40. ASTM D1765-99, Standard classification system for carbon blacks used in rubber products
41. Ajayan PM, Schadler LS, Braun PV (2006) Nanocomposite science and technology. Wiley, New Jersey
42. Thakur S, Karak N (2014) Ultratough, ductile, castor oil-based, hyperbranched, polyurethane nanocomposite using functionalized reduced graphene oxide. ACS Sustain Chem Eng 2 (5):1195–1202
43. Kausar A, Rafique I, Muhammad B (2016) A review on applications of polymer/carbon nanotube and epoxy/CNT composites. Polym Plast Technol Eng 55(11):1167–1191
44. Al-Hartomy OA, Al-Ghamdi AA, Al-Salamy F, Dishovsky N, Slavcheva D, El-Tantawy F (2012) Properties of natural rubber-based composites containing fullerene. Int J Polym Sci 2012:1–8
45. Bera M, Maji PK (2017) Graphene-based polymer nanocomposites: materials for future revolution. MOJ Poly Sci 1(3):00013. https://doi.org/10.15406/mojps.2017.01.00013
46. Jaleh B, Sodagar S, Momeni A, Jabbari A (2016) Nanodiamond particles/PVDF nanocomposite flexible films: thermal, mechanical and physical properties. Mater Res Express 3 (8):085028
47. Bera M, Maji PK (2017) Effect of structural disparity of graphene-based materials on thermo-mechanical and surface properties of thermoplastic polyurethane nanocomposites. Polymer 119:118–133
48. Li Y, Shimizu H (2009) Toward a stretchable, elastic, and electrically conductive nanocomposite: morphology and properties of poly [styrene-b-(ethylene-co-butylene)-b-styrene]/multiwalled carbon nanotube composites fabricated by high-shear processing. Macromolecules 42(7):2587–2593
49. Dhakate SR, Chaudhary A, Gupta A, Pathak AK, Singh BP, Subhedar KM, Yokozeki T (2016) Excellent mechanical properties of carbon fiber semi-aligned electrospun carbon nanofiber hybrid polymer composites. RSC Adv 6(43):36715–36722

50. Zhang RL, Zhang JS, Zhao LH, Sun YL (2015) Sizing agent on the carbon fibers surface and interface properties of its composites. Fibers Polym 16(3):657
51. Jiang D, Liu L, Wu G, Zhang Q, Long J, Wu Z, Huang Y (2017) Mechanical properties of carbon fiber composites modified with graphene oxide in the interphase. Polym Compos 38 (11):2425–2432
52. Zhang RL, Huang YD, Li N, Liu L, Su D (2012) Effect of the concentration of the sizing agent on the carbon fibers surface and interface properties of its composites. J Appl Polym Sci 125(1):425–432
53. Ding X, Wang J, Zhang S, Wang J, Li S (2016) Carbon black-filled polypropylene as a positive temperature coefficient material: effect of filler treatment and heat treatment. Polym Bull 73(2):369–383
54. Liang JZ, Yang QQ (2009) Mechanical properties of carbon black-filled high-density polyethylene antistatic composites. J Reinf Plast Compos 28(3):295–304
55. Performance Improvement of Natural Rubber/Carbon Black Composites by Novel CouplingAgents, Yasuo UEKITA Yousuke WATANABE Hironobu IYAMA, Orhan OZTURK
56. Ao G, Hu Q, Kim MS (2008) Properties of activated carbon blacks filled SBR rubber composites. Carbon Lett 9(2):115–120
57. Mochalin VN, Shenderova O, Ho D, Gogotsi Y (2012) The properties and applications of nanodiamonds. Nat Nanotechnol 7(1):11–23
58. Naebe M, Abolhasani MM, Khayyam H, Amini A, Fox B (2016) Crack damage in polymers and composites: a review. Polym Rev 56(1):31–69
59. Jee AY, Lee M (2011) Thermal and mechanical properties of alkyl-functionalized nanodiamond composites. Curr Appl Phys 11(5):1183–1187
60. Zhang Q, Naito K, Tanaka Y, Kagawa Y (2007) Polyimide/diamond nanocomposites: microstructure and indentation behavior. Macromol Rapid Commun 28(21):2069–2073
61. Behler KD, Stravato A, Mochalin V, Korneva G, Yushin G, Gogotsi Y (2009) Nanodiamond-polymer composite fibers and coatings. ACS Nano 3(2):363–369
62. Neitzel I, Mochalin V, Knoke I, Palmese GR, Gogotsi Y (2011) Mechanical properties of epoxy composites with high contents of nanodiamond. Compos Sci Technol 71(5):710–716
63. Morimune S, Kotera M, Nishino T, Goto K, Hata K (2011) Poly (vinyl alcohol) nanocomposites with nanodiamond. Macromolecules 44(11):4415–4421
64. Li ZH, Zhang J, Chen SJ (2008) Effects of carbon blacks with various structures on vulcanization and reinforcement of filled ethylene-propylene-diene rubber. Express Polym Lett 2(10):695–704
65. Singh ROHIT, Shah M, Jain S, Shit S, Giri RADHASHYAM (2013) Elastomeric composite: mechanical and thermal properties of styrene butadeine rubber (SBR) based on carbon black and nanoclay. J Inf Knwl Res Mech Eng 2:515–521
66. Ahmadi Shooli S, Tavakoli M (2016) Styrene butadiene rubber/epoxidized natural rubber (SBR/ENR50) nanocomposites Containing nanoclay and carbon black as fillers for application in tire-tread compounds. J Macromol Sci Part B 55(10):969–983
67. Wang J, Li Q, Wu C, Xu H (2014) Thermal conductivity and mechanical properties of carbon black filled silicone rubber. Polym Polym Compos 22(4):393
68. Sengupta R, Bhattacharya M, Bandyopadhyay S, Bhowmick AK (2011) A review on the mechanical and electrical properties of graphite and modified graphite reinforced polymer composites. Prog Polym Sci 36(5):638–670
69. Baptista R, Mendão A, Guedes M, Marat-Mendes R (2016) An experimental study on mechanical properties of epoxy-matrix composites containing graphite filler. Procedia Struct Integr 1:74–81
70. Kim H, Macosko CW (2008) Morphology and properties of polyester/exfoliated graphite nanocomposites. Macromolecules 41(9):3317–3327
71. Wang X, Tan D, Chu Z, Chen L, Chen X, Zhao J, Chen G (2016) Mechanical properties of polymer composites reinforced by functionalized graphene prepared via direct exfoliation of graphite flakes in styrene. RSC Adv 6(113):112486–112492

72. Fang M, Wang K, Lu H, Yang Y, Nutt S (2009) Covalent polymer functionalization of graphene nanosheets and mechanical properties of composites. J Mater Chem 19(38):7098–7105
73. El Achaby M, Arrakhiz FE, Vaudreuil S, el Kacem Qaiss A, Bousmina M, Fassi-Fehri O (2012) Mechanical, thermal, and rheological properties of graphene-based polypropylene nanocomposites prepared by melt mixing. Polym Compos 33(5):733–744
74. Chen Z, Lu H (2012) Constructing sacrificial bonds and hidden lengths for ductile graphene/polyurethane elastomers with improved strength and toughness. J Mater Chem 22(25):12479–12490
75. Pitchan MK, Bhowmik S, Balachandran M, Abraham M (2016) Effect of surface functionalization on mechanical properties and decomposition kinetics of high performance polyetherimide/MWCNT nano composites. Compos Part A Appl Sci Manuf 90:147–160
76. Zhang W, Picu RC, Koratkar N (2007) Suppression of fatigue crack growth in carbon nanotube composites. Appl Phys Lett 91(19):193109
77. Gavrilov AA, Chertovich AV, Khalatur PG, Khokhlov AR (2013) Effect of nanotube size on the mechanical properties of elastomeric composites. Soft Matter 9(15):4067–4072
78. Fu SY, Lauke B, Mäder E, Yue CY, Hu X (2000) Tensile properties of short-glass-fiber-and short-carbon-fiber-reinforced polypropylene composites. Compos Part A Appl Sci Manuf 31 (10):1117–1125
79. Goertzen WK, Kessler MR (2006) Creep behavior of carbon fiber/epoxy matrix composites. Mater Sci Eng A 421(1):217–225
80. Kawai M, Takeuchi H, Taketa I, Tsuchiya A (2017) Effects of temperature and stress ratio on fatigue life of injection molded short carbon fiber-reinforced polyamide composite. Compos Part A Appl Sci Manuf 98:9–24
81. Sharma S, Chandra R, Kumar P, Kumar N (2016) Mechanical properties of carbon nanofiber reinforced polymer composites-molecular dynamics approach. JOM 68(6):1717–1727
82. Kumar S, Doshi H, Srinivasarao M, Park JO, Schiraldi DA (2002) Fibers from polypropylene/nano carbon fiber composites. Polymer 43(5):1701–1703
83. Zeng J, Saltysiak B, Johnson WS, Schiraldi DA, Kumar S (2004) Processing and properties of poly (methyl methacrylate)/carbon nano fiber composites. Compos Part B Eng 35(2):173–178
84. Rafiee MA, Yavari F, Rafiee J, Koratkar N (2011) Fullerene–epoxy nanocomposites-enhanced mechanical properties at low nanofiller loading. J Nanopart Res 13(2):733–737
85. Kim JH, Noh J, Choi H, Lee JY, Kim TS (2017) Mechanical properties of polymer-fullerene bulk heterojunction films: role of nanomorphology of composite films. Chem Mater 29 (9):3954–3961
86. Savagatrup S, Makaram AS, Burke DJ, Lipomi DJ (2014) Mechanical properties of conjugated polymers and polymer-fullerene composites as a function of molecular structure. Adv Func Mater 24(8):1169–1181
87. Worries about new composite made airplane, http://www.1001crash.com/index-page-composite-lg-2.html. Accessed from Saharanpur, India, dated 20 July 2017
88. Boeing 787, from the ground up, http://www.boeing.com/commercial/aeromagazine/articles/tr_4_06/article_04_2.html. Accessed from Saharanpur, India, dated 20 July 2017
89. Monetta T, Acquesta A, Bellucci F (2015) Graphene/epoxy coating as multifunctional material for aircraft structures. Aerospace 2(3):423–434
90. Kotal M, Banerjee SS, Bhowmick AK (2016) Functionalized graphene with polymer as unique strategy in tailoring the properties of bromobutyl rubber nanocomposites. Polymer 82:121–132
91. Gong ZG (2013) Nanotechnology application in sports. In: Advanced materials research, vol 662. Trans Tech Publications, pp 186–189
92. Nanotechnology in sports equipment: the game changer, nano work, posted on 27 May 2013
93. Shenoi RA, Dulieu-Barton JM, Quinn S, Blake JIR, Boyd SW (2011) Composite materials for marine applications: key challenges for the future. In: Composite materials. Springer, London, pp 69–89

94. Deshmukh K, Joshi GM (2014) Novel nanocomposites of graphene oxide reinforced poly (3, 4-ethylenedioxythiophene)-block-poly (ethylene glycol) and polyvinylidene fluoride for embedded capacitor applications. RSC Adv 4(71):37954–37963
95. Harun FKC, Jumadi AM, Mahmood NH (2011) Carbon black polymer composite gas sensor for electronic nose. Methods 6:8
96. Santos CM, Tria MCR, Vergara RAMV, Ahmed F, Advincula RC, Rodrigues DF (2011) Antimicrobial graphene polymer (PVK-GO) nanocomposite films. Chem Commun 47 (31):8892–8894

Polymer/Carbon Composites for Sensor Application

Subhendu Bhandari

Abstract Carbon containing fillers are widely used in the preparation of polymer composites. Incorporation of fillers with higher electrical conductivity in insulating polymeric matrices imparts significant change of electrical resistivity/conductivity of the composites. Moreover, the changes of electrical properties may also be caused by external triggering actions such as changes of temperature, mechanical strain, concentration of specific gases or vapours etc. Conductivity of the carbonaceous filler, that of the resulting composite as well as their distribution and relative alignment play significant role in the dependence characteristics of electrical properties on such external stimulus. Reversibility of such responsive behaviour opens up potential applicability of such composites in specific sensor applications.

Keywords Carbon black · Carbon nanotube · Gas sensor · Graphene Fullerene · Strain sensor · Temperature sensor

1 Introduction

Application of carbonaceous filler based polymer composites in recent years is not confined merely aiming at improvement of mechanical properties, rather it has found many diversified areas like different types of sensors too. A sensor is a device meant for detection of some events which can also be quantified from the intensity of the responses of the device. Efforts have been made to prepare different types of sensors like gas sensor, chemical sensor, strain sensor etc. Vapour or gas sensing is also important for the application in "electronic nose" for detection of odours. It is noteworthy that, in contrast with the stimuli responsive polymers, the mechanism of sensing involves a great role of the fillers rather than the matrix for polymer com-

S. Bhandari (✉)
Department of Plastic and Polymer Engineering, Maharashtra Institute of Technology, Aurangabad 431010, Maharashtra, India
e-mail: subhenduonly@gmail.com

M. Rahaman et al. (eds.), *Carbon-Containing Polymer Composites*, Springer Series on Polymer and Composite Materials,
https://doi.org/10.1007/978-981-13-2688-2_14

posite system. The electrical as well as electrochemical properties of the carbonaceous fillers are considered as the key feathers for their use in sensor application. Several works are reported using carbon black, carbon nanotube (CNT), graphene etc. towards effective use in the said areas of application. Polymers may be used both as film or powder form. Polymeric component of the composites provide dimensional stability to form a self-supported film, whereas, the powdery polymeric components of the composites. Conducting polymers, e.g. polyaniline (PAni) [1], polypyrrole (PPy) [2] etc. can be used for sensor application. However, carbonaceous materials like carbon nanotube (CNT), graphene, carbon black, carbon fibre, fullerene etc. may be used individually or along with conducting polymers for improved sensor performances [3–7]. The sensing response of the composites depends on different factors like conductivity of the fillers, shape and size of the fillers, morphology of the matrix, dispersion and relative alignment of the fillers etc. [8–10].

2 Mechanisms of Sensor Activity

Several mechanisms have been proposed to investigate the sensor activity of polymer composites:

- Most of the sensor activity of filled polymer composites exclusively depend on the percolation phenomena. Carbonaceous fillers like carbon black, carbon fibre, CNT, graphene, fullerene etc. are electrically conducting in nature, and are mixed with polymer matrices with relatively higher insulation. Beyond percolation threshold of filler loading, a conducting network forms in the bulk of the composite. Conductivity of the composite beyond percolation threshold follows the relation

$$\sigma = \sigma_1(\phi - \phi_c)^t \text{ for } \phi > \phi_c, \text{ and}$$

$$\sigma = \sigma_2(\phi - \phi_c)^{-s} \text{ for } \phi < \phi_c,$$

where ϕ is the volume fraction of the conducting filler, ϕ_c is the percolation threshold, σ_1 and σ_2 are the electrical conductivity of the conducting filler and the insulating matrix respectively, t and s are the dimensional parameters [11].

- According to quantum mechanical effects, insulating matrix between two conducting particles act as virtual tunnel resistor. For the application of low level of voltage and considering quantum mechanical rectangular barrier, the tunneling resistance may be expressed as

$$R_{\text{Tunnel}} = \frac{V}{AJ} = \frac{h^2 d}{Ae^2\sqrt{2m\lambda}} \exp\left(\frac{4\pi d}{h}\sqrt{2m\lambda}\right)$$

where V is the electrical potential difference, J is the tunneling density, d is the thickness of the insulator, e is the quantum of electrical charge, m is the mass of an electron, h is the Plank constant and λ is the barrier height of the tunneling effect [12].

3 Temperature Sensor

Temperature dependent resistivity of polymer-CNT composites have led to the ample opportunity of application as temperature sensors. For such composites, the dependence of resistivity on temperature may be of two types, i.e., exhibiting positive temperature coefficient (PTC) or negative temperature coefficient (NTC). PTC behaviour may be attributed to metallic behaviour of the composite where dR/dT is positive, whereas NTC behaviour signifies non-metallic behaviour of the composite where dR/dT is negative [13]. Polymer composites with carbonaceous fillers may exhibit either of these two characteristics. Polyethylene terephthalate (PET) screen printed with micropatterns of graphene, in the work of Kong et al. [14], exhibit NTC effect (Fig. 1a) where linear dependence of ln(R) on T^{-1} signifies applicability of the

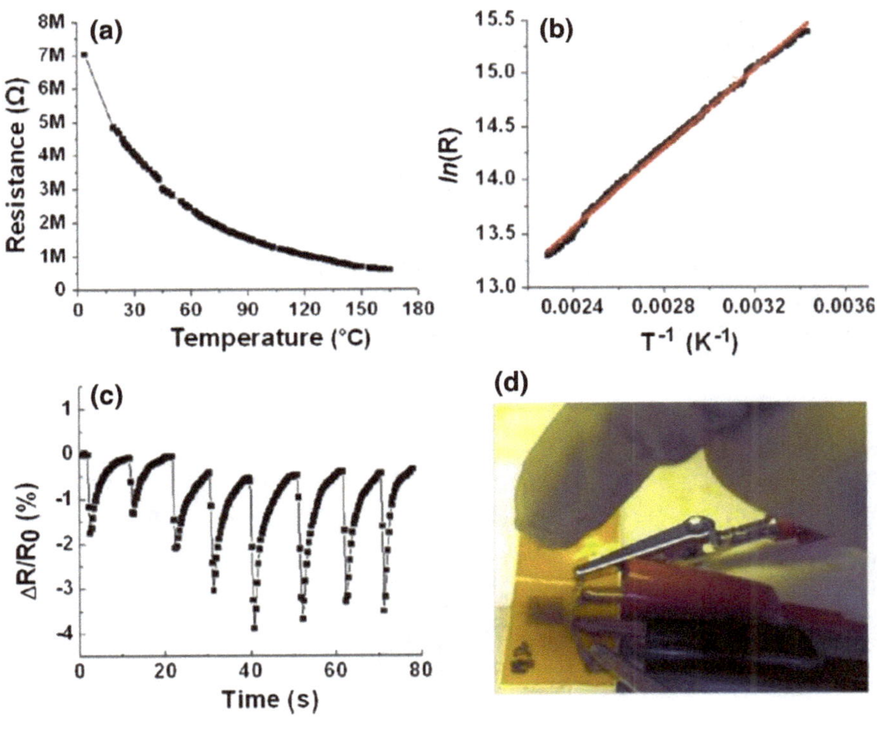

Fig. 1 **a** Temperature-dependence on electrical resistance. **b** Linear fit (red) between ln (R) versus T^{-1}. **c** Relative electrical resistance responses upon repeated fingertip tapping. **d** Experimental configuration. Reprinted with permission from [14], © 2012, American Chemical Society

material as effective temperature sensor. Moreover, the screen printed sensor also exhibited high repeatability of working after several tapping of the material with finger (Fig. 1c).

Similar temperature dependence is also observed for the resistance of filter deposited graphene oxide (GO), either metal-defused [15] or chemically reduced by hydrazine vapour [16]. Resistance of the material exhibiting NTC effect may be expressed as

$$R_{\mathrm{T}} = R_0 \exp\left(\frac{B(T_0 - T)}{T.T_0}\right)$$

where B is the material constant, T_0 is the reference temperature (298 K), R_0 and R_{T} are the corresponding resistance values at temperature T_0 and T. Matzeu et al. [17] found that, apart from NTC effect, MWCNT filled poly(styrene-*b*-(ethylene-*co*-butylene)-*b*-styrene) (SEBS) may also exhibit improved sensitivity with lower filler loading. In that experiment, sensitivity near percolation threshold of 12.5% (w/w) was found to be better (Fig. 2) compared to higher loading of 35%.

Similar trend is also reported by Oskouyi et al. [18] where graphene nano platelets (GNP) are found to be more suitable than CNT for temperature sensing application, and sensitivity decreases with the increase of volume fraction from 4.5 to 11.6%. That work also reveals that the decrease of resistivity with the increase of temperature for both CNT and GNP filled composites follow Mott's variable range hopping (VRH) model considering three-dimensional conduction:

$$\rho = \rho_0 \left(\exp\frac{T_0}{T}\right)^{1/D+1}$$

where D is dimensionality of the conduction system, and ρ_0 and T_0 are the constants.

On the other hand, PTC effect is also reported by several researchers for carbon black filled LDPE [19], MWCNT filled HDPE composites [20], short carbon fibre

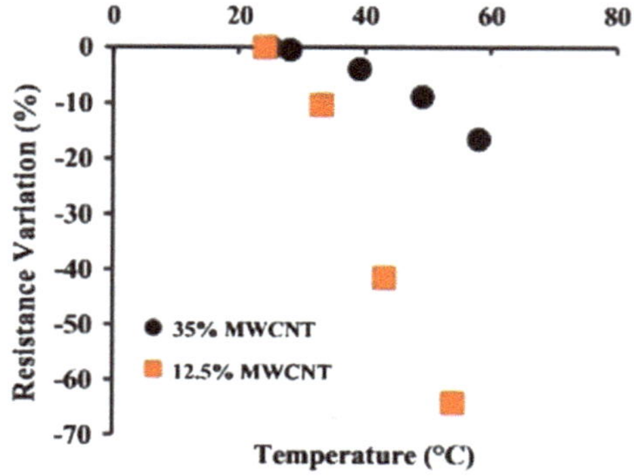

Fig. 2 Percentage variation of resistance versus temperature in sensors with MWCNT weight percentages of 12.5 and 35%. Reprinted from [17], © 2012, with permission from Elsevier

filled epoxy resin [5], graphene/polyvinylidene fluoride (PVDF) composite containing silver nanowire [21] etc. Nevertheless, the functioning of temperature sensor may also be further checked for heating and cooling cycle. For CNT deposited paper exhibiting NTC effect as observed by Karimov et al. [22], the temperature dependence of resistance follows the relation

$$\frac{R}{R_0} = \exp(-\Delta T(T/T_m)K)$$

during heating cycle, and follows the following relation

$$\frac{R}{R_0} = \exp(-\Delta T(T_m/T)K)$$

during cooling cycle, where K is the resistance-temperature factor, T_m is the maximum temperature, R_0 is the initial resistance and R is the resistance at elevated temperature T.

Fabrication of a miniaturized polymer based flexible temperature sensor filled with multiwalled carbon nanotubes has been reported by Sibinski et al. [23] which exhibited temperature sensitivity around 0.13% K^{-1}. This study reveals quasi-linear dependence of resistivity on temperature for the MWCNT filled composite (Fig. 3). For the use of graphite grains, due to the small number of connections among the separate grains, the physical contacts are lost more easily. However, for CNT composition, the very high aspect ratio (length: diameter ratio in the range of 100–1000) favours less chance to lose physical contact, which enhances stability. Thus, the sensor with CNT compositions exhibited stability during the whole range test cycles. The graphite-polymer composites also exhibited resistance change to a fair extent, however, more than 10% increase of resistance was observed beyond 200,000 cycles. A comparative study between graphite and CNT filled composites of polymethyl methacrylate reveals applicability of the composite as flexible temperature sensor at while bending it at different radius as well as for several bending cycles. Linear dependence of resistance on bending radius was observed (Fig. 3) for both cases, however, the graphite filled composite exhibits nonlinearity at very small bending radius. Resistance of the graphite filled composite increases for higher bending cycles, whereas that of the CNT filled composite almost remains unchanged even after 350,000 cycles which signifies the enhanced efficiency of the CNT filled composite as flexible temperature sensor.

Hong et al. [24] used poly(*N*-isopropylacrylamide) (PNIPAM) and fullerene (C_{60}) to prepare temperature sensor. PNIPAM exhibits reversible phase transition in water at the lower critical solution temperature (LCST). Two types of sensors were synthesized: one had two fullerene attached to the middle part of the PNIPAM chain and the pyrene based fluorophore attached to one end; and the other had fluorophore and fullerene attached to the two chain ends of PNIPAM. Thus, a combination of photophysical characteristics of fullerene as well as temperature induced coil-globule transition of PNIPAM was achieved.

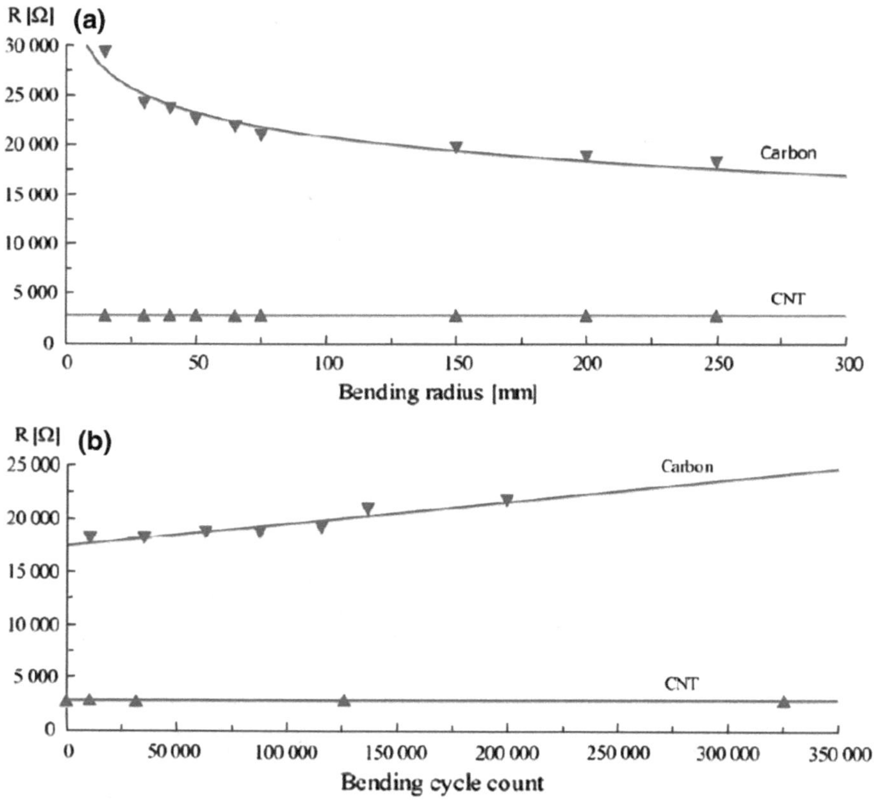

Fig. 3 Change of resistance for nanotube-polymer and graphite-polymer compositions during **a** bending cycle and **b** cyclical bending. Reprinted with permission from [23]

Pucci et al. [17, 25] have reported that solution processing of MWCNT filled composite using poly(styrene-*b*-(ethylene-*co*-butylene)-*b*-styrene) (SEBS) matrix (1:1 weight ratio) via sonication may lead to degradation of average length of MWCNT to the extent of 40% which in turn, affect the conductivity of the composite. However, nanotube degradation during sonication may be improved by alkyl functionalization of MWCNT which exhibits better dispersion efficiency and enhanced conductivity too. The sensitivity of 0.007 K^{-1} for this case is comparable to the highest values exhibited by metals 90.0037–0.006 K^{-1}). However, owing to the elastomeric behaviour of SEBS, phase stability deteriorates with the increase of temperature which results in loss of temperature sensitivity. Such problem may be overcome by using a poly(vinylbenzyl chloride) derivative with triethylamine as ionomeric surfactant, in which the cation-π interactions and weak van der Waals force between ionomer and MWCNT favours the preparation of undamaged composite [26].

4 Strain Sensor

Traditional metallic or semiconductor strain gauges are less expensive compared to polymeric strain sensors and also exhibit high sensitivity. However, they cannot be effectively embedded in the structural components, and also offer low resolution in nanoscale. Moreover, merely unidirectional effectiveness may restrict their use. Carbon nanotube based thin film sensors may be effective substitutes because of their very high resolution at nanoscale and effective use in multidirection [27].

Strain sensor behaviour is generally expressed in terms of gauge factor (G):

$$G = \frac{1}{\varepsilon}\left(\frac{\Delta R}{R_0}\right)$$

where ε is strain, R_0 is the initial resistance and ΔR is the change of resistance. For a material exhibiting strain-independent conductivity, the following relationship is generally followed [28]:

$$\frac{\Delta R}{R_0} = \varepsilon(2 + \varepsilon)$$

which signifies that, for metallic strain gauge, $G \sim 2$ at low strain. On the contrary, for polymer nanocomposite based sensors the reported value of G may be as high as 1300 for CNT/epoxy composite [29], however, mostly the reported values lie below 50.

The strain response is generally considered addressed by measuring the variation of resistance upon applying external strain. Such variation of resistance arises because of the change of local contacts between conductive fillers. Not only the type of conductive filler, the morphology of the conductive networks also plays a crucial role on the measurement of strain sensing [30].

For thermoplastic polyurethane (TPU) based conductive composites, Liu et al. [31] reported about the strain sensing characteristics, where 2D conductive network was achieved by the contacts of the planes of the flake-like graphene. For different strain patterns, high sensing stability and sensitivity were observed along with good reproducibility and recoverability after stabilization by cyclic loading.

Entangled CNT conductive network, as investigated by Zhao et al. [32] in a comparative study using CNT and carbon black individually in polypropylene matrix, a decreased response pattern was observed (Fig. 4), whereas for the conductive network using spherically shaped carbon black particles improved response pattern during cyclic strain was obtained. According to the investigation of Ferrreira et al. [33, 34], the surface functionalization and the type of CNT does not influence the piezoresistive sensitivity of the nanocomposites prepared using PVDF matrix provided the CNT content in the nanocomposites is near to the percolation threshold. The highest gauge factor (ratio of relative resistance change to strain) was reported by them as 6.2 at 0.25% strain for the 2 wt% SWCNT filled

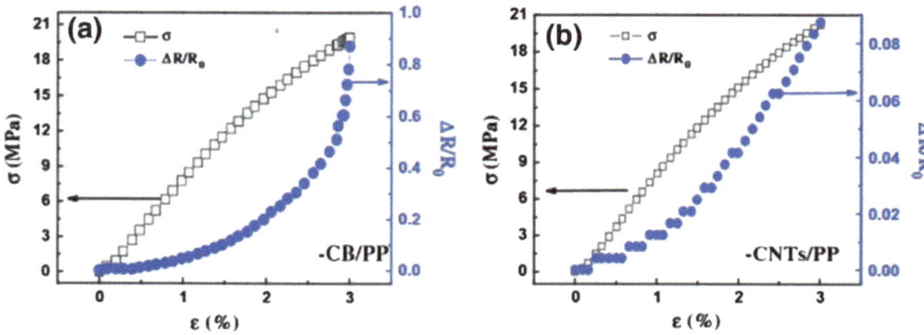

Fig. 4 Stress versus strain relationship in **a** CB (7.16 vol.%)/PP and **b** CNTs (3.41 vol.%)/PP composites during 10 extension–retraction cycles. (Inset: plot of stress vs. time). Tensile stress (left axis) and normalized change of electrical resistivity (right axis) were plotted as a function of strain for composites: **c** CB (7.16 vol.%)/PP and **b** CNTs (3.41 vol.%)/PP. Reprinted from [32], © 2013, with permission from Elsevier

nanocomposite, which was marginally higher than the of the other SWCNTs. Apart from using single filler, the use of the combination of two or more types of conductive fillers may exhibit synergistic effect in sensing behaviour and generate distinct conductive network morphology. For example, a combination of CB and CNT into thermoplastic polyurethane revealed that the addition of carbon black lowers the formation of entangled CNT network, which can be destroyed under strain easily, exhibiting higher strain sensitivity [35].

Li et al. [36] used graphene foam (GF)/polydimethylsiloxane (PDMS) composite for strain sensing application (Fig. 5). The current-voltage relationships (Fig. 5a) at different strain followed Ohm's law. The variation of relative resistance at different strain (Fig. 5b) revealed excellent reproducibility and stability at lager strain. For more than 20% strain, hysteresis was found negligible during stretching-relaxation cycle (Fig. 5c), which suggests recovery of initial resistivity after relaxation from stretched state. Gauge factor (Fig. 5d) was found to vary from 98.66% (at 5%) to 47.74 (at 20% strain).

Yasin et al. [6] fabricated an organic field effect transistor (OFET) using polymer-fullerene composite. Strain sensing behaviour was analyzed by bending the flexible device on cylinders with radii 5, 10 and 15 mm with corresponding strain values of 3.2, 1.6 and 1% respectively. The sensitivity values of the device at 3.2% strain perpendicular and parallel to the current axis was found to be 0.65 and 0.18 μA/% respectively.

Lin et al. [27] investigated the strain sensing behaviour for a polyester based thermoplastic polyurethane filled with the combination of MWCNT, nickel powder and a eutectic alloy containing bismuth and tin. The volume ratio of nickel and the eutectic alloy was maintained at 1:2, whereas the volume ratio of the combined metallic fillers and MWCNT was varied at 0:1, 10:1 and 20:1. The morphological analysis through SEM reveals that MWCNT bundles get oriented under 10% strain whereas the metallic particles exist isotropically. However, both metal particles and

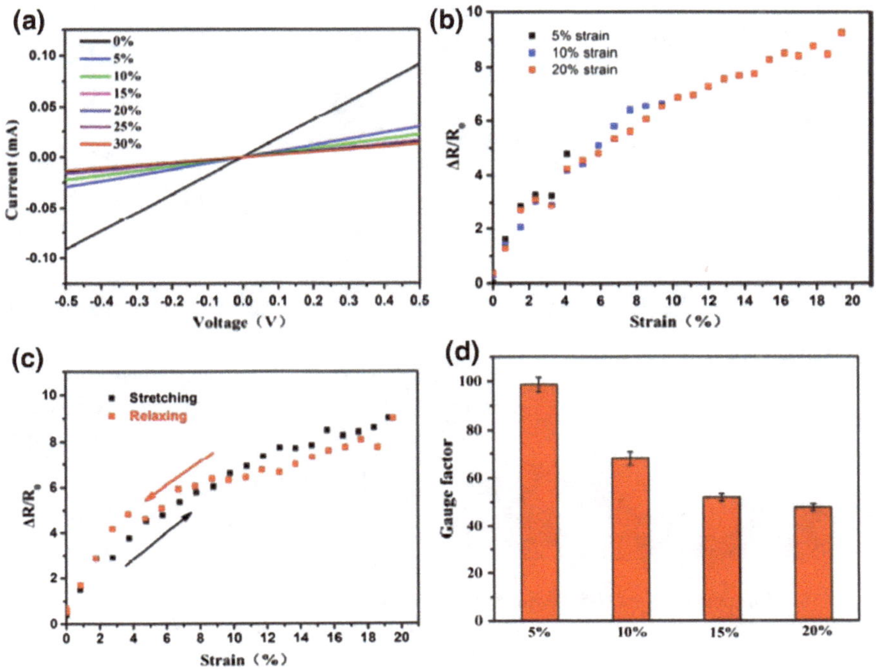

Fig. 5 **a** *I–V* curves of the strain sensor at different strains. **b** The variation of relative resistance in the stretching process. **c** The relative change of resistance in the stretching-relaxing cycles within the range of 0–20% strain. **d** Change of gauge factor under various strains. Reprinted with permission from [36], © 2016, American Chemical Society

MWCNT bundles networks get oriented at 100% strain. It is noteworthy that only the network structure gets deformed at 10% strain, while, the conductive networks get stretched apart at 100% strain (Fig. 6).

Mechanical deformation may change the electronic characteristics of CNT. While strain is applied, the bond length of CNT is altered by shifting the periodicity of the quantization which results in change of band gap. A metallic CNT does not exhibit any band gap in unstrained state. However, for a semiconducting CNT the band gap is inversely proportional to the diameter and follows the relation [8]:

$$E^0_{\text{Gap}} = \frac{2\gamma a}{\sqrt{3}d}$$

and for small-gap semiconducting CNT, the band gap is:

$$E^0_{\text{Gap}} = \frac{\gamma a^2}{4d^2}$$

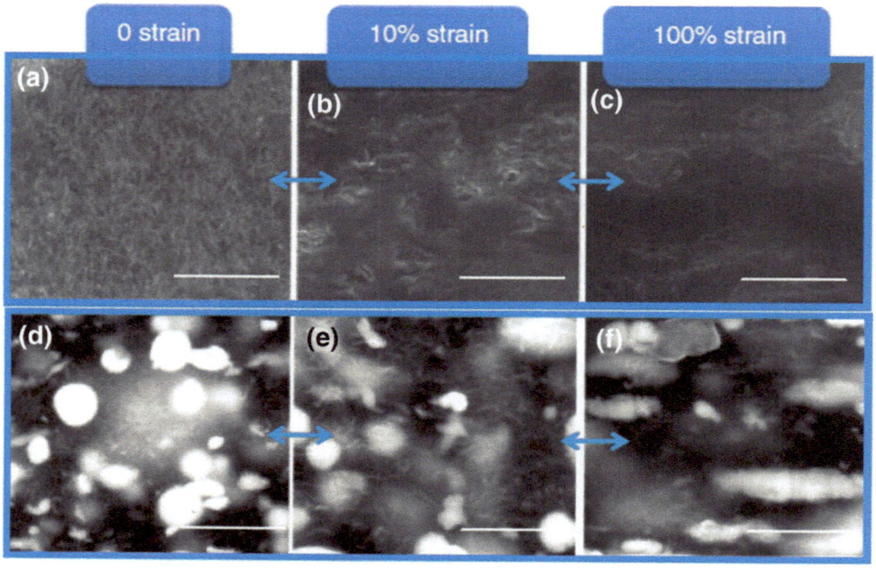

Fig. 6 SEM images of **a–c** PU-5C (with 5 vol.% MWCNT) and **d–f** PU-1.5C-15A (with 1.5 vol.% MWCNT and 15 vol.% metallic particles) at (**a**, **d**) at zero strain, **b**, **e** 10% strain and **c**, **f** 100% strain. The samples were drawn along the arrow direction. The scale bars are 1 μm. Reproduced from [27] with permission, © 2012, John Wiley and Sons

where d is the diameter, a is the unit vector of graphene, for an average semi-conducting CNT the tight binding overlap integral γ is the band gap E^0_{Gap} is 2.6 eV and the band gap E^0_{Gap} is ~0.1–2 eV.

Kang et al. [37] reported the use of polymethylmethacrylate (PMMA) composites filled with single-walled carbon nanotube (SWCNT) thin films (buckypapers) and single-walled carbon nanotube for strain sensing. Higher sensitivity in the linear bending range was observed for the strain response of the buckypaper sensors. However, the saturated strain behavior for more than 500 microstrains under tension occurred because of the slippage among the carbon nanotubes bundles. No slippage of individual carbon nanotubes was found during compression of the buckypaper sensor, which resulted in a lack of saturation in comparison with the tension case. In spite of exhibiting improved sensitivity, buckypaper may not be successfully used for measuring strain in the whole elastic range. Although the SWCNT composite sensors offer less sensitivity compared to the buckypaper, linearity in the linear symmetric strain response was observed during application of compression or tension. The bonds between SWCNT and the polymer restricted slippage, which in turn improved the strain in the composite sensor effectively. Chen et al. [38] achieved a lower percolation threshold using hybrid fillers of carbon black and CNT in the weight ratio 1:1 in the matrix of a 60/40 blend of polyamide 6 and acrylonitrile butadiene styrene (ABS) while compared with the individual use of the fillers. Ma et al. [39] reported that the percolation threshold in nanocomposites of epoxy filled with 0.2 wt% CNT in combination with various proportion of carbon black is lower than that of individually loaded with single filler. Socher et al. [40] reported about higher bulk conductivity for the use of hybrid of MWCNT and carbon black in comparison with either of the filers while using polyamide 12 as matrix of the melt processed composites.

The investigation of Parmar et al. [10] revealed that, during the preparation of 5 wt% MWCNT polycarbonate (PC) composite, the shear force in injection molding facilitates in partial alignment of MWCNTs in the direction of flow of the matrix. Therefore, injection molding results in much higher sensitivity factor ($\sim$20%) compared to the compression molding method ($\sim$0.3%). Moreover, the gauge factor for injection molded composite was also found 1.5–2 times higher in contrast to the compression molded composite.

Ke et al. [41] explored the correlation between the sensing behaviour and morphological characteristics (Fig. 7) for the nanocomposite of PVDF filled with both non-functionalized MWCNT and conductive carbon black (CCB) at different proportions. It was found that 0.5CNT-1CCB (i.e., 0.5 wt% CNT + 1 wt% CCB) combination exhibits the highest rate of the increase of $\frac{\Delta R}{R_0}$ (Fig. 7a), for which case string-like CNT-CCB-CNT or CCB-CNT-CCB bridging are formed (Fig. 7b). For 0.75CNT-0.75CCB, the amount of CNT is sufficient to form conductive path (Fig. 7c), whereas addition of CCB strengthen the contact points of CNTs. Again, for the sample with higher proportion of CNT:CCB, i.e. 1CNT-0.5CCB, much denser network structure was observed (Fig. 7d) which favours more robustness against deformation together with $\frac{\Delta R}{R_0}$ in spite of exhibiting higher conductivity.

Tuukkanen et al. [42] used solution processed CNT and graphene nanocomposite films on rubber substrates for investigation of their strain sensing behaviour. The blend of natural rubber (NR) and polybutadiene rubber (BR) as well as individual rubbers like chlorosulphonated polyethylene (CSM) and acrylonitrile butadiene rubber (NBR) were used as the substrate on which CNT and graphene ink were deposited by blade coating. Relaxation of rubber after stretching affects CNT composite less compared to the graphene composite. Moreover, the change of resistance is more for the CNT composite which suggests the its superiority as strain sensor in contrast to the graphene composite.

Jang et al. [43] took the strategy of improvement of dielectric constant of the composite to achieve enhanced performance in force sensing. They functionalized MWCNT with alkylamine which exhibited improved dispersion in polydimethylsiloxane, improved dielectric constant by threefold and capacitance by 1.8 fold while using only 0.75 wt% of filler loading. Eswaraiah et al. [44] prepared PVDF/graphene composite by solid state mixing of PVDF and reduced graphite oxide which was exposed to focused solar electromagnetic radiation using a 130 mm converging lens. Percolation was achieved at >2 wt% use of graphene, and the composite exhibited good strain sensing behaviour with gauge factor of 12.1.

Hu et al. [12] predicted the gauge factor values through mathematical modeling as 6, 21, 62 and 117 for conductivity of the filler to the extent of 103, 104, 105 and

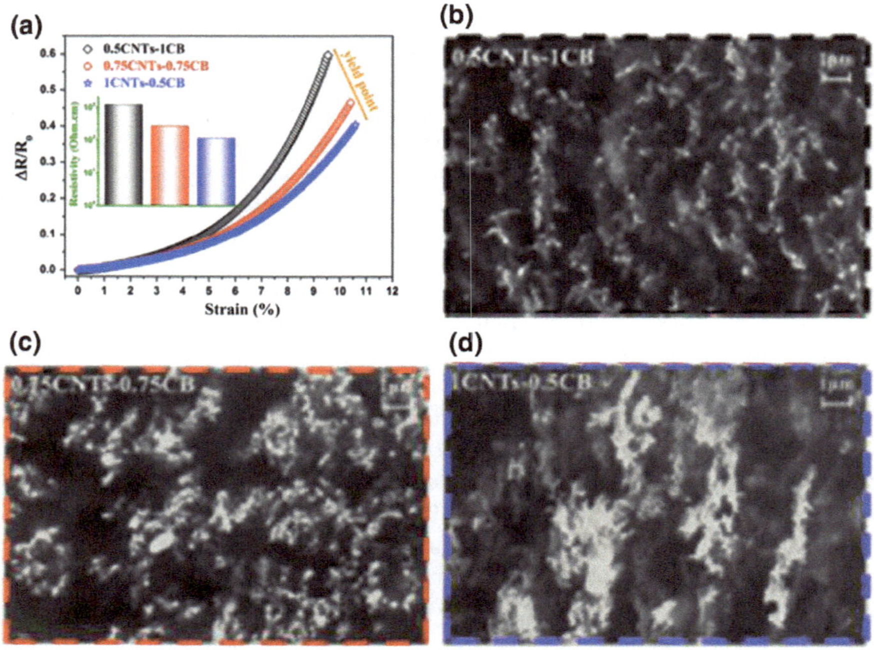

Fig. 7 **a** $\frac{\Delta R}{R_0}$ with strain, initial resistivity together with charge contrast imaging (CCI)-SEM images illustrating conductive network structure (before stretching) of nanocomposites such as **b** 0.5CNT-1CCB, **c** 0.75CNT-0.75CCB and **d** 1CNT-0.5CCB. Reprinted with permission from [41], © 2016, American Chemical Society

106 S/cm respectively. Moreover, they also experimentally reported in the same work that, for epoxy/CNT composite the gauge factor increases with the increase of stirring rate and decreases with the increase of curing temperature. The reproducibility restrictions on strain sensing behaviour for SWCNT films are higher than MWCNT because of the influence of impurities, chirality and electrical properties [45].

Dinh et al. [46] have reported a comparative study of the piezoresistive properties of polyethylene oxide (PEO)/MWCNT composite as well as the films based on MWCNT dispersed in the surfactants deoxycholic acid (DOC) and sodium dodecyl sulphate (SDS). Films were fabricated drop casting of solutions on rectangular tensile specimens. The resistance versus strain characteristics exhibit a quadratic behavior. The strain sensitivity of the samples SDS, DOC and PEO has been measured as 4, 7.5 and 12.5 respectively, whereas the SDS exhibits the most stable mechanical behavior. The maximum gauge factor reported by different researchers for polymer-CNT nanocomposites prepared by different methods has been shown below (reproduced with permission from Kanoun et al. [47]; Table 1).

Table 1 Comparison of different polymer-carbon composites and the maximum gauge factor values as reported by the researchers

Type of filler	Polymeric matrix	Maximum gauge factor	Reference
SWCNT	Polymethylmethacrylate	5.3	[48]
Graphite oxide	Polyvinylidene fluoride	12.1	[48]
MWCNT	Polycarbonate	2.457	[49]
MWCNT	Polymethylmethacrylate	4.59	[50]
MWCNT	Vinyl ester	2.6	[51]
MWCNT	Polyurethane	4	[52]
MWCNT	Epoxy	5.8	[53]
MWCNT	Polyethylene oxide	50	[54]
MWCNT	Epoxy	22.4	[55]
MWCNT	Polysulfone	2.68	[56]

5 Chemical Sensor

Chemical sensors are the sensors which respond in presence of some specific chemicals, and the response signal can be used to detect those chemicals. The mechanism of the chemical sensors may depend on chemical reactions, charge carrier concentrations, electroactive properties of the materials etc. Because of extensive applicability in environmental, biomedical or industrial monitoring of presence of specific gases or vapours, gas sensors have drawn attraction of researchers for further improvement in their efficiencies as well as extending their applicability for different gases and vapours. The target gas or vapour may be toxic in nature which needs to be detected with good precision to examine its presence in the permissible limit. For example, allowable concentration limit of ammonia at industrial workplaces for 8 h exposure is 25 ppm, and for 10 min is 35 ppm [57] where the olfactory limit of NH_3 for human body is 55 ppm [58]. However, for practical applications, the sensors should be capable of detection of NO_2 with concentration <3 ppm, NH_3 with concentration <25 ppm etc. [59, 60]. Electroactive polymers, porous silicon, metal oxides etc. may be used for this purpose. Some of the important performance criteria for an effective gas sensor are discussed below in brief [61].

- *High sensitivity of the sensor to the target gas or vapour*:

Generally, sensitivity follows the mechanism of reversible redox reactions where resistance changes during oxidation or reduction. Sensitivity for p- and n-type metal oxide or polymeric semiconductors towards reducing or oxidizing gases may be expressed as

$$S^{\mathrm{p}}_{\mathrm{red(oxd)}} = \frac{R_{\mathrm{gas(air)}}}{R_{\mathrm{air(gas)}}}$$

and

$$S^{n}_{red(oxd)} = \frac{R_{air(gas)}}{R_{gas(air)}}$$

where R_{air} is the resistance exhibited by the sensor in air as reference.

- *High degree of selectivity*:

Selectivity of a sensor signifies the effectiveness of selectively target gas specific response among the mixtures of a number of gases or vapours. Selectivity of a sensor toward the gas i among the mixture of n number of gases may be expressed as

$$\mathrm{Sel}_i = \frac{S_i}{\sum_i^n S_i}$$

- *Low detection threshold*:

The minimum concentration of the target gas should be as minimum as possible, i.e. the detection threshold should be low so as to detect any minute change of concentration with low tolerance value.

- *Quick response and recovery*:

An effective sensor should exhibit fast response and recovery time which are measured from the time required to attain 90% of the saturated resistance value upon exposure to the gas, or the time needed to attain 90% of the initial baseline value of resistance.

- *Good life cycle*:

Because of reversible redox reaction dependent detection mechanism, gas sensors can detect both increase and decrease of the gas concentration and may be used repeatedly. The maximum number of cycles of detection with allowable deviation from initially calibrated baseline response denotes life cycle of a gas sensor.

Yun et al. [62] studied the effect of oxyfluorination of polyaniline (PAni)-coated MWCNT for ammonia sensing. In their work, polyaniline was synthesized by oxidative chemical synthesis method and oxyfluorination was carried out with varying fluorine:oxygen ratio at 2:8, 5:5 and 8:2. It was observed that oxyfluorination improved the interfacial affinity as well as charge transfer between PAni and MWCNT which plays an important role to improve the gas sensing performance. Moreover, gas sensing also improves with the decrease of PAni layer thickness on MWCNT. The highest sensing performance was exhibited for the use of fluorine: oxygen in the ratio of 8:2.

Synergistic effect of reduced graphene oxide (RGO) and MWCNT in natural rubber composite was studied by Ponnamma et al. [63] for detection of xylene, toluene and benzene. Graphene oxide was separately treated at 200 and 600 °C and MWCNT was mixed with the RGOs. The mixed particles were added with natural rubber using melt mixing process. The electrical conductivity was recorded during the breakdown of conductive network while swelling in different solvents. The weaker NR-RGO interactions and stronger RGO-RGO interactions resulted exhibited higher swelling behaviour. The use of RGO, formed at 200 °C, exhibited more solvent swelling compared to the other nanocomposites.

Liao et al. [7] proposed a disposable sensor to detect low concentrations of NH_3 and HCl (100 ppb). In their work, 1 wt% carboxyl functionalized SWCNT was used during the chemical synthesis of polyaniline (PAni) in presence of *N*-phenyl-*p*-phenylenediamine and the sensor was fabricated by electron beam lithography. PAni/SWCNT composite exhibited much faster detection of the vapours compared to the use of pristine PAni. For NH_3 vapour detection, pristine PAni sensor took 1000 s for 20 fold increase in resistance, whereas the composite responded to the same extent only in 120 s. Carbon black may be used along with different polymers like polyethylenimine, polystyrene etc. for preparation of sensor electrodes by ink-jet printing [64].

Wei et al. [65], developed at US Air Force Research Laboratory (AFRL), the sensors with excellent environmental stability, high sensitivity and good selectivity offered at low power consumption for the detection of different chemical vapours. Perpendicularly aligned carbon nanotube arrays were partially coated along their tube length with polyvinylacetate (PVAc) thin film (Fig. 8). Changes in the inter-tube distance occurs during the absorption and desorption of chemical vapours, consequently, the surface resistance also changes. The flexible composite films may be potentially used as multifunctional sensors.

Xie et al. [66] used poly(4-vinyl phenol) and poly(ethylene oxide) individually as the matrix materials for the composites with carbon black. Fabrication of electrode was carried out by micromachining. Analyte vapours of methanol, ethanol and acetone were used in different concentrations, i.e. 300, 600, 900, 1200 and 1500 ppm. Nearly good linearity of resistance ratio was achieved for using poly (ethylene oxide) to ethanol and acetone in the concentration range of 300–1500 ppm. The linear dependence of resistance ratio was found to follow the relation:

$$\frac{\Delta R}{R} = 0.0007c + 0.4733 \text{ (for ethanol), and}$$

$$\frac{\Delta R}{R} = 0.0005c + 0.4545 \text{ (for acetone)}$$

Since –OH group of ethanol interacts with the hydrogen bond of poly(4-vinyl phenol), its ethanol absorption capacity as well as the sensing performance is superior than poly(ethylene oxide).

Hernández-López et al. [67] used 10 wt% of carbon black to acrylamidomethyl cellulose acetate butyrate (ACAB) and analyzed the vapour sensing behaviour

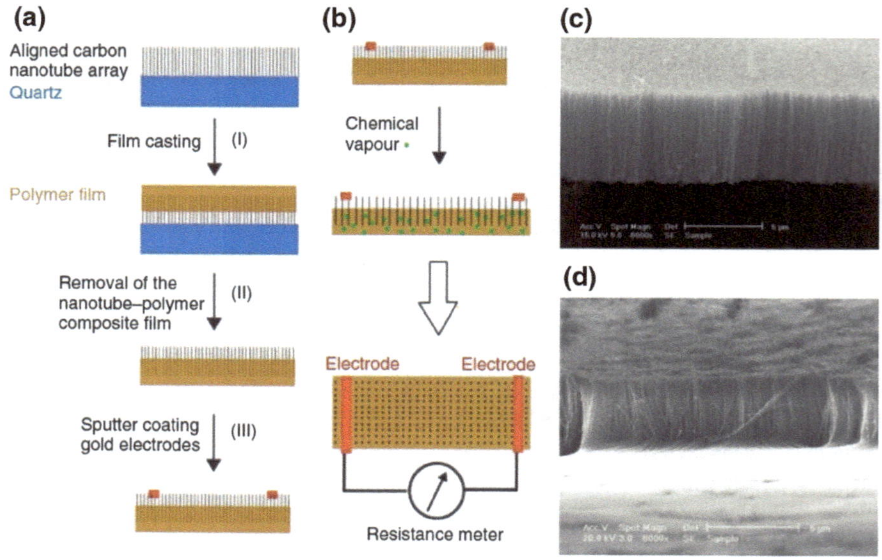

Fig. 8 Schematic representation of the procedures for **a** fabrication and **b** morphological characterization of the aligned carbon nanotube–polymer composite vapour sensor. Scanning electron micrographs of the aligned carbon nanotube array **c** before and **d** after being partially coated with PVAc on top and turned upside down (scale bars: 5 μm). Reprinted with permission from [65], © 2006, American Chemical Society

towards tetrahydrofuran. 2–24 s of response time was recorded. The durability of the sensor decreased from 20 to 10 pulses while the amount of the analyte was increased from 0.1 to 0.4 ml. However, the reproducibility was found to the level of 65%.

Holmes et al. [68] used the composites of buckyball (C_{60}) at 0.01–10% loading in epichlorohydrin using both spin coating and spray-coating technique. The sensors were used for detection of dimethyl methyl phosphonate (DMMP), methyl salicylate (MS), water and methanol (MeOH) at different time intervals. The responses of the sensors are shown in Fig. 9.

Pede et al. [69] prepared a photodiode array using the composite of fullerene (C_{60}) with conjugated polymer polythiophene. Polymer coating was done for both the lower and upper layers of electrodes, which were separated by an insulating layer at the crossover points. The fabricated device was used as chemical sensor for detection of iodine [70].

Zheng et al. [71] successfully used the composite of graphene oxide with PAni for the detection of ammonia, formaldehyde and cyclohexanone vapour.

Jang et al. [72] used the composite of polypyrrole (PPY) with MWCNT for the detection of ammonia. In their work, in situ chemical synthesis of PPY was carried out in presence of MWCNT at different wt% (0, 0.5, 1, 3 and 5) and the dispersion of the composite in acetone was drop cast on silicon wafer. Addition of 5 wt%

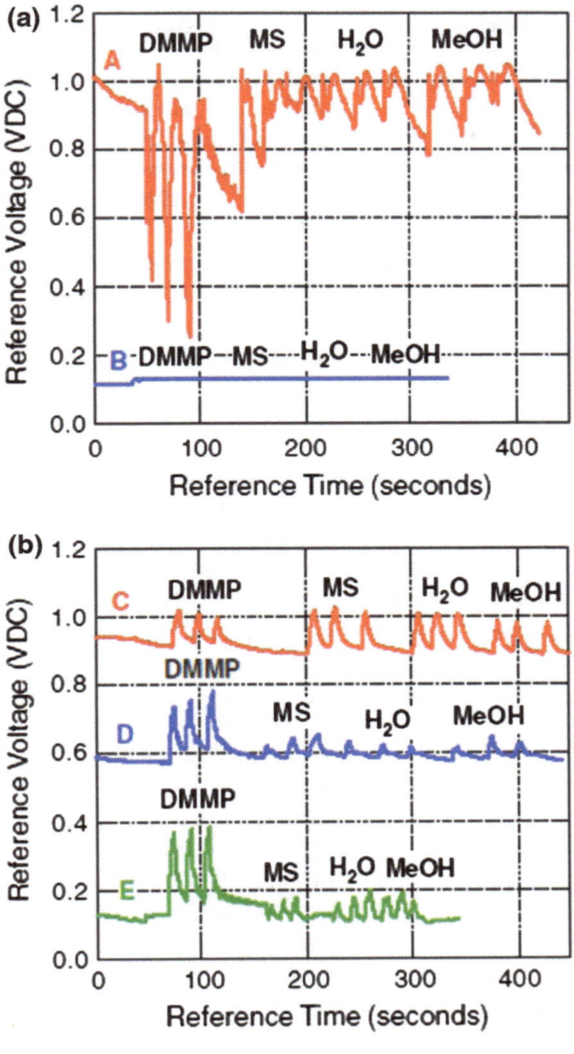

Fig. 9 Response curves of the sensor. **a** Line 2 coated with pure polyepichlorohydrin. (A) Response immediately after spray-coating. (B) Response after 1 day. **b** Line 2 coated with polyepichlorohydrin with 10% fullerenes by mass. (C) Response immediately after spray-coating. (D) Response after 3 days. (E) Response after spraying more solution 4 days after first spraying. Reprinted by permission from [68], © 2009, Springer Nature

MWCNT exhibited the best performance in terms of quick detection as well as excellent reproducibility of more than 90% recovery of resistive response for cyclic testing. Apart from pristine CNTs, functionalized CNTs are also being investigated in several applications, however, controversial results are reported regarding the electrical properties of functionalized MWCNTs [73].

Kumar et al. [74] prepared a sensor by spraying 1 wt% of CNT/polycarbonate (PC) on a clean interdigitated printed circuit board (PCB). The sensing behaviour was found coherent with the respective Flory-Huggins polymer-solvent interaction parameter, i.e., in the increasing order of water < methanol < toluene.

A similar observation about the correlation between gas sensing behaviour and polymer-solvent interaction was also reported by Kobashi et al. [75]. They achieved percolation threshold below 0.5% addition of MWCNT for a melt processed composite of polylactic acid (PLA) which was used for the detection of water, toluene, chloroform, *n*-hexane, ethanol, dichloromethane and tetrahydrofuran. Some researchers reported that functionalization favours improvement of dispersion, which in turn enhances conductivity; whereas the others reported about the reduction of conductivity arising due to disruption of π-conjugated system [76].

Philip et al. [77] prepared surface functionalized MWCNT by oxidization in presence of potassium dichromate and a phase transfer agent. Composite of polymethylmethacrylate (PMMA) with functionalized MWCNT exhibited improved detection response and reversibility for dichloromethane, chloroform and acetone. The COOH and OH groups, generated on CNTs because of oxidation, can physically interact with the PMMA matrix by hydrogen bonding which results in stronger interfacial adhesion and better dispersion of functionalized CNTs in contrast to the pristine CNTs. A small swelling of matrix results in significant increase of resistance because of large volume of conducting channel in the polymeric matrix. Moreover, adsorption of solvent molecules like methanol, even though which is not a good solvent for PMMA, may also increase polar groups on the CNT surface.

Sensing of a wide range of organic solvent vapours using polymer optical fiber sensing array fabricated with hydrophilic graphene oxide (GO) and hydrophobic reduced graphene oxide (rGO) was reported by Some et al. [78]. The comparative performance of solvent detection by GO, rGO and their combination has been shown in Fig. 10. They prepared two-headed GO/rGO, GO/GO and rGO/rGO polymer optical fiber sensors with either rGO or GO, or one head was coated with rGO and the other head with GO. Sensitivity to organic solvent vapours was found to be higher for GO compared to rGO. The as-prepared GO at pH = 5 exhibited high sensitivity toward THF (RH 90%), nitromethane (RH 50%), diethylamine (RH 60%), acetone (RH 80%), methanol (RH 90%), ethanol (RH 90%), and hydrazine (RH 60%) at 500 ppb vapour concentration.

Screen printed gas sensor was prepared by Seekaew et al. using graphene-poly (3,4-ethylenedioxythiophene): poly (styrenesulfonate) (G-PEDOT:PSS) composite [79]. Graphene was gradually added to PEDOT:PSS solution to prepare the composite which was further used for inkjet printing on interdigitated Ag electrode on a flexible and transparent substrate. The fabricated sensor was used for selective sensing of toluene, methanol, ethanol and diethylamine vapours. G-PEDOT:PSS sensor exhibited faster detection (3 min) as well as higher response (9.6%) towards the detection of 500 ppm of NH_3 while compared with pristine graphene (2.4%) or PEDOT:PSS (9.4%).

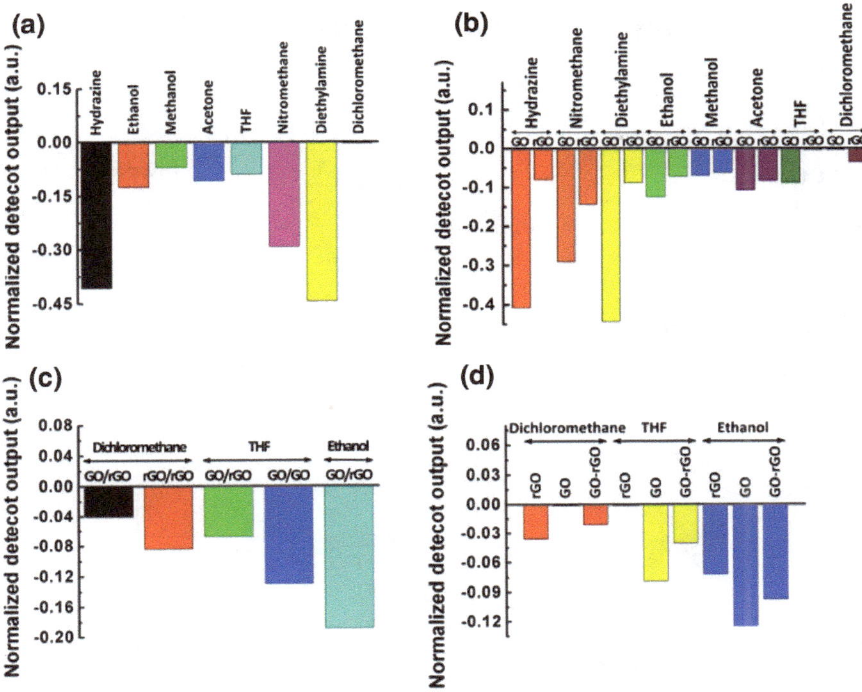

Fig. 10 Comparative sensitivity response plots toward different vapours at concentration of 500 ppb for the polymer optical fiber with **a** only GO, **b** only GO and only rGO. **c** Plot of selective sensing responses of rGO and GO polymer optical fiber to THF, dichloromethane, and ethanol with a two-headed polymer optical fiber sensor. **d** Selectivity plot of one-headed GO/rGO polymer optical fiber to THF, dichloromethane, and ethanol. Reprinted by permission from [78], © 2009, Springer Nature

Yang et al. [80] used the composite of PEDOT and RGO for sensing of NO_2 gas. Because of the π–π interactions between PEDOT and rGO sheets the charge carrier concentration increases. During the adsorption of NO_2 molecules, the number of charge carriers increases which in turn favours the decrease of the resistance of the composite.

Chen et al. performed a comparative study (Fig. 11) of the vapour sensing activities of neat carbon black (CB) and its composite with waterborne polyurethane (WPU) using 3.5% CB content in the composite [81]. It was found that the rise in vapor pressure slightly enhances the resistance increase of pristine CB. The improved transport of the organic solvent (carbon tetrachloride, chloroform, acetone, hexane and ethanol) molecules into the matrix plays an important role in the increase of the maximum responsivity of the composites. In order to respond to osmotic swelling, the polymer chains need to undergo relaxation so as to accommodate the penetrant molecules in the matrix. While measuring at equilibrium

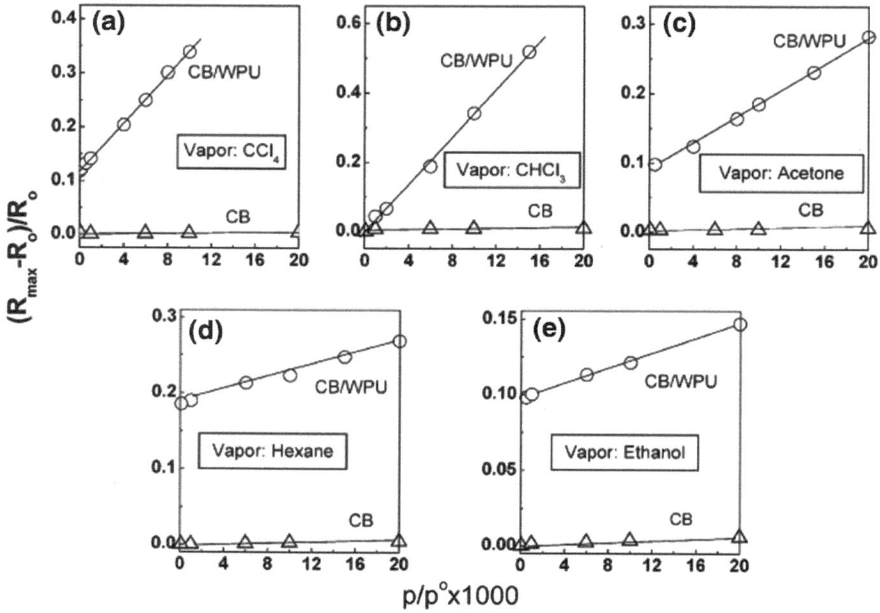

Fig. 11 Dependence of maximum responsivity on vapor pressure for carbon black filled waterborne polyurethane (CB/WPU) composites (CB content 3.5%) in comparison with pristine CB against various vapors. Reprinted from [81], © 2005, with permission from Elsevier

swelling condition, differences in solvent-matrix interactions originating from different solvents were found difficult to be seen.

Yang et al. [80] investigated the influence of porous morphology of PEDOT on the gas sensing performance of its composite with RGO based gas sensors (Fig. 12). It was observed that the RGO/porous PEDOT composite sensor with the holes of larger size and uniform distribution, exhibits enhanced sensing performance. During the long exposure of 5 ppm of NO_2 gas, the sensor device prepared by depositing the composite on an interdigitated electrode, shows fast response and recovery to the initial state rapidly after cutting off the analyte gas. It was proposed that the uniform distribution of the nanoporous structures of PEDOT facilitates the improvement of the adsorption and desorption speed of gas molecules tremendously during the course of gas sensing. However, for the porous PEDOT layer, less number of holes and their unordered distribution was more like a common PEDOT layer. The common RGO/PEDOT composite sensor exhibited long response as well as recovery time because of the inferior surface structure for gas molecule adsorption and desorption. Moreover, they have also indicated π–π interaction between porous PEDOT and RGO may increase charge carriers, which in turn results in the decrease of resistance, thus facilitating improvement of sensitivity of the composite.

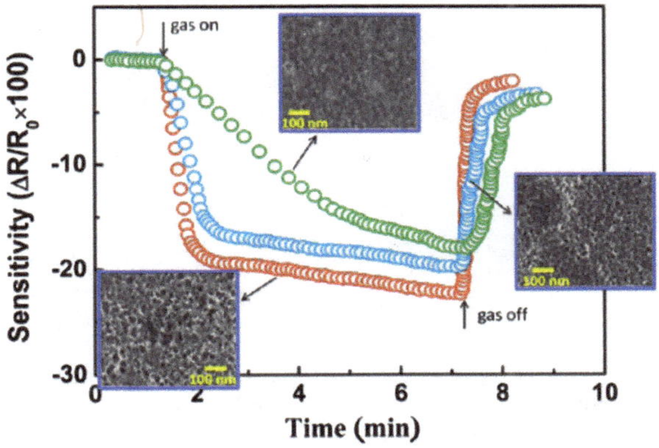

Fig. 12 Effect of porous structure on gas sensitivity for the gas sensor prepared with the composite of porous PEDOT with RGO for the detection of 5 ppm of NO_2. Reprinted with permission from [80], © 2014, American Chemical Society

Several examples of three dimensional graphene or RGO based gas sensors have been reported so far reporting good performance; however complexity of fabrication process and high cost is also a major concern for effective use. For example, a macroscopic foam-like three dimensional graphene network fabricated by chemical vapor deposition (CVD) process may be used for highly sensitive gas detection [82, 83]. However, the CVD process involves elevated temperature, expensive equipment and vacuum environment, which increases fabrication cost. Similarly, complexity of fabrication and increase of cost is applicable for the use of nickel foam template or lithography too [59, 84].

As an effective substitute of such costlier processes, Wu et al. [85] used the aqueous dispersion of reduced graphene oxide hydrogel (RGOH) by drop casting on interdigitated electrode, attached with a microheater on the back side of it. They synthesized RGOH by treating 2 mg/ml of homogeneous aqueous dispersion of GO in an autoclave and heating at 180 °C for 10 h followed by natural cooling to room temperature. For NO_2 gas sensing, the resistance of the sensor decreased considerably. The decrease of concentration of NO_2 from 10 to 2 ppm at the exposure time of 600 s led to a monotonic decrease of the response from 3.74 to 0.19%. Furthermore, a decrease of the response from 3.74 to 1.5% was also observed for the reduction of exposure time of 10 ppm of NO_2 from 600 to 300 s signifying that sufficient time was required for the gas adsorption process to reach saturation. The use of microheater improved signal recovery from 45% at 22 °C to 92% at 38 °C. On the other hand, the signal recovery for 1000 ppm of NH_3 was found to decrease from 0.67 to 0.25% for the increase of temperature from 22 to 38 °C which might be because of desorption of NH_3 molecules at elevated temperature. Due to the adsorption of p-type dopant NO_2, the hole conduction improves which leads to a remarkable decrease in resistance, while the adsorption of n-type dopant NH_3 decreases hole conduction, causing increase in resistance [59, 86]. However, as an indication of good selectivity, the RGOH sensor exhibited little response to other various gases (Fig. 13). Apart from surface adsorption, NO_2 and NH_3 molecules can also diffuse between the layers of RGOH because of relatively high adsorption

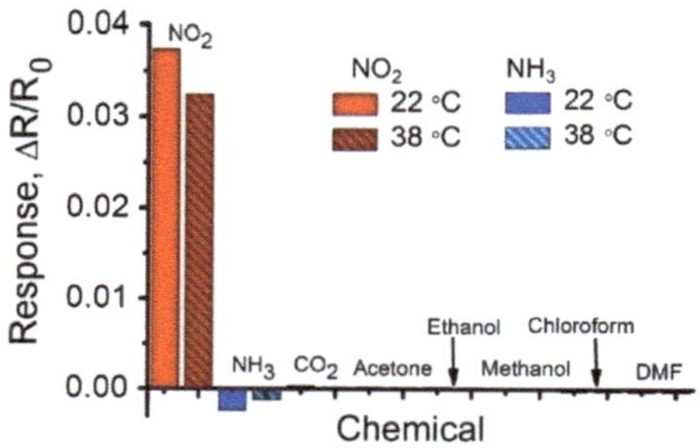

Fig. 13 The sensing response of the reduced graphene oxide hydrogel (RGOH) sensor to 600 ppm CO_2, 400 ppm NH_3, 10 ppm NO_2 and the vapours of some other organic chemicals at different temperatures. The lined and solid bars correspond to the responses at 38 and 22 °C respectively. The negative and positive values of the response correspond to increased and decreased resistance upon exposure to these gases respectively. Reprinted with permission from [85], © 2015, American Chemical Society

energy, whereas, the molecules of alcohols may interact only with the top few layers, thus resulting in relatively smaller change of resistance change [87, 88].

A SWCNT deposited cellulose paper as well as glass was used by Han et al. [89] for the investigation as humidity sensor. The schematic representation of the mechanism of sensing behaviour of the sensors is shown in Fig. 14. SWCNT was filtered using the cellulose paper and the resulting composite paper exhibited improved sensing performance at 10% relative humidity in contrast to the glass counterpart. The arrangement of CNT and cellulose fibres (CF) are depicted schematically in Fig. 14 a, b. The porosity of paper facilitates soaking of SWCNT which facilitates the charge carrier conduction through the unconnected nanotubes (Fig. 14c). The overall resistance of the SWCNT deposited cellulose paper was expressed as:

$$\frac{1}{R_{\text{Paper}}} = \frac{1}{R_{\text{CNT}}} + \frac{1}{R_{\text{CF}}} + \frac{1}{R_{\text{CNT,CF}}}$$

where, the resistance of the paper (R_{Paper}) was considered as the equivalent resistance of the resistance of CNT (R_{CNT}), cellulose fibre (R_{CF}) and CNT-cellulose fibre system ($R_{\text{CNT,CF}}$) in parallel connection. Moreover, each of these individual resistances was also considered to be connected in series:

$$R_{\text{CNT}} = \sum r_{\text{CNT}}$$

$$R_{\text{CF}} = \sum r_{\text{CF}} \text{ and}$$

$$R_{\text{CNT,CF}} = \sum r_{\text{CNT,CF}}$$

where r is the resistance of individual CNT or cellulose fibre.

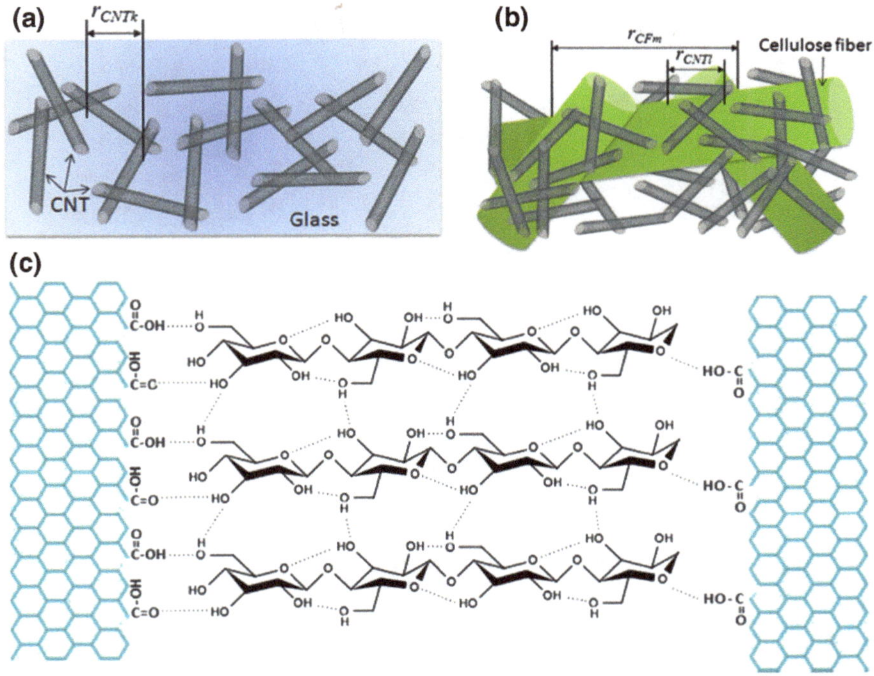

Fig. 14 Schematic representation of humidity sensing mechanism for **a** the CNT sensor on glass substrate and **b** CNT sensor on cellulose paper substrate. **c** Schematic representation of carrier conduction between unconnected carbon nanotubes assisted by cellulose fiber. Reprinted with permission from [89], © 2012, American Chemical Society

Iwaki et al. [90] reported about the model with improved signal processing for the application of thermally modulated polymer composite sensor. The composite of polyvinylpyrrolidone (PVP) was prepared with 20 wt% of carbon black followed by deposition on to a micro-hotplate using a commercial air brush through a mask made by microstereolithography. Temperature modulation at 25, 35, 45 and 55 °C was carried out by controlling voltage. The experimental temperature dependence of the sensing behaviour was found in well correlation with the theoretical prediction. Moreover, they also studied the sensing performance towards the mixture of water and methanol. Figure 15 represents the comparative results of the sensing responses towards water, methanol and the mixture of 1000 ppm of water with varying concentrations (910, 1810 and 2710 ppm) of methanol. The respective responses for individual vapours were superimposed to predict the response for the mixture of vapours. The trend of the predicted and experimental sensing towards the vapour mixtures were in well correlation, however, the experimental signal was somewhat weaker than the predicted one.

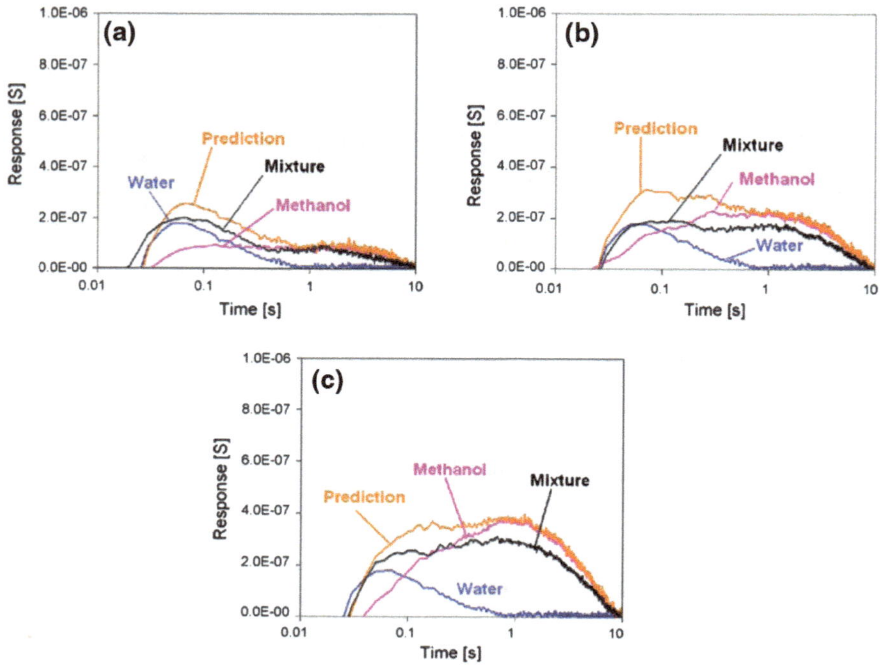

Fig. 15 Sensing performance towards the mixture of 1000 ppm water and methanol with the concentration of **a** 910 ppm, **b** 1810 ppm and **c** 2710 ppm. Reprinted from [90], © 2009, with permission from Elsevier

Hossain et al. [3] used polymeric composite with carbon black for sensing of the low concentration (2% of vapour pressure) of 1,4-benzoquinone, 1-octanol, acetone in presence of water vapour and 400 ppm of CO_2 as background. Slightly variation in the responses were observed compared to in the absence of background gases. The responses were reduced by on an average of 6–15% for poly(vinyl butaryl), poly(bisphenol-A-carbonate) and poly(vinyl acetate) based sensor while exposed to 50% water vapour.

Homer et al. [91] used 20 wt% of carbon black in six different polymers, i.e., poly 4-vinyl phenol-*co*-2-hydroxyethylmethacrylate (P4VPhcHEMA), poly 2-vinyl pyridine (PZVPy), poly *N*-vinyl pyrrolidone (PVPyr), ethyl cellulose (EC), polyvinylidene chloride-acrylonitrile (*N*) and poly methylvinylether-*alt*-maleic acid (PMVe-MAZ) for studying the gas sensing behaviour of the composites. Sensor response was found to be adversely affected by the increase of temperature. The matrix materials were ranked from greatest to least decrease in response to 100 ppm of methanol as: PVPyr, P4VPhcHEMA, PMVe-MA2, EC, PZVPy, and N.

6 Concluding Remarks

Various carbonaceous fillers like carbon black, single walled and multiwalled carbon nanotube, graphene, fullerene etc. have been extensively investigated for potential application in sensors. The matrix-filler combination played a major role in the sensing activities. Apart from pristine fillers, chemical modifications resulted in improved sensing performance in several cases. Moreover, the combination of two fillers can be used effectively, and in some cases may exhibit synergistic improved sensing behaviour. However, further researches aim at the improvement of sensitivity, reversibility and selectivity. Since sensing of mechanical strain depends on percolation phenomenon, achieving satisfactory level of strain sensing at lower loading of filler is a challenge. Similarly, the scope of further improvement of the range of temperature sensing, or the selectivity of vapours or gases are of prime interest for improvement of the performance of temperature or gas sensors. Nevertheless, the synthesis or preparation procedure may also influence significantly the sensing behaviour. Therefore, the future research in the field of the polymer composites of carbonaceous fillers may be oriented towards the optimization of the material selection, composite preparation, surface modification as well as the use of combination of fillers at different proportions to achieve enhanced response, selectivity and reversibility of the sensing behaviour.

References

1. Pramanik S, Das G, Karak N (2013) Facile preparation of polyaniline nanofibers modified bentonite nanohybrid for gas sensor application. RSC Adv 3:4574–4581
2. Al-Mashat L, Debiemme-Chouvy C, Borensztajn S, Wlodarski W (2012) Electropolymerized polypyrrole nanowires for hydrogen gas sensing. J Phys Chem C 116:13388–13394
3. Hossain ME, Freund MS, Jayas DS, White NDG, Shafai C, Thomson DJ (2012) Carbon black polymer sensor array for incipient grain spoilage monitoring. Agric Res 1:87–94
4. Wu Z, Chen X, Zhu S, Zhou Z, Yao Y, Quan W, Liu B (2013) Enhanced sensitivity of ammonia sensor using graphene/polyaniline nanocomposite. Sens Actuators B Chem 178:485–493
5. Chekanov Y, Ohnogi R, Asai S, Sumita M (1998) Positive temperature coefficient effect of epoxy resin filled with short carbon fibers. Polym J 30:381–387
6. Yasin M, Tauqeer T, Rahman HU, Karimov KS, San SE, Tunc AV (2015) Polymer-fullerene bulk heterojunction-based strain-sensitive flexible organic field-effect transistor. Arab J Sci Eng 40:257–262
7. Liao Y, Zhang C, Zhang Y, Strong V, Tang J, Li X-G, Kalantar-zadeh K, Hoek EMV, Wang KL, Kaner RB (2011) Carbon nanotube/polyaniline composite nanofibers: facile synthesis and chemosensors. Nano Lett 11:954–959
8. Obitayo W, Liu TA (2012) Review: carbon nanotube-based piezoresistive strain sensors. J Sens 2012:652438/1–652438/15
9. Laird ED, Li CY (2013) Structure and morphology control in crystalline polymer-carbon nanotube nanocomposites. Macromolecules 46:2877–2891

10. Parmar K, Mahmoodi M, Park C, Park SS (2013) Effect of CNT alignment on the strain sensing capability of carbon nanotube composites. Smart Mater Struct 22:075006/1–075006/12
11. Nan CW, Shen Y, Ma J (2010) Physical properties of composites near percolation. Ann Rev Mater Res 40:131–151
12. Hu N, Karube Y, Arai M, Watanabe T, Yan C, Li Y, Liu Y, Fukunaga H (2010) Investigation on sensitivity of a polymer/carbon nanotube composite strain sensor. Carbon 48:680–687
13. Bartolomeo AD, Sarno M, Giubileo F, Altavilla C, Iemmo L, Piano S, Bobba F, Longobardi M, Scarfato A, Sannino D, Cucolo AM, Ciambelli P (2009) Multiwalled carbon nanotube films as small-sized temperature sensors. J Appl Phys 105:064518/1–064518/6
14. Kong D, Le LT, Li Y, Zunino JL, Lee W (2012) Temperature-dependent electrical properties of graphene inkjet-printed on flexible materials. Langmuir 28:13467–13472
15. Zhuge F, Hu B, He C, Zhou X, Liu Z, Li R (2011) Mechanism of nonvolatile resistive switching in graphene oxide thin films. Carbon 49:3796–3802
16. Sahoo S, Barik SK, Sharma GL, Khurana G, Scott JF, Katiyar RS (2012) Reduced graphene oxide as ultra fast temperature sensor. arXiv:1204.1928v1 [cond-mat.mes-hall]
17. Matzeu G, Pucci A, Savi S, Romanelli M, Di Francesco F (2012) A temperature sensor based on a MWCNT/SEBS nanocomposite. Sens Actuators A Phys 178:94–99
18. Oskouyi AB, Sundararaj U, Mertiny P (2014) Effect of temperature on electrical resistivity of carbon nanotubes and graphene nanoplatelets nanocomposites. J Nanotechnol Eng Med 5:044501/1–044501/6
19. Hou YH, Zhang MQ, Rong MZ (2003) Carbon black-filled polyolefins as positive temperature coefficient materials: the effect of in situ grafting during melt compounding. J Polym Sci B Polym Phys 41:127–134
20. He XJ, Du JH, Ying Z, Cheng HM (2005) Positive temperature coefficient effect in multiwalled carbon nanotube/high-density polyethylene composites. Appl Phys Lett 86:062112/1–062112/3
21. He L, Tjong S-C (2014) Electrical behavior and positive temperature coefficient effect of graphene/polyvinylidene fluoride composites containing silver nanowires. Nanoscale Res Lett 9:375/1–375/8
22. Karimov KS, Chani MTS, Khalid FA (2011) Carbon nanotubes film based temperature sensors. Phys E 43:1701–1703
23. Sibinski M, Jakubowska M, Sloma M (2010) Flexible temperature sensors on fibers. Sensors 10:7934–7946
24. Hong SW, Kim DY, Lee JU, Jo WH (2009) Synthesis of polymeric temperature sensor based on photophysical property of fullerene and thermal sensitivity of poly (N-isopropylacrylamide). Macromolecules 42:2756–2761
25. Calisi N, Giuliani A, Alderighi M, Schnorr JM, Swager TM, Francesco FD, Pucci A (2013) Factors affecting the dispersion of MWCNTs in electrically conducting SEBS nanocomposites. Eur Polym J 49:1471–1478
26. Giuliani A, Placidi M, Francesco FD, Pucci A (2014) A new polystyrene-based ionomer/MWCNT nanocomposite for wearable skin temperature sensors. React Funct Polym 76:57–62
27. Lin L, Deng H, Gao X, Zhang S, Bilotti E, Peijs T, Fu Q (2013) Modified resistivity–strain behavior through the incorporation of metallic particles in conductive polymer composite fibers containing carbon nanotubes. Polym Int 62:134–140
28. Window AL (1992) Strain gauge technology. Springer, Berlin
29. Anand SV, Mahapatra DR (2009) Quasi-static and dynamic strain sensing using carbon nanotube/epoxy nanocomposite thin films. Smart Mater Struct 18:045013/1–045013/13
30. Liu H, Gao J, Huang W, Dai K, Zheng G, Liu C, Shen C, Yan X, Guo J, Guo Z (2016) electrically conductive strain sensing polyurethane nanocomposites with synergistic carbon nanotubes and graphene bifillers. Nanoscale 8:12977–12989
31. Liu H, Li Y, Dai K, Zheng G, Liu C, Shen C, Yan X, Guo J, Guo Z (2016) Electrically conductive thermoplastic elastomer nanocomposites at ultralow graphene loading levels for strain sensor applications. J Mater Chem C 4:157–166

32. Zhao J, Dai K, Liu C, Zheng G, Wang B, Liu C, Chen J, Shen C (2013) A comparison between strain sensing behaviors of carbon black/polypropylene and carbon nanotubes/polypropylene electrically conductive composites. Compos A Appl Sci Manuf 48:129–136
33. Ferrreira A, Rocha J, Ansón-Casaos A, Martínez M, Vaz F, Lanceros-Mendez S (2012) Electromechanical performance of poly (vinylidene fluoride)/carbon nanotube composites for strain sensor applications. Sens Actuators A Phys 178:10–16
34. Ferreira A, Martínez M, Ansón-Casaos A, Gómez-Pineda L, Vaz F, Lanceros-Mendez S (2013) Relationship between electromechanical response and percolation threshold in carbon nanotube/poly (vinylidene fluoride) composites. Carbon 61:568–576
35. Lin L, Liu S, Zhang Q, Li X, Ji M, Deng H, Fu Q (2013) Towards tunable sensitivity of electrical property to strain for conductive polymer composites based on thermoplastic elastomer. ACS Appl Mater Interfaces 5:5815–5824
36. Li J, Zhao S, Zeng X, Huang W, Gong Z, Zhang G, Sun R, Wong C-P (2016) Highly stretchable and sensitive strain sensor based on facilely prepared three-dimensional graphene foam composite. ACS Appl Mater Interfaces 8:18954–18961
37. Kang I, Schulz MJ, Kim JH, Shanov V, Shi D (2006) A carbon nanotube strain sensor for structural health monitoring. Smart Mater Struct 15:737–748
38. Chen J, Du X-C, Zhang W-B, Yang J-H, Zhang N, Huang T, Wang Y (2013) Synergistic effect of carbon nanotubes and carbon black on electrical conductivity of PA6/ABS blend. Compos Sci Technol 81:1–8
39. Ma P-C, Liu M-Y, Zhang H, Wang S-Q, Wang R, Wang K, Wong Y-K, Tang B-Z, Hong S-H, Paik K-W (2009) Enhanced electrical conductivity of nanocomposites containing hybrid fillers of carbon nanotubes and carbon black. ACS Appl Mater Interfaces 1:1090–1096
40. Socher R, Krause B, Hermasch S, Wursche R, Pötschke P (2011) Electrical and thermal properties of polyamide 12 composites with hybrid fillers systems of multiwalled carbon nanotubes and carbon black. Compos Sci Technol 71:1053–1059
41. Ke K, Pötschke P, Wiegand N, Krause B, Voit B (2016) Tuning the network structure in poly (vinylidene fluoride)/carbon nanotube nanocomposites using carbon black: towards improvements of conductivity and piezoresistive sensitivity. ACS Appl Mater Interfaces 8:14190–14199
42. Tuukkanen S, Hoikkanen M, Poikelispää M, Honkanen M, Vuorinen T, Kakkonen M, Vuorinen J, Lupo D (2014) Stretching of solution processed carbon nanotube and graphene nanocomposite films on rubber substrates. Synth Met 191:28–35
43. Jang H, Yoon H, Ko Y, Choi J, Lee S-S, Jeon I, Kim J-H, Kim H (2016) Enhanced performance in capacitive force sensors using carbon nanotube/polydimethylsiloxane nanocomposites with high dielectric properties. Nanoscale 8:5667–5675
44. Eswaraiah V, Balasubramaniam K, Ramaprabhu S (2012) One-pot synthesis of conducting graphene-polymer composites and their strain sensing application. Nanoscale 4:1258–1262
45. Li X, Levy C, Elaadil L (2008) Multiwalled carbon nanotube film for strain sensing. Nanotechnology 19:045501/1–045501/8
46. Dinh TN, Steitz J, Bu L, Kanoun O (2009) Influence of the composition of MWCNTs layers on the properties of strain gauges. In: Proceedings of the 9th IEEE Conference on Nanotechnology, Genoa, Italy, 26–30:477–480
47. Kanoun O, Müller C, Benchirouf A, Sanli A, Dinh TN, Al-Hamry A, Bu L, Gerlach C, Bouhamed A (2014) Flexible carbon Nanotube films for high performance strain sensors. Sensors 14(6):10042–10071
48. Loh KJ, Lynch JP, Shim BS, Kotov NA (2008) Tailoring piezoresistive sensitivity of multilayer carbon nanotube composite strain sensors. J Intell Mater Syst Struct 19:747–764
49. Loh KJ, Kim J, Lynch JP, Kam NWS, Kotov NA (2007) Multifunctional layer-by-layer carbon nanotube–polyelectrolyte thin films for strain and corrosion sensing. Smart Mater Struct 16:429–438
50. Pham GT, Park Y-B, Liang Z, Zhang C, Wang B (2008) Processing and modeling of conductive thermoplastic/carbon nanotube films for strain sensing. Compos B Eng 39:209–216

51. Ku-Herrera JJ, Avilés F (2012) Cyclic tension and compression piezoresistivity of carbon nanotube/vinylester composites in the elastic and plastic regimes. Carbon 50:2592–2598
52. Slobodian P, Riha P, Saha P (2012) A highly-deformable composite composed of an entangled network of electrically-conductive carbon-nanotubes embedded in elastic polyurethane. Carbon 50:3446–3453
53. Yin G, Hu N, Karube Y, Liu Y, Li Y, Fukunaga H (2011) A carbon nanotube/polymer strain sensor with linear and anti-symmetric piezoresistivity. J Compos Mater 45:1315–1323
54. Park M, Kim H, Youngblood JP (2008) Strain-dependent electrical resistance of multi-walled carbon nanotubes/polymer composite films. Nanotechnology 19:055705/1–055705/7
55. Hu N, Itoi T, Akagi T, Kojima T, Xue J, Yan C, Atobe S, Fukunaga S, Yuan W, Ning H, Surina, Liu Y, Alamusi (2013) Ultrasensitive strain sensors made from metal-coated carbon nanofiller/epoxy composites. Carbon 51:202–212
56. Oliva-Avilés AI, Avilés F, Sosa V (2011) Electrical and piezoresistive properties of multi-walled carbon nanotube/polymer composite films aligned by an electric field. Carbon 49:2989–2997
57. Christie S, Scorsone E, Persaud K, Kvasnik F (2003) Remote detection of gaseous ammonia using the near infrared transmission properties of polyaniline. Sens Actuators B Chem 90:163–169
58. Bendahan M, Lauque P, Lambert-Mauriat C, Carchano H, Seguin JL (2002) Sputtered thin films of CuBr for ammonia microsensors: morphology, composition and ageing. Sens Actuators B Chem 84:6–11
59. Duy LT, Kim D-J, Trung TQ, Dang VQ, Kim B-Y, Moon HK, Lee N-E (2015) High performance three-dimensional chemical sensor platform using reduced graphene oxide formed on high aspect-ratio micro-pillars. Adv Funct Mater 25:883–890
60. ACGIH (2005) In 2005 TLVs and BEIs based on the document of the threshold limit values for chemical substances and physical agents & biological exposure indices. American Conference of Governmental Industrial Hygienists (ACGIH): Cincinnati, Ohio pp 8–29
61. Berber MR, Hafez IH (eds) (2016) Carbon nanotubes—current progress of their polymer composites. In Tech, p 460
62. Yun J, Im JS, Kim H-I, Lee Y-S (2012) Effect of oxyfluorination on gas sensing behavior of polyaniline-coated multi-walled carbon nanotubes. Appl Surf Sci 258:3462–3468
63. Ponnamma D, Sadasivuni KK, Strankowski M, Guo Q, Thomas S (2013) Synergistic effect of multi walled carbon nanotubes and reduced graphene oxides in natural rubber for sensing application. Soft Matter 9:10343–10353
64. Loffredo F, Del Mauro ADG, Burrasca G, Ferrara VL, Quercia L, Massera E, Francia GD, Della Sala D (2009) Ink-jet printing technique in polymer/carbon black sensing device fabrication. Sens Actuators B Chem 143:421–429
65. Wei C, Dai L, Roy A, Tolle TB (2006) Multifunctional chemical vapor sensors of aligned carbon nanotube and polymer composites. J Am Chem Soc 128:1412–1413
66. Xie H, Wu J, Huang P, Ji X, Hunag Y (2006) The study of the gas microsensors based on polymer-carbon black composites. In: 2006 8th international conference on solid-state and integrated circuit technology proceedings, Shanghai, pp 637–639
67. Hernández-López S, Vigueras-Santiago E, Mora MM, Mancill JRF, Contreras EAZ (2013) Cellulose-based polymer composite with carbon black for tetrahydrofuran sensing. Int J Polym Sci 2013:381653/1–381653/7
68. Holmes MA, Mackay ME, Giunta RK (2007) Nanoparticles for dewetting suppression of thin polymer films used in chemical sensors. J Nanopart Res 9:753–763
69. Pede D, Smela E, Johansson T, Johansson M, Inganäs O (1998) A general-purpose conjugated-polymer device array for imaging. Adv Mater 10:233–237
70. Dai L, Soundarrajan P, Kim T (2002) Sensors and sensor arrays based on conjugated polymers and carbon nanotubes. Pure Appl Chem 74:1753–1772
71. Zheng J, Ma X, He X, Gao M, Li G (2012) Preparation, characterizations, and its potential applications of PANi/ graphene oxide nanocomposite. Procedia Eng 27:1478–1487

72. Jang WK, Yun J, Kim H-I, Lee Y-S (2012) Improvement in ammonia gas sensing behavior by polypyrrole/multi-walled carbon nanotubes composites. Carbon Lett 13:88–93
73. Moniruzzaman M, Winey KI (2006) Polymer nanocomposites containing carbon nanotubes. Macromolecules 39:5194–5205
74. Lu J, Kumar B, Castro M, Feller J-F (2009) Vapour sensing with conductive polymer nanocomposites (CPC): polycarbonate-carbon nanotubes transducers with hierarchical structure processed by spray layer by layer. Sens Actuators B Chem 140:451–460
75. Kobashi K, Villmow T, Andres T, Pötschke P (2008) Liquid sensing of melt-processed poly(lactic acid)/multi-walled carbon nanotube composite films. Sens Actuators B Chem 134:787–795
76. Salavagione H, Díez-Pascual AM, Lázaro E, Vera S, Gómez-Fatou MA (2014) Chemical sensors based on polymer composites with carbon nanotubes and graphene: the role of the polymer. J Mater Chem A 2:14289–14328
77. Philip B, Abraham JK, Chandrasekhar A, Varadan VK (2003) Carbon nanotube/PMMA composite thin films for gas-sensing applications. Smart Mater Struct 12:935–939
78. Some S, Xu Y, Kim Y, Yoon Y, Qin H, Kulkarni A, Kim T, Lee H (2013) Highly sensitive and selective gas sensor using hydrophilic and hydrophobic graphenes. Sci Rep 3:1868/1–1868/8
79. Seekaew Y, Lokavee S, Phokharatkul D, Wisitsoraat A, Kerdcharoen T. Wongchoosuk C (2014) Low-cost and flexible printed graphene-PEDOT:PSS gas sensor for ammonia detection. Org Electron 15:2971–2981
80. Yang Y, Li S, Yang W, Yuan W, Xu J, Jiang Y (2014) In situ polymerization deposition of porous conducting polymer on reduced graphene oxide for gas sensor. ACS Appl Mater Interfaces 6:13807–13814
81. Chen SG, Hu JW, Zhang MQ, Rong MZ (2005) Effects of temperature and vapor pressure on the gas sensing behavior of carbon black filled polyurethane composites. Sens Actuators B Chem 105:187–193
82. Yavari F, Chen Z, Thomas AV, Ren W, Cheng HM, Koratkar N (2011) High Sensitivity gas detection using a macroscopic three-dimensional graphene foam network. Sci Rep 1:166/1–166/5
83. Yang G, Lee C, Kim J (2013) Three-dimensional graphene network-based chemical sensors on paper substrate. J Electrochem Soc 160:B160–B163
84. Wu J, Yu CH, Li SZ, Zou BH, Liu YY, Zhu XQ, Guo YY, Xu HB, Zhang WN, Zhang LP, Liu B, Tian DB, Huang W, Sheetz MP, Huo FW (2015) Parallel near-field photolithography with metal-coated elastomeric masks. Langmuir 31:1210–1217
85. Wu J, Tao K, Miao JM, Norford L (2015) Improved selectivity and sensitivity of gas sensing using 3D reduced graphene oxide hydrogel with integrated microheater. ACS Appl Mater Interfaces 7:7502–7510
86. Fowler JD, Allen MJ, Tung VC, Yang Y, Kaner RB, Weiller BH (2009) Practical chemical sensors from chemically derived graphene. ACS Nano 3:301–306
87. Lipatov A, Varezhnikov A, Wilson P, Sysoev V, Kolmakov A, Sinitskii A (2013) Highly selective gas sensor arrays based on thermally reduced graphene oxide. Nanoscale 5:5426–5434
88. Lipatov A, Varezhnikov A, Augustin M, Bruns M, Sommer M, Sysoev V, Kolmakov A, Sinitskii A (2014) Intrinsic device-to-device variation in graphene field-effect transistors on a Si/SiO_2 substrate as a platform for discriminative gas sensing. Appl Phys Lett 104:013114/1–013114/5
89. Han J-W, Kim B, Li J, Meyyappan M (2012) Carbon nanotube based humidity sensor on cellulose paper. J Phys Chem C 116:22094–22097
90. Iwaki T, Covington JA, Udrea F, Gardner JW (2009) Identification and quantification of different vapours using a single polymer chemoresistor and the novel dual transient temperature modulation technique. Sens Actuators B 141:370–380
91. Homer ML, Lim JR, Manatt K, Kisor A, Lara L, Jewell AD, Shevade A, Yen S-PS, Zhou H, Ryan MA (2003) Using temperature effects on polymer-composite sensor arrays to identify analytes. Sensors, P IEEE 1:144–147

The Use of Polymer–Carbon Composites in Fuel Cell and Solar Energy Applications

Aniruddha Chatterjee, D. P. Hansora, Purabi Bhagabati and Mostafizur Rahaman

Abstract Energy conversion devices such as solar cells and fuel cells are considered as highly efficient devices that are capable of transforming and conserving energy directly from the chemical energy into electricity. These are the devices which do not cause evolution of any noxious pollutants to the environment. The different carbon/polymer composites have concerned immense interest as functional components in solar cells and fuel cells as energy conversion applications. In this chapter, the point here is to give initiative about basic perceptive of different types of carbon structured polymer nanocomposites, their characteristic properties, and applications for the fuel cell and solar energy devices.

Keywords Polymer–carbon composites · Fuel cells · Solar energy

A. Chatterjee (✉)
Maharashtra Institute of Technology, Aurangabad 431010, Maharashtra, India
e-mail: aniruddha.chatterjee@mit.asia

D. P. Hansora
University Institute of Chemical Technology, North Maharashtra University, Jalgaon 425001, Maharashtra, India
e-mail: 2568dharmesh@gmail.com

P. Bhagabati
Chemical Engineering Department, Indian Institute of Technology Guwahati, Guwahati 781039, India
e-mail: purabi.bhagabati08@gmail.com

M. Rahaman
Department of Chemistry, College of Science, King Saud University, Riyadh 11451, Saudi Arabia
e-mail: mrahaman@ksu.edu.sa

M. Rahaman et al. (eds.), *Carbon-Containing Polymer Composites*, Springer Series on Polymer and Composite Materials,
https://doi.org/10.1007/978-981-13-2688-2_15

1 Introduction

Major global issues related to energy and environments induce the pollution problems related to the environment. Energy production from conventional sources such as fossil fuels is recognized as the primary reason for the degradation of our environment. Considerable efforts have been made since the past 25 years to utilize the excellent characteristics of carbon nanomaterials, such as fullerenes, carbon nanotubes (CNTs), and graphene, as a source to produce energy in different forms. A remarkable progress has been attained for the development of high-performance energy conversing and storing devices (e.g., fuel cells and solar cells). Nanoscience and technology play a crucial role in revolutionizing the energy conversion devices. Hybrid nanocomposites made from CNTs possess peculiar electronic and mechanical properties, and as a result, they have been enjoying immense attention from scientists and technologists toward its applications in the diversified field. Integrating and exploring the carbon/polymer nanocomposites in solar and fuel cells devices has increased the energy conversion efficiency of these devices, which can serve as the future sustainable energy resources. Carbon–polymer nanocomposites materials have been also used successfully in energy storage devices such as electrochemical batteries [1–8].

CNTs have been used as excellent materials for absorption of sunlight and for generation of photo-carriers. There are mainly two types of CNTs: one is known as single wall (SWCNTs) and another is known as multiwall (MWCNTs). Both are in the range nanometric range of its diameter and compatible for organic solar cells due to their properties. Diameter up to 20 nm of CNTs is preferred for OPV device applications, and typically the diameter of SWCNTs and MWCNTs lies within a range of 2–10 and 5–100 nm, respectively. On the other hand, both CNTs and conducting polymers like polyaniline possess conjugated π-systems, which are capable of conducting electrons from one end to the other. Their behavior of electronic interaction is also projected to take place through π–π electron stacking. Surface functionalized CNTs are capable to enhance the compatibility in water and organic solvents. Moreover, functionalization of CNTs can improve the interfacial compatibility with polymer matrix to avoid CNT aggregation in CNT/polymer nanocomposites before their applications in energy conversion devices [9, 10].

Fuel cell efficiently transforms chemical energy into the electrical energy in hydrogen-rich materials. Conducting polymer composites-based electrolyte membranes are superior to the membrane made from standard materials, which is because of ease handling, high energy density and low operating temperature of fuel except for hydrogen. The contribution of CNTs is reviewed and reported due to its proton conductivity and improved mechanical strength of PEMFC. Their role in electrocatalysis has been studied with respect to facilitating the use of noble metal catalysts (such as platinum) and also exploring the use of platinum-free catalysts [11, 12]. Even though many researchers have reported about conductive polymer composites (CPCs) system as bipolar plates for the application of PEMFCs and PV devices, in-depth research is necessary to study the combined system of hybrid fillers bipolar plate materials for the better electrical conductivity of the conducting polymer composites. Therefore, some carbon-based conductive fillers are required, which include GO, CNTs, RGO, graphite, carbon fibers, and carbon black that can be used as reinforced filler to improve the overall performance of CPCs for the application in bipolar plates [1]. Yuan et al. [6] reported redox flow batteries and lithium ion flow batteries developed using various polymer/carbon nanocomposites (such as PEDOT/CNTs, Nafion/CNTs, PEG/CNTs, PI/CNTs, PANI/CNTs, PPY/CNTs, and PEDOT/CNTs). Tan et al. [8] reviewed CNT-based composites for fuel cell and solar cell applications.

Solar energy emerged as a great potential for sustainable and clean energy source because of its plenty and more regular distribution in nature compared to other competing renewable sources. Solar cells with improved efficiencies have been a keen interest due to the need for green energy sources and most commercially used cells today. Development of various solar cells has shown considerable attention since the last two decades, which includes first-generation silicon (Si) based solar cells, second-generation semiconductor thin films-based solar cells and recent development of third generation dye-sensitized solar cell (DSSC) also known as hybrid cells, and organic semiconductor solar cells or organophotovoltaic (OPV) cells. New generation of solar cells including thin films based DSSC, quantum DSSC, and polymer/organic solar cells possesses the excellent capacity to be used as PV devices due to their high energy conversion performance and low fabrication cost. One of the developing areas of these solar cells is the addition of CNTs to the various components of the devices. Comprehensive research has been conducted in order to attain high photoconversion efficiency of solar cells after addition of CNTs into the device like nanoarchitecture on the working electrode surface, CNTs composite as counter electrode, and addition of CNTs to the electrolyte [8, 10].

Dai et al. [13] reviewed carbon nanomaterials (fullerenes, CNTs, graphene) for energy conversion devices as solar cells made from based on polymer/carbon nanocomposites. Alturaif et al. [10] reviewed solar cells based on polymer/carbon nanocomposites. Composites of CNTs with small molecules and polymers are most widely studied structures used in solar cell design, where CNTs took the role of an electron acceptor and light is getting absorbed through the CNT flattering element. Polymer and composites materials were also studied for energy applications [14].

Polymer/carbon nanocomposites are promising potential materials for energy conversion and storage devices due to their merits such as mechanical strength, conductivity, flexibility, and cost. In this chapter, different conducting polymer/carbon nanocomposites made from different carbon materials like graphite, carbon blacks, carbon fibers, CNTs and graphene, GO, and RGO are discussed for fuel cell and energy device applications.

2 Carbon/Polymer Composites in Fuel Cell Applications

Carbonaceous nanomaterials are very important for the better performance of PEMFC irrespective of their use in cathode, anode and in bipolar plates [1, 3, 4, 11, 15–17]. Polymer/carbon nanocomposites have been used in fuel cell such as PEMFC, alkaline fuel cell, molten carbonate fuel cell, phosphoric fuel cell, and solid oxide fuel cell [1, 3, 4, 11, 15–17]. The use of nitrogen-doped functionalized CNTs refers to the possibility of total platinum removal from membrane electrolyte assembly in the near future. Some of these recent advances in the application of CNTs and related materials in the design of PEMFC have been reviewed and discussed by Kannan et al. [4]. The use of polymer nanocomposites-based electrolytes with the help of selectively functionalized CNTs opens prospects to combine the nanostructure-tuning capabilities of CNTs with the unique ionic transport properties of polymeric membrane electrolytes to enable key benefits although the Nafion–sulphonated CNTs are in areas of hydrogen–oxygen fuel cells exploiting the nanostructured electrolyte membrane's ionic transport and improved mechanical properties [4].

Bairan et al. [1] studied the bipolar plate made from polypropylene (PP)/carbon nanomaterials (graphite, carbon black, CNTs) based composites for PEM fuel cell applications. They found that CNTs as third filler (6 wt% loading) in a G/CB/PP composite showed the density, shore hardness, and in-plane electrical conductivity of 1.64 g/cm^3, 81.5 (SH), and 158.32 S/cm, respectively. Moreover, the optimum value of flexural strength was reported to be 29.86 MPa for 5 wt% of CNT [1].

Singh et al. [3] developed polymer–metal nanoparticle-carbon micro-nanofiber-based nanocomposites are efficient material for airborne cathode in a membraneless microbial fuel cell having only one cell. The polymer with polyvinyl alcohol (PVA) and poly methyl vinyl ether-alt-maleic anhydride (PMVEMA) was used as a monomer and a cross-linking agent, respectively. The polymer composites were coated over a multi-scale web of activated carbon fibers (ACF) and carbon nanofibers (CNF) where alumina (AA) and Ni nanoparticles (NPs) were dispersed. Figure 1 shows the SEM micrographs of the ACF in micrometer size with smooth surface with uniform pores on the surface (Fig. 1a). A uniform dispersion of metal NPs was also observed at both places inside the pores as well as on the surface of the ACFs (Fig. 1b). SEM images of the samples with low magnification can also be observed in the insets of Fig. 1a–d [3].

Novel cross-linked polymer (PVA-PMVEMA) composites were prepared and coated on the ACFs/CNFs using transition metal (AA/Ni) NPs. These polymer–metal–carbon nanocomposites were used as an efficient material for air-cathode in SMFC without any requirement of proton exchange membrane. The polymer composites permitted the passage of only cations (Hþ), except the anions. Different CNT/polymer-based nanocomposites like CNT/Nafion and CNT/SPEEK possess excellent physical properties (e.g., mechanical strength, chemical stability, ion selectively, good cell performances) and have been successfully used in fuel cell for energy application. For obtaining the good properties of CNT and SPEEK matrix and also eliminating their individual disadvantages, short-carboxylic MWCNTs were well

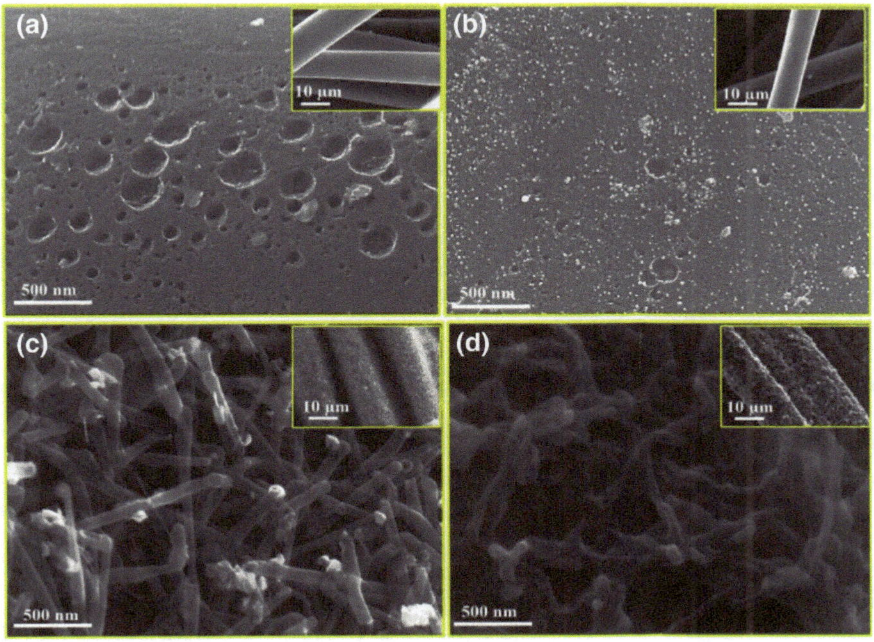

Fig. 1 SEM images of the **a** activated carbon fiber, **b** alumina: nickel-activated carbon fiber, **c** alumina: nickel-activated carbon fiber/carbon nanofiber, and **d** polymer–alumina: nickel-activated carbon fiber/carbon nanofiber (inset images for all cases are at low magnification). Reprinted with permission from Ref. [3]

dispersed in the traditional SPEEK membrane for better interaction. Besides, proper dispersion of CNTs and van der Waals interactions, good electrostatic interactions and π–π interactions between the aromatic groups of SPEEK and the sidewall of CNTs enhanced the mechanical strength drastically (61.8 MPa) as compared to a pristine SPEEK membrane (39 MPa) [3]. In situ chemically oxidative synthesized carbon/poly(3,4-ethylene dioxythiophene) (C/PEDOT) composites were also developed for carbon-supported catalysts used in PEMFC [18]. Zhang et al. [19] reviewed carbon nanohorns/polymer composites as catalysts in fuel cells applications. Zhu et al. [20] also examined about various graphene–polymer nanocomposites and reported as a potential material for building the ideal anodes in fuel cells which possess excellent electrical conductivity and mechanical properties, profuse surface functionality, and exceptional electrochemical properties. Polyaniline (PANI)-graphene nanosheets (GNS) modified cathodes were prepared by in situ polymerization of aniline in uniformly dispersed GNS solution. The PANI-GNS electrode was used as the cathode of MFCs to improve the capacity of electricity generation and ORR efficiency. Similarly, PEDOT/graphite, PANI/RGO and polypyrrole (PPy)/RGO, PPy/GO, and PPyERGO nanocomposites were also reported to be used as anode catalysts in fuel cells [20].

Huyen et al. [11] reviewed the future research trend for development of CPs/CNTs systems. Literature indicates that CNTs have great potential to be used as multi-functional materials for better PEMFC performance [12]. The PANI/CNTs nanocomposites have been used as proficient electrocatalyst in cathode–anode fuel cell reactions like oxygen reduction and methanol oxidation. Generally, CNTs provide high surface area and show better electronic conductivity, while PANI facilitates the appropriate electron transfer throughout the resulting conducting polymer matrix. MWCNTs-*g*-PANI composites showed approximately 610 mV more positive current onset potential for the two-electron oxygen reduction reaction. A poly(*o*-phenylenediamine)–MWCNTs composite was reported to have a polymer redox-mediated electrocatalytic effect for oxygen reduction reaction with a fivefold enhancement in current and a positive potential shift of 130 mV compared to the values acquired at a pure poly(*o*-phenylenediamine) electrode. The PANI/CNTs composites are outstanding host matrices for metal NPs as resulting composite materials can be used as electrode materials in methanol oxidation reaction for fuel cell application. The size of the metal nanoparticles embedded on the composite matrix was reported to be smaller than that on pristine PANi material. As a result, a higher dispersion can be achieved and hence metal nanoparticles-impregnated composites can be used in an optimized way for better and high performance and stability. The PANi/CNTs/metal nanoparticles-based composites have shown a reduced poisoning effect from adsorbed carbon monoxide other than enhanced electrocatalytic activity effect. PANI/CNTs composites-based anode in microbial fuel cell, in 0.1 M phosphate buffer consisting of 5.5 mM of glucose, *Escherichia coli* bacteria as the microbial catalyst, and 2-hydroxy-1,4-naphthoquinone as mediator, have been reported with a capability of cell voltage of 450 mV and a power density of 42 mW/m^2. The polymer nanocomposites-based membrane exhibited higher thermal and ionic conductivity associated with the lower fuel drag than that of bare polymeric materials due to the synergetic effect of the ingredient properties. In near future, the combined effort of both industrial and academic research activities can be triggered by the development of polymer nanocomposites-based electrolyte membranes for their commercialization in fuel cells [11, 12]. Sharma et al. [17] summarized various CNT–polymer nanocomposites based on cathode catalysis in PEM fuel cell as well as challenges and progress in the application of various carbon nanomaterials (CNTs and graphene) owing to their promising properties such as corrosion resistance, high electronic conductivity, and large surface area.

3 Carbon/Polymer Composites in Solar Energy Applications

Dai et al. [13] reviewed carbon nanomaterials (fullerenes, CNTs, graphene) for solar cells based on polymer/carbon nanocomposites. The observation of PV effects arises from the photo-induced charge transfer at the interface between C_{60} film as an acceptor and conjugated polymers as donors. This suggests attractive openings for enhancing the energy conversion productivities of conjugated CPC-based PV cells. Solar cells with bilayer heterojunction, interpenetrating networks consist of two semiconducting polymers are ideal as PV materials for a high efficiency of PV conversion. The CPC structures with interpenetrating network provide the spatially dispersed interfaces essential for both a resourceful photogeneration as well as a facile collection of the electrons and holes. The thermal annealing technique improves the efficiency of polymer solar cells. It was observed that the photoluminescence, photoconductivity, and electroluminescence can be improved by enhancing the wettability between the surface of CP and MWCNTs. Bilayered polymer (poly(3- hexylthiophene) (P3HT)/PCBM) based solar cells with a concentration gradient exhibited an improved power-conversion efficiency and photocurrent density as compared to those of the conventional heterojunction PV cells. The C_{60} layer in bilayered devices generally provides an additional protection to the polymer layer from degradation due to oxygen and humidity, which overall improves the stability of device. The efficiency of P3HT/P3OT-based PV cells was strongly influenced after addition of functionalized graphene as an acceptor, which indicates an efficient charge transfer to graphene acceptor from the polymer as donor. However, the device performance depends strongly on the concentration of graphene, annealing time, and temperature, respectively. The graphene quantum dots (GQDs) as electron acceptors, electrochemically synthesized with a uniform size of 3–5 nm, have been developed in application of P3HT-based solar cells. The power-conversion capacity of P3HT devices based on GQDs as acceptors was reported to be improved greatly due to the effective charge transport and separation [13].

Kymakis et al. [21] reported spin-casted composite films made from poly (3-octylthiophene) (P3OT) and SWCNTs. The film was formed onto a substrate made of indium tin oxide (ITO), which was coated on quartz. According to his observations, Al/P3OT–SWCNT/ITO structure-based diodes (Fig. 2A) possess PV characteristic of an open circuit voltage of 0.7–0.9 V at very low CNT concentration (<1%). The short circuit current (I_{sc}) was reported to increase by twofold compared to the neat polymer-based devices (as shown in Fig. 2B), but an increasing fill factor was observed within the range 0.3–0.4 for the CNT–polymer-based solar cell. This improvement in performance of the device was ascribed due to the better electronic properties and large surface areas of CNTs. When dye molecule was added into the polymer–SWCNT based solar cell, there was a substantial improvement of the light absorption capability especially in the UV as well as red region that led to five times increment in the short circuit current (I_{sc}) [13, 21].

The one-atom thick 2D carbon networks of graphene have a larger interface in a polymer matrix than that of CNTs. Hence, there was a lesser chance of short circuit through the PV. Similar to 0D-fullerenes and 1D-CNTs, the 2D-graphene nanosheets are new class of transport materials and have been used as acceptors in polymer solar cells (PSCs) [13, 22].

Yu et al. [22] have reported the preparation of PV devices based on soluble P3HT grafted graphene (G–P3HT) those are bilayered and heterojunction in nature.

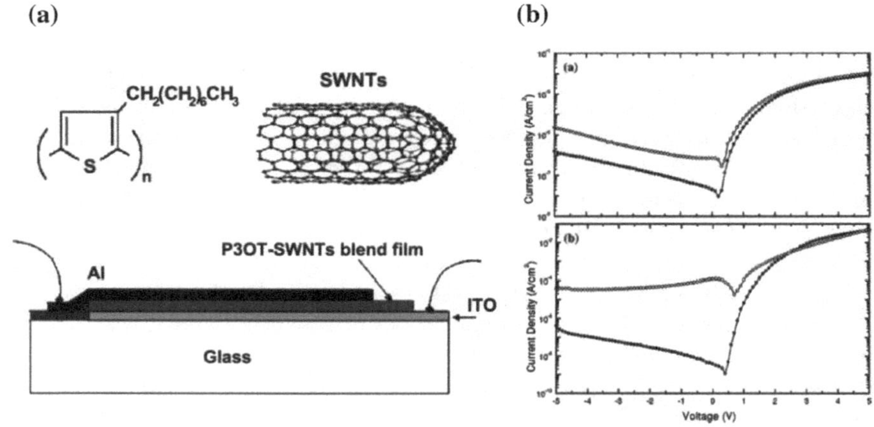

Fig. 2 **A** Illustrative example of the P3HT/SWCNT PV cell. **B** (a) current–voltage (*I–V*) curves of ITO/P3OT/Al device in dark (lower one) and (b) *I–V* curves for ITO/P3OT–SWNT/Al device. Reprinted with permission from Ref. [21]

Chemical grafting was done by esterification reaction to graft CH_2OH-terminated regioregular P3HT onto carboxylic acid functionalized GO (Fig. 3A). It was reported that the resultant product G–P3HT was found to be soluble in almost all common organic solvents. It is also suitable for fabricating devices through solution mixing process. A strong electronic interaction and covalent linkage was shown between P3HT and graphene moiety in G–P3HT composites, which was proved by both spectroscopically and electrochemically. This played a vital role in improving the charge transport and hence better device performance. A solution-casted G–P3HT/C_{60} heterostructures-based bilayer photovoltaic (PV) device showed a 200% increment in the power-conversion efficiency (η = 0.61%) compared to the P3HT/C_{60} counterpart under illumination of 100 mW/cm^2 (Fig. 3B). This considerably enhanced solar device performance was due to the good bandgap matching and strong electronic interaction [Fig. 3B(b)] between the chemically bonded P3HT and graphene. This work also indicated that graphene is an efficient hole transporting material in CPC-based solar cell devices [13, 20, 22].

Alturaif et al. [10] reviewed solar cells based on polymer/carbon nanocomposites. Different organic materials such as conjugated polymers, organic molecules, and known carbon materials such as amorphous carbon, CNTs, fullerenes, and graphene, respectively, were used for both organic as well as organic–inorganic hybrid solar cell devices. Besides, CNTs exhibit interesting physical, chemical, and optoelectronic properties required for many feasible and potential applications. Moreover, polymer–SWCNT-based solar cell devices have been reported as new alternatives, flexible and lightweight devices for space power applications. They also summarized the performance of CNT/conducting polymer-based solar cell devices.

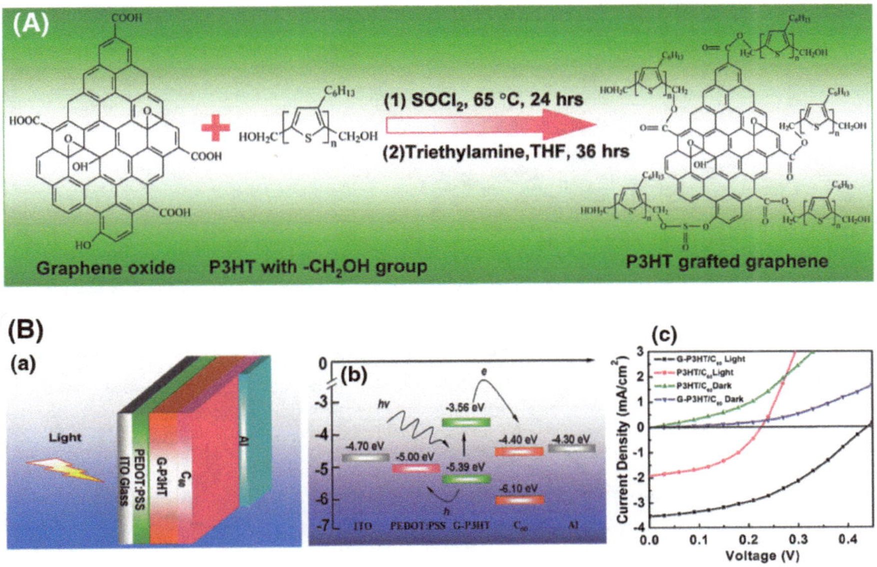

Fig. 3 **A** A scheme representing chemical grafting process onto graphene. **B** (a) A scheme representing structures of different cells, (b) energy levels of G–P3HT/C60 PV device, and (c) *I–V* curve of both PV devices. Reprinted with permission from reference [22]

MWCNT-tetrasulfonate copper phthalocyanine-based nanocomposites have been studied for development of bulk heterojunction devices. These devices are known as a hole extracting material and possess high work function. These devices have been reported with the efficiency of 1.25% [23–26]. Different approaches have been reported for reinforcement of CNTs in the whole harvested or conductive component for a polymer solar cell device, as shown in Fig. 4.

Alturaif et al. [10] reviewed transparent MWCNT sheets for the application as an interlayer anode electrode for tandem OPV devices. According to the report, thin films of PEDOT:PSS/P3HT:PCBM nanocomposites on top of an ITO coated glass substrate acted as a front cell, and the Al/Bathocuproine (BCP)/C60/CuPc:C60/CuPc/PEDOT:PSS nanocomposites structure as back cell. The MWCNT layer was sandwiched between two front cells and back cells, which resulted in a device efficiency of approximately 0.31%. However, adding a thin CNT film at the interface of P3HT:PCBM layer showed the promising efficiency of 4.9%. This is attributing due to the thin CNT film can be easily inserted either between the PEDOT-PSS layer and the active layer or between the ITO and PEDOT-PSS layers. Poly [(2-methoxy,5-octoxy)-1,4-phenylenevinylene](MO-PPV)/SWCNT-based nanocomposites are also accounted with nonstandard solar simulation and low efficiencies of device up to 0.004%. Poly-hexylthiophene derivatives are the most commonly used polymer with SWCNTs because of their resulting composites-based donor–acceptor solar cells exhibited a better efficiency of 0.52% [10].

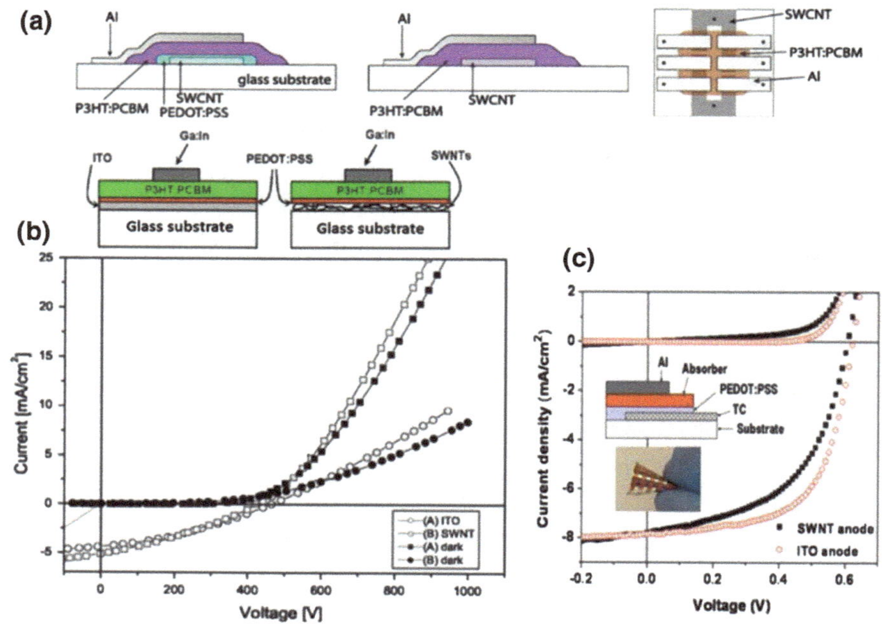

Fig. 4 Different approaches for incorporating carbon nanotubes in the whole harvested or conductive component for a polymeric solar cell. **a** First diagram showed the replacement of ITO by CNT layer; second diagram showed the replacement of PEDOT:PSS as well as ITO by CNT layer; and third diagram showed device's top view where black dots represented contacts points. **b** Device schematically representing anodes made with ITO/SWCNT thin film showed at the upper panel. The corresponding *I–V* characteristics measured at 100 mW/cm^2 in the presence of halogen white light and in dark using a reference solar cell on ITO-glass substrate, whereas the best solar cell using an SWCNT–glass current collector is presented at the lower panel. **c** The potential flexibility where SWCNTs and polymer substrate were used. Reprinted with permission from Ref. [24] for Fig. **a**; from Ref. [25] for Fig. **b**; and from Ref. [26] for Fig. **c**

Zhang et al. [19] reviewed DSSC made from carbon nanohorns/polymer composites. Design and progress of graphene/polymer based PV are one of the promising routes to connect the extraordinary properties for the development of efficient solar-to-power conversion devices. Graphene and functionalized GO and rGO are the smart and potential materials for PV cells with specific functions depending upon their fascinating qualities [20]. They reviewed graphene filled polymers and their application in PV solar devices. They discussed the multi-functional applications of graphene-based composites as the whole transport components counter electrode and electron acceptor, and hybrid polymer solar cells. Carbon materials have also a major role in the development of alternative clean and sustainable energy technologies [14, 27, 28]. Candeleria et al. [27] summarized carbon nanostructured materials and their application in energy conversion application such as solar cell and fuel cell devices. Hosseini et al. [29] reviewed the properties of CNT–polymer nanocomposites and their application in solar cell PV devices. He explained the essential aspects and the physical mechanisms of CNT–polymer nanocomposites which can control the photo-conversion, light–matter interaction, charge transfer, photo-generated charge-carrier transport, and energy transfer in the CNT–polymer nanocomposites. The influence of the processing conditions on the electronic properties of PV solar cell device-related performances was also discussed. Adler et al. [30] prepared polystyrene/MWCNT based nanocomposites for solar cell application.

4 Summary

The fundamental interface chemistry and ionic transportation generally control the development of energy storage devices. Here, various carbon nanomaterials-based conducting polymer composites for solar cells and fuel cells applications have been discussed. Both cells have a keen interest for the development of next-generation energy production and conversion devices. The conducting polymer/CNT based hybrid solar cells are still keeping little bit behind the other solar cell technologies. In terms of cost, the conducting polymer/carbon-based energy devices would be much cheaper. On the other hand, the stability of conducting polymer/carbon composites may influence the cycling stability which requires an in-depth research.

References

1. Bairan A, Selamat MZ, Sahadan SN, Malingam SD, Mohamad N (2016) Effect of carbon nanotubes loading in multifiller polymer composite as bipolar plate for PEM fuel cell. Proced Chem 19:91–97
2. Wang Z, Feng K, Li Z, Lu F, Huang J, Wu Y, Chu PK (2016) Self-passivating carbon film as bipolar plate protective coating in polymer electrolyte membrane fuel cell. Int J Hydrog Energy 41:5783–5792
3. Singh S, Modi A, Verma N (2016) Enhanced power generation using a novel polymer-coated nanoparticles dispersed-carbon micro-nanofibers-based air-cathode in a membrane-less single chamber microbial fuel cell. Int J Hydrog Energy 41:1237–1247
4. Kannan R, Pillai VK (2009) Applications of carbon nanotubes in polymer electrolyte membrane fuel cells. J Indian Inst Sci 89(4):425–436
5. de Oliveira PN, Mendes AMM (2016) Preparation and characterization of an eco-friendly polymer electrolyte membrane (PEM) based in a blend of sulphonated poly(vinyl alcohol)/chitosan mechanically stabilised by nylon 6,6. Mater Res 19(4):954–962
6. Yuan D, Zhu YG, Jia C (2016) Chapter 17: Carbon nanotube-polymer composites for energy storage applications. In: Berber M (ed) Book edition: Carbon nanotubes-current progress of their polymer composites. InTech Publications, pp 439–457
7. Vandiver MA (2014) Effect of hydration on the mechanical properties of anion exchange membranes. Ph.D. theses, Colorado School of Mines, Colorado
8. Tan CW, Tan KH, Ong YT, Mohamed AR, Zein SHS, Tan SH (2012) Energy and environmental applications of carbon nanotubes. Environ Chem Lett 10:265–273
9. Liu YL (2016) Effective approaches for the preparation of organo-modified multi-walled carbon nanotubes and the corresponding MWCNT/polymer nanocomposites. Polym J 48:351–358

10. Alturaif HA, ALOthman ZA, Shapter JG, Wabaidur SM (2014) Use of carbon nanotubes (CNTs) with polymers in solar cells. Molecules 19:17329–17344
11. Huyen DN (2011) Chapter-22 Carbon nanotubes and semiconducting polymer nanocomposites. In: Carbon nanotubes-synthesis, characterization, applications. InTech Publications, pp 469–486
12. Zhang W, Ravi S, Silva P (2011) Application of carbon nanotubes in polymer electrolyte based fuel cells. Rev Adv Mater Sci 29:1–14
13. Dai L, Chang DW, Baek JB, Lu W (2012) Carbon nanomaterials for advanced energy conversion and storage. Small 8(8):1130–1166
14. Rahman A, Ali I, AL Zahrani SM, Eleithy RH (2011) A review of the applications of nanocarbon polymer composites. Nano 6(3):185–203
15. Kim DJ, Jo MJ, Nam SY (2015) A review of polymer–nanocomposites electrolyte membranes for fuel cell application. J Ind Eng Chem 21:36–52
16. Kumar GG, Nahm KS (2011) Chapter-27 Polymer nanocomposites-fuel cell applications. In: Advances in nanocomposites-synthesis, characterization and industrial applications. InTech Publications, pp 639–660
17. Sharma R, Cherusseri J, Kar KK (2015) Polymer electrolyte membrane fuel cells: role of carbon nanotubes/graphene in cathode catalysis. In: Kar KK, Pandey JK, Rana S (eds), Handbook of polymer nanocomposites. Processing, performance and application. Volume B: carbon nanotube based polymer composites. Springer, pp 361–390
18. Memioğlu F, Bayrakçeken A, Öznülüer T, Ak M (2014) Conducting carbon/polymer composites as a catalyst support for proton exchange membrane fuel cells. Int J Energy Res 38:1278–1287
19. Zhang Z, Han S, Wang C, Li J, Xu G (2015) Single-walled carbon nanohorns for energy applications. Nanomaterials 5:1732–1755
20. Zhu J, Liu F, Mahmood N, Hou Y (2015) Graphene polymer nanocomposites for fuel cells. In: Sadasivuni KK, Ponnamma D, Kim J, Thomas S (eds), Graphene-based polymer nanocomposites in electronics. Springer, pp 90–131
21. Kymakis E, Amaratunga GAJ (2002) Single-wall carbon nanotube/conjugated polymer photovoltaic devices. Appl Phys Lett 80:112
22. Yu D, Yang Y, Durstock M, Baek JB, Dai L (2010) Soluble P3HT-grafted graphene for efficient bilayer-heterojunction photovoltaic devices. ACS Nano 4(10):5633–5640
23. Moniruzzaman M, Winey KI (2006) Polymer nanocomposites containing carbon nanotubes. Macromolecules 39:5194–5205
24. Van De Lagemaat J, Barnes TM, Rumbles G, Shaheen SE, Coutts TJ, Weeks C, Levitsky I, Peltola J, Glatkowski P (2006) Organic solar cells with carbon nanotubes replacing In_2O_3: Sn as the transparent electrode. Appl Phys Lett 88:233503
25. Pasquier AD, Unalan HE, Kanwal A, Miller S, Chhowalla M (2005) Conducting and transparent single-wall carbon nanotube electrodes for polymer-fullerene solar cells. Appl Phys Lett 87:203511
26. Rowell MW, Topinka MA, McGehee MD, Prall HJ, Dennler G, Sariciftci NS, Hu L, Gruner G (2006) Organic solar cells with carbon nanotube network electrodes. Appl Phys Lett 88:233506
27. Candelaria SL, Shao Y, Zhou W, Li X, Xiao J, Zhang JG, Wang Y, Liu J, Li J, Cao G (2012) Nanostructured carbon for energy storage and conversion. Nano Energy 1:195–220
28. Shearer CJ, Cherevan A, Eder D (2014) Application and future challenges of functional nanocarbon hybrids. Adv Mater 26:2295–2318
29. Hosseini T, Kouklin N (2016) Chapter-4 Carbon nanotube–polymer composites: device properties and photovoltaic applications. In: Carbon nanotubes-current progress of their polymer composites. InTech Publications, pp 95–123
30. Adler Q, Gerhardt RA, Muhlbauer RL (2011) Carbon nanotube-Polymer composites for solar cell application. Santa Clara University

Polymer-Carbon Composites as Anti-corrosive Materials

Sanjay Pal, Tuhin Chatterjee, Partheban Manoharan and Kinsuk Naskar

Abstract The objective of this chapter is to focus on the recent developments of polymer-carbon composites in terms of utilizing various forms of carbon, i.e., graphene or graphene-like sheets, carbon nanotube, along with polymer binder for high-performance anti-corrosive paints. The attributes to final properties of dried film coating can be resolved into two major parts, i.e., attributes due to binder and additives. The role of binder and additives has been discussed separately in the subsequent sections. A general idea has also been discussed as a premise to give the right context before the objective of this chapter is discussed, i.e., Polymer-carbon composites as anti-corrosive material.

Keywords Corrosion · Anti-corrosive materials · Graphene · Carbon nanotube

1 Introduction

Initially, the methods and technologies to protect the metal against corrosion were rooted to local area or country only. During the colonization period from seventeenth to nineteenth century, scientific methods and technologies transcended all geographical boundary, and the major rapid development was observed. Metals are the engineer's first choice for structural applications because they are mechanically strong and exhibit excellent load bearing capacity. But unfortunately most metals those used for structural applications, i.e., iron, aluminium, copper, etc., are quite

S. Pal · T. Chatterjee · P. Manoharan · K. Naskar (✉)
Rubber Technology Centre, IIT Kharagpur, Kharagpur 721302, India
e-mail: naskark73@gmail.com

S. Pal
e-mail: sanjay.cusat.psrt@gmail.com

T. Chatterjee
e-mail: papuchem@gmail.com

P. Manoharan
e-mail: parthevlr@gmail.com

M. Rahaman et al. (eds.), *Carbon-Containing Polymer Composites*,
Springer Series on Polymer and Composite Materials,
https://doi.org/10.1007/978-981-13-2688-2_16

reactive to oxygen and other corrosive species, which leads to impairment or deterioration of mechanical properties. Therefore, there is a need for some measures which can protect these metals from corrosion. In this era of growing competitiveness, economy plays a major role in the material selection for any structural application. Some metals in its pure form may qualify the required physical properties and desired inertness against corrosion simultaneously, limited availability and high cost of such materials would rather push our engineers to look for some other ways to protect readily available metal from the corrosion. The requirement of inertness to corrosion and desired mechanical properties are contrary to each other, and therefore, an additional step, called 'coating of the material' is required in the structure building process. Undoubtedly the technologies for corrosion prevention has become well evolved, the basic mechanism working behind these measures is still the same-covering metal surface with some impermeable material to cut-off the contact between corrosive species and the metal surface. Corrosion is an undesirable phenomenon that leads to the formation of a stable compound from a pure form of metal. The process of corrosion is slow, spontaneous and complex. Its study involves several observations often driven by the environment-substrate system.

2 Various Forms of Corrosion

There are various types of corrosions which are shown in Fig. 1.

2.1 General Attack Corrosion

This is the most common type of corrosion, and it is also named as uniform attack corrosion shown in Fig. 2. In general, it is caused by a chemical and electrochemical reaction which results in the deterioration of the entire surface of the metal exposed to the environment.

2.2 Localized Corrosion

Unlike uniform attack corrosion, localized corrosion specifically attacks only one area of the metal structure. It can be classified into three categories shown in Fig. 3.

Pitting

Pitting corrosion results the formation of a small hole or cavity in the metal usually as a result of de-passivation of the small area shown in Fig. 4. This area becomes anodic, and the remaining metal surface becomes cathodic which produces the

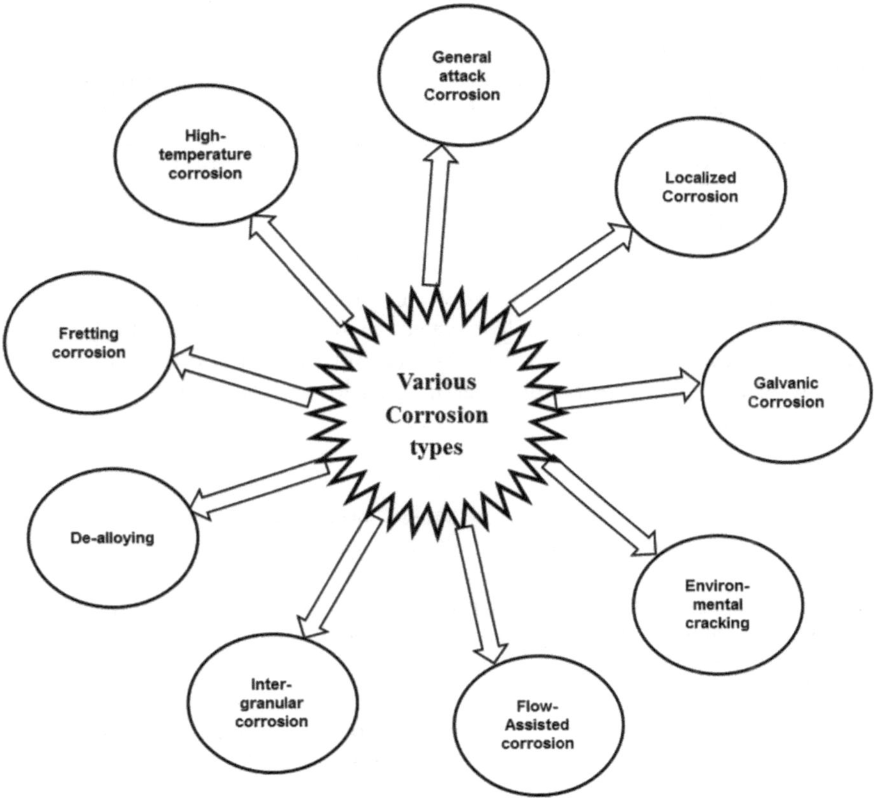

Fig. 1 Various types of corrosion

Fig. 2 General attack corrosion

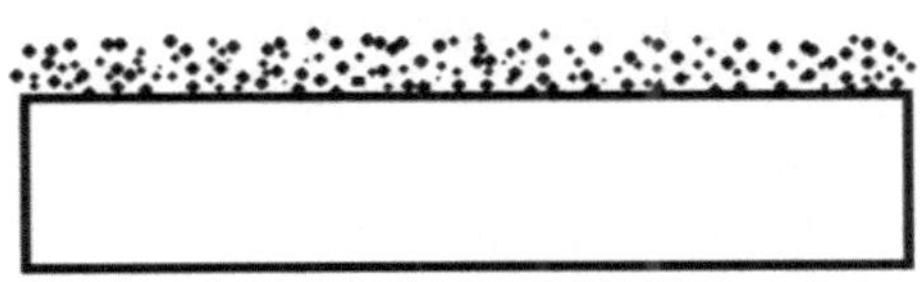

Fig. 3 Different types of localized corrosion

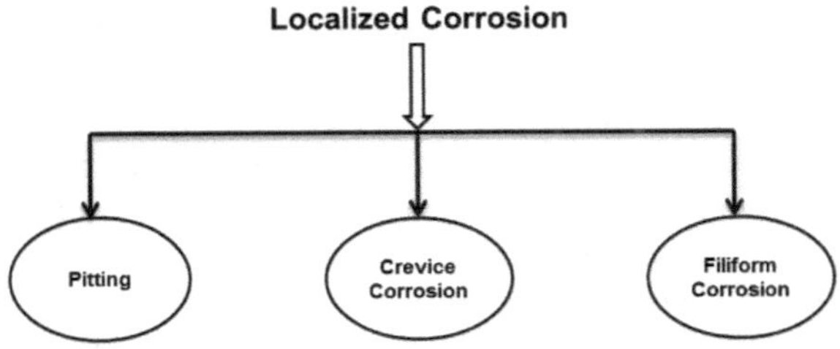

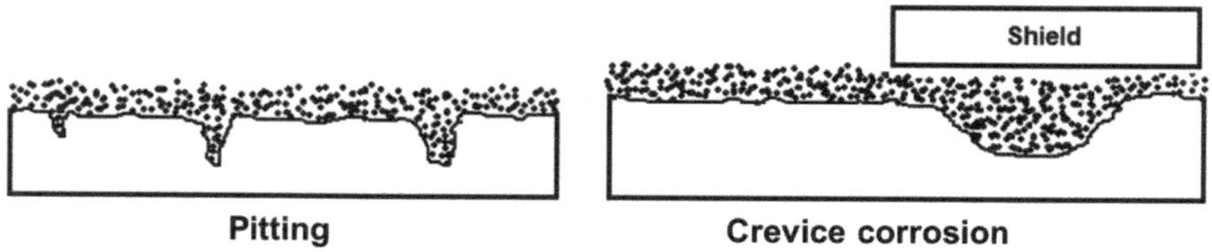

Fig. 4 Pitting and crevice corrosion

localized galvanic reaction. This form of corrosion is considered to be more dangerous because it is often difficult to detect or predict due to the fact that it is relatively small and it is mostly covered and hidden by the corrosion products. A small, narrow pit with minimal overall metal loss can lead to the failure of an entire engineering system.

Crevice Corrosion

Like pitting, it also occurs at a specific location. Crevice corrosion is usually associated with a stagnant solution on the micro-environmental level. Those stagnant microenvironments tend to occur in crevices (shielded areas) such as those formed under gaskets, washers, insulation material, fastener heads, surface deposits, disbonded coatings, threads, lap joints, and clamps shown in Fig. 4.

Filiform Corrosion

This type of corrosion occurs under painted or plated surfaces. When water breaches the coating, filiform corrosion begins at small defects in the coating and spreads to cause structural weakness.

2.3 Galvanic Corrosion

It is also termed as dissimilar metal corrosion, and it happens when two different metals are electrically connected together in a corrosive electrolyte. As a result, a galvanic couple forms between the two metals in which one metal act as anode and the other as a cathode which is shown in Fig. 5.

In the presence of environment, the most active metal (anode) will corrode at a faster rate than it would alone while the less active metal (cathode) deteriorates slowly. Galvanic corrosion occurs because different elements have different chemical activities and different alloys with different chemical compositions will have different chemical and electrochemical properties in the same environment. Apart from, several factors including the ratio of the areas of each alloy exposed to the environment, the potential difference between the alloys, the electrical resistivity of the joint and the conductivity of the environment are responsible for the galvanic corrosion.

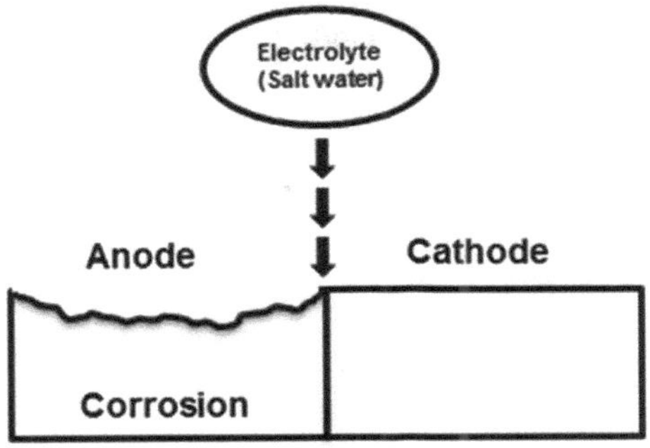

Fig. 5 Galvanic corrosion

2.4 *Environmental Cracking*

Environmental cracking results from the combination of various environmental conditions such as temperature, a chemical environment that can affect the metal. It can be of different types

(a) Hydrogen-induced cracking
(b) Corrosion fatigue
(c) Stress corrosion cracking
(d) Liquid metal embrittlement.

2.5 *Flow Assisted Corrosion*

It is also termed as flow-accelerated corrosion. In the presence of wind or air, if the protective layer of oxide on top of the metal surface is removed or dissolved flow assisted corrosion takes place. As a result, the underlying metal surface is exposed to the environment to deteriorate further and corrode. It can be various categories such as

(a) Impingement
(b) Cavitation
(c) Erosion-assisted corrosion.

2.6 *Intergranular Corrosion*

The microstructure of metals or alloys is made up of grains which are separated by the grain boundaries. Intergranular corrosion is the localized attack near to the grain boundaries or immediately adjacent to the grain boundaries while the bulk of the grains remain unaffected which is shown in Fig. 6.

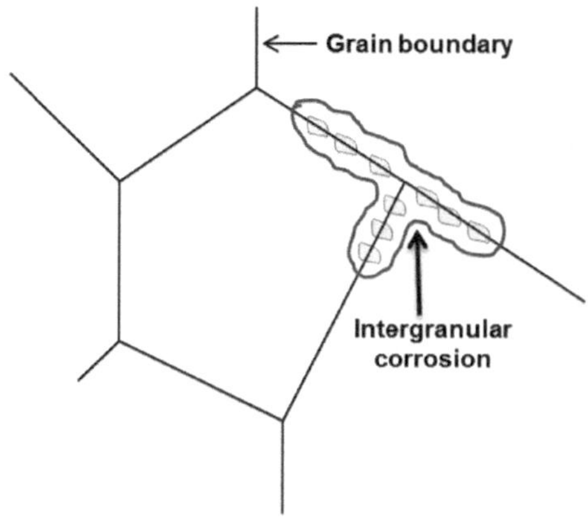

Fig. 6 Intergranular corrosion

This occurs due to the presence of the impurities in the metal. Impurities often have the tendencies to be present in large quantity in the vicinity of the grain boundaries and therefore these grain boundaries are more vulnerable to corrode than the bulk of the metal.

2.7 Dealloying or Selective Leaching

This is also termed as demetalification, parting or selective corrosion. This is the process in which selective removal or selective corrosion of an element takes place from an alloy shown in Fig. 7.

Dezincification of the unstabilized brass is the most common example of dealloying which results the formation of porous copper structure.

2.8 Fretting Corrosion

Fretting corrosion refers to corrosion damage at the asperities of contact surfaces. It happens as a result of repeated wearing, weight or vibration on an uneven, rough

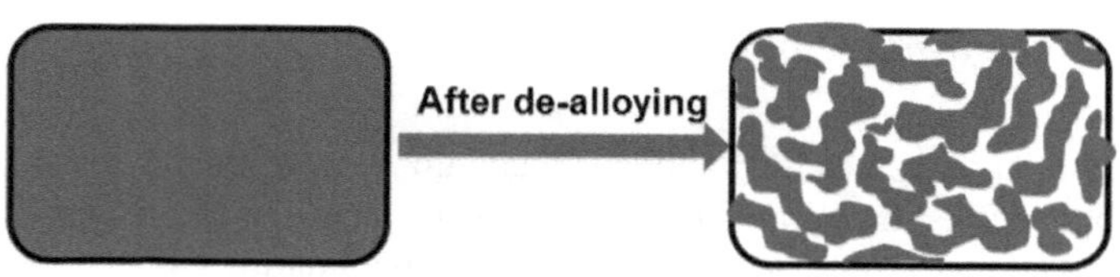

Fig. 7 De-alloying

surface. Contact surfaces exposed to vibration during transportation are exposed to the risk of fretting corrosion.

3 Factors Affecting the Corrosion

Corrosion of metal depends upon various factors. These factors can be divided into two categories: Primary factors that are related to metal and the secondary factors that are related to the environment. Let us first consider the primary factors.

3.1 Primary Factors

Nature and Reactivity of the Metal

Metals with lower electrode potential (e.g., zinc, magnesium, aluminium) corrodes at a much faster rate than the metals with higher electrode potential (gold, silver, platinum). As a result, when the two metals are in contact with each other, higher the difference in electrode potential between the two metals higher will be the corrosion rate.

Ratio of Cathodic to Anodic Region:

The ratio of cathodic and anodic region is an important factor that governs the corrosion rate. If a metal has a small anodic region and large cathodic region, the corrosion rate is very high. The free electrons in the anode are easily consumed in the cathodic region. Therefore, the free electrons of the anode are rapidly consumed at the cathode, if the cathodic region is larger. This results in further enhancement of the anodic reaction which leads to overall increase of the corrosion rate.

The Surface State of the Metal

As corrosion is a surface phenomenon, larger the surface area or finer the grain size of the metal.

Presence of Impurities

The rate of corrosion enhances in presence of impurities (even in small amount) in the metal also. Presence of impurities helps to set up a voltaic cell that promotes the corrosion.

Nature of the Corrosion Products:

Sometimes corrosion products (e.g., metal oxide) formed on top of metal acts as a protective barrier. It prevents the corrosion by acting as a barrier between the metal surface and the corrosion medium. The corrosion products must have to be stable, non-porous, and insoluble. On the other side, if the corrosion products are porous, unstable, and soluble, it further enhances the rate of corrosion.

Hydrogen Overvoltage

Higher the hydrogen overvoltage of metals on its surface lesser is the corrosion. Metals with lower hydrogen overvoltage on its surface are more susceptible to corrosion. At lower hydrogen overvoltage, the liberation of hydrogen gas is very easy which in turn makes the cathodic reaction very fast. As a result, the anodic reaction becomes faster and thus increases the rate of corrosion.

3.2 Secondary Factors

Temperature

In general, corrosion rate increases with the rise of temperature. With increase of temperature, conductance of the medium increases and hence the diffusion rate also increases. Therefore corrosion progresses faster at higher temperature.

pH of the Medium

Rate of corrosion is higher in acidic pH rather than the neutral and alkaline pH. For iron, at very high pH protective coatings of iron oxide formation takes place which prevents the corrosion. On the other hand at low pH several corrosion takes place.

Presence of Electrolyte

Presence of electrolytes in normal water also increases the corrosion rate. In sea water iron corrodes at a much faster rate than the normal distilled water. This is due to the presence of salts (e.g., sodium chloride) that can acts as an electrolyte.

Presence of Carbon Dioxide in Water

Carbon dioxide (CO_2) present in normal water also increases the rate of corrosion. CO_2 acts as an electrolyte and increases the flow of electron from one side to another side and thus enhances the rate of corrosion.

Effect of oxygen
Humidity

Humidity plays a major role in promoting and accelerating the corrosion rate. The more the humidity of the environment more will be the corrosion.

Apart from, acid rain (a chemical by-product from the industries), chlorides (in costal area) also promotes the corrosion.

4 Corrosion Prevention

Considering the above matter of fact, since hundred years corrosion and its prevention has become the serious concern for human being. To meet the demands of common citizens of any nation, continuous work is being done to develop a good infrastructure system, especially in developing countries like India, China, etc. Therefore, the production of metals for structural application is growing each year, and so the expenditure towards corrosion prevention. Corrosion related expenditure indeed constitutes a considerable part of the total economy of a country. However, such values may vary from country to country, but the common goal for many countries is to keep their infrastructure.

In state of the art development, the metals substrates are made corrosion resistant in the first place by alloying or diffusion process to maximize the metal's service life [1]. Once the suitable metal is chosen for structural construction considering the inherent resistance of metal towards corrosion, surface of metal is treated via various techniques [2, 3]. The purpose of surface treatment is to increase the surface energy so that the anti-corrosive coating adheres well with the substrate to inhibit the contact between corrosive species and metal. The next step is the application of anti-corrosive coating. Coating is applied to meet the requirements. Figure 8 exhibits the critical step involved in the process of protecting metal against corrosion.

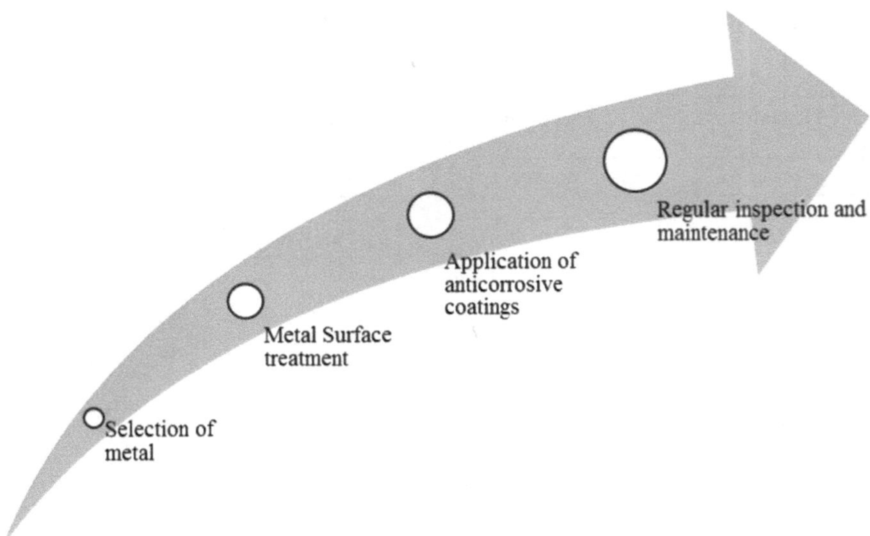

Fig. 8 Critical steps involved in the process of corrosion prevention

4.1 Various Coatings Used for Corrosion Prevention

Figure 9 shows the classification of coatings used for corrosion prevention: organic, inorganic, conversion, and metallic. All these types of coatings are better in one way or another. It is important to understand that no single coating is effective enough to protect metals against corrosion under all kind of environment for prolong time. Therefore, a combination of coatings is often required to ensure good protection against corrosion. However, most organic-based coatings will be discussed in this chapter.

Paint is one of the most important and widely used commodities in various applications. Paint is applied to metal surfaces to prevent corrosion. The paint industry for protection against corrosion is one of the longest running industries for the production of goods of basic need. However the technologies have evolved dramatically over past few decades, the underlying mechanism for corrosion prevention remains the same, i.e., barrier, sacrificial, and inhibitive.

In recent time, environmental concern over the emission of volatile organic compounds (VOCs) from the applied paint has caused a significant change in the anti-corrosive coating technology. International and national legislation aimed at reducing the emission of VOC, which led to significant change in the formulations and development of new VOC-compliant anti-corrosive paint, and commercialization of several alternatives to organic solvent-borne paint. The thorough

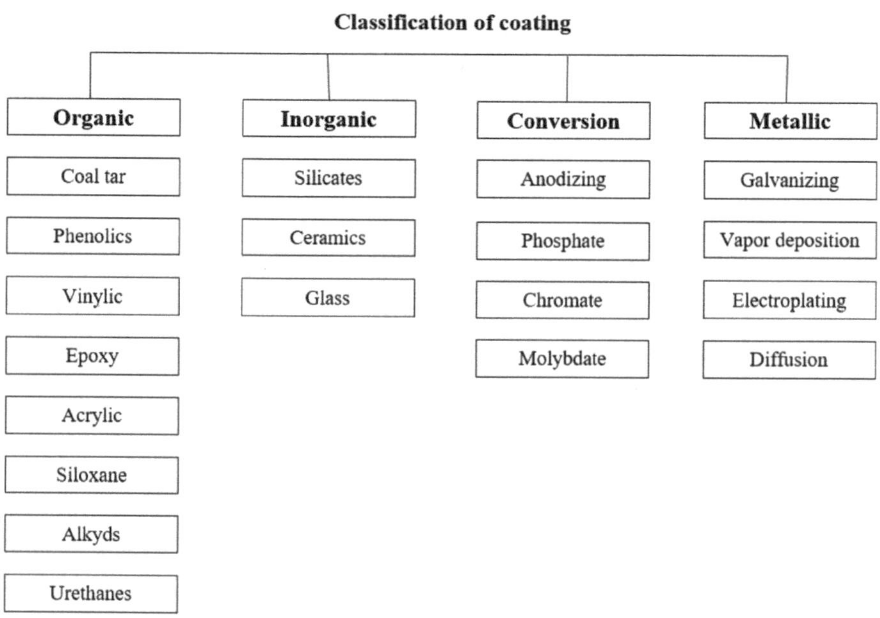

Fig. 9 Classification of various coatings

understanding of all components in the anti-corrosive paint, substrate environment interaction and detailed coating failure analysis are the key aspect in the development of VOC compliant, high-performance anti-corrosive coatings. The anti-corrosive coating companies are now compelled to produce high solid content or water-borne coating with a low amount of organic solvent to comply with the current norms on restricting VOCs level up to a certain limit. However, in a harsh environment, these water-borne coating may not be effective as compared to that of the organic solvent-borne coating. Undoubtedly for general purpose applications or where the requirements are not very stringent, the manufacturing procedure of anti-corrosive paint is well established and commercialized, but the high-performance anti-corrosive coating is yet being studied at laboratory scale. The complexity of paint-substrate and paint-environment system affecting the efficacy and durability of the paint is one of the reasons why there is less number of high-performance anti-corrosive coating- surface type, surface pre-treatment, curing, coating thickness, compounding ingredients and adhesion of paint with substrate also decides the performance of anti-corrosive paints. In this chapter, the role of environment and compounding ingredient in anti-corrosive coatings is discussed in the subsequent sections.

4.2 Environmental Effect of Anti-corrosive Paint

Since environment plays a critical role in the corrosion process, the important aspect in the efficacy of coating/paint technology is the understandings of environment for the development of advanced high-performance anti-corrosive coating with prolong service life. An anti-corrosive paint may be exposed to, i.e., water, soil, atmosphere or UV radiation, etc. The type of environment and conditions during the whole service life sets the specific requirements for the development of anti-corrosive coating. According to ISO 12944 "coating and varnishes—Corrosion protection of steel by protective coating system" the environment can be classified into three categories, i.e., immersion, atmospheric, and splash zone [4] that is shown in Table 1.

Table 1 Classification of environment experienced by the anti-corrosion coatings

	Soil
Immersion	Fresh water
	Sea water
	Chemicals
Splash zone	Costal area
	Reverine
Atmospheric	Marine
	Industrial
	Rural

The applied coatings in the atmospheric environment are exposed to several potential factors, i.e., UV radiation, moisture, heat, gas, pollutants, soot, salt, etc. Such factors significantly affect the corrosivity of the atmospheric environment and the requirements for the high-performance anti-corrosive coating. Further, the corrosivity level of atmospheric environment can be divided into six categories from very low (C1) to very high corrosivity (C5-I and C5-M), as listed in Table 2.

The corrosivity level in the rural area is low since the content of solid particle, soot and sulfate salts in the rural atmosphere is less as compared to the industrial and marine atmosphere. The high corrosivity level of industrial and urban atmosphere and the combination of sulfur based pollutant and rain results in acid rain which consequently let the coating exposed to the acidic environment. Like the industrial atmosphere, the marine atmosphere is also highly corrosive, containing a high content of chloride ion, thus making the marine environment very aggressive, causing corrosion like pitting. The environment impact of marine and industrial atmosphere is very different from that of rural one. Therefore, proper care should be taken while designing the formulation and fabrication as there will be very different conditions subjected on anti-corrosive coatings.

The anti-corrosive coating applied on structures installed near the offshore area is frequently exposed to ion rich atmosphere and a continuous splash of electrolyte from the sea. Besides the mechanical stress due to the combination of alternating moisture and dryness, the ultraviolet radiation also accelerates the failure of the coating applied to the substrate.

The requirements for anti-corrosive paints, when exposed to the environment of continuous immersion are more stringent and specific. For many applications, the structures are intended to work underneath the soil or water for prolonged period. Therefore, the correct choice of compounding ingredients becomes a very important aspect of the development of high-performance anti-corrosive paint and ensuring good durability and effectiveness. The soil type, humidity, bacteria, salts, dissolved oxygen and pH are some of the important factors to be considered while designing the paints for structures buried in soil. The combination of salt and dissolved gases in water is more detrimental and corrosive to substrate then atmospheric exposure.

Table 2 Division of corrosivity level and impact (reproduced after permission granted from Ref. [6])

Environmental Corrosivity level	Impact	Examples
C1	Very low	Indore dry room (relative humidity <60%)
C2	Low	Indoor non-heated and ventilated room
C3	Medium	Rural environment far from industrial area
C4	Heavy	Urban or industrial areas
C5-I	Very heavy industry	Industrial area with high relative humidity
C5-M	Very heavy marine	Coastal and offshore areas

Apart from good mechanical properties and thermal stability, an anti-corrosive paint should exhibit resistance to microbial action and biofouling in sea water which contains a very high concentration of the electrolytes.

4.3 Various Anti-corrosive Coatings

An anti-corrosive coating consists of several layers of thin coating each having different purpose and characteristics; the numbers of coat obviously will depend on the requirements of an application. Typically an anti-corrosive system consists of three layers of coating: a primer, an intermediate and a top coat. This individual layer can be metallic, inorganic, or organic. The primary purpose of a coat regardless of its type is to create barriers between the surface of a substrate and the environment surroundings. The primer protects the surface against corrosion and ensures good adhesion between the substrate and the intermediate coat. Primer coat generally consists of metallic zinc or some inhibitive pigments. Next, the intermediate coat is applied to build up the required thickness of the anti-corrosive coating. The intermediate coat should impede the diffusion of corrosive species and ensure good adhesion between the top coat and the primer. The top most layer (topcoat) of an anti-corrosive coating is exposed to the external hostile environment, and therefore, it should be resistant to alternating whether, impact from sharp objects, UV radiation, moisture, salts, bacterial action, temperature and such other factors. Providing color code, aesthetic appearance, gloss and good surface finish are some other objectives behind using the topcoat in the anti-corrosive coating.

(a) *Epoxy*

Epoxy resin exhibits excellent resistance to heat, water and chemicals, and excellent adhesion to the metal substrate. Epoxy resin has been traditionally used as a binder in paint applications due to relatively cheap initial materials and stable carbon–carbon and an ether bond. However, the application of epoxy binder is limited to maintenance work, primer and intermediate coatings due to yellowing of epoxy coating when exposed to UV radiation. Although epoxy resin has good resistance to heat and chemicals, it provides fair protection against corrosion, and often need a topcoat made from superior polymers such as polyurethane or polysiloxane.

Epoxy resins are prepared by the condensation reaction between the diphenyl propane derivative and epichlorohydrin, i.e., bisphenol A and epichlorohydrin. The solvent bore epoxy resins are more suitable for applications under aggressive environment. The molecular weight of unreacted epoxy resin molecule plays a crucial role in the development of a tailor-made coating for the specific application. Several properties can be tuned to the requirement with the alteration in the molecular weight of unreacted epoxy resin molecule. As the molecular weight of unreacted epoxy resin molecule increases, properties such as crosslink density, hardness and solvent resistance decreases. Other properties, i.e., flexibility, impact strength, substrate wetting, and adhesion also increases The functional groups present in the epoxy resin and the curing agent reacts together to form a three-dimensional crosslinked structure. The curing agent used for crosslinking the epoxy resin is aliphatic or cyclic amine. Apart from the molecular weight of unreacted epoxy resin molecule, the degree of resistance to corrosion will depend on the type of curing agent used in the paint and coating formulation. Although Polyamine and amide are slow in reaction but offer enhanced flexibility and less toxicity compared to another curing agent, i.e., aliphatic or cyclic amine. Water-borne epoxy is an emulsion made from diglycidyl ether of bisphenol A and water-soluble amine curing agent which also serve the purpose of emulsifier. While the film formation process, the curing agent migrates from aqueous phase to the epoxy particle and form the covalent bond between the epoxy particles to form a three-dimensional network, though, there is a risk of irregular film formation. The traditional water-borne epoxy coatings are somewhat less tactful in protecting the substrate from corrosion than a solvent-borne epoxy coating. However, cationic electrodeposited coating exhibit improved performance when used as a primer, and provide excellent adhesion to the substrate. The combination of zinc rich primer and the intermediate multifunctional epoxy resin cured with suitable curing agent lead to the formation of high-performance anti-corrosion coating suitable for the industrial atmospheric environment.

(b) *Acrylic*

Acrylic polymer and copolymers exhibit a unique set of properties: very good chemical inertness, resistance to UV radiation, environment stability. Acrylic can also be modified with epoxy, silane or carbamate to develop an anti-corrosion coating for a wide range of applications. Acrylic resin is an acrylate ester which can be formulated as water-borne emulsion latex, thermosetting and thermoplastic resin. Poly(methyl methacrylate) and Poly(ethyl acrylate) are the most commonly used binder in high-performance paint application. There are various types of acrylic monomers used in the manufacturing of acrylic paints, the type of monomers and their proportions set the properties of final film product.

The solvent-borne acrylic coatings can be dried quickly, exhibit durability and improved adhesion with the substrate. The curing of acrylic resin involves the use of isocyanate or amine to form a thermosetting film on the substrate. The thermosetting acrylic resin can be used as lacquers, baking enamels and emulsions. Lacquers and enamels are widely used in automotive and appliances finishing as a topcoat. Acrylic resins are brittle and have poor impact strength, but exhibit excellent weathering resistance, excellent adhesion, good abrasion resistance, and chemical resistance.

Methacrylate based acrylic binder does not have hydrogen atom in the polymer backbone chain. Therefore, they are resistant to acid, base, mild oxidizing agent and hence suitable for corrosive industrial gases and fumes environment. However the side group is resistance to hydrolysis, acrylic paints are not suitable for constant immersion or the strong chemical environment.

Acrylic coatings exhibit excellent properties: light and UV stability, gloss and color retention, good chemical resistance, excellent weathering resistance, resistance to chemical fumes and occasional chemical splash and spillage and better mar resistance than those of alkyd anti-corrosive coatings. Acrylic resin is also often mixed with other resin to achieve some specific properties, since it is compatible with wide variety of resin, i.e., epoxy, alkyd, etc. Thermoplastic acrylic coatings are obtained through simple solvent evaporation method. Water-borne acrylic paints, however, are not used as a topcoat for immersion and substantial acid or alkaline chemical exposure.

(c) *Siloxane*

Siloxane is a polymeric material, consist of Si–O bond as the main chain, and pendent functional group, i.e., –CH3, etc., polysiloxane coating exhibit superior chemical resistance, durability, gloss and color retention properties. The bond strength of Si–O bond is 443 kJ/mol, significantly higher than the C–C bond strength, i.e., 360 kJ/mol, therefore, polysiloxane coating are more stable at elevated temperature exhibit excellent resistance to oxidation. The siloxane polymers are cured by organo-metallic compounds, involving the reaction with the side group of the polymer chain. Research is being carried to enhance the performance of siloxane-based coating by modifying it with epoxy or acrylic polymer to meet the requirements for wide range of applications.

(d) *Alkyd resin*

Alkyd resin based paints are widely used for general purpose application because of its good durability at the low coat. Alkyd resin has been in use as a binder in the paint industry for more than 80 years, which stopped the use of nitrocellulose and oleoresions coating in almost all applications. Until the 1960s and commercialization of acrylic paints, the alkyd resin was used extensively in automotive industry. Alkyd resin exhibit inferior performance with regard to outdoor durability, color, and gloss retention when compared to the coatings of acrylic based paint.

Alkyd resin is a reaction product of three components: polybasic acid, polyol, and fatty acid. The proportion and nature of this three component govern the final properties of film product. Table 3 demonstrates the classification of the alkyd resin based on the dosage of fatty acid used in the formulation.

Table 3 Classification of alkyd resin based on fatty acid weight percentage

Alkyd resin type	Weight percentage
Oil free resin	0
Short oil resin	<45
Medium oil resin	45–55
Long oil resin	≥ 60

Long oil alkyd resin is suitable for exterior thin wall coatings, marine, metal maintenance paint, and clear lacquers. Long oil alkyd resin is soluble in aliphatic solvent and often uses a brush for coating over the substrate. On the other hand, short alkyd resin is soluble in aromatic solvents and used as air drying type for baking primer and enamals.

Table 4 shows the classification of alkyd resin based on the level of unsaturation present in the fatty acid.

Table 5 show the comparative study on the properties of final film product of alkyd paint when the different dosage of fatty acid is used in the formulation.

The alkyd paints as a group can be characterized as good substrate adherent paint with good drying properties as relatively low cost. Therefore, alkyd paints are still being used for common application, i.e., metal finish and wood product under less severe chemical environment. However, the exterior durability of alkyd paints is only fair; alkyd paints are best suitable for indoor applications considering a cost factor. The dry type alkyd resin exhibits good weather resistance, good adhesion and ability to withstand occasional chemical splashing. The alkyd paints ca be used as a primer and intermediate coating for a wide range of substrates. Further modification can improve the final properties of the coating with another type of resins, i.e., acrylic, epoxy, chlorinated rubber.

(e) *Polyurethane*

Isocyanate and material with hydroxyl group react together to form a urethane linkage, Fig. 10.

The material with hydroxyl groups can be ester, epoxy, acrylic, drying oil, or moisture (water). Urethane coating formulated form epoxy or polyester polyol

Table 4 Classification of alkyd resin based on degree of unsaturation in the fatty acid

Alkyd resin type	Iodine number
Drying resin	>140
Semidrying resin	125–140
Nondrying resin	<125

Table 5 Effect of fatty acid dosage on the paint and coating properties

Properties		Comparison(>/<)	
Color and gloss retention	Short oil alkyd	>	Long oil alkyd
Flexibility		<	
Adhesion		<	
Coating hardness		>	
Pigment dispersion		<	
Rheological properties		<	
Storage stability		<	

Fig. 10 Formation of urethane linkage

exhibit more chemical and moisture resistance than acrylic polyol. The isocyanate can be aromatic aliphatic, but a mixture of monomer can be used to formulate the urethane coating for a specific application.

The urethane coating is relatively expensive than alkyd resin based coating but cheaper than an epoxy coating. Urethane coating can be formulated as oil modified, moisture curing, blocked, two component, and lacquer.

Other resins are also used to modify certain properties with the intention to meet requirements for some specific applications. Acrylic urethane resin based coating exhibit excellent outdoor and weathering properties. Moisture curing of urethane coating need humidity and temperature, prolong UV exposure cause yellowing of coating. Two component urethane coating, consist of a catalyst to initiate the curing the process, exhibit good chemical resistance. However, they are not recommended for immersion services or exposure to strong acid/base environment. Special care should be taken while paint formulation due to high reactivity of isocyanate towards moisture in under humid condition and light, which lead to bubble formation and pin holes due to the evolution of carbon dioxide gas.

Urethane coatings are extensively used in transportation industry: aircraft, automobile, railroads, cars, truck, ship, etc. Urethane coating is resistant to biological and radiological warfare. Therefore, urethane coatings are frequently applied to military vehicle.

4.4 Working Principle of Anti-corrosive Coatings

Figure 11 shows the three mechanistic roots to protect the substrate from corrosion by different types of coatings.

Fig. 11 Anti-corrosion mechanism

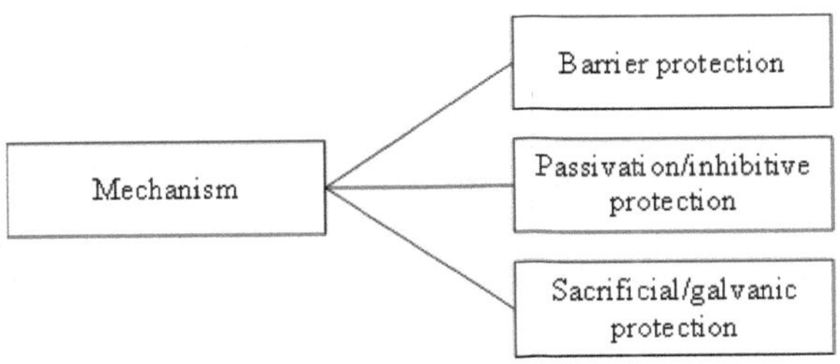

4.4.1 Barrier Protection

In barrier protection mechanism, the protection to the substrate is ensured by restricting or impeding the diffusion of corrosive species through the section of the applied coat. The binder's Impermeability to gases, liquid or ion is the essential requirement of the anti-corrosive coating either applied as primer, intermediate or topcoat. The lower dosage of pigment volume is taken in the formulation for better cohesion and low permeability. Depending on the requirement of an application other materials like Inert fillers, titanium dioxide, lamellar alumina, micaceous iron oxide, carbon black, graphene or graphene-like sheets, glass flake, etc., are also incorporated in the paint formulation to increase the tortuosity. Some recent development in the field of the polymer-carbon composite as anti-corrosive materials has been discussed in the later section of this chapter. For improved performance, it is always recommended to build anti-corrosive coat from multiple thin layer coat to the required thickness, but labor cost and potential revenue loss restricts to fewer and thicker coat. The barrier coating is used for all type of environment. However, the required type of binder used is always application specific. The type of binder and suitable application area is briefly discussed in the previous section.

4.4.2 Inhibitive Protection

The inhibitive coating protects the substrate from corrosion through the formation of the stable inorganic compound on the metal surface. The inhibitive coating is primarily applied as primers since they are alone effective and it is often exposed to the atmospheric environment. However, the barrier coating is used in combination of inhibitive protection for protecting structures working underneath water or soil. Inhibitive coating layers consist of inorganic inhibitive pigments such as phosphate, chromate, molybdate, borates, nitrates, and silicates. These inhibitive pigments in the presence of moisture react with the metal substrate and form a passive layer. Therefore high dosage of such inhibitive pigment is needed to ensure adequate leaching from the coating to the metal surface. However, the excess dosage of inhibitive pigment causes other problem like blisters, mud cracking due to internal stress, etc. The two opposite attributes, i.e., barrier properties of the coating and leaching out of sufficient quantity of inhibitor, has to be properly balanced for improved performance of the anti-corrosive paint.

4.4.3 Sacrificial Protection

In sacrificial protection mechanism, the substrate is protected against the corrosion at the cost of galvanically more active metal. The mechanism involves the oxidation of sacrificial metals and protecting the substrate, unlike barrier coatings the sacrificial coating is only applied as a primer. The sacrificial coating consists of active metallic power just below the critical pigment volume concentration to ensure direct current transfer with the substrate. Typically a dry coat of sacrificial paint consists 92–95% of active metallic powder and 5–8% binder by weight. Metallic zinc powder has been extensively used as a primer for protecting the steel substrate against corrosion. In sacrificial coating, the active metal acts as an anode and sacrifice itself to protect the substrate. The performance of zinc-based primer depends on the galvanic current transfer between zinc and steel substrate, which means, the substrate is galvanically protected as long as there is sufficient zinc to be sacrificed. Zinc rich sacrificial coatings are resistant to heat and mild chemicals. There is organic sacrificial coating as well, which are suitable for marine and industrial splash zone and offer good protection against corrosion.

5 Additives

The Improvement of anti-corrosive properties of a paint and its performance under the different environmental condition is the utmost priority of the manufacturers and scientist. The most common way to improve the performance of paints is to use some dopant, i.e., pigments, heat stabilizer, antifouling agent, fillers, etc., which improve paint's ability to protect the substrate against corrosion. Adding opacity to the paint is the widely deployed approach to inhibit the corrosion led by Ultraviolet radiation. The paint is made opaque with the help of pigments, most frequently called "hiding pigments". Titanium dioxide, lead oxide, iron oxide, carbon-based pigment, phthalocyanine blue are some of the hiding pigments. Due to the toxicity of lead, the use of lead pigment in the paint for residential application is prohibited in many nations. Therefore, other alternatives to lead pigment have been developed and deployed in paint application.

Fillers, on the other hand, are used to enhance the durability and hiding power. Fillers are broadly divided into two categories, i.e., reinforcing and inert filler. Whereas the reinforcing filler enhances or supports the structural integrity, the inert fillers are primarily used for the cost-reduction purpose. Clay, carbon black, talc, barite, lime, etc., are some examples of filler used in paint application. Depending on the application requirement, suitable fillers are chosen. Factors such as filler particle size and particle structure (i.e., aspect ratio) have a huge impact on various properties such as, the viscosity, volume and covering power of a paint over the substrate. Fillers can serve multiple purposes in paint application. Filler enhances the anti-corrosive ability, depending on size and structure of the filler particle, the permeability to the corrosive element can be reduced to a greater extent. The lamellar structure has overlapping plates and high aspect ratio than that of spherical filler particle; consequently high aspect ratio filler impede more the path of corrosive elements [5]. Figure 12 shows the mechanism of impediment by filler particle; corrosive elements can migrate relatively easily through paint coating filled with low aspect ratio filler i.e., spherical filler particle, and lamellar filler particle offers more hindrance and subsequently lower the diffusion process [6].

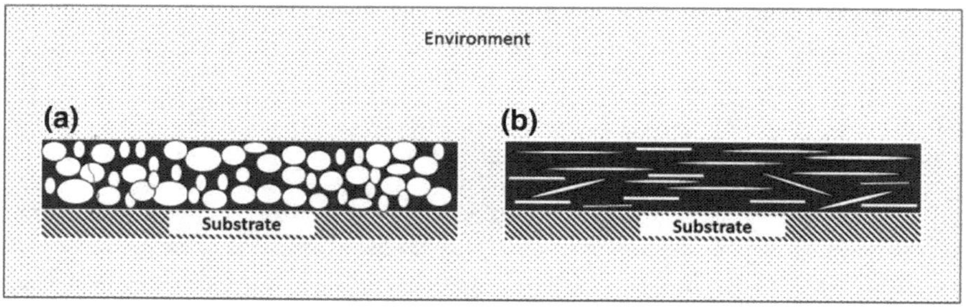

Fig. 12 Impediment mechanism by filler with different aspect ratio; **a** low aspect ratio filler containing coating and **b** high ratio filler containing coating

Fillers can also be employed as a pigment: carbon black, titanium dioxide is also referred as special pigments; often an outermost layer of water storage tanks is colored black because carbon black filler provides good protection against ultra-violet radiation and weatherability. Similarly, titanium dioxide improves the thermal stability of applied paint coating and hence inhibits the metal properties impairment.

There are some special purpose additives whose function is to add significant value to the product in paint application: surface tension modifiers, thickeners, catalyst, UV stabilizer, antifouling agent, adhesion promoter, antioxidant, etc.

6 Role of Carbon in Advance Anti-corrosive Paint

As discussed in the previous section, carbon serves multiple roles in paint application, and several major developments has been achieved in the past few years. Carbon can exist in many forms, i.e., diamond, graphite, graphene, carbon black, fullerene, carbon fiber, carbon nanotube. This different architecture of carbon has a huge effect on the performance of the paint.

Graphene, graphene oxide (GO) and reduced graphene oxide (rGo) have received considerable attention in the development of many commercially viable and energy efficient products due to their peculiar physio-chemical properties. A study done on graphene by several research groups reveals that graphene whether single layer or multiple layers coated onto metal surface significantly reduces the diffusion of corrosive species, i.e., O_2, H_2O, hence impede the environmental oxidation process [7–9]. Graphene is now rapidly being established as a potential candidate for corrosion inhibitor in paints and coating application. Graphene exhibit superior inertness than any other material which would degrade faster at the hostile condition. Graphene does not only provide good oxidation resistance but also has been used in the manufacturing of airtight rubber bladder, indicating it is very low impermeable property to air. Graphene can also affect the electrochemical reaction of metals in electrolytic solutions. A recent study has proven the ability of graphene to prevent metal dissolution in the electrolytic solutions, indicating graphene's impact on the anode-cathode reaction of metal underlying the graphene layer. More work is being done on exploring the possibility of using graphene or carbon layer for passivating the metals [10].

7 Recent Developments in the Polymer-Carbon Coating for Anti-corrosion Applications

7.1 Graphene Oxide as a Corrosion Inhibitor for the Aluminium Current Collector in Lithium Ion Batteries

R. Prabhakar et al., have shown the use of GO in the coating for protection against corrosion of a current aluminium collector used in lithium ion batteries [11]. In power storage devices, Al is established as a material for primary current collector due to its stable surface oxide layer, i.e., Al_2O_3. However, at high-voltage Al_2O_3 layer can undergo pitting corrosion, this may result in many problems such as; cell impedance loss, more self-discharge, loss of capacity, and mechanical integrity. Since GO has the unique characteristics, i.e., very high surface area, the presence of active functional groups (carboxylic, ketone), high conductivity GO is suitable for the development of efficient lithium ion batteries. Figure 13, shows the idealized mechanism of GO preventing the corrosion of Al metal from PF_6^- ion.

7.2 Graphene Oxide Based Nanopaint

Functional coating is yet another area where GO is proving its ability and competence as a material for cost effective solutions. Functional coatings are high-performance coating and serve multiple functions, i.e., anti-corrosion, antibacterial and antifouling properties [12–14]. K. Krishnamoorthy et al. have reported preparation and study on the practical application of GO based nanopaint using ball-milling process [15]. Typical formulation GO-based nanopaint is listed in Table 6.

All ingredients except solvent is taken in tungsten carbide bowl. The ball-to-powder weight ratio is set as 10:1 and 300 rpm to ensure effective milling. After 5 h of milling the required solvent is added and milling is continued for another 30 min. The finished product is stored in an airtight container. The values of several properties of GO nanopaint is summarized in Table 7, [16–19].

Depending on the solid content a paint can be divided into two categories, i.e., low and high solid content. The prepared nanopaint has 64.2% solid content which is on the higher side. The high solid paints are more advantageous than low solid

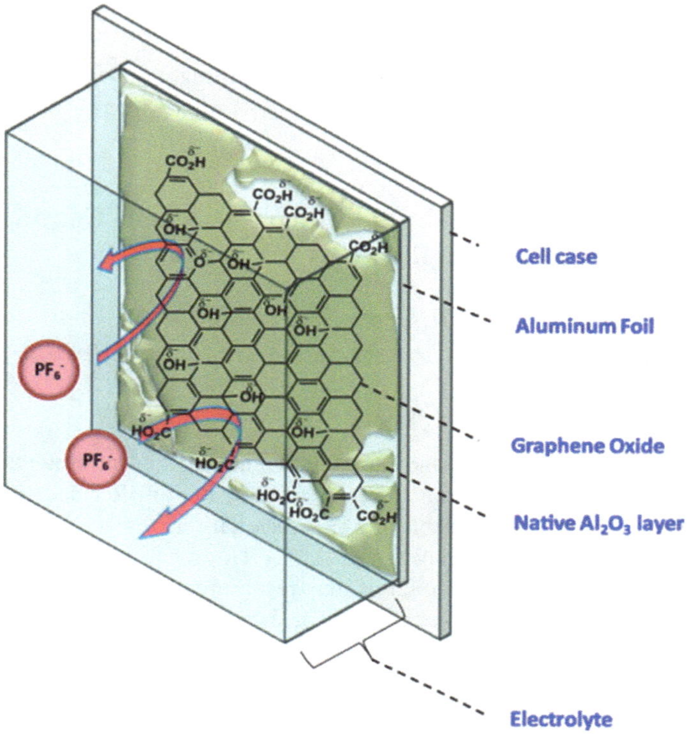

Fig. 13 The scheme showing the root to prevent corrosion, the deposited GO layer repels PF_6^- ion flux which leads to improved cyclability over bare Al (reproduced after permission granted from Ref. [11])

paint due to the merits including: low emission of volatile organic compounds, and high paint layer thickness which are important factors for reducing the application cost and labor times [20]. The GO nanopaint has high-gloss level and good surface finish, i.e., 75 GU at 60 °C, which is classified as high-gloss surface as per ASTM standard [16]. GO nanopaint also showed good corrosion resistance in acidic condition proving the prospective utility of GO in anti-corrosion application. GO based nanopaint has very fewer hazards and environmental impact. GO has significant potential for the development of next generation protective coatings.

7.3 Polyaniline Graphene Composite: Advance Anti-corrosion Coating for Metals

Inherently conducting polymers (ICP) are special class of polymeric materials having high electronic conductivity due to the conjugated electronic pattern.

Table 6 GO-based nanopaint formulation (reproduced after permission granted from Ref. [15])

Additive	Material	Purpose	Weight percentage (%)
Pigment	GO	Providing the color of the paint	16.0
Binder	Linseed alkyd resin	Film-forming component of paint	64.0
Stabilizer	Nanosized ZnO	Reducing the color fading effect	0.6
Anti-setting agent	Aluminum stearate	Preventing pigment and binder settling	0.5
Thickener	Thickener A	Improving the viscosity and preventing coagulation	0.1
Wetting agent	Soya lecithin	Wetting the pigment in the binder for uniform dispersion	0.6
Inner coat drier	Nanosized zirconia	Chemical crosslinking agent of unsaturated fatty acids	0.6
Outer coat drier	Cobalt napthenate	Active catalyst for the lipid autoxidation process	0.6
Solvent	Mineral turpentine oil	Dispersing agent	17.0

Table 7 Properties of GO nanopaint

Properties	Values
Solid content[a] (%)	64.2
Surface dry time (h)	1
gloss level[b] (GU at 60 °C)	75
Corrosion inhibition efficiency[c] (%)	88.7

[a]Solid content = $[W_{dry}/W_{initial}] * 100$; where $W_{initial}$ is the initial weight of the paint and W_{dry} is the weight of paint after dry

[b]1 g of paint coated on drawdown card and gloss level measured using Rajdhani digital gloss meter (RSPT-200 model, India)

[c]Weight loss measured and compared between coated and uncoated galvanized iron (GI) substrate submersed in 0.1 N HCL solution for 24 h. corrosion inhibition efficiency = $[(W_{uncoated}-W_{coated})/W_{uncoated}] * 100$

Polyacetylene, polydiacetylene, polyaniline, polythiophene, poly(*p*-phenylene), poly(*p*-phenylene sulfide), etc., are some of the examples of ICP. ICP show metal-like properties, i.e., electrical, electronic, magnetic, and optical properties and yet ICP can be processed like conventional polymers. ICP has received huge interest due to its several attractive properties, i.e., high electronic conductivity, light weight, adequate mechanical flexibility and low cost. Polyaniline is one of the widely studied polymeric material belonging to the class of ICP. It is a commercially successful polymeric material, and find its utility in many applications. Polyaniline and its derivatives are widely used in research and development activities for high-performance materials.

Polyaniline has some special attributes, such as relatively low cost, easily synthesized, good environmental stability, distinct color, and oxidation state. Polyaniline also exhibits the electromagnetic interference (EMI) shielding effect and shows pH-dependent response, which makes is suitable for use in many areas, such as, actuators, supercapacitor, chemical vapor sensor, biosensor, electrically conducting yarns, etc. [21, 22].

Polyaniline has also been studied as a prospective material for the high-performance coating for protection against corrosion. Polyaniline offers superior protection against corrosion compared to other polymers due to high corrosion potential and the catalytic ability for metal passivation [23]. The approach to further improve the performance of an anti-corrosive paint is to use a nanostructured filler with high aspect ratio, i.e., ~200 [24]. As discussed in the previous section, the mechanism of enhanced impediment of corrosive species through the lamellar nanostructured filler coating, high aspect filler increases the pathway for diffusion of reactive elements and improves the performance of paint against the corrosion significantly. A similar concept is applied to develop polymer/graphene composites for numerous applications. Graphene is having a higher aspect ratio, i.e., ~500, offers superior specific properties and enhanced performance at low cost than other nano-fillers [25].

The effect of functionalization of graphene on the barrier properties of anti-corrosion coatings is also being investigated. C. H. Chang et al. have studied the graphene functionalization and its effect on corrosion protection. The process of exfoliation of graphite to yield a graphene or graphene-like sheets is well studied and established. Figure 14, shows one of the methods and schematic reaction mechanism of electrophilic substitution on graphite, resulting in the formation of functionalized graphene sheet.

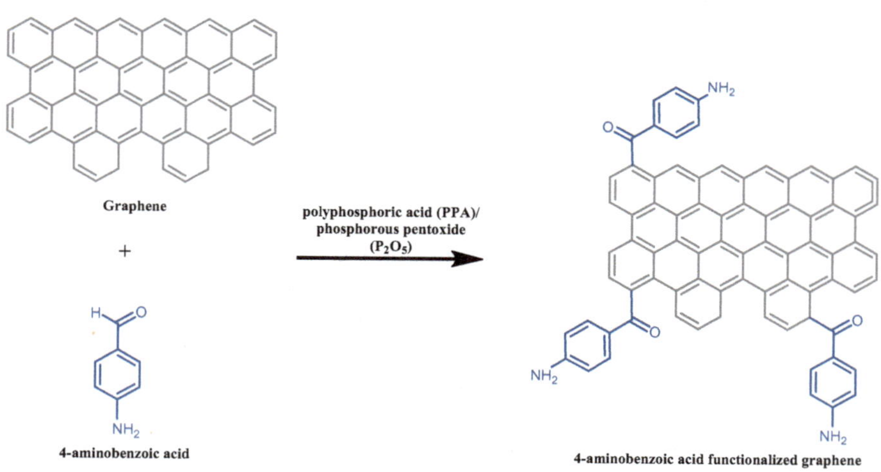

Fig. 14 Functionalization of graphene or graphite

Aminobenzoic acid and graphite are mixed vigorously in PPA/P^2O^5 medium under an inert atmosphere at elevated temperature. After 3 days the precipitate is filtered, washed thoroughly and freeze dried to yield aminobenzoic acid functionalized graphene [26].

Figure 15 shows the schematics of in situ dispersion of aminobenzoic acid functionalized graphene in polyaniline matrix of polyaniline/graphene composite.

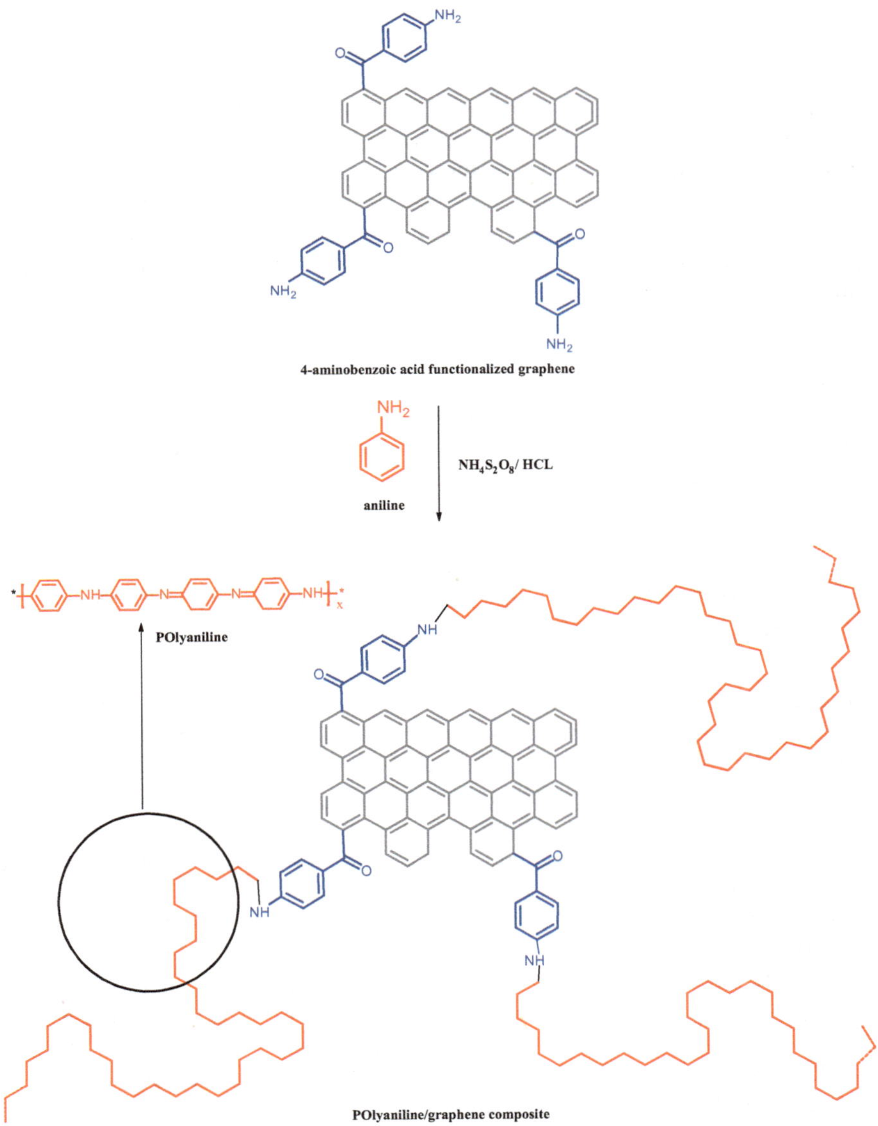

Fig. 15 In situ synthesis of polyaniline/graphene composite

Table 8 Oxygen permeability and vapor permeability rate

Sample	Permeability O_2 (barrer)	Vapor permeability rate (g/h-m^2)
Polyaniline	0.74	173
Polyaniline/Graphene—0.1 wt%	0.33	70
Polyaniline/Graphene—0.25 wt%	0.22	47
Polyaniline/Graphene—0.5 wt%	0.1	20
Polyaniline/clay—0.5 wt%	0.3	64

Aniline, functionalized graphene and ammonium persulfate is taken in a reaction vessel and stirred in HCL solution for six hours at ~4 °C temperature. The residue product is retreated with base to neutralize the pH of the polyaniline/graphene composite. The composite is finally dissolved in NMP solvent to yield the anti-corrosion coating [18]. The polyaniline/graphene composite is coated over steel substrate, and the solvent is dried. The performance of graphene based anti-corrosive coating is compared with the clay-based paint. Table 8 comprises the data on permeability of oxygen and vapor permeability rate of polyaniline/graphene and polyaniline/clay paint.

The permeability of oxygen and other corrosive species decreases significantly with increase in the dosage of graphene. Graphene or graphene-like sheets are more cost effective and superior in term of corrosion barrier performance than other type of fillers, i.e., clay.

7.4 Graphene Barrier Properties to Dissolution of Nickel and Copper Metal

The chemical vapor deposit method for growing graphene is rapidly becoming very popular because it leads to the deposition of mostly a single graphene layer over the substrate [27]. N.T. Kirkland et al. grew predominantly single-layer graphene film on ~1 mm thick Nickel (99.9% Ni) and Copper (99.9% Cu) metal sheet. Following the surface-catalyzed process, the graphene film gets deposited by exposing the metal sheets in the hot furnace filled with hydrogen and methane gas for a fixed period [9].

Figure 16 and 17 show the optical microscopic image, Raman mapping at ~2700 cm^{-1} peak and Raman spectra of graphene deposited Ni and Cu sheets respectively. Typically the single-layer graphene structure shows a sharp G (1580 cm^{-1}), 2D (~2650–2700 cm^{-1}) peaks and a small G/2D ratio, whereas, the multi-layer graphene structure display higher G/2D ratio (≥ 1) and broad 2D peak. The Raman and Scans study shows that approximately 80 and 60% area of the surface is consistent with single or few layer graphene for Cu and Ni metal sheets respectively.

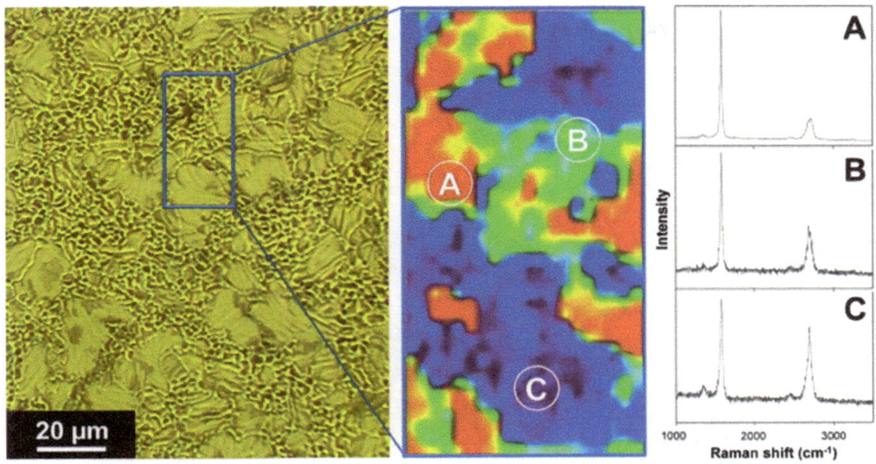

Fig. 16 Optical microscopic image, Raman mapping at ~2700 cm peak and Raman Spectra of Graphene deposited Nickel metal sheet (reproduced after permission granted from Ref. [9])

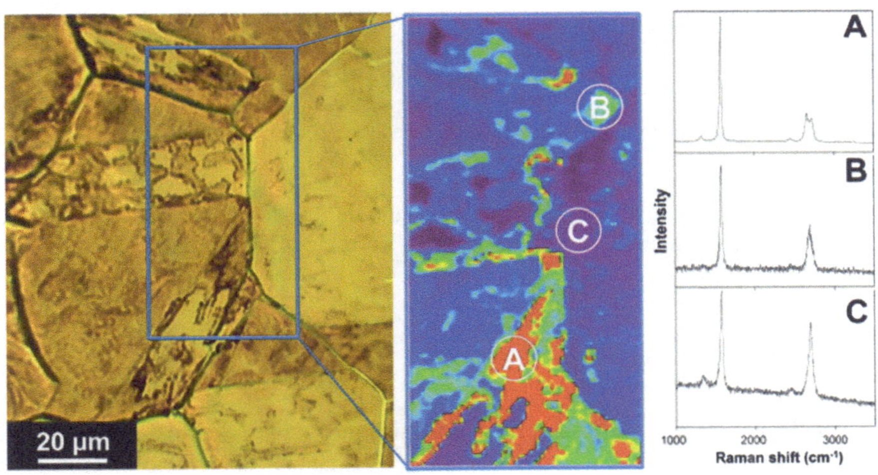

Fig. 17 Optical microscopic image, Raman mapping at ~2700 cm peak and Raman spectra of graphene deposited copper sheet (reproduced after permission granted from Ref. [9])

Figure 18 shows the electrochemical reaction scan of graphene coated Ni and Cu metal sheet with their predecessor metals, i.e., bare Ni and bare Cu metal sheets.

Ni coated with graphene layer exhibits decrease in both anode and cathode reaction rate, and also there is an increase in the corrosion potential by ~0.3 V than that of bare Ni metal sheet. In contrast to Ni, the Cu metal covered with graphene shows a decrease in the corrosion potential by ~0.15 V than that of bare Cu metal.

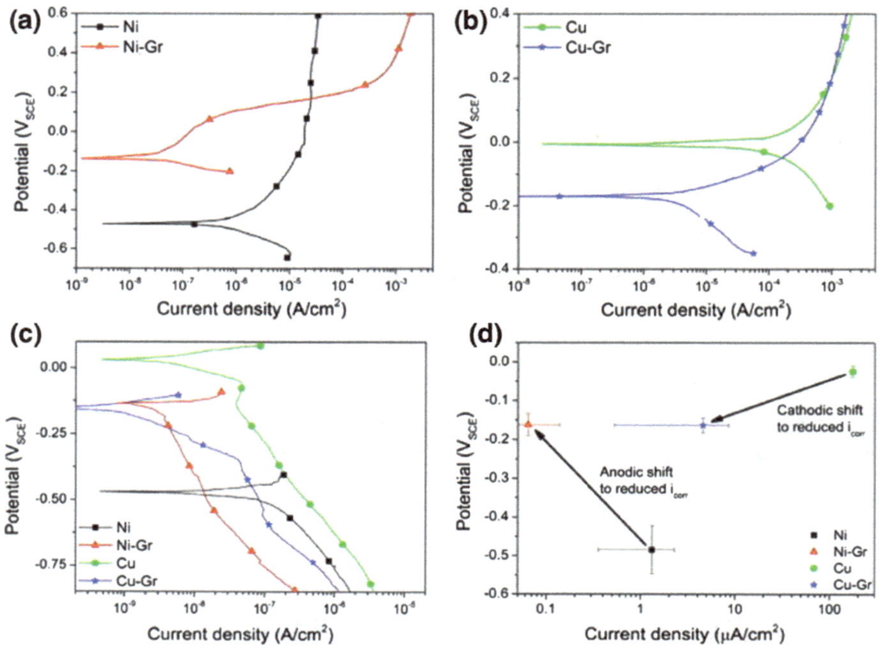

Fig. 18 Anode polarization reaction for graphene coated Ni and bare Ni metal (**a**), graphene coated Cu and bare Cu metal (**b**), cathode polarization scan for all samples (**c**), and shift in current density and corrosion potential for all samples (**d**) (reproduced after permission granted from Ref. [9])

It means the graphene acts as an inhibitor of anodic reaction to Ni and as a cathodic barrier for the Cu metal.

8 Conclusion

In conclusion, Graphene is indeed a potential material for corrosion protection, however, the reason for this strange behavior of graphene specific to metal type is still not clear. Probably, deeper insight into the effect of graphene orientation, layer thickness, surface defects, etc., could help in finding more effective and economical solution to protect corrosion.

References

1. Hashimoto K, Asami K, Naka M, Masumoto T (1979) The role of alloying elements in improving the corrosion resistance of amorphous iron base alloys. Corros Sci 19:857–867

2. Sathish Kumar P, Rajasekar R, Pal SK, Nayak GC, Syed Ismail SMR (2016) Paints and coating of multicomponent product. Springer, The Netherlands, pp 157–226
3. Schweitzer PA (2006) Paint and coatings: applications and corrosion resistance. CRC Press, New York
4. ISO-12944-2-1998
5. El-Fattah MA, El Saeed AM, Dardir MM, El-Sockary MA (2015) Studying the effect of organo-modified nanoclay loading on the thermal stability, flame retardant, anti-corrosive and mechanical properties of polyurethane nanocomposite for surface coating. Prog Org Coat 89:212–219
6. Sørensen PA, Kiil S, Dam-Johansen K, Weinell CE (2009) Anti-corrosive coatings: a review. J Coat Technol Res 6:135–176
7. Sutter E, Albrecht P, Camino FE, Sutter P (2010) Monolayer graphene as ultimate chemical passivation layer for arbitrarily shaped metal surfaces. Carbon N Y 48:4414–4420
8. Dedkov YS, Fonin M, Laubschat C (2008) A possible source of spin-polarized electrons: the inert graphene/Ni(111) system. Appl Phys Lett 92:52506
9. Kirkland NT, Schiller T, Medhekar N, Birbilis N (2012) Exploring graphene as a corrosion protection barrier. Corros Sci 56:1–4
10. Haslam GE, Sato K, Mizokawa T, Chin X-Y, Burstein GT (2012) Charge transfer effects in electrocatalytic Ni–C revealed by X-ray photoelectron spectroscopy. Appl Phys Lett 100:231601
11. Richard Prabakar SJ, Hwang Y-H, Bae EG, Lee DK, Pyo M (2013) Graphene oxide as a corrosion inhibitor for the aluminum current collector in lithium ion batteries. Carbon N Y 52:128–136
12. Akhavan O, Ghaderi E (2010) Toxicity of graphene and graphene oxide nanowalls against bacteria. ACS Nano 4:5731–5736
13. Kholmanov IN et al (2012) Nanostructured hybrid transparent conductive films with antibacterial properties. ACS Nano 6:5157–5163
14. Krishnamoorthy K, Veerapandian M, Zhang L-H, Yun K, Kim SJ (2012) Antibacterial efficiency of graphene nanosheets against pathogenic bacteria via lipid peroxidation. J Phys Chem C 116:17280–17287
15. Krishnamoorthy K, Jeyasubramanian K, Premanathan M, Subbiah G, Shin HS, Kim SJ (2014) Graphene oxide nanopaint. Carbon N Y 72:328–337
16. Sward GG (1972) Paint testing manual: physical and chemical examination of paints, varnishes, lacquers, and colors, 13th edn
17. Tang J, Shao Y, Zhang T, Meng G, Wang F (2011) Corrosion behaviour of carbon steel in different concentrations of HCl solutions containing H_2S at 90 °C. Corros Sci 53:1715–1723
18. Chang C-H et al (2012) Novel anti-corrosion coatings prepared from polyaniline/graphene composites. Carbon N Y 50:5044–5051
19. Akhavan O et al (2011) $CuO/Cu(OH)_2$ hierarchical nanostructures as bactericidal photocatalysts. J Mater Chem 21:9634
20. Díaz I, Chico B, de la Fuente D, Simancas J, Vega JM, Morcillo M (2010) Corrosion resistance of new epoxy–siloxane hybrid coatings. A laboratory study. Prog Org Coat 69:278–286
21. Wijewardane S, Goswami DY (2012) A review on surface control of thermal radiation by paints and coatings for new energy applications. Sustain Energy Rev 16:1863–1873
22. Ghosh P (2011) Polymer science and technology: plastics, rubbers, blends and composites. McGraw-Hill Education LLC
23. Kumar SA, Bhandari H, Sharma C, Khatoon F, Dhawan SK (2013) A new smart coating of polyaniline-SiO_2 composite for protection of mild steel against corrosion in strong acidic medium. Polym Int 62:1192–1201
24. Yeh J-M, Liou S-J, Lai C-Y, Wu P-C, Tsai T-Y (2001) Enhancement of corrosion protection effect in polyaniline via the formation of polyaniline—clay nanocomposite materials. Chem Mater 13:1131–1136

25. Compton OC, Kim S, Pierre C, Torkelson JM, Nguyen ST (2010) Crumpled graphene nanosheets as highly effective barrier property enhancers. Adv Mater 22:4759–4763
26. Choi E-K et al (2010) High-yield exfoliation of three-dimensional graphite into two-dimensional graphene-like sheets. Chem Commun 46:6320
27. Brownson DAC, Varey SA, Hussain F, Haigh SJ, Banks CE (2014) Electrochemical properties of CVD grown pristine graphene: monolayer-vs. quasi-graphene. Nanoscale 6:1607–1621

FSC
www.fsc.org
MIX
Papier | Fördert
gute Waldnutzung
FSC® C083411

Kolibri 360 GmbH
Ferdinand-Jühlke-Straße 7
99095 Erfurt, Deutschland
produktsicherheit@kolibri360.de